HUMAN REPRODUCTIVE PHYSIOLOGY

HUMAN REPRODUCTIVE PHYSIOLOGY

EDITED BY

Rodney P. Shearman

M.D. D.G.O. F.R.C.O.G.

Professor of Obstetrics and Gynaecology
The University of Sydney
Honorary Obstetrician and Gynaecologist
King George V Memorial Hospital, Sydney

BLACKWELL SCIENTIFIC PUBLICATIONS
OXFORD LONDON EDINBURGH MELBOURNE

© 1972 by Blackwell Scientific Publications
Osney Mead, Oxford
3 Nottingham Street, London W.1.
9 Forrest Road, Edinburgh
P.O. Box 9, North Balwyn, Victoria, Australia

All rights reserved. No part of this publication may be reproduced, stored in a retrieval system, or transmitted, in any form or by any means, electronic, mechanical, photocopying, recording or otherwise without the prior permission of the copyright owner.

ISBN 0 632 08680 7

First published 1972

Distributed in the U.S.A. by
F.A. Davis Company, 1915 Arch Street,
Philadelphia, Pennsylvania

Printed in Great Britain by
Alden & Mowbray Limited
at the Alden Press,
Osney Mead, Oxford
and bound by the Kemp Hall Bindery, Oxford

CONTENTS

Contributors	vii
Preface	ix
1 Cytogenetics and Genetic Mechanisms C.B.KERR	1
2 Ovarian Function and its Control RODNEY P.SHEARMAN	45
3 The Physiology and Function of the Testes BRYAN HUDSON & HENRY G.BURGER	91
4 Endocrinology of the Foeto-Maternal Unit G.C.LIGGINS	138
5 Immunology WARREN R.JONES	198
6 Liquor Amnii D.R.ABRAMOVICH	258
7 Myometrial and Tubal Physiology CARL WOOD	324
8 Foetal Cardio-Respiratory Physiology E.D.BURNARD	376
9 The Adrenal Cortex A.W.STEINBECK	415
10 Thyroid Function H.K.IBBERTSON	478
11 Carbohydrate and Lipid Metabolism During Pregnancy JOHN R.TURTLE	525
12 Cardiovascular Function in Pregnancy WILLIAM A.W.WALTERS	554
13 Renal Physiology—Normal Renal Physiology and Changes in Pregnancy RANJIT S.NANRA & PRISCILLA KINCAID-SMITH	594
14 Respiratory Physiology in Pregnancy ANN J. WOOLCOCK & JOHN READ	639
15 Haemopoiesis and Coagulation P.A.CASTALDI & DOUGLAS R. HOCKING	664
Appendix A Steroid Biosynthesis and Metabolism BRYAN HUDSON	716
Appendix B Gonadotrophins—Chemistry and Measurement HENRY G.BURGER	729
Appendix C Measurement of Adrenal Steroids A.W.STEINBECK	752
Index	770

CONTRIBUTORS

D.R.ABRAMOVICH, M.B., B.S., D.G.O., PH.D., M.R.C.O.G. Senior Lecturer in Obstetrics and Gynaecology, University of Aberdeen.

HENRY G. BURGER, M.D., F.R.A.C.P. Executive Director, Medical Research Centre, Prince Henry's Hospital, Melbourne. Senior Lecturer (Part-time), Monash University, Melbourne.

E.D.BURNARD, M.B., M.R.C.P., F.R.A.C.P. Senior Research Fellow, Children's Medical Research Foundation, Royal Alexandra Hospital for Children, Sydney. Paediatrician, The Women's Hospital, Crown Street, Sydney.

P.A.CASTALDI, M.D., F.R.A.C.P. Director of Haematology, Haematology Department and Departments of Medicine and Pathology, Austin Hospital, University of Melbourne.

DOUGLAS R. HOCKING, M.B., B.S., M.R.A.C.P. Haematologist, Geelong Hospital, Geelong. Physician, Haematology, Austin Hospital, Melbourne.

BRYAN HUDSON, M.D., PH.D., F.R.A.C.P., F.R.C.P. Professor and Chairman, Monash University Department of Medicine, Melbourne. Honorary Director, Medical Research Centre, Prince Henry's Hospital, Melbourne.

H.K.IBBERTSON, M.B., F.R.C.P., F.R.A.C.P. Professor of Endocrinology, Auckland School of Medicine, Auckland.

WARREN R. JONES, M.B., B.S., PH.D., D.G.O., M.R.C.O.G. Senior Lecturer, Department of Obstetrics and Gynaecology, The University of Sydney. Honorary Obstetrician and Gynaecologist, King George V Memorial Hospital, Sydney.

C.B.KERR, M.B., D.PHIL., M.R.A.C.P. Professor of Preventive and Social Medicine, University of Sydney. Honorary Medical Geneticist, Royal Alexandra Hospital for Children, Sydney.

PRISCILLA KINCAID-SMITH, M.D., B.SC., F.R.C.P., F.R.A.C.P., D.C.P. Reader in Medicine, University of Melbourne. Physician in Charge, Renal Unit, Royal Melbourne Hospital, Melbourne.

G.C.LIGGINS, M.B., CH.B., PH.D., F.R.C.O.G., F.R.C.S. Associate Professor, Postgraduate School of Obstetrics and Gynaecology, University of Auckland, Auckland.

RANJIT S.NANRA, M.B., B.S., M.R.A.C.P. Assistant Physician, Medical Renal Unit. Associate in Medicine, University of Melbourne, Royal Melbourne Hospital.

JOHN READ, M.D., F.R.A.C.P., Professor of Medicine, University of Sydney.

RODNEY P.SHEARMAN, M.D., D.G.O., F.R.C.O.G. Professor of Obstetrics and Gynaecology, The University of Sydney. Honorary Obstetrician and Gynaecologist, King George V Memorial Hospital, Sydney.

A.W.STEINBECK, M.D., B.S., PH.D., F.R.C.P., F.R.A.C.P. Associate Professor of Medicine, University of New South Wales, Physician and Head of Division of Endocrinology and Metabolism, The Prince of Wales and Prince Henry Hospitals and Consulting Endocrinologist to the Department of Paediatrics. Consultant Endocrinologist to the Royal Hospital for Women, Sydney.

JOHN R.TURTLE, M.D., M.R.A.C.P., Associate Professor in Medicine, University of Sydney. Physician, Royal Prince Alfred Hospital, Sydney.

WILLIAM A.W.WALTERS, M.B., B.S., PH.D., M.R.C.O.G. Associate Professor of Obstetrics and Gynaecology, Monash University, Melbourne. Honorary Assistant Obstetrician and Gynaecologist, Queen Victoria Memorial Hospital, Melbourne.

CARL WOOD, M.B., B.S., F.R.C.S., F.R.C.O.G. Chairman and Professor, Department of Obstetrics and Gynaecology, Monash University, Melbourne. Honorary Obstetrician and Gynaecologist, Queen Victoria Memorial Hospital, Melbourne.

ANN J.WOOLCOCK, M.D., M.R.A.C.P. Senior Research Fellow, Department of Medicine, University of Sydney.

PREFACE

Obstetrics and gynaecology embraces now a great deal more than the midwifery and operative gynaecology that the editor of this book learned 20 years ago as an undergraduate. The questions asked then and the answers given, were largely mechanical. In the area of human reproduction it has become clear that questions to which there were no valid answers two decades ago would remain unanswered without rational physiological study of the human.

Reproductive biology is at the core of man's modern paradox—the rightful demand that each individual born should have the best prospect of survival and the realization that man left to indiscriminate fertility will cause the most horrifying plague of all, that of man himself.

The subjects reviewed here were not chosen at random but rather because either of their central importance in reproductive physiology, or because of the profoundly modifying effect exerted by pregnancy on a particular body system. The book had its genesis in the belief that no person practising obstetrics and gynaecology as it should be today, can properly do so without greater physiological insight than that received as a medical student. It is aimed first towards those working for higher professional qualifications in obstetrics and gynaecology, secondly at those engaged in practice of the speciality. Finally it is directed at some medical students as a source of reference, particularly for those whose undergraduate curriculum allows the exercise of options or electives.

The best physiological laboratory for the human, is the human. It is therefore, not by coincidence nor meant as a reflection on non-clinical physiologists that each of the contributors to their book is, at least in part, a clinician. It is hoped that what offence, if any, this gives the more fundamental physiologist will be outweighed by the relevance of the material presented to a better understanding of human reproduction.

Rodney P. Shearman,
Sydney, June 1972

1
Cytogenetics and Genetic Mechanisms

C.B.KERR

CYTOGENETICS OF MAN

Human cytogenetics commenced in 1956 with the correct assignment of 46 chromosomes to the somatic cells of man. This discovery stimulated intense activity which has contributed new knowledge to human biology, the scientific basis of medicine and clinical practice. The chromosomal mechanism of sex determination was clarified; established syndromes like Turner's and Klinefelter's became better understood and new syndromes were recognized, for example, the serious conditions in childhood due to trisomy-13 and trisomy-18. Attention turned to elucidating mechanisms whereby chromosome complements became disordered. The chromosomal theory of neoplasia was revived. Human population cytogenetics developed as a systematic approach to measuring the surprisingly large load of chromosomal variation in mankind. More recently there has been much work on physiological correlations, including disorders of behaviour, associated with chromosomal anomalies. In the relatively new field of somatic cell genetics, cytogenetic techniques are used for studying fundamental aspects of gene effect at the molecular level.

The explosive and exponential increase of knowledge in human cytogenetics yielded problems, especially during its early golden years of discovery when a vast array of chromosomal variation was uncritically associated with an equally vast range of human variation—most of it harmful. There were additional difficulties with a new and confusing terminology together with having to adapt for man the lessons from an older and distinguished cytogenetic experience with lesser organisms. The situation is somewhat clearer now despite many gaps and loose ends especially in the area to be emphasized here, that of sexual development and its abnormalities. As in so many fields of human physiology our present imperfect insights come from a study of the exceptional. We have, for instance, some cytogenetic markers of *abnormal* growth and behaviour but as yet very few ideas on the genetic control of the comparable *normal* systems.

NORMAL HUMAN CHROMOSOMES

Situated within the cell nucleus, chromosomes harbour the genetic code in the form of deoxyribonucleic acid (DNA) molecules. The latter are organized along with three other macromolecules—ribonucleic acid, histones and acidic residual protein, in a manner not clarified for mammalian chromosomes, whose ultrastructure so far eludes useful analysis by electron microscopy.

Chromosomes cannot be visualized for most of the cell cycle. They are seen only during cell division which commences immediately after DNA has undergone replication. Preparations of somatic cell chromosomes are made at the stage of cell-division (mitosis) called metaphase when each chromosome is about to split longitudinally into two daughter chromosomes. Here the halves (chromatids) remain joined at a region known as the centromere. In this way the 46 human chromosomes are recognized; 44 paired autosomes in each sex with XY sex chromosomes in the male and XX in the female. Chromosome complements (karyotypes) of cells are classified on the characteristics of chromosomes engaged in mitosis; the specialized cell-division in germ cells (meiosis) reveals information of a different nature which in practice is related more to the origins of certain chromosomal patterns detected or suspected in somatic cells.

Chromosome preparation can be obtained directly from actively dividing cells such as are found in bone marrow aspirate or some tumours. However, the standard method of obtaining human karyotypes involves culturing small lymphocytes from peripheral blood for 3–4 days under the influence of the mitotic stimulant, phytohaemagglutinin. Micromethods based on 3–4 drops of capillary blood are widely used for routine purposes and are especially valuable for infants. Sometimes, and especially during a search for chromosome mosaicism (two or more different karyotypes in the cells from one individual), it is necessary to obtain chromosome preparations from material other than peripheral lymphocytes. Therefore fibroblasts from fragments of skin or occasionally other fibrous tissue are cultured.

The actual preparation of human chromosomes has become a straightforward routine procedure but expert analysis of karyotypes remains a lengthy and tedious business. Most cytogenetic laboratories base a standard investigation on the chromosome count of 20–30 cells with 5–10 cells analysed in detail. Automatic chromosome analysis using computers, although well advanced in principle (Perry 1969), is not yet sufficiently reliable or economical for routine karyotyping purposes.

Identification of individual metaphase chromosomes is based both on total length and the relative length of each pair of arms above and below the centromere. Chromosomes can readily be arranged into seven groups, A–G, but a more precise classification into numbered pairs is difficult. Expert cytogeneticists exploit further structural features that may be present, for example, secondary constrictions in the long arms of chromosome pairs 1, 9 and 16. Classification of human chromosomes is summarized in Fig. 1.1.*

Incorporation of tritiated DNA precursors, most usually thymidine, into newly synthesized DNA can be recorded autoradiographically as grain patterns over the chromosomes. Unhappily early hopes that individual chromosomes would consistently show characteristic labelling have been dimmed by variability in DNA replication patterns (Steele 1969). However, one of the two X-chromosomes in female cells constantly yields an out-of-phase pattern by becoming densely labelled after other chromosomes have ceased to incorporate isotope. This is known as a 'late-labelling' or 'hot' X-chromosome.

By international agreement human karyotypes are nowadays recorded in a descriptive manner designed for ease of processing information by digital computers; the system is summarized with examples in Table 1.1 and further examples of usage follow in the text.

CELL DIVISION

Mechanisms of cell division are the cornerstone of cytogenetics. Apart from fundamental implications, practically all significant distortions of the genetic code detected in the form of chromosome abnormalities

* Histochemical techniques have been developed recently which give distinct banding patterns on chromosome arms. These methods involve use of DNA-binding fluorescent agents and heat-dependent Giemsa staining. Because banding patterns are specific and consistent for each chromosome, classification is much facilitated especially with regard to the origins of components in composite chromosome figures. A full account of the new methods will appear during 1972 in the Proceedings of the Fourth International Congress of Human Genetics to be published by Excerpta Medica, Amsterdam.

originate from some error in the dividing process. Even if chromosome damage results from an external insult like ionizing radiation the critical factor is whether the induced defect can survive cell division and so be perpetuated in the individual.

The main features of mitosis and meiosis will be summarized here; full details are given by Swanson, Merz & Young (1967).

Group	Appearance	Centromere position	Numerical classification
A		Median/submedian	1–3
B		Submedian	4,5
C		Submedian	6–12,X
D		Subterminal (acrocentric)	13–15
E		Median/submedian	16–18
F		Median	19,20
G		Subterminal (acrocentric)	22,22,Y

FIG. 1.1. Classification of human chromosomes.

Mitosis

In somatic cell division the longitudinal splitting of each chromosome yields two chromatids each of which contains an identical sequence of DNA which was replicated before division commenced. Mitosis begins when chromosomes migrate to the equator of the nucleus and a spindle of fibres develops with attachments at opposite nuclear poles. The spindle fibres fan outwards to meet the separating chromatids of each longitudinally dividing chromosome. Then separation of chromatids occurs—each one traversing pole-wards so that two new sets of 46 chromosomes assemble at opposite ends of the old nucleus ready for incorporation into each of two daughter cells as the also-dividing nuclear

TABLE 1.1. Symbolic designation of human karyotypes system adopted by Chicago Conference (1966)

Principles of karyotype recording
Write in sequence:
(1) Total number of chromosomes
(2) Sex chromosomes, qualified with symbols if necessary
(3) Autosomes only if qualification required (use alphabetical grouping of autosomes, A–G)

Symbols
+ An additional chromosome present
− Absence of a chromosome or part of a chromosome
p Short (petit) arm of chromosome
q Long arm of chromosome
t Translocation of chromosome
i Isochromosome
mar Marker chromosome
r Ring chromosome
dic Dicentric
h Secondary constriction
inv Inversion (pericentric)

Examples

46,XY	46 chromosomes, sex chromosomes XY (normal male).
45,X	45 chromosomes, only one X-chromosome (formerly written XO for karyotype in Turner's syndrome).
47,XYY	47 chromosomes, additional Y in sex chromosomes.
47,XY, G+	47 chromosomes, sex chromosomes XY, extra chromosome in G group (male mongol).
45,XX, D−, G−, t (Dq Gq)+	45 chromosomes, sex chromosomes XX, one chromosome absent from D group and one from the G group; an additional translocation chromosome is present and is formed from the long arm of a D-chromosome and the long arm of a G-chromosome.
45,X/46,XX/47,XXX	A mosaic individual with three cell lines containing respectively, one, two and three X-chromosomes.
46,XXpi	46 chromosomes, one normal X-chromosome, the other an isochromosome of the short arm of an X.

membrane circumvents each set. The net result of mitosis is to provide two exact replicas of the parent cell, each daughter cell having a complete (diploid) set of 23 chromosome pairs which is identical to the parent set.

Meiosis

The specialized mechanism for gamete formation involves two cell divisions for one division of the chromosomes. It takes place in two stages. During the first stage each of the 46 chromosomes in a spermatocyte or oocyte assemble in side-to-side homologous pairs excepting the X- and Y-chromosomes in the male which merely become adjacent end-to-end (Sasaki & Makino 1965). Each member of a pair duplicates itself into two chromatids (a bivalent). An exchange of material then takes place between a chromatid of each pair. Interchange of DNA is effected via crosslinking sequences of chromosomes, known as chiasmata. Each chiasma can be visualized in meiotic preparations. No interchange has been observed to take place between the adjacent ends of an X- and Y-chromosome in meiotic preparations from male testes (Sasaki & Makino 1965). This crossing over of DNA segments between homologous chromatids results in new combinations of genes, the process known as genetic recombination or re-assortment. Subsequently there is a cell division which yields two daughter cells each with an unpaired or half (haploid) set of 23 chromosomes.

The second stage of meiosis involves division of each cell produced at the end of the first stage. It must be recalled that meiosis in an oocyte yields one ovum and three polar bodies. Fusion of two gametic nuclei after fertilization restores the normal diploid set in the zygote and unless the latter is one of two identical twins the meiotic process results in a uniquely coded individual. This follows because, excepting monozygotic twin pairs, no two persons are ever genetically identical.

Preparations of spermatocytes in meiosis can readily be obtained from biopsied testicular material (Fig. 1.2).

Meiosis in spermatogonia is a continuing process in post-pubertal males and the full cycle of spermatogenesis occupies about 2 months. By contrast, oocytes have completed the early part of the first meiotic division during the latter stages of intra-uterine life. Then the oocytes remain in a state of suspension (the dictyate stage). This persists until the cells are activated to complete their maturation by luteinizing hormone. Meiosis resumes and the first stage ends with extrusion of the first polar body. The second meiotic division then begins only to halt at mid-stage while the egg is ovulated. Fertilization provides the stimulus for completion of division and extrusion of the second polar body. The meiotic process can be effected *in vitro* on oocytes artificially liberated from Graafian follicles (Edwards 1965, Edwards, Bavister & Steptoe 1969) or by inducing ovulation by administration of luteinizing hormone to women about to undergo laparotomy (Jagiello, Karnicki & Ryan 1968).

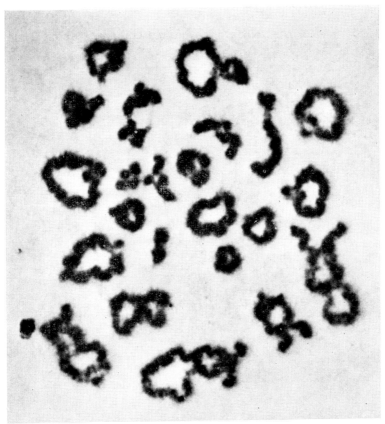

FIG. 1.2. Meiotic chromosomes in the normal male. Elongated figure at top right consists of end-to-end alignment of X and Y chromosomes. Preparation by P.L.Pearson.

CHROMOSOMAL BASIS OF SEX

The fundamental genetic difference between sexes resides in the sex chromosome dimorphism. The X- and Y-chromosomes in mammals represent the culmination of an evolutionary refinement which accumulated genes controlling the heterogametic sex into one chromosome, the

Y. Ohno (1967) has marshalled an impressive array of evidence to demonstrate the evolution of mammalian sex chromosomes from their origins as undifferentiated homologous autosomes in lower animals. Yet the actual sex-determining mechanism remains an enigma. It appears essentially a question of genetic manipulation of steroid hormone pathways based on a decision to differentiate the gonadal ridge one way or other before the human embryo is about 2 months old.

The fundamental chromosomal contribution to sex determination in man is that normally the presence of two X-chromosomes leads to female development whereas an X- and a Y-chromosome result in a male. The Y-chromosome is a powerful male determinant whose presence is required for testicular development. Its strength in this regard is demonstrated by the maleness of 47,XXY, 48,XXXY and 49,XXXXY individuals, in all of which the effects of more than one X-chromosome do not overcome the male determining effects of a single Y-chromosome.

Sex-chromosome dimorphism in man is readily detected by demonstrating the presence or absence of a sex-chromatin (or Barr) body in cell nuclei between mitotic divisions. When cells from normal 46,XX females are examined, 25–60% of nuclei will show a sex-chromatin body as a deeply staining semi-circular mass about 1 μm in diameter situated against the nuclear membrane. By contrast, cells from normal 46,XY males contain no sex-chromatin bodies. Any convenient preparation of cells can be examined but those obtained by scraping the buccal surface of the cheek are generally used for routine purposes. Sex-chromatin analysis of foetal cells desquamated into amniotic fluid permits the antenatal diagnosis of sex (Fig. 1.3).

The sex-chromatin body is an X-chromosome lying tightly condensed (heteropycnotic) and deeply stained with basic dyes (heterochromatic). If there are more than two X-chromosomes in a nucleus all except one will be rendered into a sex-chromatin body. This gives the general rule that there is one less sex-chromatin body in the nucleus than the number of X-chromosomes. Accordingly sex-chromatin analysis is a useful method for screening sex-chromosome anomalies in man.

Structural abnormalities of the X-chromosome may be reflected in the size of the sex-chromatin body which is larger in the case of an isochromosome of the long arms or smaller if there is a partial deletion. A specialized variety of sex-chromatin is the drumstick, a lobed appendage attached to the nuclei of polymorphonuclear leucocytes in females.

It is now possible to demonstrate directly the presence of Y-chromosomes in resting nuclei (Fig. 1.4). This is achieved by detecting fluorescence over the Y when the cell is stained with a quinacrine compound (Pearson, Bobrow & Vosa 1970). Moreover, Y-carrying spermatozoa

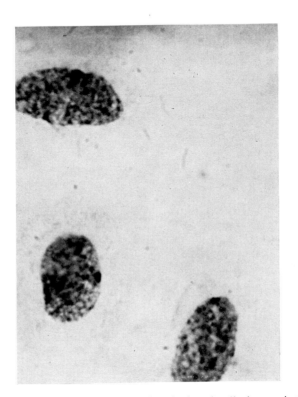

FIG. 1.3. Sex-chromatin body from female foetal cells in amniotic fluid. Obtained during third trimester; foetal sex confirmed at term. Preparation by S.G.Purvis-Smith.

can be identified by the same method (Barlow & Vosa 1970). Identification of a Y-chromosome in this manner (Fig. 1.4) is a powerful adjunct to screening populations for sex-chromosome anomalies because karyotypes with additional Y-chromosomes can be directly detected. Utilizing only sex-chromatin analysis for screening, 47,XYY males will be overlooked because like normal 46,XY males they have no sex-chromatin body.

GENE DOSAGE COMPENSATION; X-CHROMOSOMAL INACTIVATION

The term 'dosage compensation' was introduced in 1932 to describe the phenomenon whereby females with two X-chromosomes and males with one X-chromosome displayed similar phenotypes for normal characteristics determined wholly or in part by X-linked (X-chromosomal) genes.

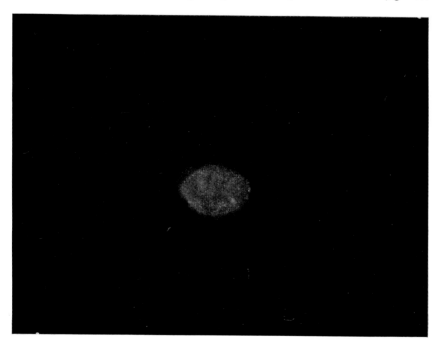

FIG. 1.4. Fluorescent Y-chromosome in a buccal mucosa cell from a normal male. Preparation by P.L.Pearson.

Some mechanism existed which equated the effects of one and two doses of X-linked genes. Various hypotheses were developed to account for a mechanism of gene dosage compensation, largely from work on Drosophila. There were two main theories; the first based on differences in the developmental physiology of males and females so that the two-gene dose in females was rendered functionally similar to the one-gene dose in males (Goldschmidt 1954). The other theory postulated a compensating system of modifying genes which equated the X-linked gene dosage in either sex (Muller 1950). Although both theories explained to some extent observations in Drosophila, there were great difficulties in accommodating findings in mammals. In fact the developmental theory proved virtually untestable and was abandoned; the modifying gene theory foundered on the findings in human X-chromosomal anomalies. Muller (1950) emphasized that the compensating system had to be adjusted so that a double dose of modifying genes would reduce the effect of a compensated gene to half the effect that would be present with a single dose of modifiers. Stern (1960) showed as follows that such an hypothesis would not result in dosage compensation in 47,XXX females. If the effect of a single X-linked gene dose was equal to 4 arbitrary units its compensated effect in individuals with one X-chromosome would be $\frac{1}{2} \times (1 \times 4)$ or 2, in XX individuals it would be $(\frac{1}{2})^2 \times (2 \times 4)$ or 2, but with XXX the effect would be $(\frac{1}{2})^3 \times (3 \times 4)$ or 1·5. This latter finding was difficult to explain on a theory of genetic modifiers.

The advent of modern cytological techniques provided the background for a new theory of gene-dosage compensation. Several investigators concluded independently that the sex-chromatin body in nuclei of female cells was a functionally inert X-chromosome which did not participate in metabolic activities of the cell. In 1961, Mary Lyon incorporated the above observation into a theory to explain actions of X-linked genes in the mouse. She postulated that the genetically inactive X-chromosome in female cells could be derived from one or other parent; in other words, it was a random matter for each cell whether the functioning X-chromosome originated from the father or the mother. Moreover, to account for findings in females heterozygous for an X-linked gene the decision on which X was to be inactivated had to be made early in embryogenesis, at a stage when there were relatively few precursor cells. This theory, which became widely known as the Lyon hypothesis, explained two sets of observations on the mouse. First, mice with only one sex chromosome (XO, comparable to 45,X in man) were normal fertile females—indicating that only one active X-chromosome was required for normal development. Secondly, the theory suggested a plausible mechanism for the mosaic appearance in female mice heterozygous for an X-linked mutation affecting fur colour. The mechanism was that pigment cells descended from stem-cells in which the X-chromosome carrying the mutant gene was inactivated would cause a normally coloured fur patch; conversely when the normal gene was on the inactivated X, the patches would have coloration determined by the mutant gene (Fig. 1.5).

The Lyon hypothesis provided an elegant explanation for dosage compensation in man by postulating that any X-chromosome in excess of one per cell was rendered genetically inert. Thus 46,XX females would be identical to 46,XY males with regard to normal X-linked alleles. When there was a mutant gene on one female X, the resulting effect (phenotype) in a female carrier (heterozygote) would be intermediate between the full effect seen in affected males and the normal effect in

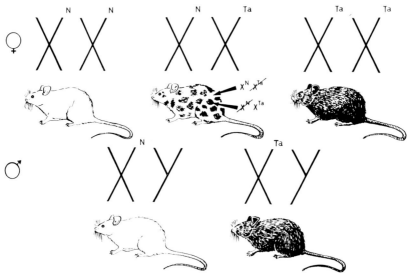

FIG. 1.5. Schematic interpretation of tissue mosaicism in terms of the Lyon (inactive X-chromosome) hypothesis. X-chromosomal loci are denoted by Ta, the X-linked gene Tabby conferring a dark fur colour, and N, its normal allele.

normal males. Direct evidence of a mosaic cell population has been obtained for several X-linked mutations in man—notably in women heterozygous for a glucose-6-phosphate dehydrogenase variant where enzyme-deficient and enzyme-normal cells can be demonstrated in cultured fibroblasts from a female carrier (Davidson, Nitowsky & Childs 1963).

One problem with the Lyon hypothesis has been to explain phenotypes of human sex-chromosome anomalies. If, as the theory predicts, only one X-chromosome is active in each cell then the 45,X female with Turner's syndrome should be equivalent to a normal 46,XX female.

Likewise, a 47,XXY male with Klinefelter's syndrome should not differ on theoretical grounds from a normal 46,XY male. These variant observations will be considered later during discussion of phenotype in disorders of sex chromosomes.

The evidence on which the Lyon hypothesis was originally based has been greatly strengthened by subsequent work. Autoradiographic studies of human chromosomes support the assumption that only one X-chromosome in a female cell participates along with autosomes in biosynthetic activity during the interphase stage of the somatic cell cycle. Randomness of X-chromosomal inactivation has been directly demonstrated for one hybrid animal, the female mule. Because horse and donkey X-chromosomes are morphologically distinct it was possible to show that individual cells from a female mule contained one or other parental X-chromosome (Mukerjee & Sinha 1963). Such randomization of inactivation is, of course, implicit in the finding of a mosaic cell population in female heterozygotes. If this were not the case, phenotypes in heterozygous females would segregate into two distinct classes representing each parental phenotype. An exception to random inactivation is when there is a structural alteration of an X-chromosome. In man (but not always in the mouse) a structurally abnormal X-chromosome undergoes inactivation to form a sex-chromatin body.

The final theoretical postulate—that inactivation takes place early during embryogenesis—has been difficult to determine due to technical problems with identifying sex-chromatin in early foetuses. However, Park (1957) concluded that the earliest appearance of a sex-chromatin body in human females was on the sixteenth day of gestation although he suggested that the time for detection varied in different embryonic tissues.

CHROMOSOMAL ABNORMALITIES; FREQUENCY AND CAUSES

Chromosome abnormalities usually arise from some disturbance of cell division. The exceptions are when there is direct structural damage due to some external agent such as ionizing radiation or a chemical compound.

Harmful effects result from a deficiency or excess of chromosomal material in all or part of an individual's cells. Generally the consequences are more serious with autosomal anomalies; trisomy (triplication of a chromosome) of a small autosome, 47,G +, yields the grossly abnormal person with mongolism but the comparable X-chromosome trisomy, 47,XXX is quite compatible with normal fertile womanhood. Conversely, monosomy of the X-chromosome, 45,X, results in the short infertile female with Turner's syndrome, whereas with few not absolutely certain exceptions (e.g. Al-Aish, De la Cruz *et al* 1967) monosomy for an autosome appears to be lethal.

It has been estimated that at least 4% of all conceptuses carry an abnormal chromosome complement that will cause the majority (over 80%) to be eliminated before birth (Polani, 1970). Approximately 1 in 150 liveborn children carry a chromosomal anomaly that will do them some degree of harm. About the same proportion again will have a visible variant of the normal karyotype yet suffer no obvious ill effects. It can be assumed in such people that despite an unusual arrangement of chromosomes the genetic code is distributed in correct sequences for normal coding function.

Harmful chromosomal anomalies are divided about equally between errors of autosomes and sex chromsomes. However, the latter have been more adequately surveyed in numerous sex-chromatin studies of different populations. Some frequencies for specific chromosomal syndromes are given in Table 1.2.

Although quite a lot is known about the mechanism of abnormal chromosome formation relatively little is understood about what actually causes the disordered mechanisms in man. Nevertheless a number of probable or possible causes can be considered in turn.

Parental age

Penrose (1939) demonstrated the significance of advanced maternal age in the genesis of mongolism, and at the same time eliminated the father's age and sibship size as irrelevant. Subsequently a similar maternal age effect was found for other forms of autosomal trisomy, 47,XXY males and 47,XXX females but not for 45,X females (Court-Brown & Smith 1969).

The increased risk to offspring of mothers past the age of 30 has initiated much research into factors which may disrupt meiosis in an ageing ovum. Circumstantial supporting evidence has been noted in animal ova. Austin (1967) found evidence of chromosomal degeneration in over-ripe rabbit ova. Studies on mouse oocytes indicated more abnormalities of bivalent association during the first stage of meiosis in the eggs of older animals (Edwards, Bavistock & Steptoe 1970). If this holds for man then one would expect the aged mother to have a higher incidence of trisomies. As yet there are no comparable data on human oocytes.

A theory that delayed fertilization associated with decreased frequency of coitus due to advancing age was put forward by German (1968) to account for the chromosomal error in mongolism. However, this theory has not been supported. It has proved impossible to construct any model

TABLE 1.2. Approximate frequencies of some chromosomal abnormalities at birth

47,G+	1·6 per 1000 liveborn[1]
47,D+ 47,E+	0·14 per 1000 liveborn[1]
47,XXY	1·4 per 1000 males[1]
47,XYY	1·8 per 1000 males[2]
45,X	0·5 per 1000 females[1]
47,XXX	0·8 per 1000 females[3]

[1] Polani (1970).
[2] Ratcliffe, Stewart, Melville, Jacobs & Keay (1970).
[3] Court-Brown & Smith (1969).

based on average frequency of coitus at various ages and known survival rates of sperm and ova that fits the observed age-dependent risk for mongolism with its near-exponential increase after the maternal age of 30 (Matsunaga & Murayama 1969). Neither was there found any increased detriment among the offspring of illegitimate unions for whom delayed fertilization associated with sporadic coitus could reasonably be assumed (Fabia 1969). Moreover, Shaver & Carr (1967) could not demonstrate in the rabbit any rise in chromosomally abnormal embryos after fertilization had been delayed.

Radiation

The findings in man are equivocal but some careful studies suggest that irradiation of women should be regarded with suspicion. In an extensive retrospective study of radiation histories in mothers, Sigler, Lilienfeld *et al* (1965) detected a slightly increased (P <0·05) exposure in the mother of mongols thus reversing the negative findings of Schull & Neel (1962) in an earlier and comparable study. Uchida, Holunga & Lawler (1968) undertook a prospective analysis and concluded that abdominal radiology increased the risk of meiotic error especially in older women.

Drugs and chemicals

Many chemicals including a large number of pharmaceutical agents are known to damage chromosomes. Shaw (1970) lists 200 compounds which cause known damage to *in vitro* chromosome preparations or which are strongly suspected of adverse cytological effects. Much attention has centred on the psychotropic drugs, notably lysergic acid diethylamide (LSD). Infants born to mothers who took LSD during pregnancy have had evidence of persisting damage in the form of chromosome breaks—so far with no proven relationship to any harmful effect. Cohen & Mukherjee (1968) found meiotic changes in the gonads of LSD-fed mice. This gives rise to the possibility of complex chromosome arrangements which may cause no obvious detriment to offspring but which in a subsequent meiosis would undergo further rearrangement ending with an unbalanced karyotype and harmful effects. But the evidence for LSD-induced meiotic damage in man remains conflicting and there are methodological factors related to tissue culture techniques which render difficult any interpretation based on *in vitro* experiments (Dorrance, Janiger & Teplitz 1970).

Viruses

Again, there are conflicting findings on the direct effects of viruses and chromosomes. During the viraemic stages of certain infections, increased chromosome breakage has been observed by some workers and not by others (Harnden 1964). McDougall (1970) found that one of the adenoviruses (type 12) when added to cell cultures appeared to cause damage localized to E-group chromosomes. By contrast, other adenoviruses had a random effect throughout the entire karyotype.

A viral influence on female meiosis was suspected by Collmann & Stoller (1962) who found in Victoria an increased birth-rate of mongols about 9–10 months after an epidemic of infectious hepatitis. This epidemiological approach suggested that the hepatitis virus in some way influenced the dictyate oocytes so that subsequent meiosis yielded chromosomally abnormal ova. However, several subsequent studies have failed to confirm any definite relationship between hepatitis epidemics and the later birth of mongols.

Genes

There are three rare genetic disorders with autosomal recessive inheritance in which chromosomes cultured from an affected person's lymphocytes show an increased frequency of structural abnormality. These conditions are: Fanconi's anaemia, Bloom's disease and ataxia-telangiectasia. Apart from possessing excessive chromosome-breakage these disorders carry an increased risk of malignancy. As to any genetic predisposition to meiotic disturbance—no real evidence is available. Kwiterovich, Cross & McKusick (1966) studied mongolism in an inbred religious isolate and concluded that no genetic factor was responsible for meiotic errors which resulted in visible detriment.

Autoimmunity

It has been suggested that autoimmune disturbance in some persons may cause them to produce chromosomally unbalanced offspring. In sex-chromosomal conditions a significant association has been found between the possession of thyroid autoantibodies and various forms of ovarian dysgenesis and an anomalous karyotype (Doniach, Roitt & Polani 1968). Fialkow (1966) discovered an increased frequency of thyroid autoantibodies among the parents of mongols. Such findings suggest a relationship between production of autoantibodies and chromosomally abnormal children—the latter tending to inherit their parents' immunological variant. However, any actual mechanism for such an association remains unknown.

NUMERICAL ABNORMALITIES OF CHROMOSOMES

Any abnormal number of chromosomes in a cell which is not a multiple of the normal haploid number of 23 is known as aneuploidy. Where there exists some multiple of 23 other than a diploid set of 46, the karyotype is polyploid. The latter occurs sometimes in tumour cells presumably by a process of nuclear division without simultaneous division of the cytoplasm. So one can find cells in malignant tissue with, say, tetraploidy (92 chromosomes) or even octoploidy (184 chromosomes). Polyploidy, usually in the form of triploidy (69 chromosomes), is occasionally found in human abortuses; a few triploids with co-existent normal cell-lines have survived after birth as retarded and malformed individuals. Mechanisms for triploid formation are uncertain but probable explanations include digyny where an ovum has become diploid due to a meiotic error before fertilization or alternatively, diandry where a normal haploid ovum has been fertilized by two spermatozoa.

Aneuploidy usually results from non-disjunction (non-separation) of a chromosome or chromatid pair during cell division so that an unequal number of chromosomes enters the daughter nuclei. This happens most frequently during the second stage of meiosis. If a pair of chromosomes do not part company then, instead of each daughter cell receiving a total of 23 chromosomes, there will be 24 in one cell and 22 in the other. When the gamete carrying 24 chromosomes combines with a normal haploid gamete at fertilization the resulting zygote will have 47 chromosomes. Moreover one chromosome will be represented three times in the zygote, that is, trisomy for that chromosome. Such is the most common origin for the trisomy-21 karyotype in mongolism, now written 47,G+.

In the converse situation where a gamete has lost a chromosome through non-disjunction, the zygote will have a total of 45 chromosomes. In practice this occurs only with sex chromosomes yielding the classical karyotype for Turner's syndrome 45,X (formerly XO). A comparable lack of one chromosome (monosomy) involving autosomes is generally regarded as incompatible with life although there has been at least one recorded exception, 45,G− in a retarded girl (Al-Aish *et al* 1967).

Non-disjunction during oogenesis can result in zygotes with either 47,XXY; 45,X; 47,XXX or 45,Y. The latter has never been observed in man; presumably it is inviable like its extensively studied counterpart in Drosophila. Non-disjunction during spermatogenesis can result in zygotes with either 47,XXY; 45,X; 47,XXX or 47,XYY. Because of the two divisions of meiosis it is possible for more complicated numerical consequences. Such aberrations could account for individuals with two or three additional sex chromosomes in their cells. The exploitation of X-linked marker genes has helped to throw light on the origins of these complex karyotypes. This principle is illustrated in Fig. 1.6 where the X-linked Xg blood group system was tested in a family containing a 49,XXXXY male. In this instance the blood groups did not segregate in an informative manner but as an example of how interpretations of

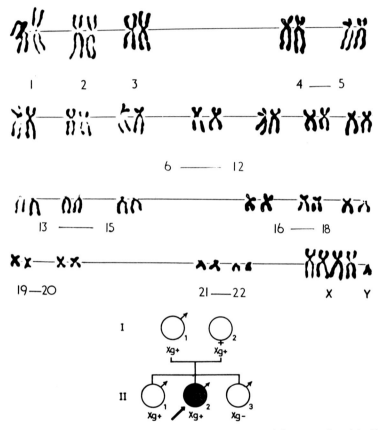

FIG. 1.6. Use of Xg blood groups on a family containing a male with 49, XXXXY karyotope. Phenotypes Xg(a+) and Xg(a−) are marked on the pedigree in abbreviated form.

Xg phenotypes are used to elucidate the origins of aneuploidy, consider the following situation: A 49,XXXXY male who is Xg (a−) has a father Xg (a+) and mother Xg (a−). The affected male cannot have received the paternal X-chromosome because it contains the allele Xg^a which would make his red cells react with anti-Xg^a serum to give an Xg (a+) phenotype.

The patient's Xg (a −) phenotype indicate that all four X-chromosomes must possess the silent allele *Xg*. Thus the patient's X-chromosomes are maternal in origin—a conclusion which fits with the mother's Xg (a −) phenotype reflecting the presence of an *Xg* allele on each of her X-chromosomes. Because a 49,XXXXY karyotype cannot be obtained by a single non-disjunctional event the conclusion is that in this particular instance the defect originated from non-disjunction at both the first and second meiotic divisions of oogenesis.

The Xg blood grouping of families containing a person with abnormal sex chromosomes has been undertaken mainly by Race & Sanger (1969). Analysis of family units containing 47,XXY males indicates that in about 40% of cases the additional X is paternal in origin, thus implying a defect in the first meiotic division of spermatogenesis to yield an XY-bearing sperm. With 45,X females the causal meiotic mishap occurs more frequently in male germ cells resulting in a gamete lacking an X- or Y-chromosome. This follows from deducing the presence of a maternal X in 76% of 45,X females and a paternal X in 24%. Additional information is available on some of the more unusual X-chromosomal anomalies (Race & Sanger 1969).

Non-disjunction during meiosis results in a single aberrant cell-line in the resulting zygote. But if chromosomes fail to separate at an early cleavage division of the embryo then more than one cell-line can occur. This is chromosomal mosaicism and persons with two or more karyotypes in their cells are known as mosaics. Usually only two cell-lines are detected as in the 46,XX/45,X Turner's syndrome mosaic. Occasionally both aberrant lines survive in the zygote from a mitotic non-disjunction and co-exist with normal cells, e.g. 46,XX/47,XXX/45,X. Almost all individuals who possess a structurally abnormal X-chromosome are mosaics with an additional 45,X or 46,XX cell-line.

The maintenance of mosaicism in an embryo depends on the viability of blastomeres with abnormal chromosomes. It is theoretically possible for certain numerical anomalies to arise post-zygotically and persist as a single cell-line; for instance non-disjunction of sex chromosomes in a 46,XY zygote at the first cleavage division would yield one blastomere with XXYY sex chromosomes and the other with none. The latter blastomere would perish, leaving an embryo that could develop into a 48,XXYY individual. Alternatively, two cell-lines may develop but one may be favoured to successively outgrow the other. Taylor (1968) followed over 4 years the relative proportions of cells in eight children who were normal/47,G+ mosaics. In four children the proportion of normal cells increased progressively, in two there were fluctuations but no apparent trend and in the remaining two instances there was, surprisingly, an increase in the proportion of abnormal cells.

The above and similar findings suggest that certain individuals with a single abnormal cell-line originating from a meiotic error might be converted into mosaics through post-zygotic non-disjunction and selective proliferation of the 'new' cell-line. Some indirect evidence to support this notion comes from the Xg blood grouping of 46,XY/47,XYY mosaics. If such individuals arose from a mitotic aberration at an early cleavage division of a 46,XY zygote one would expect them to have the normal male distribution of Xg blood groups. But they show a highly significant deviation from the expected male distribution (Race & Sanger 1969). Thus one is forced to conclude that these individuals originated as 47,XYY due to meiotic error and subsequently were rendered mosaic by a mitotic non-disjunction yielding a line which had lost a Y and so reverted to normal 46,XY.

Another mechanism for mosaicism is a disturbed segregation of chromosomes into daughter cells due to the lagging behind and subsequent elimination of one chromosome in the later stages (anaphase) of mitosis. This anaphase lagging is a well-known cause of aneuploidy in plants and its occurrence in man is supported by a few exceptional instances where monozygotic (identical) twins differ in karyotype. Where, for example, one twin is a 45,X female and the other a 46,XY male (Turpin, Lejeune *et al* 1961) the most plausible explanation is that the first cleavage division of a 46,XY zygote gives one blastomere which develops as a normal male but due to lagging of a Y-chromosome on the mitotic spindle the second blastomere lacks one sex chromosome and thus yields a 45,X female.

A specialized form of mosaicism is chimaerism. A chimaera is usually defined as an organism whose cells are derived from two or more zygotes whereas the conventional mosaic has cells derived from only one individual zygote. In other words the chimaera differs by containing chromosomes and hence, genes from more than one zygotic source. Ford (1969) extends the definition of chimaerism to include the products of dispermic fertilization and reviews two examples found in men who were 46,XX/46,XY chimaeras with mosaic blood groups and intersexual features. In one instance there was evidence that sperm from two different males were involved in fertilization. Human chimaeras formed from two independent zygotes have arisen from placental cross-circulation between dizygotic twins, maternal-foetal transplacental exchange or artificially via blood transfusion or bone marrow grafting (Ford 1969).

Aneuploidy is a consistent feature of neoplasia. The variation in chromosome numbers ranges from a simple loss or excess readily explainable by mitotic non-disjunction, through polyploidy to highly irregular counts resulting from sequential mitosis of nuclei with abnormal chromosome complements. Jones, Woodruff *et al* (1970) describe the progressive changes in chromosome number (and structure) that accompany histopathological progression of cervical cancer.

STRUCTURAL ABNORMALITIES OF CHROMOSOMES

Although structurally abnormal chromosomes can result from exposure to ionizing radiations or various drugs and exist as a feature of neoplasia, most instances arise without apparent reason during meiosis. In each case the origin of a structural variant can be explained by an incorrect re-joining of broken chromosomal sequences during chiasma formation at the first meiotic division.

There are five main groups of structural abnormality known or suspected to cause harmful effects in man.

Translocation

This involves the transfer of a segment from one chromosome to a site on a different chromosome. There is usually a mutual or reciprocal exchange which results in two composite figures each containing part of the other. The minor product of the exchange may be lost from the cell because it cannot survive subsequent divisions. Even if this loss occurs there may be no harmful consequences because the major product contains all the genetic information essential for coding purposes. This state is known as balanced translocation.

Clinically normal people who carry a balanced translocation risk having children with an unbalanced karyotype. For instance a female may carry a translocation consisting of the long arms of a middle-sized D-chromosome and a small G-chromosome. She has only 45 chromosomes, lacking a normal D and G, each of which is replaced by the translocated chromosome. Her karyotype is written 45,XX, D−, G−, t(Dq Gq)+. At meiosis the composite chromosome will pair with the remaining normal D-chromosome. There are four possible types of gamete yielding at fertilization four possible types of zygote: (i) 46,D−, t(Dq Gq)+ with mongolism, because having received a G-chromosome from each parent, the additional G material in the translocated chromosome renders the zygote trisomic for G; (ii) 45,D−, G−, t(Dq Gq)+, a balanced translocation as in the mother; (iii) a normal child with 46 chromosomes for whom the mother supplied an ovum with a D- and a G-chromosome; and (iv) 45,G− which is practically certain to be non-viable. Although each type of zygote does not necessarily occur with the same frequency, risks of a mongol child being born to a balanced translocation-carrying female are about 1 in 6. If the father carries a translocation the risk of an abnormal child is much less—at most 1 in 20 presumably because a sperm bearing the composite chromosome has a reduced chance of effecting fertilization.

There are many varieties of translocation involving the autosomes (but not the sex chromosomes). In fact, Court-Brown & Smith (1969) estimate that about 0·5% of the population carry a balanced chromosome translocation.

Deletion

This refers to a visible loss of chromosome substance where the terminal part of an arm has been deleted or alternatively a segment within the arm has been lost. A well-known example is deletion of the short arm of a B-chromosome (Bp−) associated with the *cri-du-chat* syndrome, so called on account of the plaintive, mewing cry of affected children. Another example is the marker chromosome found in the blood and marrow cells of persons with chronic myeloid leukaemia. This, the Philadelphia chromosome, is a G-chromosome with its long arm deleted (Gq−). The Philadelphia chromosome is the most consistent of a wide variety of structurally abnormal chromosomes found in neoplastic tissue.

Duplication

The most definite example of gene duplication in man is an iso-chromosome. Each arm of an iso-chromosome has an identical sequence of genes and arises from a transverse instead of longitudinal splitting of the centromere during cell division. Thus in the case of the X-chromosome there can be either an iso-chromosome of the short arm, 46,XXpi, or of the long arm, 46,XXqi.

Ring chromosome and other abnormal configurations

A ring chromosome results when there are two breaks in a chromosome, each one near an end. The terminal fragments are lost and the two ends of the middle portion join together producing a circular chromosome. Other structural abnormalities which can be reproduced by mitosis (i.e. are stable and not immediately lost from the cell) include fragments of chromosomes which sometimes do not have a centromere (acentric) or alternatively larger figures which have two centromeres (dicentric). These latter configurations are encountered most frequently after breaks in chromosomes have been induced by agents such as radiations, certain drugs or viral infections.

Inversion

This refers to an inverted sequence of DNA so that the code lies in the reverse of its original order. A few instances of inversion have been strongly suspected in man where there appears to have been a shift of chromosome segment to the opposite arm in the region of the centromere (a pericentric inversion) and which is therefore detectable by a disturbance in arm-ratio. A good case has been made for a pericentric inversion involving the Y-chromosome (Jacobs 1969).

PHENOTYPIC CORRELATIONS WITH CYTOGENETIC VARIATION

A tremendous range of human variation, much of it harmful, has been described in association with deviant karyotypes (Turpin & Lejeune 1969). Some general principles have been evolved from these observations.

Abortion

Between 20–25% of first trimester abortions have a detectable abnormal karyotype (Benirschke *et al* 1966). The percentage may in fact be greater because of the difficulty in culturing tissue from early abortuses (Larson & Titus 1970). Trisomies are the most frequent anomaly (40% of the total) followed by monosomies (20%)—of which almost all are 45,X. The remainder are polyploids (20%), triploidy most commonly and then tetraploidy, mosaics (7%), translocations and some unusual configurations. For different categories of chromosomal aberration it is estimated that of all foetuses with 45,X less than 10% survive to be liveborn; for trisomic foetuses the figure is 20% and for polyploids, less than 1% (Edwards 1966). There is little doubt that intra-uterine inviability is a powerful method of natural selection whereby the genetically disadvantaged are eliminated. Although information is scanty on development of chromosomally abnormal foetuses, not all have gross evidence of physical malformation and so more subtle consequences of genetic imbalance presumably take their toll.

Another finding from studies on chromosomes of abortuses is that earlier assumptions were erroneous in concluding that there was an excess of males in the primary sex ratio. It seems that there were misinterpretations of anatomical studies based on identifying the foetal phallus (which is relatively well developed in the female). In addition, sex-chromatin bodies are difficult to identify in foetuses so that males were most probably overestimated. From karyotype studies, the primary sex ratio is approximately 1·0 (Stevenson & Bobrow 1967).

Autosomal monosomy has very rarely been detected among karyotyped material. The loss of one autosome may operate at one of several stages; the abnormal gamete may not survive, the zygote may die before implantation, early embryonic rejection may occur—or it may be that a monosomic cell-line fails to grow in tissue culture.

Autosomal anomalies

Numerical abnormalities of autosomes include three well-defined syndromes: mongolism, trisomy-13 (a D-group chromosome) and trisomy-18 (an E-group chromosome) endowed also with eponyms, respectively, of Down, Patau and Edwards. The possession of an extra chromosome in each case is accompanied by gross physical and mental maldevelopment. As the size of the extra chromosome increases so does the magnitude of adverse effects. With trisomy-13 or -18 about 75% of affected infants are dead within 6 months; the prognosis in mongolism is somewhat better so that currently persons with trisomy-21 constitute the largest single diagnostic category of those requiring institutional care for mental retardation.

Each syndrome is characterized by a constellation of clinical features (summarized for trisomy-13 and -18 by Polani 1969 and for mongolism by Penrose & Smith 1966). Gross disturbances of development indicate the adverse consequences of mis-coding in the trisomic state; one indication of maldevelopment which may be useful diagnostically is a characteristic pattern of dermatoglyphic abnormalities for each syndrome.

All three syndromes may be associated with structural anomalies of autosomes due to interchanges of the translocation variety. In each case the affected individual has an excess chromosome material which renders him effectively trisomic for the autosome in question.

A number of autosomal deletions have syndromal associations (Polani 1969). All are characterized by markedly retarded development and frequent physical malformation although in the *cri-du-chat* syndrome (Bp−) the defects may be relatively mild. Deletions of an E-group chromosome either of the long arm (Eq−) or the short arm (Ep−) generally have gross anatomical lesions and a poor survival. A deficiency of a G-group autosome results in a syndrome sometimes known as anti-mongolism because of an anti-mongoloid slant of the palpebral fissure and hypertonus instead of the hypotonicity of mongols.

Deletions appear almost exclusively to involve specific arms of relatively few chromosomes. This may indicate some peculiarity of meiotic breakage at such sites and also raises the possibility that some sequences of DNA may be lost without conferring inviability whereas losses from other chromosomes are lethal during early foetal development and so remain undetected. Autosomal deletions have considerable fundamental interest as one approach to mapping the genetic constitution of man. The principle is to detect anomalous inheritance of a genetic character by testing family units containing an individual with a chromosomal deficiency. Absence of a homologous chromosome segment will alter expression of the unpaired genetic locus. So far a large amount of negative information has been collected on loci that cannot be assigned to sites which have been deleted in investigated persons.

Sex-chromosomal anomalies

By contrast to autosomal anomalies, studies of abnormal sex chromosomes have provided somewhat more precise information on the correlations between karyotype and phenotype. This is because sex chromosomes can be readily identified and their imbalance often leads to less disastrous effects on the individual. Specific areas of interest are sexual development, stature and physical development and mental development.

Sexual development. Much has been learnt about the role of X-chromosomes in ovarian development from studying females with a 45,X sex-chromosome complement or some variant of it. A 45,X karyotype is associated with Turner's syndrome in which females have short stature, infantile sexual development, streak gonads and some congenital malformations. The latter, which can collectively be termed 'Turner stigmata' include webbing of skinfolds at the neck, a scaphoid chest, increased carrying angles at the elbow joints, peripheral lymphoedema at birth, shortened fourth metacarpals and a high frequency of coarctation of the aorta.

It was against this classical phenotype of Turner's syndrome that comparisons were subsequently made with other phenotypes associated with different varieties of sex-chromosome anomalies including mosaicism and structurally altered X-chromosomes. Such comparisons revealed that two intact X-chromosomes are required for normal ovarian development. If genes are deficient on either the short or long arms of the X-chromosome, as in females with deletions 46,XXp − or 46,XXq −, then functioning ovaries do not develop and are replaced by streak gonads.

Hamerton (1968) pointed out that although 45,X girls were almost invariably sterile and possessed only stromal remnants of ovaries, 45,X foetuses during early gestation have relatively normal gonads containing germ cells in similar numbers to 46,XX foetuses. Follicles have even been observed in the ovaries of 45,X infants. Thus the presence of only one X-chromosome does not prevent formation and migration of primordial ridge cells into the germinal ridge nor subsequent organization into a functional ovary. But the follicular atresia and stromal replacement which follows indicate that a second X-chromosome is needed to avert a breakdown in ovarian development.

Females with a 47,XXX complement are as a group normally fertile and, as far as is known, the possession of an additional X-chromosome has no detectable effect on their sexual development.

The presence of a Y-chromosome results in the development of a male phenotype consequent to testis formation. An exception is the intersexual syndrome of testicular feminization where 46,XY individuals have an external female phenotype despite the possession of intra-abdominal or inguinal testes. This syndrome is inherited in a manner which is either autosomal dominant with only males affected or alternatively X-linked. Target-cell unresponsiveness to testosterone has long been suspected in testicular feminization; a probable enzymatic defect has been located by Northcutt, Island & Liddle (1969) which prevents the conversion of testosterone to its active intracellular form, dihydrotestosterone.

Although relatively few persons have been discovered with a structural anomaly of the Y-chromosome the evidence so far suggests that genes responsible for testicular development are situated on the short arm of the Y-chromosome (Jacobs 1969). Males with very small Y-chromosomes due to a deletion of the long arms (46,XYq−) have had demonstrably normal testicular tissue.

The presence of an additional Y-chromosome in 47,XYY males does not affect their sexual development or fertility. Many have produced offspring; with one exception (Sundequist & Hellstrom 1969) all were chromosomally normal due apparently to a meiotic mechanism in primary spermatocytes whereby the additional Y-chromosome seems to be selectively eliminated (Melnyk, Thompson *et al* 1969).

By contrast, males with additional X-chromosomes are sterile due to testicular hypoplasia. Presumably the sequence of events is that testes development is initiated by genes on the Y-chromosome but subsequent progress is inhibited in some way by an abnormal metabolic influence from excess X-chromosomes.

Stature and physical development. There are strong correlations between the height of individuals and their sex-chromosome complements. Males with a 47,XXY karyotype and Klinefelter's syndrome are taller than average. This is characteristic also of 48,XXXY males and is due to proportionately greater growth of limbs than of the trunk.

An excess of Y-chromosomes confers an overall greater height with proportionate growth. Males with a 47,XYY complement are mostly over 6 feet in height; 47,XYY boys have heights nearly always at or above the 90th percentile for age.

The short stature and Turner stigmata of 45,X females do not occur if only the long arm of one X-chromosome is deficient. Females with ovarian dysgenesis and 46,XXq− or 46,XXpi karyotypes have a mean height not different from normal 46,XX females. On the other hand shortness (commonly less than 5 feet tall) is found when the short arm of one X-chromosome is missing from cells as in 46,XXqi or 46,XXp− females. It is apparent from these observations that normal development of stature and of some physical features malformed in Turner's syndrome, requires the presence of genes on both short arms of two X-chromosomes.

Mental development. Mental subnormality is found among 47,XXY males with Klinefelter's syndrome, 47,XXX females and in individuals of either sex with more than three sex chromosomes. Of the latter, 48,XXXY males, 49,XXXXY males and 48,XXXX females are all severely retarded. As a general rule, the more X-chromosomes a person has, the greater seems to be the intellectual impairment. It must be noted that not all 47,XXY males or 47,XXX females are retarded; many are intellectually normal. In addition there are complicating personality factors in Klinefelter's syndrome with its eunuchoid body proportions and other features of sexual maldevelopment.

It is generally held that 45,X females have normal intelligence. However, Polani (1969b) records a very slight degree of global intellectual impairment in a group of 45,X subjects. But there is certainly no excess of these females in institutions for the mentally subnormal.

Schizophrenia or a schizophrenia-like illness occurs more frequently among 47,XXY males with Klinefelter's syndrome and 47,XXX females than would be suspected by chance (Polani 1969b). Aggression and violence have been established as not infrequently occurring in Klinefelter's syndrome. There is a spectacular association of violent aggressive behaviour with the 47,XYY karyotype in males. These individuals are over-represented in special-security institutions for the dangerously antisocial (Bartlett, Hurley *et al* 1968). Apart from being unusually tall, XYY males may have a high frequency of neurological abnormalities (Daly 1969). As yet there are no data on any cerebral or other disturbance that reveal any objective cause for the behavioural consequences of an additional Y-chromosome.

Production of phenotypic effects by sex-chromosome imbalance. The manner by which sex-chromosome abnormalities produce their effects is not known, although *a priori* it is assumed that an absence or excess of genetic information in cells must disrupt a carefully adjusted and regulated process. There is a problem with reconciling the predictions of the inactive X-chromosome (Lyon) hypothesis with observed phenotypic effects. On theoretical grounds and assuming that only one X-chromosome is genetically active in a female cell, additional or deficient X-chromosomes in a nucleus should have effects identical with the 46,XX situation. But it is clear that additional X-chromosomes are not genetically inert because their presence is manifest although, by comparison to autosomes, not to a necessarily detrimental degree. It is possible that genes on multiple X-chromosomes are available for coding purposes during the brief period before all but one X-chromosome are rendered genetically inactive. Alternatively inactivation may not be complete and some gene loci on additional X-chromosomes participate in the metabolic activities of cells. Theoretical implications are considered further by Hamerton (1968) and Polani (1969b).

Although there are conflicting opinions on the nature of X- and Y-chromosome association in meiosis it is possible that a homologous segment exists on either chromosome. Ferguson-Smith (1966) has proposed this mechanism to account for almost all true hermaphrodites (possessing both ovarian and testicular tissue) having a 46,XX karyotype. He assumes that a Y-chromosome segment has become translocated to the X-chromosome the hermaphrodite has received from its father. Moreover, if the translocated chromosome was involved in X-inactivation then the male-determining function of Y-chromosomal genes would be active in some cells and not in others.

GENETIC MECHANISMS IN MAN

Modern genetics is based on Mendel's original concept of the gene as a distinct particulate unit of inheritance. Quite early on it was recognized that the gene must have certain fundamental properties so as to function specifically in each cell. The gene had to be capable of duplicating itself exactly during cell division so that descendant replicas of functional cells could be produced in their millions during the life of a multicellular organism such as man. And in view of this almost infinite capacity for producing an exact copy of cells with only an occasional error, the gene had to be extremely stable. But because errors did occur and were manifest as some variant of normal gene action there was obviously a need to allow for a sudden change (mutation) which would result in a new altered gene differing in function from the original one but still capable of self-replication in its new form.

The notion that genes were responsible for the synthesis of proteins dates back for more than 50 years (Garrod first proposed the idea in 1908 when describing the first four inborn errors of metabolism). However, studies on the genetic control of metabolic pathways in micro-organisms were necessary to establish the first firm concept of protein synthesis as formulated in the one gene-one enzyme hypothesis. Following the discovery that genes were composed of deoxyribonucleic acid, DNA, the stage was set for Watson and Crick who in 1952 accurately predicted the macromolecular organization of DNA in the form of the now famous double helix. Subsequent study of DNA proved that the molecule fulfilled all criteria necessary for carrying genetic information. Moreover, knowledge of the interactions of DNA with related nucleic acids and enzymes has provided insights into genetics at its fundamental molecular level.

MOLECULAR BASIS OF HEREDITY

Being a nucleic acid, DNA consists of three main constituents: phosphoric acid, a sugar (pentose) and the nitrogenous bases, pyrimidines (cytosine, C and thymine, T) and purines (adenine, A and guanine, G). These constituents are organized first into nucleotides composed of one sugar, one phosphate and one base. Many of these nucleotides polymerize to form a long polynucleotide chain, two of which are twisted around each other as a double-stranded helix. The backbone of each strand consists of a regular alternation of phosphate and sugar groups and to each sugar is attached a nitrogenous base. The bases are bonded in a critical manner to form base-pairs and a pyrimidine always pairs with a purine, i.e. G–C, A–T.

Although it is not yet resolved exactly how DNA is duplicated during each cell-division, the mechanism must provide for separation of the two strands so that one strand acts as a template for synthesis of a new chain for the daughter cell.

The linear order of bases in DNA comprises the genetic code and acts as a blueprint for specifying protein products of a cell. A typical gene which codes for a polypeptide is a sequence of DNA containing 100 or more bases. The essence of the code is that a triplet of three bases in DNA is responsible for inserting one amino acid into a polypeptide chain. Because human proteins are composed of only 20 different amino acids there is excess information in the genetic code for with 4 bases there are 64 possible combinations of triplets. Accordingly different triplets code for the same amino acid and a few are used for 'punctuation'—to begin or terminate synthesis of a polypeptide chain.

Genetic code and polypeptide synthesis

DNA molecules are situated in chromosomes within cell nuclei; polypeptides are synthesized at ribosomes in the cytoplasm. The process whereby information encoded in base sequences of DNA is utilized for protein synthesis involves two mechanisms. First, the information in DNA is copied into complementary messenger ribonucleic acid (mRNA) molecules—the process of *transcription*. Then the message is translated into the amino-acid sequence of a polypeptide chain—the process of *translation*.

RNA is structurally similar to DNA but differs in being single-stranded, possessing a different sugar (ribose) and one different base uracil (U) instead of thymine. A copy of the nucleotide sequence is taken by mRNA from one of the two strands of DNA following the base-pairing rules so that A links with T (or with U in mRNA) and G with C. In this manner the mRNA chain is synthesized along the length of a gene and transcription ends with an exact copy of the genetic code embodied in an mRNA template. Then mRNA migrates from the nucleus to the ribosomes and translation of the code commences.

Another variety of RNA is concerned with translation—transfer RNA (tRNA). Its function is to form an attachment to amino acids which are then assembled sequentially as directed by base triplets (codons) on mRNA. Once again the linked relationships of bases on mRNA and tRNA determines selection of amino acid. In this fashion a polypeptide chain is built up, with the number and order of amino acids ultimately determined by the original message in DNA transcribed onto mRNA, and translated through the agency of tRNA.

Once the polypeptide chain has been formed it is joined with other chains and folded into the appropriate form for a functional protein.

Definition of a gene

The definition most widely used is a functional one—the linear sequence of DNA bases that specify the amino-acid composition of one polypeptide chain. This is a refinement of the classical 'one gene-one enzyme' relationship which became necessary because of the realization that genes code for all proteins not just enzymes and many proteins are composed of several polypeptide chains each coded by different genes. Human adult haemoglobin (HbA) is a good illustration. The molecule has a tetramer protein section (globin) which contains 534 amino acids arranged in dimer pairs, written $\alpha_2 \beta_2$. Each α polypeptide contains 141 amino acids, each β chain, 146. The sequence of amino acids in each pair of chains is controlled by a pair of genes. Each pair of genes is made up of two alleles. An allele or allelic gene is one of two genes, each of which occupies an identical site (or locus) on both members of a chromosome pair. For haemoglobin, each allele of a pair codes for a polypeptide chain and as there are two alleles each with identical DNA base sequences there will be two identical polypeptide chains. If these are α chains then the other pair of allelic genes codes for two identical β chains. The resulting four chains are arranged together with the haem moiety to form a haemoglobin A molecule.

A gene defined in the above manner to code for a polypeptide chain is often known as a cistron. There are other definitions at the molecular level such as the smallest coding unit—a triplet of three bases coding for one amino acid, which is called a codon.

Gene control

Genes which produce mRNA for polypeptide synthesis are known as structural genes. Such genes are present in all body cells but their activity varies enormously; cells in liver, tooth buds and brain tissue have widely differing biological functions. Some controlling system is obviously required to switch on and off structural genes so that proteins are synthesized in required amounts depending on the needs of embryonic differentiation and the subsequent specific functions of specialized cells. A theoretical system of controlling genes has been evolved by Jacob & Monod (1961) to account for a controlling mechanism in unicellular organisms. This, the operon theory of genetic control, has to some extent been validated for bacteria.

Essentially, there are two categories of controlling genes; operator genes and regulator genes. The operator gene is located immediately adjacent to the structural gene and its function is to influence the latter to initiate synthesis of protein. The operator gene is in turn controlled by the regulator gene situated at some distance from the operator and structural gene unit. The regulator gene determines synthesis of a specific product known as the repressor. The latter (now isolated in *E. coli*) combines chemically with the operator gene and blocks its initiating function so that no protein synthesis can be undertaken by the structural gene. In order to get the structural gene working, the mechanism of enzyme induction is brought into play. Induction is the process by which the presence of a specific substrate of an enzyme stimulates the synthesis of that enzyme. Presumably the substrate interacts with the repressor neutralizing the effect of the latter so that it no longer inhibits the operator gene. With the operator gene derepressed, protein synthesis is commenced by the 'switched-on' structural gene.

Although the operon theory explains the regulatory mechanisms controlling gene expression in bacteria there are other factors influencing the vastly more complicated cellular organization in higher animals. For instance, hormones are known to affect protein synthesis possibly through a selective regulation of mRNA synthesis. And then there are the enigmatic histones—highly basic nuclear molecules that can attach themselves to the acidic phosphate groups in the DNA backbone. Despite speculation that histones can act as gene repressors their true function remains unknown.

Gene mutation

There are different types of mutation. The profound disturbances of the genetic code associated with abnormal karyotypes represent one form of mutation. Chromosomal deletions too small to be visualized by current microscopy may be another variety. However, the most common variety of gene mutation involves a mistake in a triplet of bases (codon) that comprises a structural gene. The mistake generally involves the substitution of one base for another and evidence for this is available from human haemoglobinopathies. Most of these can be shown to have one different amino acid at a given position in an α or β chain. Because the coding specificity of base triplets in DNA is known, any alteration in the sequence of amino acid can be related back to the substitution of a single base which has resulted in a different amino acid being inserted into the polypeptide chain.

Structural gene mutations result in structurally different molecules. In many cases these altered molecules function quite normally and are not associated with clinical disease. Under current evolutionary theory it is held that these different varieties of molecules represent normal biochemical variations that have arisen within a species during the long course of adaptation to environmental conditions. Such genetic variation or polymorphism is seen with many serum proteins or enzymes, for instance with the 19 varying forms of transferrin or the 28 varieties of glucose-6-phosphate dehydrogenase listed by Giblett (1969).

Some mutations in man may affect a control gene. It is likely though not yet proven (Weatherall 1965) that thalassaemia is due to a defect in a controlling operon because there is no structural change in the haemoglobin molecule and the condition results from a genetically determined reduced rate of polypeptide chain synthesis.

INHERITANCE OF SINGLE GENES WITH LARGE EFFECTS

Classical Mendelian genetics applied to man is concerned with genetic variations in individuals, families and populations which arise from mutation of a single gene and which have a detectable effect. Most interest centres around the harmful inherited conditions of man which have been catalogued by McKusick (1968). It is customary to consider these conditions under the operational terms of dominant, recessive or X-linked (sex-linked).

Dominant inheritance

A dominant condition is defined as one in which a person shows the effect of a single mutant gene situated at one locus of a chromosome. On the corresponding locus of the other paired chromosome is a normal allele. Such a person is a heterozygote. So the essence of dominant inheritance is that one dose of a mutant gene results in a detectable variation from normal.

Genes responsible for dominant conditions are carried on autosomes. Accordingly persons of either sex may be affected. The usual situation is that the first case in a family results from a fresh mutation and thus appears sporadically. Both parents, any siblings and other relatives are normal. But if the affected person has children it could mark the start of a pedigree as shown in Fig. 1.7 because there is a 50% chance that the chromosome carrying the mutant gene will enter a gamete concerned with fertilization. Thus each offspring has a 1 in 2 risk of becoming affected.

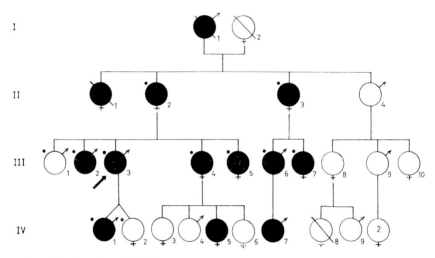

FIG. 1.7. Dominant inheritance. A typical pedigree for a relatively harmless condition, in this case a disorder of tooth formation, dentinogenesis imperfecta.

Dominant conditions are frequently marked by considerable variation between different affected persons. With osteogenesis imperfecta, for instance, the effect (or expressivity) of the abnormal gene can vary from extremely severe, causing intra-uterine fractures, to mild, with a blue coloration of sclerae as the only sign of gene effect. Sometimes, the mutant gene may be undetectable—a situation known as failure of penetrance. Even so, the use of refined tests may detect signs of mutant gene effect in a clinically normal person—as with faecal protoporphyrin in porphyria variegata.

Recessive inheritance

A recessive condition becomes apparent when a person possesses a mutant gene at comparable loci on each chromosome of a pair. Such a person is a homozygote with two doses of the mutant gene. A heterozygote with a normal allele matching the gene mutation is generally quite without obvious genetic effects at a clinical level. However, if the underlying metabolic system is understood, detection of the heterozygous state can usually be made.

The family pattern of recessive inheritance is that, with few exceptions, affected persons are confined to the same sibship. Parents, almost always both heterozygotes, are clinically normal (Fig. 1.8).

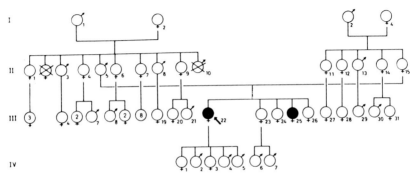

FIG. 1.8. Recessive inheritance. Typically, the affected persons are either sporadic cases or, as illustrated here, confined to one sibship. This pedigree, prepared by Dr R.S.Wells, contains two cases of a rare recessive dermatological condition, ichthyosiform erythrodermia.

Two heterozygous parents can produce three types of zygote: affected homozygotes possessing both abnormal alleles, normal homozygotes containing both normal alleles and heterozygotes like each parent. These are formed with the frequency on average of two heterozygotes to one of each type of homozygote. Because the two heterozygotes are clinically normal, the risk of an affected child being born to such a union is 1 in 4 for each conception.

An important feature of recessive inheritance relates to consanguinity. If a person carrying a recessive gene mates with a first cousin or other related person the chances are much increased that both will carry the same gene. Say the gene in question is for albinism which has an incidence of about 1 in 10,000 persons. If a heterozygote marries an unrelated person the chances that the latter is also a heterozygote come to about 1 in 50 (approximately twice the square root of the incidence). But if both are first cousins and one carries the gene, the chances that the other is also heterozygous are increased to 1 in 8. As a general rule, the rarer a recessive condition the more frequently will consanguineous marriages be found among the parents of affected children.

Consanguinity is nowadays unusual in countries with significant Anglo-Saxon origins; probably no more than 1 in 1000 marriages are between related people. However, certain population isolates have a tendency to become inbred; the isolation may be geographical as with islands or remote communities, racial especially with reference to minorities, religious where sectarian rules favour inter-sect mating or social as with the deaf or visually handicapped.

X-linked inheritance

When a mutant gene is situated on the X-chromosome there is a family pattern characterized by absence of male-to-male transmission, because males transmit only their Y-chromosomes to sons. Almost all the 80 X-linked conditions known in man are recessive in the sense that clinical manifestations are infrequently detectable in heterozygous females. Full effects are seen only in males with the mutant gene on their sole X-chromosome and the condition is carried by clinically normal females (Fig. 1.9). Nevertheless by employing refined investigations some evidence of gene effect can be found in many females heterozygous for X-linked mutations.

The risk to offspring of a heterozygous female is 1 in 2 for each son to be affected and 1 in 2 for each daughter to be a heterozygote. All the daughters of affected males must be heterozygotes because they receive the paternal X-chromosome with the mutant gene upon it.

A proportion of males with serious X-linked conditions have arisen from fresh mutations and appear as sporadic cases. This can result from a mutation in a maternal ovum or farther back in the mother's lineage where there has been neither the opportunity or an appropriate segregation of X-chromosomes into gametes to yield other affected males.

POLYGENIC INHERITANCE

Genetic disorders resulting from chromosomal abnormalities and gene mutations are distinguished as abrupt changes from normality. So are many consequences of environmental effects such as malnutrition, infections or injury. But most normal characteristics and processes

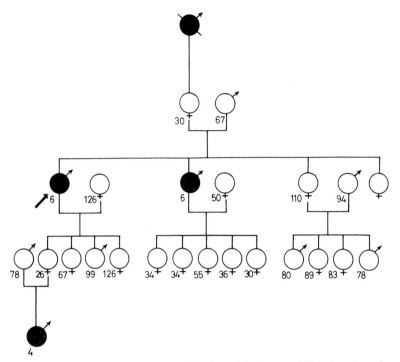

Fig. 1.9. X-linked inheritance. A family with haemophilia showing the typical pattern of inheritance via clinically normal females. Factor VIII levels (percentage of the average normal concentration) are marked on the pedigree and illustrate the wide variation in gene effect found among female carriers of an X-linked mutation.

cannot be described in this fashion. Developmental parameters like stature or intelligence and normal physiological variables (e.g. blood pressure and blood sugar levels) do not exist as a series of discrete events but rather as a continuous distribution of values. With the appropriate statistical treatment such a distribution fits, or approximates to, a normal (Gaussian) curve.

It is known from twin-studies and family investigations that a considerable genetic component is involved in the causation of characteristics with a continuous pattern of variation. Essentially, conclusions are based on assessing the degree of resemblance between persons of known relationship and relating this to the proportion of genes shared between different classes of relative. Considering relatives in pairs, then the regression coefficient (which is equivalent to the correlation coefficient) of one relative upon the other will depend on the number of genes common to both. Observations on measurements can then be compared with those expected on theoretical grounds; for instance, each member of a monozygotic twin-pair has identical genes (correlation coefficient, $r = 1 \cdot 0$) whereas each twin of a dizygotic pair is no more alike, genetically, than are siblings and share half their genes in common ($r = 0 \cdot 5$). A parent transmits half his genes to each child ($r = 0 \cdot 5$); first cousins share one-eighth of their genes ($r = 0 \cdot 125$) and an unrelated husband and wife should have no genes in common ($r = 0 \cdot 0$).

When the above approach is applied to continuously distributed human measurements it is possible to obtain some estimate of the relative contributions made by heredity and environment to a given characteristic. A human developmental trait largely unaffected by environmental factors is the distribution of finger-print ridges. Ridge-counts observed for different degrees of relationship agree extremely closely with theoretical predictions based on the number of shared genes. This is to be expected because the finger-print ridges are laid down at about the third month of intra-uterine life under almost entirely genetic control. But with characteristics significantly influenced by environmental factors, the fit of observed theoretical values is not so close; height and intelligence come into this category.

A polygenic basis implies that many genes are involved in creating a genetic predisposition with which environmental factors interact to produce a given characteristic. There is evidence for a polygenic threshold effect in the causation of many common diseases or malformations (Carter 1969). Beyond the genetic threshold there is a progressively greater probability that harmful consequences will be induced by environmental factors.

Very little is known of the individual genes that make up the polygenic component in causation of specific conditions or natural characteristics. However, it has long been known that relatives of patients with duodenal ulcer have an increased risk of being similarly affected. Formal studies on families and twins have confirmed a strong genetic liability. In addition there is the well-known association of duodenal ulcer with persons of blood group O. Therefore for one condition at least a genetic locus has been identified which, apart from its major role in determining a blood group antigen, participates as one component in the polygenic set that predisposes to duodenal ulcer.

REFERENCES

AL-AISH M.S., DE LA CRUZ F., GOLDSMITH L.A., VOLPE J., MELLA G. & ROBINSON J.C. (1967) Autosomal monosomy in man. *New Eng. J. Med.* **277**, 777–784.

AUSTIN C.R. (1967) Chromosome deterioration in ageing eggs of the rabbit. *Nature* **213**, 1018.

BAIRD P.A. & MILLER J.R. (1968) Some epidemiological aspects of Down's syndrome in British Columbia. *Brit. J. prev. soc. Med.* **22**, 81–85.

BARLOW P. & VOSA C.G. (1970) The Y chromosome in human spermatozoa. *Nature* **226**, 961–962.

BARTLET D.J., HURLEY W.P., BRAND C.R. & POOLE E.W. (1968) Chromosomes of male patients in a security prison. *Nature* **219**, 351–354.

BENIRSCHKE K., EDWARDS J.H., GROPP A., DE GROUCHY J., HILL R.T., KIRK R.L., KLINGER H.P., MILLER O.J., POLANI P.E., PROKOFJEVA-BELGOVSKAJA A., SASAKI M., SCHWARZACHER H.G., STANDLEY C.C. & STEFFEN J. (1966) Standardization of procedures for chromosome studies in abortion. *Bull. Wld. Hlth. Org.* **34**, 765–782.

CARTER C.O. (1969) Genetics of common disorders. *Brit. med. Bull.* **25**, 52–57.

Chicago Conference (1966) *Standardization in Human Cytogenetics.* Birth defects: Original series, 2, No. 2. New York, The National Foundation.

COHEN M.M. & MUKHERJEE A.B. (1968) Meiotic chromosome damage induced by LSD-25. *Nature* **219**, 1072–1074.

COLLMANN R.D. & STOLLER A. (1962) A survey of mongoloid births in Victoria, Australia, 1942–1957. *Amer. J. publ. Hlth.* 52, 813–829.

COURT-BROWN W.M. & SMITH P.G. (1969) Human population cytogenetics. *Brit. med. Bull.* **25**, 74–80.

DALY R.F. (1969) Neurological abnormalities in XYY males. *Nature* **221**, 472–473.

DAVIDSON R.G., NITOWSKY H.M. & CHILDS, B. (1963) Demonstration of two populations of cells in the human female heterozygous for glucose-6-phosphate dehydrogenase variants. *Proc. nat. Acad. Sci. (Wash.)* **50**, 481–485.

DONIACH D., ROITT I.M. & POLANI P.E. (1968) Thyroid antibodies and sex-chromosome anomalies. *Proc. roy. Soc. Med.* **61**, 278–280.

DORRANCE D., JANIGER O. & TEPLITZ R.L. (1970) *In vivo* effects of illicit hallucinogens on human lymphocyte chromosomes. *J. Amer. med. ass.* **212**, 1488–1491.

EDWARDS, J.H. (1966) Chromosomal abnormalities and prenatal selection. *Ann. hum. Genet.* **29**, 320 (Abst.)

EDWARDS, R.G. (1965) Maturation *in vitro* of human ovarian oocytes. *Lancet* **ii**, 926–929.

EDWARDS R.G., BAVISTER B.D., STEPTOE P.C. (1969) Early stages of fertilization *in vitro* of human oocytes matured *in vitro*. *Nature* **221**, 632–635.

FABIA J. (1969) Illegitimacy and Down's syndrome. *Nature* **221**, 1157–1158.

FERGUSON-SMITH M.A. (1966) X-Y chromosomal interchange in the aetiology of true hermaphroditism and of XX Klinefelter's syndrome. *Lancet* **ii**, 475–476.

FORD C.E. (1969) Mosaics and chimaeras. *Brit. med. Bull.* **25**, 104–109.

GERMAN J. (1968) Mongolism, delayed fertilization and human sexual behaviour. *Nature* **217**, 516–518.

GIBLETT E.R. (1969) *Genetic Markers in Human Blood*. Oxford, Blackwell.

GOLDSCHMIDT R. (1954) Different philosophies of genetics. *Science* **119**, 703–710.

HAMERTON J.L. (1968) Significance of sex chromosome derived heterochromatin in mammals. *Nature* **219**, 910–914.

HARNDEN D.G. (1964) Cytogenetic studies on patients with virus infections and subjects vaccinated against yellow fever. *Amer. J. Hum. Genet.* **16**, 204–213.

JACOB F. & MONOD J. (1961) Genetic regulatory mechanisms in the synthesis of proteins. *J. Molec. Biol.* **3**, 318–356.

JACOBS, P. (1969) Structural abnormalities of the sex chromosomes. *Brit. med. Bull.* **25**, 94–98.

JAGIELLO G., KARNICKI J. & RYAN R.J. (1968) Superovulation with pituitary gonadotrophins. Method for obtaining metaphase figures in human ova. *Lancet* **i**, 178–180.

JONES H.W., WOODRUFF J.D., DAVIS H.J., KATAYAMA K.P., SALIMI R., PARK I-J., TSENG P-Y. & PRESTON E. (1970) The evolution of chromosomal aneuploidy in cervical atypia, carcinoma *in situ* and invasive cancer of the uterine cervix. *Johns Hopkins med. J.* **127**, 125–135.

KWITEROVICH P.O., CROSS H.E. & McKUSICK V.A. (1966) Mongolism in an inbred population. *Bull. Johns Hopk. Hosp.* **119**, 268–275.

LARSON S.L. & TITUS J.L. (1970) Chromosomes and abortion. *Mayo Clin. Proc.* **45**, 60–72.

LYON M.F. (1961) Gene action in the X-chromosomes of the mouse (*Mus musculus* L.). *Nature* **190**, 372.

MATSUNAGA E. & MARUYAMA T. (1969) Human sexual behaviour delayed fertilization and Down's syndrome. *Nature* **221**, 642–644.

McDOUGALL J.K. (1970) Effects of adenoviruses on the chromosomes of normal human cells and cells trisomic for an E chromosome. *Nature* **225**, 456–458.

McKUSICK V.A. (1968) *Mendelian Inheritance in Man*; 2nd ed. Baltimore, Johns Hopkins.

MELNYK J., THOMPSON H., RUCCI A.J., VANASEK F. & HAYES S. (1969) Failure of transmission of the extra chromosome in subjects with 47,XYY karyotype. *Lancet* **ii**, 797–798.

MUKHERJEE B.B. & SINHA A.K. (1963) Single-active-X hypothesis. Cytological evidence for random inactivation of X-chromosomes in a female mule complement. *Proc. nat. Acad. Sci. (Wash.)* **51**, 252–259.

MULLER H.J. (1950) Evidence of the precision of genetic adaption. *Harvey Lectures*, Series 63, vol. 1. Springfield, Thomas.

NORTHCUTT R.C., ISLAND D.P. & LIDDLE G.W. (1969) An explanation for the target organ unresponsiveness to testosterone in the testicular feminization syndrome. *J. clin. Endocr. Metab.* **29,** 422–425.

OHNU, S. (1967) *Sex Chromosomes and Sex-linked genes.* New York, Springer-Verlag.

PARK W.W. (1957) The occurrence of sex-chromatin in early human and macaque embryos. *J. Anat.* **91,** 369–373.

PEARSON P.L., BOBROW M. & VOSA C.G. (1970) Technique for identifying Y-chromosomes in human interphase nuclei. *Nature* **226,** 78.

PENROSE L.S. (1939) Maternal age, order of birth and developmental abnormalities. *J. Ment. Sci.* **85,** 1141–1150.

PENROSE L.S. & SMITH G.F. (1966) *Down's Anomaly.* London, Churchill.

PERRY J. (1969) System for semi-automatic chromosome analysis. *Nature* **224,** 800–803.

POLANI P.E. (1969a) Autosomal imbalance and its syndromes, excluding Down's. *Brit. med. Bull.* **25,** 81–93.

POLANI P.E. (1969b) Abnormal sex chromosomes and mental disorder. *Nature* **223,** 680–686.

POLANI P.E. (1970) The incidence of chromosomal malformations. *Proc. roy. Soc. Med.* **63,** 14–16.

RACE R.R. & SANGER R. (1969) Xg and sex-chromosome abnormalities. *Brit. med. Bull.* **25,** 99–103.

RATCLIFFE S.G., STEWART A., MELVILLE M., JACOBS P. & KEAY A.J. (1970) Chromosome studies on 3500 new-born male infants. *Lancet* **i,** 121–122.

SASAKI M. & MAKINO S. (1965) The meiotic chromosomes of man. *Chromosoma (Berl.)* **16,** 637–651.

SCHULL W.J. & NEEL J.V. (1962) The effect of inbreeding on mortality and morbidity in two Japanese cities. *Proc. nat. Acad. Sci. U.S.A.* **48,** 573–582.

SHAPIRO L.R. (1970) Hormones and the XYY syndrome. *Lancet* **i,** 623.

SHAVER E. & CARR D.H. (1967) Chromosome abnormalities in rabbit blastocysts following delayed fertilization. *J. Reprod. Fert.* **14,** 415–420.

SHAW M.W. (1970) Human chromosome damage by chemical agents. *Ann. rev. med.* **21,** 409.

STEELE M.W. (1969) Autoradiography may be unreliable for identifying human chromosomes. *Nature* **221,** 1114–1116.

STERN C. (1960) Dosage compensation. Development of a concept and new facts. *Canad. J. Genet. Cytol.* **2,** 105.

STEVENSON A.C. & BOBROW M. (1967) Determinants of sex proportions in man, with consideration of the evidence concerning a contribution from X-linked mutations to intrauterine death. *J. med. Genet.* **4,** 190–221.

SUNDEQUIST U. & HELLSTROM E. (1969) Transmission of 47,XYY karyotype? *Lancet* **ii,** 1367.

SWANSON C.P., MERZ T. & YOUNG W.J. (1967) *Cytogenetics.* New Jersey, Prentice-Hall.

TAYLOR A.I. (1968) Cell selection *in vivo* in normal G. trisomic mosaics. *Nature* **219,** 1028–1030.

TURPIN R. & LEJEUNE J. (1969) *Human Afflictions and Chromosomal Aberrations.* London, Pergamon Press.

Turpin R., Lejeune J., Lafourcade J., Chigot P.L. & Salmon C. (1961) Présomption de monozygotisme en dépit d'un dimorphisme sexuel: subjet masculin XY et sujet neutre, hapto X. *C.R. Acad. Sci. (Paris)* **252**, 2945–2946.

Uchida I.A., Holunga R., & Lawler C. (1968) Maternal radiation and chromosomal aberrations. *Lancet* **ii**, 1045–1049.

Weatherall, D.J. (1965) *The Thalassaemia Syndromes*. Oxford, Blackwell.

2
Ovarian Function and its Control

RODNEY P. SHEARMAN

INTRODUCTION

The purpose of this chapter is to present material relating to the human. It must be accepted that the human is understandably and quite properly a reluctant experimental subject. One may hypophysectomize a quail in the pursuit of knowledge, insert electrodes into the hampster hypothalamus, place chronic catheters in the carotids of rabbits, render a rat foetus decerebrate, a piglet anosmic or expose a mouse to continuous light. The human will have none of this. The clinician is often reluctant to accept that findings from other species produced under these circumstances have relevance to the clinical problems of a particular patient. Yet it is one of the continuing joys of gynaecology that endocrinological disorders once thought obscure—and because of this hiding behind euphonious eponyms such as the syndromes of Chiari-Frommel or Ahumada-del Castillo—ultimately are explicable in terms of disturbed physiology. Almost always the basic physiological knowledge has come from these experimental animals. Caution should always be used in applying data from other species to the human but, perhaps because reproduction is biologically a fairly old practice, an enormous amount of information, found first in the rodent, has ultimately been applicable to man.

Where information is available on the control of ovarian function in the human it will be presented here. Of necessity much of the evidence will be from animal studies. As far as possible these differences will be made clear; but it must be stressed that animal data should not be dismissed or ignored because human knowledge lags behind.

This chapter will deal in some detail with human ovarian function. The role of the pineal and hypothalamus will be presented followed by pituitary control with some discussion of the mechanisms of action of pituitary gonadotrophins. Current concepts of steroid biogenesis are presented in Appendix A but the secretion patterns and the biological activities of ovarian hormones will be dealt with here. Some of these subjects have been reviewed elsewhere (Shearman 1971). Finally, a theoretical concept of ovarian control will be presented.

PUBERTY AND MENARCHE

There is anecdotal evidence that the age of female puberty during the last millennium was much the same as it is now. Ignoring the anecdotes, puberty has appeared at a progressively earlier age in those countries that have reliable data going back to the last century (Fig. 2.1). A significant age reduction in menarche from one generation to the next is well documented (Damon *et al* 1969).

The operative factors responsible for this change remain obscure. Climate and geography, once thought to be important, are now excluded. Improved nutrition accounts for only part of this advance: other factors—the effect of urbanization, the effect of normal light perception and sense of smell are important. Polish data have shown an earlier age of menarche for urban girls than rural girls since 1880 (Tanner 1968) and while some of these differences may be attributed to nutritional variables, there remain areas of psycho-social environment that appear to influence the onset of puberty in an unidentified manner. It is interesting that the laboratory rat, like the urban girl, has earlier gonadal development than her free-ranging counterpart.

In 1964, Zacharias and Wurtman indicated that blind human females had a significant acceleration of menarche when compared with normally sighted girls. Although this finding did not meet with universal initial acceptance there is now no doubt that it is true (Magee *et al* 1970). The fact that removal of the olfactory bulbs from piglets will make them forever sexually infantile may do little more than raise a tired shrug when considered in isolation. But when it is realized that the congenitally anosmic human infant has the same sexual destiny as the piglet this piece of otherwise inconsequential knowledge becomes of real but still unexplained importance. It does, however, give potential human relevance to the pheromones of arthropods and rodents.

It is not clear how these many factors—nutrition, domestic environment, health, light perception, the sense of smell—operate to affect the age of menarche but there is increasing evidence that the pineal gland may be the mediator.

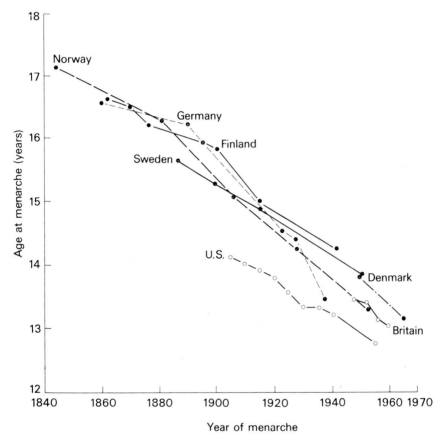

FIG. 2.1. Showing progressive reduction in the age of menarche. From J.M. Tanner (1968) *Scientific American* **218**.

PINEAL GLAND

The possible function of the pineal gland has been debated for years. Its attributed actions in the past have varied from nothing, to a vestigial third eye, or an atavistic osmoreceptor. Only one clear correlation has been recognized for nearly 100 years; that pineal tumours in male children are often associated with sexual precocity. The sporadic interest in the possible relationship of the pineal gland to sexual maturation has been concentrated by the demonstration of the effect of indoles on gonadotrophic function. A specific indole, melatonin, is one of a family of hormones, the methoxyindoles, produced by the pineal gland (Wurtman 1969).

Wurtman has pointed out that the pineal is not a gland, and for its function he has introduced the phrase 'a neuroendocrine transducer', terminology shared by the adrenal medulla and some of the centres in the hypothalamus. 'The typical gland (i.e. the thyroid, adrenal cortex, ovary) responds to instructions which are delivered to it from the circulation (i.e. TSH, ACTH, FSH) by secreting a different set of instructions into the blood stream (i.e. its hormone: thyroxine, cortisol, oestradiol). Neuro-endocrine transducers also secrete coded messages into the circulation; however, their input of instruction comes not from the blood stream but from nerves.'

Melatonin (5-methoxy,N-acetyl tryptamine) was first described as a specific pineal hormone in 1960, by Lerner et al. The synthetic sequence was shown to be tryptophan→5-hydroxytryptophan→serotonin→N-acetylserotonin→melatonin. The enzyme necessary for this last step is unique to the pineal-hydroxyindole-O-methyltransferase (HIOMT).

In the mammal, the main afferent stimulus to the pineal is light and melatonin synthesis has a circadian rhythm that might be expected because of this. The systemic administration of melatonin to rats suggests that the secretion of LH but not FSH is inhibited (Martini et al 1968). The same inhibitory effect has been demonstrated after local placement of melatonin or pineal tissue into the median eminence (Clementi et al 1969). In rats it is known that continuous light exposure reduces the pineal activities of HIOMT, a response abolished by blinding (Alexander et al 1970a). The role of the pineal in human sexual maturation remains to be determined, but it is of more than passing interest, in view of its photosensitivity, that—as mentioned earlier—blind humans show a significant acceleration of menarche (Zacharias & Wurtman 1964). Finally, it may add to one's sense of universal order to reflect that while pinealectomy will delay sexual maturation in the Japanese quail (Sayler & Wolfson 1968), exposure to continuing light has no effect on the HIOMT content of this curious bird's pineal (Alexander et al 1970b).

Clearly the pineal has a very real though possibly differing role in gonadal control of rodents and birds. Unravelling its true physiological role in the human is an exciting, but still distant prospect.

HYPOTHALAMIC RELEASING FACTORS

The concept of neural control of the anterior pituitary advanced by Marshall in 1937 was removed from the realm of hypothesis with the demonstration of the hypophyseal portal circulation by Green and Harris 10 years later. Concrete evidence of specific releasing factors was forthcoming in the 1960s and currently rapid progress is being made in characterization and synthesis of the specific releasing factors controlling the various pituitary trophic hormones. Acting as neuroendocrine transducers, the hypothalamic centres produce releasing factors from nerve endings which pass via the portal circulation to the pituitary. Seven of these are now well established in mammals. The terminology of the releasing compounds remains equivocal, some referring to them as hormones, others as foctors. Because luteinizing hormone release factor (LRF) is less repetitive than luteinizing hormone release hormone (LRH) the former terminology will be used here.

Only two of these neurohumors warrant discussion in a section dealing with the ovary—Luteinizing Hormone Release factor and Follicle Stimulating Hormone release factor. The possible relevance of a third—Prolactin Inhibitory Factor—awaits further evidence for its existence in the human and further elucidation of the role of prolactin in human biology. This will be discussed shortly.

LH RELEASING FACTOR (LRF)*

LRF was first identified in the hypothalamus of rats (McCann *et al* 1960) and its presence confirmed in man 7 years later (Schally *et al* 1967). *In vitro* and *in vivo* evidence suggests that the hypothalamic transmitter for LRH release is dopamine (Schneider & McCann 1970). LRH is probably secreted in the suprachiasmatic and arcuate ventromedial regions and stored in the median eminence (Martini *et al* 1968, Crighton *et al* 1970).

* There is very recent evidence that FRF and LRF are identical with the structure (pyro) Glu-Trp-Ser-Tyr-Gly-Leu-Arg-Pro-Gly-NH$_2$ (Schally, A. V., Arimura, A., Kastin, A. J. *Res. in Reprod.* (1971) **3**, No. 6, 1).

The structural secrets of the releasing hormones are yielding very quickly. Thyrotrophin releasing factor consists of three amino acids—glutamic acid, histidine and proline—in the sequence (Pyro) Glu-His-Pro (NH2) and has been synthesized (Bowers *et al* 1970). It is probable that LRF is also a very small molecule and that human and animal materials are very similar structurally and biologically (Schally *et al* 1970).

Little surprise should be evoked by the statement that the suggested controlling mechanisms for LRF become more complex as more is known about them. The older belief in a long negative servo of gonadal steroids on either the hypothalamus or pituitary is no longer tenable, and now long and short positive and negative feedbacks are invoked.

Specific receptors for oestradiol have been demonstrated in the hypothalamus and pituitary (Kahwanago *et al* 1970) but they appear to be receptors for different purposes. Schneider & McCann (1970) have shown that oestradiol has a direct effect on the pituitary to stimulate LH release while acting paradoxically on the hypothalamus to inhibit the dopamine-dependent induction of LRF release. As will be discussed more fully below when dealing with the effect of oestradiol on 'target cells', it seems likely that oestradiol acts on both pituitary and hypothalamic tissue to stimulate RNA synthesis, which in turn stimulates synthesis of mediator proteins or peptides.

The short servo mechanism is a relatively recent concept. These are controls independent from the hormones secreted by the peripheral target gland and in which the signal is provided by pituitary hormones themselves (Martini *et al* 1968). LH implanted into the median eminence will cause a decrease in pituitary LH (short negative feedback). The specific receptors for and the mode of action of this inhibition are not yet clear.

There is much more to LRF than physiology. It is clear that the releasing hormones are much simpler molecules than the giant trophic hormones. Some already are, and most will shortly be synthesized. It is equally clear that whereas the pituitary hormones have a real degree of species specificity, the releasing hormones do not, sheep, pig, rat and human responding alike to the same compound. The probable availability of LRF in the near future is an exciting prospect, as there is no doubt that porcine LRF has a profound effect on human LH secretion (Kastin *et al* 1970).

FSH RELEASING FACTOR (FRF)

This was first clearly established as a separate hormone from LRF by Igrashi & McCann (1964). It is probably synthesized in a single centre in the hypothalamus—the paraventricular area—but like LRF is stored in the median eminence (Martini *et al* 1968). Also once thought to be a polypeptide, it is now thought that FRF is a much smaller molecule, probably a polyamine (Schally *et al* 1968).

As with LRF, long and short feedback controls have been demonstrated. The short receptors for both LH and FSH are highly specific and there is no cross over in specific centres between FSH and LH feedback.

Again there may be great therapeutic potential in FRF. This was first shown by Igarashi *et al* (1967) who induced ovulation with bovine FRF in previously anovulatory women. It is now suggested that FRF may be effective when given orally (Schally & Kastin 1969).

THE GONADOTROPHINS

Since the brief for this chapter is human ovarian function and control it is difficult to know what to do with prolactin. It seems clear that three gonadotrophins relate to control of the rodent ovary—Follicle Stimulating Hormone (FSH), Luteinizing Hormone (LH) and Luteotrophin or Prolactin (LTH)—although there is some doubt that prolactin is truly luteotrophic even in that species (Savard *et al* 1965). While there remains doubt about its function, it seems reasonable to present briefly some of the current concepts of prolactin in the human, and then to follow this with a more detailed discussion of FSH and LH. The chemistry and problems of assay are presented in Appendix B.

PROLACTIN (LUTEOTROPHIN, LTH)

Argument has raged for years about the existence of a purely prolactational hormone in the human, many believing it to be the exclusive property of Human Growth Hormone (HGH). On purely clinical grounds, study of patients with secondary amenorrhoea and inappropriate lactation provides strong evidence for the existence of such a hormone distinct from growth hormone (Shearman & Turtle 1970).

Morphological study of a pituitary tumour in a patient with Forbes Albright syndrome suggests that prolactin is a separate hormone produced by cells different from those secreting HGH (Peake *et al* 1969) and study of the rhesus monkey provides further supporting evidence (Nicolle *et al* 1970). Quantitatively and qualitatively different patterns of prolactin secretion compared with GH secretion have been demonstrated. An unexplained but interesting observation is that the female rhesus monkey produces much more prolactin than the male, an observation made earlier in the rat. To inject a note of nihilism it should be pointed out that other responsible workers are still unable to identify a prolactin from human pituitary tissue that is separate from HGH (Nicholson 1970).

Proof of the evidence of prolactin as a separate hormone in the human is at the moment circumstantial, but very strong. The mechanisms controlling its secretion, and its function in the non-pregnant female remain clouded in obscurity.

An acceptable concept is that there is no specific releasing factor for prolactin in the hypothalamus but probably a specific prolactin inhibiting factor (PIF) the absence of which allows prolactin secretion. It has been suggested that a corollary of decreased PIF release and/or prolactin secretion is inhibition of LH and FSH. Although the Scottish verdict of 'not proven' applies here, this hypothesis provides a rational explanation for the amenorrhoea seen in most conditions associated with lactation, whether physiological or pathological.

The function of prolactin in the human can only be the subject of speculation, but not necessarily loose speculation. The hypophysectomized human female may be made to ovulate without prolactin and by all objective criteria, including ensuing normal pregnancy, induced ovarian function is quite normal. While the effect of prolactin, if any, on the human ovary is not yet known, and may ultimately prove to be negligible, because of its widespread effects in lower animals it appears quite clear that there will be a place somewhere for prolactin in the human economy.

The interdependence of the gonads and the pituitary has been known since Cushing's classical experiments early in this century. Following the initial demonstrations of gonadotrophic activity by Zondek, Fevold *et al* (1931) indicated the probable existence of two such substances. Many years of uncertainty and dispute about the existence of two separate gonadotrophins followed; many workers believed and showed

OVARIAN FUNCTION AND ITS CONTROL

to their own satisfaction that there was one gonadotrophin, with two actions. There is now indisputable evidence for two separate trophic hormones controlling the ovary, FSH, and LH.

FOLLICLE STIMULATING HORMONE (FSH)

While real progress can only come from ideas, in the field of biology ideas without the techniques to test them will often be fruitless. Although it is now more than 40 years since the idea of FSH was first proposed, difficulties in isolating biologically pure FSH have been so formidable that the physiology of this substance untouched by other trophic hormones is still conjectural. The continuing uncertainty about differences between pituitary and urinary FSH continues (Appendix B). Studying the effect of porcine FSH in rats, Greep's work (Chow *et al* 1942, Greep *et al* 1942) remained the foundation of FSH physiology for many years. His graphic description of its action in the rat is worth quoting: '... it will not lead to stimulation of the uterus; the ovaries will exhibit many mature but not cystic follicles and no follicles are in pre-ovulatory swelling; the preparation will not produce any trace of luteinization or thecal swelling in the hypophysectomized female rat nor any enlargement of the prostate in males' (Greep 1959).

Implicit in these observations is that FSH alone will have no effect on ovarian steroid production, although preparing all the props for LH to do so (*vide infra*). Using highly purified human material Eshkol & Lunenfeld (1967) supported the view that FSH alone had no effect on steroidogenesis. As so often happens to disturb what appeared to be a tidy situation, the Stockholm group have cast grave doubt on this (Petrusz *et al* 1970).

They have reached the fundamentally important and different conclusion that pure FSH does have an effect in inducing oestrogen synthesis from the ovary of both intact and hypophysectomized female rats. Careful examination of their data and of earlier suggestive fragments from other sources leaves little reason to doubt their conclusions. So that whereas the observations of Greep quoted above have been taken until now as virtually an *ex cathedra* statement permitting FSH by itself no part in oestrogen synthesis, the revision by Petrusz, Robyn and Diczfalusy (1970) should now be quoted in its place as the currently approved version.

'... the biological effects of *human urinary* FSH in hypophysectomized rats can be described as follows: in females it induces ovarian weight increase and growth and maturation of the Graafian follicles, without repair of the interstitial cells and without any sign of luteinization. These ovarian changes are associated with an increase in uterine weight and with vaginal cornification. The above described biological effects of human urinary FSH in hypophysectomized rats agree rather closely with those of swine pituitary FSH reported more than 25 years ago by Greep et al (1942). The only difference between these two types of FSH preparations seem to be the capacity of human urinary FSH to induce uterine weight increase and vaginal cornification.'

Vive la différence!

LUTEINIZING HORMONE (LH)

LH is produced by different cells in the pituitary from those secreting FSH; the specific basophil cells producing LH may be demonstrated by immunofluorescent techniques (Bain & Ezrin 1970). It is still unclear if LH is a single hormone or a conglomerate (Peckham & Parlow 1969).

If LH is given alone to an immature or hypophysectomized animal, no effect is seen, but in the mature animal, or a hypophysectomized animal treated concurrently with FSH, many of its effects are visible to the naked eye—superovulation, luteinization and the effect of secreted steroids on target organs.

There is no evidence in the human that LH has any effect on the life span of the corpus luteum (Neill et al 1967). Other factors that may affect the life of the corpus luteum will be discussed in the next section.

From what has been said earlier it must be accepted that FSH does have some effect on ovarian oestrogen production. Despite this it seems that LH plays the dominant role in ovarian steroid production as well as triggering ovulation and causing luteinization of the ruptured follicle.

LH will cause an increased secretion of steroids from the corpus luteum *in vitro*. These *in vitro* findings are supported *in vivo* by the finding that the corpus luteum removed early in the luteal phase, when it is presumably under the influence of considerable LH stimulation, has a higher rate of incorporation of labelled precursor than a corpus luteum removed later in the cycle (Savard et al 1965).

Some general concepts of the mechanisms of protein hormone action

This is probably as good a place as any to introduce some concepts of the mechanisms of trophic hormone action in general. These will then be followed further in discussing the specific mechanisms of LH. In general terms a hormone could have a selective effect on a target organ either because the target organ had a specific method of binding that hormone not found in other tissue, or because the target organ had certain metabolic pathways capable of responding to the hormone, metabolic pathways not shared by non-target tissue. The majority of current evidence suggests that specific binding is the usual mechanism.

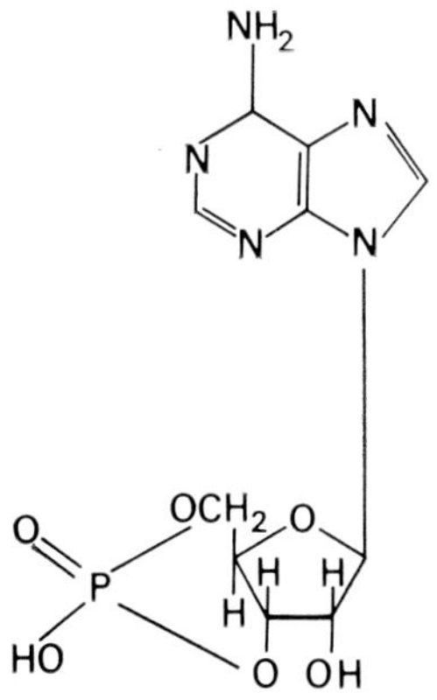

FIG. 2.2. Structure of cyclic AMP.

The initial work of Sutherland now supports an immense and rapidly growing superstructure indicating that activation of cyclic AMP (adenosine 3', 5' cyclic monophosplate) is of central importance in the mediation of the activity of most peptide and protein hormones (Sutherland et al 1968). Because of its importance, cyclic AMP (Fig. 2.2) has been called 'the Second Messenger', the first being the hormone itself. In sequence the hormone is bound to a specific receptor on the cell membrane of the target cell; it then reacts with adenyl cyclase to produce cyclic AMP from adenosine-triphosphate (ATP). The other main enzyme in the regulation of cyclic AMP levels is a phosphodiesterase which degrades cyclic AMP to 5'-AMP. The latter is biologically inactive (Editorial comment 1970, Catt 1970).

These sequences are shown schematically in Fig. 2.3.

The cyclic AMP formed within the cell in this way then triggers a sequence of events specific for that cell which leads to the physiological response associated with the hormone (Editorial comment 1970).

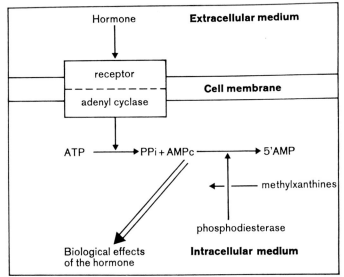

ATP: adenosine triphosphate
AMPc: cyclic adenosine 3'–5' monophosphate
5'AMP: adenosine monophosphate
PPi: pyrophosphate

FIG.2.3. Diagram of mechanism of action of a hormone eliciting intracellular production of the 'second messenger'. From F.Morel (1969) *Triangle, Sandoz Journal of Medical Science* **9**, No. 4.

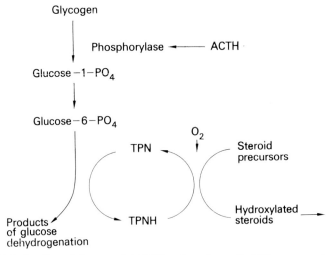

FIG. 2.4. Outline of theory of ACTH action. From R.C.Haynes *et al* (1960) *Rec. Progr. Horm. Res.* **16**, 129.

One of the earlier suggested models based on the second messenger concept was for the mode of action on corticotrophin (Fig 2.4).

LH is no exception to the statement that trophic hormone action is mediated by stimulation of cyclic AMP activity in the specific target cell. There is some evidence that the ubiquitous prostaglandins—in this case

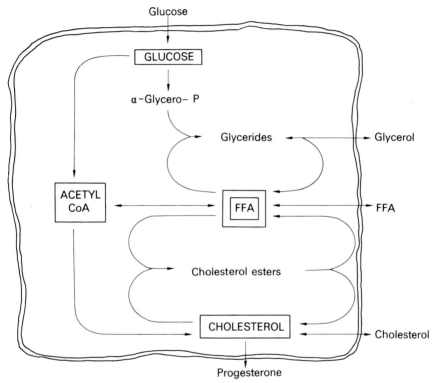

FIG. 2.5. Some possible relationship between glycolysis and dynamics of ovarian lipid metabolism. From D.T.Armstrong (1968) *Rec. Progr. Horm. Res.* **24**, 255.

prostaglandins E and E_2 (PGE_1, PGE_2) are the specific LH receptors, at least in the mouse ovary (Kuetil *et al* 1970). These findings indicate that LH acting with PGE_1 stimulates adenyl cyclase. A review of current concepts of the role of prostaglandins in the mediation of hormone action will be found in Ramwell & Shaw (1970).

The possible mechanisms by which LH influences ovarian steroid production have been reviewed fully by Armstrong (1968). He has shown that LH increases glucose utilization, via glycolysis in the ovarian cell. The possible relationships between glycolysis and the dynamics of ovarian lipid depletion are shown in Fig. 2.5. If reference is made to this and to Fig. 2.4 it will be noted that the second messenger would probably provide the stimulus for phosphorylation of glucose to glucose phosphate. It is probable that LH increases the conversion of cholesterol to progesterone by inhibiting esterification of cholesterol (Fig. 2.6). This decrease in esterification of cholesterol in the ovary would increase the rate of incorporation of cholesterol precursors into pregnenolone and thence progesterone.

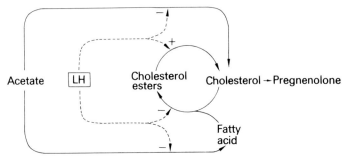

FIG. 2.6. Some possible explanations for the observed effects of LH upon ovarian cholesterol dynamics. From D.T.Armstrong (1968) *Rec. Progr. Horm. Res.*, **24**, 255.

The mechanism(s) by which LH also increases oestrogen synthesis is not yet clear. To the writer the 'two cell' theory of Short (1962, 1964) which is based on the differential ability of thecal and granulosal cells to produce progesterone or oestradiol because of their differing enzyme content remains an attractive hypothesis. This will be discussed in more detail when describing ovarian steroid production (*vide infra*).

OTHER FACTORS IN OVARIAN CONTROL

As already mentioned, there is no evidence that LH, at least in the human, affects the life span of the corpus luteum. It is stretching credibility too far to believe that a structure, that in the absence of pregnancy recurs each month in the sexually mature woman, has the seeds of self destruction sown within itself. Some signal or signals very probably control the inexorable 12 to 14 days' life span of the normal corpus luteum of the menstrual cycle.

In 1923, Loeb made the surprising observation that hysterectomy during the luteal phase in the guinea pig prolonged the life of the corpus luteum. It was suggested that in some animals the uterus produced a substance that normally caused regression of the corpus luteum. The unknown factor was named 'uterine luteolysin'. Caldwell *et al* (1967) placed the Syrian hampster 'among those mammals in which a definite uterine antiluteal effect has been demonstrated (including) the laboratory rat, guinea pig, rabbit, pig, cow, and sheep while the possum, ferret and primate fail to show the uterus–corpus luteal relationship'.

These relationships in the sheep have been studied carefully by Moor (Moor & Rowson 1964, Moor *et al* 1970). The uterus appears 'to be essential both for the initiation and the continuation of those processes that lead to the death of the lutein cells'. Further regression may be prevented by hysterectomy even after functional and morphological regression has commenced. Because grafting of endometrial tissue into the animal after hysterectomy will re-initiate luteal regression it is assumed that the luteolysin is in the endometrium.

It has been suggested that a prostaglandin may be the responsible luteolysin in these animals. $PGF_2\alpha$ is an abundant uterine prostaglandin. While *in vitro* PGE and $PGF_2\alpha$ increase luteal steroid production (Speroff & Ramwell 1970) *in vivo* $PGF_2\alpha$ induces luteal regression in the sheep (McCracken *et al* 1970). As already mentioned the steroidogenic action is probably dependent on the local sequence LH-Prostaglandin-adenyl cyclase, whereas the luteolytic action may be due to an effect on ovarian blood flow.

From the information available, the situation in the human seems to be different. Careful study of women with congenital absence of the uterus and vagina indicates normal ovarian function in every respect. Beling *et al* (1970) have studied the effect of hysterectomy on the human corpus luteum. They could find no evidence that hysterectomy affected the corpus luteum either in the operative or subsequent cycles.

The subject of luteal control in the human is by no means closed. Because such control offers hope of an effective method of contraception a great deal more information will be forthcoming in the very near future. The observation that progesterone production in the corpus luteum of the rhesus monkey is reduced by $PGF_2\alpha$ (Kirton *et al* 1970) is already extended to the use of this compound as a contraceptive in clinical trials in the human, where it appears to be effective by the vaginal, anal or even oral route—although it tastes vile. It should soon be clear whether this action is luteolytic, abortifacient or both.

OVARIAN STEROIDS

Steroid biogenesis is discussed in Appendix A and only those elements necessary for local continuity of text will be referred to here. Three important items should be remembered:

(1) For all practical purposes, oestradiol is *the* biologically active oestrogen produced by the ovary, with smaller quantities of oestrone. Except for the possible secretion of small amounts of oestriol in the luteal phase, the vast bulk of oestriol and the multitude of other known human urinary oestrogens are metabolic breakdown products of secreted oestradiol/oestrone.

(2) Oestradiol may be produced through the Δ^4 pathway via progesterone→17 hydroxy progesterone→androstenedione→oestradiol or via the Δ^5 pathway, pregnenolone→17 hydroxy pregnenolone→dehydroepiandrosterone→oestradiol.

(3) All oestrogens secreted by the ovary are produced from androgenic precursors.

It is a mistake to think only of follicular and luteral cells in the dynamics of ovarian steroid production. It is better to think of three separate steroidogenic compartments, each responsive to gonadotrophins (Savard *et al* 1965). These are as follows:

(a) The stroma, which may synthesize androstenedione, dehydroepiandrosterone, testosterone and small quantities of oestradiol and progesterone.

(b) The follicle, synthesizing mainly oestradiol with trace amounts of progesterone and androgens.

(c) The corpus luteum, producing progesterone as the dominant hormone, but also significant quantities of oestradiol.

There are very few data on the changes in steroidogenesis coincidental with cellular changes during the human menstrual cycle. Short (1962, 1964) has taken advantage of the large amount of follicular and luteal fluid present in the mare ovary to study steroid production *in vivo*. On the basis of his results he puts forward the two cell theory which 'postulates that the theca interna cells have all the enzyme systems necessary for the synthesis of oestradiol 17β from cholesterol, whereas the granulosal cells have only a weak 17-hydroxylase ability and little or no 17-desmolase activity'. Short suggests that the synthetic potential of the granulosal cells is not evident in the follicular phase because of its avascularity. After ovulation with a rapid ingrowth of blood vessels, the capacity of the lutein-transformed granulosa becomes dominant. He concludes that in the mare, changes in steroid secretion so clearly evident after ovulation are 'a direct result of the change in cell type', the thecal cells of the follicle producing oestrogens and androstenedione, the granulosal-derived luteal cells producing predominantly progesterone.

Savard (1967) emphasized the species variability in production of steroids by the corpus luteum. *In vitro*, the human corpus luteum produces significant quantities of oestrogen and the urinary excretion pattern and plasma levels of oestrogens in the luteal phase of the cycle indicates that this also occurs *in vivo*. It may be relevant that whereas in the mare—studied by Short—the thecal cells almost entirely disappear after ovulation, in the human they remain prominent as the theca lutein cells.

Ryan & Petro (1966) have studied the steroidal potential in human granulosal and thecal cells *in vitro*. They conclude that in the human the differences are quantitative rather than qualitative. Both cells convert precursor pregnenolone to oestrogens, but this is of much greater degree in thecal cells. Granulosal cells showed a striking preference for the conversion of pregnenolone to progesterone.

Although the evidence for production of ovarian androgens is unequivocal, the physiological roles of androgens in the human female are not clear. Their potential for mischief in pathological circumstances is undoubted. The two classical 'female' hormones were, and remain, oestradiol and progesterone.

OESTRADIOL

In this section the general effects of oestradiol both genital and extragenital will be presented, followed by a discussion of the early current knowledge of the effects of oestradiol on cellular function.

General effects

One of the most comprehensive reviews of the general actions of oestrogens remains Burrows' monograph (1949). The growth of the vulva seen in all normal adolescent girls and the pigmentation seen in some of them appears to be a direct action of oestrogens.

The cyclic changes in the vaginal epithelium during the menstrual cycle remain one of the more spectacular effects of target organ response to oestradiol. The thin atrophic epithelium of the pre-pubertal girl or the post-menopausal woman is transformed by oestrogens. Initially, in a morphological sense, oestrogens cause an increase in the number of cells and a rapid transformation of the cells into typical squames. This epithelium is free of leucocytes. The cells, although nucleated, are pyknotic. These changes are due to a direct action on the vaginal epithelium, the earliest ultrastructural signs being visible very quickly after topical administration (see below). The oestrogenic response is not restricted to the epithelium—there is general hyperaemia and some degree of hypertrophy of the muscular wall.

The glycogen content of vaginal epithelium is oestrogen dependent. Glycogen granules can be demonstrated in cornified epithelium and the glycogen increases as cornification progresses. Vaginal pH decreases *pari passu*. In children and post-menopausal women vaginal pH is 7 or greater; during sexual maturity the pH is between 4·5 and 5·0. Administration of oestrogens to castrated women causes a rapid fall in pH.

The earliest structural effect of oestrogen on a uterus not previously exposed to oestrogens is stromal oedema and hyperaemia. This increase in water content is seen during the normal cycle—at least in the monkey—when it has been noted that uterine water content is highest in the follicular phase.

Initial oedema and hyperaemia is followed by mitosis in the endometrial epithelium, glands and muscle, but muscular hyperplasia follows that seen in the epithelium. Concurrently there is an increase in actomyosin content of the muscle cells which show spontaneous contractibility of low amplitude and high frequency—previously termed A waves (see chapter 7). The glands of the human uterine cervix are moved by oestrogens to produce a copious, thin, elastic and clear cervical mucus with a very high electrolyte content. The physical characteristics of this mucus have been examined fully by Odeblad (1968) and its ultrastructural characteristics by Singer & Reid (1970).

The effects of oestradiol on the human breast are curiously poorly documented. In passing it is interesting that while very obvious effects can be produced on the normal breast and nipple in the human, accessory tissue responds much less.

Pigmentation of the areola and growth and mobility of the nipple are oestrogen dependent; the effects can be observed after local application.

The main initial effect of oestrogen on the monkey breast proper is to cause development of both ducts and alveoli (Benson *et al* 1959). The effects in the human are much less clear. Certainly in girls with gonadal agenesis, it is possible to secure primary breast bud development with low doses of oestrogen alone and to effect reasonable gross breast development with larger doses of oestrogens.

It is a well-known fact that women tend to be shorter than men of the same ethnic group. It is also known that girls stop linear growth earlier than boys and that hypogonadotrophic eunuchoid girls are taller after the age of normal puberty than sexually mature age matched controls. Because of this it is reasonable to conclude that the ovaries normally exert a significant effect on linear growth.

Having said this the facts start to run very thin. Every clinician knows that the bone age of hypogonadal girls shows a significant lag after the age of normal puberty. Every clinician knows that when therapeutic doses of oestrogens are given to these girls there is rapid cessation of linear growth accompanied by a very rapid increase in bone age with concurrent epiphyseal closure.

The most reliable data on human bone growth come from anatomical studies (Burrows 1949). Carpal ossification centres appear several months earlier in girls than boys while epiphyseal fusion occurs from 6 months earlier (lower radial) to 2 to 3 years earlier (metatarsal) in girls.

Animal data suggest that this effect is specifically ovarian, and within the ovary, specifically oestrogenic.

The effect of oestrogen on bone calcification is unresolved. While there is good evidence that large doses of oestrogens may increase bone density in some laboratory animals, the role in the human is much less clear. The evidence that post-menopausal osteoporosis is due to oestrogen deficiency is no more cogent than other evidence suggesting that the changes are related to calcium of fluoride intake. At the moment there is no very good reason to regard post-menopausal osteoporosis as an entity different from the loss of bone that occurs with age in either sex (Hall *et al* 1969).

Although there are hormonal factors at work in the shaping of the true bony pelvis, it now seems clear that differences in the pelvic brim, once thought to be hormonally determined, are in fact influenced by environment. Comparing the pelvic brim in men and women from the same and different socio-economic groups, there is very little difference apart from that due to social class (Thoms & Greulich 1940).

The effects of oestrogens on fat storage are not clear. The characteristic distribution of fat in the female appears to be oestrogen dependent but the mechanisms controlling this distribution are not known.

The haemodynamic effects of oestradiol are complex and ill understood. There does appear to be a specific effect causing dilation of small vessels and increased capillary permeability. The effect of oestrogens on plasma volume and the renin-angiotensin system is discussed by Walters in chapter 12. The effects on the cellular content of blood and bone marrow are described by Castaldi and Hocking in chapter 15.

All of the transporting globulins are effected by oestrogen stimulation. An oestrogen-dependent increase in transcortin, transferrin and thyroid binding globulin is well documented.

General metabolic effects

It is increasingly clear that oestrogens have protean effects on many metabolic pathways. Because of the clinical relevance, most recent efforts have been aimed at the two synthetic oestrogens in the commercially available oral contraceptives (ethinyl oestradiol and mestranol). Regrettably no definite statement can be made as to whether oestradiol shares all of the properties of these synthetic oestrogens and the following summary relates mainly to these synthetic compounds. A full review of the problem will be found in Sandler & Billing (1970). Where further references are not indicated in the text, this reference should be used.

In chapter 11 Turtle summarizes the evidence that, at least in pregnancy, oestrogens may induce end organ resistance to glucose transport in peripheral tissues.

In experimental studies much work has been done with patients taking combined oestrogen progestin pills. There is fairly good evidence now that most of these metabolic effects are related to the oestrogen rather than the progestin (Editorial comment 1971).

Synthetic oestrogens cause a dose related increase in total, free and protein bound cortisol. It is only, however, with the very high dose of oestrogens given to patients with prostatic carcinoma that there is a real increase of urinary free cortisol excretion—the best index of increased tissue exposure to free cortisol throughout the day. It has been suggested that the impairment of carbohydrate tolerance seen in some women as a result of ingestion of ethinyl oestradiol or mestranol is in turn due to changes in free cortisol levels, but at the doses used clinically this view remains unsubstantiated.

The increase in serum cholesterol, serum triglycerides, plasma insulin and growth hormone found in many patients taking oral contraceptives appears to be oestrogen related. Once more it must be emphasized that it is not known whether or not these metabolic characteristics are shared by oestradiol.

The effects of pregnancy on coagulation factors are discussed by Castaldi and Hocking in chapter 15. Brief reference must be made here to the effects of oestrogens on these parameters in non-pregnant women although the effects remain ill defined. Exposure to ethinyl oestradiol and mestranol causes an increase in factors VII and X and increased platelet aggregation. Since these effects are not produced by progestins and do occur in pregnancy, it is probable that natural oestrogens would have the same effect. The changes in platelet aggregation are accompanied by striking changes in platelet electrophoretic activity.

The emotional stresses of the pre-menstrual tension syndrome suggest that ovarian hormones may affect mood profoundly. The emotional vagaries of the normal menstrual cycle have been the subject of many theories, endless speculation but an almost total absence of facts. There is good evidence that a feature of depressive illnesses is an abnormally low conversion of tryptophan to 5-hydroxytryptophan. Oestrogens have been shown to increase the conversion of tryptophan to nicotinic acid derivatives. If experience with the rat holds good in the human this would be at the expense of 5-hydroxytryptophan synthesis and could lead to depression.

The cellular action of oestradiol

Although oestradiol has many extra-genital effects, the most dramatic effects are on the sexual target organs. There is no longer any doubt that the secret of this specificity is selective binding. The less dramatic but very important extra-genital effects may be based on a different mechanism and will be discussed shortly.

The first evidence of specific hormonal binding came from the experiments of Jensen & Jacobson (1962). They showed that if labelled oestradiol was given in physiological quantities to castrated rats, the pattern of incorporation seen in responsive tissues was different from that seen in non-responsive organs. Non-responders (such as liver, kidney and muscle) showed tissue concentrations closely resembling blood concentrations, with an early maximum concentration, followed by a rapid decrease. Responsive tissue, such as the uterus, vagina and pituitary, continued to incorporate and retain a high concentration of oestradiol for a much longer period (Fig. 2.7).

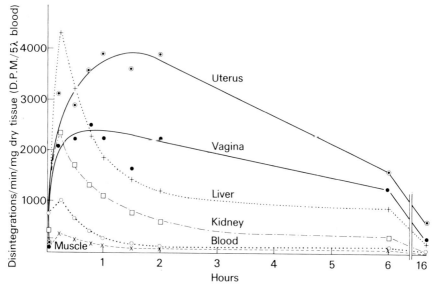

FIG. 2.7. Concentration of radioactivity in rat tissues after single subcutaneous injection of oestradiol-6,7-H^3. From Jensen & Jacobson (1962) *Rec. Progr. Horm. Res.* **18**, 390.

In addition, in non-reactive tissues, such as liver, there is rapid conjugation of oestradiol, with significant conversion to oestrone and probably 2-methoxy-oestrone. In the uterus, however, the steroid remains predominantly as unconjugated oestradiol and appears to be effective without itself being metabolized.

Oestrone differs chemically from oestradiol only in the presence of an oxo instead of a hydroxyl group at C17 but has biologically only one-tenth its potency. The reasons for this are emerging. The uterus does not display the same ability to bind oestrone as it does oestradiol, the concentration gradients for oestrone in the uterus being somewhat similar to those found for oestradiol in non-reactive tissue. About one-tenth of administered oestrone is reduced to oestradiol and the uterine

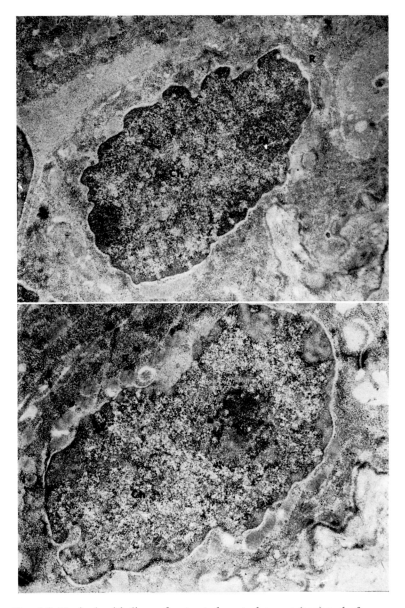

FIG. 2.8. Vaginal epithelium of untreated control mouse (top) and of mouse 30 min after oestradiol treatment (bottom) stained with uranyl acetate, EDTA and lead nitrate. Note heavier contrast in the nucleolus (N), nuclear chromatic region (Ch), and cytoplasmic ribosomes (R) in the section from the hormone-treated animal. Magnification ×14,000. From Irina Pollard (1970) *J. Endocrin.* **47**, 143.

behaviour to this converted portion produces the pattern of native oestradiol.

Gorski et al (1968) have isolated a binding component for oestradiol in the reproductive tract, while Rochefort & Baulieu (1969) indicate two types of binders, one having 'great affinity, limited capacity and high specificity', the other having smaller affinity and an apparently unlimited capacity. Gorski et al (1968) suggest as a model, that oestrogens are bound by this target-tissue specific 9·5 S receptor and that this large receptor-oestrogen complex migrates into the nucleus.

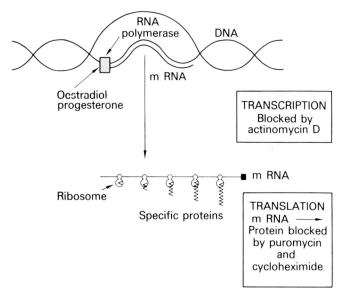

FIG. 2.9. Site of action of steroid hormones. Oestradiol and progesterone have been shown to act by modifying gene expression (transcription). From K.J.Catt (1970) *Lancet* **i**, 763.

Morphological evidence has shown that the earliest observable ultrastructural change in vaginal epithelium after local application of oestradiol occurred in the nucleolus. It was suggested that these changes were due to RNA synthesis (Pollard et al 1966). RNA produced in this way will reproduce the stimulating effects of oestradiol itself (Unhjem et al 1968). The elegant chemical work of the Glasgow group (Billing et al 1969a and b) has demonstrated clearly the effect of oestradiol on the transport of RNA precursors into the cell and the subsequent synthesis of RNA, while Pollard (1970) has provided eloquent ultrastructural confirmation that this ferment, although dominant in the nucleus, also occurs in the nuclear chromatic region and ribosomes (Fig. 2.8). A summary of the suggested sequence is shown in Fig. 2.9 (Catt 1970).

It has been suggested (Morel 1969) that hormones can be classified because of their mode of action into two groups; those that modify the differentiation of receptor cells in a qualitative sense (for example steroids) and exert their effect without invoking cyclic AMP and those that quantitatively modify the function of receptor cells and act via cyclic AMP (for example the pituitary trophic hormones) the hormone simply regulating the cell's intensity. Although this may have been a reasonable division there is now cogent work to indicate that oestradiol too affects the cyclic AMP of its receptor cells while cyclic AMP can increase intracellular RNA synthesis (Talwar 1970).

In addition to these actions on specific target cells the observations of Billing *et al* (1969a and b) together with the older observation of the very early increase in water content in oestradiol stimulated tissue, provide suggestive evidence by which to explain some of the extra-genital effects of oestrogens—that is a non-specific effect on membrane permeability.

PROGESTERONE

General actions

Progesterone has little effect on the genital tract if administered alone, but striking effects if given after oestradiol. Under its influence the typical oestrogen-dependent vaginal squames of oestrogen stimulation disappear and cells with larger nuclei appear together with a leucocytic infiltration (see below).

Although progesterone if given alone to a castrated animal will produce some hyperaemia and stromal mitoses, these effects are negligible compared with those seen after prior oestrogen administration. Progesterone causes the typical secretory endometrium of the luteal phase of the cycle with evident glycogen formation. Uterine motility under the influence of progesterone is characterized by spontaneous contractions of low frequency and high amplitude—previously B waves and discussed more fully by Wood (chapter 7).

Cervical mucus characteristically becomes thicker, viscous, opaque and infiltrated with leucocytes. The ultrastructural characteristics are very different from those seen in oestrogen stimulated mucus (Singer & Reid 1970).

The effects of progesterone on the human breast remain obscure. Although widely stated to cause glandular development, the evidence for this is slender (Benson et al 1959).

It is also frequently stated that progesterone causes smooth muscle relaxation in the gut and ureter. Examination of the data suggests that much of it is anecdotal or based on unphysiological experiments (Hytten & Leitch 1964).

Metabolic effects

The best documented of these effects is on basal body temperature, seen clinically to occur in the luteal phase of the cycle. The mechanism of this action is not clear, but it is a steroid-dependent phenomenon shared by etiocholanolone.

Galletti & Klopper (1962) have shown that freely fed rats treated with progesterone store more fat than untreated controls. Unfortunately it is not known if this is due to increased calorie intake, decreased calorie expenditure or altered metabolism. The central sedative action of progesterone—seen clinically in early pregnancy and reproducible in treated males (Hytten & Leitch 1964) may explain part of the increased fat storage.

The reduction in alveolar and arterial P_{CO_2} caused by progesterone is discussed by Read and Woolcock in chapter 14.

Other metabolic effects of progesterone have been reviewed by Fotherby (1964). It will diminish the sodium retaining effects of aldosterone, and also increase aldosterone secretion rate. In normal subjects progesterone in a dose of 50 mg daily causes a prompt increase in urinary sodium excretion.

Because of the widely stated belief that some synthetic progestogens are anabolic and others are not it is important to emphasize that progesterone itself is catabolic increasing the excretion of urinary nitrogen. This effect may also be observed in the luteal phase of normal ovulatory cycles but cannot be seen in anovulatory cycles.

The cellular actions of progesterone

There is much less known of progesterone binding than of oestradiol binding. Following in the wake of Jenson and Jacobson, Falk & Bardin (1970) showed specific uptake of tritiated progesterone by the uterus when compared with non-target tissue such as heart or diaphragm. While their work indicates that there was no competitive binding by cortisol, concurrent work by Milgrom & Baulieu (1970) showed a highly specific but low capacity binding site in the rat myometrium very similar to corticosteroid binding globulin. It may be expected that this dilemma will be resolved before long, and while specific binding of progesterone to the myometrium is very probable, the nature and site of the binding protein remains to be determined. The French workers appear confident that the binding protein is intranuclear.

As noted above, oestradiol is effective in the target cell without itself being metabolized. Bryson & Sweat (1969) have produced good evidence that progesterone pays a higher price to exact its metabolic toll in the cell, being broken down to pregnenolone, pregnanolone, pregnenedione and pregnanedione.

Although the intracellular binding and metabolism of progesterone are clearly different from oestradiol it has one important effect in common—action at transcriptional and then translational level to increase synthesis of messenger RNA (O'Malley & Maguire 1969).

Ovarian hormones and human sexuality

'Investigators of the role of gonadal hormones in human eroticism are more restricted by the modes of our culture than are students in other areas of biologic and medical research' (Money 1961).

Human libido, and the direction in which it will be aimed is modified by environmental and hormonal factors. The detailed studies of the Johns Hopkins group on the psychosexual orientation of intersexes leaves little doubt that general psychosexual orientation is dominated by the sex of rearing. Although the hypothalamus and subsequent behaviour of female foetuses of many species may be 'masculinized' by exposure to androgens *in utero* or shortly after birth, the evidence for this in the human is not so clear. The closest common clinical situation to the experimentally androgenized animal foetus is a female foetus with congenital adrenal hyperplasia. These children, providing they are reared as females are essentially feminine in psychosexual outlook, although they do show more aggression than normal female controls. It could be argued that at a foetal level, feminine orientation as with feminine internal and external genital sex is the neutral norm, unaffected by oestrogens, requiring exposure to a competent testis to be modified along male lines (see Liggins, chapter 4).

It is even difficult to assess the effect of ovarian hormones on behavioural maturation at the time of puberty. It is clearly fallacious to assess psycho-sexual behaviour in a group of 'normal' female adolescents and a group of hypogonadal girls of the same age and then attribute differences in emotional maturation to the presence or absence of ovarian function. 'When a teen-ager with sex-hormonal failure is responded to as a juvenile from all quarters, the chances are very great that she will respond as a juvenile and lag behind in psychologic and behavioural development as she gets older' (Money 1961).

It is a common clinical experience that castration in the mature female—or other causes of secondary hypogonadism in an otherwise healthy female—does not usually affect sexual behaviour in those women who have acquired the learned habit of pleasurable sexual response. Equally, in otherwise normal women complaining of frigidity, it has not been possible to demonstrate any abnormalities of ovarian hormone production.

Ovarian hormones and vaginal cytology

Reference has already been made to the response of vaginal epithelium to oestrogens and progesterone. In normal women the cyclicity of hormonal secretion produces a predictable response in the vaginal epithelium which many people find useful clinically. The operative word in the last sentence is 'normal'. Women who are producing excessive amounts of androgens, or who are taking an anti-oestrogen such as clomiphene are difficult subjects for interpretive hormonal cytology.

Different descriptive terms are used by different workers to quantitate these changes—karyopyknotic index, maturation index, eosinophilic count. There is an underlying theme common to them all. A useful review will be found in Wied (1968). The various classifications are based on the proportions of three cell types. The superficial cells are those cells exfoliated from the most highly proliferated layers of epithelium. The cells are mature, polygonal squames with a pyknotic nucleus and eosinophilic or cyanophilic cytoplasm. The intermediate cells are somewhat less mature but still well differentiated. The cytoplasm is eosinophilic or cyanophilic and the nucleus non-pyknotic. The Basal-Parabasal cells come from the least mature and deeper layers of the epithelium. They are small oval or round immature squamous cells. The general appearances may be further modified by cell folding, leukocytic infiltration and of course in pathological conditions, abnormal constituents.

A summary of the changes throughout the normal cycle is shown diagrammatically in Fig. 2.10 (Wied 1968).

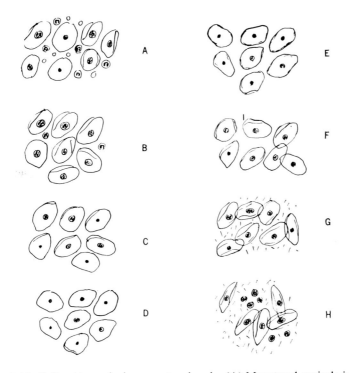

FIG. 2.10. Cell patterns during menstrual cycle. (A) Menstrual period: intermediate and superficial cells, leukocytes, and erythrocytes. (B) Early proliferative phase: mostly intermediate cells, decrease in number of leukocytes. (C) Middle proliferative phase: increase in number of superficial cells, present cells demonstrating decreased folding. (D) Time of ovulation: maximal proliferation, flattening of cells, some cells exhibiting extremely fine outlines, maximal number of superficial cells with eosinophilic cytoplasm. (E) Post-ovulatory phase: more prominent cell borders as observed at time of ovulation, beginning decrease of the KPI. (F, G, H) Luteal phase (early, middle, late): gradual decrease in number of superficial cells, increased number of folded and crowded intermediate cells, increased amount of *Bacillus vaginalis* Doderlin, moderate to marked cytolysis. From G.L.Wied (1968) in Jay Gold (ed.) *Textbook of Gynecologic Endocrinology*. New York, Hoeber.

THE MENOPAUSE

This term means the final cessation of menstruation and post-menopausal implies the portion of a woman's life from the time of physiological cessation of menstruation until death. The term relates only to menstrual bleeding, climacteric being used more correctly to cover the broader group of symptoms occurring at this time in a woman's life.

The menopause is a physiological event and warrants discussion in a text of this type. The mean age of the menopause in Australia is 51. Cessation of ovulation and cessation of ovarian hormonal function are not usually concurrent, the former preceding the latter. The decline in ovarian hormones production causes the expected increase in gonadotrophin secretion. During the normal climacteric these events occur relatively slowly and are poorly documented in an endocrine sense. A telescoped version is seen after ovarian irradiation and is shown in Fig. 2.11 (Brown *et al* 1958).

The cardinal symptom of the menopause is the hot flush or heat flash. This has been attributed both to high gonadotrophin levels or oestrogen withdrawal. Since treatment with oestrogens clearly increases oestrogen levels, controls the flushes and depresses gonadotrophin secretion it is difficult to dismiss oestrogen deprivation as an important contributory factor. However, further thought indicates that the position is more complicated. Many patients with idiopathic hypothalamic secondary amenorrhoea have endogenous oestrogen production at the same level as post-menopausal women, yet never complain of hot flushes. Most patients with gonadal agenesis or dysgenesis have 'post-menopausal' levels of gonadotrophins, yet neither do they complain of hot flushes.

The human female is the only example of a species outliving its reproductive life. Why this is, and what factors control the onset of physiological ovarian failure are not known.

SECRETION AND EXCRETION PATTERNS OF GONADOTROPHINS AND OVARIAN STEROIDS IN OVULATORY CYCLES

GONADOTROPHINS

The problems of measurement of gonadotrophins are discussed in Appendix B. The data presented here appear to be the best available at

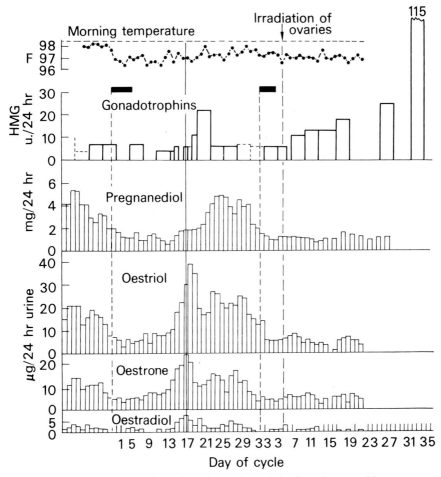

FIG. 2.11. Urinary steroid and total gonadotrophins in a 41-year-old woman with breast cancer. Estimations are made during a normal cycle and following ovarian irradiation. From J.B.Brown et al (1958) J. Endocrin. 17, 401.

the time of writing but are not meant to be regarded as definitive statements.

It is easy to understand why information about the FSH and LH content of the human pituitary at various stages in the menstrual cycle is very scanty. In the monkey, the pituitary content of FSH and LH is highest between the ninth and eleventh days of the cycle (Simpson et al 1956). The most complete pituitary study in the human is from the autopsy studies of Bischoff et al (1969). These workers showed a high content of FSH in the early menstrual phase while it was lowest in the luteal phase. There was a marked depletion of LH in the immediate post-ovulatory phase.

Once we leave the pituitary and move to biological fluids such as blood and urine, more information is available.

The classic criteria of reliability applied to hormone assays are those of accuracy, precision, specificity and sensitivity. It is only in the last half of the 1960s that assays for specific gonadotrophins in blood and urine have satisfied these fundamental requirements.

Much of the older literature, and an unfortunate amount of current literature, refers to measurement of 'total gonadotrophins'. These results are without much biological meaning, as the end point for quantitation is not very sensitive and the methods used usually measure an unspecified and inconstant mixture of FSH and LH. The development of immunoassay and then radioimmunoassay has cleared up much of this confusion, but there is still conflict in attempts to correlate bioassay with radioimmunoassay.

In 1966, Bagshawe *et al* convincingly documented levels of LH in plasma and urine throughout the cycle. These results showed an unequivocal mid-cycle peak of LH in both fluids, with low levels throughout the rest of the cycle. This basic observation has been confirmed by others (Odell *et al* 1967, Neill *et al* 1967) and further refinement has been added by the studies of Cargille *et al* (1969). In essence, LH levels are lower in the luteal phase than in the follicular phase, and the only clear peak is seen at mid cycle, coinciding in most women with the second and lesser peak of FSH (*vide infra*). After this peak, levels drop rapidly and remain low throughout the luteal phase. During menstruation they rise slowly, reaching the follicular plateau by day four, which remains relatively constant until the mid-cycle peak starts to develop about day ten. (All of these changes refer to an idealized 28-day cycle.)

The development of a satisfactory radioimmunoassay for FSH presented greater problems than with LH, and earlier bioassays had produced a confusing picture. In 1967, Faiman and Ryan using radioimmunoassay indicated that there were two cyclic peaks of FSH, one early in the proliferative phase and the second at or near the time of ovulation. Despite disagreement by some groups, the work of Cargille *et al* (1969) has confirmed and refined these observations. They describe FSH levels increasing rapidly from the onset of menstruation and reaching a follicular peak on day four. There is then a rapid drop, the lowest levels occurring just before mid cycle, followed by a lesser, but clear peak, coinciding with the mid-cycle LH peak described above. The levels then drop abruptly in the luteal phase and continue to decline until several days before the onset of the next period. There is then a slow rise which continues throughout menstruation to reach the peak on day four of the subsequent cycle described above.

The final word has not yet been said about gonadotrophin levels in the normal cycle. A reasonable and current view is shown in Fig. 2.12 where the results of radioimmunoassay and bioassay in the same patients are plotted (Stevens 1969).

The precise relationships of these mid-cycle peaks of FSH and LH to the climax of ovulation in the human have been subject to dispute. The very careful work of Yussman & Taymor (1970) offers the best evidence yet available. They had the considerable advantage of being able to study plasma levels at 8-hour intervals over a 5-day period in women subject to laparotomy early in the postovulatory phase. LH showed a significant rise 24 hours before ovulation and peaked 16 hours prior to ovulation (Fig. 2.13). The mid-cycle peak of LH is therefore pre-ovulatory.

OVARIAN STEROIDS

Progesterone

There is general agreement that plasma progesterone is low in the follicular phase and high during the luteal phase (Neill *et al* 1967, Cargille *et al* 1969). As with the gonadotrophins, there has been real difficulty in relating mid-cycle changes to the actual time of ovulation and the problems of obtaining multiple blood samples from volunteers have contributed materially to the uncertainty. Because of the intricate relationship of progesterone secretion and gonadotrophin production clarification is very important.

The work of Yussman & Taymor (1970) confirms that there is a rise in plasma progesterone 16 hours *before* ovulation but following the first rise in plasma LH (Fig. 2.13). An important corollary of this is that progesterone production begins to rise before luteinization of the follicle. Another is that the initiation of the mid-cycle LH peak is not progesterone dependent although the very sharp following surge may be.

For historical and technical reasons, urinary assays of pregnanediol have been used for many years as an index of progesterone production.

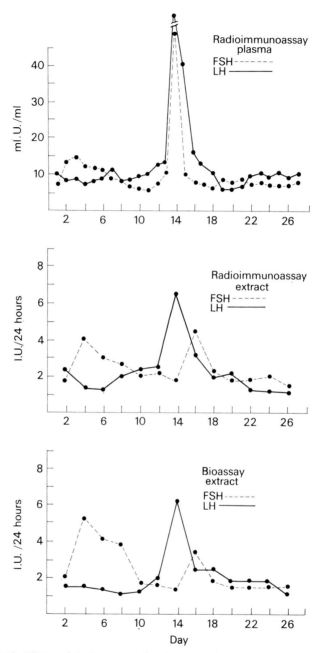

FIG. 2.12. FSH and LH patterns in plasma and urinary extract by radio-immunoassay and in urinary extract by bioassay during the menstrual cycle. From V.C.Stevens (1969) *J. clin. Endocrin.* **29**, 904.

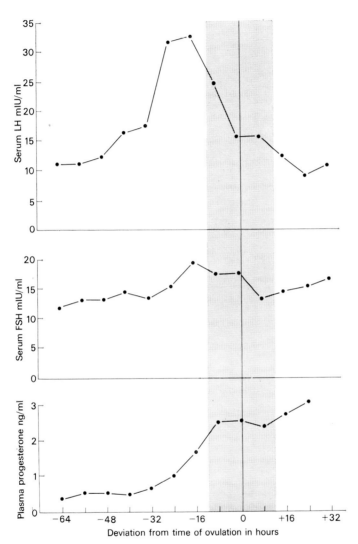

FIG. 2.13. Mean 8-hour serum LH, serum FSH and plasma progesterone in five subjects related to ovulation. From M.A.Yussman & M.L.Taymor (1970) *J. clin. Endocrin.* **30**, 396.

Although it must be accepted that the correlation between plasma progesterone and urinary pregnanediol is neither constant nor close in a quantitative sense, qualitatively, there is good correlation. The original observations of Klopper (1957) based on the first adequate chemical method for urinary assay have been amply confirmed. Throughout the follicular phase, pregnanediol excretion is low, usually less than 1 mg daily. This level is not dissimilar to that seen in men. The parent progesterone for this urinary offspring comes from the adrenal cortex and it is only after ovulation that ovarian progesterone secretion begins to influence urinary pregnanediol. Peak levels of between 2·5 and 8 mg per 24 hours are reached before the pre-menstrual decline.

Oestrogens

The behaviour of plasma oestradiol throughout the normal ovulatory cycle, and its relationship to the pre-ovulatory peak of LH has been the subject of careful study (Mikhail *et al* 1970, Abraham & Klaiber 1970). These studies show a mid-cycle peak of oestradiol and a secondary increase during the luteal phase. *Vis-à-vis* LH, the mid-cycle rise in oestradiol precedes the LH peak and may therefore play an important role in the LH surge (Fig. 2.14).

The striking contributions made by J.B.Brown to knowledge of urinary oestrogen patterns during the cycle (Brown 1955, Brown & Matthew 1962) provide one of the watersheds in gynaecological endocrinology. His development of a suitable chemical method for assay of oestradiol, oestrone and oestriol in urine made possible remarkable advances in this area. His original observations during the normal cycle have been fully confirmed, and the mean values and normal ranges are shown in Fig. 2.15. Typically the pattern is biphasic. During the first 2 or 3 days of menstruation, oestrogen excretion is low. In a 28-day cycle there is then a fairly rapid rise to a well-defined peak at mid cycle, termed the ovulation peak. This is followed by a fall in oestrogen excretion and then a second broader rise—the luteal maximum. In the last few days of the cycle there is a decrease in oestrogen excretion and menstruation follows.

Other steroids

As might be expected from the biosynthetic pathways, steroids other

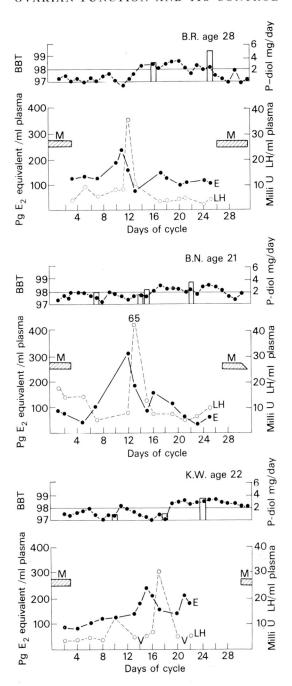

FIG. 2.14. Plasma patterns of immunoreactive oestrogens (E_2) and LH during three normal menstrual cycles. From G.E.Abraham & E.L.Klaiber (1970) *Amer. J. Obstet. Gynec.* **108,** 528.

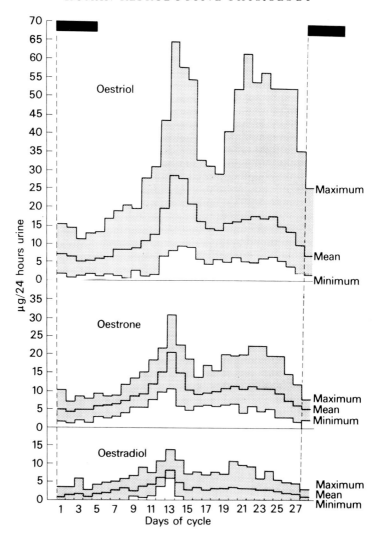

FIG. 2.15. Urinary oestrogens during the normal cycle. Cumulative data from sixteen subjects, aged 18–41 years. From J.B.Brown & G.D.Matthew (1962) *Rec. Progr. Horm. Res.* **18,** 337.

than the classical ovarian hormones of oestrogens and progesterone show cyclic changes throughout the menstrual cycle.

There is a distinct cyclicity in pregnanetriol and testosterone. The former might be expected because of the important intermediary role of

17-hydroxyprogesterone and 17-hydroxypregnenolone in oestrogen synthesis. It is interesting, therefore, that an increase in pregnanetriol excretion is only seen during the luteal phase of the cycle, more or less following the pattern of pregnanediol excretion. In plasma, a rise in 17-hydroxyprogesterone can be demonstrated coincidental with the LH peak and occurring before ovulation (Strott *et al* 1969).

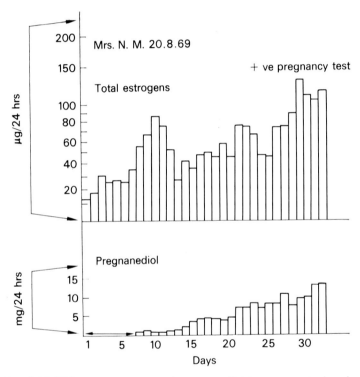

FIG. 2.16. Urinary oestrogens and pregnanediol in a conceptual cycle.

Pregnanetriol is biologically inert. Testosterone is, on the other hand, the most potent androgen known. Changes in plasma levels related to ovulation are quite clear, levels being higher in the luteal phase than in the follicular phase (Loraine & Bell 1966). This has clinical relevance when one remembers the cyclic fluctuations of acne in many adolescent girls—and the dramatic improvement seen in many of them when this and other ovarian androgens are inhibited by oral contraceptives.

CONCEPTUAL CYCLES

It might be reasonable to expect that a conceptual cycle should show the same gonadotrophin and steroid profile as an ovulatory cycle in which pregnancy did not occur, until the impact of the trophoblast becomes apparent with implantation about 7 days after fertilization. This reasonable expectation is valid. With suitable methods, a rise in plasma HCG can be detected about 11 days after fertilization and it is the advent of this unique hormone that transforms the pattern of a conceptual cycle. Unlike LH, HCG is luteotrophic and the effect on both the life span and steroid output of the corpus luteum is visible before the first missed period (Fig. 2.16). It will be noted that instead of abating about day 24, there is a continued increase in oestrogen and pregnanediol excretion. Very probably some of this increase comes from the truly infant trophoblast; the various contributions of corpus luteum and trophoblast to steroid output at this very early stage in pregnancy remains to be determined.

SUMMARY

It might now be reasonable to attempt a brief synthesis and summary of hypothalamic–pituitary–ovarian interactions. It should be clearly understood however, that much of this summary is speculative.

It is assumed that of the two hypothalamic centres controlling LH release one of these is a tonic centre responsible for the background, small, but vital levels of LH seen at all stages of the cycle. Starting at the time of menstrual bleeding, the falling levels of oestrogen, by long feedback, cause an increase in FSH levels via FRF, giving the early follicular rise in FSH. The consequent maturing follicle(s) produce increasing quantities of oestrogens and both these, via long feedback and FSH itself via short feedback, cause rapid inhibition of FRF. As the follicle(s) mature there is a rapid increase in oestrogens which release a postulated inhibitor of the second LRF centre causing a sudden flood of LH which triggers ovulation. The sudden post-ovulatory drop in oestrogens causes a short surge of FSH, while the LH peak is directly inhibited by short LH feedback on the releasing centres and then the long feedback of progesterone. As luteal oestrogens and progesterone rise further, FSH falls again and then as the pre-menstrual decline in luteal function reduces steroid output, FSH commences to rise again just before menstrual bleeding and the cycle starts again.

For those who prefer the precision of mathematical models to serve as a basis for computer simulation studies of the ovarian cycle, the reviews of Schwartz (1969) and Vande Wiele *et al* (1970) are recommended.

REFERENCES

ABRAHAM G.E. & KLAIBER E.L. (1970) Plasma immunoreactive estrogens and LH during the menstrual cycle. *Amer. J. Obstet. Gynec.* **108,** 528.

ALEXANDER B., DOWD A.J. & WOLFSON A. (1970a) Effects of prepubertal hypophysectomy and ovariectomy on hydroxyindole-O-methyltransferase. Activity in the female rat. *Endocrinology* **86,** 1166.

ALEXANDER B., DOWD A.J. & WOLFSON A. (1970b) Effect of continuous light and darkness on hydroxyindole-O-methyltransferase and 5-hydroxytryptophan decarboxylase activities in the Japanese quail. *Endocrinology* **86,** 1441.

ARMSTRONG D.T. (1968) Gonadotropins, ovarian metabolism, and steroid biosynthesis. *Rec. Progr. Horm. Res.* **24,** 255.

BAGSHAWE K.D., WILDE C.E. & ORR A.H. (1966) Radio immuno assay for human chorionic gonadotrophin and luteinizing hormone. *Lancet* **i,** 1118.

BAIN J. & EZRIN C. (1970) Immunofluorescent localization of the LH cell of the human adenohypophysis. *J. clin. Endocr.* **30,** 181.

BELING C.G., MARCUS S.L. & MARKHAM S.M. (1970) Functional activity of the corpus luteum following hysterectomy. *J. clin. Endocr.* **30,** 30.

BENSON G.K., COWIE A.T., FOLLEY S.J. & TINDAL J.S. (1959) Recent developments in endocrine studies on mammary growth and lactation, in C.W. Lloyd (ed.) *Recent Progress in the Endocrinology of Reproduction,* pp. 457. New York, Academic Press.

BILLING R.J., BARBIROLI B. & SMELLIE R.M.S. (1969a) The mode of action of oestradiol. I. The transport of RNA precursors into the uterus. *Biochem. Biophys. Acta* **190,** 52.

BILLING R.J., BARBIROLI B. & SMELLIE R.M.S. (1969b) The mode of action of oestradiol. II. The synthesis of RNA. *Biochem. Biophys. Acta* **190,** 60.

BISCHOFF K., BETTENDORF G. & STEGNER H.E. (1969) FSH and LH content of the human pituitary during the menstrual cycle. *Arch. Gynäk.* **208,** 44.

BOWERS C.Y., SCHALLY A.V., ENZMANN F., BOLER J. & FOLKERS K. (1970) Porcine thyrotrophin releasing hormone is (pyro)glu-his-pro (NH_2). *Endocrinology* **86,** 1143.

BROWN J.B. (1955) Urinary excretion of oestrogens during the menstrual cycle. *Lancet* **i,** 320.

BROWN J.B., KLOPPER A. & LORAINE J.A. (1958) The urinary excretion of oestrogen, pregnanediol and gonadotrophins during the menstrual cycle. *J. Endocrin.* **17,** 40.

BROWN J.B. & MATTHEW G.D. (1962) The application of urinary oestrogen measurements to problems in gynaecology. *Rec. Progr. Horm. Res.* **18**, 337.

BRYSON M.J. & SWEAT M.L. (1969) Metabolism of progesterone in human myometrium. *Endocrinology* **84**, 1071.

BURROWS H. (1949) *Biological Actions of Sex Hormones*, 2nd ed. Cambridge University Press.

CALDWELL B.W., MAZER R.S. & WRIGHT P.A. (1967) Leuteolysin as affected by uterine transplantation in the Syrian hampster. *Endocrinology* **80**, 477.

CARGILLE C.M., ROSS G.T. & YOSHIMI T. (1969) Daily variations in plasma follicle stimulating hormone, luteinizing hormone and progesterone in the normal menstrual cycle. *J. clin. Endocr.* **29**, 12.

CATT K. J. (1970) Hormones in general. *Lancet* **i**, 763.

CHOW B.F., VAN DYKE H.I.P., GREEP R.C., ROTHEN A. & SHELOVSKY T. (1942) Gonadotrophins of the swine pituitary. III. Preparation and biological and physiochemical characterization of a protein apparently identical with Metakentrin (ICSH). *Endocrinology* **30**, 650.

CLEMENTI F., DE VIRGILUS G. & MESS B. (1969) Influence of pineal gland principles of gonadotrophin producing cells of the rat anterior pituitary gland; an electron microscopic study. *J. Endocr.* **44**, 241.

CRIGHTON D.B., SCHNEIDER H.P.G. & MCCANN S.M. (1970) Localization of LH releasing factor in the hypothalamus and neurohypophysis as determined by an *in vitro* method. *Endocrinology* **87**, 323.

CUSHING H. (1912) *The Pituitary Body and its Disorders. Clinical States produced by Disorders of the Hypophysis cerebri*. Philadelphia and London, Lippincott.

DAMON A., DAMON S.T., REED R.R. & VALADIAN I. (1969) Age at menarche of mothers and daughters with a note on accuracy of recall. *Hum. Biol.* **41**, 162.

Editorial comment (1970) Cyclic A.M.P.: the second messenger. *Lancet* **ii**, 1119.

Editorial comment (1971) Progestagen—only contraception. *Lancet* **i**, 25.

ESHKOL A. & LUNENFELD B. (1967) Purification and separation of follicle stimulating hormone (FSH) and luteinizing hormone (LH) from human menopausal gonadotrophin (HMG). *Acta Endocr.* **54**, 91.

FAIMAN C. & RYAN R.Y. (1967) Serum follicle stimulating hormone and luteinizing hormone concentrations during the menstrual cycle as determined by radioimmunoassays. *J. clin. Endocr. & Metab.* **27**, 1711.

FALK R.J. & BARDIN C.W. (1970) Uptake of tritiated progesterone by the uterus of the ovariectomized guinea pig. *Endocrinology* **86**, 1059.

FEVOLD H.L., HISAW F.L. & LEONARD S.L. (1931) The gonad stimulating and the luteinizing hormone of the anterior lobe of the hypophysis. *Amer. J. Physiol.* **97**, 291.

FOTHERBY K. (1964) The biochemistry of progesterone. *Vitamins & Hormones* **22**, 153.

GALLETTI F. & KLOPPER A. (1962) Effect of administration of progesterone on the quantity, and distribution of body fat in the female rat. Excerpta Medica. International Congress Series No. 51, 253.

GORSKI J., TOFT D., SHYAMALA G., SMITH D. & NOTIDES A. (1968) Hormone receptors: studies on the interaction of estrogen with the uterus. *Rec. Progr. Horm. Res.* **24**, 45.

GREEN J.D. & HARRIS G.W. (1947) The neurovascular link between the neurohypophysis and adenohypophysis. *J. Endocr.* **5**, 136.
GREEP R.O., VAN DYKE H.B. & CHOW B.F. (1942) Gonadotrophins of the swine pituitary. I. Various biological effects of purified thylakentrin (FSH) and pure metakentrin (ICSH). *Endocrinology* **30**, 635.
GREEP R.O. (1959) Discussion. *Rec. Progr. Horm. Res.* **15**, 139.
HALL R., ANDERSON J. & SMART G.A. (1969) *Fundamentals of Clinical Endocrinology*, p. 353. Pitman, London.
HAYNES R.C., SUTHERLAND E.W. & RALL T.W. (1960) The role of cyclic adenylic acid in hormone action. *Rec. Progr. Horm. Res.* **16**, 121.
HYTTEN F.E. & LEITCH I. (1964) *The Physiology of Human Pregnancy*, p. 147. Oxford, Blackwell.
IGARASHI M. & McCANN S.M. (1964) A hypothalamic follicle stimulating hormone-releasing factor. *Endocrinology* **74**, 446.
IGARASHI M., YOKATA N.J., EHARA, Y., MAYUZUMI R. & MATSUMOTO S. (1967) Induction of human ovulation with purified beef hypothalamic FSH-releasing factor, in C. Wood (ed.) *Proceedings of Vth World Congress of Gynaecology and Obstetrics, Sydney, Butterworth*, p. 349.
JENSEN E.V. & JACOBSON H.I. (1962) Basic guides to the mechanism of oestrogen action, in Gregory Pincus (ed.) *Recent Progress in Hormone Research: The Proceedings of the 1961 Laurentian Hormone Conference*, vol. XVIII. New York, Academic Press.
KAHWANAGO I., HEINRICHS W.L. & HERRMANN W.L. (1970) Estradiol 'Receptor' in hypothalamus and anterior pituitary gland: Inhibition of estradiol binding by SH-group blocking agents and clomiphene citrate. *Endocrinology* **86**, 1319.
KASTIN A.J., SCHALLY A.V., GUAL C., MIDGLEY A.R., BOWERS C.Y. & GOMEZ-PEREZ F. (1970) Administration of LH-releasing hormone in selected subjects. *Amer. J. Obstet. Gynec.* **108**, 177.
KIRTON K.T., PHARRISS B.B. & FORBES A.D. (1970) Luteolytic effects of prostaglandin F_2a in primates (34464). *Proc. Soc. Exp. Biol. Med.* Quoted by Speroff Ramwell. **133**, 1, 314.
KLOPPER A.I. (1957) Excretion of pregnanediol during the normal menstrual cycle. *J. Obstet. Gynec. Brit. Emp.* **64**, 504.
KUEHL F.A., HUMES, J.L., TARNOFF J., CIRILLO Y.J. & HAM E.A. (1970) Prostaglandin receptor site: Evidence for an essential role in the action of luteinizing hormone. *Science* **169**, 883.
LERNER A.B., CASE J.D. & TAKAHASHI Y. (1960) Isolation of melatonin and 5-methoxyindole-3-acetic acid from bovine pineal glands. *J. biol. Chem.* **235**, 1992.
LOEB L. (1923) The mechanism of the sexual cycle with special reference to the corpus luteum. *Amer. J. Anat.* **32** 304.
LORAINE J.A. & BELL E.T. (1966) The clinical application of testosterone assays in blood, in *Hormone Assays and their Clinical Application*, 2nd ed.
McCANN S.M., TALEISNI S. & FRIEDMAN H.M. (1960) LH-releasing activity in hypothalamic extracts. *Proc. Soc. Exp. Biol. Med.* **104**, 432.
McCRACKEN J.A., GLEW M.E. & SCARRAMUZZI R.J. (1970) Corpus luteum regression induced by prostaglandins F_2a. *J. clin. Endocr.* **30**, 544.

MAGEE K., BASINSKA J., QUARRINGTON B. & STANCER H.C. (1970) Blindness and menarche. *Life Sciences* **9**, 7.

MARSHALL F.H.A. (1937) On the change over in the oestrous cycle in animals after transference across the equator, with further observations on the incidence of the breeding seasons and the factors controlling sexual periodicity. *Proc. roy. Soc.* Series B, **122**, 413.

MARTINI L., FRASCHINI F. & MOTTA M. (1968) Neural control of anterior pituitary functions. *Rec. Progr. Horm. Res.* **24**, 439.

MIKHAIL G., WU C. H., FERIN M. & VANDE WIELE R.L. (1970) Radioimmunoassay of plasma estrone and estradiol. *Steroids* **15**, 333.

MILGROM E. & BAULIEU E.E. (1970) Progesterone in uterus and plasma. I. Binding in rat uterus 105,000 g supernatant. *Endocrinology* **87**, 276.

MONEY J. (1961) Sex hormones and other variables in human eroticism, in W.C. Young (ed.) *Sex and Internal Secretions*, vol. II, p. 1383. Baltimore, Williams & Wilkins.

MOOR R.M. & ROWSON L.E.A. (1964) Influence of the embryo and the uterus on luteal function in the sheep. *Nature* **201**, 522.

MOOR R.M., HAY M.F., SHORT R.W. & ROWSON L.E.A. (1970) The corpus luteum of the sheep: effect of uterine removal during luteal regression. *J. Reprod. Fert.* **21**, 319.

MOREL F. (1969) Cyclic adenosine monophosphate, intracellular mediator of the action of numerous hormones. *Triangle* **9**, 119.

NEILL J.D., JOHANSSON E.D.B., DATTA J.K. & KNOBIL E. (1967) Relationship between the plasma levels of luteinizing hormone and progesterone during the normal menstrual cycle. *J. clin. Endocr.* **27**, 1167.

NICOLL C.S., PARSONS J.A., FIORINDO R.P., NICHOLS C.W. & SAKUMA M. (1970) Evidence of independent secretion of prolactin and growth hormone *in vitro* by adenohypophyses of rhesus monkeys. *J. clin. Endocr.* **30**, 512.

NICHOLSON P.M. (1970) A study of prolactin-like activity in individual human pituitary glands. *J. Endocr.* **48**, 639.

ODEBLAD E. (1968) The functional structure of human cervical mucus. *Acta obstet. gynec. scand.* **67**, Suppl. 1.

ODELL W.D., ROSS G.T. & RAYFORD P.L. (1967) Radioimmunoassay for luteinizing hormone in human plasma or serum: physiological studies. *J. clin. Invest.* **46**, 248.

O'MALLEY B.W. & MCGUIRE W.L. (1969) Progesterone-induced synthesis of a new species of nuclear RNA. *Endocrinology* **84**, 63.

O'MALLEY B.W., MCGUIRE W.L., KOHLER P.O. & KORENMAN S.G. (1969) Studies on the mechanism of steroid hormone regulation of synthesis of specific proteins. *Rec. Progr. Horm. Res.* **25**, 105.

PEAKE G.T., MCKEEL D.W., JARETT L. & DAUGHADAY W.H. (1969) Ultrastructural, histologic and hormonal characterization of a prolactin-rich human pituitary tumour. *J. clin. Endocr.* **29**, 1383.

PECKHAM W.D. & PARLOW A.F. (1969) Isolation from human pituitary glands of three discrete electrophoretic components with high luteinizing hormone activity. *Endocrinology* **85**, 618.

PETRUSZ P., ROBYN C. & DICZFALUSY E. (1970) Biological effects of human urinary follicle stimulating hormone. *Acta Endocr.* **63**, 454.

POLLARD I., MARTIN L. & SHOREY C.D. (1966) The effects of intravaginal oestradiol-3:17β on the cell structure of the vaginal epithelium of the ovariectomized mouse. *Steroids* **8**, 805.

POLLARD I. (1970) Ultrastructural evidence for the stimulation of nuclear ribonucleic acid synthesis by oestradiol-17B in the vaginal epithelium of the ovariectomized mouse. *J. Endocr.* **47**, 143.

RAMWELL P.W. & SHAW J.E. (1970) Biological significance of the prostaglandins. *Rec. Progr. Horm. Res.* **26**, 139.

ROCHEFORT H. & BAULIEU E.E. (1969) New *in vitro* studies of estradiol binding in castrated rat uterus. *Endocrinology* **84**, 108.

RYAN K.T. & PETRO Z. (1966) Steroid biosynthesis by human ovarian granulosa and thecal cells. *J. clin. Endocr.* **26**, 46.

SANDLER M. & BILLING B. (1970) The pill. Biochemical consequences. *J. clin. Path.* **23**, Suppl. 3 (Ass. Clin. Path.)

SAVARD K., MARSH J.M. & RICE B.F. (1965) Gonadotropins and ovarian steroidogenesis. *Rec. Progr. Horm. Res.* **21**, 285.

SAVARD K. (1967) In E.T. Bell & J.A.Loraine (eds.) *Recent Research on Gonadotrophic Hormones*, p. 170. Edinburgh, Livingstone.

SAYLER A. & WOLFSON A. (1968) Influence of the pineal gland on gonadal maturation in the Japanese quail. *Endocrinology* **83**, 1237.

SCHALLY A.V., MULLER E.E., ARIMURA A., BOWERS C.Y., SAITO T., REDDING T.W., SAWANO S. & PIZZOLATO F. (1967) Releasing factors in human hypothalamic and neurohypophysial extracts. *J. clin. Endocr.* **27**, 755.

SCHALLY A.V., ARIMURA A., BOWERS C.Y., KASTIN A.J., SWANO S. & REDDING T.W. (1968) Hypothalamic neurohormones regulating anterior pituitary function. *Rec. Prog. Horm. Res.* **24**, 497.

SCHALLY A.V. & KASTIN A.J. (1969) The present concept of the nature of hypothalamic hormones stimulating and inhibiting the release of pituitary hormones. *Triangle* **9**, 19.

SCHALLY A.V., ARIMURA A., BOWERS C.Y., WAKABAYASHI I., KASTIN A.J., REDDING T.W., MITTLER J.C., NAIR R.M.G., PIZZOLATO P. & SEGAL A.J. (1970) Purification of hypothalamic releasing hormones of human origin. *J. clin. Endocr.* **31**, 291.

SCHNEIDER H.P.G. & MCCANN S.M. (1970) Mono- and indolamines and control of LH secretion. *Endocrinology* **86**, 1127.

SCHNEIDER H. & MCCANN S.M. (1970) Estradiol and the neuroendocrine control of LH release *in vitro*. *Endocrinology* **87**, 330.

SCHWARTZ N.B. (1969) A model for the regulation of ovulation in the rat. *Rec. Progr. Horm. Res.* **25**, 1.

SHEARMAN R.P. (1970) Physiology of ovarian control and steroid biosynthesis, in R.R.Macdonald (ed.) *Scientific Basis of Obstetrics and Gynaecology*. London, J. & A. Churchill.

SHEARMAN R.P. & TURTLE J.R. (1970) Secondary amenorrhoea with inappropriate lactation. *Amer. J. Obstet. Gynec.* **106**, 818.

SHORT R.V. (1962) Steroids in the follicular fluid and the corpus luteum of the mare. A 'two-cell type' theory of ovarian synthesis. *J. Endocr.* **24**, 59.
SHORT R.V. (1964) Ovarian steroid synthesis and secretion *in vivo*. *Rec. Progr. Horm. Res.* **20**, 303.
SIMPSON M.E., VAN WAGENEN G. & CARTER F. (1956) Hormone content of anterior pituitary of monkey (*Macaca mulatta*), with special reference to gonadotrophins. *Proc. Soc. Ex. Biol. Med.* **91**, 6.
SINGER A. & REID B.L. (1970) Effect of the oral-contraceptive steroids on the ultrastructure of human cervical mucus—a preliminary communication. *J. Reprod. Fert.* **23**, 249.
SPEROFF L. & RAMWELL P.W. (1970) Prostaglandin stimulation of *in vitro* progesterone synthesis. *J. clin. Endocr.* **30**, 345.
STEVENS V.C. (1969) Comparison of FSH and LH patterns in plasma, urine and urinary extracts during the menstrual cycle. *J. clin. Endocr.* **29**, 904.
STROTT C.A., YOSHIMA T., ROSS G.T. & LIPSETT M.B. (1969) Ovarian physiology: relationship between plasma LH and steroidogenesis by the follicule and corpus luteum; effect of HCG. *J. clin. Endocr.* **29**, 1157.
SUTHERLAND E.W., ROBINSON G.A. & BUTCHER R.W. (1968) Some aspects of the biological role of adenosine $3'$, $5'$-monophosphate (cyclic AMP). *Circulation* **37**, 279.
TALWAR G.P. (1970) *Res. in Reprod.* **2**, 1.
TANNER J.M. (1968) Earlier maturation in man. *Sci. Am.* **218**, 21.
THOMS H. & GREULICH W.W. (1940) A comparative study of male and female pelves. *Amer. J. Obstet. Gynec.* **39**, 56.
UNHJEM O., ATTRAMADEL A. & SOLNA J. (1968) Changes in uterine RNA following stimulation with 17B-oestradiol. *Acta Endocr.* **58**, 227.
VANDE WIELE R.L., BOGUMIL J., DYRENFURTH I., FERIN M., JEWELEWICZ R., WARREN M., RIZKALLAH T. & MIKHAIL G. (1970) Mechanisms regulating the menstrual cycle in women. *Rec. Progr. Horm. Res.* **26**, 63.
WEID G.L. (1968) Evaluation of endocrinologic condition by exfoliative cytology, in J.J.Gold (ed.) *Textbook of Gynecologic Endocrinology*, p. 133. New York, Hoeber.
WURTMAN R.J. (1969) The pineal gland in relation to reproduction. *Amer. J. Obstet. Gynec.* **104**, 32.
YUSSMAN M.A. & TAYMOR M.L. (1970) Serum levels of follicle stimulating hormone and luteinizing hormone and of plasma progesterone related to ovulation by corpus luteum biopsy. *J. clin. Endocr.* **30**, 396.
ZACCHARIAS L. & WURTMAN R.J. (1964) Blindness: its relation to age of menarche. *Science* **144**, 1154.
ZONDEK B. (1930a) Über die hormone des hypophysenvorderlappens. *Klin. Wschr.* **ix**, 245.
ZONDEK B. (1930b) Hypophysenvorderlappen. *Arch. Gynäk.* **134**, 133.

3

The Physiology and Function of the Testes

BRYAN HUDSON & HENRY G. BURGER

INTRODUCTION

Normal testicular function confers upon the adult male two important attributes—virility and fertility—which represent the separate functions of the Leydig or interstitial cells and the seminiferous epithelium respectively. The purpose of this chapter is to provide the reader with a simple outline of the physiology and biochemistry of the testes with respect to these two functions, so far as these are known.

Because the biosynthesis of steroids is a function not only of the Leydig cells, but also of steroid secreting cells of the adrenal cortex and ovary, a separate Appendix A has been devoted to a discussion of this topic. Although pathways of steroid biosynthesis peculiar to the testis are outlined in this Appendix, they will be discussed in slightly more detail in this chapter. Likewise, the gonadotrophins—follicle stimulating hormone (FSH), luteinizing hormone (LH) or interstitial cell stimulating hormone (ICSH)—are important regulators of both testicular and ovarian function; for this reason, a discussion of the chemistry, the standardization and methods for the measurement of these hormones will be discussed separately in Appendix B. It is important to point out that the term 'luteinizing hormone' is used by those who deal with the female, while this same hormone is commonly termed 'interstitial cell stimulating hormone' by those who work with the male. For those who work with both males and females, these terms tend to be used interchangeably. As far as is known, these two hormones (LH and ICSH) have chemical identity. For a fuller discussion on the problems created by this nomenclature, the reader is referred to a reivew by Hall (1970). In this chapter LH will be used to describe the pituitary gonadotrophin that stimulates the interstitial cells of the testis.

EMBRYOLOGY AND DEVELOPMENT OF THE TESTIS

Over the past two decades, advances in cytogenetics and experimental embryology have helped clarify our understanding of the factors that determine the development and differentiation of the gonads from primordial germ cells. These cells, which can usually be identified between the 25th and 30th days of embryonic life, lie in the dorsal endoderm of the yolk sac. After increasing in number by mitotic division they migrate between the 5th and 6th weeks to the urogenital ridges which lie on either side of the dorsal mesentery. These primitive indifferent gonads are usually well demarcated by the 40th day of embryonic life, but at this stage it is not possible to identify them either as testes or ovaries. The factor which determines the destiny of this indifferent gonad is the sex-chromatin content of these cells.

At this stage, in the urogenital ridges two separate groups of cells can be identified, those derived from the coelomic epithelium and those of mesenchymal origin. Sheets of coelomic epithelial cells migrate into this mesenchyme, become condensed and form the primary sex cords. These coelomic and mesenchymal cells form the cortex and medulla of the developing gonad respectively, and from this time onwards the dominant development of one or other group of cells determines the differentiation of the gonad into testis or ovary. In the formation of the testis, it is the dominant development of medullary elements. In either instance primitive germ cells are carried into the mesenchyme with the migrating coelomic cells, ultimately to become the spermatogonia or oogonia. These cells do not arise in the gonad but from the endoderm of the yolk sac; their course can be traced histochemically by their content of alkaline phosphatase (McKay *et al* 1953).

If the primordial gonad is to develop into a testis, the primary sex cords, which contain spermatogonia, proliferate to form the seminiferous tubules, the cells of the cords becoming the Sertoli cells. Between the 7th and 8th weeks of intra-uterine life, the innermost portions of the proliferating primary sex cords join with the mesonephric tubules to form the rete testis and epididymis. A little later, usually between the 8th and 9th weeks, the mesenchymal medullary cells mature to form

Leydig cells, the secretions of which play an important role in subsequent male development (see Fig. 3.1).

The differentiation of the primitive gonad into an ovary usually occurs between 10 and 14 days later than testicular development. The principal difference between the two gonads is that in the formation of the ovary, cortical development dominates that of the medulla. Primary sex cords recede toward the hilar region and further cortical ingrowths form the secondary sex cords in which the oogonia proliferate to give rise to primary oocytes which, with the secondary sex cords become the primordial follicles.

DEVELOPMENT OF THE MALE DUCT SYSTEM AND GENITALIA

At the time the primitive gonad is developing to testis or ovary, no difference can be seen in the external genitalia which, at this stage, are capable of differentiation in either direction. The urogenital slit is bounded ventrally by the genital tubercle consisting of glans and corpora cavernosa, laterally by urethral folds outside which are the labioscrotal swellings. In the male, the urethral folds fuse to form the urethra which becomes enclosed in the corpus spongiosum; there is also fusion of the labioscrotal swellings to form the scrotum into which the testes ultimately migrate. The descent of the testes from the mid-abdomen begins in the 7th month of gestation and is complete by the 6th to 8th week of extra-uterine life.

At the time of the development of the primitive gonad two duct systems can be recognized—the Müllerian and Wolffian. In the female, the Müllerian ducts form the Fallopian tubes, uterus and upper third of the vagina, while in the male the Wolffian duct system forms the epididymis, vas deferens, seminal vesicles and ejaculatory ducts. If the primitive gonad develops into a testis there is regression of the Müllerian ducts and development of the Wolffian ducts. These occur simultaneously and are complete by the 3rd month. The development of the male ducts and external genitalia, and the involution of the female ducts, is influenced significantly but not entirely by the secretions of the foetal Leydig cells. These cells develop rapidly after the 9th or 10th week of foetal life and show ultrastructural characteristics similar to those described for adult cells by Fawcett & Burgos (1960). They also show a well-developed agranular endoplasmic reticulum which is a typical feature of active steroidogenesis in other endocrine tissues

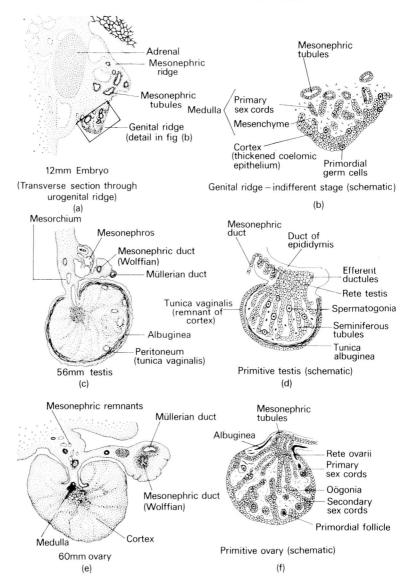

Fig. 3.1. A schematic representation of early gonadal differentiation. (A & B) Transverse sections of the indifferent gonad in the region of the urogenital ridge. (C & D) Transverse sections through the foetal testis at the 56 mm stage. Note that the medullary elements dominate in the development of the testis. (E & F) A transverse section through the foetal ovary at the 60 mm stage. In the developmental of the ovary the cortical elements are dominant. See text for details. From Van Wyk & Grumbach 1968; redrawn from Arey & Witchi.

(Niemi *et al* 1967). The capacity of the foetal testis to synthesize testosterone from acetate at this stage has also been shown (Serra *et al* 1970). It seems likely that the stimulus for the differentiation and maintenance of foetal interstitial tissue is placental gonadotrophin which has an excretory peak at the time of the development of Leydig cells (Venning 1955) and has been shown to be present in foetal tissues (Bruner 1951). Whether chorionic gonadotrophin is entirely responsible for the maintenance of Leydig cell function is uncertain; although the foetal pituitary does not become differentiated until after the twelfth week, it is quite probable that it also plays a role in the maintenance of Leydig cell activity, because in the anencephalic foetus with a presumptive lack of gonadotrophin secretion, interstitial cell function is diminished (Jost 1960, 1970).

At birth these cells can be clearly identified as interstitial cells between the immature tubules. During the neonatal period Leydig cells regress to immature mesenchymal forms, and remain in this state until the onset of puberty.

DEVELOPMENT OF THE TESTIS BETWEEN BIRTH AND PUBERTY

Following the regression of Leydig cells after birth, the testis remains dormant until puberty. Although there is a gradual increase in testicular size with age, this is only in proportion to the increase in the size of other organs with general body growth, and results from gradual but progressive changes in the seminiferous tubules until the time of puberty. Until the age of 5 or 6 years, seminiferous tubules appear as solid cord-like structures in which primitive spermatogonia can be recognized; these show occasional mitoses. In the latter years of childhood, the seminiferous cords acquire a lumen and the epithelium becomes arranged in a characteristic pattern with two or three layers of cells. At the time these changes are taking place in the seminiferous epithelium, the Leydig cells remain unchanged and unrecognizable as such, being represented by fibroblast-like cells in the loose myxomatous interstitial tissue.

With the onset of puberty, striking and dramatic changes can be observed, at first within the tubules, and somewhat later in the interstitial cells. The tubular diameter increases, and in the enlarged tubules Sertoli cells can be seen, along with mitotic and meiotic activity associated with spermatogenesis, with the ultimate formation of spermatozoa.

With the maturation of the seminiferous tubules, Leydig cells become differentiated from the precursor fibroblast-like cells, and ultimately aggregate into the characteristic islands of cells in the angles between the tubules (Albert *et al* 1953).

THE SECRETION AND METABOLISM OF TESTICULAR STEROIDS

STEROID PRODUCTION BY THE TESTIS

Loss of virility is one of the obvious and most ancient observations concerning the effects of castration in man. These changes result from failure of androgen secretion following removal of the testes. Testosterone and androstenedione are the biologically significant androgenic hormones from the testes of man and other vertebrates; of these, testosterone possesses the more potent androgenic activity. Although a number of other steroids are known to be produced by testicular tissue *in vitro*, or to be secreted by the testes, none possesses more potent biological effects than testosterone.

STEROID BIOSYNTHESIS BY THE TESTIS

There is good experimental evidence that testicular tissue forms testosterone from acetate, with cholesterol as an intermediate. The early steps of this biosynthesis are probably the same as those in other steroid secreting tissues. Thus, the formation of squalene from acetate has been shown, and a number of recognized intermediates have been isolated, which would suggest that within the testis the synthesis of squalene from acetate follows much the same pathway as in liver (Tsai *et al* 1964, Ying *et al* 1965, Nightingale *et al* 1967). Testicular tissue has also been shown to form lanosterol from acetate and cholesterol from lanosterol. One of the problems about the interpretation of such data is that the results have been derived from *in vitro* studies in which whole testicular tissue rather than isolated Leydig cells have been used. Because the testis is rich in cholesterol, much of which is associated with germinal cells, it is not possible to be certain that these conversions necessarily take place only in Leydig cells; germinal cells could also be involved.

Because little is known about the sizes or the turnover rates of cholesterol pools within the testis, difficulties arise in the interpretation of data relevant to the formation of steroids from cholesterol. It seems likely, however, that the cholesterol pool for steroid biosynthesis is small and most experiments would suggest that cholesterol is the precursor of all steroid hormones and that this cholesterol is formed locally.

Pregnenolone is the first compound formed in the transformation of cholesterol to steroids. This reaction, which occurs in mitochondria involves the oxidative cleavage of the six carbon side-chain and the conversion of a C-27 sterol to a C-21 steroid. These steps in biosynthesis have been extensively studied; it seems likely that trophic hormones act on one of the steps in this conversion, the mechanisms of which have recently been reviewed by Hall (1970).

With the formation of pregnenolone, numbers of pathways exist for the subsequent biosynthesis of testosterone. It is probable that all these pathways may contribute to the production of testosterone; certainly there is experimental evidence to suggest that this is so, but it is not possible to say with certainty which is the preferred pathway. This uncertainty arises because most of the studies undertaken to elucidate these points are unphysiological. *In vitro* studies suffer from the defects that enzymes leak into the incubation medium from cells that are cut or damaged, that only informed guesses can be made as to the co-factor requirements for a particular study, and that individual cells or organelles used in the preparation are severely injured or moribund. Another problem is that of product accumulation, for it has been shown that end-product enzyme inhibition can occur (Mahajan & Samuels 1962, Kowal *et al* 1964). Problems of interpretation also arise from studies *in vivo*; although the cells are alive, have a normal circulation, utilize co-factors already present in the tissues and the products are removed, there are the difficulties of the effects of anaesthesia, the penetration of infused substrates into cells which may have differing permeabilities to individual substrates, and the different pool sizes and turnover rates of products—those with rapid turnover rates may either not enter the effluent or be present in such minute concentrations as to escape detection—problems that have been fully discussed by Eik-Nes (1970).

Notwithstanding these difficulties, it seems likely that the following are the major pathways of testosterone biosynthesis in the Leydig cell:

98   HUMAN REPRODUCTIVE PHYSIOLOGY

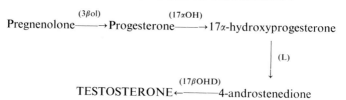

Alternate pathways are the hydroxypregnenolone-hydroxyprogesterone pathway:

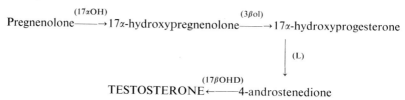

and the DHEA-4-androstenedione pathway:

Pregnenolone —(17αOH)→ 17α-hydroxypregnenolone —(L)→ DHEA

↓ (3βol)

TESTOSTERONE ←(17βOHD)— 4-androstenedione

and the 5-androstenediol pathway:

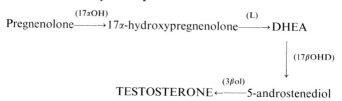

The enzyme systems necessary for these conversions have all been isolated from testicular tissue, namely, the 3β-hydroxysteroid dehydrogenase complex (3βol), the 17α-hydroxylase (17αOH), the lyase (L) and the 17β-hydroxysteroid dehydrogenase (17βOHD).

Until recently most studies directed toward the biosynthesis of androgens by the testis had been concentrated upon the Leydig cell which has long been accepted as the principal source of these hormones. Many years ago, McCullagh (1932) suggested that the Sertoli cell may be the source of a hormone, the nature of which was by no means clearly defined. Howard et al (1950) have also suggested that the testis may produce two hormones, one from the Leydig cells and another from the tubules—possibly the Sertoli cells. More recently, Lacy and his colleagues have produced experimental evidence which suggests that the Sertoli cell may be a site of steroid biosynthesis, both in the rat and man. They undertook *in vitro* studies on the seminiferous tubules of rats whose testes had been subjected to heat treatment which destroys the germinal epithelium but leaves the Sertoli cells intact. With both progesterone and pregnenolone as precursors, they showed that steroid biosynthesis took place in isolated tubules and suggested that the Sertoli cell was the site of this biosynthesis. In addition to identifying 17α-hydroxyprogesterone, 4-androstenedione and testosterone, they showed that progesterone metabolites were also formed, including 20α--hydroxyprogesterone, allopregnanedione, allopregnanolone and pregnanediol. The significance of the formation of the progesterone metabolites remains to be elucidated, but Lacy suggests that testosterone biosynthesis by the Sertoli cell is one of the factors responsible for the maintenance of normal spermatogenesis (Lacy & Pettit 1970).

Most earlier studies of the steroid biosynthetic capacities of endocrine tissue ignored the possibility that steroid conjugates might be biosynthetic products. Baulieu (1962) showed that DHEA sulphate was a secretory product of the adrenal and Saez *et al* (1967) have shown testosterone sulphate is a secretory product of the testis. Although DHEA sulphate is present in high concentrations in the testicular tissue and spermatic vein blood of certain animals (e.g. boar and dog), it is not known in man whether this is a secretory product or a precursor for testosterone or its sulphate.

THE SECRETION OF OESTROGENS BY THE TESTIS

Although both androstenedione and testosterone are known to be precursors of oestrogens in the ovary and the placenta (Ryan & Engel 1953, Ryan & Smith 1965, Siiteri & MacDonald 1966), and oestrogens have been isolated from human testicular tissue (Goldzieher & Roberts 1952) the testis normally secretes only small amounts of oestrone and insignificant amounts of oestradiol. In the male, plasma oestrone is also derived from the adrenal cortex and by peripheral conversion from androstenedione; plasma oestradiol is mainly derived by peripheral interconversion from testosterone and oestrone (Fishman *et al* 1966, Baird *et al* 1969).

THE SECRETION OF ANDROGENS

The secretion of testosterone by the testis is the most significant contribution to the maintenance of the normal androgenic stimulus in the adult male. Normally the adrenal cortex secretes only very small amounts of testosterone. The secretion rate of testosterone in the normal adult male is between 5 and 10 mg per day, and at these secretion rates the normal concentration of testosterone in peripheral plasma is between 350 and 1200 ng/100 ml, with a mean concentration of about 700 ng/100 ml. Androstenedione is secreted both by the testis and the adrenal cortex; the secretion rate from these two sources is between 1 and 2 mg/day, and the level in peripheral plasma between 80 and 120 ng/100 ml. The concentrations of testosterone and androstenedione in peripheral plasma have been determined by a number of different methods in many laboratories. The principal methods that have been used to make these measurements include double isotope dilution derivative analysis, gas-liquid chromatography and competitive protein binding; for a review of the results obtained by these methods and of the methods themselves, see Hudson *et al* (1970) and van der Molen (1970).

The procedure initially contrived to measure the secretion rate of testosterone depended upon the principle of isotope dilution. By this technique a known tracer dose of radioactive testosterone is injected and the specific radioactivity in urinary testosterone glucuronide is measured after all the radioactivity has been excreted. The results obtained by this technique have been shown to be fallacious, particularly in women and in prepubertal or hypogonadal males in whom the secretion rates of testosterone are low. The reason for this fallacy is that urinary testosterone glucuronide is not uniquely derived from secreted testosterone. This was shown experimentally in two ways: following the injection of labelled precursors such as DHEA or androstenedione, radioactivity was found in urinary testosterone glucuronide, indicating that these steroids had contributed to this urinary metabolite; also, if after an infusion or injection of isotopic testosterone, the specific radioactivities of testosterone in blood and of testosterone glucuronide in urine are measured, the former is found to be higher than the latter. Because testosterone glucuronide is formed in the liver, it is postulated that testosterone is produced in the liver, not only from blood testosterone but also from precursors such as DHEA and androstenedione, and that the testosterone formed in this way neither circulates nor mixes with secreted testosterone.

These observations led to a much closer scrutiny of testosterone secretion and metabolism, and resulted in a series of new approaches to this problem. One of these approaches is to measure the blood production rate (P_B^T) of testosterone. This is possible if the plasma level and the rate at which the steroid is cleared from plasma are known. This latter value, the metabolic clearance rate (MCR), is the amount of blood (or plasma) irreversibly cleared of steroid in a given time. It can be measured experimentally by infusing labelled testosterone at a constant rate and measuring the level of radioactivity, specifically as the steroid, when a steady state has been reached. For testosterone, values of about 1100 litres of plasma/day for men, and 650–700 for women were found, a difference that was not eliminated if a correction was made for body surface area; for androstenedione, the corresponding values were 2300 and 2000 l/day respectively. The rate of entry of new steroid into blood can be calculated as the product of the MCR and the peripheral level of the steroid in question. If this new steroid entering the blood is entirely secreted, then this value is the same as the secretion rate of testosterone (S_B^T); however, it can be shown that during an infusion of testosterone, radioactivity is present in androstenedione which indicates that androstenedione in blood is not derived entirely from secreted androstenedione; the same applies for blood testosterone during an infusion of androstenedione or DHEA; radioactivity is present in testosterone, indicating that not all blood testosterone is secreted testosterone.

Measurements of the ratios of radioactivity in precursor and product have shown that between 3 and 4% of blood androstenedione is converted to blood testosterone, and that conversion in the reverse direction is between 7 and 10%. These conversion ratios are not significantly different between the sexes, nor have they been shown to differ in a variety of disease states. The question as to where these conversions take place cannot be answered with complete certainty; while it is clear that some testosterone produced in the liver escapes into the blood, and thus contributes to blood testosterone, it is also likely that conversions occur in peripheral tissues. The relative contributions of the liver and peripheral tissues to these conversions have not been measured precisely.

If the precursor-product conversion ratio is known, it is possible to calculate the overall contribution (transfer constant) of precursor to product, and thus arrive at a figure for the secretion rate for the steroid in question. Thus:

$$S_B^T = P_B^T - P_B^{A \to T} - P_B^{D \to T}$$

where $P_B^{A \to T}$ and $P_B^{D \to T}$ are the blood production rates of testosterone derived from androstenedione and DHEA respectively. These values can be calculated knowing the conversion ratios and the MCRs of the steroids concerned.

Using this approach, it can be shown that for the adult male in excess of 95% of blood testosterone is secreted testosterone. For androstenedione about 60% is secreted, the remainder being derived by peripheral conversion from testosterone and DHEA. In the female, in whom the blood production rate of testosterone is of the order of 200 μg/day, only about 40% of this is secreted, the remainder coming by conversion from androstenedione (50%) and DHEA (10%).

In pre-pubertal boys in whom plasma levels of testosterone are in the female range, and in whom androstenedione levels approximate those of the adult male, it is probable that much of the testosterone is derived from precursors rather than from secreted testosterone. There is some evidence suggesting that this may also be true for hypogonadal males in whom testicular function is reduced or absent.

In elderly males plasma levels of testosterone are frequently found to be in the normal range, although mean levels are lower than those found in young adults. If blood production rates are measured in these subjects, they are found to be significantly reduced, and this reduction in blood production—and presumably in secretion of testosterone—results from a lowering of the MCR which approaches that found in the female (Kent & Acone 1966).

Many laboratories have contributed to the concepts and data cited in this section; for a more comprehensive account of the hypotheses and experimental work the reader is referred to the following papers and reviews on this subject. These contain many source references; Baird *et al* (1968 and 1969), Baulieu & Robel (1970), Horton & Tait (1966), Lipsett *et al* (1966), Tait & Burstein (1964), Vande Wiele *et al* (1963).

THE TRANSPORT OF ANDROGENS

The method by which androgens, once synthesized in the Leydig cells, enter the spermatic venous plasma is not known. Christensen (1965) has suggested that the transport from the Leydig cell may involve the proteins synthesized by the cell, but at present this is speculative. Once in plasma, most of the testosterone is bound to plasma proteins, one of which is a β-globulin specific for testosterone and other 17β-hydroxysteroids. This protein has a high affinity but relatively low capacity for testosterone; it also binds oestradiol but the affinity for this steroid is probably about one-third that for testosterone. Corticosteroid binding globulin (CBG) also binds testosterone with a moderate affinity (about one-twentieth that of cortisol) but with relatively high capacity. Androstenedione is also bound to CBG, and both steroids are weakly bound to serum albumin which has a high capacity for these and other steroids.

At physiological concentrations and at 4°C, in excess of 98% of testosterone is bound to plasma proteins (Pearlman & Crepy 1967), so that in the adult male the concentration of free (unbound) testosterone is approximately 12 ng/100 ml, and for the female this value is less than 1 ng/100 ml. The testosterone binding capacity of plasma (TBC) can be measured (moles/litre) using equilibrium dialysis or other techniques. These have shown that female plasma has a higher TBC ($7-8 \times 10^{-8}$ moles/litre) than does male plasma ($4-5 \times 10^{-8}$ moles/litre), and that the administration of sex hormones can alter these binding capacities. Following the administration of oestrogens the TBC increases greatly, whereas treatment with testosterone will reduce this (Vermeulen *et al* 1969). While it is not known with certainty whether the hormonal effects of testosterone are related to the degree to which this steroid is bound, it would seem likely that the binding of the steroid to a specific transport protein retards its metabolism by the liver. Thus, steroids that are not bound specifically to a transport protein but only to albumin have been shown to have a high splanchnic extraction (mainly liver) whereas for those which are bound to specific proteins (e.g. cortisol and testosterone) the splanchnic extraction is low. For testosterone in the male the splanchnic extraction is about 50%, whereas for the female this is between 35 and 40%. These figures are almost certainly reflected by the difference between the metabolic clearance that can be induced by treatment with oestrogens, which reduce the MCR, or by testosterone which increases it (Bird *et al* 1969). Thus, in the pre-pubertal male, TBC is increased and the MCR reduced (with corresponding lowering of the blood production rate), a situation that is also found in the hypogonadal male.

The physiological signfiicance of these observations is that with an increased binding capacity, not only may the splanchnic extraction be reduced but the extraction of testosterone at a tissue level may be similarly affected.

THE FURTHER METABOLISM OF ANDROGENS

The common pathway for both testosterone and androstenedione is reductive. In ring A the double bond is reduced to give dihydro compounds (5α or 5β); 3α and 3β-hydroxysteroid reductases cause a further reduction of the 3-oxo group to yield 3α or 3β-hydroxysteroids. The compounds that are commonly formed are androsterone (5α-androstane-3αol-17-one), aetiocholanolone (5β-androstane-3αol-17-one) iso-androsterone (5α-androstane-3βol-17-one), and 5α and 5β-androstanediols. Androsterone and aetiocholanolone are quantitatively the most important of these metabolites. Having undergone reductive catabolism in this way, these metabolites are conjugated in the liver by glucuronyl and sulphuryl transferases to form steroid glucuronides and sulphates. From the liver these conjugates enter the blood in which they are mainly found as sodium salts; they are cleared by the kidney and appear in the urine. Sulphates, which are albumin bound, are cleared only slowly by the kidney with clearance rates of 10 ml/min or less, whereas glucuronides are cleared at rates approaching those of creatinine clearance. Some of the conjugates are excreted by the bile into the intestine, whence they are mainly absorbed. In man, faecal excretion accounts for only a minute amount of the total excretion of steroid metabolites.

It is important to appreciate that this mixed group of urinary conjugates has a diverse origin (some are also derived from DHEA), and that varying proportions of the 5α- and 5β-androstanes are formed depending on the precursors and the sex of the individual. For a more detailed account of this metabolism, the reader is referred to Baulieu *et al* (1965), and Baulieu & Robel (1970).

Measurement of the urinary 17-oxosteroids has been used for many years to assess androgen function in both men and women. It should be appreciated that quite significant changes in testosterone secretion can occur without any appreciable—certainly any significant—alteration in the rate of excretion of this group of urinary steroid metabolites. Thus, in a female with a normal 17-oxosteroid excretion rate of 10 mg per 24 hours and a daily secretion rate of testosterone 0·1 mg, a ten-fold increase in the rate of secretion of testosterone would only contribute 1·0 mg, at the most, to the urinary 17-oxosteroid level. Differences of this order are not significant by the methods used to make this measurement. Such an increase would certainly be associated with significant biological effect.

ACTION OF MALE SEX HORMONES

The most potent androgen secreted by the testis is testosterone; other steroids produced in the testis such as androstenedione and DHEA are also androgenic, but less strongly so. Any description of the actions of these hormones should take into account not only their actions on the whole organism, but also on androgen-sensitive tissues at both the cellular and molecular levels.

IN THE FOETUS

Testosterone starts to exert its effects on the developing foetus. Jost (1953) has demonstrated that the secretions of the foetal testis play a determinant role in the development of genital ducts. The presence of functional testes causes an involution of the Müllerian duct system, and permits the Wolffian duct system to develop completely. In the absence of the testes Müllerian duct systems mature and Wolffian structures are resorbed, changes which occur independently of the presence of the ovary. All evidence points to this being a local effect, because the early unilateral removal of the testis is associated with Müllerian duct development on the side from which the testis is removed, while male duct development proceeds normally on the side of the intact testis. The role of male sex hormones in promoting these changes is not entirely clear. If testosterone is given systemically to the developing embryo, these effects are not reproduced; however, when testosterone is applied in high doses locally, Wolffian duct stimulation occurs but Müllerian ducts fail to regress. Studies by Jost (1970) have shown that if cyproterone acetate, an antagonist of testosterone, is administered to pregnant rabbits, there is a suppression of the male duct development in the male foetus, but the Müllerian ducts retrogress. It would seem clear that the presence of a testis in the developing foetus is essential for male duct development, and that testosterone plays a role in this development;

however, for the complete organization of the duct system the testis secretes another substance, as yet unidentified, which is probably the major determinant of duct structure, and this, with testosterone, enables complete and unambiguous male duct development.

IN THE WHOLE ORGANISM

The post-natal effects of testosterone on the whole organism are most evident in the genital tract, but are not confined to it. Each region of the genital tract in the sexually immature or castrate male is stimulated by androgens—the penis, the scrotum, the seminal vesicles, the prostate and accessory glands. To a lesser degree, androgens also affect the female genital tract; in the adult female they may produce some features of masculinization, particularly clitoral hypertrophy and a redistribution of pubic hair. Androgens also affect non-genital structures. They are responsible for the growth of the beard, the distribution of body hair, the development of the larynx and for an increase in muscle mass. They exert effects on the distribution of fat, bone growth and epiphyseal closure, and almost certainly on the development of male psyche and sexual drive.

THE TIMING OF PUBERTY IN BOYS

In a study of 228 British boys using the stages of puberty as defined by Tanner (1962), Marshall & Tanner (1970) have shown that in 95% of this population genital development had commenced between the ages of $9\frac{1}{2}$ and $13\frac{1}{2}$ years, and reached maturity between the ages of 13 and 17 years. Peak height velocity was also measured in a small group of the subjects in this study; this was shown to occur between the third and fourth stages of puberty, at a mean chronological age of about 14 years. The data from this study provide a reasonable basis for determining whether pubertal changes are premature or abnormally delayed. Thus, puberty cannot be regarded as being early unless it commences before the age of 9, nor abnormally delayed unless it commences beyond the age of 15 years.

The observed correlations between the adolescent growth spurt and the stage of puberty are also helpful in predicting the ultimate height of the individual; thus, a boy who is abnormally short and shows well-developed pubertal changes is likely to become a small adult, whereas normal adult height is likely to be achieved if, at the age of 14 or 15, genital development is just commencing.

At puberty androgens play an important role in the adolescent growth spurt, and may be responsible for initiating and accelerating growth at this time. It is not clear whether growth hormone also plays the major role in this adolescent growth spurt or whether the two hormones act synergistically (Marshall & Tanner 1970). There is now good evidence in at least some subjects that if puberty is delayed because of failure of gonadotrophin and androgen secretion, the secretion of growth hormone may also be impaired, an impairment that can be corrected by the administration of androgens (Illig & Prader 1970, Deller *et al* 1970).

ON THE CENTRAL NERVOUS SYSTEM

In the higher vertebrates both the central nervous and endocrine systems are finely integrated to regulate reproductive activity. Although androgens may exert their effect on many parts of the brain and thus contribute to the psychological aspects of male sex drive, the most clearly identifiable effects of androgens are upon the hypothalamus. Axones from the hypothalamic nuclei terminate in the region of the median eminence; secretions from these axones enter the capillaries of a primary plexus which forms the hypophyseal-portal system by which blood is delivered to the sinusoids of the anterior lobe of the pituitary. These neurohumoral secretions have been termed releasing factors or hormones, and are directly responsible for the regulation of the secretion of the gonadotrophins.

Because much of the information about the neurohumoral control of anterior pituitary-gonadal function has been obtained in rodents, only general extrapolations can be made to man. It is clear, however, that the pattern of secretion of gonadotrophin releasing hormones by hypothalamic nuclei depends upon the sex of the individual. In the male, after puberty there is a tonic release of gonadotrophins without evidence of cyclical activity (Burger *et al* 1968) while in the female there are surges of LH and FSH secretion that result in ovulation. In rats and mice this intrinsic activity of the hypothalamus can be modified by treatment with androgens. The female pattern of hypothalamic function can be converted into a male-type pattern by the administration of testosterone in the first 4 or 5 days of life. Harris (1964) has shown that female animals so treated will not mate, and show no oestrus cycles.

It would seem unlikely that the human hypothalamus can be similarly sensitized; females with pseudohermaphroditism due to congenital adrenal hyperplasia often have abnormally high levels of circulating androgens at birth, but following appropriate treatment even after months or years of androgen exposure, normal ovulatory patterns usually become established. Despite this, there is some evidence that a male-type brain may exist in these subjects; thus, it has been shown that treated female patients with congenital adrenal hyperplasia may be erotically aroused by visual stimuli that would normally provoke the adult male.

Androgen secretion by the testis is directly regulated by circulating levels of pituitary LH which, in turn, is controlled by the secretion of a luteinizing releasing factor (LRF) from the hypothalamus. The hypothalamic nuclei producing LRF are involved in a feedback control by gonadal steroids. The existence of this feedback loop has been established by the implantation of minute pellets of gonadal steroids (oestrogens or androgens) into the hypothalamus, following which gonadal atrophy occurs in the male, while in the female there is ovarian and uterine involution along with an abolition of the oestrous rhythm.

The releasing factors are dealt with in greater detail in chapter 2.

ACTION OF ANDROGENS AT A MOLECULAR LEVEL

If isotopically labelled testosterone is administered to an adult male, radioactivity is concentrated in specific areas, particularly but not exclusively in the accessory sexual glands—the prostate, seminal vesicles, etc. It is on these tissues that testosterone is known to have its most important biological effects. It would be difficult to construct any theory about the mechanism of action of androgens without postulating that there are primary receptors for these (and other) hormones in the responsive cells. It is further postulated that the receptors are macromolecules on the surface of or within the sensitive cells, and that the hormones become attached to these receptors by noncovalent forces.

Following the infusion or injection of testosterone *in vivo* and examination of prostatic tissue, a number of unconjugated metabolites can be isolated. These include androstenedione, dihydrotestosterone, androstane-$3\alpha,17\beta$-diol, androsterone and small amounts of saturated diketones and other diols. The significance of these findings was not entirely apparent until it was shown that although these metabolites were identifiable within the cytoplasm of prostatic cells, only testosterone and dihydrotestosterone were demonstrable within prostatic cell nuclei. Although radioactive metabolites are demonstrable in the cytoplasm within minutes of injection, only testosterone and dihydrotestosterone are associated with cell nuclei for as long as 2 hours after the injection. It has also been shown that this reduction of testosterone- to dihydrotestosterone occurs principally in androgen-sensitive tissues —prostate, seminal vesicles and the preputial glands—whereas no dihydrotestosterone is recovered from the liver, gut or lungs. It has also been shown that prostatic cell nuclei promote the NADH-dependent reduction of testosterone to dihydrotestosterone, and that the 5α-reductase appears to be bound to nuclear chromatin. These observations have led Bruchovsky & Wilson (1968) to conclude that dihydrotestosterone was the active form of testosterone. This work has been independently confirmed by Anderson & Liao (1968) who have shown that when radioactive testosterone is injected into rats, dihydrotestosterone is the principal radioactive product within the cell nuclei from the prostate. They also have shown that a number of other radioactive products can be isolated from the soluble (cytoplasmic) fraction; but of these less than 5% is dihydrotestosterone. Nuclei of other tissues such as brain, liver or thymus, do not possess this capacity.

Although this evidence may strongly suggest that dihydrotestosterone is the active metabolite of testosterone at the tissue level, it is possible that this may not be the only active metabolite. If testosterone and dihydrotestosterone are given at the same dose level to castrate male rats, there is a restoration of normal sexual activity in those rats treated with testosterone but not in those treated with dihydrotestosterone, despite the fact that dihydrotestosterone was almost as potent for the restoration of accessory sex glands. It may thus be that in the hypothalamus either testosterone itself or another metabolite may be responsible for the restoration of male behavioural characteristics (McDonald *et al* 1970). Ito & Horton (1970) have described a method for the measurement of dihydrotestosterone in peripheral plasma. They have found values of about 55 ng and 15 ng/100 ml in male and female plasma respectively. Dihydrotestosterone was not detected in plasma from prepubertal boys and was found to be elevated in some female patients with disorders thought to result from androgen excess. They suggest that in the male most, if not all, plasma dihydrotestosterone originates from plasma testosterone.

These interesting observations do not explain the mechanism of action of androgens; however, they do carry our knowledge one step further forward so that there is now good reason to believe that dihydrotestosterone may be the active form of testosterone at a cellular level. It does not explain the dramatic effects of androgens on sensitive and responsive tissues. The biochemical events in the epithelial cells of these tissues have been extensively studied in order to determine the primary site of action of these hormones, but the primary event is as yet unknown. What does seem clear is that the androgens stimulate protein synthesis and that an initial event in this stimulus is an increase in DNA-mediated RNA synthesis. Prolonged treatment with testosterone is associated with a significant increase in DNA polymerase activity and the incorporation of tritiated thymidine into DNA. For a more detailed account of the action of androgens at a molecular level, the reader is referred to the review by Williams-Ashman (1970).

ACTIONS OF ANDROGENS ON THE TESTIS

Many reviews of testicular function and physiology tend to separate the endocrine and exocrine (gametogenic) functions of the testis. While such an approach may simplify the understanding of the separate functions of the interstitial cells and the seminiferous epithelium, it overlooks the close and presumably important topographic relationship between these structures. It has often been assumed that the spermatic venous blood is the only medium by which testosterone and other secretions of the interstitial cells are carried from the testis into the general circulation. Testosterone is also secreted into the interstitial fluid and the lymph of the testis; Lindner (1963, 1966) has shown that testicular lymph from the ram contains both testosterone and androstenedione, and that the ratio between these two steroids is much the same in lymph as it was in spermatic venous blood. However, the contribution from this lymphatic drainage to the overall androgen effects in the whole animal is negligible owing to the low rate of lymph flow from the testis.

The importance of the passage of interstitial cell secretions into the lymph and interstitial fluid should not be overlooked. Within the testis, blood vessels do not appear to penetrate the basement membrane of the seminiferous tubules which are nourished by the contents of the interstitial fluid of the peritubular space, and by a transfer of fluid across the walls of the capillaries that surround the tubular basement membrane. It seems highly probable, therefore, that androgens are delivered in high concentrations to the seminiferous epithelium and that this has an important direct role not only on the maintenance of spermatogenesis but also on the integrity of the epididymis and the survival of sperm within this accessory structure.

Hypophysectomy in man or experimental animals is associated ultimately with a complete regression of spermatogenesis which can be restored by treatment with FSH and LH (MacLeod *et al* 1966). In the rat, Clermont & Harvey (1967) have shown that following hypophysectomy spermatogenesis can be maintained by testosterone if given in sufficiently large doses—between 30 and 40 mg of testosterone proprionate/kg body weight/day; this is equivalent to the administration of between 2 and 3 g of testosterone proprionate to an adult male each day. When given in smaller doses, sufficient to maintain accessory reproductive organs—and presumably to maintain a normal peripheral blood level—spermatogenesis is not maintained. Woods & Simpson (1961) have also shown that the administration of LH in doses that failed to maintain accessory structures was sufficient to sustain spermatogenesis. These observations support the idea that the close anatomical relationship between the interstitial cells and the seminiferous epithelium is of fundamental importance. The finding of androgens in seminal plasma (White & Hudson 1968) in concentrations similar to that in peripheral blood is additional evidence of the passage of androgens from the interstitial fluid into the seminiferous tubule and would support the view that within the testis testosterone is essential for the maintenance of the seminiferous epithelium, for the secretory activity of the epididymis and the survival of sperm.

SPERMATOGENESIS

The production of spermatozoa by the testes of normal adult animals is a complex, continuous and dynamic process involving cellular differentiation and division—both mitotic and meiotic. Although the morphology of the different types of cells that populate the seminiferous epithelium has been recognized for many years, to the casual observer microscopic examination of this epithelium may present a picture of apparent disorder and chaos. Studies, such as those by Roosen-Runge & Giesel (1950), Le Blond & Clermont (1952), and Heller & Clermont

(1964), have shown that this casual impression is not justified. An important contribution to understanding the dynamics of spermatogenesis is that in separate parts of the seminiferous tubule, different generations of germ cells are regularly grouped together as 'cell associations'.

Approximately twenty different types of cells can be recognized within the seminiferous epithelium, nearly all of which are germ cells at different stages of development. Because very adequate descriptions of the morphology of these cells already exist, only a general outline of the different cell types that are encountered will be given here. Three groups of maturing sperm cells can be recognized within the seminiferous epithelium; spermatogonia, spermatocytes and spermatids.

Lining the wall of the seminiferous tubule are the spermatogonia which are the least mature of the germ cells. These are of three morphologic types: dark type A, pale type A and type B, each derived from the other by mitotic division. The next generation of cells are the spermatocytes. The primary spermatocytes are produced from the last mitotic division of the spermatogonia. These primary (also called resting or preleptotene spermatocytes) pass through the well-recognized stages of meiosis, and this is the last time that DNA is synthesized during spermatogenesis. The prophase of this first maturation division is long, and the cells derived from the different steps of this phase can usually be easily recognized, these cells being termed leptotene, zygotene and pachytene spermatocytes. Following completion of this division, primary spermatocytes give rise to secondary spermatocytes, which soon proceed to the second maturation division to yield spermatids. The different phases of this development are depicted in Fig. 3.2.

In the human, six characteristic steps can be recognized in the maturation of spermatids, a process which is known as spermiogenesis; this is one of cell differentiation rather than of cell division. There is a loss of cytoplasm and a condensation of nuclear chromatin. In the later stages of spermiogenesis the nucleus flattens, a flagellum appears and a mid-piece is formed, so that finally a cell with most of the features of a mature spermatozoon is produced. This cell, upon release from the seminiferous epithelium, becomes embedded in the cytoplasm of an adjacent Sertoli cell.

The role of the Sertoli cell in the ecology of the seminiferous epithelium is still somewhat uncertain. These cells are spaced at regular intervals along the wall of the tubule, and have an abundant cytoplasm

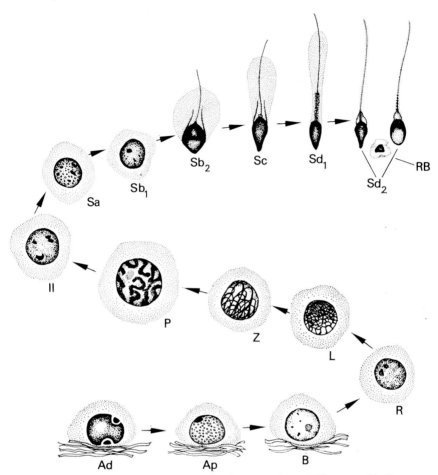

FIG. 3.2. Individual cell types seen in the normal seminiferous epithelium. These depict the main steps of spermatogenesis in man. Ad, dark type A spermatogonium; Ap, pale type A spermatogonium; B, type B spermatocyte; R, resting spermatocyte; L, leptotene spermatocyte; Z, zygotene spermatocyte; P, pachytene spermatocyte; II, secondary spermatocyte; Sa, Sb, Sb_2, Sc, Sd_1 and Sd_2, spermatids at different stages of spermiogenesis; RB, residual body. From C.G. Heller & Y. Clermont (1964) *Rec. Prog. Horm. Res.* **20**, 548.

that infiltrates between the germ cells. The Sertoli cell has always been regarded as possessing a supporting role in the germinal epithelium, but it is quite possible that these cells are the site of hormone production.

It has already been mentioned that generations of germ cells are grouped together along the walls of the tubule to form cell associations of relatively fixed composition. The existence of these cell associations were first recognized by Le Blond & Clermont (1952) in the rat, in which there are fourteen such groupings. In seeking for similar groupings in

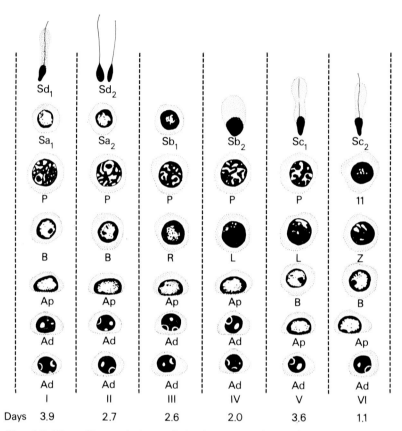

FIG. 3.3. The cell associations of the human testis. Each vertical column represents one cell association. The cells at the top of each column are those closest to the lumen and those at the bottom are located at the basement membrane. Generally all the spermatogonia lie on the basement membrane. The lettered cells correspond to those shown in Fig. 3.2 (p. 113). The figures indicate the duration of each cell association. From C.G. Heller et al (1969) Progress in Endocrinology, Series 184. Amsterdam, Excerpta Medica.

man, Heller & Clermont (1964) were able to show that there were six such cell associations, which have been interpreted as being the stages of a cycle of the seminiferous epithelium in man. In the rat and other mammals, one cell association can usually be observed in one tubular cross-section, but in man, on the average three cell associations can be seen in any cross-section, and these are placed without apparent order in the section. Heller & Clermont, however, have been able to reconstruct the areas occupied by cell associations along the wall of the tubule. They have also shown that there is no orderly arrangement of these cell associations so that in man there does not appear to be the wave of the seminiferous epithelium which can be seen in the rat and other mammals.

DURATION OF SPERMATOGENESIS

The recognition of these cell associations enabled Heller & Clermont (1964) to make calculations about the duration of the seminiferous cycle in man. These estimates are based upon the assumption that the relative duration of each stage of spermatogenesis is directly proportional to the area that an individual cell association occupies within the tubule, and from observations made by the use of ^{3}H-thymidine injected directly into the testes of normal volunteers. Using this technique they were able to show that one cycle, e.g. from mid-stage III to mid-stage III cell association, lasted approximately 16 days. However, the time taken for the development of sperm, that is from the appearance of pale type A spermatogonia in mid-stage V cell association to the release of spermatozoa in mid-stage II is about four and a half cycles. On this basis the average duration of spermatogenesis is 73·7 days, with a range of from 69·6–77·8 days. Further, they were able to show that treatment with a steroid such as norethandrolone, which depresses spermatogenesis or with human chorionic gonadotrophin did not affect the rate of development of germ cells which would appear to be a specific biological constant.

SPERM TRANSPORT AND STORAGE

The accessory reproductive structures in the male subserve two functions: to supply the constituents of seminal plasma and to act as a collection and transport system to convey spermatozoa from the testis to the ejaculatory duct. These separate functions enable the storage of spermatozoa, motionless, often for long periods of time prior to ejaculation, while at the same time enhancing their capacity for fertilization.

Whereas spermiation, the shedding of the sperm from the testis, is

116   HUMAN REPRODUCTIVE PHYSIOLOGY

probably under hormonal control (Burgos & Vitale-Calpe 1968), sperm migration within the male reproductive ducts is, from the point of view of the sperm, essentially a passive process. The mechanisms of

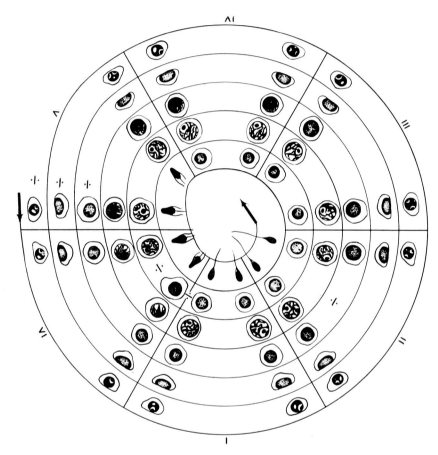

FIG. 3.4. This is intended to depict the helical arrangement of the six cell associations. The development commences with the Ad spermatogonia and is complete with the Sd_2 spermatid. Spermatogenesis, beginning with the division of the Ad spermatogonia, requires 4·6 helices. Reproduced from C.J.Heller *et al* (1969) *Progress in Endocrinology*, Series 184. Amsterdam, Excerpta Medica.

transport are not completely understood, but fluid currents, cilial and muscular activity probably all play a role. Another factor may be the smooth muscle found in the tunica albuginea of the testis which is sensitive to catecholamines (Davis & Langford 1970).

Transport through the epididymis in man takes about 3 weeks, and is mainly the result of muscular activity. While in the epididymis sperm acquire the capacity for motility; thus, sperm removed from the head of the epididymis rarely show motility, while those removed from the tail immediately become active when suspended under aerobic conditions in physiological saline. The reason why sperm become motile and acquire a capacity for fertilization within the epididymis is not known; the presence of glycerylphosphorylcholine in the epididymal fluid may indicate that this substance has a function in the further maturation of the sperm.

The largest accessory sex glands in the male are the prostate and seminal vesicles. Their principal function is in the formation of the seminal plasma. In man there is a continuous secretion of prostatic fluid, probably of the order of 1–2 ml per day. This rate of secretion is increased by sexual and nervous (parasympathetic) secretion, and is androgen dependent. The seminal vesicles, which vary considerably in size from animal to animal, and which are also androgen dependent, secrete a fluid that is rich in fructose (Mann 1946). It is probable that the role of fructose is to provide energy for spermatozoa. At the time of ejaculation when sperm come into contact with secretions which are rich in fructose they become highly motile, an activity for which energy is required. In man and laboratory animals, spermatozoa are not normally found in the seminal vesicles, which are thus not a storehouse for spermatozoa.

Although spermatozoa are stored in the epididymis, it seems likely that this is not the only accessory structure in which such storage occurs. It has been reported that following bilateral vasectomy sperm can be observed in quite abundant numbers in the first two or three ejaculates following this procedure, and have been shown to persist for as many as 8–10 ejaculates (Hanley 1968). This observation suggests that in addition to storage within the epididymis, the ampulla of the vas is an important storage site for sperm prior to ejaculation.

EVALUATION OF SPERMATOGENESIS

Most commonly the need to assess spermatogenesis arises because of the problem of infertility. The usual procedures for assessing male fertility are the examination of the seminal fluid and testicular biopsy.

EXAMINATION OF SEMINAL FLUID

Seminal fluid must be examined no more than two hours after it has been collected. In evaluating the potential of the seminal fluid with respect to its capacity to fertilize, attention should be directed to the following parameters:

  (1) Volume
  (2) Concentration and motile activity of spermatozoa
  (3) Sperm morphology
  (4) Chemistry

VOLUME

Normally the ejaculate collected after 3 days of sexual abstinence has a volume of between 2 and 5 ml, but it may be as little as 0·5 or as much as 10 ml. Although evidence on this point is fragmentary, it is probable that extremes of volume (less than 1 or greater than 7 ml) may be associated with impaired fertility.

THE CONCENTRATION AND MOTILE ACTIVITY OF SPERMATOZOA

The normal, healthy fertile seminal fluid contains between 50 and 100 million spermatozoa/ml; in excess of 60 or 70% of these are actively motile. While the sperm concentration can be measured with a reasonable degree of precision, most observers agree that it is difficult to define the term 'actively motile' because sperm motility within any specimen is, at the best, a subjective impression. Methods have been proposed by which the speed of movement of cells across ruled microscopic gratings or haemocytometer rulings can be measured, but there are certain technical difficulties with such procedures, and, as a consequence, there is still no common language by which motile activity can be described. For the present, the description of motile activity must be in semi-quantitative terms, such as 'good', 'average', 'poor' or 'indifferent'— terms which only have meaning if they have been applied by an experienced observer. Likewise, the assessment of the percentage of motile sperm can only be an approximation.

Even with reasonably precise methods for the assessment of the motile activity of spermatozoa, interpretation of the results obtained may present some difficulty in the evaluation of spermatogenesis and fertility. Quite clearly the fertility of a specimen of average volume with a sperm concentration in excess of 100 million/ml, and with 80% of sperm actively motile, is hardly in dispute. Likewise, a specimen that contains less than 10 million/ml of which less than 10% are motile, is almost certainly to be from an infertile person. The problems arise with sperm counts of between 20 and 40 million/ml, in which between only 40 and 50% of sperm are of average or indifferent motility. Significant contributions to the study of this problem have been made by MacLeod (1964), who, on the basis of his analyses of large populations, believes that there is a sharp difference between fertile and infertile seminal fluid once the percentage of active cells drops below 40%, even at sperm concentrations of 20–30 million/ml. It should be stressed, however, that data are not yet available to enable the construction of a probability nomogram with respect to fertility based on sperm concentration, total sperm count, motile activity and the proportion of motile sperm in the specimen.

SPERM MORPHOLOGY

Nearly 50 years ago, Williams & Savage (1925) reported a significant correlation between the breeding records and the morphology of the spermatozoa from bulls—those with normal sperm morphology having good breeding records. There is some controversy as to what constitutes an abnormal form and also to the proportion of abnormal forms that may appear within the seminal fluid and be consistent with fertility. MacLeod (1964, 1970) has produced convincing evidence that the sperm morphology of any individual in normal health shows a remarkable degree of stability, and that changes in the structural profile of the sperm of such an individual can be correlated with overt disease or psychologic stress. These changes are associated either with the appearance of less mature or abnormal forms of spermatozoa or spermatids within the ejaculate. In the examination of the infertile male the finding of large numbers of immature or abnormal forms predicates that the seminal fluid is likely to be infertile. In this area, also, it is not possible to prescribe in precise terms what proportion of abnormal forms are consistent with fertility.

It seems important to stress that sperm morphology should be used as an index of the activity of the seminiferous epithelium in the same way that an examination of the peripheral blood film is used as an index of bone marrow function. Quite clearly, there is a need for more studies which will correlate sperm morphology with the cytology of the seminiferous epithelium.

SEMINAL FLUID CHEMISTRY

Although the spermatozoa originate from a single site—the seminiferous epithelium—the seminal plasma is a composite mixture of secretions from many glandular tissues—the testis, epididymis, vas deferens, ampulla, seminal vesicles, prostate, bulbourethral and urethral glands. In the human it is not known how important it is for the function of all these individual glands to be preserved to maintain a fertile seminal fluid. In the past the activities of these accessory glandular tissues could only be assessed by direct (anatomical) examination, but now it is possible to make an assessment more simply and precisely by the quantitative chemical determination of substances such as fructose, citric acid, ergothioneine, acid phosphatase, glycerylphosphorylcholine and prostaglandins. Each of these represent products of secretion characteristic of individual accessory reproductive glands.

Chemical analyses of seminal plasma are frequently undertaken as a routine in the evaluation of spermatogenesis. These analyses may serve as a useful guide to regions in the reproductive tract that are the site of dysfunction. Thus, measurement of the citric acid or acid phosphatase concentrations reflect the adequacy of prostatic secretions while the measurement of fructose concentration is an index of the secretory activity of the seminal vesicles. The biochemistry of seminal fluid has been comprehensively reviewed by Mann (1964).

TESTICULAR BIOPSY

Testicular biopsy is frequently of great value in the diagnosis of abnormalities of spermatogenesis, in the intelligent application of therapeutic procedures and in the assessment of the results of such therapy. Unless it is undertaken properly, interpretation of the biopsy may be unreliable and the information provided may be misleading.

It is not within the realm of this presentation to describe the surgical techniques involved in testicular biopsy, other than to reaffirm that the amount of testicular tissue obtained should be adequate (a wedge about 1 cm long and 0·3 cm deep is satisfactory), it should be obtained by a clean excision and should immediately be fixed in a fluid that will prevent artefactual shrinking of the specimen. In order that this will not occur, specimens must under no circumstances be fixed in ordinary solutions of formaldehyde; the most satisfactory fixative for general use is Bouin's fluid; this, followed by the usual histological techniques, combined with haemotoxylin and eosin for staining is adequate for most purposes of interpretation. For a review on the interpretation of testicular biopsy the reader is referred to Albert *et al* (1953) and Nelson (1953).

In patients whose testes are immature from the lack of a gonadotrophic stimulus, the defect is usually obvious; appearances similar to those of the prepubertal testis are seen, the tubules are small, the Sertoli cells undifferentiated, and the germinal epithelium immature. Some spermatogonia and few, if any, primary spermatocytes are seen. The intertubular tissues are devoid of mature Leydig cells and contain numerous fibroblast-like precursors.

In the patient with infertility the biopsy may show no characteristic pattern but a combination of a number of defects, the examination of the specimen should be focused on the following criteria:

Size and maturation of the tubules. The normal adult male testis presents a picture of fairly uniform tubular size with clearly defined lumina in most of the tubules.

Spermatogenic activity. The seminiferous epithelium of the normal testis shows a progressive and orderly activity in which the various stages of spermatogenesis can be recognized. There should be little or no sloughing of the seminiferous epithelium into the lumen.

The peritubular connective tissue. The normal tubule has a thin, homogeneous limiting membrane immediately outside the basal layers of the germinal epithelium. The basement membrane is surrounded by a tunica propria which comprises a few lamellae of connective tissue fibres. Hyalinization of the peritubular tissue is frequently encountered in patients with testicular disorders, particularly and characteristically in Klinefelter's syndrome.

The intertubular areas. Unless the biopsy specimen has been improperly fixed, the intertubular tissue should not be more than moderately evident. In the angles between the tubules, Leydig cells can be recognized as rounded polyhedral cells with rounded nuclei and nucleoli.

It is not possible to describe all the types of lesions that may be found in patients with infertility other than to mention abnormal biopsy patterns that may be encountered.

Complete fibrosis. This is a condition in which all or most of the tubules are either completely obliterated, or so seriously damaged that spermatogenesis is no longer seen. It is not uncommon in Klinefelter's syndrome in which the Leydig cells are commonly clumped or are in a nodular arrangement and give the appearance of being increased in numbers (Paulsen *et al* 1968).

Germinal cell aplasia (Sertoli-cell only syndrome). The tubules are usually smaller than normal and populated entirely with Sertoli cells. As a rule few, if any, germinal cells are present. Peritubular fibrosis is usually absent and Leydig cells are commonly normal.

Germinal cell arrest. In many or all of the tubules, stages of spermatogenesis can be seen to be proceeding to only one of the less mature stages of the process. Peritubular fibrosis may be present, but this is not uniformly distributed. Leydig cells are normal.

Incomplete germinal cell arrest. This is similar to germinal cell arrest except that the arrest of spermatogenesis does not involve all spermatocytes. Some of the spermatocytes show fragmentation or pyknotic changes but others show the changes involved in the normal process of maturation. Spermatozoa may be detectable within the tubules but these are rarely in large numbers. Peritubular fibrosis is not common and the Leydig cells are usually normal.

Sloughing and disorganization. This is not an uncommon feature in biopsies from oligospermic men in which a somewhat disorganized appearance of the seminiferous epithelium is characteristic. This would appear to result from a disturbance of the normal orderly sequence of spermatogenesis, and a sloughing of the immature cells into the tubules, or a combination of both these defects. Peritubular fibrosis of a varying degree may be seen in this condition, but Leydig cells are usually present in normal numbers.

Methods have been devised by which testicular biopsy specimens can be more precisely quantitated. The technique of a testicular biopsy score count has been described by Johnsen (1970a) which is claimed to be a convenient and rapid method for registering the state of the germinal epithelium. Use of this technique has been reported to provide a more meaningful correlation between sperm count and other parameters, thus removing more subjective and general impressions commonly used to describe testicular biopsies.

CONTROL OF TESTICULAR FUNCTION

GONADOTROPHINS

More than forty years ago, Smith (1930) demonstrated quite conclusively that the pituitary was essential for the maintenance of normal testicular function. Involution of both the seminiferous epithelium and interstitial cells followed hypophysectomy in rats. Some years later, Greep & Fevold (1937) showed that in order to maintain the separate functions of the testis, both pituitary gonadotrophins (LH and FSH) were necessary. Subsequently, Smith (1942) showed that following hypophysectomy in the primate (monkey), there was a regression of the germinal epithelium to the stage of spermatogonia; Leydig cells became atrophic and disappeared. Treatment of these animals with equine gonadotrophins would prevent this regression and in those animals in which the regression had occurred there was a partial restoration of spermatogenesis.

The opportunity to study electively the effects of pituitary ablation in the human testis has rarely been possible, but it is well known that if pituitary function is interrupted prior to puberty, normal testicular development does not occur, the germinal epithelium remains immature and Leydig cells do not develop. Should failure of pituitary function occur in adult life, there is an involution of the germinal epithelium and Leydig cells disappear. Whether pre- or post-pubertal, treatment with human gonadotrophins will bring about some restoration of the atrophic testis to normal. Thus, patients with prepubertal gonadotrophic failure have been rendered both virile and fertile; similar effects have also been produced in patients with gonadotrophic failure in adult life (Gemzell & Kjessler 1964, Lytton & Kase 1966 and MacLeod *et al* 1966).

Notwithstanding the clear experimental and therapeutic evidence of the effects of pituitary ablation and gonadotrophic therapy on the separate functions of the testis, there is still no real consensus as to the role played by the gonadotrophins or testicular hormones in the maintenance of spermatogenesis. For instance, Cutuly & Cutuly (1940) have stated that the pre-meiotic phase of spermatogenesis can occur independently of hormonal action, that meiotic divisions appear to be controlled by pituitary gonadotrophin, and that spermiogenesis is either independent of hormonal control or may be under the influence of testicular hormones. The Steinbergers (1969) propose that testicular hormones are required for the division of gonocytes to spermatogonia, and possibly for the early maturation steps of the spermatid, that FSH is required for the final maturation of spermatids while all steps from pale type A spermatogonia to pachytene spermatocytes are not under direct hormonal influence. A different view is expressed by Albert (1961) who believes that spermatogenesis is regulated entirely by the gonadotrophins which directly influence the rates of meiotic and mitotic activity, and indirectly influence spermiogenesis by their effects on Leydig cell function.

Other opinions on the influence of gonadotrophins on spermatogenesis and spermiogenesis could be cited; all they might do is to confuse rather than clarify this still rather grey area. These difficulties are compounded because of the widely different states of purification of the pituitary hormones that have been used experimentally or therapeutically, because of species differences and because some experiments have studied the effect of gonadotrophins upon the maintenance of spermatogenesis, while others have been directed towards the study of the repair of the already involuted or regressed epithelium. These problems are accentuated even further by the complexities of spermatogenesis which, because of the intricate nature of the process, does not easily permit systematic study.

On one phenomenon there seems to be no doubt; the interstitial cells are controlled by LH. It would seem preferable, therefore, to examine the separate control of the two components of the testis, and to attempt to integrate these on the basis of current information.

CONTROL OF LEYDIG CELL FUNCTION

Reference has already been made to the abundance of Leydig cells in the neonatal testis, presumably as the result of the stimulus by placental gonadotrophin, and of their complete involution within the first few weeks of life. If appropriately stimulated before puberty, these quiescent Leydig cells are capable of secreting testosterone. Thus, the administration of LH or HCG, over a 3- or 4-day period, will produce elevations in the levels of testosterone in peripheral plasma which will return to pre-treatment levels after cessation of therapy. Histological examination of testes stimulated with HCG shows the development of interstitial cells, provided the stimulus has been of sufficient duration. In children with precocious puberty due to LH excess, examination of the testes show intense Leydig cell stimulation; measurements of the levels of testosterone in plasma in such patients show these to be elevated, not infrequently into the adult male range.

During the pre-pubertal years when mature Leydig cells cannot be identified within the testis, low levels of LH can be detected in plasma and urine by sensitive radioimmunoassay procedures. Both FSH and LH can be measured by bioassay in pooled urine samples from pre-pubertal children. With sexual maturation, urinary levels of both hormones increase, but the increase in the rate of excretion of LH is about five times greater than that of FSH (Rifkind *et al* 1967). These changes are also observed in the levels of plasma LH and FSH which together with the levels of testosterone in plasma gradually increase and correlate well with the observed stages of puberty (Frasier & Horton 1966, Johanson *et al* 1969, Raiti *et al* 1969, Lee *et al* 1970, Yen & Vicic 1970, Burr *et al* 1970), testicular biopsies taken at this stage show Leydig cell development which can be similarly correlated. More direct evidence of the effect of HCG on Leydig cell function is shown by the rapid increases (within minutes) in the levels of testosterone in the spermatic vein following the infusion of HCG into the spermatic artery (Eik-Nes 1964, 1967). In the adult with normal Leydig cell function the administration of HCG further increases the levels of testosterone in plasma, and this can be usefully employed as a test of Leydig cell function (Hudson *et al* 1966, Lipsett *et al* 1966).

The administration of testosterone or oestrogens to normal adult males is associated with a fall in the levels of both LH and testosterone in plasma; these values return to normal once the treatment is discontinued. It is of interest that suppression of LH secretion by this treatment is relatively slow (days) and incomplete as compared to the suppression of ACTH by corticosteroids, but the significance of this is not entirely understood (Alder *et al* 1968). In contrast to the administration of gonadal steroids, castration is followed by an elevation in plasma LH and FSH.

There is good evidence that the feedback mechanism between testosterone and LH is mediated within the hypothalamus. This can be abolished by hypothalamic lesions or by transplantation of the pituitary away from the hypothalamus. Further, it has been shown that crude or purified hypothalamic extracts will result in the secretion of LH by the pituitary. The hypothalamic materials that are responsible for the release of LH and other trophic hormones from the anterior pituitary are termed releasing factors (or hormones). Chemically, these releasing factors have been shown to be small peptides each specific for the pituitary hormone it releases. Unlike some of the pituitary protein hormones, e.g. growth hormone, LH and FSH, these releasing factors appear not to be species-specific. They are highly potent, minute (nanogram) amounts administered systemically resulting in trophic hormone release. They are also quickly destroyed following incubation with plasma which is of interest in view of the uniquely short system of portal vessels by which these peptides are transported from the hypothalamus to the pituitary.

The regulation of testosterone secretion from Leydig cells depends upon the secretion of LH from the anterior pituitary which, in turn, is controlled by hypothalamic LH-releasing factor (LRF). The usual long loop feedback system between gonad and hypothalamus is a significant regulating mechanism; this, however, is not the only factor regulating gonadotrophin secretion. As far as FSH is concerned, Martini (1970) has shown that FSH-releasing factor is present in a higher concentration in the hypothalami of rats that have been castrated and hypophysectomized than in those animals which have been either castrated or hypophysectomized.

In addition to the direct effects of gonadal steroids on the hypothalamus, numerous afferent pathways enter this region of the brain from the cerebral cortex and basal ganglia. The areas of the brain particularly involved are the hippocampus, amygdala and globus pallidus; the hypothalamus also has connections with the mid-brain, particularly with the reticular activating substance. It has been shown that gonadal steroids, implanted locally or administered systemically can alter the threshold of cerebral activity. Thus, synthetic steroids, used as antifertility compounds, when administered to experimental animals can block the secretion of ovulating hormones from the pituitary by preventing the limbic-hypothalamic stimuli reaching the area of the median eminence and thus preventing the release of LRF.

Recent experiments on the possible neurotransmitter substances responsible for the secretion of releasing hormones would suggest that dopamine is probably involved; thus, the injection of dopamine into the third ventricle of male rats has been shown to be followed by an increase in the concentration of LRF and FRF (FSH-releasing factor) in hypophyseal-portal blood (Kamberi *et al* 1970).

The administration of the anti-oestrogen, clomiphene, has also been shown to be associated with LH and FSH release, presumably as a direct result of the action of this compound on the hypothalamus. This action is of value in diagnosis and possibly in treatment (Odell *et al* 1967, Peterson *et al* 1968).

The roles of gonadotrophin release and of testosterone secretion at times of sexual activity have also been studied. Taleisnick *et al* (1966) have shown that acute depletion of pituitary LH occurs immediately following copulation in rabbits when acute elevations in the levels of testosterone in plasma have also been observed (Saginor & Horton 1968, Haltmeyer & Eik-Nes 1969). In the human male, evidence for the reflex release at the time of sexual activity is more fragmentary. Changes in beard growth, known to be under androgenic control, have been reported after periods of prolonged sexual abstinence. It was noted that with the anticipation of the resumption of sexual activity there is a significant increase in beard growth (Anonymous 1970).

Although the relationships between the secretion of LH and the Leydig cells seem fairly clearly defined, there is still considerable doubt as to whether a feedback mechanism exists between spermatogenesis and FSH. From an extensive study based on his technique of the testicular biopsy score count and the measurement of urinary total gonadotrophin, Johnsen (1970b) concludes that it is the last stage of spermiogenesis, involving the maturation of spermatozoa from spermatids which is the important and only stage in the seminiferous epithelium-hypophyseal feedback mechanism. Because the technique of gonadotrophin estimation does not specifically measure FSH, it is not possible to say whether this feedback involves LH or FSH. Others (Leonard *et al* 1970) claim that there is a feedback between the pituitary and the seminiferous epithelium, but make no specific claims as to the site in the epithelium involved in this mechanism.

THE ACTION OF LH

All available evidence points to a direct effect of LH on androgen biosynthesis in the testis, and to its localization mainly, if not exclusively, in the Leydig cells. The problem of how LH exerts this effect is only part of the broader problem of how trophic hormones in general stimulate steroidogenesis. Most experimental evidence would suggest that the main site of action of LH on androgen biosynthesis is between the stage of cholesterol and pregnenolone which involves 20α-hydroxylation of the cholesterol side-chain which is then cleaved. The mechanism by which LH exerts this effect, for example by stimulating protein synthesis or by increasing the levels of $3',5'$-AMP in the cell, remains to be elucidated. For a complete review of this problem the reader is referred to Hall (1970).

THE ACTION OF FSH

Most evidence would point to FSH having a direct effect on the germinal epithelium; the precise mechanism by which this hormone exerts this effect is still unknown. Means & Hall (1968) have shown that when FSH is administered to rats *in vivo* there is an increased incorporation of amino acids into testicular proteins *in vitro*. It is of interest that the type of effect observed depends upon the age of the animal; thus FSH will show this effect on rats up till 22–24 days old, but not on older rats. However, if adult rats are hypophysectomized this effect is again observed within 18 hours of hypophysectomy. It has been suggested that the failure of FSH to stimulate protein synthesis in adult rats results from the fact that they are already maximally stimulated by FSH. It is not possible to be certain on which cells within the germinal epithelium FSH exerts its principal effects. Studies by Lostroh (1963) indicate that these are on the early steps of spermatogenesis, but more definitive experiments are needed before the exact site of action of this hormone is known.

THE CONTROL OF SPERMATOGENESIS

In this rather difficult area, a number of questions remain to be answered; what initiates spermatogenesis? How is spermatogenesis maintained? And, what regulates spermatogenesis once it is initiated and maintained?

The factors responsible for the initiation of spermatogenesis are not well understood. It seems likely that those primordial germ cells carried into the gonadal mesenchyme with the primary sex cords are the progenitors of the gonocytes from which pale type A spermatogonia

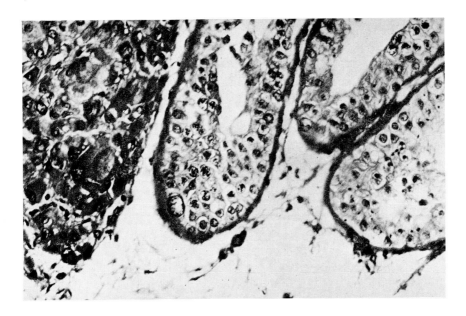

FIG. 3.5. Photomicrograph of one margin of an interstitial cell tumour of the testis of a young boy. To the left of the section are the solid tumour cells; in the centre and to the right are three seminiferous tubules in cross-section. It may be seen that spermatogenesis is proceeding to the stage of secondary spermatocytes. Reproduced by courtesy of R.H.Vines (1969).

develop. The stimulus for this development is not known with certainty, but it could well be the result of testosterone secretion. In this context it is of interest that in a few of the reported cases of Leydig cell tumours occurring in children, the testicular tissue close to the tumour shows evidence of development of the seminiferous tubules with cellular proliferation to the stage of secondary spermatocytes or spermatids; in those patients where the opposite (non-affected) testis was examined, the testis was in its normal prepubertal state. This would certainly support the view that high local concentration of androgens may initiate spermatogenesis (Stewart *et al* 1936, Cook *et al* 1952, Vines 1969).

Once spermatogenesis is initiated, its maintenance probably depends upon the combined effects of a high local concentration of androgens from the interstitial cells and possibly the Sertoli cells and the activity of FSH. It is suggested that androgens stimulate the proliferation of spermatogonia; that these cells, in the presence of androgens are able to mature to the stage of secondary spermatocytes or early spermatids, and that the final stages of spermiogenesis require FSH. The evidence for this suggestion comes from a number of sources; if normal males are treated with testosterone proprionate there is a fall in the level of plasma LH to subnormal levels and the disappearance of sperm from the ejaculate, while plasma levels of testosterone and FSH remain normal; continuing treatment with testosterone but with the addition of HCG shows a return of the sperm count to normal. This is interpreted as demonstrating the importance of the local effects of high androgen concentrations, absent during treatment with parenteral testosterone, but restored during HCG therapy (Heller *et al* 1970). Case reports by Gemzell & Kjessler (1964) and Lytton & Kase (1966) have shown that treatment with FSH (as menopausal gonadotrophin) alone was insufficient to restore spermatogenesis; treatment with HCG, apart from stimulating Leydig cells and having the expected effect on secondary sexual characteristics stimulated the germinal epithelium to the spermatid stage; the addition of FSH to this therapeutic regime promoted the maturation to the stage of spermatozoa.

CONCLUSION

A rational approach to the diagnosis and treatment of testicular disorders can only be made by an understanding of the physiological basis of testicular function. Patients with testicular disorders may present with the problem of an apparent failure of androgen secretion, of seminiferous tubular function or a combination of these defects; sexual precocity as the result of androgen excess is a much less common problem.

Techniques are now available which can give a clear indication of the cause of the failure of androgenic function, and will show whether the defect is primarily testicular in origin or secondary to a deficiency of gonadotrophin secretion. These techniques include the measurement of plasma testosterone before and after gonadotrophic stimulation with HCG; the measurement of gonadotrophins—FSH and LH—in plasma and urine; and the use of agents that may stimulate or suppress these secretions. It would not be unreasonable to anticipate that within the next decade hypothalamic releasing factors will be isolated and synthesized, an achievement that will provide further information about the capacity of the anterior pituitary to secrete gonadotrophins. Treatment of the individual whose problem is primarily one of androgen deficiency can be rationally planned and adequately monitored. These techniques may also permit a more detailed description of patients who present with the problem of impotence and of the possible neuroendocrine disturbances in these patients. At present these are far from clearly defined.

In the field of male fertility much yet remains to be achieved. At present some relief of this problem can be offered to those men whose infertility results from a failure of gonadotrophic secretion—already many conceptions have resulted from the therapeutic use of gonadotrophins in such patients. For patients with primary testicular disorders as a cause of infertility, there is still a need for more comprehensive studies by combining endocrine and cytogenetic procedures in order to determine in which of these patients a restoration of fertility is possible by the use of gonadotrophic hormones (Lunenfeld *et al* 1970).

REFERENCES

ALBERT A. (1961) The mammalian testis, in W.C. Young (ed.) *Sex and Internal Secretions*, 3rd ed., pp. 305–365. Baltimore, Williams & Wilkins Co.

ALBERT A., UNDERDAHL L.O., GREENE L.F. & LORENZ N. (1953) Male hypogonadism: I. The normal testis. *Proc. Staff Meet. Mayo Clin.* **28**, 409–422.

ALDER A., BURGER, H.G., DAVIS J., HUDSON B., SARFATY G.A. & STRAFFON W. (1968) Carcinoma of the prostate: response of plasma luteinizing hormone and testosterone to oestrogen therapy. *Brit. med. J.* **1**, 28–30.

ANDERSON K.M. & LIAO S. (1968) Selective retention of dihydrotestosterone by prostatic nuclei. *Nature* **219**, 277–279.

ANONYMOUS (1970) Effects of sexual activity on beard growth in man. *Nature* **226**, 869–870.

BAIRD D., HORTON R., LONGCOPE C. & TAIT J.F. (1968) Steroid prehormones. *Perspect. Biol. Med.* **11**, 384–421.

BAIRD D.T., HORTON R., LONGCOPE C. & TAIT J.F. (1969) Steroid dynamics under steady-state conditions. *Rec. Progr. Horm. Res.* **25**, 611–656.

BAULIEU E.E. (1962) Studies of conjugated 17-ketosteroids in a case of adrenal tumor. *J. clin. Endocr.* **22**, 501–510.
BAULIEU E.E., CORPÉCHOT C., DRAY F., EMILIOZZI R., LEBEAU M-C., MAUVIS-JARVIS P. & ROBEL P. (1965) An adrenal-secreted 'androgen': dehydroisoandrosterone sulfate. Its metabolism and a tentative generalization on the metabolism of other steroid conjugates in man. *Rec. Progr. Horm. Res.* **21**, 411–494.
BAULIEU E.E. & ROBEL P. (1970) Catabolism of testosterone and androstenedione, in K.B. Eik-Nes (ed.) *The Androgens of the Testis*, pp. 49–71. New York, Marcel Dekker Inc.
BIRD C.E., GREEN R.N. & CLARK A.F. (1969) Effect of administration of estrogen on the disappearance of ^{3}H-testosterone in the plasma of human subjects. *J. clin. Endocr.* **29**, 123–126.
BRUCHOVSKY N. & WILSON J.D. (1968) The conversion of testosterone to 5α-androstane-17β-ol-3-one by rat prostate *in vivo* and *in vitro*. *J. biol. Chem.* **243**, 2012–2021.
BRUNER J.A. (1951) Distribution of chorionic gonadotropin in mother and fetus at various stages of pregnancy. *J. clin. Endocr.* **11**, 360–374.
BURGER H.G., BROWN J.B., CATT K.J., HUDSON B. & STOCKIGT J.R. (1968) Physiological studies on the secretion of human pituitary luteinizing hormone and gonadal steroids, in M.Margoulies (ed.) *Protein & Polypeptide Hormones*, pp. 412–414. International Congress Series 161. Amsterdam, Excerpta Medica Foundation.
BURGOS M.H. & VITALE-CALPE R. (1969) Gonadotrophic control of spermiation, in C.Gual (ed.) *Progress in Endocrinology*, pp. 1030–1037. International Congress Series 184. Amsterdam, Excerpta Medica Foundation.
BURR I.M., SIZONENKO P.C., KAPLAN S.L. & GRUMBACH M.M. (1970) Hormonal changes in puberty. I. Correlation of serum luteinizing hormone and follicle stimulating hormone in the stages of puberty, testicular size, and bone age in normal boys. *Pediat. Res.* **4**, 25–35.
CHRISTENSEN A.K. (1965) The fine structure of the testicular interstitial cells in guinea pigs. *J. Cell Biol.* **26**, 911–935.
CLERMONT Y. & HARVEY S.C. (1967) Effects of hormones on spermatogenesis in the rat, in G.E.W. Wolstenholme & M. O'Connor (eds.) *Endocrinology of the Testis, Ciba Foundation Colloquia on Endocrinology* **16**, 173–196.
COOK C.D., GROSS R.E., LANDING B.H. & ZYGMUNTOWICZ A.S. (1952) Interstitial cell tumor of the testis. Study of a 5-year-old boy with pseudo-precocious puberty. *J. clin. Endocr.* **12**, 725–734.
CUTULY E. & CUTULY E.C. (1940) Observations on spermatogenesis in rats. *Endocrinology* **26**, 503–507.
DAVIS J.R. & LANGFORD G.A. (1970) Pharmacological studies on the testicular capsule in relation to sperm transport, in E. Rosemberg & C.A. Paulsen (eds.) *The Human Testis*, pp. 495–512. New York and London, Plenum Press.
DELLER J.J., BOULIS M.W., HARRISS W.E., HUTSELL T.C., GARCIA J.F. & LINFOOT J.A. (1970) Growth hormone response patterns to sex hormone administration in growth retardation. *Amer. J. Med. Sci.* **259**, 292–297.

EIK-NES K.B. (1964) Effects of gonadotrophins on secretion of steroids by the testis and ovary. *Physiol. Rev.* **44**, 609–630.

EIK-NES K.B. (1967) Factors influencing the secretion of testosterone in the anaesthetized dog, in G.E.W. Wolstenholme & M. O'Connor (eds.) *Endocrinology of the Testis, Ciba Foundation Colloquia on Endocrinology* **16**, 120–136.

EIK-NES K.B. (1970) Synthesis and secretion of androstenedione and testosterone, in K.B. Eik-Nes (ed.) *The Androgens of the Testis*, pp. 1–47. New York, Marcel Dekker Inc.

FAWCETT D.W. & BURGOS M.H. (1960) Studies on the fine structure of the mammalian testis. II. The human interstitial tissue. *Amer. J. Anat.* **107**, 245–269.

FISHMAN L.M., SARFATY G.A., WILSON H. & LIPSETT M.B. (1966) The role of the testis in oestrogen production, in G.E.W. Wolstenholme & M. O'Connor (eds.) *Endocrinology of the Testis, Ciba Foundation Colloquia on Endocrinology* **16**, 156–172.

FRASIER S.D. & HORTON R. (1966) Androgens in the peripheral plasma of prepubertal children and adults. *Steroids* **8**, 777–784.

GEMZELL C. & KJESSLER B. (1964) Treatment of infertility after partial hypophysectomy with human pituitary gonadotrophins. *Lancet* **i**, 644.

GOLDZIEHER J.W. & ROBERTS I.S. (1952) Identification of estrogen in the human testis. *J. clin. Endocr.* **12**, 143–149.

GREEP R.O. & FEVOLD H.L. (1937) The spermatogenic and secretory function of the gonads of hypophysectomized adult rats treated with pituitary FSH and LH. *Endocrinology* **21**, 611–618.

HALL P.F. (1970) Gonadotrophic regulation of testicular function, in K.B. Eik-Nes (ed.) *The Androgens of the Testis*, pp. 73–115. New York, Marcel Dekker Inc.

HALTMEYER G.C. & EIK-NES K.B. (1969) Plasma levels of testosterone in male rabbits following copulation. *J. Reprod. Fert.* **19**, 273–277.

HANLEY H.G. (1968) Vasectomy for voluntary male sterilization. *Lancet* **ii**, 207–209.

HARRIS G.W. (1964) Sex hormones, brain development and brain function. *Endocrinology* **75**, 627–648.

HELLER C.G. & CLERMONT Y. (1964) Kinetics of the germinal epithelium in man. *Rec. Progr. Horm. Res.* **20**, 545–575.

HELLER C.G., HELLER G.V. & ROWLEY M.J. (1969) Human spermatogenesis; an estimate of the duration of each cell association and of each cell type, in C. Gual (ed.) *Progress in Endocrinology*, pp. 1014–1015. International Congress Series, 184. Amsterdam, Excerpta Medica Foundation.

HELLER C.G., MORSE H.C., SU M. & ROWLEY M.J. (1970) The role of FSH, ICSH and endogenous testosterone during testicular suppression by exogenous testosterone in normal men, in E. Rosemberg & C.A. Paulsen (eds.) *The Human Testis*, pp. 249–256. New York and London, Plenum Press.

HORTON R. & TAIT J.F. (1966) Androstenedione production and interconversion rates measured in peripheral blood and studies on the possible site of its conversion to testosterone. *J. clin. Invest.* **45**, 301–313.

HOWARD R.P., SNIFFEN R.C., SIMMONS F.A. & ALBRIGHT F. (1950) Testicular deficiency; a clinical and pathologic study. *J. clin. Endocr.* **10**, 121–186.

HUDSON B., COGHLAN J.P. & DULMANIS A. (1966) Testicular function in man, in

G.E.W. Wolstenholme & M. O'Connor (eds.) *Endocrinology of the Testis, Ciba Foundation Colloquia on Endocrinology* **16**, 140–155.

HUDSON B., BURGER H.G., DE KRETSER D.M., COGHLAN J.P. & TAFT H.P. (1970) Testosterone plasma levels in normal and pathological conditions, in E. Rosemberg & C.A. Paulsen (eds.) *The Human Testis*, pp. 423–436. New York and London, Plenum Press.

ILLIG R. & PRADER A. (1970) Effect of testosterone on growth hormone secretion in patients with anorchia and delayed puberty. *J. clin. Endocr.* **30**, 615–623.

ITO T. & HORTON R. (1970) Dihydrotestosterone in human peripheral plasma. *J. clin. Endocr.* **31**, 362–368.

JOHANSON A.J., GUYDA H., LIGHT C., MIGEON C.J. & BLIZZARD R.J. (1969) Serum luteinizing hormone by radioimmunoassay in normal children. *J. Pediat.* **74**, 416–424.

JOHNSEN S.G. (1970a) Testicular biopsy score count—a method for registration of spermatogenesis in human testes: normal values and results in 335 hypogonadal males. *Hormones* **1**, 1–24.

JOHNSEN S.G. (1970b) The stage of spermatogenesis involved in the testicular-hypophyseal feed-back mechanism in man. *Acta Endocr.* **64**, 193–210.

JOST A. (1953) Problems of fetal endocrinology: the gonadal and hypophyseal hormones. *Rec. Progr. Horm. Res.* **8**, 379–418.

JOST A. (1960) The role of fetal hormones in prenatal development. *Harvey Lectures* **55**, 201–226.

JOST A. (1970) Hormonal factors in the development of the male genital system, in E. Rosemberg & C.A. Paulsen (eds.) *The Human Testis*, pp. 11–18. New York and London, Plenum Press.

KAMBERI I.A., MICAL R.S. & PORTER J.C. (1970) Follicle stimulating hormone releasing activity in hypophyseal portal blood and elevation by dopamine. *Nature*, **227**, 714–715.

KENT J.R. & ACONE A.B. (1966) Plasma testosterone levels and aging males, in *Androgens in Normal & Pathological Conditions*, pp. 31–35. International Congress Series 101. Amsterdam, Excerpta Medica Foundation.

KOWAL J., FORCHIELLI E. & DORFMAN R.I. (1964) The Δ^5-3β-hydroxysteroid dehydrogenases of corpus luteum and adrenal. II. Interaction of C_{19} and C_{21} substrates and products. *Steroids* **4**, 77–100.

LACY D. & PETTIT A.J. (1970) Sites of hormone production in the mammalian testis and their significance in the control of fertility. *Brit. med. Bull.* **26**, 87–91.

LE BLOND C.P. & CLERMONT Y. (1952) Definition of the stages of the cycle of the seminiferous epithelium in the rat. *Ann. New York Acad. Sci.* **55**, 548–573.

LEE P.A., MIDGLEY A.R. & JAFFE R.B. (1970) Regulation of human gonadotrophins: VI. Serum follicle stimulating and luteinizing hormone determinations in children. *J. clin. Endocr.* **31**, 248–253.

LEONARD J.M., LEACH R.B. & PAULSEN C.A. (1970) Interrelationship of follicle stimulating hormone (FSH) and spermatogenesis. *Clin. Res.* **18**, 169 (Abstract).

LINDNER H.R. (1963) Partition of androgen between lymph and venous blood of the testis of the ram. *J. Endocr.* **25**, 483–494.

LINDNER H.R. (1966) Participation of lymph in the transport of gonadal steroid hormones. *Proc. 2nd. Int. Congr. Hormonal Steroids*, pp. 821–827. International Congress Series 132. Amsterdam, Excerpta Medica Foundation.

LIPSETT M.B., WILSON H., KIRSCHNER M.A., SARFATY G.A. & BARDIN C.W. (1966) Studies on Leydig cell physiology and pathology: secretion and metabolism of testosterone. *Rec. Progr. Horm. Res.* **22**, 245–271.

LOSTROH A.J. (1963) The effect of follicle stimulating hormone and interstitial cell-stimulating hormone on spermatogenesis in Long-Evans rats hypophysectomized for six months. *Acta Endocr.* **43**, 592–600.

LUNENFELD B. & SHALOVSKY-WEISSENBERG R. (1970) Assessment of gonadotropin therapy in male infertility, in E. Rosemberg & C.A. Paulsen (eds.) *The Human Testis*, pp. 613–628. New York and London, Plenum Press.

LYTTON B. & KASE N. (1966) Effects of human menopausal gonadotrophin on a eunuchoidal male. *New Eng. J. Med.* **274**, 1061–1064.

MACLEOD J. (1964) Human seminal cytology as a sensitive indicator of the germinal epithelium. *Int. J. Fertil.* **9**, 281–295.

MACLEOD J., PAZIANOS A. & RAY B. (1966) The restoration of human spermatogenesis and of the reproductive tract with urinary gonadotropins following hypophysectomy. *Fertil. & Steril.* **17**, 7–23.

MACLEOD J. (1970) The significance of deviations in human sperm morphology, in E. Rosemberg & C.A. Paulsen (eds.) *The Human Testis*, pp. 481–492. New York and London, Plenum Press.

MCCULLAGH D.R. (1932) Dual endocrine activity of the testes. *Science* **76**, 19–20.

MCDONALD P., BEYER C., NEWTON F., BRIEN B., BAKER R., TAN H.S., SAMPSON C., KITCHING P., GREENHILL R. & PRITCHARD D. (1970) Failure of 5α-dihydrotestosterone to initiate sexual behaviour in the castrated male rat. *Nature* **227**, 964–965.

MCKAY, D.G., HERTIG A.T., ADAMS E.C. & DANZIGER S. (1953) Histochemical observations on the germ cells of human embryos. *Anat. Record.* **117**, 201–219.

MAHAJAN D.K. & SAMUELS L.T. (1962) Inhibition of steroid desmolase by progesterone. *Federation Proc.* **21**, 209 (Abstract).

MANN T. (1946) Studies on the metabolism of semen. III. Fructose as a normal constituent of seminal plasma. Site of formation and function of fructose in semen. *Biochem. J.* **40**, 481–491.

MANN T. (1964) *The Biochemistry of Semen and of the Male Reproductive Tract*. London, Methuen.

MARSHALL W.A. & TANNER J.M. (1970) Variations in the pattern of pubertal changes in boys. *Arch. Dis. Childh.* **45**, 13–23.

MARTINI L. (1970) Hypothalamic control of gonadotropin secretion in the male, in E. Rosemberg & C.A. Paulsen (eds.) *The Human Testis*, pp. 187–203. New York and London, Plenum Press.

MEANS A.R. & HALL P.F. (1968) Protein biosynthesis in the testis: I. Comparison between stimulation by FSH and glucose. *Endocrinology* **82**, 597–602.

NELSON W.O. (1953) Interpretation of testicular biopsy. *J.A.M.A.* **151**, 449–454.

NIEMI M., IKONEN M. & HERVONEN A. (1967) Histochemistry and fine structure of the interstitial tissue of the human foetal testis, in G.E.W. Wolstenholme &

M. O'Connor (eds.) *Endocrinology of the Testis, Ciba Foundation Colloquia on Endocrinology* **16**, 31–35.

NIGHTINGALE M.S., TSAI S.C. & GAYLOR J.L. (1967) Testicular sterols. IV. Incorporation of mevalonate into squalene and sterols by cell-free preparations of testicular tissue. *J. biol. Chem.* **242**, 341–349.

ODELL W.D., ROSS G.T. & RAFORD P.L. (1967) Radioimmunoassay for luteinizing hormone in human plasma or serum: physiological studies. *J. clin. Invest.* **46**, 248–255.

PAULSEN C.A., GORDON D.L., CARPENTER R.W., GANDY H.M. & DRUCKER W.D. (1968) Klinefelter's syndrome and its variants: A hormonal and chromosomal study. *Rec. Progr. Horm. Res.* **24**, 321–363.

PEARLMAN W.H. & CREPY O. (1967) Steroid-protein interaction with particular reference to testosterone binding by human serum. *J. biol. Chem.* **242**, 182–189.

PETERSON N.T., MIDGLEY A.R. & JAFFE R.B. (1968) Regulation of human gonadotropins. III. Luteinizing hormone and follicle stimulating hormone in sera from adult males. *J. clin. Endocr.* **28**, 1473–1478.

RAITI S., JOHANSON A., LIGHT C., MIGEON C.J. & BLIZZARD R.M. (1969) Measurement of immunologically reactive follicle stimulating hormone in serum of normal male children and adults. *Metabolism* **18**, 234–240.

RIFKIND A.B., KULIN H.E. & ROSS G.T. (1967) Follicle stimulating hormone (FSH) and luteinizing hormone (LH) in the urine of prepubertal children. *J. clin. Invest.* **46**, 1925–1931.

ROOSEN-RUNGE E.C. & GIESEL L.O. (1950) Quantitative studies on spermatogenesis in the albino rat. *Amer. J. Anat.* **87**, 1–30.

RYAN K.J. & ENGEL L.L. (1953) The interconversion of estrone and estradiol by human tissue slices. *Endocrinology* **52**, 287–299.

RYAN K.J. & SMITH O.W. (1965) Biogenesis of steroid hormones in the human ovary. *Rec. Progr. Horm. Res.* **21**, 367–409.

SAEZ J.M., SAEZ S. & MIGEON C.J. (1967) Identification and measurement of testosterone in the sulfate fraction of plasma of normal subjects and patients with gonadal and adrenal disorders. *Steroids* **9**, 1–14.

SAGINOR M. & HORTON R. (1968) Reflex release of gonadotropin and increased plasma testosterone concentration in male rabbits during copulation. *Endocrinology* **82**, 627–630.

SERRA G.B., PEREZ-PALACIOS G. & JAFFE R.B. (1970) De novo testosterone biosynthesis in the human fetal testis. *J. clin. Endocr.* **30**, 128–130.

SIITERI P.K. & MACDONALD P.C. (1966) Placental estrogen biosynthesis during human pregnancy. *J. clin. Endocr.* **26**, 751–761.

SMITH P.E. (1930) Hypophysectomy and replacement therapy in the rat. *Amer. J. Anat.* **45**, 205–273.

SMITH P.E. (1942) Effects of equine gonadotropin on testes of hypophysectomized monkeys. *Endocrinology* **31**, 1–12.

STEINBERGER E. & STEINBERGER A. (1969) The spermatogenic function of the testes, in K.W. McKerns (ed.) *The Gonads*, pp. 715–737. Amsterdam, North-Holland Publishing Co.

STEWART C.A., BELL E.T. & ROEHLKE A.B. (1936) An interstitial-cell tumour of the testis with hypergenitalism in a child of five years. *Amer. J. Cancer* **26**, 144–150.

TAIT J.F. & BURSTEIN S. (1964) *In vivo* studies of steroid dynamics in man, in G. Pincus, K.V. Thimann & E.B. Astwood (eds.) *The Hormones* 5, pp. 441–445. New York, Academic Press.

TALEISNIK S., CALGARIS L. & ASTRADA J.J. (1966) Effect of copulation on the release of pituitary gonadotropins in male and female rats. *Endocrinology* **79**, 49–54.

TANNER J.M. (1962) *Growth at Adolescence*, 2nd ed. Oxford, Blackwell Scientific Publications.

TSAI S.C., YING B.P. & GAYLOR J.L. (1964) Testicular sterols. I. Incorporation of mevalonate and acetate into sterols by testicular tissue from rats. *Arch. Biochem. Biophys.* **105**, 329–338.

VAN DER MOLEN H.J. (1970) Estimation of androstenedione and testosterone by physico-chemical methods, in K.B. Eik-Nes (ed.) *The Androgens of the Testis*, pp. 145–215. New York, Marcel Dekker Inc.

VANDE WIELE R.L., MACDONALD P.C., GURPIDE E. & LIEBERMAN S. (1963) Studies on the secretion and interconversion of the androgens. *Rec. Progr. Horm. Res.* **19**, 275–310.

VAN WYK J.J. & GRUMBACH M.M. (1968) Disorders of sex differentiation, in R.H. Williams (ed.) *Textbook of Endocrinology*, 4th ed., pp. 551. Philadelphia, W.B. Saunders Company.

VENNING E.H. (1955) Clinical value of hormone estimations. *Brit. med. Bull.* **11**, 140–144.

VERMEULEN A., VERDONCK L., VAN DER STRAETEN M. & ODRIE N. (1969) Capacity of the testosterone-binding globulin in human plasma and influence of specific binding of testosterone on its metabolic clearance rate. *J. clin. Endocr.* **29**, 1470–1480.

VINES, R.H. (1969) Interstitial tumour of the testis. *Aust. Paediat. J.* **5**, 248–249.

WHITE I.G. & HUDSON B. (1968) Testosterone and dehydroepiandrosterone concentration in fluids of the mammalian male reproductive tract. *J. Endocrinol.* **41**, 291–292.

WILLIAMS-ASHMAN H.G. (1970) Biochemistry of testicular androgen action, in K.B. Eik-Nes (ed.) *The Androgens of the Testis*, pp. 117–143. New York, Marcel Dekker Inc.

WILLIAMS W.W. & SAVAGE A. (1925) Observations on the seminal micropathology of bulls. *Cornell Vet.* **15**, 353–375.

WOODS M.C. & SIMPSON M.E. (1961) Pituitary control of the testis of the hypophysectomized rat. *Endocrinology* **69**, 91–125.

YEN, S.S.C. & VICIC, W.J. (1970) Serum follicle-stimulating hormone levels in puberty. *Amer. J. Obstet. Gynec.* **106**, 134–137.

YING B.P., CHANG Y.J. & GAYLOR J.L. (1965) Testicular sterols. III. Effect of gonadotrophins on the biosynthesis of testicular sterols. *Biochim. Biophys. Acta* **100**, 256–262.

4
Endocrinology of the Foeto-Maternal Unit

G.C.LIGGINS

NON-STEROIDAL HORMONES OF THE PLACENTA

HUMAN CHORIONIC GONADOTROPHIN (HCG)

Chorionic gonadotrophin is remarkable amongst the hormones in that its function remains as uncertain today as it was when Halban (1905) correctly deduced from clinical observations its presence in the placenta. Furthermore, nothing is known of the factors that regulate the release or production of HCG. Nevertheless, despite these deficiencies in knowledge, the assay of HCG for the diagnosis of pregnancy (Ashcheim & Zondek 1927) represented the first routine clinical application of a hormone assay.

There is no doubt now that HCG is correctly named (that is, it is chorionic in origin), but the cellular site of production is uncertain. Tissue culture, histochemistry and immunofluorescent studies (reviewed by Brody 1969) have located the hormone in both cytotrophoblastic cells and syncytiotrophoblastic cells. Information available from electron-microscopic investigations suggests that the syncytiotrophoblast is derived by differentiation from the cytotrophoblast and it is reasonable, therefore, to suppose that HCG can be synthesized in both types of cell but more predominantly, perhaps, in the syncytiotrophoblast. Most of the hormone is secreted into the maternal circulation and the concentration in foetal plasma is much lower than that in maternal plasma. But the low foetal concentrations may be due in part to rapid destruction in the foetal circulation, a possibility that is supported by the presence of a higher concentration of HCG in the umbilical vein than in the umbilical artery (Lauritzen & Lehmann 1967).

Chemistry of HCG

The complete purification and structural analysis of HCG has not been achieved, the hormone being either unstable or a hormone-complex. It is a glycoprotein showing physico-chemical similarities as well as immunological cross-reactivity with the pituitary gonadotrophins, luteinizing hormone (LH) and follicle stimulating hormone (FSH). Bahl (1969) has isolated two chromatographically homogeneous glycopeptides and has also determined the sequence of monosaccharides in the carbohydrate part of the molecule. He also reported that HCG has no N-terminal and no C-terminal amino acids, but contains instead N-acetylneuraminic acid and fructose as non-reducing terminal units. After enzyme treatment with neuraminidase, biological activity is lost, suggesting that neuraminic acid is essential for activity. The polypeptide chains of HCG are linked to each other by disulphide bridges.

Measurements of the molecular weight of HCG vary with the technique used. By gel filtration its weight is approximately 59,000 but by ultracentrifugation, particularly after treatment of HCG, estimates of the order of 30,000 are usually obtained. The wide variation in estimates of the molecular weight of HCG are the result either of polymerization or of heterogeneity and both of these possibilities have received support. Bell *et al* (1969) observed that HCG is a dimer made up of identical and chromatographically homogeneous sub-units, each having a molecular weight of 23,000–28,000. On the other hand, Hamashige *et al* (1967) observed differences in electrophoretic, ion-exchange, chromatographic, immunological and biological properties which led them to conclude that HCG represents a complex of hormones composed of at least three different types of HCG molecule.

Structural similarities of HCG, LH, FSH and TSH are probably responsible for the similarities in immunological properties that create problems in the radioimmunoassay of these hormones. An additional problem in the radioimmunoassay of HCG is the existence of biologically inactive, but immunologically active, forms of the protein that arise spontaneously or result from chemical or enzymatic treatment of HCG (Butt 1967).

Secretion of HCG

A rough estimate of the secretion rate of HCG can be made by dividing the excretion rate in pregnancy urine by the percentage recovery of HCG in non-pregnant women after intravenous injection of known amounts. Such estimates yield figures of the order of 500,000–1,000,000 I.U./day in the third month of pregnancy, and 80,000–120,000 I.U./day in late pregnancy. Factors responsible for regulating the secretion are unknown. There is no doubt that secretion rate varies with trophoblastic mass, being highest when trophoblastic mass is abnormally increased by multiple pregnancy, diabetes, erythroblastosis and trophoblastic tumours and being lowest when trophoblastic mass is reduced by damage or retarded growth. However, the determinants of the changing pattern of secretion throughout normal pregnancy are uncertain.

Less than 10% of circulating HCG is excreted in the urine in a biologically active form, the remaining 90% being metabolized in the body. The renal clearance remains constant throughout pregnancy and thus the urinary excretion of HCG is probably a good indication of its secretion rate. The concentration in maternal tissues, blood and urine runs parallel throughout pregnancy, following a curve reaching a sharp peak between the 7th and 12th weeks, followed by an equally steep drop to levels that remain fairly constant at 1/5 to 1/10 of the peak value from the 15th week until term (Loraine & Bell 1966).

Physiological role of HCG

An intact corpus luteum is necessary for the continuation of human pregnancy for the first few weeks, but after that period the endocrine function of the ovaries can be dispensed with. It is generally accepted that regression of the corpus luteum is prevented by the luteotrophic effect of HCG which can be detected in the plasma within 9–12 days after ovulation (Wide 1969). Evidence for a luteotrophic role of HCG comes from the report of several investigators (Brown & Bradbury 1947, Bradbury *et al* 1950, Segaloff *et al* 1951), who succeeded in postponing menstruation by administering HCG to women during the menstrual cycle. However, Short (1969) has critically re-evaluated these studies and considers the evidence to be weak. He postulates that menstruation was postponed because large doses of HCG may induce the formation of new corpora lutea. On the other hand, Savard *et al* (1965) have shown that HCG is capable of stimulating progesterone synthesis by corpus luteum slices *in vitro*. At the present time, the precise role of HCG as a stimulus to steroidogenesis by the corpus luteum in early pregnancy remains to be defined.

HCG may stimulate placental steroid biogenesis by accelerating the aromatization of neutral steroids. Perfusion of the human placenta with HCG increases the rate of conversion of dehydroepiandrosterone or testosterone to oestradiol (Cédard *et al* 1964, Varangot *et al* 1965). However, it is not known whether HCG plays any part in the regulation of placental steroidogenesis under normal conditions.

Morphological differentiation of the Leydig cells of the interstitial tissue of the foetal testis starts at about 8 weeks, reaches maximal development at about 15 weeks and declines after the 17th week (Niemi *et al* 1967). The synthesis of luteinizing hormone (LH) by the foetal pituitary begins too late to act as a source of a gonadotrophic stimulus (Gitlin & Biasucci 1969) but the developmental course of Leydig cell differentiation fits well with the pattern of secretion of HCG. Furthermore, there is a relationship between the sex of the foetus and maternal HCG levels (Brody & Carlstrom 1965). These observations have prompted the suggestion, as yet unsubstantiated, that HCG serves an important function in sexual differentiation (Fig. 4.1).

Growth and function of the foetal zone of the foetal adrenal undoubtedly depend on ACTH secreted by the foetal pituitary. However, it is necessary to postulate the presence of an additional trophic stimulus to explain the normal development of the adrenal gland of anencephalic foetuses during the first half of pregnancy (Benirschke 1956) and the marked involution that is seen after birth. Villee (1969) has suggested that HCG may stimulate growth of the foetal adrenal. In non-pregnant women the administration of large doses of HCG induces a rise in excretion of androgens but not in that of 17-hydroxycortico-steroids, and it is possible that the same action occurs in the foetal adrenal (Pauerstein & Solomon 1966). Indeed, Lauritzen *et al* (1969) found that injections of HCG into male infants stimulated the excretion of dehydroepiandrosterone in four of seven infants studied.

Another possible function of HCG is in contributing to altered immunological reactivity in pregnancy, particularly in the early stages, by participating in a local hormonal immuno-suppression of maternal lymphocytes in the vicinity of the developing trophoblast (Kaye & Jones 1971). HCG depresses phytohaemagglutin-induced transformation of lymphocytes in culture and may also be responsible, together with cortisol, for the lymphopoenia that accompanies normal pregnancy and is particularly marked in trophoblastic tumours (see also chapter 5).

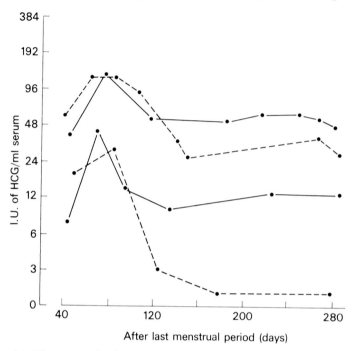

FIG. 4.1. The range of values of urinary excretion of HCG in normal pregnancies with male (broken line) or female (continuous line) foetuses. From S. Brody & G. Carlström (1965) *J. clin. Endocr.* **25**, 792.

Clinical significance of HCG assays

The urinary excretion and plasma concentration of HCG is generally increased when either the mass or the activity of the trophoblastic tissue is abnormally high. Thus, raised levels are seen in multiple pregnancy, maternal diabetes and trophoblastic neoplasms (Klopper 1970). Rather surprisingly, in view of the extent of placental damage which may be present, levels of HCG may also be increased in severe pre-eclamptic toxaemia. Histological examination of such placentas commonly reveals trophoblastic proliferation which may explain the increase in HCG excretion. It is also possible that leakage of protein through the glomerular membrane of damaged kidneys may increase renal clearance of the glycoprotein. Large lutein cysts in association with massive increases in

HCG excretion have been observed in otherwise uncomplicated pregnancies. The cause is unknown.

In the rare examples of foetal triploidy in which foetal survival has continued into late pregnancy the placentas show hydatidiform changes, HCG excretion is high and lutein cysts are present (Paterson *et al* 1971).

HCG excretion may be abnormally low in threatened abortion, but assays do not discriminate clearly between those patients who are going to abort and those who are not (Klopper 1970). Nevertheless, in conjunction with assays of urinary oestrogen and pregnanediol, HCG excretion may be useful in predicting the outcome. It is as well to remember, however, that in missed abortion detectable levels of HCG may persist for several weeks.

The most important uses of HCG assays are in the diagnosis of normal pregnancy and in the recognition and follow-up of trophoblastic neoplasm. Ellegood *et al* (1969) suggest that although high HGC levels are not always conclusive evidence of the presence of trophoblastic tumour, the persistence of high levels beyond 15 weeks of pregnancy strongly supports this diagnosis. But as pointed out by Halpin (1970) abnormally high HCG titres may be present in multiple pregnancy and since the clinical features are similar, misdiagnosis can occur. Appropriate investigations to exclude multiple pregnancy should always precede evacuation of a suspected hydatidiform mole. Hobson & Wide (1968) observed that urine from pregnant women with hydatidiform moles has a higher ratio of bioassayable HCG to immunoassayable HCG than urine from women with normal pregnancies. Determination of this ratio may enhance the reliability of diagnosis of trophoblastic tumours. It may be particularly important in distinguishing chorion carcinoma from early pregnancy when HCG assays become positive after follow up tests have been negative in women who have been successfully treated for a trophoblastic neoplasm. Occasionally, assay of HCG in cerebrospinal fluid can be useful if intracranial metastases of chorion carcinoma are suspected. Rushworth *et al* (1968) found that the ratio of plasma:CSF concentration of HCG was 100:1 in patients without intracranial metastases and 35:1 in those in whom intracranial metastases had occurred.

The value of HCG assays in the management of pregnancy complications such as diabetes, toxaemia and Rh isoimmunization has been thoroughly investigated (for review see Klopper 1970) and the consensus of opinion is that they are not helpful.

HUMAN CHORIONIC SOMATOMAMMOTROPHIN (HCS)

The human placenta was found by Fukushima (1961) to contain a somatotrophic substance and by Ito & Higashi (1961) to contain a prolactin-like material. About the same time, Josimovich & MacLaren (1962) reported the isolation of a substance that was lactogenic and that showed a partial cross-reactivity with antisera to human pituitary growth hormone. They designated this hormone 'human placental lactogen' (HPL). Subsequently, Kaplan & Grumbach (1964) introduced the term 'Chorionic Growth Hormone-Prolactin' (CGP). The confusion in terminology has been more or less resolved by general adoption of the term 'Human Chorionic Somatomammotrophin' (HCS).

Chemistry of HCS

HCS is a polypeptide with a molecular weight between 19,000 and 30,000. It has chemical, immunological and biological similarities to HGH that appear to depend on similarities of amino-acid composition. Sherwood (1967) noted not only that there was identity in the amino-acid content of about 40% of tryptic digest fragments of the two hormones but also that there were striking similarities in the location of the disulphide bridges. The physical properties of human prolactin have yet to be characterized but it is possible that prolactin, too, has close chemical similarity to HCS.

Secretion of HCS

The half-life of HCS in the maternal blood stream is about 20 minutes (Beck & Daughaday 1967). From this it can be calculated that the daily secretion rate near term is 0·5–3·0 g. Little is known of the control of the rate of secretion of HCS; it is unaffected by maternal exercise, by the ingestion of food and by maternal blood sugar levels and there is no circadian rhythm (Spellacy *et al* 1966). Trophoblast in tissue culture readily secretes HCS and the rate of production is unaffected by insulin, progesterone, cortisol and cyclic AMP (Suwa & Friesen 1969).

Like HCG, most HCS is secreted into the maternal rather than the foetal circulation; the concentration in cord blood is only about 1% of that in maternal blood. It is detectable in plasma at the 6th week of pregnancy and thereafter rises progressively to reach a plateau after 34 weeks. The curve of plasma HCS concentration follows very closely the curve of placental weight, suggesting that the secretion rate per unit mass of placenta is more or less constant throughout pregnancy (Fig. 4.2). This is further supported by the finding that the concentration of HCS in the placenta varies little in relation to the duration of pregnancy. The relationship of plasma concentration of HCS to placental weight is preserved in a wide variety of complications such as erythroblastosis, multiple pregnancy and diabetes that may cause placental weight to deviate markedly from the normal (Josimovich 1971).

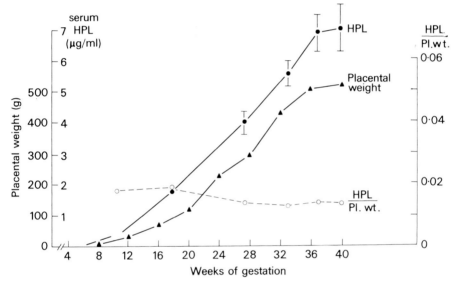

FIG. 4.2. Serum HCS concentrations in normal pregnancies and their correlation with placental weight. The concentration of serum HCS per unit weight of placenta is constant throughout pregnancy. From H.A.Selenkow et al (1969) in A.Pecile & C.Finzi (eds.) *The Foeto-Placental Unit*. Amsterdam, Excerpta Medica.

Little HCS is secreted in the maternal urine and, as already noted, the concentration in foetal blood is low. However, the concentration in amniotic fluid is approximately 20% of that in maternal plasma but the route of entry into the amniotic fluid is unknown.

Physiological role of HCS

The major function of HCS is thought to be related to a glucose-sparing effect in the mother which diverts glucose from maternal metabolism towards the foetus. The energy requirements of the foetus are met almost exclusively by glucose, and it is important that the supply of glucose to the foetus is maintained even when the intake of food by the mother is restricted. The biological activity of HCS is only about 10% of that of HGH but the large amount secreted leads to physiological changes similar to those induced by excess HGH secretion. In addition, a synergistic effect of HCS with HCG has been demonstrated (Murakawa & Raben, 1968). Much of the information about physiological actions of HCS has been obtained in animals and the results of experiments in humans have been rather equivocal. Nevertheless, it is likely that the changes in carbohydrate metabolism occurring in human pregnancy are due in part to HCS. Other factors including oestrogen and placental insulinase also contribute (see also chapter 11).

In normal pregnancy there is hypersecretion of insulin by the β cells of the pancreas but at the same time there is increased resistance to the action of insulin. Thus, the fasting blood sugar levels of a normal pregnant woman remain within normal limits and the increase in blood glucose following an intravenous load of glucose is unchanged by pregnancy. But a glucose load leads to a considerably greater secretion of insulin in the pregnant woman than it does when she is non-pregnant. In normal subjects given HCS infusions, insulin secretion is increased but no carbohydrate intolerance occurs after a glucose load (Beck & Daughaday 1967). Women who have gestational diabetes respond to HCS infusion with smaller increases in insulin secretion and with evidence of carbohydrate intolerance when given an intravenous load of glucose. These observations are consistent with HCS being one of the factors responsible for the diabetogenic stress of pregnancy.

If HCS were identical in its somatotrophic properties with HGH it would be expected that, in addition to the above changes in insulin secretion, administration of HCS would lead to increased nitrogen retention and also to mobilization of free fatty acid. In fact, both these effects can be demonstrated when large doses of HCS are given to hypopituitary subjects. There is retention of nitrogen, potassium, and phosphorus as well as an increase in the urinary excretion of calcium and hydroxyproline; the concentration of free fatty acids in the plasma rises and there is increased resistance to insulin-induced hypoglycaemia (Grumbach *et al* 1968).

Little is known of the lactogenic properties of HCS in human pregnancy. Injections of HCS into rats increases the weight of the mammary glands and leads to development of mammary alveoli. In rhesus monkeys, HCS causes lactogenesis after pre-treatment with progesterone and oestrogens. None of these effects has so far been demonstrated in humans. Nevertheless, in view of its undoubted prolactin-like activity it is likely that HCS contributes to the mammary changes of human pregnancy.

Clinical significance of HCS assays

In general, assays of serum HCS levels in various pregnancy complications yields similar information to that obtained by urinary pregnanediol estimations. In threatened abortion, HCS levels may be low in patients who subsequently abort and levels are also low in the presence of foetal growth retardation, particularly if the latter occurs as a complication of toxaemia or maternal hypertension (Saxena *et al* 1969). Falling levels of HCS in pregnancies complicated by diabetes, toxaemia or postmaturity are indicative of a serious degree of foetal jeopardy. Spellacy *et al* (1970) found that foetal death occurred almost invariably in women with hypertension when plasma HCS levels fell below 4 μg/ml. Assays of HCS are of no prognostic value in rhesus isoimmunization (Samaan *et al* 1969).

The abnormal trophoblastic tissue of hydatidiform moles apparently produces little HCS. This observation may be useful in the differential diagnosis of molar pregnancy in which high levels of HCG are coupled with low levels of HCS (Saxena *et al* 1968).

HUMAN CHORIONIC THYROTROPHIN (HCT)

There is strong evidence that extracts of human placenta contain a thyrotrophic hormone but it is as yet unproven that the hormone is elaborated by the placenta. However, increased thyroid function, sometimes amounting to thyrotoxicosis, has been described in patients with chorion carcinoma or hydatidiform moles (Kock *et al* 1966), suggesting that in these pathological states, at least, the trophoblast can secrete HCT.

HCT is a glycoprotein with physico-chemical properties similar to those in TSH. It has immunological properties that are related to TSH, particularly TSH of bovine and porcine origin (Hennen *et al* 1969). The concentration in maternal plasma is highest during the first 2 months of pregnancy and thereafter declines progressively to reach a minimum at term.

The physiological role of HCT during pregnancy is unknown. It is possible that it may be partly responsible for the increase in maternal thyroid activity that occurs in normal pregnancy. It has also been suggested that it may be responsible for the early development of the foetal thyroid gland. Thyroid follicles start forming at about the 8th week of pregnancy and iodination begins at about the 11th week (Shepard 1967) but the foetal pituitary does not begin to synthesize TSH until about the 14th week (Gitlin & Biasucci 1969).

OTHER NON-STEROIDAL HORMONES ON THE PLACENTA

Several hormones have been isolated from the placenta but there is no evidence to show that they are synthesized there. They include adrenocorticotrophic hormone, melanocyte stimulating hormone, oxytocin, vasopressin, catecholamines, relaxin and a uterotrophic hormone.

SEX DETERMINATION

Sex is expressed in a variety of ways, each of which is the result of specific influences at certain stages of development. A normal adult male (and conversely for a female) behaves like a male, has the general body configuration of a male, has male external genitalia, has male internal genitalia, has male gonads and has the sex chromosomes of the male. Such concordance of sexual characteristics is not always present and any one or more may be at variance thus giving rise to a wide spectrum of anatomical and psychological deviants. In any discussion of sex determination it is essential, therefore, that each type of sexual characteristic is clearly specified. The categories to be used in this chapter are shown in Table 4.1.

CHROMOSOMAL SEX

The somatic cells of the human body contain 46 chromosomes arranged in 22 pairs of autosomes and one pair of sex chromosomes. In the female the sex chromosomes are alike and are designated XX; in the male the paired sex chromosomes are unequal in size and are designated

X and Y. In the meiotic division of the germ cells that leads to formation of the gametes the paired chromosomes are parted, one going to each of two cells. The female (XX) produces only one kind of gamete (X) but the male (XY) produces two kinds of gamete (X and Y). At fertilization the diploid number of chromosomes is restored by the union of two gametes each containing a haploid complement (see chapter 1).

The translation of genetic sex into phenotypic sex is achieved in the human by the gonad and the steroidal sex hormones that it secretes. The genetic sex of all other somatic tissues is apparently irrelevant to their subsequent sexual differentiation and development. The designation of X-chromosomes as sex chromosomes can be misleading if it is

TABLE 4.1 Categories of sexual characteristics

Category	Characteristics
1. Chromosomal sex	XX or XY
2. Gonadal sex	Ovary or testis
3. Sex-duct sex	Müllerian or Wolffian duct development
4. External genital sex	Clitoris and vulva or penis and scrotum
5. Phenotypic sex	Breasts, body build, hair distribution, voice, etc.
6. Psychological sex	Feminine or masculine behaviour
7. Legal sex	Birth registration

taken to imply a functional association with sexual development. In fact, the X-chromosome carries genes whose functions are apparently quite unconnected with sex and include genes concerned with haemophilia, muscular dystrophy, colour blindness, skin diseases, and the Xg blood group, amongst many others (McKusick 1962).

Differentiation of the primitive gonad into testis or ovary is determined by the Y-chromosome. The indifferent primitive gonad is destined to form an ovary unless a Y-chromosome is present to switch development into that of a testis. The action of the Y-chromosome may be exerted through an organizer substance or by stimulating DNA synthesis (Mittwoch 1970). Because the Y-chromosome is smaller than the X and does not carry corresponding genes the question arises how the potentially different amounts of X-linked gene products in the two sexes are integrated with a constant amount from the autosomal genes. The solution to this problem appears to be that whatever number of X-chromosomes are present in a somatic cell, only one remains genetically active and forms messenger RNA. The others, normally one in a female and none in a male, become inactive (Lyon 1961). The inactive chromosome forms the sex-chromatin body found lying against the nuclear membrane. It is usually assumed that the male determining gene is carried on the Y-chromosome. However, Hamerton (1968) has proposed that all the male-determining genes are in fact located on the X-chromosome and so are present, but latent, in normal female individuals. He postulates that the Y-chromosome of the male acts merely as a controlling centre, 'switching on' the male-determining genes of the X-chromosome.

GONADAL SEX

The primitive gonads develop on the mesonephros and consist of coelomic epithelium, an underlying mesenchyme (probably originating from the mesonephros), and germ cells which migrate from a site in the yolk sac. The germ cells lie in relation to the coelomic epithelium and together they comprise the cortex (Fig. 4.3). The inner layer of mesenchyme together with some mesonephric tubules form the medullary component. Downgrowths of coelomic epithelium into the medulla form the primary sex cords. Until the 6th week of embryonic life the primitive gonad appears identical in the two sexes but differentiation into a testis starts at this time and is followed a little later by differentiation of the ovary. In a gonad differentiating into an ovary the cortex continues growing but the rete cords stop proliferating and become progressively hollowed. In a gonad differentiating to a testis the rete cords continue growing and they branch and attract the germ cells; the cortex is thus emptied and reduced to a thin coelomic epithelium. The germ cells become oocytes if they remain located in the cortex, and they become spermatogonia if they are embedded in the medulla; their fate apparently depends upon the somatic cells which enclose them (Jost 1970). Witschi (1967) developed the concept that cortex and medulla synthesize a dual system of antagonistic inductors which govern not only the differentiation of the germ cells but also the sex differentiation of the whole gonad. Cortexine and medullarine are postulated as inductive substances respectively formed under genetic control by the cortex or the medulla and which compete to impose either ovarian or testicular development.

The germ cells arise in the yolk sac and migrate via the hind-gut endoderm to the mesentery of the gut, thence to the mesonephric folds

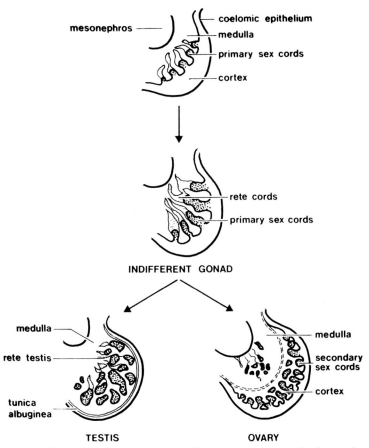

FIG. 4.3. Differentiation of the gonads. The cortex regresses in the testis to form the tunica albuginea but persists in the ovary. After M.M.Grumbach & M.L.Barr (1958).

and finally they enter the germinal ridges. They actively migrate along this pathway by amoeboid movement. Multiplication of the germ cells occurs not only during migration but also after their arrival in the primitive gonad so that the few hundred germ cells that leave the yolk sac give rise to hundreds of thousands of daughter cells in the gonad.

Mitotic division of the germ cells of the ovary (oogonia) continues for about 5 months and the oogonia then undergo meiotic division to become oocytes. Waves of atresia subsequently reduce the numbers of oocytes until only about 5% of the peak number of germ cells survive in the gonad at birth (Baker 1963). In the testis, the germ cells give rise to the spermatogonia which continue mitotic division until puberty when meiotic division initiates spermatogenesis.

Meiosis in spermatogonia is rapidly completed but in oocytes it is arrested in prophase; meiosis in the oocyte is completed only years later, shortly before ovulation.

SEX-DUCT SEX

In the female, the Müllerian ducts develop to form the Fallopian tubes, uterus and upper vagina whereas the Wolffian ducts regress. In the male, the Wolffian ducts develop into the epididymis, vas deferens and seminal vesicles whereas the Müllerian ducts regress, leaving only vestiges in the form of the appendix of the testis and the utriculus masculinus. The mechanism controlling differentiation of the sex ducts was established by the classical studies of Jost (1969). He showed that in rabbit foetuses castrated before the initiation of sexual differentiation of the genital tract, the whole internal tract becomes feminine whatever the genetic sex of the foetus. On the other hand, grafts of testis cause masculinization regardless of genetic sex. Many observations on humans congenitally deprived of gonads show that in the absence of gonads the genital tract becomes feminine whatever the chromosomal sex. It appears that the testes are the body-sex differentiators; they impose masculinity on the whole genital sphere which would become feminine in their absence; the presence or absence of ovaries is of no significance. The testicular influence causes the Wolffian ducts to develop and the Müllerian ducts to regress. The mode of action of the testis upon the ducts remains uncertain but it appears that two distinct testicular secretions are involved, both acting locally. One of the secretions is an androgen that causes development of the Wolffian duct but does not inhibit the Müllerian duct. The inhibitor of the Müllerian ducts, although unidentified, is not an androgen (Fig. 4.4).

The conclusions reached by Jost in his early investigations have been supported more recently by observations made in male rabbits exposed to the anti-androgen, cyproterone acetate, in early foetal life. The rabbits show absence of masculinization of the Wolffian ducts but the Müllerian ducts are nevertheless inhibited, presumably because the action of the non-androgenic-inhibitor is not affected by anti-androgens (Elgar 1966). In the human syndrome of testicular femininization which seems to result from lack of sensitivity of tissues to androgens, the whole body, including the genitalia, becomes feminine but the Müllerian ducts are inhibited. This suggests that the testicular inhibitory action takes place despite the absence of response to androgens.

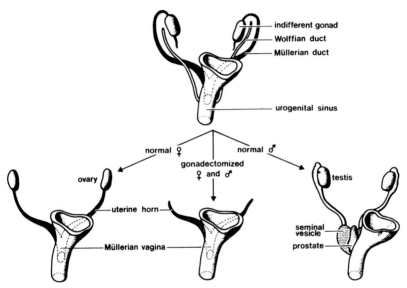

FIG. 4.4. Differentiation of the Müllerian and Wolffian ducts in male and female foetuses. After A. Jost (1959).

EXTERNAL GENITAL SEX

At the 8th week of foetal life the external genitalia consist of a urogenital slit that is bounded by the urethral folds and, more laterally, by the labioscrotal swellings. Anterior to the urogenital sinus is the genital tubercle which may subsequently develop into the corpora cavernosa and glands of the penis or remain small as the clitoris. By the 14th week of foetal life it is possible to distinguish the male external genitalia from those of the female; the latter shows no major change from the appearance of an 8-week foetus but in the male the urethral folds are forming the perineal and penile urethra and the corpus spongiosum. The scrotum forms by fusion of the labioscrotal folds and the penis develops by hypertrophy of the genital tubercle.

Differentiation of the neuter external genitalia of the 8-week foetus to form the external genitalia of the male depends upon androgenic stimulation before the 12th week. In the absence of androgen the external genitalia are those of a female irrespective both of the chromosomal sex and even of the gonadal sex.

PHENOTYPIC SEX

The body form and the development of secondary sexual characteristics depend upon the hormones secreted from the gonads and have nothing to do with the sex-chromosome complement of the cells of tissues other than the gonad. The characteristic features of the male phenotype that are induced by androgen include relatively wide shoulder span as compared to the narrower hips, greater muscle bulk, conspicuous facial hair, pubic hair that extends to the umbilicus and a greater amount of body hair particularly on the chest, the limbs and the back of the fingers. Fat tends to be distributed mainly on the trunk and the voice is deep. The female phenotype is determined largely by the absence of androgen and is not hormone dependent. In addition, however, the female shows changes induced by oestrogen. The latter changes are most obvious as mammary development and a tendency to accumulate fat around the hips and thighs. However, normal development of pubic and axillary hair in the female is dependent on androgen (of adrenal origin) as well as on oestrogen.

PSYCHOLOGICAL SEX

Work over the last 10 years gives a strong indication that the embryonic brain may be sexually differentiated in a manner very similar to that of the reproductive tract. The critical period during which gonadal hormones exert an inductive influence on neural mechanisms occurs later than the critical period during which the tract is influenced. This may be related to the fact that the reproductive tract is affected by a local diffusion of hormones from the developing gonad while the central nervous system has to await formation of efficient general and local circulatory systems for the effect to be mediated (Harris 1970).

In the rat, experimental investigation of the effects of androgens on subsequent neurological functions is facilitated by the fact that sexual differentiation of the brain occurs during the first 3 days after birth. Pfeiffer (1936) showed that female rats into which testes were implanted at birth failed to show oestrus cycles when they became adult. Harris & Levine (1962) found that a similar effect could be obtained by a single injection of testosterone into newborn female rats. They also noted a loss of normal female sexual behaviour pattern. Similar observations have been made in other mammals, including the rhesus monkey, but there is no clear evidence that androgens can cause permanent imprinting on the hypothalamus of the human female foetus. In congenital adrenal hyperplasia, evidence of long-standing intra-uterine effects of androgen can be seen in the female neonate with masculinized external genitalia. When untreated, girls with adrenal hyperplasia are boyish and aggressive but when successfully treated by cortisone their behaviour becomes feminine, they menstruate regularly, they make successful marriages and have children. Nor have behavioural problems been seen in girls with masculinized genitalia due to androgens administered to the mother during pregnancy unless the wrong sex was designated to the child at birth. Furthermore, there is no evidence that sexual differentiation of the human hypothalamus is determined by the sex chromosomes of the nervous tissue. In the testicular feminization syndrome, for example, in which there is a biochemical disorder that causes complete lack of response to androgen, sexual behaviour is that of a normal woman despite the presence of XY-chromosomes in every cell.

In humans, it seems that the sex of rearing plays the main part in moulding sexual behaviour in later life. Children of ambiguous sex who are designated to the wrong sex and are reared as such will develop the normal behaviour of the designated sex and will remain well-adjusted providing that pubertal changes conform to the designated sex and ill-advised attempts to revert to the true sex are avoided.

None of the psychosexual disorders including homosexuality, transvestism and masochism has a hormonal basis as far as is known. Treatment with sex hormones and even castration are without beneficial effects. However, it should be noted that Lorraine *et al* (1970) have reported that the blood testosterone levels of male homosexuals may be abnormally low and that of female homosexuals abnormally high. It is clear, however, that their findings have no therapeutic implications and probably little aetiological significance.

LEGAL SEX

The legal sex of an individual is that shown on the birth certificate and is decided by the parents at the time of registration of the birth. In most instances the registered sex is not at variance with chromosomal, phenotypic, etc., sex but on occasions difficulties arise from incorrect designation of sex at birth (usually in pseudohermaphrodites) or in psychological transexualism when medical opinion favours a change in the sex of upbringing. The Birth and Death Registration Acts of many countries contain provisions for redesignation of the registered sex and usually require statutory declarations from persons having full knowledge of the facts. The parents and the doctor in charge of the patient will usually be those called on to make such declarations.

THE FOETO-PLACENTAL COMPLEX

The designation of the human foetus and the placenta as an endocrine unit or complex recognizes the concept developed by Diczfalusy (1962) that the foetus and placenta form a functional unit to carry out together steroid biosynthetic reactions which the placenta *per se* or the foetus *per se* are incapable of completing. The secretion by the placenta of HCG, HCS and progesterone occurs independently of a foetus and may continue unchanged after foetal demise or in the absence of a foetus. On the other hand, oestrogen biosynthesis requires the combined efforts of both foetus and placenta since neither contains the full range of enzymes required for the conversion of acetate or cholestrol to oestrogen.

THE BIOSYNTHESIS AND METABOLISM OF OESTROGEN IN HUMAN PREGNANCY

Oestrogens in body fluids

Throughout pregnancy there is a progressive increase in the urinary excretion of oestrogens of the order of a thousand-fold in the case of oestriol and almost one hundred-fold in the case of oestrone (Brown 1956). In all, over 20 oestrogens have been isolated from late pregnancy urine and it is likely that further remain to be identified. The oestrogens present in highest concentration are those substituted at carbon atom 16, a reflection of the high activity of 16-hydroxylating enzymes in the foetal liver and adrenal; among these 16-hydroxylated oestrogens, oestriol accounts for some 90% of the total.

The mean oestriol excretion curve is characterized by relatively low values during the first trimester, increasing values after the 10th–12th week, a levelling off after the 24th–26th week and a sharp rise during

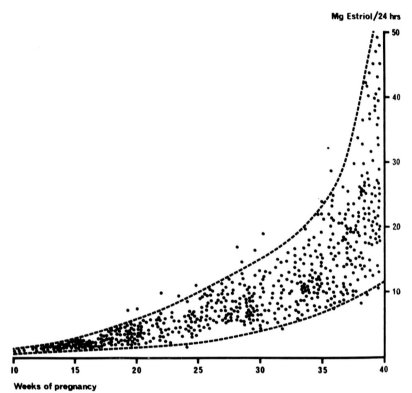

FIG. 4.5. Urinary oestriol excretion throughout pregnancy in thirty-one normal women. From C.F.Beling (1971) in G.Fuchs & A.Klopper (eds.) *Endocrinology of Pregnancy*. New York, Harper & Row.

the last 3–4 weeks of pregnancy. Variability from one woman to another tends to obscure these features when data from large groups of patients are plotted so that the mean values and 95% confidence limits in published graphs usually show a steady rise in the rate of increase of urinary excretion (Fig. 4.5). Govaerts-Videtzky (1965) found that the increase in the logarithmic means of oestriol values fitted to regression lines with different slopes before and after the 24th week of pregnancy. No satisfactory explanation for the changed pattern of oestriol excretion after the 24th week has been given but it seems likely to be related to the level of activity in the foetal zone of the adrenal cortex. Oestrone and oestradiol curves are similar to those of oestriol but at a much lower level. At term, approximately 1·5 mg of oestrone, 0·5 mg of oestradiol and 25 mg of oestriol are present in the urine.

Amniotic fluid contains oestrogens totalling less than 1 mg per litre; the oestrogen is almost entirely oestriol, only minute quantities of oestrone and oestradiol being present (Diczfalusy & Magnusson 1958). Oestriol similarly predominates in foetal urine and blood. By way of contrast, maternal blood contains almost as much oestrone and oestradiol as oestriol.

Sources of oestrogen in pregnancy

Urinary excretion of oestrogen in late pregnancy is unaffected by removal of the ovaries. Moreover, bilaterally adrenalectomized women also excrete large amounts of oestrogen during pregnancy although in some cases the mean values are lower than normal (Harkness *et al* 1966). Thus maternal endocrine organs can be excluded as a major site of oestrogen production in the foeto-placental complex.

Unlike the ovary which readily synthesizes oestrogens from precursors such as acetate, cholesterol and pregnenolone the placenta can utilize only C-19 steroids as precursors for oestrogen biosynthesis. The inability of the placenta to convert steroids with a side-chain at the 17-carbon position depends on the virtual absence from the placenta of the enzymes necessary for side-chain cleavage. Placental preparations incubated with pregnenolone, 17α-hydroxypregnenolone, progesterone or 17-hydroxyprogesterone do not form oestrogen (Ryan 1959). Furthermore no oestrogens are found when C-21 steroids are perfused through normal mid-term placentas *in situ* (Pion *et al* 1965). On the other hand, there is no doubt that the placenta has a very potent aromatizing enzyme system that is capable of converting a wide variety of neutral C-19 steroids into the corresponding oestrogens. Not only testosterone, androstenedione, and dehydroepiandrosterone but also many substituted neutral C-19 steroids, particularly 16-oxygenated steroids, are aromatized in good yield (Magendantz & Ryan 1964).

It can be concluded from the above observations that the biosynthesis of oestrogens by the placenta during pregnancy depends on C-19 precursors coming either from the maternal compartment or from the foetus (Fig. 4.6). As mentioned above, removal of the maternal ovaries or adrenals is compatible with continued production of large amounts of oestrogen in pregnancy suggesting that the major source of oestrogen precursors is the foetus. That such is the case is now supported by a large body of evidence, both direct and indirect. Cassmer (1959) ligated

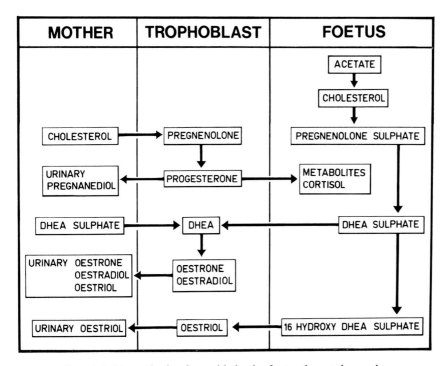

FIG. 4.6. Biosynthesis of steroids in the foeto-placental complex.

the umbilical cord in patients undergoing legal abortions and left the placenta and dead foetus *in situ* for 3 days. Interruption of the foeto-placental circulation resulted in a very marked and immediate drop in urinary excretion of the three 'classical' oestrogens, oestrone, 17β-oestradiol and oestriol, whereas pregnanediol excretion showed only a slight decrease. Frandsen & Stakemann (1961) noticed that the urinary oestrogen excretion in women pregnant with anencephalic monsters is low and they surmized that the defective production of oestrogens might be the result of low production by the abnormal foetuses of oestrogen precursors. A characteristic feature of anencephalic monsters is their hypoplastic adrenals, an association which led Frandsen and Stakemann to suggest that in normal pregnancy the foetal adrenals produce substantial quantities of a precursor for oestrogen biosynthesis in the placenta. The precursor is now known to be dehydroepiandrosterone sulphate (DHAS).

Dehydroepiandrosterone sulphate is present in cord blood in high concentration (Colás et al 1964) and there is no doubt that it originates in the foetal adrenals. DHAS is formed by slices of human foetal adrenal incubated with acetate or cholesterol (Bloch & Benirschke 1959) and, in addition, little DHAS is found in cord blood of anencephalic foetuses or in foetuses whose adrenal activity has been suppressed by exogenous corticosteroids. Although *in situ* perfusion of placentas shows that androstenedione, testosterone, dehydroepiandrosterone, and dehydroepiandrosterone sulphate are all readily aromatized, the great preponderance in cord blood of DHAS compared with the other androgens is consistent with it being the major foetal precursor for placental production of oestrone and 17β-oestradiol. However, studies of the production rate of 17β-oestradiol show that the placenta uses approximately equal quantities of DHAS from the maternal compartment (maternal adrenal) and foetal compartment (Siiteri & MacDonald 1966). DHAS is further metabolized by the placenta to androstenedione before conversion to oestrone and thence to 17β-oestradiol. It should be emphasized that the foetal zone of the foetal adrenal cortex has highly specialized enzyme activities that permit it to secrete enormous quantities (75 mg/day) of DHA sulphate without at the same time secreting excesses of more potent androgens or corticosteroids. The conversion of pregnenolone to progesterone and to corticosteroids and the conversion of DHA to androstenedione and testosterone requires the presence of the enzyme 3β-hydroxysteroid dehydrogenase; this enzyme is absent from the foetal zone. Further protection from unwanted biological activity is provided by the extremely active sulphurylating enzymes in foetal tissues including the adrenal cortex. Steroids, whether neutral or phenolic, are found in the foetal circulation almost entirely in the conjugated form, mainly sulphoconjugates; glucuronides are present in relatively low concentration. In contrast to foetal tissues, the placenta has a most active steroid sulphatase system associated with very little steroid sulphotransferase activity (Warren & Timberlake 1962). Consequently, the emphasis in the placenta is on hydrolysis of conjugates; the liberated free steroids are then further metabolized or are secreted into the maternal circulation. The remarkable distribution of sulphotransferase and sulphatase activity between the foetus and the placenta has certain advantages to the foetus. First, oestrogens and oestrogen precursors can circulate in large quantities in the foetus without exerting undesirable biological effects. Secondly, hydrolysis of the conjugates in the placenta permits rapid conversion of oestrogen precursors to oestrogen. In addition, free steroids pass through the placenta several times more readily than the sulphoconjugates (Levitz et al 1967); thus placental sulphatase activity enhances clearance of steroidal metabolites in the foetal circulation.

The predominance of oestriol over oestrone and 17β-oestradiol is attributable to the high level of activity of 16α-hydroxylating enzymes in the foetus, particularly in the foetal liver. The placenta on the other hand has no 16α-hydroxylating activity (Bolté et al 1966). The concentration of 16α-hydroxydehydroepiandrosterone sulphate in cord blood is two or three times that of DHAS (Colás et al 1964) and it is now considered that the major pathway to oestriol is by a neutral pathway, i.e. oestriol is formed from 16α-hydroxylated C-19 neutral steroids rather than by 16α-hydroxylation of oestradiol. Perfusion of *in situ* mid-term placentas or of *in vitro* term placentas has confirmed that 16α-hydroxylated C-19 steroids are readily converted to oestriol. 16α-hydroxylation in the maternal compartment is small compared to that in the foetus and most of the DHAS secreted by the maternal adrenal reaches the placenta unchanged. Siiteri & Macdonald (1966) showed that most of the oestriol produced during normal pregnancy is derived from foetal DHAS and that the maternal contribution is minimal. The contribution of maternal DHAS to the oestriol pool is less than 10% whereas its contribution to the pool of oestrone and oestradiol is approximately 50%. These observations account not only for the relatively high ratio of oestrone and oestradiol to oestriol in cases of anencephaly but also for the disordered maternal urinary oestrogen excretion observed by Coyle (1962) in association with congenital cirrhosis of the foetal liver. In this rare disease total oestrogen excretion is normal but the oestriol fraction is abnormally low, no doubt because 16α-hydroxylation by the foetal liver is defective.

It has been thought that the foetal adrenal synthesizes DHAS from circulating pregnenolone that is synthesized in the placenta from cholesterol. However, there is now good evidence that androgens secreted by the foetal adrenal are synthesized from pregnenolone formed within the foetus itself, and that most of the cholesterol used by the foetal organs to synthesize pregnenolone is of foetal, rather than maternal, origin. The studies of Hellig *et al* (1970) in which radioactive cholesterol was infused into pregnant women showed that only 18% of cholesterol in the foetal circulation came from the maternal pool of cholesterol. The remaining cholesterol in foetal plasma is formed from acetate by the foetal liver. Telegdy *et al* (1970) infused labelled acetate into mid-term foetuses and found that large quantities of acetate were converted to cholesterol. A low conversion of labelled acetate to pregnenolone by the normal foetal adrenal was demonstrated by Solomon (1966) but definitive experiments on foetal production of pregnenolone sulphate from foetal precursors have yet to be performed. Nevertheless, it seems likely that the foetal adrenal is able to convert cholesterol to pregnenolone sulphate and that the latter is partly secreted into the foetal circulation and partly metabolized further by the adrenal to androgens. France (1971) observed that the levels of pregnenolone sulphate in the plasma of anencephalic foetuses near term were extremely low. This finding strongly favours the foetal adrenals as the source of pregnenolone in foetal plasma in normal pregnancies for it is known that the placentas of anencephalic pregnancies are capable of normal rates of synthesis of pregnenolone and progesterone. These observations also indicate that the foetal contribution to placental progesterone production must be of limited significance, a conclusion further supported by the absence of altered levels of progesterone in maternal plasma after foetal death (Lurie *et al* 1966).

The rate-limiting steps in the biosynthesis of oestrogens by the placenta are not yet known. In normal pregnancy, the administration of DHAS to either the foetus or to the mother is followed by an increase in urinary excretion of oestrogen (Lauritzen 1967), suggesting that hydrolysis and aromatization by the placenta of DHAS and 16-hydroxy DHAS is not normally rate-limiting. It follows that the production rate of oestrogen by the placenta is determined by the availability of precursors, particularly those in the foetal circulation. Little is known of the mechanisms regulating foetal adrenal androgen production but it is clear that ACTH plays an important part. When ACTH secretion is deficient because of congenital absence of the foetal hypothalamus, or when ACTH secretion is inhibited by passage into the foetus of corticosteroids administered to the mother (Simmer *et al* 1966) low oestrogen production is consequent upon low production of DHAS by the foetal adrenal. Furthermore, abnormally high urinary excretion of oestrogens is observed in pregnancies associated with foetal congenital adrenal hyperplasia due to an inherited deficiency of 11β-hydroxylase or 21α-hydroxylase. It is postulated that in such pregnancies inability to synthesize cortisol at a normal rate causes increased ACTH secretion which leads, in turn, to increased secretion of DHAS. The homeostatic mechanisms regulating ACTH release in normal circumstances are unknown but it is of interest that the placenta contains an active 11β-hydroxysteroid dehydrogenase which converts cortisol to its inactive metabolite, cortisone (Osinski 1960). Because of the activity of this enzyme the concentration of cortisone in foetal plasma exceeds that of cortisol in marked contrast to the situation in maternal plasma where the concentration of cortisol normally exceeds by several times that of cortisone. The presence of 11β-hydroxysteroid dehydrogenase gives the placenta the potential to influence the secretion of ACTH from the foetal pituitary by means of an effect on plasma levels of free cortisol. Chorionic gonadotrophin also might serve as a means by which the placenta could influence the function of the foetal adrenal; not only does the intra-amniotic administration of HCG at mid-pregnancy lead to ultrastructural changes in the foetal adrenal cortex consistent with increase in activity (Johannisson 1968) but also its administration to newborn infants causes an increase in urinary excretion of DHA (Lauritzen 1966).

Decreased urinary oestrogen excretion in certain pathological states usually reflects either a diminished production by the foetal adrenal of precursors (see above) or inability of a placenta damaged by one of the various hypertensive complications of pregnancy to convert precursors to oestrogen. Only one example of a specific enzyme defect in the pathway from acetate to oestrogen is known. France and Liggins (1969) have described a patient in whom a defect of placental sulphatase present in two successive pregnancies was associated with levels of urinary oestrogen that did not rise above 1 mg/24 hr at any stage of pregnancy. That the biosynthetic pathway was otherwise intact was demonstrated by normal pregnanediol excretion, by normal levels of DHAS and 16-hydroxy DHAS in cord blood and by normal placental conversion of DHA to oestrone and oestradiol both *in vivo* and *in vitro*. In connection with possible functions of the large quantities of oestrogens synthesized in human pregnancy it is interesting to note that the exceptionally low oestrogen production resulting from placental sulphatase deficiency has no deleterious effects on foetal growth and development.

Interest in oestrogen metabolism in pregnancy has centred mainly on the classical oestrogens, oestrone, oestradiol and oestriol. However, certain other metabolites are now attracting attention since they may be formed only within the foetal compartment and thus may reflect more accurately the state of foetal metabolic activities. In particular, oestetrol (15α-hydroxyoestriol) is excreted in the urine of infants injected with 17β-oestradiol in quantities comparable to that of oestriol (Hagen et al 1965). Oestetrol is present in pregnancy urine in readily measurable quantities and it is possible that it may prove more useful than oestriol in the management of various complications of pregnancy.

CLINICAL APPLICATION OF OESTROGEN ASSAYS

Qualitative assessment of placental function can be made by measuring any one of a number of the products of placental biosynthetic activity. These include both endocrine and non-endocrine materials. Of the non-endocrine placental secretions the enzymes heat-stable alkaline phosphatase, diamine oxidase and oxytocinase (leucine aminopeptidase), all of which are readily measured in the maternal plasma, have attracted most interest. Plasma concentrations rise progressively throughout pregnancy and may be lower in conditions associated with major placental dysfunction but the range of normal values is wide and little change may occur after foetal death. The maternal urinary excretion of pregnanediol and the plasma levels of HCS have also been used for assessing placental function. It will be appreciated that all of these tests involve the assay of materials that are synthesized by the placenta without the need for foetal participation and thus reflect placental rather than foeto-placental function. The excretion of oestrogen, on the other hand, depends on metabolic processes in the foetus as well as the placenta and assays of urinary oestrogen have proved to give a sensitive assessment of foetal viability.

Nevertheless, the fact that a number of foetal endocrine organs as well as the placenta take part in the biosynthesis of oestrogens gives rise to greater opportunity for misinterpretation of results if urinary excretion of oestrogens is unquestioningly accepted as an overall index of foeto-placental function. If the best use of oestrogen excretion is to be made a detailed knowledge of the various factors influencing the rate of oestrogen synthesis is essential. Due consideration must be given to the various possibilities that exist for abnormal excretion to be unrelated to placental function and, indeed, to foetal well-being. Although it is usual to describe results of urinary assays in terms of oestriol excretion many of the simplified methods in common use (not including gas chromatography) actually measure total oestrogens. However, because oestriol predominates, the usefulness of the test is not impaired by lack of specificity. Assays of oestriol in blood are difficult compared with urinary assays and this disadvantage probably outweighs the advantage of avoiding a delay of 24 hours while urine is collected. In general, assay of oestriol in blood gives similar information to that obtained from assays of 24-hour urine samples. As yet, assay of oestriol in amniotic fluid has not yielded sufficiently useful data to justify the need for repeated amnioceniteses.

The significance of abnormally low urinary oestriol excretion

It is helpful in discussing the significance of abnormally low urinary oestriol excretion to dissect the foeto-placental unit into a foetal component and a placental component then to further subdivide the foetal component into the various endocrine organs playing a part in oestrogen biosynthesis. It will be apparent from perusal of a simplified scheme of the pathway involved in the production and excretion of oestriol (Fig. 4.7) that the organs upon whose normal function the normal excretion of oestriol depends are the foetal hypothalamus and pituitary, the foetal adrenals, the foetal liver, the foetal trophoblast (placenta) and the maternal kidneys. The contribution of the maternal adrenals and maternal liver is small and can usually be ignored.

The foetal hypothalamus and pituitary. The extremely low oestriol excretion associated with anencephalic foetuses first described by Frandsen & Stakemann (1961) is often attributed to absence of the foetal pituitary. However, the pituitary is almost invariably present (Angevine 1938) and it is more reasonable to attribute the functional abnormality to the absence of the hypothalamus. Although it can only be assumed that hypoplasia of the anencephalic adrenal cortex results from insufficient release of ACTH from the foetal pituitary there is no doubt of the functional inadequacy of the foetal zone of the adrenal cortex since the levels of DHAS and 16-hydroxy DHAS in cord plasma of anencephalics is very low (Easterling et al 1966). Furthermore in their studies of oestrogen production rates MacDonald & Siiteri (1965) found that what small amount of oestrogen was formed by the placenta came almost entirely from maternal DHAS. Defective oestrogen biosynthesis in anencephalic pregnancy can thus be attributed to lack of the foetal precursor of oestrogen consequent upon adrenal hypoplasia which, in turn, is the result of lack of hypothalamic stimulation of ACTH release.

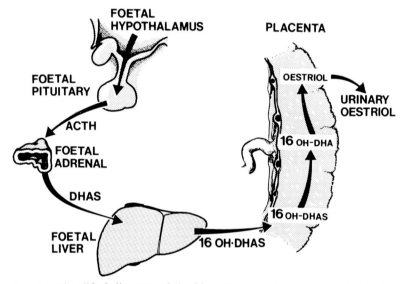

FIG. 4.7. Simplified diagram of the biosynthetic pathway of oestriol in the foeto-placental complex to illustrate the factors influencing secretion rate.

Not only may the hypothalamus be defective for anatomical reasons as in anencephaly, hydranencephaly and occasionally in hydrocephaly but it may also have defects of a functional nature. Administration of relatively large daily doses of corticosteroids (more than 75 mg of cortisone) depresses urinary oestriol levels to those observed in anencephaly (Oakey 1970; Simmer et al 1966). No concomitant fall in urinary pregnanediol is seen, suggesting that the cause of diminished oestrogen production lies not in impaired placental function but in passage of the corticosteroid into the foetal circulation where it acts on the hypothalamus and pituitary to inhibit release of ACTH. Clearly, depressed values of oestriol excretion in such circumstances in no way reflects foetal well-being and this is borne out in studies of women treated throughout pregnancy with large doses of prednisone in whom perinatal mortality is not increased.

Foetal adrenal cortex. Congenital absence of the foetal adrenal is extremely rare and oestrogen excretion in a patient with such a complication has yet to be described. However, an isolated example of congenital adrenal hypoplasia reported by Cathro & Coyle (1966) was associated with very low urinary oestrogen levels. Congenital deficiency of an adrenal enzyme in the pathway from cholesterol to DHAS could cause a low production rate of oestrogens but this theoretical possibility has not been recognized in practice.

In foetal dysmaturity, hypofunction of the adrenals is suggested not only by low oestriol excretion but also by low concentrations of DHAS and 16-hydroxy DHAS in the cord blood. In addition, liver glycogen deposition is abnormally low, possibly due to reduced secretion of corticosteroids.

The foetal liver. The foetal liver being the main site of 16-hydroxylation, liver disorders might cause reduced production of oestriol; the synthesis of oestrogens that are not 16-hydroxylated would, on the other hand, be unaffected by liver disease. A pregnancy complicated by foetal liver cirrhosis described by Coyle (1962) was associated with a low oestriol fraction and a high oestrone/oestradiol fraction. The foetal liver is also the main site of cholesterol synthesis which, if impaired, might limit the ability of the foetal adrenal to synthesize DHAS.

The placenta. A variety of pathological states may affect the ability of the placenta to synthesize oestrogens at a normal rate. It is probably true that pathological states of the placenta invariably affect the ability of the foetus to produce oestrogen precursors and that function of the placenta can only rarely be considered in isolation. However, little is known of the ways in which the foetus and placenta interact in disease and it is usually more convenient to discuss the effects of various complications of pregnancy under a heading such as disorders of the foetoplacental complex.

The small placenta. The clinical syndrome of foetal dysmaturity includes a number of ill-defined entities which at the present time cannot be easily categorized during pregnancy. In one form of the disorder, so-called idiopathic dysmaturity, the foetus and placenta are small and signs of intra-uterine foetal malnutrition are present but there are no predisposing causes such as pre-eclamptic toxaemia or essential hypertension. The obstetrician's attention is attracted mainly by the underweight baby but retardation of growth of the placenta is usually even more marked and foeto-placental weight ratios of 8:1 or 10:1 are not uncommon. It is not known whether the primary cause of the failure of normal growth of the conceptus lies in the foetus or in the placenta or, as seems most likely, involves both from the outset. In any case, urinary oestriol excretion is usually low and successive values follow a course that is horizontal. It is probable that there is little likelihood of foetal death before term in such patients but it is difficult to be sure that the placenta is small yet undamaged and that the life of the foetus is not in serious danger from placental insufficiency. For this reason, delivery before the 38th week is commonly practised and it is only in retrospect when the placenta is examined that the diagnosis of idiopathic growth retardation can be made.

The damaged placenta. Robertson et al (1967) have given a detailed account of the vascular lesions in the maternal blood vessels of the placenta in pregnancies complicated by hypertension. The lesions found in placental biopsies from women with pre-eclamptic toxaemia or eclampsia resemble the vascular lesions of malignant hypertension and consist mainly of fibrinoid necrosis in both the basal arteries and the spiral arterioles. This may be accompanied by occlusive thrombosis with infarction of the decidua. In chronic hypertension, whether essential or associated with chronic renal disorders, the arteriolar lesions are usually less marked and take the form of a hyperplastic arteriosclerosis. As a result of the vascular lesions there is slowing of blood flow in the intervillous space (Dixon et al 1963) which presumably is the immediate cause of impaired function of the trophoblast. This is reflected in diminished production of oestrogens but many other vital functions of the placenta are also affected. It should be borne in mind when assessing the significance of urinary oestriol excretion that one is measuring only a single function of the placenta but making the assumption that this applies equally well to other functions such as gas exchange and transport of nutrient materials. Fortunately, this assumption is usually correct and it is for this reason that assays of urinary oestriol are widely used when placental insufficiency is suspected.

Most investigators agree with Taylor et al (1958) who found a correlation between the severity of the hypertension and the extent of reduction in urinary oestriol excretion. In mild toxaemia or in chronic hypertension with a diastolic pressure of less than 100 mm Hg, reduction in oestriol excretion is minimal unless it is accompanied by foetal growth retardation (Klopper 1965). In more severe toxaemia and chronic hypertension urinary oestriol values are reduced to levels that are approximately 50% of normal. A further fall, usually to levels below 5 mg per 24 hours, occurs before foetal death.

Two abnormal components may be present in graphs of oestriol values from women with either toxaemia or some form of chronic hypertension. First, there may be loss of the marked increase in oestriol excretion that normally occurs in the last few weeks of pregnancy. Thus, the graph tends to lie below the mean and to be flatter than normal. The trend of successive values can usually be obtained by weekly assays providing that they have been started sufficiently early in pregnancy. Second, there may be a rapid fall in values. To reliably detect the latter pattern, assays must be done at intervals of no more than 3 days so that a trend of values can be established. Obstetrical intervention should not be based on the dubious significance of a single low result.

The valuable contributions that urinary oestriol estimations have made to the rational management of pregnancies complicated by hypertension is widely recognized. In many countries, obstetricians place considerable reliance on oestriol excretion in deciding the optimal time for delivery in patients with toxaemia of pregnancy. Most authors are in agreement with McLeod et al (1967) who concluded that there was little danger of foetal death in women excreting more than 12 mg of oestriol daily, particularly when the levels were rising. Indeed, according to Beling (1967) the foetus can apparently survive in the uterus with a very low oestriol level (even as low as 2–3 mg per 24 hours) and it seems that as long as the level is horizontal or shows a small increase the foetus is not in any immediate danger. In any event, there seems to be good evidence that interruption of pregnancy on foetal indications is not justified in women with toxaemia of pregnancy when successive values show no fall and are above 12 mg per 24 hours. On these criteria, most cases of mild and moderately severe toxaemia will progress safely to term. It has been pointed out (Liggins & Evans 1963) that the greatest benefits from urinary oestriol estimations may derive from the avoidance of unnecessary immaturity and the morbidity and mortality that it entails. Nevertheless, although various authors differ in what they regard as levels indicating impending foetal death there is general agreement that a sustained fall in levels is associated with a high risk of foetal death in the immediate future. Not uncommonly, meconium is found in the amniotic fluid when induction is performed on the basis of falling oestriol levels, lending support to the diagnosis of foetal jeopardy. Even more commonly, foetal bradycardia occurs in early labour and caesarean section is necessary to avoid intra-partum death from hypoxia.

On occasions, a horizontal curve of oestriol excretion in toxaemic pregnancy does not provide any useful information about the optimal time for delivery since, if the values are low, the opportunity to observe falling values does not arise. In these circumstances, resort is made to other tests of placental function. Urinary pregnanediol excretion, and plasma levels of HCS, heat stable alkaline phosphatase, or oxytocinase may be helpful. A test which shows more promise of usefulness is based on the dynamic response to the administration of a load of DHAS. The author uses a single dose of 100 mg of DHAS administered intravenously to the mother. Urinary oestriol, oestradiol and oestrone are measured during the 24 hours before and after administration. A substantial increase in urinary oestrogen excretion suggests that reserves of placental function are not entirely exhausted and that the cause of low oestrogen excretion may lie as much in diminished production of DHAS by the foetal adrenal as in placental insufficiency (Fig. 4.8). Foetal survival for a further 2 weeks has invariably followed DHAS loading tests in which the increase in excretion of oestrone and oestradiol totalled more than 1 mg per 24 hours regardless of how low the basal oestriol excretion may be at the time of the test.

Maternal diabetes. The high perinatal mortality of diabetic pregnancies arises in three different ways. First, intra-uterine death may occur before term and, as shown by Farquhar (1962), is most common in babies whose birth weight lies at or below the normal mean for gestational age. The mothers are usually insulin-dependent diabetics and it is likely that vascular disease is present in many. The second group comprises the overweight babies who die neonatally, often from respiratory distress. The mothers of these babies may be poorly controlled diabetics or

ENDOCRINOLOGY OF THE FOETO-MATERNAL UNIT 171

unrecognized latent diabetics. The third group includes overweight babies who die *in utero* beyond term; the mothers are commonly unrecognized latent diabetics. Because of the wide variation in foetal weight and in the extent of pathological changes in both the placenta and the foetus in diabetes it is not surprising that the range of values

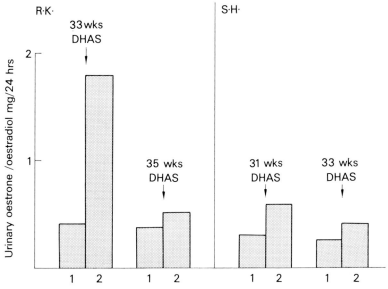

FIG. 4.8. The responses of two pregnant patients suffering from severe pre-eclamptic toxaemia to intravenous loads of dehydroepiandrosterone sulphate. Both woman excreted 5-6 mg of oestriol daily. Patient R.K. showed a considerable increase in excretion of oestrone and oestradiol at 33 weeks but little response at 35 weeks, 5 days before delivery. Patient S.H. who showed minimal response at 31 weeks and again at 33 weeks, was delivered at 34 weeks. Both babies were very dysmature but survived.

of urinary oestriol is larger than normal. Mean values are slightly lower than those of normal pregnancies (Greene *et al* 1965) but there is a correlation between foetal size and oestriol excretion, diabetics with large babies having oestriol values in the upper part of the normal range. High oestriol excretion may draw attention to the possible presence of an abnormally large baby but otherwise has little correlation with the cause of perinatal death which is most likely to be from respiratory distress in the first few days of life. On the other hand, oestriol values that are low should be viewed with some concern for they suggest that there is an element of foetal growth retardation which is a sinister sign in diabetics. Likewise, low values in the presence of toxaemia in a pregnant diabetic are of serious significance and should be given at least the same prognostic importance that the values would have if the mother were not diabetic. Again, falling oestriol levels are a strong indication for delivery providing that a reasonable maturity has been reached. Beling (1971) considers that urinary oestriol excretion plays an important part in deciding when delivery should be effected but other authors have not shared his enthusiasm. There is no doubt that intra-uterine death of a large foetus can occur at or beyond term in the presence of normal or even high oestriol levels. Nevertheless, the significance of low or falling oestriol levels does not differ from that of other complications of pregnancy and can help avert intra-uterine death of the foetus in advance of the planned date for premature delivery. On the other hand, normal oestriol values should not be an encouragement to prolong a pregnancy beyond the time when usual clinical indications point to the desirability of delivery.

Rhesus incompatibility. Severe foetal anaemia or hydrops seems not to impair either the ability of the foetal adrenal to secrete DHAS or the ability of the placenta to utilize DHAS in the synthesis of oestrogens. Most reports of studies of oestriol excretion in rhesus incompatibility describe instances of normal excretion in the presence of foetal hydrops and within 1 or 2 days of foetal death. There is a marked fall in urinary oestriol excretion immediately after intra-uterine transfusion for reasons that are unknown. Usually the oestriol excretion returns to normal a few days later but in some, values remain low and the foetus succumbs. In general, however, urinary oestriol assays have not been helpful in the management of pregnancies complicated by rhesus incompatibility (Klopper & Stephenson 1966).

Prolonged pregnancy. The question of whether or not prolongation of pregnancy without any pathological features is dangerous to the foetus continues to be argued. Even so, there is no doubt at all that degrees of placental insufficiency that are consistent with foetal survival to term and which may go unrecognized can prove lethal 2 or 3 weeks beyond term. Whatever other value urinary oestriol excretion may have in assessing the risks of prolonged pregnancy it is valuable in unmasking occult placental damage which may have been present for some time before term. Opinions vary on what constitutes an abnormally low oestriol output in these circumstances. Smith *et al* (1966) and Lundwall & Stakemann (1967) recommend 12 mg per 24 hours as a lower limit of normal in prolonged pregnancy and observed no mortality attributable to prolonged pregnancy when excretion was above this figure. The safest course to follow may well be to screen all patients at, say, the 42nd week, to induce those with oestriol output below 12 mg and to monitor the remainder twice weekly. In the latter group, Klopper (1969) recommends induction when two successive falls have occurred, whatever the absolute values.

Antepartum haemorrhage. Urinary oestriol excretion can be valuable in the management of antepartum haemorrhage (Beischer *et al* 1967). Levels are usually unaffected by bleeding from placenta praevia but in patients who happen to be in hospital at the time of minor placental abruption transient, though marked, falls in excretion may be observed. In others, oestriol output may remain low after antepartum haemorrhage and although there may be no other clinical evidence to indicate that the foetus is endangered this observation suggests that more of the placenta has been damaged by abruption than might be suspected from the modest nature of the haemorrhage. Estimation of oestriol excretion is well worth while in all patients admitted to hospital with antepartum haemorrhage; those who do not show a prompt return to normal levels are best kept under observation until delivered.

Abnormal oestriol excretion that does not reflect foeto-placental well-being. Reference has already been made to abnormally low levels of oestriol output that may be observed in certain congenital abnormalities such as anencephaly and sulphatase defect, during corticosteroid therapy and in conditions causing impaired clearance of oestriol by the maternal kidneys. Other possibilities arise and in particular consideration must always be given to the possibility that a 24-hour urine collection is incomplete. Suspicions should be aroused when a low oestriol value is coupled with a small 24-hour urine volume and the urinary creatinine output should be determined. Artefactual values may also be obtained in patients taking certain types of medication. Urine containing methenamine mandelate given in the treatment of urinary tract infection destroys oestriol during determination and the values as low as 3–4 mg/24 hr may be obtained in normal patients (Touchstone *et al* 1965).

Ampicillin therapy in usually employed dosage may depress oestriol output by 50% and if interpreted incorrectly could lead to unnecessarily premature delivery (Willman & Pulkinnen 1971). Neither ampicillin nor its metabolites interferes with accurate determination of urinary oestrogens and since plasma, as well as urine, concentrations of oestriol fall it is likely that ampicillin depresses the synthesis of oestriol in the foeto-placental unit. However, there is no evidence that ampicillin has other adverse effects on foetal metabolism.

THE BIOSYNTHESIS AND METABOLISM OF PROGESTERONE IN HUMAN PREGNANCY

PROGESTERONE IN BODY FLUIDS AND TISSUES

The concentration of progesterone in plasma rises rapidly from the time of conception to reach a peak of about 50 ng/ml 4 weeks after ovulation. During the next 4 weeks, levels fall but from the 8th week they begin to rise again and continue to do so throughout the remainder of pregnancy until at term the concentration reaches approximately 150 ng/ml. Plasma 17α-hydroxyprogesterone levels show a similar peak at 4 weeks but fall after the 8th week in contrast to the rising levels of progesterone (Fig. 4.9). The placenta has little capacity for 17α-hydroxylation and thus it is likely that the source of 17α-hydroxyprogesterone is the corpus luteum and that the rise and fall in concentration of this steroid is an index of corpus luteum function. If this is so, corpus luteum production of progesterone is largely replaced by placental production from the 7th week of pregnancy, a conclusion that is supported by the many observations that show that abortion does not occur when the corpus luteum is removed after the 6th week of pregnancy.

Progesterone in plasma is almost entirely bound to protein, only 1–2% being free. Protein-bound progesterone is distributed between albumin which has a large capacity but low affinity, orosomucoid, a glycoprotein with a high affinity but small capacity and transcortin, an α_1-globulin with a very high affinity for both cortisol and progesterone. The concentration of transcortin rises three-fold during pregnancy as a result of stimulation by oestrogen of its synthesis in the liver (Seal & Doe 1966) and this, together with an increase in the production rate of progesterone, is responsible for the high levels of progesterone found in the second half of pregnancy.

Serial measurements of plasma progesterone before and during labour show no evidence of a fall in concentration until late in the first stage and lend no support to the idea that withdrawal of progesterone is responsible for the onset of labour. However, women with low levels of progesterone at term respond more readily to oxytocin than women with high levels (Johansson 1968), suggesting that progesterone plays some

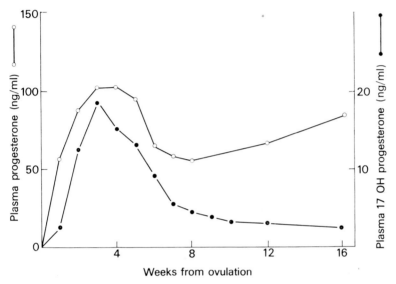

FIG. 4.9. Plasma levels of progesterone (○—○) and 17α-hydroxyprogesterone (●—●) in early pregnancy. Ovulation was induced by human menopausal gonadotrophin and chorionic gonadotrophin. Redrawn from T. Yoshimi et al (1969).

part in regulating uterine activity at the time of onset of labour. According to Zander et al (1969) the concentration of progesterone in myometrium remains more or less constant throughout pregnancy and there is no evidence that local changes in concentration precede labour.

Progesterone is rapidly removed from the blood and is metabolized in a wide variety of tissues including liver, lungs, kidneys, brain and heart. In all these tissues the concentration of progesterone is very low but in the myometrium relatively large amounts are present because of the presence in the cytoplasm of a specific binding protein with a high affinity for progesterone. Amniotic fluid contains a similar level of progesterone to that in maternal blood but levels in foetal blood are much higher. Mean concentrations in umbilical vein and artery are respectively 720 ng/ml and 440 ng/ml (Harbert *et al* 1964). Such a large arteriovenous difference indicates that a considerable amount of progesterone is removed by the foetus and it has been estimated that as much as a third of the total production of progesterone by the placenta is metabolized within the foetus.

SOURCES OF PROGESTERONE IN PREGNANCY

From the 7th week of pregnancy the ovarian contribution to the progesterone pool is insignificant and the placenta is almost the exclusive source of progesterone. Biosynthesis of progesterone from acetate is not a very efficient process in the human placenta and it is probable that cholesterol is the prime precursor. Most of the cholesterol is of maternal origin but a small amount comes from the foetus and this, together with some foetal pregnenolone, may provide up to 15% of the total precursor pool (France 1971). However, foetal death is not usually associated with an immediate fall in progesterone production, indicating that the placenta can dispense with foetal precursors. In this respect, the synthesis of progesterone contrasts sharply with that of oestrogen for which the combined activities of foetus and placenta are essential.

METABOLISM OF PROGESTERONE

Progesterone secreted by the placenta into the maternal circulation is metabolized mainly to pregnanediol but, in addition, a number of other metabolites including pregnanolone, 5α-pregnane-3, 20-dione and 3α-hydroxy-5α-pregnane-20-one are formed. These metabolites are conjugated with glucuronic acid in the liver and kidneys and are excreted in the urine and faeces.

Metabolism in the foetus is extensive and differs in several ways from that in the maternal compartment. Reduction occurs not only at the 5β, 3α and 20α positions as in the mother but also at the 6β, 14α, 15α, 16α and 17α positions. In particular, the metabolite 15α-hydroxy progesterone appears to be uniquely formed by the foetus and may, like the 15α-hydroxylated oestrogen, oestetrol, have potential in the assessment of foetal metabolic status. Injection of radioactive progesterone into mid-trimester foetuses has also demonstrated significant conversion to corticosteroids (Solomon & Fuchs 1971). However, information is not yet available on the quantitative details of the relative importance of progesterone, acetate, pregnenolone and cholesterol as precursors for corticosteroid synthesis in the foetal adrenal cortex.

Conjugation of progesterone metabolites in the foetus is predominantly with sulphuric acid rather than with glucuronic acid. The sulphoconjugates of progesterone metabolites including corticosteroids are not hydrolysed in the placenta but enter the maternal circulation as the conjugates and are excreted unchanged in the urine.

The urinary excretion of pregnanediol during pregnancy reflects progesterone production in the placenta and as such it can be used as an index of placental function. The test has found less favour in clinical practice than oestriol determinations, partly because it provides information on placental, rather than foeto-placental, function and partly because the flat slope of urinary levels in late pregnancy gives a less sensitive index of changing function than does the more steeply rising curve of values for urinary oestriol excretion. Nevertheless, it may be important to examine pregnanediol excretion in women excreting abnormally low levels of oestriol since this may provide the clue to the presence of a specific defect of oestrogen metabolism such as foetal anencephaly or an enzyme defect that renders oestriol excretion an unreliable index of foetal well-being.

THE FOETAL ADRENAL GLANDS

The human foetal adrenal is unusual because of its large size relative to body weight due to the huge development of the internal, or foetal, zone of the cortex which regresses after birth. Normal foetal adrenal weight at term is 7–9 g which is comparable to that of a child of 10–12 years. Apart from the foetal zone the anatomy of the foetal adrenal is similar to that of the adult, there being a clearly defined medulla, an outer glomerular zone, a fascicular zone and a reticular zone, the major part of which comprises the foetal zone.

THE FOETAL ADRENAL MEDULLA

Little is known of the functions of the adrenal medulla in the human foetus. Chromaffin tissue is recognizable in the adrenal primordium when the embryo has a crown–rump length of 8 mm and catecholamines are present by the 12th week of pregnancy. In late pregnancy both noradrenaline and adrenaline are present in almost equal amounts, unlike the chromaffin tissue of the organs of Zuckerkandl which contain only noradrenaline (West et al 1953). The factors causing release of catecholamines from the adrenal medulla in the human foetus are unknown but have been studied in foetal lambs and calves by Comline & Silver (1966). They found in these foetal animals that the effective stimulus is a direct action of hypoxia on the gland itself. The adult type of response to hypoxia, that mediated via the splanchnic nerves, appears only near term. Release of catecholamines in the foetal lamb also differs from the adult type of response in that a high proportion of noradrenaline is released which bears no relationship to the catecholamine content of the gland. In general, noradrenaline comprises the major part of the catecholamine content of the medulla early in gestation but with increasing age there is a gradual increase in the proportion of adrenaline possibly related to the stimulation of methylation of noradrenaline by corticosteroids. The conversion of noradrenaline to adrenaline is catalysed by phenylethanolamine-N-methyl transferase, an enzyme which is induced by cortisol from the surrounding cortical tissue (Margolis et al 1966).

Dawes et al (1959) have concluded that maintenance of the circulation is of prime importance for survival of foetal and newborn animals after prolonged asphyxia, and the pressor amines released from the adrenal medulla under these conditions may be essential for this purpose. On the other hand, foetal lambs subjected to bilateral adrenalectomy are capable of continued survival in utero up to and beyond term (Drost & Holm 1968) but this may be no more than a reflection of the functional capacity of the extramedullary chromaffin tissue. In addition, the catecholamines may stimulate heat production after birth in a manner similar to that described for kittens (Moore & Underwood 1960); it may be significant that noradrenaline is a more calorigenic agent than adrenaline.

THE FOETAL ADRENAL CORTEX

Control of adrenocortical function

The various peculiarities of foetal zone development, such as its prominence in primates, its abrupt postnatal involution, and its absence in anencephaly, has made it difficult to find a trophic stimulus which explains all these observations. However, there is no doubt that the normal functioning of the human foetal adrenal depends upon ACTH and the release of ACTH is controlled by the hypothalamus which is sensitive to inhibitory effects of circulating cortisol. In anencephalic foetuses, growth of the adrenal is normal until the 20th week but thereafter little further growth occurs and at term the combined weight of adrenal tissue may be less than 1 g. Administration of metyrapone to pregnant women with normal foetuses causes an increase in urinary excretion of oestrogens but when the foetus is anencephalic there is no increase in oestrogen excretion despite a rise in maternal ketosteroid levels (Oakey & Heys 1970). This suggests that metyrapone crosses the placenta and stimulates release of ACTH in the normal foetus by inhibiting cortisol production. Lanman (1962) found more direct evidence of responsiveness of the adrenal cortex to ACTH. He administered long-acting ACTH to newborn infants with congenital malformations and found greatly enlarged adrenals when death occurred after 2–4 weeks. Although it has not been possible to measure cortisol production from the foetal adrenal during intra-uterine life in humans this has been done in foetal sheep. Near term, cortisol production in lambs doubles following the acute administration of ACTH (Fairclough & Liggins 1971). The human placenta is permeable to maternal corticosteroids and permits their entry into the foetal circulation where they inhibit ACTH release. Within 24 hours of administration of large doses of corticosteroids to pregnant women depressed adrenal function becomes evident as a fall in oestrogen excretion (Simmer *et al* 1966).

The glomerular zone of the foetal cortex is relatively undeveloped and has ultrastructural characteristics that suggest a low level of activity. The zone does not appear to be responsive to ACTH since it is not hypertrophic in newborn infants who receive ACTH injections. The foetal zone, on the other hand, has a level of activity and of development that cannot be explained in terms of ACTH alone. HCG has been suggested as an additional trophic stimulus and this attractive possibility would readily explain the regression of the adrenal zone after birth. *In vitro* studies have shown that HCG will stimulate the rate of synthesis by placental slices of steroids from acetate (Bloch & Benirschke 1959) but otherwise there is little evidence to support this hypothesis. Lanman (1962) failed to preserve a foetal zone by administration of HCG postnatally to an infant with a congenital malformation. Growth hormone, which is present in the foetal circulation in high concentration (Kaplan

& Grumbach 1967), has also been suggested as a trophic stimulus to the foetal zone since it can be synergistic with ACTH in adults. For the same reason HCS might also be an adrenocorticotrophic factor, but there is no satisfactory evidence to support either of these suggestions. Another possibility that would explain regression of the foetal zone after birth depends, not on a second trophic hormone, but on the presence of an enzyme inhibitor. The foetal zone is peculiar both in its hypertrophy and in its absence of 3β-hydroxysteroid dehydrogenase and these characteristics could result from the combined influence of ACTH and an inhibitor of 3β-hydroxysteroid dehydrogenase. Goldman (1968) induced congenital adrenal hyperplasia by administration of oestradiol to pregnant rats and showed that this effect depended on the ability of oestradiol to inhibit 3β-hydroxysteroid dehydrogenase whose activity is necessary for the synthesis of cortisol. If oestradiol has the same action on this enzyme in human foetal adrenals oestrogens secreted by the placenta into the foetal circulation might contribute to the development of the foetal zone. But Lanman (1962) failed to preserve a foetal zone in a newborn infant to whom oestrogen was administered post-natally.

Functions of the adrenal cortex

The glomerular zone is not known to have any important functions during foetal life. The need for regulation of electrolyte concentrations in foetal body fluids is minimal because of free exchange of sodium and potassium across the placenta. Children suffering from hereditary hypoaldosteronism who have a deficiency of 18-hydroxylase and who thus are unable to synthesize aldosterone are normal at birth and develop signs of a salt-losing syndrome only after 3 or more weeks (Visser & Cost 1964).

The function of the foetal zone has been described in the previous section on the foeto-placental complex. It will not be discussed further except to remind the reader that the main product of the foetal zone is DHAS and other $\Delta 5$ steroids; the absence of 3β-hydroxysteroid dehydrogenase largely precludes the synthesis of $\Delta 4$ steroids (including cortisol).

The definitive zone of the foetal adrenal contains the same array of enzymes as the fascicular and reticular zones of the adult cortex. Technical limitations have largely precluded *in vivo* studies of corticosteroid production in the human foetal adrenals. Cortisol production rates have been measured in newborn infants by Kenny et al (1966). The mean value for normal infants was 12·1 mg per sq cm per 24 hours. Surprisingly, the production rate in anencephalic infants was within the normal range, although babies with growth retardation and neonatal hypoglycaemia had abnormally low production rates. During foetal life it is usually assumed that the rate of cortisol secretion is low and that cortisol present in the foetal circulation is largely of maternal origin. Such opinions are based on the observations of Migeon et al (1957) who injected labelled cortisol into pregnant women shortly before delivery and identified labelled steroid in the cord blood within the next hour. However, the steroid carrying the label in the foetal blood was not identified and the authors concluded that cortisol and/or a steroid metabolite was transferred across the placenta to the foetal circulation. Subsequent investigations that have demonstrated the high activity of 11β-hydroxysteroid dehydrogenase in the human placenta (Osinski 1960) casts some doubt on the passage of cortisol into the foetus in an unchanged form. It is not unlikely that maternal cortisol is partially converted to biologically inactive cortisone during passage through the placenta. This seems even more possible in the light of the demonstration by Hillman & Giroud (1965) that the ratio of cortisone to cortisol in foetal blood may be as high as 10:1. Foetal cortisol levels are approximately one-third that of the mother, this ratio being preserved when maternal levels are raised; this constant gradient can be explained equally well either by increased transplacental passage of cortisol from mother to foetus or by reduced placental clearance of cortisol from the foetal circulation. If corticosteroid sulphates present in maternal urine during pregnancy are of exclusively foetal origin as suggested by Klein et al (1969) assay of cortisol sulphate and cortisone sulphate in maternal urine may give some indication of foetal adrenal secretory activity. Of course, the steroid conjugates may have arisen from corticosteroids transferred into the foetal circulation from the maternal blood stream. Moreover, the foetal adrenal may secrete cortisol in the sulphoconjugated form which makes no contribution to the free cortisol pool.

In other species the placenta is relatively impermeable to corticosteroids, at least until late in pregnancy, and corticosteroids in the foetal blood are more unequivocally of foetal adrenal origin. In the foetal lamb, for example, when labelled cortisol is infused into the foetal circulation there is minimal transfer of label into the maternal circulation and the rate of secretion of cortisol by the foetal adrenal is 1–2 mg per 24 hours (Beitins *et al* 1970). The concentration of cortisol in foetal and maternal circulations is similar until near term when the foetal concentration rises well above that of the mother.

Bloch & Benirschke (1959) found a very small conversion of labelled acetate to cortisol in incubations of slices of foetal adrenal glands and Solomon *et al* (1967) found very little conversion of acetate and cholesterol to C-21 steroids during perfusions of previable human foetuses. On the other hand, a large number of metabolites were obtained when labelled progesterone was perfused or injected into pre-viable mid-term foetuses. It is likely that the main substrate for cortisol biosynthesis in the foetal adrenal cortex is the abundant supply of progesterone that is secreted into the foetal circulation by the placenta.

The metabolism of cortisol in the human foetus differs markedly from that of the adult in whom loss of biological activity depends largely on A-ring reductase activity which leads to the excretion of reduced products such as tetrahydrocortisol. In the foetus, however, the pattern of circulating corticosteroids is dominated by the activity of two types of enzymes, namely, 11β-hydroxysteroid dehydrogenase and 21-hydroxysteroid sulphokinase. The former enzyme is present in the placenta (Osinski 1960) and the latter is present in both the adrenal and the liver (Klein *et al* 1965). Corticosteroid sulphates are transferred to the maternal circulation without prior hydrolysis and are excreted in the maternal urine as such. In late pregnancy the maternal urine contains sulphates of cortisol and corticosterone that are increased several-fold above the values observed in non-pregnant subjects and represent up to 15% of the excretion of tetrahydrocortisol plus tetrahydrocortisone glucuronides. It is likely that the corticosteroid sulphates in maternal urine are the product of foetal metabolism and may prove useful in the assessment of foetal adrenocortical activity. It is as yet unknown what proportion of corticosteroid sulphates are secreted by the foetal adrenal as such, are formed by the sulphation of cortisol entering the foetal circulation from the mother or represent sulphoconjugation in the foetal liver of cortisol secreted by the foetal adrenal. It is also unclear whether the cortisol from the maternal circulation or from the foetal adrenal serves as the major substrate for placental 11β-hydroxysteroid dehydrogenase which gives rise to the relatively high levels of cortisone circulating in the foetus. In any event, it is clear that the foetus has effective means for disposing of cortisol and maintaining its homeostasis. As judged by the maternal excretion of urinary oestrogens the foetus is able to maintain its homeostasis in the face of elevated levels of corticosteroid activity in the maternal circulation since depression of oestrogen excretion is seen only when corticosteroids are administered in doses that are equivalent to several times the normal adult secretion rate (Oakey 1970).

Functions of cortisol in the foetus

In general terms the actions of cortisol in the human foetus, as is also the case in the adult, depend on an ability of cortisol to increase the activity of certain enzymes. In the foetus, the role of enzyme induction in the timing of important developmental events imparts a special significance to the function of cortisol. In some instances, specific cortisol-inducible enzymes have been identified and extensively studied; in others, cortisol-related developmental events have been observed but the exact nature of the enzymes involved remains obscure. Some of the more important functions of cortisol are related to activities that can be best described as preparations for birth. These include maturation of neurological function, storage of glycogen, maturation of lung function and, at least in some species, the initiation of both lactation and parturition.

Cortisol-inducible enzymes. Precocious maturation of function in the small intestine of newborn rats and mice can be induced by the administration of corticosteroids. Treatment is followed by a large increase in the alkaline phosphatase activity of the duodenum (Moog & Kirsch 1955) which coincides with a sharp decrease in the ability to absorb antibody from the gut (Halliday 1959). Other enzymes that are induced by cortisol in foetal or neonatal rats include intestinal invertase and the liver enzymes, tyrosine aminotransferase and phosphopyruvate carboxylase (Holt & Oliver 1968). Reference has already been made to induction of phenylethanolamine-N-methyltransferase in the adrenal medulla of foetal rats. Cortisol-inducible enzyme systems in human foetal tissues have not been extensively investigated but it is probable that cortisol has similar effects to those observed in experimental animals.

Control of glycogen storage. Whereas the foetus received a continuous supply of glucose from its mother, the newborn faces a period of variable length during which all the exogenous supplies have been cut off. During the latter part of gestation the foetus of many species accumulates large amounts of glycogen not only in the liver (Fig. 4.10) but also in the skeletal muscles and heart (Shelley & Neligan 1966). The concentration of glycogen in the liver falls rapidly after birth and is almost depleted within 24 hours. During this period, maintenance of blood

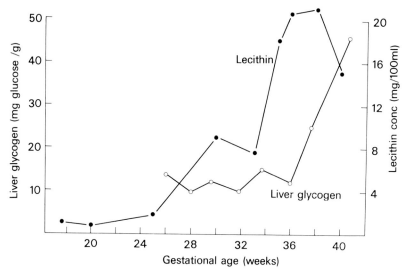

FIG. 4.10. The concentrations of glycogen (○) in the liver of human foetuses and the concentrations of lecithin (●) in amniotic fluid according to gestational age. After H.J.Shelley & G.A.Neligan (1966) and L.Gluck *et al.* (1971).

glucose and the supply of glucose to the brain is largely dependent on liver glycogen stores. The duration of asphyxia that can be withstood by the foetus and newborn is directly related to the amount of glycogen stored in the cardiac muscle. The hormonal control of glycogen storage is not fully understood but there is little doubt that cortisol plays a major part. One of the earliest studies in the field of foetal endocrinology was reported by Jost & Hatey (1949) who observed that the livers of decapitated rat foetuses contain twelve times less glycogen than those of litter-mate controls. Similar observations were made in rat foetuses and in addition it was found that liver glycogen could be restored to normal levels by the administration of corticosteroids. In foetal rabbits, however, corticosteroids do not restore liver glycogen unless an extract of rat placenta or prolactin is administered at the same time. Jost & Picon (1970) propose that glycogen deposition in the foetal liver obeys a dual hormonal control effected by corticosteroids and by a metabolic hormone. The latter hormone could be growth hormone, prolactin or, in the case of the human foetus, chorionic somatomammotrophin. In many species a surge of corticosteroids in the foetus occurs in late pregnancy and it is at this time that most of the glycogen in the liver is deposited. Premature glycogen storage can be induced by the administration of corticosteroids to foetal animals. The human newborn who has suffered intra-uterine growth retardation has a small liver containing little glycogen and hypoglycaemia is a common complication of the early neonatal period. Abnormally low oestrogen excretion in association with foetal growth retardation suggests that adrenocortical insufficiency may be part of the syndrome and may contribute to impaired glycogen storage.

Maturation of lung function. To be capable of functioning as an organ of gaseous exchange, foetal lung must attain not only anatomical maturity but also the alveolar lining membrane must synthesize and release into the alveolus adequate quantities of a surface active material. Unless this surfactant is present, the alveoli are unstable and deflate completely when a distending pressure is withdrawn. In most species surfactant concentration in the alveoli rises rapidly and the lung attains functional maturity about seven-eighths of the way through pregnancy. Before this time, inflation of the lungs is poorly maintained, atelectasis is widespread and the newborn suffers increasingly severe respiratory distress.

In foetal lambs infused either with ACTH or with corticosteroids Liggins (1969) found that functional maturation of the lungs was greatly accelerated enabling premature lambs to breathe without distress and to maintain inflation of the lungs 7–10 days before surfactant normally appears. Similarly accelerated appearance of pulmonary surfactant in response to corticosteroids has been observed in the foetal rabbit (Kotas & Avery 1971).

In the human foetus, functional maturity of the lungs is attained at about 35 weeks of pregnancy when there is a surge of lung surfactant. Before 35 weeks surfactant activity is relatively low and falls progressively with diminishing maturity until at 26 weeks surfactant activity is almost absent. Between 26 weeks and 35 weeks the newborn infant maintains alveolar distension with difficulty and respiratory distress is common. Gluck *et al* (1971) have shown that surfactant activity in the lung fluid is reflected in the lecithin concentration of the amniotic fluid and it is now possible to predict accurately whether or not respiratory distress will develop in prematurely born infants by examination of a sample of amniotic fluid obtained by amniocentesis. It is probably significant that the lecithin surge and the glycogen surge occur almost simultaneously in the human foetus. In sheep and rabbit foetuses also these surges are synchronous and can be induced prematurely by corticosteroids. This suggests the possibility of the therapeutic use of corticosteroids before premature delivery of human foetuses in an attempt to avoid the often-fatal neonatal consequences of surfactant deficiency. Preliminary results of a clinical trial are promising and suggest that respiratory distress syndrome is an avoidable disease.

Control of parturition. Appreciation of the vital role that the foetus of a number of species plays in the mechanisms controlling the onset of labour has led to renewed interest in this problem. It now seems likely that the intense activity of the foetal adrenal cortex that occurs near term is causally related not only to metabolic processes such as glycogen deposition and lung maturation but also to the initiation of parturition. In the human, foetal anencephaly has long been known to be associated with extreme prolongation of pregnancy (Rea 1898).

More recent studies of anencephalic pregnancies have shown that extended duration of pregnancy is almost invariably present when complications such as polyhydramnios are excluded. Furthermore, the extent of prolongation can be correlated with the degree of hypoplasia of the foetal pituitary and adrenals (Anderson *et al* 1969). Although this evidence in human pregnancy for the role of the foetal adrenal is circumstantial, in other species the results of experimental work is unequivocal. Ablation of the foetal hypothalamus, pituitary or adrenals by intra-uterine surgical procedures in sheep causes failure of labour at term and pregnancy is prolonged for many weeks (Liggins, Kennedy & Holm 1967). When the foetal lamb adrenals are stimulated by ACTH or when cortisol is infused into the lamb premature delivery occurs within a few days. Ablation of maternal endocrine organs or administration of ACTH and corticosteroids to the ewe have no effect on pregnancy except close to term. Other species in which administration of corticosteroids or ACTH to the foetus induces premature labour include cows, rabbits, and goats.

The mechanism by which corticosteroids of the foetal adrenal control the activity of the uterus is not completely understood but in the sheep two major endocrine changes are known to be related to increased

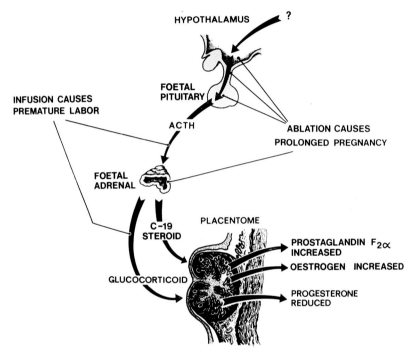

FIG. 4.11. Diagram of the endocrine changes involved in the mechanism by which the foetal lamb controls the onset of labour. The various experimental procedures that prolong or shorten pregnancy are shown. After G.C. Liggins (1969).

cortisol secretion from the foetal adrenal and both occur not only in spontaneous labour at term but also when premature delivery is induced by administration of ACTH to the foetus. First, there is a profound fall in the rate of secretion of progesterone from the placenta. Secondly, there is an increase in the rate of synthesis and release of prostaglandin $F_{2\alpha}$ from the tissues of the maternal placenta followed by a secondary rise in the myometrial concentration of this prostaglandin. It is thought

G

that the onset of labour may result from the combined effects of withdrawal of the inhibitory influence of progesterone and the oxytocic effects of $PGF_{2\alpha}$ (Fig. 4.11).

It is of interest that foetal corticosteroids can start labour not only in species (sheep) in which the hormones maintaining pregnancy are secreted by the placenta but also in species (goats and rabbits) in which maintenance of pregnancy depends upon corpus luteum activity. In the latter mammals increased activity of the foetal adrenals causes the corpora lutea to degenerate and progesterone levels to fall (Thorburn et al 1971). $PGF_{2\alpha}$ is known to be luteolytic in non-pregnant animals and it is possible that it provides the link between foetal adrenal activity and corpus luteum activity in pregnancy.

Although the human foetal adrenal is connected in some way with the control of labour it is uncertain whether the definitive zone or the foetal zone is the more important. In anencephaly, hypoplasia is most evident in the foetal zone and the striking endocrine abnormality that is observed is reduced oestrogen production. Moreover, Turnbull et al (1967), found that the pattern of maternal oestrogen excretion at the 34th week of pregnancy correlated (though not strongly) with gestation length. On the other hand, there is no evidence that the human foetal definitive zone is involved in the control of labour. Maternally administered corticosteroids are thought to cross the placenta into the foetus yet large doses of potent corticosteroid analogues have not been observed to cause premature labour. But the interpretation of such observations is confused by the effect of exogenous corticosteroids in depressing oestrogen production. Further studies of the functions of the human foetal adrenal will be necessary before these issues can be resolved.

REFERENCES

ANDERSON A.B., LAURENCE K.M. & TURNBULL A.C. (1969) The relationship in anencephaly between the size of the adrenal cortex and the length of gestation. *J. Obstet. Gynaec. Brit. Cwlth.* **76**, 196.

ANGEVINE D.M. (1938) Pathologicanatomy of hypophysis and adrenals in anencephaly. *Arch. Path.* **26**, 507.

ASCHHEIM S. & ZONDEK B. (1927) Hypophysenvorderlappen, hormon und ovarial hormon im tarn von Schwangeren. *Klin. Wschr.* **6**, 1322.

BAHL O.P. (1969) Human chorionic gonadotropin. I. Purification and physico-chemical properties. *J. biol. Chem.* **244**, 567.

BAKER T.G. (1963) A quantitative and cytological study of germ cells in human ovaries. *Proc. roy. Soc.* B. **158**, 417.
BECK P. & DAUGHADAY W.H. (1967) Human placental lactogen; studies of its acute metabolic effects and disposition in normal man. *J. clin. Invest.* **46**, 103.
BEISCHER N.A., BROWN J.B., MACLEOD S.C. & SMITH M.A. (1967) The value of urinary oestriol estimation in patients with antepartum vaginal bleeding. *J. Obstet. Gynaec. Brit. Cwlth.* **74**, 51.
BEITINS I.Z., KOWARSKI A., SHERMETA D.W., DE LEMOS R.A. & MIGEON C.J. (1970) Foetal and maternal secretion rate of cortisol in sheep: Diffusion resistance of the placenta. *Pediat. Res.* **4**, 129.
BELING C.G. (1967) Estriol excretion in pregnancy and its application to clinical problems, in S.C.Marcus & C.C.Marcus (eds.) *Advances in Obstetrics and Gynecology*, p. 88. Baltimore, Williams & Wilkins Co.
BELING C.G. (1971) Estrogens, in F. Fuchs & A. Klopper (eds.) *Endocrinology of Pregnancy*, p. 57. New York, Harper & Row.
BELL J.J., CANFIELD R.E. & SCIARRA J.J. (1969) Purification and characterization of human chorionic gonadotropin. *Endocrinology* **84**, 298.
BENIRSCHKE K. (1956) Adrenals in anencephaly and hydrocephaly. *Obstet. Gynec.* **8**, 412.
BLOCH E. & BENIRSCHKE K. (1959) Synthesis *in vitro* of steroids by human fetal adrenal gland slices. *J. biol. Chem.* **234**, 1085.
BOLTÉ E., WIQVIST N. & DICZFALUSY E. (1966) Metabolism of dehydroepiandrosterone and dehydroepiandiosterone sulphate by the human foetus at mid pregnancy. *Acta endocr. (Kbh.)* **52**, 583.
BRADBURY J.T., BROWN W.E. & GRAY L.A. (1950) Maintenance of the corpus luteum and physiologic actions of progesterone *Rec. Progr. Horm. Res.* **5**, 151.
BRODY S. (1969) Protein hormones and hormonal peptides from the placenta, in A. Klopper & E. Diczfalusy (eds.) *Foetus and Placenta*, p. 299. Oxford, Blackwell.
BRODY S. & CARLSTRÖM G. (1965) Human chorionic gonadotropin pattern in serum and its relation to the sex of the fetus. *J. clin. Endocr.* **25**, 792.
BROWN J.B. (1956) Urinary excretion of oestrogens during pregnancy, lactation, and re-establishment of menstruation. *Lancet* **i,** 704.
BROWN W.E. & BRADBURY J.T. (1947) Study of physiologic action of human chorionic hormone; production of pseudopregnancy in women by chorionic hormone. *Amer. J. Obstet. Gynec.* **53**, 749.
BUTT W.R. (1967) Chemical properties of gonadotrophins, in W.R.Butt (ed.) *The Chemistry of Gonadotrophins*, p. 58. Springfield, Charles C. Thomas.
CASSMER O. (1959) Hormone production of the isolated human placenta. *Acta endocr. (Kbh.)* Suppl. 45, 1.
CATHRO D.M. & COYLE M.G. (1966) Adrenocortical function in newborn infants of low birthweight, in Excerpta Medica International Congress Series 132, p. 688. Amsterdam, Excerpta Medica Foundation.
CÉDARD L., VARANGOT J. & YANNOTTI S. (1964) Influence des gonadotrophines chorioniques sur le métabolisme des stéroides dans les placentas humains perfusés in vitro. *C. R. Acad. Sci. (Paris)* **258**, 3769.
COLÁS A. & HEINRICHS L. (1964) Pettenkofer chromogens in the maternal and fetal

circulations: anencephalic pregnancies, cesarean sections, and tentative identification of 3β, 17β-dihydroxyandrost-5-en-16-one in umbilical cord blood. *Steroids* **5**, 753.

COMLINE R.S. & SILVER M. (1966) The development of the adrenal medulla of the foetal and newborn calf. *J. Physiol. (Lond.)* **183**, 305.

COYLE M.G. (1962) The urinary excretion of oestrogen in four cases of anencephaly and one case of foetal death from cirrhosis of the liver. *J. Endocr.* **25**, viii.

DAWES G.S., MOTT J.C. & SHELLEY H.J. (1959) The importance of cardiac glycogen for the maintenance of life in foetal lambs and newborn animals during anoxia. *J. Physiol. (Lond.)* **146**, 516.

DICZFALUSY E. & MAGNUSSON A.M. (1958) Tissue concentration of oestrone oestradiol-17β and oestriol in the human foetus. *Acta endocr. (Kbh.)* **28**, 169.

DICZFALUSY E. (1962) Endocrinology of the foetus. *Acta obstet. gynec. scand.* **41**, Suppl. 1, 45.

DIXON H.G., BROWNE J.C. & DAVEY D.A. (1963) Choriodecidual and myometrial blood-flow. *Lancet* **ii**, 369.

DROST M. & HOLM L.W. (1968) Prolonged gestation in ewes after adrenalectomy. *J. Endocr.* **40**, 293.

EASTERLING W.E., SIMMER H., DIGNAM W.J., FRANKLAND M.W. & NAFTOLIN F. (1966) Neutral C-19 steroids and steroid sulfates in human pregnancy. *Steroids* **8**, 157.

ELGER W. (1966) Die Rolle der Fetalen Androgene in der sexual Differenzierung des Kaninchens und ihre somatische Faktoren durch anwendung eines starken Antiandrogens. *Arch. Anat. Micr. Morph. Exp.* **55**, Suppl. 657.

ELLEGOOD J.O., MAHESH U.B. & GREENBLATT R.D. (1969) Chorionic gonadotropin tests for diagnosis of pregnancy and chorionic tumours. *Post Grad. Med.* **46**, 105.

FAIRCLOUGH R.J. & LIGGINS G.C. (1971) Cortisol production rates in the foetai lamb, in *Abstracts of the Fourth Asia and Oceania Congress of Endocrinology, Auckland.*

FARQUHAR J.W. (1962) Birthweight and survival of babies of diabetic women. *Arch. Dis. Childh.* **37**, 321.

FRANCE J.T. (1971) Levels of 16α-hydroxydehydroepiandrosterone, dehydroepiandrosterone and pregnenolene in cord plasma of human normal and anencephalic fetuses. *Steroids* **17**, 697.

FRANCE J.T. & LIGGINS G.C. (1969) Placental sulfatase deficiency. *J. clin. Endocr.* **29**, 138.

FRANDSEN V.A. & STAKEMANN G. (1961) The site of production of oestrogenic hormones in human pregnancy. Hormone excretion in pregnancy with anencephalic foetus. *Acta endocr. (Kbh.)* **38**, 383.

FUKUSHIMA M. (1961) Studies on somatotropic hormone secretion in gynecology and obstetrics. *Tohoku J. exp. Med.* **74**, 161.

GITLIN D. & BIASUCCI A. (1969) Ontogenesis of immunoreactive growth hormone, follicle-stimulating hormone, luteinizing hormone, chorionic prolactin and chorionic gonadotropin in the human conceptus. *J. clin. Endocr.* **29**, 926.

GLUCK L., KULOVICH M.V., BORER R.C., BRENNER P.H., ANDERSON G.G. & SPELLACY W.N. (1971) Diagnosis of the respiratory distress syndrome by amniocentesis. *Amer. J. Obstet. Gynec.* **109**, 440.

GOLDMAN A.S. (1968) Production of congenital adrenocortical hyperplasia in rats by estradiol-17β and inhibition of 3β-hydroxysteroid dehydrogenase. *J. clin. Endocr.* **28,** 231.

GOVAERTS-VIDETSKY M. (1965) Elimination urinaire des oestrogènes au cours de la grossesse normale. *Rev. Franc. Etud. Clin. Biol.* **10,** 815.

GREENE J.W., SMITH K., KYLE G.C., TOUCHSTONE J.C. & DUHRING J.L. (1965) The use of urinary estriol excretion in the management of pregnancies complicated by diabetes mellitus. *Amer. J. Obstet. Gynec.* **91,** 684.

GRUMBACH M.M., KAPLAN S.L., SCIARRA J.J. & BURR I.M. (1968) Chorionic growth hormone-prolactin (CGP) secretion, disposition, biologic activity in man, and postulated function as the 'growth hormone' of the second half of pregnancy. *Ann. New York Acad. Sci.* **148,** 501.

GRUMBACH M.M. & BARR M.L. (1958) Cytologic tests of chromosomal sex in relation to sexual anomalies in man. *Rec. Progr. Horm. Res.* **14,** 255.

HAGEN A.A., BARR M. & DICZFALUSY E. (1965) Metabolism of 17β-oestradiol-4-14-C in early infancy. *Acta endocr. (Kbh.)* **49,** 207.

HALBAN J. (1905) Die innere Secretion von Ovarion und Placenta und ihre Bedeutung für die Function der Milchdruse. *Arch. Gynäk.* **75,** 353.

HALLIDAY R. (1959) The effect of steroid hormones on the absorption of antibody by the young rat. *J. Endocr.* **18,** 56.

HALPIN T.F. (1970) Human chorionic gonadotropin titers in twin pregnancy. *Amer. J. Obstet. Gynec.* **106,** 317.

HAMASHIGE S., ASTOR M.A., ARQUILLA E.R. & VAN THIEL D.R. (1967) Human chorionic gonadotropin: a hormone complex. *J. clin. Endocr.* **27,** 1690.

HAMERTON J.L. (1968) Significance of sex chromosome derived heterochromatin mammals. *Nature (Lond.)* **219,** 910.

HARBERT G.H., MCGAUGHEY H.S., SCOGGIN W.A. & THORNTON W.N. (1964) Concentration of progesterone in newborn and maternal circulations at delivery. *Obstet. Gynec.* **23,** 413.

HARKNESS R.A., MENINI E., CHARLES D., KENNY F.M. & ROMBAUT R. (1966) Studies of urinary steroid excretion by an adrenalectomized woman during and after pregnancy. *Acta endocr. (Kbh.)* **52,** 409.

HARRIS G.W. (1970) Hormonal differentiation of the developing central nervous system with respect to patterns of endocrine function. *Phil. Trans. roy. Soc. Lond.* B. **259,** 165.

HARRIS G.W. & LEVINE S. (1962) Sexual differentiation of the brain and its experimental control. *J. Physiol. (Lond.)* **163,** 42P.

HELLIG H., GATTEREAU D., LEFEBVRE Y. & BOLTÉ E. (1970) Steroid production from plasma cholesterol I. Conversion of plasma cholesterol to placental progesterone in humans. *J. clin. Endocr.* **30,** 624.

HENNEN G., PIERCE J.G. & FREYCHET P. (1969) Human chorionic thyrotropin: further characterization and study of its secretion during pregnancy. *J. clin. Endocr.* **29,** 581.

HILLMAN D.A. & GIROUD C.J. (1965) Plasma cortisone and cortisol levels at birth and during the neonatal period. *J. clin. Endocr.* **25,** 243.

HOBSON B. & WIDE L. (1968) Human chorionic gonadotrophin excretion in men and women with invasive trophoblast assayed by an immunological and a biological method. *Acta endocr. (Kbh.)* **58**, 473.

HOLT P.G. & OLIVER I.T. (1968) Factors affecting the premature induction of tyrosine aminotransferase in foetal rat liver. *Biochem. J.* **108**, 333.

ITO Y. & HIGASHI K. (1961) Studies on the prolactin-like substance in human placenta II. *Endocr. Japan* **8**, 279.

JOHANNISSON E. (1968) Foetal adrenal cortex in the human: Ultrastructure. *Acta endocr. (Kbh.)* Suppl. **130**, 1.

JOHANSSON, E.D.B. (1968) Progesterone level and response to oxytocin at term. *Lancet* **2**, 570.

JOSIMOVICH J.B. (1971) Placental lactogenic hormone, in F.Fuchs & A.Klopper (eds.) *Endocrinology of Pregnancy*, p. 184. New York, Harper & Row.

JOSIMOVICH J.B. & MACLAREN J.A. (1962) Presence in the human placenta and term serum of a highly lactogenic substance immunologically related to human growth hormone. *Endocrinology* **71**, 209.

JOST A. (1959) In C.R.Austin (ed.) *Sex Differentiation and Development*. Society for Endocrinology Memoirs No. 7. London, Cambridge University Press.

JOST A. (1969) In H.W. Jones & W.W. Scott (eds.) *Hermaphroditism, Genital Anomalies and Related Endocrine Disorders*, 2nd ed. Baltimore, Williams & Wilkins.

JOST A. (1970) Hormonal factors in the sex differentiation of the mammalian foetus. *Phil. Trans roy. Soc. Lond. B.* **259**, 119.

JOST A. & HATEY J. (1949) Influence de la décapitation sur la teneur en glycogène du foie du fœtus de lapin. *Compt. Rend. Soc. Biol.* **143**, 146.

JOST A. & PICON L. (1970) Hormonal control of fetal development and metabolism. *Adv. Metab. Disord.* **4**, 123.

KAPLAN S.L. & GRUMBACH M.M. (1964) Studies of a human and simian placental hormone with growth hormone-like and prolactin-like activities. *J. clin. Endocr.* **24**, 80.

KAPLAN S.L. & GRUMBACH M.M. (1967) In *Growth Hormone: International Symposium on Growth Hormone No. 98*, p. 57. International Congress Series 142. Amsterdam, Excerpta Medica Foundation.

KAYE M.D. & JONES W.R. (1971) Effect of human chorionic gonadotropin on *in vitro* lymphocyte transformation. *Amer. J. Obstet. Gynec.* **100**, 1029.

KENNY F.M., PREEYASOMABA C., SPAULDING J.S. & MIGEON C.J. (1966) Cortisol production rate IV. Infants born of steroid-treated mothers and of diabetic mothers. Infants with trisomy syndrome and with anencephaly. *Pediatrics* **37**, 960.

KLEIN G.P., BRANCHAUD C., CHAN S.K., HALL C.ST.G., SCHWEITZER M. & GIROUD C.J.P. (1969) Identification and possible origin of C-21 steroid sulfates within the foeto-placental unit, in A. Pecile & C. Finzi (eds) *The Foeto-Placental Unit*, p. 176. Amsterdam, Excerpta Medica Foundation.

KLEIN G.P. GIROUD C.J. & BROWNE J.S.L. (1965) Sulfation of corticosteroids by the adrenal of the human newborn. *Steroids* **5**, 765.

KLOPPER A. (1965) Assays of urinary oestriol as a measure of placental function,

in, C. Cassans (ed.) *Research on Steroids* vol. 1, p. 119. Rome, Il Pensiero Scientifico.
KLOPPER A. (1969) Assessment of placental function in clinical practice, in A. Klopper & E.Diczfalusy (eds.) *The Foeto-Placental Unit*, p. 471. Oxford. Blackwell.
KLOPPER A. (1970) Assessment of fetoplacental function by hormone assay. *Amer. J. Obstet. Gynec.* **170**, 807.
KLOPPER A. & STEPHENSON R. (1966) The excretion of oestriol and pregnanediol in pregnancy complicated by Rh immunisation. *J. Obstet. Gynaec. Brit. Cwlth.* **73**, 282.
KOCK H., KESSEL H., VON STOLTE L. & LEUSDEN H. VON (1966) Thyroid function in molar pregnancy. *J. clin. Endocr.* **26**, 1128.
KOTAS R.V. & AVERY M.E. (1971) Accelerated appearance of pulmonary surfactant in the fetal rabbit. *J. appl. Physiol* **30**, 358.
LANMAN J.T. (1962) An interpretation of human foetal adrenal structure and function, in A.R.Currie, T.Synington & J.K.Grant (eds.) *The Human Adrenal Cortex*, p. 547. Edinburgh, Livingstone.
LAURITZEN C. (1966) Action of HCG and steroids in physiological processes in the neonatal period, in C.Cassana (ed.) *Research on Steroids*, vol. 2, p. 109. Rome, Il Pensiero Scientifico.
LAURITZEN C. (1967) A clinical test for placental functional activity using DHEA-sulphate and ACTH injections in the pregnant woman. *Acta endocr. (Kbh.)* Suppl. **119**, 188.
LAURITZEN C. & LEHMANN W.D. (1967) Levels of chorionic gonadotrophin in the newborn infant and their relationship to adrenal dehydroepiandrosterone. *J. Endocr.* **39**, 173.
LAURITZEN C., SHACKLETON C.H. & MITCHELL F.L. (1969) The effect of exogenous human chorionic gonadotrophin on steroid excretion in the newborn. *Acta endocr. (Kbh.)* **61**, 83.
LEVITZ M., CONDON G.P., DANCIS J., GOBELSMANN U., ERIKSSON G. & DICZFALUSY E. (1967) Transfer of estriol and estriol conjugates across the human placenta perfused *in situ* at midpregnancy. *J. clin. Endocr.* **27**, 1723.
LIGGINS G.C. (1969) Premature delivery of foetal lambs infused with glucocorticoids. *J. Endocr.* **45**, 515.
LIGGINS G.C. (1969) The foetal role in the initiation of parturition in the ewe, in G.E.W.Wolstenholme & M.O'Connor (eds.) *Foetal Autonomy*. Ciba Symposium. London, Churchill.
LIGGINS G.C. & EVANS M. (1963) Patterns of oestriol excretion in abnormal pregnancy. A study of 234 cases. *New Zeal. Med. J.* **62**, 365.
LIGGINS G.C., KENNEDY P.C. & HOLM L.W. (1967) Failure of initiation of parturition after electrocoagulation of the pituitary of the fetal lamb. *Amer. J. Obstet. Gynec.* **98**, 1080.
LORAINE J.A. & BELL E.T. (1966) *Recent Research on Gonadotrophic Hormones. Proceedings of Fifth Gonadotrophin Club Meeting*, 2nd ed. Edinburgh, Livingstone.
LORAINE J.A., ISMAEL A.A.A., ADAMOPOULOS D.A. & DOVE G.A. (1970) Endocrine function in male and female homosexuals. *Brit. med. J.* **4**, 406.

LUNDWALL F. & STAKEMANN G. (1966) The urinary excretion of oestriol in postmaturity. *Acta obstet. gynec. scand.* **45**, 301.

LURIE A.O., REID D.E. & VILLEE C.A. (1966) The role of the fetus and placenta in maintenance of plasma progesterone. *Amer. J. Obstet. Gynec.* **96**, 670.

LYON M.F. (1961) Gene action in the X-chromosome of the mouse (*Mus musculus* L.) *Nature (Lond.)* **190**, 372.

MACDONALD P.D. & SIITERI P.K. (1965) Origin of estrogen in women pregnant with an anencaphalic fetus. *J. clin. Endocr.* **24**, 685.

MCKUSICK V.A. (1962) On the X chromosome of man. *Quart. Rev. Biol.* **37**, 69.

MCLEOD S.C., BROWN J.B., BEISHER N.A. & SMITH M.A. (1967) The value of urinary oestriol measurements during pregnancy. *Aust. N.Z. J. Obstet. Gynaec.* **7**, 25.

MAGENDANTZ H.G. & RYAN K.J. (1964) Isolation of an oestriol precursor 16α-hydroxydehydroepiandrosterone, from human umbilical sera. *J. clin. Endocr.* **24**, 1155.

MARGOLIS F.L., ROFFI J. & JOST A. (1966) Norepinephrine methylation in fetal rat adrenals. *Science* **154**, 275.

MIGEON C.J., BERTRAND J. & WALL P.E. (1957) Physiological disposition of 4-C14-cortisol during late pregnancy. *J. clin. Invest.* **36**, 1350

MITTWOCK U. (1970) How does the Y chromosome affect gonadal differentiation? *Phil. Trans roy. Soc. Lond.* B. **259**, 113.

MOOG F. & KIRSCH M.H. (1955) Quantitative determination of phosphatase activity in chick embryo duodenum culture in fluid media with and without hydrocortisone. *Nature (Lond.)* **175**, 722.

MOORE R.E. & UNDERWOOD M.C. (1960) Noradrenaline as a possible regulator of heat production in the newborn kitten. *J. Physiol. (Lond.)* **150**, 13P.

MURAKAWA S. & RABEN M.S. (1968) Effect of growth hormone and placental lactogen on DNA synthesis in rat costal cartilage and adipose tissue. *Endocrinology* **83**, 645.

NIEMI I., IKONEN M. & HERVONEN A. (1967) Histochemistry and fine structure of interstitial tissue in human foetal testis, in G.E.W.Wolstenholme & M.O'Connor (eds.) *Endocrinology of Testis*, p. 3. CIBA Foundation. London, Churchill.

OAKEY R.E. (1970) The interpretation of urinary oestrogen and pregnanediol excretion in pregnant women receiving cortico-steroids. *J. Obstet. Gynaec. Brit. Cwlth.* **77**, 922.

OAKEY R.E. & HEYS R.F. (1970) Regulation of the production of oestrogen precursors in the foetus. *Acta endocr. (Kbh.)* **65**, 502.

OSINSKI P.A. (1960) Steroid 11β-ol-dehydrogenase in human placenta. *Nature (Lond.)* **187**, 777.

PATERSON W.G., HOBSON B.M., SMART G.E. & BAIN A.D. (1971) Two cases of hydatidiform degeneration of the placenta with fetal abnormality and triploid chromosome constitution. *J. Obstet. Gynaec. Brit. Cwlth.* **78**, 136.

PAUERSTEIN C.J. & SOLOMON D. (1966) LH and androgenesis. *Obstet. Gynec.* **28**, 692.

PFEIFFER C.A. (1936) Sexual differences of the hypophyses and their determination by gonads. *Amer. J. Anat.* **58**, 195.

PION R., JAFFE R., ERIKSSON G., WIQVIST N. & DICZFALUSY E. (1965) Studies on the metabolism of C-21 steroids in the human foeto-placental unit. I. Formation of a β-unsaturated 3-ketones in midterm placentas perfused *in situ* with pregnenolone and 17α-hydroxypregnenolone. *Acta endocr. (Kbh.)* **48**, 234.

REA C. (1898) Prolonged gestation, acrania monstrosity and apparent placenta praevia in an obstetrical case. *J. Amer. Med. Ass.* **30**, 1166.

ROBERTSON W.B., BROZENS I. & DIXON H.G. (1967) The pathological response of the vessels of the placental bed to hypertensive pregnancy. *J. Path. Bact.* **93**, 81.

RUSHWORTH A.G., ORR A.H. & BAGSHAWE K.D. (1968) The concentration of HCG in the plasma and spinal fluid of patients with trophoblastic tumours in the central nervous system. *Brit. J. Cancer* **22**, 253.

RYAN K.J. (1959) Biological aromatization of steroids. *J. biol. Chem.* **234**, 268.

SAMAAN N.Z., BRADBURY J.T. & GOPLERUD C.P. (1969) Serial hormonal studies in normal and abnormal pregnancy. *Amer. J. Obstet. Gynec.* **104**, 781.

SAVARD K. MARSH J.M. & RICE B.F. (1965) Gonadotrophins and ovarian steroidogenesis. *Rec. Progr. Horm. Res.* **21**, 285.

SAXENA B.N. EMERSON K. & SELENKOW H.A. (1969) Serum placental lactogen (HPL) levels as an index of placental function. *New Engl. J. Med.* **281**, 225.

SAXENA B., GOLDSTEIN D.P., EMERSON K. & SELENKOW H.A. (1968) Serum placental lactogen levels in patients with molar pregnancy and trophoblastic tumors. *Amer. J. Obstet. Gynec.* **102**, 115.

SCOMMEGNA A., NEDOSS B.R. & CHATTORAI S.C. (1968) Maternal urinary estriol excretion after dehydroepiandrosterone-sulfate infusion and adrenal stimulation and suppression. *Obstet. Gynec.* **31**, 526.

SEAL U.S. & DOE R.P. (1966) Corticosteroid-binding globulin, in G.Pincus, T.Nakao & J.F.Tait (eds.) *Steroid Dynamics*, p. 84. New York, Academic Press.

SEGALOFF A., STERNBERG W.H. & CASKILL C.J. (1951) Effects of luteotropic doses of chorionic gonadotropin in women. *J. clin. Endocr.* **11**, 936.

SELENKOW H.A., SAXENA B.N., DANA C.L. & EMERSON K. (1969) Measurement and pathophysiologic significance of human placental lactogen, in A.Pecile & C. Finzi (eds.) *The Foeto-Placental Unit*, p. 352. Amsterdam, Excerpta Medica Foundation.

SHELLEY H.J. & NELIGAN G.A. (1966) Neonatal hypoglycaemia. *Brit. med. Bull.* **22**, 34.

SHEPARD T.H. (1967) Onset of function in the human fetal thyroid: biochemical and radioautographic studies from organ culture. *J. clin. Endocr.* **27**, 945.

SHERWOOD L.M. (1967) Human placental lactogen. Partial analysis of chemical structure and comparison with pituitary growth hormone, in C.Gual (ed.) *Progress in Endocrinology*, p. 394. Amsterdam, Excerpta Medica Foundation.

SHORT R.V. (1969) Implantation and the maternal recognition of pregnancy, in G.E.W.Wolstenholme & M.O'Connor (eds.) *Foetal Autonomy*, p. 1. Ciba Symposium. London, Churchill.

SIITERI P.K. & MACDONALD P.C. (1966) Placental estrogen biosynthesis during human pregnancy. *J. clin. Endocr.* **26**, 751.

SIMMER H.H., DIGNAM W.J., EASTERLING W.E., FRANKLAND M.V. & NAFTOLIN F.

(1966) Neutral C-19-steroids and steroid sulfates in human pregnancy. III. Dehydroepiandrosterone sulfate, 16α-hydroxydehydroepiandrosterone and 16α-hydroxydehydroepiandrosterone sulfate in cord blood and blood of pregnant women with and without treatment with corticoids. *Steroids* **8**, 179.

SMITH K., GREENE J.W. & TOUCHSTONE J.C. (1966) Urinary estriol determination in the management of prolonged pregnancy. *Amer. J. Obstet. Gynec.* **96**, 901.

SOLOMON S. (1966) Formation and metabolism of neutral steroids in the human placenta and fetus. *J. clin. Endocr.* **26**, 762.

SOLOMON S., BIRD C.E., LING W., IWAMIYA M. & YOUNG P.C.M. (1967) Formation and metabolism of steroids in the fetus and placenta. *Rec. Progr. Horm. Res.* **23**, 297.

SOLOMAN S. & FUCHS F. (1971) Progesterone and related neutral steroids, in F. Fuchs & A.Klopper (eds.) *Endocrinology of Pregnancy*, p. 66. New York, Harper & Row.

SPELLACY W.N., CARLSON K.L. & BIRK S.A. (1966) Dynamics of human placental lactogen. *Amer. J. Obstet. Gynec.* **96**, 1164.

SPELLACY W.N., TEOH E.S. & BUHI W.C. (1970) Human chorionic somatomammotropin (HCS) levels prior to fetal death in high-risk pregnancies. *Obstet. Gynec.* **35**, 685.

SUWA S. & FRIESEN H. (1969) Biosynthesis of human placental proteins and human placental lactogen (HPL) *in vitro*. Dynamic studies of normal term placentas. *Endocrinology* **85**, 1037.

TAYLOR E.S., BRUNS P.D., HEPNER H.J. & DROSE V.E. (1958) Measurements of placental function. *Amer. J. Obstet. Gynec.* **76**, 983.

TELEGDY G., WEEKS J.S., ARCHER D.F., WIQVIST N. & DICZFALUSY E. (1970) Acetate and cholesterol metabolism in the human foeto-placental unit at midgestation. *Acta endocr. (Kbh.)* **63**, 119.

THORBURN G.D., NICOL D. & BASSETT J.M. (1971) The foetal role in parturition in the goat. *J. Reprod. Fert.* (in press).

TOUCHSTONE J.C., STOJKEWYCZ M. & SMITH K. (1965) The effect of methenamine mandelate (Mandelamine) on determination of urinary estriol. *Clin. Chem.* **11**, 1019.

TURNBULL A.C., ANDERSON A.B. & WILSON G.R. (1967) Maternal urinary oestrogen excretion as evidence of a foetal role in determining gestation at labour. *Lancet* **ii**, 627.

VARANGOT J., CÉDARD L. & YANNOTTI S. (1965) Perfusion of the human placenta *in vitro*: study of the biosynthesis of oestrogens. *Amer. J. Obstet. Gynec.* **92**, 534.

VILLEE C.B. (1969) Development of endocrine function in the human placenta and fetus. *New Eng. J. Med.* **281**, 533.

VISSER H.K.A. & COST W.S. (1964) A new hereditary defect in the biosynthesis of aldosterone: urinary C21-corticosteroid pattern in three related patients with a salt-losing syndrome, suggesting an 18-oxidation defect. *Acta endocr. (Kbh.)* **47**, 589.

WARREN J.C. & TIMBERLAKE C.E. (1962) Steroid sulfatase in the human placenta. *J. clin. Endocr.* **22**, 1148.

WEST G.B., SHEPHERD D.M., HUNTER R.B. & MACGREGOR A.R. (1953) The function of organs of Zuckerkandl. *Clin. Sci.* **12**, 317.
WIDE L. (1969) Early diagnosis of pregnancy *Lancet* **ii**, 863.
WILLMAN K. & PULKINNEN M.O. (1971) Reduced maternal plasma and urinary estriol during ampicillin treatment. *Amer. J. Obstet. Gynec.* **109**, 893.
WITSCHI E. (1951) Embryogenesis of the adrenal and the reproductive glands. *Recent Prog. Horm. Res.* **6**, 1.
YOSHIMI T., STROTT C.A., MARSHALL J.R. & LIPSETT M.B. (1969) Corpus luteum function in early pregnancy. *J. clin. Endocr.* **29**, 225.
ZANDER J., HOLZMANN K., VON MUNSTERMANN A.M., RUNNEBAUM B. & SIEBER W. (1969) New results on the metabolism of progesterone in the foeto-placental unit, in A.Pecile & C.Finzi (eds.) *The Foeto-Placental Unit*, p. 162. Amsterdam, Excerpta Medica Foundation.

5
Immunology

WARREN R. JONES

INTRODUCTION

Mammalian pregnancy represents the highly developed end point of the evolution of viviparous reproduction. The human deciduate haemochorial placenta, in which trophoblast invades maternal tissues, provides an extremely efficient respiratory and trophic connection between mother and foetus. It also presents to the biologist the immunological paradox of the acceptance and control of trophoblastic invasion in the maternal organism.

In 1933 Grosser expressed these problems in simple and striking terms. He saw an obligatory role for the placenta in the preservation of the individuality of mother and conceptus. At the same time he warned that if orthogenetic, or predetermined as opposed to adaptive, evolution of the placenta continued, the invasive properties of trophoblast may lead to the ultimate destruction of the species. Over the last 40 years, however, it has become apparent that another highly developed and potentially destructive organ, the brain, has established a firm hold on the genesis of human extinction.

Grosser's concepts of placentation expressed in modern immunological terms form the basis of the science of reproductive immunology. The dynamic success of the foetal allograft* in its potentially hostile maternal environment offers wide and, as yet, relatively unexplored implications in the field of organ transplantation. The basic immunological

---

* Allograft: (synonyms: homograft, allogeneic graft) a graft derived from one individual to another individual of the same species but differing in genetic constitution.

mechanisms contributing to the controlled survival of trophoblast may also hold the key to many aspects of neoplastic proliferation.

As well as being of central importance in the maintenance of pregnancy it is now realized that immunological factors are involved in gamete production, fertilization and implantation, and normal embryonic and foetal development.

Clinical aspects of reproductive immunology are concerned with adverse immunological reactions in the reproductive tract and with situations involving disruption of the maternal-foetal relationship; the classical example being rhesus isoimmunization.* It also seems probable that any disruption of the normal immunological integrity of the foetus, such as viral infection, may have far-reaching effects on future development. The evolution of foetal medicine must surely include the possibility of prophylactic intra-uterine immunological manipulation.

The aim of this chapter is primarily to review in systematic fashion, the present knowledge of reproductive immunology. The field is an extensive one and references will include a large proportion of review articles.

As a background to the applied immunology of reproduction, and as an expression of the immense amount of new knowledge in the field, an initial section is devoted to a review of basic mechanisms and current concepts in general immunology.

SECTION 1—GENERAL IMMUNOLOGY

The broad spectrum of modern immunological theory contains little that is recognizable of the classical studies of communicable diseases and their prevention, which formed the cornerstone of this branch of biology. Certainly the crude reasoning which preceded the practical application of smallpox vaccination in the mid-eighteenth century could not be construed as immunological theory. However, the fact that this event antedated the elaboration of the germ theory of disease established a unique place for immunology in medical history. Since that time the study of acquired immunity to infection, and the practical application of prophylactic immunization against pathogenic organisms, has influenced profoundly the epidemiology of human disease.

The elucidation of immunological mechanisms involved in resistance

* Isoimmunization: immunization against antigens (isoantigens) present in another member of the same species resulting in the production of isoantibodies.

to infection has led to a wider biological era of immunology; an era with ramifications in genetic theory and molecular biology, and with clinical implications in every branch of medicine.

The importance of the maintenance of immunological defence (immunity) against foreign microorganisms or macromolecules has given way to the more basic concept of the body's recognition of 'self' and 'non-self' antigens. An antigen may now be regarded in general terms as something producing a 'non-self' reaction. It may be of external or internal origin, the latter relating to altered 'self' components and providing the basis of modern immunological theories of neoplasia (Green 1954) and autoimmunity (Burnet 1959a).

The term immunological surveillance has been coined to imply the control of potentially harmful altered 'self' antigens. This concept will be discussed later in relation to mechanisms of antibody formation and immunological tolerance.

In its simplest terms, the specific immunological response to antigenic stimulation may be mediated in two ways, one humoral and the other cellular. The humoral response involves the production of circulating antibodies, and the cellular response involves the stimulation and aggregation of cells of the lymphoid system—mostly small lymphocytes.

Before dealing in detail with the mechanisms underlying these responses and with their relationship to antigenic stimulation and processing, it is necessary to describe the physiological and anatomical components of immunological activity.

For supplementary information in general immunology, the reader is also referred to the following texts:

HOLBOROW E.J. (1967) *An ABC of Modern Immunology*, Lancet.
BURNET F.M. (1969) *Cellular Immunology*, Books 1 and 2. Melbourne University Press.
HUMPHREY J.H. & WHITE R.G. (1970) *Immunology for Medical Students*, 3rd ed. Oxford, Blackwell.

THE IMMUNOGLOBULINS

Although the localization of tetanus antitoxin activity in plasma was recognized as long ago as 1890, the birth of modern immunochemistry was delayed until the advent of serum protein electrophoresis in the 1930s. At that time it was established that nearly all antibodies were contained in the most slowly moving electrophoretic fraction, the gammaglobulins.

NOMENCLATURE

As more complex biochemical and biophysical methods were utilized, it was soon realized that a confusing heterogeneity could be demonstrated within the gammaglobulins. The resultant nomenclature remained a bewildering stumbling block to scientific communication until it was rationalized by international agreement (Bull. W.H.O. 1964).

The term 'immunoglobulin' was proposed as the generic name for all types of proteins with antibody activity. The various immunoglobulin classes (and sub-classes) are sub-divided on a functional basis related to their antigenic activity and denoted by the symbols Ig or γ prefixed to letters indicating their class. The nomenclature and properties of the immunoglobulins are outlined in Table 5.1.

Rowe (1970) has recently reviewed the evolution and current status of immunoglobulin biochemistry. It is of interest that many of the structural and functional characteristics of each class were established by the study, not only of normal antibodies, but of abnormal immunoglobulins (usually homogeneous or 'monoclonal') produced in excess by patients with multiple myeloma, and related conditions called paraproteinaemias.

STRUCTURE

All five classes of immunoglobulins are constructed on the same basic plan—namely four peptide chains, two short and two long, the former each composed of about 214 amino acids, the latter of about twice that number. They are now generally referred to as the light (L) and heavy (H) chains with respective molecular weights of about 25,000 and 50,000, the complete molecule being held together by disulphide bonds, and possessing a molecular weight of 150,000. Whereas most of the molecules of IgG exist in this form, the IgM molecule is made up of five such units covalently linked to each other by sulphur bonds between cystine residues. By careful chemical reduction of these bonds the IgM molecule can be dissociated into its constituent unit of about 150,000 molecular weight. IgD and IgE probably exist mainly in the monomer form (single units) like IgG, but IgA shows a marked tendency to aggregate into polymers.

The H chains are specific for each immunoglobulin class and are the basis on which the different classes are identified immunologically. The L chains, of which two forms exist, are common to all immunoglobulins.

The mutual relationship of the four constituent chains was first established by study of the fragments released by proteolytic enzymes. The basic structure of rabbit immunoglobulin was formulated into a diagram (Porter 1959) which has provided the standard model for human immunoglobulins (Fig. 5.1).

Papain digestion yields two identical Fab fragments and one Fc fragment. The Fab fragments contain specific antigen binding sites and represent the immunologically active portion of the molecule. The Fc fragment has no such activity but contains the specific portions of the H chains which distinguish one immunoglobulin from another. It is also of considerable biological significance because the properties of complement fixation,* cell surface binding capacity, and ability to cross the placenta (in the case of IgG) are inherent in its structure.

Pepsin digestion yields one $F(ab^1)_2$ fragment and splits the carboxy terminal halves of the heavy chains into small peptides. The $F(ab^1)_2$ fragment contains both combining sites of the parent molecule (one on each Fab fragment) and therefore retains the capacity of specific antigen precipitation.

Studies of amino-acid composition of both L and H chains have shown that both are composed of a constant (C) region, where the amino-acid sequence is virtually unchanged from chain to chain within the class, and a variable (V) region where the sequence varies from molecule to molecule. In more detailed terms, then, the antibody specificity of an immunoglobulin molecule is thought to be determined by the particular sequences of amino acids in the V regions of both the L and H chains. Each pair of L and H chains contributes a single antigen binding site.

Immunoglobulin allotypes

Allotypic specificities of immunoglobulin molecules are those antigenic determinants which are different in different groups of individuals within

---

* Complement fixation: complement is a group of co-factors present in fresh normal blood serum which are necessary for the full activity of antibody on cells but which are by themselves inactive. The 'fixation' of these factors in antigen–antibody reactions forms the basis of an *in vitro* test.

TABLE 5.1. The immunoglobulins

Class	Old terminology	Molecular weight	Serum concentration (adult) mg%	Function	Placental transfer	Presence in colostrum
IgM	β2M, γ, M, 19s γ macroglob.	900,000	50–200	Early primary antibody response	No	No
IgG	γ, 7s, γ2, SS	150,000	600–1500	Late primary, and secondary antibody response	Yes (active)	Yes
IgA	β2 A, γ, A	180,000 (and polymers 400,000)	20–500	Mucosal production Local and topical immunity	No	Yes
IgD	—	186,000	1–14	Unknown	No	—
IgE	—	200,000	Trace	Reaginic antibodies Immediate type sensitivity Histamine release reactions	No	—

the same species. This may be demonstrated by injecting animals or, in some clinical circumstances, human subjects, with IgG from another individual of the same species. This results in the production of isoantibodies reacting with donor IgG. A similar isoimmunization may take place naturally between mother and foetus during pregnancy. The situation here is similar to that with blood groups—the ability of one individual to respond by producing antibodies against IgG from another individual depends upon whether they have inherited the same or different antigenic variants of immunoglobulins.

These inherited variants are called allotypes and in man two allotypic systems are known for immunoglobulins—the Gm system and the InV system. The Gm factors, of which twenty-nine are now known, are present on the H chain of IgG and are therefore characteristic of that class only. The InV system on the other hand is confined to L chains and appears in all immunoglobulin classes.

FUNCTIONS

The functions of the various classes of immunoglobulins are summarized in Table 5.1. The presence in man and most mammals of specific immunoglobulin classes implies that each class may have unique properties. This is true to a certain extent, but a functional overlap is apparent particularly with IgG and IgM.

IgG forms the major part of the humoral antibody response. It is distributed equally between the blood and interstitial tissues, and IgG antibodies excel in their ability to react with soluble antigens such as toxins.

IgM, a macromolecule, is predominantly intravascular. It participates in the first phase of the humoral response, and IgM antibodies are also known to be particularly efficient in causing lysis of micro-organisms.

IgA has unique properties of great interest. Though its concentration in plasma is only one-fifth that of IgG, it is the dominant immunoglobulin in the body's secretions such as tears, saliva, colostrum and milk, and in the intestinal and respiratory tracts. This ability to pass through epithelial surfaces, often against a concentration gradient, is associated with the addition of a 'secretory piece' of 50,000 M.W. to the original serum IgA molecule. The concentration of IgA secreting plasma cells at body surfaces suggests that it has a special role in local immunological reactions. Passive local protection for the neonatal intestine via

IMMUNOLOGY 205

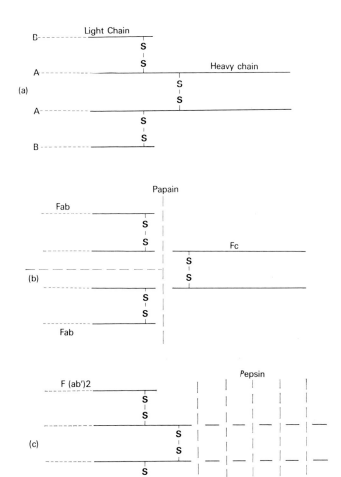

FIG. 5.1. Immunoglobulin structure. Immunoglobulin molecules may consist of one or more such units. (a) Light and heavy chains joined by disulphide (s-s) bonds. The N-terminal amino acids (at dotted ends of chains) may be the regions carrying specific combining sites. A - - - - Variable (V) regions of heavy chains. B - - - - Variable (V) regions of light chains. (b) Papain digestion yielding one Fc and two Fab fragments. (c) Pepsin digestion yielding one F(ab¹)₂ fragment and splitting the carboxy terminal halves of the heavy chains into small peptides.

IgA in colostrum is a modern vindication of breast feeding, at least in the first week of life.

The function of IgD is as yet unknown but recent interest in IgE has identified antibodies in this class as the 'reagins' which have for some time been thought to mediate anaphylactic type sensitivity reactions. IgE is capable of fixing mast cells and other tissue components and thereby mediates histamine release and immediate hypersensitivity. With the development of more sensitive assay methods the serum concentration of IgE is now known to be increased in atopic and allergic conditions.

THE LYMPHOID SYSTEM

THE LYMPHOCYTE FAMILY OF CELLS

The capacity of an individual to respond immunologically depends on the lymphocyte family of cells which includes not only lymphocytes but also plasma cells, macrophages and certain 'active' transitional cell forms called immunoblasts, plasmablasts and lymphoblasts. It is important at this stage to place the members of this family in perspective so that their functional inter-relationships may be understood. It must be remembered, however, that cell morphology as such expresses very little about the extremely heterogeneous distribution and function of these cells within the lymphoid system.

Lymphocytes themselves, distinguishable into three morphological categories, small, medium and large, are collectively the most numerous cells in lymphoid tissue. They constitute a considerable proportion of the nucleated cell population in blood, and practically the whole population of the efferent lymph. It is only in the last 10 years that the lymphocyte has been functionally characterized as the keystone of immunological activity.

Small lymphocytes

The small lymphocyte is probably the commonest mobile nucleated cell in the body. It is 6–10 microns in diameter with a dense nucleus and very scanty cytoplasm. This ubiquitous cell can find its way into all tissues and is therefore ideally suited to its role in immunology.

Small lymphocytes can synthesize new RNA and protein but may require an immunological stimulus before they can divide and make new DNA. Burnet (1969) has called them 'highly mobile carriers of genetic information with no more executive capacity than is needed to stimulate them to take the form of functioning cells'.

Our understanding of the role of small lymphocytes has evolved dramatically from the traditional view of them as a collection of inert end products. These morphologically featureless cells form a complex and dynamic population, capable in their active form of participating in both humoral and cellular immune responses. They are concerned in the initiation of the primary immune response and qualify for the label, immunologically competent cells.* Their role as the 'committed' or 'memory' cells in the mediation of the secondary immune response is now well established.

The cells which have the characteristic morphology of the small lymphocyte in the blood, bone marrow and lymphoid organs are heterogeneous with regard to life span. It is now clear that there is a short-lived population of lymphocytes and a second population which may be extremely long lived.

The former type are found mainly in thymus and bone marrow; although some are found in spleen, very few are present in the lymph nodes and thoracic duct lymph. The lymph nodes and lymph are the residence and pathway respectively of the long lived cells. The lymphocytes of the peripheral blood are composed of both types.

Plasma cells

There is no doubt now that plasma cells are the major producers of antibody. The mature plasma cell has a small, round, eccentric nucleus with a chromatin pattern producing a 'clock face' effect and is well known to the pathologist in various chronic granulomatous lesions.

The immunologist sees the plasma cell as the dynamic end point of a 'plasmacytic' differentiation which probably begins with small lymphocytes and certainly passes through an active undifferentiated immunoblast–plasmablast stage.

The presence of specific antibody can be demonstrated in the plasma cell line using immunofluorescent methods. Each clone, or family, of cells is capable of producing antibodies in only one immunoglobulin class (e.g. IgG, IgM or IgA).

* Immunologically competent cells; see p. 215.

Macrophages

The phagocytes or scavengers of the reticulo-endothelial system (RES) are the blood monocyte and the tissue macrophage. The immunological significance of the monocyte is uncertain, but the macrophage seems to have an important place amongst the immunologically active cells.

Macrophages are widespread in the RES and perform an antigen processing function in the lymph follicles of lymphoid organs. The role of these cells in antibody production is unclear but it has been suggested that, at least for particulate antigens, the macrophage processes antigenic material by complexing it with RNA and presenting it to immunologically competent cells. The macrophage may also have a duty to 'soak up' excess antigens and reduce the antigenic stimulus to a biologically acceptable level; in other words to provide a mechanism for preventing antigen overdosage and immunological paralysis. There is evidence that soluble antigens, particularly in low dosage, are capable of bypassing the macrophage and thereby inducing tolerance.

THE GENESIS OF THE LYMPHOID SYSTEM

Before considering the structure and function of lymphoid organs it is pertinent to review briefly the genesis of the tissues involved. The phylogenetic aspects* of development immunology form an extensive literature and have been the subject of several comprehensive reviews (Miller & Davies 1964, Good & Papermaster 1964, Miller 1966a, Sterzl & Silverstein 1967, Solomon 1970). The following phylogenetic preamble serves only to introduce the discussion of lymphoid development in higher mammals.

The phenomena of adaptive immunity seem to be limited to the vertebrate species. Even so, the most primitive of these, the hagfish, has no organized lymphoid or thymic tissue and no capacity for immune response. By contrast the lamprey, a fellow cyclostome, shows a distinct capacity for humoral and cellular immunity and also demonstrates the presence of lymphoid cells and a primitive counterpart of the thymus.

With the progressive evolution of organized lymphoid tissue in more highly developed fishes and higher vertebrates, more sophisticated

* Phylogenesis: the racial evolution of an animal species.

immunological mechanisms appear. This pattern is repeated in the ontogenesis* of the immune response so that the timing of immunogenesis in the developing vertebrate animal corresponds to the period at which lymphocytes first appear. Thus, for example, the tadpole develops the capacity for skin allograft rejection at 40–50 days post-hatching, the time at which small lymphocytes first appear. Similarly, in the immature opossum, antibody production coincides with the appearance of lymphocytes.

In all vertebrate species lymphocytes appear in the thymus at a very early stage of development. This is closely followed by their appearance in the tonsil in mammals and the bursa of Fabricius (a gut associated lymphoid organ) in birds. Other tissues (lymph nodes, spleen, lymphoid aggregates in the intestinal tract) become lymphoid later in foetal life or even after birth (Table 5.2).

TABLE 5.2. Lymphoid development in various species

Species	Gestation period	Thymus	Spleen	Lymph nodes	Bursa of Fabricius	Gut associated
Chick	21 days	10–12*	—	—	14	>18
Mouse	20 days	14	>20	>20	—	>20
Rabbit	32 days	20	28	36	—	40
Man	40 weeks	12	14–16	14	—	20

* Time intervals shown are from conception to the first appearance of lymphoid tissue (modified after Miller, 1966a).

LYMPHOID EMBRYOGENESIS

The ancestral undifferential cell for the immunological and haemopoietic systems appears to originate in the blood islands of the yolk sac. In the chick embryo, stem cells enter the thymus and bursa of Fabricius and this determines the subsequent lymphopoietic activity in these tissues (Moore & Owen 1967). This system also exists in the mouse and presumably in all mammals; pluripotential cells in the yolk sac of the embryo migrate to the liver, the site of foetal haemopoiesis, and later to the bone marrow, the site of adult haemopoiesis. Stem cells migrate also

* Ontogenesis: origin and development of an individual. Synonymous with embryogenesis in this context.

into the primary lymphoid organ, the thymus, at an early stage and are thence seeded to secondary lymphoid tissue (*vide infra*).

In man, small lymphocytes may be detected in blood at about 7 weeks gestation. Their concentration increases to peak levels of about 7000/mm^3 (sometimes over 10,000/mm^3) by 20–25 weeks and then declines. Playfair *et al* (1963) have postulated that this 'lymphoid surge' of lymphocytes at mid-gestation indicates a significant stage in the development of immunological potential, particularly since it coincides with the maximum rates of development of the foetal thymus and spleen (Fig. 5.2).

THE THYMUS

Embryology

It has been clearly established for some time (see Good & Papermaster 1964) that the thymus becomes a lymphoid organ early in embryogenesis, showing lymphopoiesis by the 9th week and having a well-developed lymphocyte population by 12 weeks.

The thymus develops from the endoderm of the 3rd branchial cleft and in mammals it alone fulfils the criteria for a primary or 'central' lymphoid organ. Early theories of direct thymic lymphopoiesis have been supplemented by the concept that the stem cell precursors mentioned above enter the epithelial anlage of the thymus at 8 to 9 weeks of gestation and rapidly proliferate as small lymphocytes.

Structure

The thymus differs in structure from other lymphoid organs. It has a cortex and medulla, but normally it contains neither germinal centres nor plasma cells. The two main cell-types of thymic tissue are small lymphocytes (thymocytes) and epithelial (reticular) cells. The latter are most numerous in the medulla, but also surround the closely packed masses of lymphocytes that make up the bulk of the cortex.

Cell dynamics

Thymocyte proliferation continues in the complete absence of foreign antigens but the majority of these cells die without leaving the thymus. A minority are exported via the blood to populate the secondary lymphoid organs (lymph nodes, Peyer's patches, spleen and bone marrow) where subsequent proliferation is dependent on antigenic stimulation.

Having once been disseminated from the thymus the small lymphocytes become long lived. They are concentrated in specialized areas of

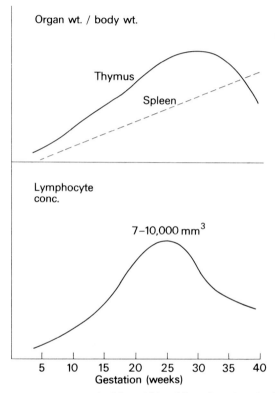

FIG. 5.2. The 'lymphoid surge' of foetal blood lymphocytes at mid-gestation compared with rates of development of foetal thymus and spleen.

lymphoid tissues, namely the para-cortical zones of lymph nodes, and a sheath surrounding the Malpighian bodies of the spleen. These are the 'thymus-dependent' areas and the lymphocytes which inhabit them are termed 'thymus-derived' or 'T-cells'. Throughout life these cells are constantly recirculated along the following pathway: blood stream, postcapillary venule of lymphoid organs, para-cortical area of lymph nodes, lymph, thoracic duct and back to blood stream (Fig. 5.3).

It is convenient to introduce here a second family of circulating small lymphocytes which are 'thymus independent'. These are the 'B cells' (bursal equivalent cells) so named since they subserve the same functions as cells derived from the bursa of Fabricius, a primary lymphoid organ peculiar to birds.

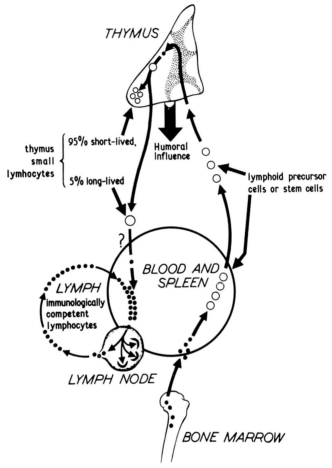

Fig. 5.3. Relationship of the thymus to other lymphoid organs, showing the entry of lymphoid precursor cells into the thymus, and the circulation of long-lived (T) lymphocytes (re-drawn after J.F.A.P.Miller 1966b).

The bursa mediates humoral immunity in birds and although there is no discrete anatomical equivalent in mammals, the bursal equivalent (B) cells are known to give rise to antibody forming cells of the plasmacyte series. These cells are short lived and their origin is presumed to be from bone marrow. They are also present in the thymus but, as mentioned above, they mostly die without leaving it. The functional significance and interrelationships of the T and B cell systems are discussed below.

Immunological significance of the thymus

The thymus in its role as a primary lymphoid organ plays a fundamental role in immunological development and function, and it has been the subject of several monographs and reviews (Good & Gabrielson 1964, Miller 1965, Wolstenholme & Porter 1966, Miller 1966b, Goldstein & Mackay 1969).

The prominence of the thymus during foetal and neonatal life suggests that a major part of its immunological influence occurs at these times. It has been shown in a number of species that neonatal thymectomy results in a suppression of immunogenesis, whereas thymectomy after immunological competence has been achieved is usually without untoward effects. For example, thymectomy in neonatal chicks, mice, rats and hamsters strongly interferes with the capacity of the mature animal to reject foreign skin grafts and to express delayed hypersensitivity. In the rabbit, and in other species such as the dog, in which the lymphoid system has reached a stage of development at birth more advanced than in chicks and rodents, thymectomy has little or no effect on the capacity to perform immune reactions in later life.

Experimental reconstitution of immunological function in neonatally thymectomized animals suggests that at least part of this effect is mediated by an ill-characterized humoral factor elaborated by the thymic epithelium, as well as by the reintroduction of thymic lymphocytes. A clinical parallel for this experimental model can be seen in the syndromes of cellular immune deficiency associated with congenital absence of the thymus in which thymic grafting repairs the immunological defects (Kay 1970).

In summary, the thymus may be said to exercise its control functions by the seeding of small lymphocytes to secondary lymphoid organs as well as by an ill-defined humoral mechanism. A wider role for the thymus in immunological surveillance and elimination of abnormal lymphocyte clones during foetal life has been proposed (see Burnet 1969) and awaits critical analysis.

SECONDARY LYMPHOID ORGANS

Embryology

The thymic initiation and control of the immunological function of secondary lymphoid organs is established. Embryologically these organs lag behind the thymus in their lymphoid development. In man, lymphocytes appear in the connective tissue around lymphatic plexuses by 7–8 weeks gestation but primary lymph node development and lymphopoiesis do not occur until at least 11–12 weeks. Lymphopoiesis is not established in the spleen until after the 16th week.

The development of gut associated lymphoid tissue in man is probably delayed until the second half of gestation (Kay *et al* 1962a); certainly the foetal appendix lacks lymphoid follicles until about 25 weeks gestation (Jones & Kaye 1971). It has been suggested that, in rabbits, the appendix may function as a primary lymphoid organ equivalent to the avian bursa of Fabricius (Archer *et al* 1963). A similar role has been postulated for gut associated lymphoid tissue in other mammalian species (Good *et al* 1966); however the delayed development of this tissue in man until after the capability for humoral antibody production is established (Jones 1969a) argues against such a role. So too does the demonstration by Dwyer & Mackay (1970) of antigen binding by thymic lymphocytes in the 20-week human foetus. These authors suggest that the thymus contributes cells mediating humoral antibody production during foetal life and subserves the function of the bursa of Fabricius.

Structure and function

The primary lymph follicle of lymph nodes or other secondary lymphoid tissue can be thought of as a spongy reticulum of supporting cells whose dendritic processes are in part directed along a light mesh of reticulum fibres. Within this framework lie the relatively closely packed small lymphocytes, in life probably a mobile mass of amoeboid cells.

The lymph nodes form a front-line immunological defence mechanism guarding both vascular and lymphatic systems. The spleen is ideally situated to process blood-borne antigen, especially when this is of particulate nature. Gut associated lymphoid tissue is accumulated in the pharynx (tonsils), the small intestine (Peyer's patches and solitary follicles) and the appendix. It functions presumably as a local immunological guardian of the intestinal tract.

Antigen processing

Examination of secondary lymphoid tissue following the administration of labelled antigen suggests that antigen is trapped in the lymph follicles by dendritic processes of the phagocytic reticular cells (Nossal 1969). Burnet (1969) proposes that this cell complex, in which macrophages are also present, provides a suitable means of stimulating any appropriately patterned cell receptor carried by small lymphocytes. As discussed earlier, the role of macrophages may well be to protect the antigen processing system from too great an uptake of individual antigen. Following stimulation, the lymphocytes undergo differentiation (to plasma cells) within the secondary lymphoid system to produce antibody.

FUNCTIONAL INTERRELATIONSHIPS

Initial exposure to an antigen activates a primary immune response in the manner described above. Subsequent exposure to the same antigen elicits a secondary response which can by-pass antigen processing mechanisms. This response involves a marked and very rapid proliferation of lymphocytes due to an 'immunological memory' mechanism which is primed by the acceptance of the primary stimulus.

The cellular basis for these events must now be considered. The initiation of a primary response involves the interaction of antigen with cells that are normally present in the body and possess potential immunological activity—immunologically competent cells.

This term was coined by Medawar in 1958 to refer to cells of the lymphoid system which have the capacity to undertake a specific immunological response. These cells, immunocytes as Burnet (1969) calls them, or members of the 'immunocyte complex' of Dameshek (1966) may be recognized in experimental models by their ability to restore immunological capability to immunologically deficient animals or to produce graft-versus-host reactions* (Medawar 1963).

ANTIGENIC 'SELECTION' OF IMMUNOLOGICALLY COMPETENT CELLS

The antigen (immunogen) 'selects' and stimulates only cells already capable of making a specific response to it, rather than instructs undifferentiated cells to respond. The specific response of a particular

* Graft-versus-host reaction: see p. 217.

clone, or family, of immunologically competent cells to an antigen is evoked by a specific surface configuration (of antigenic determinants) on the antigen. This concept assumes that the population of immunologically competent cells contains a very wide variety of individual cells each of which is able to respond only to a limited number (possibly only one) of specific determinants. It also depends on availability of the specific antigenic determinants. Availability can be obscured by pre-existing circulating antibody in some instances, a mechanism which has been proposed to explain the apparent antigenic inertness of trophoblast.

The nature of the complementary specific receptor on immunologicaly competent cells is uncertain, although there is some evidence that it could be an immunoglobulin.

CLONAL SELECTION

The antigenic 'selection' mechanisms outlined above indicate in practical terms the acceptance into basic immunology of the clonal selection theory of antibody formation first elaborated by F.M.Burnet (Burnet 1959b). This theory has provided a rationalization of many of the diverse features of the immunological response. It does, however, require the acceptance of an extraordinary genetic capacity in each individual, so that he is potentially able to make the enormous variety of antibodies which are known to be possible; and this must be achieved by individual cells expressing only one or, at most, two of these possibilities.

Implicit also in this theory is the concept of a self-recognition mechanism overlying antibody production to ensure that only foreign antigens are able to elicit immune responses. Thus the ability to recognize 'self' is due to the fact that, once the precursors of antibody-forming cells in foetal life have encountered potentially antigenic material in the tissues, they are thereafter unable to make a specific immune response. The clonal selection theory formed a vital background to the subsequent discovery of acquired immunological tolerance (Medawar 1961).

IMMUNOLOGICAL TOLERANCE

As implied above, the phenomenon of immunological tolerance is of central importance to our understanding of the ontogenesis of the immune response. Immunological tolerance occurs when an individual is rendered incapable of responding immunologically to foreign antigens. Thus, for example, a mouse of a pure line strain, inoculated during foetal life with tissue cells from a mouse of another strain, will subsequently accept an allograft of skin from this second strain as readily as an autograft of its own skin. Despite the apparent simplicity of the reaction in this model, it is now realized that acquired immunological tolerance is a very complex phenomenon which depends on dosage, processing and persistence of antigen, as well as on the chronological age of the subject.

It has been shown experimentally that tolerance may occur in two separate antigen dosage zones; a 'high zone' type of immunological paralysis with large doses of antigen, and a 'low zone' tolerance when antigen in microgram doses is given repeatedly over several weeks. With antigen in the intermediate range a standard immune response is obtained. The immunological pathways leading to tolerance instead of immunity are indistinct but there is some evidence that low zone tolerance depends on antigen escaping macrophage processing.

GRAFT-VERSUS-HOST REACTION

If the donor cells in the mouse experiment described above are immunologically competent lymphocytes from an allogeneic strain* that lacks tissue antigens present in the host, they may react against the immunologically tolerant host, producing a graft-versus-host reaction. Such a reaction can be produced experimentally in many situations where the host cannot destroy a graft of foreign lymphoid cells; its severity depends on the extent of the antigenic disparity involved, and it can lead to a fatal wasting condition—runt syndrome or homologous disease.

The mouse experimental model for immunological tolerance and graft-versus-host reaction is shown in Fig. 5.4.

'T' AND 'B' CELL SYSTEMS

Our understanding of immunological responses has been enhanced by the delineation of immunological competent lymphocytes into two functionally separate lineages, the 'T' and 'B' cell systems mentioned

---

* Allogeneic: of different genetic constitution but still within the species.

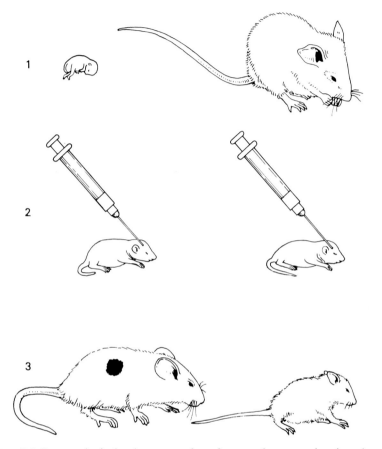

FIG. 5.4. Immunological tolerance and graft-versus-host reaction in mice. At left, spleen cells from embryo mouse of strain B(1) are injected into newborn mouse of strain A(2). Later skin graft from B takes on A(3), showing mutual tolerance of host and graft.

At right, spleen cells from adult mouse A are injected into second newborn mouse B, which develops runt disease (3) because injected cells set up a graft-versus-host reaction. (Reproduced by kind permission of Sir F. Macfarlane Burnet and the Editor. *Scientific American.*)

earlier. The role of these systems in immunological responses has been reviewed by Roitt *et al* (1969) and Mackay (1970).

Both systems contain antigen-sensitive cells probably with specific antibody-like receptors on their surface. B-cells are short lived, thymic independent and can differentiate, proliferate, and mature into plasma cells which synthesize humoral antibody. T-cells are long lived, thymic dependent, and have the following properties (Roitt *et al* 1969):

(1) On contact with antigen (which may have to be macrophage processed) they transform into large blast cells and divide both *in vivo* and in tissue culture (Fig. 5.5).

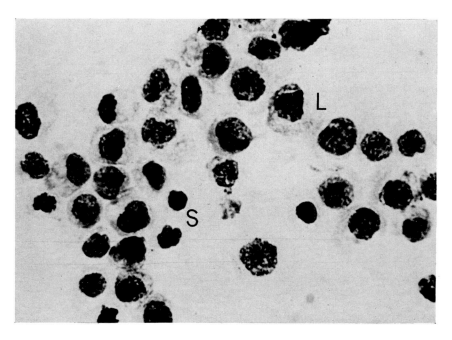

FIG. 5.5. Lymph node lymphocytes cultured for 3 days with phytohaemagglutinin, a non-specific immunological stimulant which transforms thymic dependent small lymphocytes (S) into large lymphoblast-like cells. (L)

(2) They divide further to form an expanded population of primed, antigen-sensitive cells, which provide immunological memory because of their long life span.

(3) They are 'killer' cells which are cytotoxic for graft target cells (i.e. cause graft rejection).

(4) They release a number of soluble factors called lymphokines (Dumonde 1970). These factors, once produced, are active in the absence of the cells which produced them. They include the mitogenic factor, which stimulates DNA metabolism and lymphocytic division; the cytopathic factor which damages monolayers of target cells; the

macrophage inhibitory factor (MIF) which inhibits the migration of macrophages *in vitro*; and transfer factor which confers reactivity on non-sensitized lymphocytes. These characteristics plus lymphocyte cytoxicity form the basis for the phenomena of cell-mediated (delayed) hypersensitivity and cellular immunity.

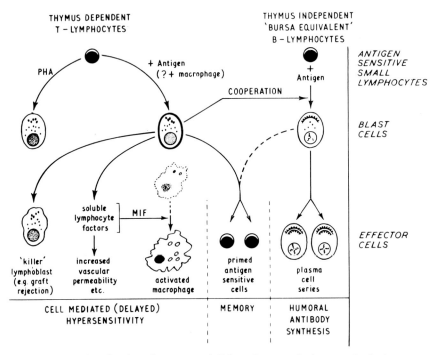

FIG. 5.6. Role of T-lymphocytes and B-lymphocytes in immunological responses. PHA: Phytohaemagglutinin, a non-specific immunological stimulant. MIF: Macrophage inhibitory factor. From I.M.Roitt *et al* (1969) *Lancet* **ii**, 367.

(5) They may co-operate during the immune response to certain ('thymus-dependent') antigens by stimulating the antigen-sensitive B-cell This co-operative function of T-cells at their lymphoblast stage would account for the relationship often observed between delayed hypersensitivity and humoral antibody formation.

The role of T- and B-cells in immunological responses is depicted diagrammatically in Fig. 5.6.

THE BASIS OF TRANSPLANTATION IMMUNOLOGY

An organ grafted from one individual to another encounters the 'histocompatibility barrier'; it is recognized as foreign, the host is sensitized, and the graft rejected. Recognition depends on cell surface histocompatibility antigens* in the grafted tissue. Histocompatibility antigens are also present on leucocytes and they are studied using specific antisera obtained from multiparous women who have been naturally isoimmunized against foetal leucocytes by repeated pregnancies. The complex system of leucocyte antigens is determined genetically at a single locus known as HL-A which represents three closely linked sub-loci with five to eight alleles on each. Despite the complexity of the HL-A system, matching of donors and recipients in renal transplantation significantly influences the survival of an organ graft.

A graft is recognized as foreign by patrolling long-lived (T) lymphocytes which sensitize or augment the predetermined capacity of lymphocytes to react with histocompatibility antigens and so mediate rejection of the graft. Rejection is effected mainly by sensitized cytotoxic lymphocytes which invade and destroy the graft. Humoral antibody may also participate since there is deposition of immunoglobulin and complement demonstrable in vessels and glomeruli of kidneys undergoing rejection (McKenzie & Whittingham 1968).

THE SPECIFIC IMMUNOLOGICAL RESPONSE

From the point of view of the maintenance or disruption of immunological integrity in the individual, the complex spectrum of immunological mechanisms reviewed above can be synthesized into the four elements which constitute the classical specific immunological response.

(1) Antibody production.
(2) Specific cell-mediated immunity.
  (a) Delayed hypersensitivity (e.g. tuberculin reaction).
  (b) Foreign tissue graft rejection.
  (c) Surveillance and control of malignancy.
(3) Immunological memory and the secondary immune response;

* Histocompatibility antigens: (synonym: transplantation antigens) certain cell-bound antigens which are concerned with the capacity of tissue grafts to evoke allograft rejection.

the accelerated immune response (humoral or cellular) occurring on re-exposure to an antigenic stimulus.

(4) Immunological unresponsiveness based on immunological paralysis or tolerance.

SECTION II—IMMUNOLOGICAL ASPECTS OF REPRODUCTION

The combined evolutionary advantages of genetic polymorphism and the development of internal fertilization and viviparity have provided a system of reproduction in which immunological factors are pre-eminent. Commenting on the importance of mammalian reproduction as an immunological model, Medawar (1953) stated that '. . . fertilization and pregnancy are almost the only natural occasions on which the antigenic diversity of the individuals of a species has any opportunity to reveal its consequences'. Reproductive immunology is the study of these consequences. More specifically, it involves the analysis and control of those components of the reproductive process which exhibit features analogous to those of immunological systems.

IMMUNOLOGICAL FACTORS AND FERTILITY

One of the earliest animal models studied by reproductive immunologists was the sea-urchin. Fertilization in this species is controlled by an antigen–antibody-like reaction between a substance called fertilisin on the egg surface and antifertilisin on the surface of the sperm. The ease with which reproduction could be influenced in this lowly animal by manipulating this single mechanism (Tyler 1949) established the rationale for the study of the immunological control of fertility in higher species.

The basic steps in the human reproductive process leading up to implantation are depicted in Fig. 5.7. Theoretically, immunological factors could operate at any stage in this scheme, since the gametes and the fertilized egg, as well as the hormones, tissues and other secretions in their environment are all potentially antigenic, and capable of eliciting an immune response. Cells and proteins with a foreign (in this context, male) genetic constitution may cause isoimmunization in the female partner; isoimmunization being immunity in one individual directed against antigens derived from another member of the same species. On the other hand, autoimmunization* against components ('self' antigens) of the reproductive tract may occur in either the male or the female.

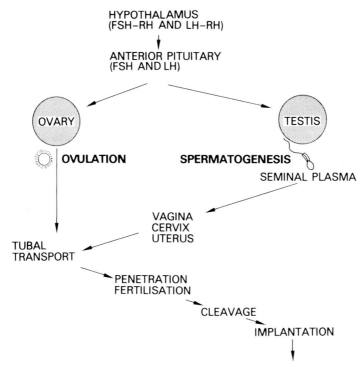

FIG. 5.7. The human reproductive process leading up to implantation showing the steps at which immunological influences might be expected to operate.

The precedent established in the sea-urchin has its counterpart in our accumulation of knowledge in this field. Experimental manipulation in animal models, the study of abnormal reproductive states in man, and the pressures for therapeutic fertility control have unlocked most of the doors to our present, albeit incomplete, understanding of physiological mechanisms.

* Autoimmunization: immunization against constituents (autoantigens) of the individual's own tissues producing autoantibodies as one manifestation.

A synthesis of current knowledge of immunological influences on fertility may be obtained in several recent reviews (Katsh 1967; Behrman 1968; Edwards 1969a, 1970).

THE HIGHER CONTROL OF REPRODUCTION

The first level at which reproduction may be influenced immunologically is at the hypothalamic-pituitary axis (Fig. 5.7). Although at present there is no evidence that naturally occurring immunological factors operate at this level, there has been a longstanding interest in the production of antisera to pituitary hormones. This has stemmed from the development of gonadotrophic hormone immunoassay techniques and from the search for methods of fertility control at a hypothalamic and pituitary level.

It has only been relatively recently that the individual pituitary gonadotrophins have been adequately characterized to allow their successful use in the production of specific antisera in experimental animals. Human Luteinizing Hormone (LH) is now available in purified form, and active or passive immunization of female rodents with this hormone produces marked effects on reproductive function. These include suppression of ovulation, as well as of oestrus and the pseudo-pregnancy reaction, and the prevention of implantation. It is noteworthy that a single dose of passive anti-LH will consistently interrupt pregnancy in the mouse. Experimental immunization against the partially purified Follicle Stimulating Hormone (FSH) preparations currently available produces similar effects.

In the male rat and rabbit, anti-LH immunization neutralizes endogenous hormone, depresses testicular weight and spermatogenesis, and causes involution of the prostate and seminal vesicles. It must be stressed that these studies have not yet progressed beyond the experimental stage and their application to human fertility control is uncertain.

The gonadotrophin releasing hormones (LH–RH and FSH–RH) secreted by the hypothalamus control the cyclic release of LH and FSH from the anterior pituitary (see chapter 2). Whilst the elucidation of their molecular structure is not yet complete, they have been characterized as polyamines of M.W. 1000–2000. Being simple molecules it is reasonable to assume that they will eventually be synthesized, and that in either their natural or synthetic state, they can be bound to protein to render them immunogenic. Active or passive immunization against such hormonal conjugates, allowing selective control of pituitary function, would be a particularly tidy therapeutic principle.

IMMUNIZATION AGAINST MALE REPRODUCTIVE TRACT COMPONENTS

The antigenicity of testicular and seminal components has been exhaustively investigated in a variety of experimental animals over the past 80 years (Behrman 1968). The results of these studies have demonstrated great variations in response to immunization against male reproductive tract components depending upon the preparation of the antigen, the animal used, the method and route of immunization, and the immunological techniques used to evaluate results. While sterility can be induced in experimental animals by heterologous* immunization with male antigens, the application of these principles to human reproductive physiology and pathology has proved frustrating.

In mammalian reproduction, exposure to potentially immunogenic male components may occur in the male himself (auto-immunity) and in the female (isoimmunity). Before considering these situations the nature of the antigenic stimulus must be considered.

THE ANTIGENICITY OF HUMAN SEMEN

Information relating to the antigenic composition of human semen is confusing and incomplete, but suggests that very complex systems exist (Edwards 1969b). Seminal plasma is known to contain at least sixteen antigens contributed by the accessory reproductive organs. Some of these are common to serum (albumin, possibly some immunoglobulins, ABO and Lewis blood groups) but others are specific to semen. Certain of the seminal antigens adhere strongly to testicular spermatozoa and are thought to constitute a major part of the antigenicity of ejaculated sperm.

One of these, called Sperm Coating Antigen, has been extensively investigated (Weil 1961) and has the following properties:

(1) It originates in the seminal vesicle.
(2) It cannot be detected on testicular or epididymal spermatozoa.

---

* Heterologous: (synonym: heterospecific) derived from another species.

(3) It is present in seminal plasma, even in the absence of spermatozoa, such as in azoospermic ejaculates.

(4) It is a β-globulin of large molecular size; it is non-dialysable; it is destroyed by boiling.

Further analysis characterized it as an iron-binding protein with the same electrophoretic mobility as transferrin and it acquired a second name, Scaferrin (Boettcher 1969). There is evidence that Scaferrin, or a closely related protein, is identical with lactoferrin in human milk (Hekman & Rumke 1969), which would help to explain the ubiquitous nature of some anti-sperm antibodies (*vide infra*).

SCA or Scaferrin is strongly antigenic after heterologous injection, but only weakly so in the same species. Along with other seminal plasma antigens, it is species specific, rather than individual specific, which accounts for the cross-reactivity of some seminal antibodies. Antibodies to SCA may agglutinate sperm but usually fail to immobilize them, the latter probably being the more significant influence on sperm function.

Spermatozoal antigens, as opposed to seminal plasma antigens, originate from the germ cell line and are thought to be individual specific and hence more important in provoking a specific immunological reaction. There are at least seven spermatozoal antigens, four of which may be common to seminal plasma. They are located in all segments of the sperm with those situated on the acrosome cap being possibly most important.

Recent evidence establishes the presence of histocompatibility (HL-A) antigens on sperm, an intriguing example of haploid expression of antigenic characteristics (Fellous & Dausset 1969). This finding foreshadows sperm tissue-typing and a more precise approach to studies of anti-sperm immunization.

AUTOIMMUNITY TO SEMINAL COMPONENTS

A well-established manifestation of autoimmunization in the male is the presence of sperm agglutinating antibodies in the serum of a small proportion of infertile men (Table 5.3). Rumke (1969) found that two-thirds of the infertile men with autoantibodies in his series had either obstruction or infection demonstrable in their genital tract.

Antibody formation would therefore seem to be largely related to some disruption of the male genital tract with consequent exposure of the normally 'secluded' testicular antigens to immunologically competent cells. This is a possible explanation for the characteristic autoimmune-like pathology seen in mumps orchitis.

Infertility in men with normal sperm counts and anti-sperm antibodies in their serum is at least partially due to intense agglutination of ejaculated sperm and to a decreased capacity for sperm penetration of cervical mucus.

TABLE 5.3. Incidence of sperm agglutinating autoantibodies in infertile males

Investigator	Infertile males		Fertile males	
	No.	Positive agglutination (%)	No.	Positive agglutination (%)
Fjallbrant (1968)	400	6·8	500	2·6
Rumke (1969)	1913	3·2	416	0·0

ISOIMMUNITY TO SEMINAL COMPONENTS

Studies of couples with unexplained infertility have demonstrated the presence of isoimmunization against seminal components in the female partner to a significantly greater degree than in fertile controls. The first comprehensive investigation of sperm agglutinating isoantibodies was that of Franklin & Dukes (1964). Using a micro-agglutination technique they detected anti-sperm activity in the sera of 79% of nineteen women with unexplained infertility. Subsequent studies using the same technique place this incidence considerably lower (Table 5.4).

The wide incidence range (14–50%) for sperm agglutinating iso-antibodies in unexplained infertility, and their presence in up to 10% of fertile women, and up to 22% of women with organic sterility has prompted the hopeful use of other immunological methods for their detection (Behrman 1968). These include micro-immunodiffusion in agar, haemagglutination, complement fixation and immunofluorescent microscopy. All have proved difficult to apply on their own, which is not surprising when the antigenic heterogeneity of semen is considered. It seems obvious that a combination of two or three tests may be necessary to provide meaningful results.

On theoretical grounds, a sperm immobilization test utilizing complement, either on its own or in combination, would be expected to correlate more closely with fertility status and there is some evidence that this is so (Isojima 1969).

TABLE 5.4. Incidence of positive micro-agglutination tests for sperm iso-antibodies in unexplained infertility

Investigator	Infertility category (primary or secondary, unexplained)	No. of patients	Incidence (%)
Malinak et al (1968)*	Primary	127	48·0
Schwimmer et al (1967a)	Primary	64	37·5
	Secondary	32	50·0
Isojima (1969)	Unstated	34	26·5
Glass & Vaidya (1970)	Primary	77	17·0
	Secondary	45	24·0
Present author (1971)	Primary	52	19·3
	Secondary	19	21·0
Boettcher & Hay (1968)	Unstated	42	19·0
Tyler et al (1967)	Unstated	43	14·0

* This series is expanded from the original study of Franklin & Dukes (1964).

The nature of the isoimmune response

The immune mechanisms involved in isoimmunization to semen have been summarized by Behrman (1968). The female genital tract is well endowed with immunologically competent cells which presumably process seminal antigens and mediate local immune responses. Mononuclear cells in the cervix are known to phagocytose sperm fragments (Fig. 5.8), and tissue antibodies to semen have been demonstrated in the vagina, cervix and uterus of rodents (Paine & Behrman 1969) and humans (Schwimmer et al 1967b).

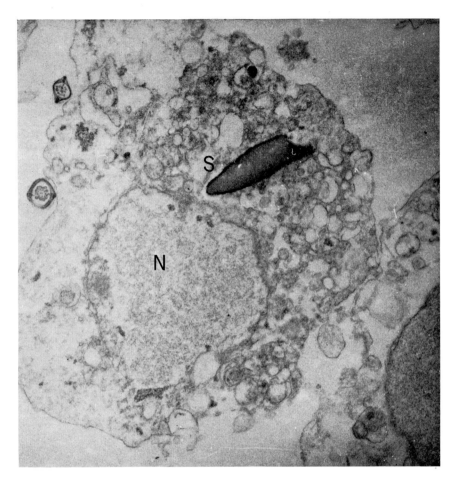

Fig. 5.8. Phagocytosed sperm head (S) in the cytoplasm of a mononuclear cell (macrophage) in the uterine wall of the mouse. Electron micrograph ×7,200. N. Nucleus. (By courtesy of Dr B.L.Reid.)

Immunoglobulins are present in normal cervical mucus and a cytotoxic IgG anti-sperm antibody has been described in the mucus and serum of an infertile woman (Parrish & Ward 1968). At present, however, the precise relationship of local immunity in genital secretions to humoral immunity and ultimately to fertility status is unresolved. Certainly there is no convincing evidence that pre-zygotic selection based on ABO iso-agglutinins in cervical mucus can account for the deficit of group A and B children of ABO incompatible parents (Solish 1969).

The immunofluorescent characterization of humoral anti-sperm antibodies indicates that they are mainly of the IgM and IgA classes (Hjort & Hansen 1971). These workers found that IgM antibody reacted primarily with the acrosome cap and IgG with the tail. It is of interest to mention at this point a dramatic case of acute anaphylaxis following sexual intercourse reported by Halpern *et al* (1967). This was associated with high serum levels of reaginic (IgE) antibodies to a seminal plasma antigen and must thankfully be regarded as a very rare occurrence.

The significance of sperm isoimmunization

Despite the technical problems involved, it is obvious that humoral isoimmunity can be detected in a significant proportion of women with unexplained infertility. That circulating anti-spermatozoal antibodies are in some way associated with infertility is indirectly confirmed by the fact that, if conversion to negative serology can be achieved by occlusion therapy (male partner using condom), then subsequent pregnancy rates of 50–60% may be anticipated (Behrman 1968).

The significance of seminal antibodies in normal women, the extreme heterogenicity of the antibody responses involved, and the precise characterization of local immunity in the genital tract, all await elucidation.

AUTOIMMUNITY IN THE FEMALE GENITAL TRACT

By contrast to the extensive investigation of isoimmunity in the female, little is known about possible intrinsic immunological influences on female reproductive function. The reciprocal relationship of the thymus with certain endocrine organs, including the gonads, is well established. A need for further investigation of the immunological control of ovarian function is suggested by a recent report of inhibition of ovarian development and germ cell maturation in neonatally thymectomized mice. (Nishizuka & Sakakura 1969).

Indeed, there is clinical evidence that some instances of premature gonadal failure (premature menopause) belong to a spectrum of autoimmune diseases involving endocrine glands. Such cases exhibit organ-specific autoantibodies* which react against ovarian granulosa and theca interna cells as well as against steroid producing cells in other organs (Irvine et al 1968). It must be admitted, however, that the majority of patients with premature menopause unrelated to extra-ovarian autoimmunity demonstrate no evidence of ovarian auto-antibodies (Irvine 1971).

No naturally occurring autoimmune reactions have been demonstrated against the proteins of follicular, oviductal and uterine secretions, or against ovarian steroid hormones. The induction of antibodies against specific antigens in these secretions for the purposes of fertility control could theoretically be expected to block the reproductive process in a highly localized and specific manner (Edwards 1970).

PRE-IMPLANTATION AND IMPLANTATION PHENOMENA

It is well known in veterinary practice that mammalian fertilized eggs will survive interspecies transfer to the uterus of a foster mother. This apparent robustness belies the very complex and somewhat precarious immunological situation which exists in the earliest stages of embryonic development.

Although in-bred animals reproduce with varying degrees of success, there is evidence that, in general terms, efficient reproduction depends on a degree of genetic disparity between the fertilized egg and the maternal organism. Notwithstanding this, the survival of the egg, with its paternally derived antigens, in the potentially hostile female genital tract, requires an immunological explanation. The possible operation, at implantation, of maternal immunological selection based on blood groups of sex-chromosome-linked antigens has been postulated but remains unresolved (Kirby et al 1967, Galton 1967, Kirby 1969).

Detailed studies in the mouse have helped to clarify at least a few of the immunological mechanisms involved in zygote transport and implantation (see Simmons & Russell 1967, Galton 1967, Kirby 1970). The salient features only of these studies will be presented here.

* Organ specific antibodies: antibodies directed against components (organ specific antigens) of specific organs. Organ specific antigens may be species specific or have an inter-species distribution on particular cell types.

PRE-IMPLANTATION PHENOMENA

(1) The very early embryo (tubal egg and uterine blastocyst) develops normally when transplanted ectopically (beneath the kidney capsule) into a maternal strain recipient, but is rejected when the recipient has been hyperimmunized against the host. At this early stage, therefore, histocompatibility antigens are expressed (albeit weakly) on the surface of the conceptus.

(2) The zona pellucida which probably exists to protect tubal eggs from mechanical disruption may also have an immunological role. The enzymatic removal of this layer from mouse blastocysts allows their *in vitro* destruction by antibodies (Heyner *et al* 1969). Kirby (1970), however, reported the survival of zona-free blastocysts when transplanted into the non-pregnant uterus of an immunized host.

IMPLANTATION PHENOMENA

(1) Although ectopically transplanted uterine blastocysts are destroyed in hyperimmunized hosts they develop normally in the pseudopregnant uterus of these same hosts. This suggests a protective role for the deciduum at the time of implantation.

(2) Embryonic tissue devoid of trophoblast is rejected when implanted ectopically into a pre-sensitized recipient but pure trophoblast and the intact post-blastocyst embryo are not. This confirms a significant role for trophoblast in the immunological protective mechanism of the embryo, at least beyond the very earliest stages of its development (*vide infra*).

With the acceptance by the mother of the implanted embryo the immunological paradox of placentation comes into full focus. The apposition of maternal and foetal tissues is obvious on light microscopy (Fig. 5.9), and electron microscopy in rodents reveals cytoplasmic interchange (hybridization) at the trophoblast–deciduum interface (Larsen 1961, Potts 1969). However, apart from the apparently non-specific accumulation of leucocytes at the interface in some animal species and on occasions in man, the establishment of the embryo is unaccompanied by manifestations of immunological reactivity.

THE CONCEPTUS AS AN ALLOGRAFT

The conceptus is an allograft since half its antigenic quota is paternal in origin. Moreover, ' . . . it is an allograft which grows, invades, synthesizes, excretes, differentiates and yet remains exempt from the laws of tissue transplantation' (Currie 1969).

The possible mechanisms involved in the survival of the genetically foreign conceptus in its potentially hostile immunological environment have been extensively reviewed elsewhere (Billingham 1964, Douglas 1965, Galton 1967, Simmons & Russell 1967, Currie 1969). The present account will be confined to an outline of their relative importance and implications (Fig. 5.10).

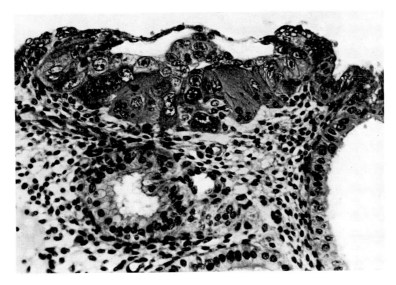

FIG. 5.9. Section through the middle of the Hertig-Rock embryo estimated to be $7\frac{1}{2}$ days old. The trophoblast is in intimate opposition with maternal decidual cells. ×168. (Reproduced by kind permission of Drs A.T.Hertig & J.Rock and the Carnegie Institute of Washington).

The immunological mechanisms are embraced in three main hypotheses:

(a) Maternal immunological unresponsiveness.
(b) Foetal antigenic immaturity.
(c) Immunological separation of mother and conceptus.

Maternal immune reactions will be reviewed below; it can be stated, however, with regard to the second hypothesis, that a general theory of antigenic immaturity in the conceptus is unacceptable. The role of trophoblast in the immunological separation of mother and conceptus has excited considerable interest, and will be considered in some detail later.

MATERNAL IMMUNE REACTIONS

Maternal immunological reactivity has been extensively investigated for a possible role in foetal allograft survival. Mechanisms of immunological unresponsiveness in the mother can be classified as either non-specific or specific in nature.

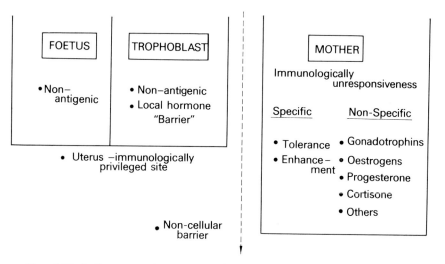

FIG. 5.10. A diagrammatic summary of possible immunological mechanisms involved in the survival of the foetus as an allograft.

NON-SPECIFIC IMMUNOLOGICAL UNRESPONSIVENESS

A general theory of non-specific immunological unresponsiveness during pregnancy is untenable since the capacity of the pregnant animal for allogenic skin graft rejection is, at most, only slightly impaired in a few of the several species investigated. Also any impairment demonstrated has been confined to the initial ('first set') graft but not for a second graft from the same donor which is rejected in the normal accelerated fashion (the 'second set' phenomenon) (Andresen & Monroe 1962).

Hormonal immunosuppression

The possibility that the profound endocrinological changes in pregnancy might influence immunological reactivity has prompted studies of the immunosuppressive properties of pregnancy hormones. It is now well established that plasma levels of both free and bound cortisol are increased in pregnancy and it is of interest to recall that the recognition of the therapeutic potential of cortisone followed the observation by Hench (1938) of what might now be recognized as the corticosteroid-mediated immunosuppressive effect of pregnancy.

Studies of the influence of other steroid hormones on skin allograft rejection in animal models have produced conflicting results. Any possible contribution of the increased oestrogen and progesterone levels to altered immunological reactivity in pregnancy is therefore unresolved.

Human Chorionic Gonadotrophin (HCG) suppresses lymphocyte activity *in vitro* (Kaye & Jones 1970), and there is an inverse relationship between HCG titres and absolute lymphocyte counts in women with malignant trophoblastic tumours (Nelson 1966). Once again, however, the effects of this hormone on graft survival in animals provide no clear cut evidence of immunosuppression.

Lymphoid tissue in pregnancy

Pregnancy is accompanied by an absolute lymphopaenia of varying degree; lymphocyte kinetic studies, however, have shown that the *in vitro* immunological reactivity of maternal lymphocytes is normal (Lewis *et al* 1966, Thiede *et al* 1968). Hypoplasia of the thymus and other lymphoid organs occurs in pregnant rodents and this effect can be reproduced by administering oestrogen, progesterone and HCG in various combinations to intact non-pregnant animals (Nelson *et al* 1967). Similar changes occur in the thymo-lymphatic system in human pregnancy and Nelson & Hall (1965) discovered that pelvic lymph lacked discrete germinal centres, implying depressed immune potential. These changes are all non-specific but probably reflect some degree of systemically mediated immunosuppression in pregnancy.

Placental hormone barrier

Although Simmons & Russell (1966) have shown that transplanted trophoblast will not prolong the survival of adjacent allografts of normal tissue, it has been suggested that the high hormone concentration at the placental interface might have a local influence on maternal cellular immunity (Zipper *et al* 1966, Nelson *et al* 1967, Watnick & Russo 1968). There is some experimental evidence in support of this hypothesis in relation to oestrogen (Zipper *et al* 1966) and oestrogen–progesterone combinations (Watnick & Russo 1968) but not in relation to HCG (Kaye *et al* 1971).

SPECIFIC IMMUNOLOGICAL UNRESPONSIVENESS

An important hypothesis relating to the survival of the foetal allograft is based on the induction in the mother of a specific incapacity to react to paternally derived antigens (Breyere & Barrett 1960). This theory has been sustained in the face of an early study in the mouse (Medawar & Sparrow 1956) which failed to demonstrate appreciable specific unresponsiveness during pregnancy to paternal skin grafts.

The hypothesis and its ramifications lean heavily on the concept that the trophoblast, although forming an immunological barrier between foetal histocompatibility antigens and the maternal immune system, continues to allow limited access of paternally derived antigens to the mother from the time of implantation. The stage is therefore set for the induction in the mother of either a type of immunological tolerance, or of immunological enhancement, both well-recognized mechanisms of immunological unresponsiveness. The experimental data bearing on these mechanisms are extensive and are derived almost exclusively from rodents (see Currie 1969). The general implications of these studies can be summarized as follows:

(1) The mother is probably exposed to small doses of foetal (or more precisely, paternal) antigens on trophoblast cells. This exposure occurs at the placental interface and also by trophoblastic deportation in man. Foetal lymphocytes are known to cross the placenta in several species including man and are another source of foetal histocompatibility antigens.

(2) This low-grade antigenic stimulation provokes a weak humoral antibody response but no true transplantation immunity. Isoantibodies to foetal components have been demonstrated in man and other animals, but their relationship to histocompatibility systems is uncertain.

(3) Immunological tolerance or enhancement are the two efferent mechanisms which may be provoked by these events.

'Parity induced' immunological tolerance

Anderson & Benirschke (1964) first proposed a hypothesis based on the specific abrogation of the maternal immune response by migrant foetal cells. They called this phenomenon, which co-exists with a humoral antibody response, 'immunological inertia', to distinguish it from true immunological tolerance, in which there is no immune response at all.

Immunological enhancement

Immunological enhancement refers to the impaired development of effective cellular immunity against tissue grafts or experimentally transplanted tumours in the presence of circulatory isoantibodies directed against tissue or tumour antigens. This leads to enhanced survival of grafts and enhanced growth of tumours.

Enhancement in relation to the placental allograft implies the localization or coating of antibodies on weak antigenic sites on trophoblast (Karliss & Dagg 1964). This low-grade antigen–antibody reaction has no cytotoxic action but rather protects the cell from attack by cellular immune reactions. This interesting hypothesis derives support from experiments correlating increased trophoblast proliferation in rodents with genetic disparity. The placental hyperplasia seen in hydrops foetalis due to rhesus isoimmunization may also be an expression of immunological enhancement.

THE ROLE OF THE UTERUS

Apart from the possible role of the deciduum at implantation and the influence of local hormone production, there is no experimental evidence that any significant immunological privilege can be ascribed to the uterus (Schlesinger 1962). The successful development of human pregnancies in extra-uterine sites supports this contention, and indeed, immunological isolation of the uterus similar to that of immunologically privileged sites such as the hamster cheek pouch, the anterior chamber of the eye and the brain seems unlikely when the rich vascular and lymphatic supply of the pelvis is considered.

THE ROLE OF THE PLACENTA

The survival of murine trophoblast in ectopic sites in hyperimmunized hosts, and its importance as a protective screen for the post-implantation embryo were mentioned earlier. In the labyrinthine haemochorial placenta of the mouse, and of course in the villous haemochorial placenta of man, the trophoblast is in direct contact with maternal blood; in man for example the area of this contact is of the order of 10 square metres. On anatomical grounds alone it is reasonable to presume that the trophoblast might represent an immunological barrier between foetus and mother. This concept has been investigated extensively and there is now widespread agreement that such a barrier exists; the precise nature and histological location of this barrier, however, remain uncertain (see Simmons & Russell 1967, Billington 1967, Bagshawe 1969, Currie 1969).

THE ANTIGENICITY OF TROPHOBLAST

While there is no doubt that foetal tissues themselves possess a full range of antigens from an early age, considerable confusion remains regarding the antigencity of trophoblast cells (Curzen 1970, Gross 1970). There is some evidence that organ specific, but not transplantation, antigens on trophoblast can elicit an immune response in normal human pregnancy and following experimental immunization in animals. Simmons & Russell (1967) consider that the placenta is immunologically inert and ascribe this to an intrinsic defect in transplantation (histocompatibility) antigen expression on trophoblast. They propose that the placenta therefore acts as a functional buffer zone between the immunological systems of foetus and mother.

PLACENTAL 'FIBRINOID' LAYER

A second explanation of the peculiar immunological properties of trophoblast stems from the demonstration of an electron dense mucoprotein layer separating foetal and maternal tissues in the mouse (Kirby et al 1964) and in man (Bradbury et al 1969, 1970). It is postulated that this layer, which has emerged from studies in the mouse with the rather loose label of 'fibrinoid', provides a pericellular immunological protection for trophoblast cells (Fig. 5.11).

The presence of 'fibrinoid' material in other placental sites (stroma, basement membrane, Nitabuch's layer*) is an unrelated phenomenon

* Nitabuch's layer: Raissa Nitabuch, a nineteenth-century German physician, described a band of fibrinoid material in the decidua basalis and decidua capsularis of the human placenta.

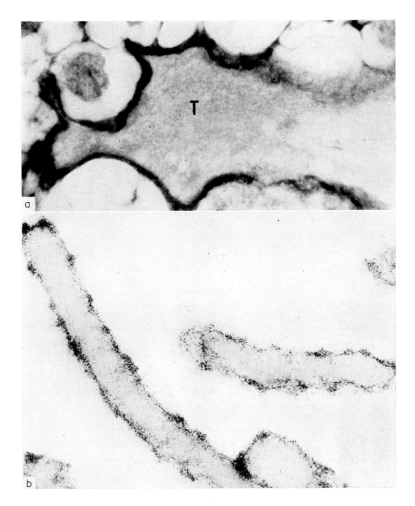

FIG. 5.11. Pericullular 'fibrinoid' (mucoprotein) in the placenta.

Mouse trophoblast. Junction of trophoblast cell (T) with maternal decidual cells showing the darkly staining cell surface coat of mucoprotein. (Electron micrograph reproduced by kind permission of Dr S.Bradbury and the Editor, *Journal Royal Microscopical Society*.)

Human trophoblast. An electron micrograph of syncitiotroblast microvilli in human term placenta. The preparation has been treated with colloidal iron which has aggregated at the surface membrane. This property, taken in conjunction with complementary histochemical staining, indicates the presence of a muco-protein layer at this site. Reproduced by kind permission of Dr W.D.Billington.

and may represent secondary manifestations of an immunological response. A general increase of 'fibrinoid' in genetically disparate mice (Kirby et al 1964) and in certain pregnancy complications in man (Fox 1969) supports this contention.

Initial studies of pericellular 'fibrinoid' in the mouse (Bradbury et al 1964, Currie 1969) and in man (Bradbury et al 1969) indicated that it had a sialic acid-rich mucoprotein structure, and that its removal from trophoblast cells by enzyme digestion rendered them immunogenic. Subsequent and more detailed histochemical characterization of the pericellular trophoblast coat in man has revealed that it is a membrane-bound basic mucoprotein rich in carboxyl and sulphate groups (Bradbury et al 1970). These workers suggest that it is the presence of these groups and not of a high sialic acid content which accounts for the proposed immunological masking effect of the layer.

The implications of these studies must be tempered by discordant data from other mammalian species with non-haemochorial placentation in whom a pericellular 'fibrinoid' layer is absent (Wynn 1967). Indeed, the presence of a complete pericellular layer in the mouse has been disputed by some workers (Simmons et al 1967, Martinek 1971) although difficulties in the preservation of mucosubstances during fixation for electron microscopy may account for some of these discrepancies.

Notwithstanding these technical problems, Kirby (1969) has formulated an interesting hypothesis which, in essence, reconciles the role of enhancement (*vide supra*) with the concept of pericellular mucoprotein as an important but incomplete immunological barrier allowing limited access of foetal antigens to the maternal immune system.

TROPHOBLASTIC PROLIFERATION

The immunological control of trophoblastic proliferation was investigated by Billington (1964) who found that antigenic dissimilarity of mother and conceptus increases placental size in mice. This effect can be augmented or diminished by making mothers immune or tolerant, respectively, to paternal strain antigens (James 1965). The results of human studies on placental size in relation to ABO blood group antigens as histocompatibility markers have been inconclusive (Jones 1968a, Seppala & Tolonen 1970, Toivanen & Hirvonen 1970).

TROPHOBLASTIC TUMOURS

Much of the interest in the immunological aspects of trophoblastic development in man has been stimulated by observations on the unusual biological behaviour of trophoblastic tumours (Bardawil & Toy 1959; Bagshawe 1969, 1970). There is some evidence that the occurrence and behaviour of these tumours is related to immunological compatibility of reproductive partners. Despite this, however, the results of immunotherapy in choriocarcinoma have been disappointing and this approach to treatment has been largely abandoned in the face of the successful results now obtained with chemotherapy. The spontaneous regression of some of these tumours, and the correlation of the initial histology of the host reaction with response to chemotherapy probably reflect immunological influences.

TROPHOBLASTIC DEPORTATION

It has long been known that trophoblastic deportation is an accompaniment of normal pregnancy in man. The magnitude of this phenomenon has been established relatively recently (Bagshawe 1969) and it has been estimated that about 100,000 cells or cell clumps enter the maternal circulation daily. The lack of cellular immunity induced by these cells when they lodge in the lung parenchyma is yet another expression of the immunological inertness of trophoblast.

The possible biological implications of trophoblastic deportation have been considered in relation to the induction of immunological enhancement (p. 237), but the restriction of deportation to man and the chinchilla amongst all the mammalian species studied modifies the immunological importance of this phenomenon.

IMMUNOLOGICAL COMPETENCE OF THE PLACENTA

Cell suspensions prepared from mouse placenta induce homologous syndrome or runt disease* in totally irradiated recipients (Douglas 1965). The precise cell type responsible for this reaction is uncertain but the concept that the placenta itself can react immunologically raises two interesting possibilities. First, it has been suggested (Douglas 1965) that an immunologically competent placenta provides protection for the

* Homologous syndrome, runt disease: see p. 217.

foetus against the influx of maternal cells and potentially injurious antigens. Secondly, Tyler et al (1967) have evolved a theory of parturition involving the rejection of the mother by the placenta—in effect, a graft-versus-host reaction. Although this theory is no more valid than any other relating to the paradoxical transplantation immunology of the placenta, it does imply a foetal role in determining the length of gestation, and is compatible with current concepts in foetal endocrinology (see chapter 4).

FOETAL IMMUNE REACTIONS

The initial emphasis in developmental immunology fell on passively transferred immunity. The mechanisms by which the foetus and neonate acquire passive protection with maternal antibodies have been elucidated in a number of species and the importance of the placental route of transfer has been established for primates, including man (Freda 1961). In the light of the considerable immunological protection conferred by passively transferred antibody, it seemed reasonable to assume that the mammalian foetus and neonate were immunologically inert and unable to initiate active immunity. The apparent inability of the human neonate to respond to the standard paediatric immunizations for up to 6 months after birth further reinforced this view. The lack of immunological response by neonatal laboratory rodents, and the ease with which these animals could be rendered immunologically tolerant of a variety of antigens, also contributed to the generally held concept of a relatively long immunologically inert period during mammalian development.

Over the past decade, studies in developmental immunology have made it necessary to revise these earlier interpretations. It has now been established that, at least in some mammalian species, if the immunological isolation of the pre-natal animal is overcome by suitable antigenic stimuli, then a wide range of immunological activity can be elicited (Good & Papermaster 1964, Miller 1966a, Sterzl & Silverstein 1967).

IMMUNITY IN THE NEONATE

Much of our knowledge of human foetal immunity has arisen by inference from evidence provided by neonatal studies. It has been established that, in a proportion of infants, adequate immunization to routine antigens can be accomplished in the first 3 months of life and often soon after birth. There may be quantitative deficiencies in this response, but since premature and mature infants can respond in equal fashion, it seems that immunological reactivity is not a consequence of chronological maturity alone. Antibody formation has also been detected in the neonatal period in response to a variety of spontaneous antigenic stimuli, including certain viruses and gram-negative enterobacteria. In the light of such evidence it has been realized that the immunological inertness ascribed to the neonatal period is at least partly due to inhibition of active immunity by passively transferred maternal antibody.

Cellular immunity can also be expressed in the newborn period. Delayed hypersensitivity can be induced in a proportion of full-term and premature infants exposed to 2,4-dinitro-fluorobenzene and poison ivy, and Mantoux conversion may occur following BCG vaccination at birth. Skin allografts were applied to newborn infants by Fowler *et al* (1960) who found a normal adult type of rejection at 12–20 days in four out of six subjects including a premature baby of 34 weeks gestation. However, they also demonstrated an impairment of rejection in babies who had previously received fresh blood exchange transfusions from their skin graft donors, suggesting that partial immunological tolerance had been induced by the initial exposure to leucocyte antigens.

It seems, therefore, that the neonate has a capacity for immunity, but that the realization of this may be incomplete in some circumstances. The variable occurrence and nature of immunity in the newborn probably reflects three factors: (a) individual variation in the time-scale of immunological development, (b) the influence and extent of antigenic stimulation, and (c) the state of the antigen processing mechanisms in the neonatal reticulo-endothelial system.

HUMORAL IMMUNITY IN THE FOETUS

The human foetus acquires passive immunity by the active placental transfer of maternal IgG (*vide infra*). Other immunoglobulin classes do not cross the placenta and their detection in foetal serum can be taken as evidence of active synthesis by the foetus. With the advent of sensitive immunochemical techniques of protein quantitation, IgM has been consistently detected in cord serum in late gestation (Jones 1969a). It first appears in foetal serum before mid-gestation, but unlike IgG shows no correlation with either foetal weight or gestational age (Fig. 5.12). The lack of association of IgM with chronological maturity implies its production in response to a variable form of intra-uterine immunological stimulation.

Evidence of humoral immunity in the foetus has also come from van Furth *et al* (1965) who found IgM in serum and cultured spleen cells as early as the 20th week of gestation. Silverstein (1962) had previously described plasma cell proliferation in second trimester foetuses with congenital syphilis.

IgA is undetectable in normal full-term foetuses (McCracken & Shinefield 1965, Johansson & Berg 1967, Jones 1969a) but has been reported in the cord sera of infants with intra-uterine infection (Mc-Cracken & Shinefield 1965). The presence of IgA in these circumstances provides further evidence for an active humoral immune response by the foetus to antigenic stimuli.

IgD is absent from cord blood (Johansson & Berg 1967, Rowe *et al* 1968) but IgE is present in low levels and is probably of foetal origin (Johansson 1968). The presence of ABO group isohaemagglutinins of foetal origin has also been reported in cord blood (Thomaidis *et al* 1969).

CELLULAR IMMUNITY IN THE FOETUS

The direct investigation of cellular immunity in the human foetus is not possible. However, certain higher mammals (e.g. pigs, cattle, sheep and monkeys) provide adequately long gestation periods and sufficiently large foetuses in the latter half of gestation for immunological processes to be initiated and studied. Results obtained in these species show patterns of immunological development which are markedly different from those in small mammals, including the common laboratory rodents.

Skin allograft rejection can be initiated by mid-gestation in the foetal lamb (Schinckel & Ferguson 1953) and monkey (Silverstein 1967). The porcine foetus has also been shown to develop an immune response to cellular antigens introduced after the 80th day of its 120-day gestation period (Binns 1967). Bangham *et al* (1962) demonstrated that 60–100-day foetal rhesus monkeys (gestation period 160 days) were not runted by the injection of nucleated adult bone marrow cells in doses which were known to cause homologous disease in irradiated adult recipients. Such foetuses also showed a capacity for allograft rejection as neonates, when challenged with bone marrow cells or skin from the original donor. These results were taken to imply that the monkey is immunologically competent towards marrow cells by 9 weeks gestation and the authors

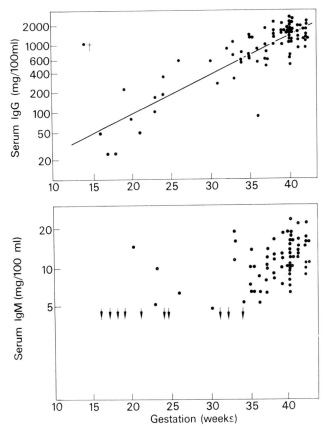

FIG. 5.12. Serum IgG and IgM levels (log scale) plotted against gestational age. (top) IgG (correlation with gestational age). (bottom) IgM (lack of correlation with gestational age). ↓ Undetected at test sensitivity, 5 mg per 100 ml. Reproduced by permission of the Editor, *Journal Obstetrics and Gynaecology British Commonwealth*.

suggest that a similar capability could be expected in the human foetus of corresponding gestational age, that is 16 weeks.

In the human foetus too, it has been found that attempts at intrauterine grafting of allogeneic haemopoetic cells in early gestation for rhesus disease have been uniformly unsuccessful (Kay et al 1962b; and see Jones 1968b). Further data have come from the investigation of the immunological effects of intra-uterine transfusion. Blood used in these transfusions can reasonably be expected to contain immunologically competent cells in the form of small lymphocytes and therefore represents a form of intra-uterine allograft. Since graft-versus-host reaction can occur following the allogeneic grafting of immunologically competent small lymphocytes in an immature animal, the possibility has been raised of the occurrence of this syndrome as a sequel to intra-uterine transfusion. The only recorded case of overt runt disease following intra-uterine transfusion was reported by Naiman et al (1966). This infant had received three intra-uterine transfusions between 27 and 31 weeks gestation. At 8 weeks of age, marked features of runt disease were present, and chromosome studies on peripheral blood, and on thymus obtained at autopsy 2 weeks later, revealed the presence of donor lymphocytes.

Scattered cases of lymphoid chimaerism* involving donor lymphocytes have been reported but in none of these was there clinical evidence of immunological abnormality (Jones 1968b). Classical graft-versus-host reaction is, therefore, a rare (and almost theoretical) complication of intra-uterine transfusion and it seems likely that lymphocytic chimaerism following this procedure is the exception rather than the rule. This suggests that in the great majority of cases the foetus is immunologically capable of eliminating the donor lymphocytes by allograft rejection.

Further evidence of a potential for cellular immune reactions has derived from *in vitro* testing of foetal lymphocytes. The response to immunological stimuli of foetal blood lymphocytes in culture parallels that of adult lymphocytes (Jones 1969b). Thymic, splenic and lymph node lymphocytes from second and third trimester foetuses are also capable of *in vitro* immunological activity (Kay et al 1962b, Kay et al 1966, Jones 1969b).

A further relevant report comes from Wallach et al (1969) who found that lymphocytes from infants born to mothers with *E. coli* bacilluria were capable of a secondary immune response when exposed to these organisms *in vitro*. This suggests that a primary immune response had been initiated *in utero* by transplacental bacterial infection of the foetus.

---

* Chimera: an individual with a double population of cells. The term usually refers to blood or lymphoid elements and implies that one of the cell populations exists under conditions of immunological tolerance.

MATERNAL–FOETAL RELATIONSHIPS

It has been stressed that the placenta performs an important biological function as an immunological barrier between two genetically distinct individuals—the mother and her foetus (Fig. 5.13). This barrier, however, is incomplete since it is known that it may be traversed by proteins, cells and sub-cellular particles with immunological activity. The clinical

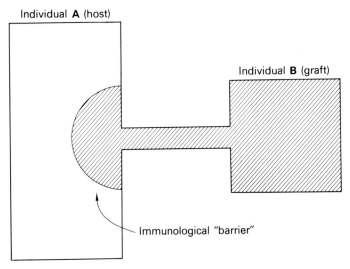

Fig. 5.13. The mother (A) and her foetus (B) separated by the incomplete immunological barrier of the placenta. Reproduced by permission of the Editor, *British Journal of Hospital Medicine*.

consequences of immunological contact between mother and foetus are reviewed elsewhere (Scott 1966; Galton 1967; Jones 1967, 1971), and are beyond the scope of this chapter. What follows here is a brief outline only of the extent and immunological implications of deficiencies in the placental barrier.

IMMUNOGLOBULIN TRANSFER

In the primate, including man, antibody transfer from mother to foetus is almost exclusively via the placenta, and the intestinal route is relatively unimportant (Freda 1962). The routes of transfer in other species are summarized elsewhere (Burnett 1968).

IgG is actively transported across the placenta in man against a concentration gradient. Transfer begins before mid-gestation and foetal levels correlate with gestational age (Fig. 5.12). The ability of IgG molecules to cross the placenta provides passive immunity for the neonate but also makes it possible for maternal isoimmunisation due to foetal antigens, and for maternal autoimmune disease, to affect adversely the foetus and neonate.

MATERNAL AUTOIMMUNE DISEASE AND THE FOETUS

Transient neonatal syndromes have been reported following transplacental passage of organ specific autoantibodies from mothers with thyrotoxicosis, myasthenia gravis, and idiopathic thrombocytopaenic purpura. Apart from the clinical importance of this phenomenon, the utilization of pregnancy as an experimental immunological system may provide important evidence concerning the pathogenesis of other diseases with a suspected or proven immunological basis. For example, in systemic lupus erythematosus, the placental transmission of antinuclear antibody, but not of the clinical manifestations of the disease, confirms the disassociation of this autoantibody from the cause of the condition.

MATERNAL ISOIMMUNIZATION

The clinical sequelae of isoimmunization involving foetal red cell antigens in the rhesus and ABO groups are well known to obstetricians. These disorders have their origin in the bidirectional breaching of the materno-foetal barrier by both cells and antibodies.

The mechanism of rhesus isoimmunization is well established and can be taken as the basic model for all forms of isoimmunization (Fig. 5.14). The sensitization of a rhesus negative mother involves the leakage of rhesus positive red cells across the placental barrier, usually at the time of delivery. There is a variable association between the amount of leakage and the degree of maternal sensitization, and one of the main modifying factors is the effect of foeto-maternal ABO blood group incompatibility, in some cases, in eliminating foetal cells before they immunize the mother. It is also apparent that factors such as antigen dosage, and the occurrence of minor episodes of ante-natal leakage of foetal red cells to the mother, may influence the incidence and extent of maternal sensitization (Clarke 1968),

In an approach based on the simulation of the natural ABO blood group protective mechanism, a method of prophylaxis of rhesus iso-immunization has been evolved which ranks as one of the most significant practical contributions made by immunology to clinical medicine. By the intramuscular administration of high titre anti-D immunoglobulin to unsensitized rhesus negative women within 72 hours of

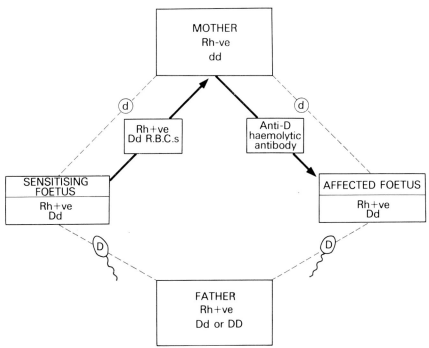

FIG. 5.14. The mechanism of production of rhesus iso-immunization. From J.S.Scott (1966) Brit. med. J., i, 1559.

delivery of a rhesus positive baby, a 95% protection against isoimmunization can be achieved. The small percentage of failures may be related to ante-natal sensitization and this has led to preliminary trials of the use of anti-D prophylaxis during pregnancy.

The historical, clinical and immunological aspects of rhesus iso-immunization prophylaxis have been comprehensively reviewed by Clarke (1968) who examines the postulated immunological mechanisms underlying the spectacular success of passive anti-D administration. The most plausible explanation is that a negative feedback system is operative in which the passive IgG antibody interferes with red cell antigen processing by macrophages, thus preventing the initiation of the immune response.

Isoimmunization may also occur following the foetal-maternal passage of leucocytes, platelets and IgG, producing leucocyte isoagglutinins, platelet isoagglutinins and anti-Gm antibodies respectively. Neonatal clinical sequelae involving these antibodies are extremely rare.

The possible role of trophoblast isoimmunization in immunological enhancement and protection of the foetal allograft has already been considered (p. 237). Experimental studies of the immunization of pregnant animals with trophoblast and other foetal tissues have revealed pathological sequelae which include abortion, toxaemia-like syndromes, and teratogenic effects (Bardawil & Toy 1959, Brent 1965). These findings have stimulated a search, largely unsuccessful, for evidence of isoimmunization as a basis for similar clinical syndromes in man.

FOETAL EXPOSURE TO FOREIGN ANTIGENS

Less well defined than the mechanisms of isoimmunization are the effects on the foetus of transplacental exposure to foreign antigenic material. Foreign material in this context may include lymphocytes, viruses, immunoglobulin, DNA fragments and neoplastic cells of maternal origin, and abnormal clones of cells arising during foetal development. The potential short and long-term effects of the influx of foreign antigens have been reviewed previously (Jones 1971) and may include chimaerism, runting, neoplasia and autoimmunity.

Viewed in this light, it is important to appreciate the role of foetal immunological integrity in the maintenance of normal intra-uterine development. The surprisingly comprehensive immunological armamentarian of the foetus represents a defence mechanism capable of activation by the influx of foreign antigens. This active immune mechanism serves the traditional function of protection from maternally transmitted infectious processes. It also probably plays a role in circumventing harmful long-term effects of exposure to, and subsequent immunological tolerance of, small doses of foreign antigens. At present the nature of the long-term effects is conjectural, but it is reasonable to postulate that normal intra-uterine immunological function will prove to be of fundamental importance to adult welfare, as well as to that of the foetus and newborn infant.

REFERENCES

ANDERSON J.M. & BENIRSCHKE K. (1964) Maternal tolerance of foetal tissue. *Brit. med. J.* **1**, 1534.

ANDRESEN R.H. & MONROE C.W. (1962) Experimental study of the behaviour of adult human skin homografts during pregnancy. *Amer. J. Obstet. Gynec.* **84**, 1096.

ARCHER O.K., SUTHERLAND D.E.R. & GOOD R.A. (1963) Appendix in the rabbit: a homologue of the bursa of the chicken. *Nature* **200**, 337.

BAGSHAWE K.D. (1969) *Choriocarcinoma.* London, Edward Arnold.

BAGSHAWE K.D. (1970) Immunological features of choriocarcinoma. *Brit. med. J.* **4**, 426.

BARDAWIL W.A. & TOY B.L. (1959) The natural history of choriocarcinoma. *Ann. New York Acad. Sci.* **80**, 197.

BANGHAM D.R., COTES P.M., HOBBS K.R. & TEE D.E.H. (1962) An attempt to determine the age at which cellular immunological maturity develops in the foetal rhesus monkey, in *Proc. Int. Symp. on Bone Marrow Therapy in Irradiated Primates,* p. 187. Netherlands, Rijswijk.

BEHRMAN S.J. (1968) The immune response and infertility, in S.J. Behrman & R.W. Kistner (eds.) *Progress in Infertility,* p. 675. Boston, Little Brown & Co.

BILLINGHAM R.E. (1964) Transplantation immunology and the maternal-foetal relation. *New Eng. J. Med.* **270**, 667 and 720.

BILLINGTON W.D. (1964) Influence of immunological dissimilarity of mother and foetus on size of placenta in mice. *Nature* **202**, 317.

BILLINGTON W.D. (1967) Transplantation immunity and the placenta. *J. Obstet Gynaec. Brit. Comm.* **74**, 834.

BINNS R.M. (1967) Bone marrow and lymphoid cell injection of the pig foetus resulting in transplantation tolerance or immunity and immunoglobulin production. *Nature* **214**, 179.

BOETTCHER B. (1969) Blood group antigens in seminal plasma and the nature of sperm-coating antigen, in R.G. Edwards (ed.) *Immunology and Reproduction.* London, IPPF.

BOETTCHER B. & HAY J. (1968) An investigation of a proposed mechanism for infertility based on ABO blood group incompatibility. *Amer. J. Obstet. Gynec.* **100**, 437.

BRADBURY S., BILLINGTON W.D. & KIRBY D.R.S. (1965) A histochemical and electron microscopical study of the fibrinoid of the mouse placenta. *J. Roy. Micr. Soc.* **84**, 199.

BRADBURY S., BILLINGTON W.D., KIRBY D.R.S. & WILLIAMS E.A. (1969) Surface mucin of human trophoblast. *Amer. J. Obstet. Gynec.* **104**, 416.

BRADBURY S., BILLINGTON W.D., KIRBY D.R.S. & WILLIAMS E.A. (1970) Histochemical characterization of the surface mucoprotein of normal and abnormal human trophoblast. *Histochem. J.* **2**, 263.

BRENT R.L. (1965) Effect of proteins, antibodies and autoimmune phenomena upon conception and embryogenesis, in J.G.Wilson & J.Warkany (eds.) *Teratology,* p. 215. Chicago, University of Chicago.

BREYERE F.J. & BARRETT M.I.C. (1960) Prolonged survival of skin homografts in parous female mice. *J. Nat. Cancer Int.* **25,** 1405.

Bull. W.H.O. (1964) Nomenclature for human immunoglobulins. **30,** 447.

BURNET F.M. (1959a) Auto-immune disease. I. Modern immunological concepts. *Brit. med. J.* **2,** 645.

BURNET F.M. (1959b) *The Clonal Selection Theory of Acquired Immunity.* Nashville, Vanderbilt University Press.

BURNET F.M. (1969) *Cellular Immunology.* Melbourne University Press.

BURNETT L.S. (1968) Development of Immune Processes, in A.C.Barnes (ed.) *Intrauterine Development,* p. 229. Philadelphia, Lea and Febiger.

CLARKE C.A. (1968) Prevention of rhesus isoimmunization. *Lancet* **ii,** 1.

CURRIE G.A. (1969) The foetus as an allograft, in G.E.W.Wolstenholme & M. O'Connor (eds.) *Foetal Autonomy. CIBA Foundation Symposium,* p. 32. London, J.&A. Churchill Ltd.

CURZEN P. (1970) The antigenicity of the human placenta. *Proc. roy. Soc. Med.* **63,** 65.

DAMESHEK W. (1966) Immunocytes and immunoproliferative disorders, in G.E.W. Wolstenholme & R.Porter (eds.) *The Thymus: Experimental and Clinical Studies. CIBA Foundation Symposium,* p. 399. London, J.&A.Churchill Ltd.

DOUGLAS G.W. (1965) The immunological role of the placenta. *Obstet. Gynec. Surv.* **20,** 442.

DUMONDE D.C. (1970) 'Lymphokines': molecular mediators of cellular immune responses in animals and man. *Proc. roy. Soc. Med.* **63,** 899.

DWYER J.M. & MACKAY I.R. (1970) Antigen binding lymphocytes in human fetal thymus. *Lancet* **ii,** 1199.

EDWARDS R.G. (1969a) (ed.) *Immunology and Reproduction.* London, IPPF.

EDWARDS R.G. (1969b) Transmission of antibodies across the membranes of the reproductive tract, in R.G. Edwards (ed.) *Immunology and Reproduction,* p. 28. London, IPPF.

EDWARDS R.G. (1970) Immunology of conception and pregnancy. *Brit. med. Bull.* **26,** 72.

FELLOUS M. & DAUSSET J. (1969) Probable haploid expression of HL-A antigens on human spermatozoa. *Nature* **225,** 191.

FJALLBRANT B. (1968) Sperm antibodies and sterility in men. *Acta obstet. gynec. scand.* **47,** Suppl. 4.

FOWLER R., SCHUBERT W.K. & WEST C.D. (1960) Acquired partial tolerance to homologous skin grafts in the human infant at birth. *Ann. New York Acad. Sci.* **87,** 403.

FOX H. (1968) Fibrinoid necrosis of placental villi. *J. Obstet. Gynaec. Brit. Comm.* **75,** 448.

FRANKLIN R. & DUKES C.D. (1964) Antispermatozoal antibody and unexplained infertility. *Amer. J. Obstet. Gynec.* **89,** 6.

FREDA V.J. (1962) Placenta transfer of antibodies in man. *Amer. J. Obstet. Gynec.* **84,** 1756.

GALTON M. (1967) Immunologic interactions between mother and foetus, in K. Bernirschke (ed.) *Comparative Aspects of Reproductive Failure,* p. 414. New York, Springer-Verlag.

GLASS R.H. & VAIDYA R.A. (1970) Sperm agglutinating antibodies in infertile women. *Fertil. Steril.* **21**, 657.
GOLDSTEIN G. & MACKAY I.R. (1969) *The Human Thymus.* London, Wm. Heinemann Ltd.
GOOD R.A. & GABRIELSON A.E. (1964) (eds.) *The Thymus in Immunology*, New York, Harper & Row.
GOOD R.A., GABRIELSON A.E., PETERSON R.D.A., FINSTAD J. & COOPER M.D (1966) The development of the central and peripheral lymphoid tissues, in G.E.W. Wolstenholme & R. Porter (eds.) *The Thymus. Clinical and Experimental Studies. CIBA Foundation Symposium*, p. 181. London, Churchill.
GOOD R.A. & PAPERMASTER B.W. (1964) Ontogeny and phylogeny of adaptive immunity. *Adv. Immunol.* **4**, 1.
GREEN H.N. (1954) An immunological concept of cancer. *Brit. med. J.* **2**, 1375.
GROSS S. (1970) The current dilemma of placental antigenicity. *Obstet. Gynec. Surv.* **25**, 105.
GROSSER O. (1933) Human and comparative placentation. *Lancet* **i**, 999.
HALPERN B.N., KY T. & ROBERT B. (1967) Clinical and immunological study of an exceptional case of reaginic type sensitization to human seminal fluid. *Immunol.* **12**, 247.
HEKMAN A. & RUMKE P. (1969) The antigens of human seminal plasma. *Fertil. Steril.* **20**, 191.
HENCH P.S. (1938) The ameliorating effect of pregnancy on chronic atrophic (infectious rheumatoid) arthritis, fibrositis and intermittent hydarthrosis. *Proc. Mayo Clinic* **13**, 161.
HEYNER S., BRINSTER R.L. & PALM J. (1969) Effect of isoantibody on preimplantation mouse embryos. *Nature* **222**, 783.
HJORT T. & HANSEN K.B. (1971) Immunofluorescent studies of human spermatozoa. *Clin. exp. Immunol.* **8**, 9.
HOLBOROW E.J. (1967) An ABC of modern immunology. *Lancet*.
HUMPHREY J.H. & WHITE R.G. (1970) *Immunology for Medical Students.* 3rd ed., Oxford, Blackwell Scientific Publications.
IRVINE W.J. (1971) Personal communication.
IRVINE W.J., CHAN M.M.W., SCARTH L., KOLB F.O., HARTOG M., BAYLISS R.I.S. & DRURY M.I. (1968) Immunological aspects of premature ovarian failure associated with idiopathic Addison's disease. *Lancet* **ii**, 883.
ISOJIMA S. (1969) Relationship between antibodies to spermatozoa and sterility in females, in R.G. Edwards (ed.) *Immunology and Reproduction*, p. 267. London, IPPF.
JAMES D.A. (1967) Some effects of immunological factors on gestation in mice. *J. Reprod. Fert.* **14**, 265.
JOHANNSON S.G. (1968) Serum Ig ND levels in healthy children and adults. *Int. Arch. Allergy* **34**, 1.
JOHANSSON S.G. & BERG T. (1967) Immunoglobulin levels in healthy children. *Acta. Ped. Scand.* **56**, 572.
JONES W.R. (1967) Immunological diseases and pregnancy. *Hosp. Med.* **1**, 788.
JONES W.R. (1968a) Immunological factors in human placentation. *Nature* **218**, 480.

Jones W.R. (1968b) Immunological aspects of intra-uterine transfusion. *Brit. med. J.* **3,** 280.

Jones W.R. (1969a) Immunoglobulins in foetal serum. *J. Obstet. Gynaec. Brit. Comm.* **76,** 41.

Jones, W.R. (1969b) *In vitro* transformation of fetal lymphocytes. *Amer. J. Obstet. Gynec.* **104,** 586.

Jones W.R. (1971) Immunological factors in pregnancy, in R.R. Macdonald (ed.) *Scientific Basis of Obstetrics and Gynaecology.* London, Churchill.

Jones W.R. & Kaye M.D. (1971) Unpublished data.

Karliss N. & Dagg M.K. (1964) Immune response engendered in mice by multiparity. *Transpl.* **4,** 416.

Katsh S. (1967) Immunological aspects of infertility and conception control, in S.L. Marcus & C.C. Marcus (eds.) *Advances in Obstetrics and Gynaecology,* p. 467. Baltimore, Williams & Wilkins Co.

Kay H.E.M. (1970) The thymus in immune deficiency. *Proc. roy. Soc. Med.* **63,** 715.

Kay H.E.M., Playfair, J.H.L., Wolfendale M.R. & Hopper P.K. (1962a) Development of the thymus in the human foetus and its relationship to immunological potential. *Nature* **196,** 238.

Kay H.E.M., Playfair J.H.L., Wolfendale M.R. & Hopper P.K. (1962b) Evidence relating to the immunological capacity of human foetal tissues. *Proc. Int. Symp. on Bone Marrow Therapy in Irradiated Primates,* p. 169. Netherlands, Rijswijk.

Kay H.E.M., Wolfendale M.R. & Playfair J.H.L. (1966) Thymocytes and phytohaemagglutinin. *Lancet* **ii,** 804.

Kaye M.D. & Jones W.R. (1971) The effect of human chorionic gonadotrophin on *in vitro* lymphocyte transformation. *Amer. J. Obstet. Gynec.* **109,** 1029.

Kaye M.D., Jones W.R., Ing R.M.Y. & Markham R. (1971) The effect of human chorionic gonadotrophin on intrauterine skin allograft survival in rats. *Amer. J. Obstet. Gynec.* **110,** 640.

Kirby D.R.S. (1969) Is the trophoblast antigenic? *Transp. Proc.* **1,** 53.

Kirby D.R.S. (1970) The egg and immunology. *Proc. roy. Soc. Med.* **63,** 59.

Kirby D.R.S., Billington W.D., Bradbury S. & Goldstein D.J. (1964) Antigen barrier of the mouse placenta. *Nature* **204,** 548.

Kirby D.R.S., McWhirter K.G., Teitelbaum M.S. & Darlington C.D. (1967) A possible immunological influence on sex ratio. *Lancet* **ii,** 139.

Larsen J.R. (1961) Electron microscopy of the implantation site in the rabbit. *Amer. J. Anat.* **109,** 319.

Lewis, J., Whang J., Nagel B., Oppenheim J.J. & Perry S. (1966) Lymphocyte transformation in mixed leukocyte cultures in women with normal pregnancy or tumours of placental origin. *Amer. J. Obstet. Gynec.* **96,** 287.

Malinak L.R., Muirford D.N., & Franklin R. (1968) An expanded study of sperm agglutinating antibodies in fertile and infertile couples. Presented at the 24th Annual Meeting of the American Fertility Society, San Francisco. Cited in R.H. Glass & R.A. Vaidya (1970) *Fertil. & Steril.* **21,** 657.

McCracken G.H. & Shinefield H.R. (1965) Immunoglobulin concentrations

in newborn infants with congenital cytomegalic inclusion disease. *Pediatrics* **36**, 933.
MACKAY I.R. (1970) Growing points of immunology. *Post Grad. Med. J.* **46**, 182.
MCKENZIE I.F.C. & WHITTINGHAM S. (1968) Deposits of immunoglobulin and fibrin in human allografted kidneys. *Lancet* **ii**, 1313.
MARTINEK J.J. (1971) Ultrastructure of the deciduotrophoblastic interface of the mouse placenta. *Amer. J. Obstet. Gynec.* **109**, 424.
MEDAWAR P.B. (1953) Some immunological and endocrinological problems raised by the evolution of viviparity in vertebrates, in *Symp. Soc. Exp. Biol.* No. VII, p. 320.
MEDAWAR P.B. (1958) The homograft reaction. *Proc. roy. Soc. Biol.* **149**, 166.
MEDAWAR P.B. (1961) Immunological tolerance. *Science* **133**, 303.
MEDAWAR P.B. (1963) Definition of the immunologically competent cell, in G.E.W. Wolstenholme & J. Knight (eds.) *The Immunologically Competent Cell. CIBA Foundation Symposium*, p. 1. London, J. & A. Churchill.
MEDAWAR P.B. & SPARROW E. (1956) The effects of adrenocortical hormones, adrenocorticotrophic hormone and pregnancy on skin transplantation immunity in mice. *J. Endocr.* **14**, 240.
MILLER J.F.A.P. (1965) The thymus and transplantation immunity. *Brit. med. Bull.* **21**, 111.
MILLER J.F.A.P. (1966a) Immunity in the foetus and newborn. *Brit. med. Bull.* **22**, 21.
MILLER J.F.A.P. (1966b) The function of the thymus in immunity. *Hosp. Med.* **1**, 199.
MILLER J.F.A.P. & DAVIES A.J.S. (1964) Embryological development of the immune mechanisms. *Ann. Rev. Med.* **15**, 23.
MOORE M.A.S. & OWEN J.J.T. (1967) Experimental studies on the development of the thymus. *J. Exp. Med.* **126**, 715.
NAIMAN J.L., PUNNETT H.H., DESTINÉ M.L. & LISCHNER H.W. (1966) Yy chromosomal chimaerism. *Lancet* **ii**, 591.
NELSON J.H. (1966) Alteration of the thymo-lymphatic system in malignant trophoblastic disease, in J.H. Holland & M.M. Hreshchysh (eds.) *Chorio Carcinoma*, p. 19. Berlin, Springer-Verlag.
NELSON J.H. & HALL J.F. (1965) Studies on the thymo-lymphatic system in humans I. Morphological changes in lymph nodes in pregnancy at term. *Amer. J. Obstet. Gynec.* **93**, 1133.
NELSON J.H., HALL J.E., MANUEL-LIMSON G., FRIEDBERG H. & O'BRIEN F.J. (1967) Effect of pregnancy on the thymo-lymphatic system. I. Changes in the intact rat after exogenous HCG, estrogen and progesterone administration. *Amer. J. Obstet. Gynec.* **98**, 895.
NISHIZUKA Y. & SAKAKURA T. (1969) Thymus and reproduction: sex linked dysgenesis of the gonad after neonatal thymectomy in mice. *Science* **166**, 753.
NOSSAL G.J.V. (1969) The cellular basis of immunity, in *Harvey Lectures*, Series 63, p. 179. New York, Academic Press.
PAINE P.J. & BEHRMAN S.J. (1968) Antibody localization in guinea pig reproductive tissues. *Int. J. Fertil.* **13**, 121.

PARRISH W.E. & WARD A. (1968) Studies of cervical mucus and serum from infertile women. *J. Obstet. Gynaec. Brit. Cwlth.* **75**, 1089.

PLAYFAIR J.H.L., WOLFENDALE M.R. & KAY H.E.M. (1963) The leucocytes of the peripheral blood in the human foetus. *Brit. J. Haemat.* **9**, 336.

PORTER R.R. (1959) The hydrolysis of rabbit gammaglobulin and antibodies with crystalline papain. *Biochem. J.* **73**, 119.

POTTS M. (1969) The ultrastructure of egg implantation. *Adv. Reprod. Physiol.* **4**, 241.

ROITT I.M., GREAVES M.F., TORRIGIANI G., BROSTOFF J. & PLAYFAIR J.H.L. (1969) The cellular basis of immunological responses. *Lancet* **ii**, 367.

ROWE D.S. (1970) Nomenclature of immunoglobulins. *Nature* **228**, 509.

ROWE D.S., CRABBE P.A. & TURNER M.W. (1968) Immunoglobulin D in serum, body fluids and lymphoid tissues. *Clin. exp. Immunol.* **3**, 477.

RUMKE P. (1969) Autoimmunity to spermatozoa in men, in R.G. Edwards (ed.) *Immunology and Reproduction*, p. 251. London, IPPF.

SCHINCKEL P.G. & FERGUSON K.A. (1953) Skin transplantation in the foetal lamb. *Aust. J. Biol. Sci.* **6**, 533.

SCHLESINGER M. (1962) Uterus of rodents as site for manifestation of transplantation immunity against transplantable tumours. *J. Nat. Cancer Inst.* **28**, 927.

SCHWIMMER W.B., USTAY K.A. & BEHRMAN S.J. (1967a) An evaluation of immunologic factors in infertility. *Fertil. & Steril.* **18**, 167.

SCHWIMMER W.B., USTAY K.A. & BEHRMAN S.J. (1967b) Sperm-agglutinating antibodies and decreased fertility in prostitutes. *Obstet. Gynec.* **20**, 192.

SCOTT J.S. (1966) Immunological diseases and pregnancy. *Brit. med. J.* **1**, 1559.

SEPPALA M. & TOLONEN M. (1970) Histocompatibility and human placentation. *Nature* **225**, 950.

SILVERSTEIN A.M. (1962) Congenital syphilis and the timing of immunogenesis in the human foetus. *Nature* **194**, 196.

SILVERSTEIN A.M. (1967) Ontogenesis of the immune response, in K. Benirschke (ed.) *Comparative Aspects of Reproductive Failure*, p. 400. New York, Springer-Verlag.

SIMMONS R.L., CRUSE V. & McKAY D.C. (1967) The immunologic problem of pregnancy II. Ultrastructure of isogeneic and allogeneic trophoblast transplants. *Amer. J. Obstet. Gynec.* **97**, 218.

SIMMONS R.L. & RUSSELL P.S. (1966) The histocompatibility antigens of fertilized mouse eggs and trophoblast. *Ann. New York Acad. Sci.* **129**, 35.

SIMMONS R.L. & RUSSELL P.S. (1967) Immunological interactions between mother and foetus, in S.L. Marcus & C.C. Marcus (eds.) *Advances in Obstetrics and Gynaecology*, p. 38. Baltimore, Williams & Wilkins Co.

SOLISH G.I. (1969) Distribution of ABO isohaemagglutinins among fertile and infertile women. *J. Reprod. Fert.* **18**, 459.

SOLOMON J.B. (1970) *Foetal and Neonatal Immunology*. Amsterdam, North Holland Publishing Co.

STERZL J. & SILVERSTEIN A.M. (1967) Developmental aspects of immunity. *Adv. Immunol.* **6**, 337.

THIEDE H.A., CHOATE J.W. & DYRE S. (1968) Pregnancy and the lymphocyte. *Amer. J. Obstet. Gynec.* **102**, 642.

THOMAIDIS T., AGATHOPOULOS A. & MATSONIOTOS N. (1969) Natural isohaemagglutinin production by the foetus. *J. Pediat.* **74,** 39.

TOIVANEN P. & HIRVONEN T. (1970) Placental weight in human foeto-maternal incompatibility. *Clin. exp. Immunol.* **7,** 533.

TYLER A. (1949) Properties of fertilization and related substances of eggs and sperm of marine animals. *Amer. Natur.* **83,** 195.

TYLER A., TYLER E.T. & DENNY P.C. (1967) Concepts and experiments in immunoreproduction. *Fertil. Steril.* **18,** 153.

VAN FURTH R., SCHUIT H.R.E. & HIJMANS W. (1965) The immunological development of the human foetus. *J. exp. Med.* **122,** 1173.

WALLACH E.E., BRODY J.I. & OSKI F.A. (1969) Foetal immunization as a consequence of bacilluria during pregnancy. *Obstet. Gynec.* **33,** 100.

WATNICK A.S. & RUSSO R.A. (1968) Survival of skin homografts *in uteri* of pregnant and progesterone-oestrogen treated rats. *Proc. roy. Soc. Exp. Biol. Med.* **128,** 1.

WEIL A.J. (1961) Antigens of the adnexal glands of the male genital tract. *Fertil. Steril.* **12,** 538.

WOLSTENHOLME G.E.W. & PORTER R. (1966) (eds.) *The Thymus, Experimental and Clinical Studies, CIBA Foundation Symposium.* London, J.&A. Churchill.

WYNN R.M. (1967) Feto-maternal cellular relations in the human basal plate: An ultrastructural study of the placenta. *Amer. J. Obstet. Gynec.* **97,** 832.

ZIPPER J., FERRANDO G., SAIZ G. & TCHERNITCHIN A. (1966) Intrauterine grafting in rats of autologous and homologous skin. *Amer. J. Obstet. Gynec.* **94,** 1056.

6
Liquor Amnii

D. R. ABRAMOVICH

The origin and disposal of liquor has long excited much interest but most information on liquor has been gathered by the clinical and pathological examination of the abnormal states of poly- and oligo-hydramnios. With the increasing use of liquor as an aid in the diagnosis of foetal conditions there is urgent need for a proper understanding of the physiology of this fluid. With new techniques being used to investigate intra-uterine functions much new information will be available; it is expected that both clinical and experimental evidence will need re-evaluation in the light of this new information.

THE VOLUME OF LIQUOR

Sufficient figures are now available to indicate the volume of liquor throughout pregnancy though more data are needed for the interval 20 to 28 weeks.

In early and midpregnancy the volume of liquor has been obtained accurately at hysterotomy by removing the sac intact and measuring the liquor obtained. At this stage of gestation the liquor volume correlates well with foetal crown-rump length and weight. The correlation is linear only when the logs of the liquor figures are used (Figs. 6.1, 6.2 and 6.3); when the actual values are plotted, the graph is curved (Fig. 6.4).

Figs. 6.1 and 6.2 show the correlation between the log of the liquor volume and the foetal crown-rump length and weight respectively. The correlation coefficient, r, is 0·94 with the foetal length and 0·88 with

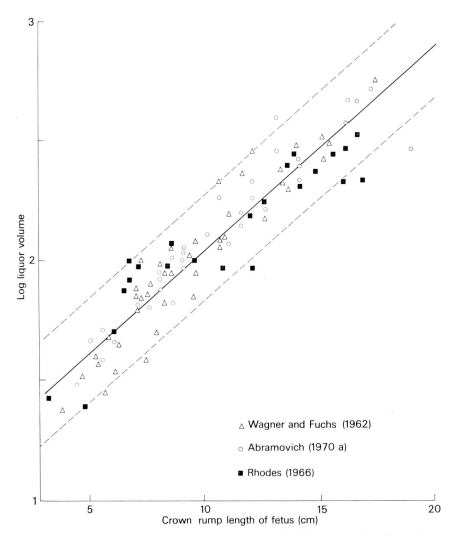

FIG. 6.1. Correlation between crown-rump length of foetus (cm) and log of liquor volume. The equation for the regression curve is $0.078x + 1.18$ where x = crown-rump length in cm and y = log of liquor volume. The correlation coefficient $r = 0.94$. The 95% confidence limits are indicated as broken lines. Modified after D.R.Abramovich (1970a). Data compiled from D.R. Abramovich (1970a), G.Wagner & F.Fuchs (1962) and P.Rhodes (1966).

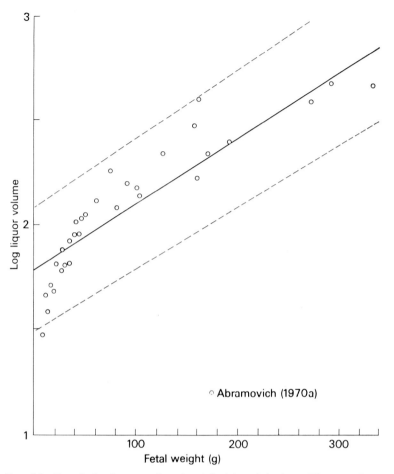

Fig. 6.2. Correlation between foetal weight (g) and the log of liquor volume. The equation for the regression curve is $y = 0.0032x + 1.78$ where x = foetal weight (g) and (y) = log of the liquor volume. The correlation coefficient is 0.88. The 95% confidence limits are indicated as broken lines. Modified after D.A.Abramovich (1970a).

foetal weight. Both these figures are significant (P <0.001 in both cases) and show that the liquor volume increases with increasing foetal size in mid-pregnancy.

Fig. 6.3 shows the correlation between the log of the liquor volume and the menstrual age of the foetus. The correlation coefficient, though still significant (P <0.001) is lower at 0.78 than the other coefficients, suggesting that there is a wider spread of liquor values at each menstrual age. While less accurate than correlation with foetal size, the correlation between duration of amenorrhoea and liquor volume is more valuable clinically. In early pregnancy there is a wide range of reported

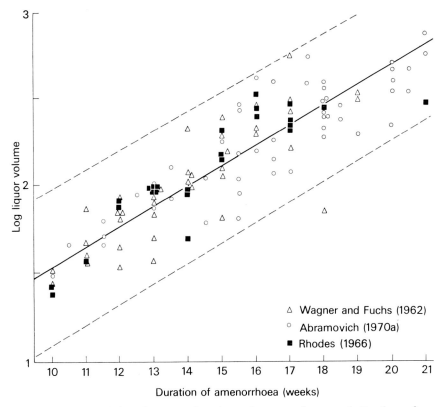

FIG. 6.3. Correlation between duration of amenorrhoea and the log of liquor volume. The equation for the regression curve is $y = 0.11x + 0.36$ where x = duration of amenorrhoea in weeks and y = log of the liquor volume. The correlation coefficient $r = 0.78$. The 95% confidence lines are indicated as broken lines. Modified after Abramovich (1970a). Data compiled from D.A.Abramovich (1970a), G.Wagner & F.Fuchs (1972) and P.Rhodes (1966).

values for each week of pregnancy. At 12 weeks the average value is 58 ml (range 35–103 ml), at 16 weeks, 171 ml (range 159–342 ml) and 20 weeks, 500 ml* (range 226–515 ml). More figures need to be collected before these average values can be regarded as reliable.

* This figure, though derived from the regression equation of Fig. 6.3, may be too high as insufficient figures are available for liquor volume in pregnancies from 20–28 weeks.

The influence of the liquor volume on uterine size in later pregnancy is well recognized and with the wide range of liquor volumes noted in mid-pregnancy it would be surprising if uterine size was not similarly influenced. An example of this influence is seen in Table 6.1.

In case 1, the dates, liquor volume and uterine size (by palpation) were consistent, but in case 2, the uterine size was nearly 1 month ahead of dates. As the liquor volume (among other factors) can influence uterine size in mid-pregnancy, care must be taken when uterine size at first visit is used to assess the gestational age of the foetus (especially if this visit has been delayed beyond the 12th week).

Where gestation is greater than 20–22 weeks the liquor volume is more commonly estimated by dye dilution techniques using either Coumassie Blue (Elliott & Inman 1961) or sodium aminohippurate. The volume in pregnancies of 21–30 weeks is poorly documented but ranges from 300–900 ml (Charles *et al* 1965). More figures are available for pregnancies from 30 weeks till term and later. Gadd (1966) found that the volume ranges from 500–1100 ml reaching a peak at 37 weeks and declining to between 0–600 ml at the 43rd week (average 100 ml). Elliott & Inman (1961) found that the volume is maximal at 38 weeks, averaging 1 litre and that it declines by 145 ml per week till by 43 weeks it is about 250 ml. Beischer *et al* (1969) confirmed these volumes, quoting figures of 484 ml at 42 weeks, 332 ml at 43 weeks and 162 ml at 44 weeks. In a series of 17 cases measured serially after 42 weeks the volume decreased in 14 but in 3 patients the volume increased, the average gain being 80 ml per week. Though these figures confirm the association of oligohydramnios and prolonged pregnancy, Beischer *et al* (1969) feel that the clinical diagnosis of oligohydramnios (and polyhydramnios) is highly subjective and frequently erroneous. Fig. 6.4 shows the suggested average volume of liquor throughout pregnancy.

The volume of liquor does not appear to vary with the type of foetal presentation. Gadd (1966) found that in cases of persistent breech presentation (where version was unsuccessful) the volume of liquor was within the normal limits for the period of gestation.

In pre-eclamptic toxaemia the amount of fluid is decreased, the mean volume being only half that of normal pregnancies of the same gestation (Elliott & Inman 1961). Gadd (1966) divided his series of cases into mild, moderate and severe groups and found a diminution of volume only in the severe toxaemias. There was no association of polyhydramnios and pre-eclamptic toxaemia. In essential hypertension Elliott &

Inman (1961) found a diminished liquor volume while Gadd, in a much smaller number of cases, showed the volume to be normal. More information is needed on this point.

There is conflicting information available on correlation of liquor volume with fetal and placental weights at term. Jeffcoate & Scott (1959) showed that large infants and polyhydramnios were associated when the gestational age was greater than 34 weeks, while Elliott & Inman (1961) found that birth weights of 3·75 kg were associated with the largest volumes and that increase or decrease of 1 kg in the birth weight was accompanied by a considerable decrease in the liquor volume. The correlation of liquor volume with placental weight at term

TABLE 6.1. Influence of liquor volume on uterine size

Amenorrhoea (weeks)	Crown-rump length (cm)	Liquor volume collected (ml)	Estimated liquor volume (ml) (fig. 6.1)	Uterine size (weeks) (by palpation)
16½	11	119	137	16
16½	13	400	204	20

was described by Elliott & Inman (1961) who showed that placentae of 725 g were associated with the smallest volumes while an increase or decrease of 300 g in placental weight was associated with a doubling of the amount of liquor.

Clinically polyhydramnios is associated with:
(a) anencephaly and other abnormal foetuses,
(b) hydrops foetalis, idiopathic or that associated with rhesus iso-immunization,
(c) oesophageal or other gut atresias,
(d) twin pregnancies,
(e) diabetes mellitus, and
(f) chorioangioma of the placenta.

Those cases traditionally associated with lack of foetal swallowing will be discussed in the next section.

There is little information available on the cause of hydramnios in twins. The excess fluid is more commonly present in one sac of uniovular twins, the affected foetus often being abnormal (Hibbard 1959). Gadd (1966) measured the volume of binovular twins at 33 weeks gestation and found 640 ml of liquor surrounding the presenting twin and 855 ml around the second foetus. While the volume of each sac lay within the normal limits of a single pregnancy the total volume of amniotic fluid in that pregnancy closely approached abnormal levels.

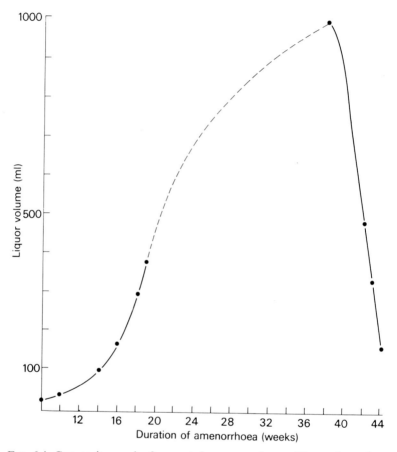

FIG. 6.4. Composite graph of suggested average volume of liquor throughout pregnancy. Volume from 10–20 weeks obtained from regression equation of Figure 3. Maximum value at 38 weeks obtained from P.M.Elliott & W.H.W. Inman (1961) while volumes for prolonged pregnancies obtained from N.A. Beischer et al (1969).

Polyhydramnios is commonly seen in cases of maternal diabetes mellitus, different authors quoting an incidence varying between 1 and 66%. Pedersen & Jorgensen (1954) found excess fluid in 20% of their cases. Though the mean liquor glucose level is higher in diabetics than in normals, Pedersen (1954) found no direct relationship between the concentration of glucose in the liquor and its volume. Similarly though close control of the diabetic state appeared to diminish the incidence of hydramnios, the lack of close association between liquor glucose levels and the amount of fluid led Pedersen and Jorgensen to believe that factors other than those which kept blood glucose levels normal were responsible for the hydramnios. Factors influencing the development of hydramnios in diabetes mellitus are unknown.

Hydramnios is associated with chorioangioma of the placenta, especially if the tumour is large (Siddal 1924). The true cause of the excess fluid is unknown but it is possible that a disturbance of the foeto-placental circulation is implicated in the aetiology.

ORIGIN OF LIQUOR

The origin of liquor in early pregnancy is unknown. The close similarity in composition of liquor and maternal plasma in early pregnancy suggests that the fluid is a transudate of maternal plasma. In an interesting experiment Berman *et al* (1967) removed foetal monkeys from the gestational sac at various stages during pregnancy, leaving the placenta *in situ* and examined the fluid formed 13–79 days later. They showed that the amnion secreted fluid which was isotonic with maternal plasma. Their results were consistent with the hypothesis that early liquor is a transudate.

However, it has been shown by Parmley & Seeds (1970) that foetal skin is permeable to water in early pregnancy and so it can be suggested that while early liquor is a transudate of maternal plasma it can just as easily be a transudate of foetal plasma. The origin of liquor is further discussed under the section 'Turnover of Liquor'.

FOETAL FACTORS WHICH INFLUENCE LIQUOR VOLUME

FOETAL SWALLOWING

It has for long been believed that the foetus swallows *in utero*. Experimental evidence to uphold this belief was provided by early workers, Needham, Haller and Doederlein (quoted by Becker *et al* 1940) who found epithelial cells, lanugo hairs and vernix caseosa in the gut and attributed this to swallowing of amniotic fluid.

Wiener injected calcium ferrocyanide into the amniotic cavity of animal foetuses in 1883 and calculated that the foetuses were regularly swallowing *in utero*. Evidence that the human foetus swallowed *in utero* was provided by Menees et al in 1930 and Ehrhardt (1937).

Rosa (1951) injected inulin into the amniotic cavity and measured its removal from the sac. He estimated that the foetus near term swallowed 500 ml/day. Confirmation of this figure for term foetuses was provided

TABLE 6.2. Foetal swallowing in mid-pregnancy. (Reproduced with permission of the Editor of the *Journal of Obstetrics and Gynaecology of the British Commonwealth*.)

Foetus	Duration of amenorrhoea (weeks)	Foetal crown-rump-length (cm)	Liquor volume (ml)	Volume swallowed in 24 hours (ml)
1	17	10	131	2·0
2	18	12·5	218	4·0
3	18	13	261	4·5
4	18	13·5	232	8·0
5	17	14	223	7·8
6	18	14·5	421	8·4
7	18	15	376	10·7
8	19	15	212	8·5
9	20	16	375	13·2
10	20	16·5	510	11·5

by Pritchard (1965) who was able to measure accurately intra-uterine swallowing using chromium^{51} labelled red cells. He found that the volume of liquor swallowed per 24 hours ranged from 210 to 760 ml. In 1966 he provided figures for foetal swallowing in mid-pregnancy, showing that in 24 hours a 16-week foetus swallowed 7 ml, a 20–21-week foetus 16 ml, and a 28-week foetus 120 ml.

Abramovich (1970) using radioactive colloidal gold, investigated swallowing in ten foetuses aged between 17 and 20 weeks (Table 6.2). The results obtained were in broad agreement with those of Pritchard (1966).

The volume of liquor swallowed per kilogram foetal weight per hour ranged from 1·2 to 2·2 ml/kg/hr while the foetuses of similar age examined by Pritchard swallowed 2 and 2·1 ml/kg/hr.

Examination of gut contents 24 hours after administration of the isotope indicates that foetal peristalsis is established by 18 weeks. The great majority of activity (80%) was found in the lower gut, suggesting that liquor was actively passed down the gut while water was being absorbed.

It was formerly believed that anencephalic foetuses did not swallow. Taussig (1927) was unable to find epithelial cells and lanugo hairs in meconium from anencephalic foetuses while Scott & Wilson (1957), who used amniography in three cases were unable to show any swallowing. Jeffcoate & Scott (1959) whilst believing that the majority of anencephalics did not swallow felt that in those cases where there was no polyhydramnios, the lesion was incomplete and that the foetus was swallowing. James (1961) also felt that the presence of polyhydramnios in anencephaly depended on the level of the brain lesion. Pritchard (1965, 1966) investigated five anencephalics and found no radioactivity in the foetal gut and also concluded that foetal swallowing was absent.

Positive evidence that some anencephalics could swallow was provided by Nichols & Schrepfer (1966) when they performed amniography on three anencephalic pregnancies and showed concentration of radio-opaque dye in the gut. Gadd (1970) also showed that four anencephalics with proven hydramnios were swallowing liquor by injecting barium sulphate into the amniotic sac and X-raying the infants on delivery when barium was seen concentrated in the gut.

Karchmer et al (1969) injected opaque dye into the liquor of normal and anencephalic foetuses and followed its progress through the foetal gut by X-ray. In the majority of cases of anencephaly the appearance of the dye in the large bowel was delayed when compared with normals. Three anencephalics with polyhydramnios showed the contrast material in the gut at the expected time, suggesting that these foetuses were swallowing normally.

Abramovich (1970b) using radioactive colloidal gold measured the volume of liquor and amount swallowed by anencephalic infants and one microcephalic (Table 6.3). The volume of liquor in these pregnancies ranged from nearly 5 litres down to 1·2 litres. Swallowing was evident in three foetuses; two of these swallowed amounts which were in the lower normal range, yet both were from hydramniotic sacs. There was no correlation between the presence or absence of swallowing and the volume of liquor. Because of this Abramovich (1970b) suggested that the presence or absence of swallowing plays little part in the aetiology of polyhydramnios in this condition.

Autopsies were performed in all cases mentioned in Table 6.3. The extent of the brain lesion did not seem to affect swallowing as all the foetuses appeared to have similar defects. The microcephalic foetus was an exception but the presence of a hypothalamus made little difference.

TABLE 6.3. Foetal swallowing in anencephaly. (Reproduced with permission of the Editor of the *Journal of Obstetrics and Gynaecology of the British Commonwealth*.)

Foetus	Period of amenorrhoea (weeks)	Foetal weight (g)	Estimated liquor volume (ml)	Liquor swallowed in 24 hours (ml)
1	43	2444	3732	237
2*	41	2724	2906	240
3	32	908	1225	10
4	31	821	3110	9
5	36	995	2851	5·5
6	38	2386	1978	6
7	38	1758	3290	2
8	36	1594	4913	22
9	36	1816	4156	79

* Microcephaly with encephalocoele.

Examination of foetuses associated with polyhydramnios has led to the current clinical concept that when the foetus cannot swallow or absorb liquor then excess fluid results. Jeffcoate & Scott (1959) qualify this concept by stating that the other mechanisms of production and absorption of liquor must be operating normally before lack of swallowing causes the hydramnios.

There are many published series of cases of hydramnios where the fetal abnormalities have been examined. Of interest are the following two series, though most figures appear to have a regional bias. The figures to be quoted have a high incidence of anencephaly while, for example, Queenan & Gadow (1970) have a high incidence of diabetes mellitus in their cases.

Jeffcoate & Scott (1959) examined 169 consecutive cases of polyhydramnios (clinically diagnosed). Of these there were 54 cases (32%)

where the authors suggested that the foetus was unable to swallow or absorb liquor from the gut. These cases were: anencephaly 32, iniencephaly 2,* gross hydrops foetalis 4, oesophageal and duodenal atresia 15, diaphragmatic hernia 1. If the presence or absence of swallowing in anencephaly is not associated with hydramnios, then there are only 15 cases (9%, the gut atresias) where hydramnios and lack of swallowing could be associated.

Gadd (1970) examined 100 consecutive confirmed cases of hydramnios. In 45 cases mother and infant were normal. There were 9 multiple pregnancies, 7 cases of diabetes mellitus, and 39 cases of foetal abnormality, 24 of which were anencephalic. The remaining foetal abnormalities were 5 cases of oesphageal or upper small bowel atresia or obstruction and 11 other cases (hydrops foetalis 3, hydrocephalus 3, achondroplasia 2, spina bifida and meningocoele 2 and a skin condition 1). In this series there were only 5 cases (5%)—the gut obstructions—in which it can be argued that hydramnios and failure to swallow or absorb fluid are associated.

Not all cases of gut obstruction are associated with hydramnios. Lloyd & Clatworthy (1958) examined a series of infants with congenital obstruction of the gut. There were 53 infants with oesphageal atresia, 7 (14%) of which were associated with hydramnios. There were 35 cases of atresia of upper small bowel, 23 (66%) of which had hydramnios. There is no doubt about the association of hydramnios and gut obstruction yet 86% of the cases of oesophageal atresia and 34% of the cases of small bowel obstruction had normal volumes of liquor.

This type of evidence can be criticized as it is likely that the number of cases with hydramnios was underestimated and the type of oesophageal atresia was not differentiated.

Carter (1960) did distinguish the various types of tracheo-oesophageal fistula when he discussed 228 cases of this abnormality. There were 203 infants where liquor could have reached the stomach, 27 (8%) of which had hydramnios. Of the 25 cases with no communication between mouth and stomach 19 (76%) had excess fluid. There were however, 6 cases (24%) in which liquor could not have reached the gut, yet the volume of liquor was normal. Again this series can be criticized in that the number of cases with hydramnios may have been underestimated.

Nevertheless all this evidence indicates that foetal swallowing plays

* Abramovich (unpublished) examined one case of iniencephaly which was shown to be swallowing normally.

a much smaller part in the control of liquor volume than previously believed.

Pritchard (1966) examined foetal swallowing in three cases of polyhydramnios with normal infants. While one foetus with a liquor volume of nearly 5·5 litres was not swallowing, the two others with volumes of 4 litres each were swallowing 730 ml and 840 ml respectively. These figures are at and just above the suggested normal limits of swallowing and indicate further that the amount swallowed does not influence liquor volume.

The arguments brought forward above to challenge established clinical concepts can themselves be criticized. In anencephaly, conditions other than lack of swallowing have been advanced as the cause of excess fluid. Gadd (1970) suggests that a transudation from the exposed meninges may be a factor but Potter (1961) thought this explanation unsatisfactory. Benirschke & McKay (1953) pointed out that the rudimentary or distorted brain of anencephalics is almost always covered with a collagen membrane. In none of the infants investigated by Abramovich (1970b) was there an obvious vascular network or choroid plexus at the base of the brain. Benirschke & McKay (1953) suggested that foetal polyuria may contribute to the hydramnios as anencephalic foetuses lack antidiuretic hormone but there is evidence that the foetal kidney does not respond to this hormone (Vernier & Smith 1968).

However, it is still difficult to argue that foetal swallowing is of little relevance in the control of liquor when the foetus swallows between 210–760 ml/day with a liquor volume of 500–1500 ml. It may be also claimed that the published investigation of amounts swallowed by foetuses *in utero* only lasted 20–96 hours, a small unit of time in the length of the pregnancy. It can also be suggested that if a foetus swallowing say 200 ml/day subsequently swallows 100/ml day for a fortnight, then polyhydramnios would develop, providing all other factors remained constant.

The explanation of these rather confusing facts may lie in an imbalance of the exchange rates of water between mother, foetus and liquor. This concept will be discussed in a later section.

Foetal voiding

It has been believed since Hippocrates that the foetus voided *in utero* and evidence has been slowly accumulating to confirm this point.

Hewer (1924) after differentially staining proximal and distal renal tubules concluded that the human kidney functions by 12 weeks while Gersh (1937) believed that tubular function begins at 9 weeks. Cameron & Chambers (1938) showed tubular secretory activity in a 14-week foetus.

There is little doubt that the foetus can void *in utero*. The presence of urine in the foetal bladder was noted by Makepeace *et al* (1931) and Wagner & Fuchs (1962) in foetuses aged 3–5 months and by Jeffcoate (1931) in term foetuses. Abramovich (1968) found urine in the bladder of normal foetuses aged from 11 weeks while Kjellberg & Rudhe (1949) and Tahti (1966) showed renal activity in foetuses of 12–20 weeks when they injected contrast medium into these foetuses and found that the kidneys were outlined by the opaque medium.

Many foetuses pass urine at delivery and Tausch (1936) found up to 44 ml in the bladder of neonates. None of this evidence proves that the foetus voids *in utero* but analogy from animal investigations makes it highly likely. Alexander *et al* (1958) measured the urine flow in the foetal lamb. They calculated that it voided 0·14 ml/min at 61 days gestation, 0·64 ml/min at 117 days and 0·14 ml/min near term. Chez *et al* (1964) catheterized a rhesus monkey foetus *in utero* and noted that it voided 5 ml/kg/hr.

Few figures are available for the volumes of urine voided by the human foetus *in utero*. Abramovich (1970) calculated that the 18-week foetus voided 7–17 ml/24 hr. These figures were based on a measured bladder capacity of 0·3–0·7 ml and an estimated frequency of voiding averaging once per hour. These figures remain speculative and need confirmation. Rosa (1951) calculated that the term foetus in labour produces 43 ml/hr but again the figures need confirmation.

Urine produced by the foetus *in utero* is hypotonic. Table 6.4 shows the osmolality, electrolyte and urea concentrations for urine from 12 foetuses aged from 17–21 weeks (Abramovich 1970b) and from 12 foetuses at term (McCance & Widdowson 1953). The urine is hypotonic compared with maternal and foetal plasma and liquor at similar gestation, the sodium and chloride concentrations are lower while the average urea concentration is higher. The urine stays hypotonic till term but there is a drop in the sodium concentration from 68 to 44 mEq/l and that of chloride from 66 to 41 mEq/l from mid-pregnancy till term, this drop being significant ($P < 0·01$ in each case). This suggests more effective reabsorption of sodium and chloride at term. The concentration of urea in urine is low in mid-pregnancy but rises as the foetus matures suggesting more effective renal tubular function.

As the foetal urine is hypotonic throughout pregnancy, many writers have argued that it is the addition of this hypotonic urine to a liquor which is isotonic (or nearly so) with plasma in early pregnancy, which converts the liquor to its hypotonic state at term. It is furthermore argued that the bulk of the liquor in the third trimester is composed of this hypotonic urine (Lind *et al* 1969) but the position is complicated by the dynamic turnover of the constituents of liquor.

TABLE 6.4. Osmolality and electrolyte concentration of 12 foetal urines in mid-pregnancy (Abramovich 1970b) and at term (McCance & Widdowson 1953). (Reprinted with permission of the Editor of the *Journal of Obstetrics and Gynaecology of the British Commonwealth*.)

	Osmolality (m Osm/kg)	Na+	K+ (mEq/l)	Cl−	Urea (mg/100 ml)
Mid-pregnancy (17–21 weeks)					
range	121–204	44–96	1·4–5·0	53–90	11–68
average	144	68	3·2	66	28
term					
range	97–232	13–66·5	1·2–16·8	20–61	48–252
average	137	44	4·7	41	102

Clinical findings have further strengthened this belief of the importance of foetal urine in the formation of liquor. The association of oligohydramnios and malformation of the foetal urinary system (renal agenesis or absent or non-patent urethra) was reviewed by Schiller & Toll (1927) when they noted 15 cases of urinary malformation out of 57 reported cases of oligohydramnios.

Jeffcoate & Scott (1959) collected data from 295 cases of renal agenesis and severe dysplasia, including 232 cases of Davidson & Ross (1954). There was firm or presumptive evidence of oligo- or anhydramnios in 100 cases; in the majority of other cases there was no statement about the volume of liquor. While there is no doubt about the association of renal agenesis and oligohydramnios, this association does not prove that foetal urine provides the bulk of liquor in the third trimester.

Normal volumes of liquor can be present even when foetal kidneys are absent (Schiller & Toll 1927, Gowar 1935, Sylvester & Hughes

1954). Nichols (personal communication) has had 3 cases, one each of phimosis, penile agenesis and renal agenesis, each with a normal volume of liquor. Goodlin & Lloyd (1968) described a further case which when delivered by caesarean section, appeared to have a normal liquor volume. Post-mortem showed renal agenesis but the lungs were not hypoplastic.

Excess fluid can be present with renal agenesis. Jeffcoate & Scott (1959) described one such case with renal agenesis and iniencephaly while Bain & Scott (1960) described two cases of renal agenesis in iniencephalic foetuses, both with polyhydramnios, and one case of renal cystic dysplasia, hydrocephalus and a normal volume of liquor.

It should be remembered that foetuses with renal agenesis are abnormal. They have a characteristic Potter facies which is unlikely to be caused by uterine pressure (Hibbard 1962); they have hypoplastic lungs (Potter 1961) and the foetal membranes show *amnion nodosum* (Jeffcoate & Scott 1959). These abnormalities may play as large a part in causing liquor deficiency as the absence of the kidneys.

Whilst the foetus voids *in utero* from early in pregnancy, it is unlikely that foetal urine contributes the bulk of the water which comprises the liquor near term. However, there are insufficient figures available for the volume of urine voided by the foetus at various stages of pregnancy and insufficient knowledge on the pathways used in the exchange of the liquor components to confirm this belief.

Respiratory tract

There is confusion over the importance of the lungs as a source of liquor amnii and indeed over the physiological functions of the foetal lung in utero (see also chapter 8).

A fluid-filled trachea is normal during foetal life. The source of this fluid is in doubt. It may be present because the foetus breathes *in utero* or because the alveolae and lungs produce the fluid; the presence of tidal flow in the trachea and its relevance is uncertain.

Ahlfeld (1888) and Reifferscheidt (1911) believed that chest movements occurred *in utero* in near-term infants while Bufe (1936) was unable to demonstrate them in mid-term foetuses. Reifferscheidt & Schmiemann (1939) described tidal flow in the bronchial tree together with inhalation into the lungs following injections into the liquor with fluoroscopy pre-operatively and microscopy post-operatively. Other workers have used radio-opaque substances to study respiration; Windle *et al* (1939) were unable to show chest movements or tidal flow using diodrast while Davis and Potter (1946) who used thorotrast found this substance in the lungs of both mid-term and term foetuses following its injection into the liquor; in contrast Liley (1963) used urographin and failed to detect due in the respiratory tract.

The cause of these conflicting results is uncertain but may depend on such factors as type of substance injected into uterus, degree of foetal oxygenation, and length of pregnancy.

Dawes *et al* (1970) investigating foetal lambs, showed that irregular rhythmic respiratory movements were always present in the last third of pregnancy, associated with changes in the EEG. There were tidal movements of fluid in the trachea but these appeared to be too small to wash out the dead space.

Evidence has been produced by Reynolds (1953) and Dawes (1954) using foetal lambs, and Jost (1954) with foetal rabbits, that fluid is secreted by the respiratory tract. Suzucki & Plentl (1968) using foetal monkeys which were kept *in utero*, criticized these results suggesting the conditions used were unphysiological.

Adams *et al* (1967) and Towers (1968) have demonstrated in foetal lambs near term that there is a sphincter at the laryngeal outlet which periodically opens, allowing fluid to be discharged to the nose and pharynx where it is swallowed. There is a positive pressure of 10–15 mm H_2O in the trachea and while the source of the fluid is unknown, at least some of it comes from the alveoli (Adams *et al* 1967).

The amount of tracheal fluid produced by the foetal lamb is 50–80 ml/day (Goodlin, personal communication) but it is not known how much of this fluid reaches the amniotic fluid. Dawes (personal communication) does not feel that it influences the liquor volume.

It should also be realized that most of this work has been performed on animals. However, it is likely that the broad biological principles apply to both man and animals though the details may differ.

A consensus would favour the view that fluid is produced by the lungs and while a small amount may reach the amniotic sac, it is unlikely to influence greatly the liquor volume.

Turnover of liquor

Vosburgh *et al* (1948) showed that liquor was a pool of rapidly exchanging water and electrolytes. They found that the water of the amniotic fluid was completely replaced every 2·9 hours and that water was exchanged from mother to foetus in large quantities, reaching a maximum of 3·6 litres/hour (Hellman et al 1948).

These findings were confirmed by Hutchinson et al (1955, 1959) who also suggested that at term the quantity of water transferred between liquor and mother was constant at 26 mols/hr (468 ml of water) and was independent of the volume of the liquor. This means that while the same amount of water is exchanged per unit time in patients with polyhydramnios and oligohydramnios, the time taken for half the water in the sac to be exchanged is different. Just as water is exchanged, so are sodium and potassium at rates of 12 mEq and 0·5 mEq/hr respectively.

Hutchinson et al (1955) believed that the cause of abnormal liquor volumes lay in an imbalance of the exchange rates, as an accumulation of only a few millilitres per hour over a period of weeks would produce polyhydramnios. This explanation appears reasonable in cases of anencephaly with polyhydramnios where foetal swallowing (or the lack of swallowing) plays little part in the aetiology of the hydramnios.

The differing exchange rates of water and electrolytes suggest that amniotic fluid is not a transudate or ultrafiltrate of maternal plasma. If liquor was an ultrafiltrate the water and electrolytes would be exchanged *en bloc* but they exchange at their own characteristic rates. In early pregnancy electrolytes are found in concentrations compatible with the Gibbs-Donnan equilibrium theory suggesting that at this stage liquor is a dialysate of maternal plasma (Seeds 1968).

Hutchinson et al (1959) thought that the foetus plays an important role in the transfer of water from the liquor to the maternal system. They suggested that at least 25% and probably more than 50% of the water transferred out of the liquor was accomplished via the foetus. In pregnant monkeys in which samples of foetal blood can be obtained without disturbing the foeto-placental circulation, at least 75% of the water leaving the amniotic sac was transmitted via the foetus.

In hydramnios in the human, the foetus plays a much smaller role in the water exchange. The situation is similar to early pregnancy in that only 16% of the water leaves via the foetus though the total amount of water leaving the sac is the same in normal and hydramniotic pregnancies (Hutchinson et al 1959).

These workers provided figures for the water exchange between mother, foetus and liquor. They found that at term water exchanged between:

(1) Liquor and foetus at 150–174 ml/hr and 0–298 ml/hr in the reverse direction.

(2) Liquor and mother at 235–259 ml/hr and 97–433 ml/hr in the reverse direction.

(3) Mother and foetus at 3540–3700 ml/hr and between foetus and mother at 3650–3720 ml/hr.

Tervila (1964) confirmed some of the findings of Hutchinson et al (1955, 1959) when he noted that the cord circulation and live foetus normally accounted for about half of the total transfer of water from maternal circulation to the amniotic fluid. Gillibrand (1969a) measured the transfer rate of water out of the amniotic sac. The rate of increase was linear between 14 and 26 weeks, appeared to reduce as term approached and there was a reduction in the transfer rate in prolonged pregnancy.

Hutchinson et al (1959) suggested that a net flow of water from mother to foetus to liquor and back to mother must occur. Scoggin et al (1964) confirmed this concept when they measured the total solute concentration in maternal and foetal blood and in liquor.

This work on the exchange of water between mother, foetus and liquor might provide an explanation for an apparently confusing situation; it has been suggested that foetal swallowing is unimportant in the control of liquor volume yet it is known that the foetus swallows between 210 and 760 ml/day in a liquor volume of 500–1500 ml.

The foetus swallows about 20 ml/hr and it can be seen that an imbalance in the exchange rates between liquor and foetus and liquor and mother could easily outweigh the 20 ml per hour of foetal swallowing. However, swallowing removes liquor in bulk while any imbalance in the water exchange would need to be finely matched by a similar imbalance in electrolyte exchange. Hibbard (1962) has pointed out that with a turnover rate of water in the liquor of 10 litres per day, a positive imbalance of 1% would result in the accumulation of 100 ml/day and polyhydramnios would ensue rapidly.

Work on water exchange measured by isotopes has been criticized on clinical (Jeffcoate & Scott 1959), technical (Tervila 1964) and biophysical grounds (Garby 1959, Seeds 1968). Seeds (1970) has suggested that isotopic water measures only diffusion and that osmosis causes a much greater rate of water transfer. Seeds (1968) states that transfer of water across the membranes *in vivo* occurs by a process of bulk flow, which he defines as a transfer of solvent water across a semi-permeable membrane by either hydrostatic or osmotic force.

While osmotic and hydrostatic forces play some part in water exchange, the foetus, as suggested by the use of isotopic water, seems to play a central role. But no theory linking all these biophysical forces and allowing for foetal swallowing and voiding has yet been put forward.

FLUID FLOW *IN UTERO* IN MID-PREGNANCY

Figures are available for foetal swallowing and voiding in mid-pregnancy while the daily increase in liquor can be calculated from the regression equation of Fig. 6.3. Table 6.5 shows the average liquor volume with the daily increase from 17–19 weeks.

The daily increase in liquor volume at 18 weeks is 10 ml (range 7–13 ml) while the foetus swallows 4–11 ml/day. Thus the net volume that is added to the liquor lies between 11 and 24 ml/day.

The main part of this fluid is provided by foetal urine as the foetus is thought to void between 7–17 ml/day. The source of the rest of the fluid, which can be from 4–7 ml/day may be the lungs, skin, amnion or umbilical cord or it may come from an imbalance of the exchange of water and electrolytes as suggested by Hutchinson *et al* (1959). No figures are yet available for movement of fluid inside the uterus at other stages of pregnancy.

PATHWAYS OF EXCHANGE AND THEIR MORPHOLOGY

D.R.ABRAMOVICH and E.W.PARRY[*]

Many experiments have shown water and electrolyte exchange between mother, foetus and liquor but the pathways by which these exchanges occur are less well defined.

Possible routes of exchange are:

(1) Placental and reflected membranes.
(2) Umbilical cord.
(3) Foetal epidermis.

While most experimental evidence points to the reflected amnion and chorion serving merely as a pathway for exchange processes, some histological and clinical evidence suggests a secretory function. Taussig

* E. W. Parry, Senior Lecturer in Histology, University of Liverpool.

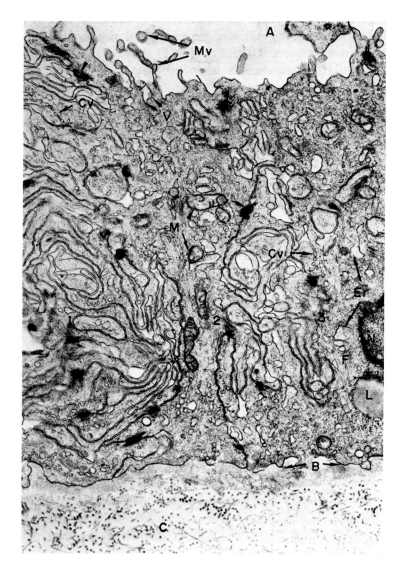

FIG. 6.5. Placental amnion of 23-week foetus. Irregular microvilli (Mv) are seen projecting into the amniotic cavity (A). Epithelial cells 1, 2 and 3 are separated laterally by complex interdigitations. Only small degrees of separation of cells to form intercellular spaces are seen. Desmosomes (arrows) are frequently seen, but there are no zonulae occludens. One nucleus (N), mitochondria (M), occasional cisternae of rough endoplasmic reticulum (Er), bundles of fine intracytoplasmic filaments (F) and a lipid inclusion (L) are seen. Small, smooth vacuoles (V) are present at the cell membranes; occasionally they form chains of vacuoles (Cv), leading into the cell. Basement membrane (B), and underlying collagen (C), are also shown. ×13,500.

THE AMNION AT TERM (Figs. 6.6 and 6.7)

The epithelium has been well described by Thomas (1965) who found two types of cell, a 'Golgi' type and a 'fibrillar' type. The 'Golgi' type, as its name implies, is characterized by the presence of an extensive

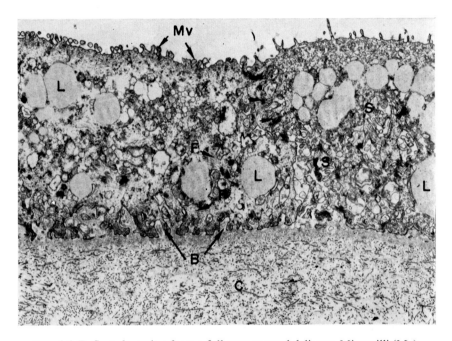

FIG. 6.6. Reflected amnion from a full-term normal delivery. Microvilli (Mv) are present at the free surface. Laterally, two cells from complex adjacent surfaces with desmosomes (arrows), and with formation of intercellular spaces (S) into which microvilli project. The cells are of the 'fibrillar' type, and fine filaments (F) are seen in the cytoplasm. Normal cytoplasmic organelles are scanty, but numerous lipid inclusions (L) are present. The irregularity of the basal aspect of the cells is well shown. The basement membrane (B) and underlying collagen (C), are also shown. × 6250.

Golgi apparatus, spreading throughout the cell. A second striking feature is the abundance of dilated rough endoplasmic reticulum* cisternae with finely fibrillar or granular contents of moderate electron density. This morphological picture is now generally considered to

* Rough endoplasmic reticulum: a system of cell membranes studded with polyribosomes which constitute the mechanism for synthesis of export protein.

indicate a capacity for synthesis of protein for use outside the cell, although in this case no evidence of specific secretory activity is seen.

The second or 'fibrillar' cell type is characterized by an abundance of intracellular fine filaments, by the absence or small size of the Golgi

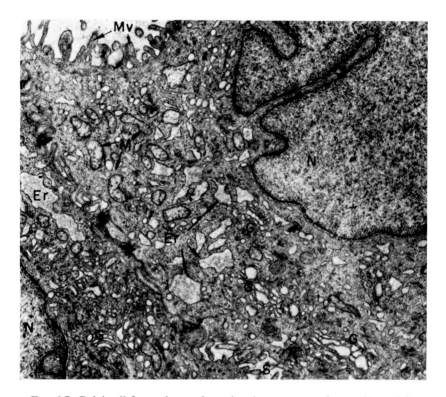

FIG. 6.7. Golgi cell from placental amnion (caesarean section at 40 weeks). Part of two Golgi cells is shown, separated by moderately complex lateral plasma membranes; an intercellular space (S) with microvilli is shown. Microvilli (Mv) are also present at the free surface. Two nucleii are shown (N), as are parts of the extensive Golgi apparatus (G). Mitochondria are numerous (M), and irregularly dilated rough endoplasmic reticulum cisternae with medium density content (Er), are also seen. × 12,500.

apparatus, by a paucity of rough endoplasmic reticulum and by fewer mitochondria than in the Golgi cell type. Thomas suggests that the 'fibrillar' cell may be a resting state of the 'Golgi' cell.

Lipid is a prominent intracellular component at term. Armstrong *et al* (1968) maintain that the lipid is limited by a trilaminar membrane, thus suggesting that these are lipid vacuoles (in contrast to lipid droplets which do not possess a limiting membrane). They further suggest that this lipid may be secreted into the amniotic fluid. This claim needs further proof.

Bevis (1968) suggested a secretory function for the amnion. He cultured amnion cells from near the insertion of the cord, near the edge of the placenta and from the reflected amnion and showed that protein was produced by the amnion near the insertion of the cord in greater quantities than from the other two sites.

Though microvilli increase in number as the pregnancy advances there are cells present at term in which the microvilli are sparse or absent. Although microvilli are usually thought to be involved in exchange, Thomas (1965) believes that at term they may have a ciliary action.

The complex appearance of the lateral cell borders is maintained at term; often the cells are closely apposed but following normal delivery intercellular canals may be widely distended (Thomas 1965). Desmosomes are present along the lateral cell borders.

In early pregnancy the basal aspects of the amniotic cells are smooth or only moderately irregular. At term the basal surface consists of numerous clefts separating foot-like processes.

Bourne (1962) suggests that the amnion cells are involved in exchange processes but show no evidence of secretory activity. It can be seen from the above descriptions that synthesis of protein and perhaps secretion of lipid does occur, while the fine structure is also compatible with exchange processes. It may also be suggested that the complex lateral cell borders play some part in controlling water and electrolyte exchange. The failure to demonstrate occlusive junctions near the surface of this epithelium and the confirmed presence of many direct passages from amniotic cavity into intercellular spaces appears to exclude surface and near surface zones from the control of the passage of water and electrolytes from the amniotic cavity to intercellular spaces.

UMBILICAL CORD

MID-TERM EPITHELIUM (Fig. 6.8)

Hoyes (1969d) has studied the fine structure of cord epithelium from 8 weeks to term. Between 8 and 20 weeks the epithelium consists of a

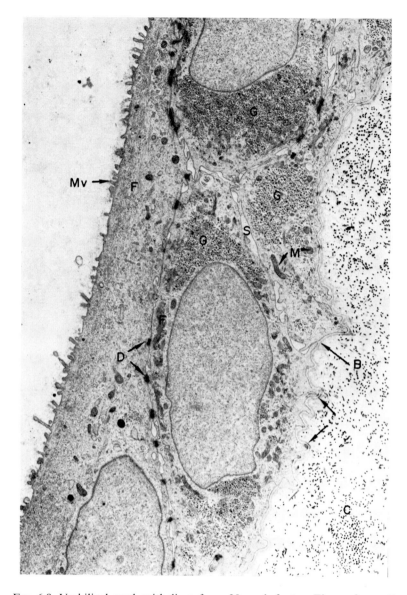

Fig. 6.8. Umbilical cord epithelium from 20 week foetus. The surface cell bears small microvilli (Mv). Cells are attached to each other by desmosomes (D), and basally to the basement-membrane (B), by hemidesmosomes (arrows). Intercellular spaces (S) are small and relatively simple. Glycogen (G) is prominent, and bundles of fine filaments (F) are present in all the cells. Mitochondria (M) are seen in all the cells. Collagen of Wharton's jelly is shown (C). × 6750.

single layer of cells. In the 12th week the epithelium becomes bilaminar and the superficial cells show a collection of organelles above the nucleus. In the 6th and 7th months the epithelium is composed of three or more layers.

Before the 6th month of pregnancy *zonulae occludens* are invariably present between the superficial layers of the epithelium. Their presence suggests that any flow of fluid must take place through the cytoplasm of these superficial cells.

The superficial layer of the epithelium in mid-pregnancy consists of cells with surface microvilli. Numerous medium-sized vesicles are present in the cytoplasm underlying these microvilli. The cytoplasm is characterized by the presence of a large quantity of glycogen distending the cell and pushing aside the normal cellular organelles. Hoyes (1969d) has suggested that some of the vesicles in the superficial epithelial layer are similar to those seen in the periderm of foetal skin, vesicles which are interpreted as being secretory in nature.

TERM EPITHELIUM (Fig. 6.9)

At term the epithelium consists of between one and five layers of rather flattened cells. There are two definite cell types at the surface with intermediate forms (Parry & Abramovich 1970). The predominant cell, a 'fibrillar' type, has a thickened plasma membrane, vacuoles limited by similar thick membrane and a striking abundance of fine intracellular filaments. The second cell type, less in number, possesses a plasma membrane of normal thickness, fewer filaments, a Golgi apparatus, endoplasmic reticulum and easily identified mitochondria, these latter features suggesting synthetic and metabolic abilities.

Only occasional *zonulae occludens* are seen at the surface layer, and as in amnion, direct communications between amniotic cavity and intercellular channels exist (Hoyes 1969d, Parry & Abramovich 1970).

Experimental evidence that the umbilical cord is important in the exchange of water was provided by Hutchinson *et al* (1959) who found a strikingly high concentration of administered tritium in Wharton's jelly. They concluded that the major part of water exchange takes place through the cord at term. In 1961, Plentl, working *in vitro*, showed that the cord exchanges 40–50 ml of water per hour. Feliks (1969), again *in vitro*, showed that the umbilical arteries are impermeable to water

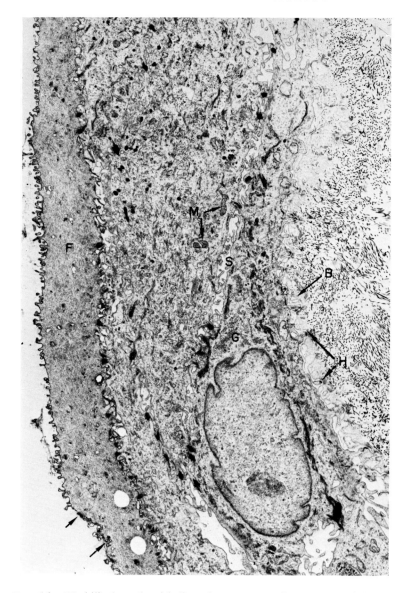

FIG. 6.9. Umbilical cord epithelium (caesarean section at 40 weeks). The surface is formed by a fibrillar cell (F) which possesses small surface microvilli; organelles are absent from the cytoplasm and the plasma membrane is thick. Remnants of a superficially attached cell are shown (arrows). Intermediate and basal cells possess scattered fine filamentous bundles, possess mitochondria (M) and a Golgi apparatus (G). Cells are attached to one another by desmosomes, and again hemidesmosomes (H), attach the cells to the basement membrane (B). Intercellular spaces (S) often contain microvilli. × 6750.

whereas the cord vein reabsorbs sodium, chloride and potassium in quantities suggesting a dialysis mechanism in the vein walls. However, as there are such great morphological changes in the cord when it has been cut, results from experiments *in vitro* must be interpreted carefully.

The morphology of the cord suggests that if there is any control of the exchange of water and electrolytes, then this must be mediated by the plasma membrane of the intercellular spaces or perhaps by the endothelium of the umbilical vessels.

FOETAL EPIDERMIS

In early pregnancy skin plays a role in the exchange of water and electrolytes. At 2 months of age foetal epidermis is composed of two layers, a basal layer and superficial cells (periderm). During the 3rd month intermediate layers of cells appear while in the 4th month globular structures, called 'bladder cells', form an apparent second layer of peridermal cells on the epidermal surface. By 17–20 weeks the periderm begins to disappear and the underlying epithelium begins to keratinize. Before keratinization occurs, the peridermal cells appear functionally active with microvilli, Golgi apparatus, mitochondria, ribosomes and endoplasmic reticulum. Membrane-bound vesicles are present in relation to the Golgi apparatus and also in clusters beneath the surface membrane (Hoyes 1968c).

These morphological changes may be interpreted as showing that the cells are capable of exchange in both directions, the water and electrolytes entering the intercellular channels via the cells (Lind *et al* 1969). Secretion may occur between 10 and 16 weeks (Hoyes 1968c). It is believed that once keratinization begins, the exchange of water and solutes diminishes (Lind *et al* 1969).

Experimental confirmation of this exchange via the skin has been provided by Parmley & Seeds (1970) who measured the permeability value for diffusion of isotopic water across foetal skin from the back and scalp of foetuses ranging from 12 to 22·5 cm crown-rump length. For foetuses of c.r.l. 12 to 18 cm (16–20 weeks) similar permeability values were found to those measured across chorion and amnion throughout pregnancy. Partially and well-keratinized skin from 18–22·5 cm c.r.l. foetuses showed little or no permeability. By 25 weeks foetal skin is no longer a site of exchange.

CONSTITUENTS AND PROPERTIES OF LIQUOR AMNII

OSMOLALITY AND ELECTROLYTE CONCENTRATION

Early workers (Zangemeister & Meissl 1903) measured the osmotic pressure (by freezing point depression) of liquor and maternal plasma in mid-pregnancy and at term. They showed that the liquor became hypotonic as the foetus matured. Their results were confirmed by Guthmann & May (1930) and Makepeace *et al* (1931) who also suggested that the decreasing tonicity of the liquor was caused by the addition of hypotonic foetal urine to a fluid which was originally isotonic with maternal plasma.

There are few figures for osmolality or electrolyte concentration in very early pregnancy liquor (from 8 to 10 weeks) but information is available from pregnancies of 10 weeks and older. Lind *et al* (1969) measured the osmolality of 9 liquors and maternal plasmas in early pregnancy, 1 at 8 weeks, the remainder aged from 11 to 16 weeks. Gillibrand (1969b) examined 60 liquors and maternal plasmas at all stages of gestation. There were 15 cases in the group of 10–16 weeks gestational age. Table 6.6 shows that the osmolality of liquor is lower than that of maternal plasma but whether the differences are significant is not known.

Gillibrand (1969b) examined fluids from 11 pregnancies of 17–30 weeks. By this time the difference in osmotic pressure was more marked, the average of the maternal plasma being 278 mOsm/kg while that of the liquor was 267 mOsm/kg. Abramovich (unpublished) in 10 cases aged from 17 to 21 weeks, found that the average osmolality of the liquor was 10 mOsm less than that of the maternal plasma.

Data are available for bulk flows inside the uterus and exchange of water (by diffusion) between liquor, mother and foetus, so the concept of the addition of a hypotonic foetal urine to a nearly isotonic liquor can be examined. At 18 weeks of age the foetus swallows between 4–11 ml/day, and voids as a maximum 7–17 ml/day (probably less) into a liquor volume of nearly 300 ml. The urine the foetus voids is hypotonic, the average osmolality being 144 mOsm/kg. There is the complicating factor that in a 20-week foetus water is exchanging between liquor and mother at 2·5 litres/day and between liquor and foetus at nearly 2 litres/day (Hutchinson *et al* 1959). Electrolytes are also exchanging between liquor and foetus and mother and it can be argued that the volumes of hypotonic urine may not be large enough to influence the liquor osmotic pressure in the face of all the exchange that undoubtedly occurs.

As the foetus matures the liquor osmotic pressure continues to drop. At term (38–41 weeks) it averages 244 mOsm/l (range 222–269 Huntingford & Marsden 1966) or 250 mOsm/kg (range 231–261 Gillibrand 1969b) compared with maternal plasma levels of 276–278 mOsm/kg. There is no correlation between liquor volume and the osmotic pressure of the fluid (Huntingford & Marsden 1966).

TABLE 6.6. Osmolality of liquor amnii and maternal plasma in pregnancies of 10–16 weeks gestation

Author and no. of cases	Osmolality (mOsm/kg)	
	Liquor amnii	Maternal plasma
Lind et al (1969) (9)	278	280
Gillibrand (1969b) (15)	273	276

Miles & Pearson (1969) suggest that the osmolality continues to decrease past term and that a level of 250 mOsm/l or less is suggestive of foetal maturity. Examination of Gillibrand's figures shows no difference between the average levels at 38–41 weeks or 42 and 43 weeks. O'Leary & Feldman (1970) pointed out that an individual reading of liquor osmolality is useless in determining foetal maturity as the normal range is so wide. These authors found no difference in the liquor values in cases of hypertension, diabetes, rhesus iso-immunization or toxaemia when compared with normal values.

Why there is a decrease in liquor osmolality and what controls this drop is unknown. Gillibrand (1969b) has suggested that the decreasing osmolality may be a factor in controlling foetal body water and electrolyte changes in pregnancy.

Electrolyte concentrations in 10–16 week pregnancies have been measured by Gillibrand who found that the sodium and potassium values were similar in maternal plasma and liquor but that liquor chloride values were higher than plasma levels. Lind *et al* (1969) measured the concentration of sodium in 9 cases from 8–16 weeks and found the plasma values greater than liquor (Table 6.7).

These results for liquor chloride confirmed those presented by Makepeace *et al* (1931). Seeds (1968) suggested that the Gibbs-Donnan equilibrium was responsible for this phenomenon.

TABLE 6.7. Concentration of electrolytes in liquor and maternal plasma in pregnancies of 10–16 weeks gestation

Author and no. of cases	Na+ (mEq/l)		K+ (mEq/l)		Cl− (mEq/l)	
	Liquor	Maternal plasma	Liquor	Maternal plasma	Liquor	Maternal plasma
Gillibrand (15)	137	136	4·0	3·9	104	100
Lind *et al* (9)	134	136	—	—	—	—

The concentration of potassium in liquor remains steady during pregnancy while that of sodium and chloride falls at term (38–41 weeks). Gillibrand showed that the average sodium value in 13 cases was 128 mEq/l (117–138 mEq/l) with the average chloride value 99 mEq/l (95–104 mEq/l). Huntingford and Marsden (1966) found the average sodium value in 21 cases was 122 mEq/l with a range of 108–135 mEq/l. Gillibrand stated that the difference in values of both sodium and chloride in liquor from early pregnancy compared with levels at term was significant ($P < 0.001$ in both cases).

Lind *et al* (1969) found a good correlation between osmolality and liquor sodium levels ($r = 0.88$) but Gillibrand while agreeing that the osmolality of liquor is largely dependent on sodium concentration, stated that the relationship between the fall in these two factors was not constant or direct. He felt that there are other unmeasured osmotically active constituents involved in keeping the liquor hypotonic.

The osmotic pressure and concentration of sodium and chloride in the liquor fall as the foetus matures. The changes in these values are not definite enough to be of value clinically but it is probable that there is a physiological reason, as yet undescribed, to account for them.

UREA, URIC ACID AND CREATININE

The levels of the so-called urinary constituents of liquor have been measured by Guthmann & May (1930), Friedberg (1955) and Lind *et al* (1969, Table 6.8).

TABLE 6.8. Levels of urea, creatinine and uric acid in liquor throughout pregnancy

Author	Gestational age (weeks)	Urea mg%	Creatinine mg%	Uric acid mg%
Guthmann & May (1930)	10	34	—	3·41
	14	37·7	—	3·99
	30	40·1	—	4·49
	term	44·3	—	5·06
Friedberg (1955)	8–12	20·7	1·04	2·78
	16–20	19·4	1·06	3·6
	28–32	31·1	2·2	5·6
	36–term	36·1	3·7	6·8
Lind *et al* (1969)	20	20·6	0·54	—
	30	19·5	0·82	—
	34	—	1·29	—
	35	22·8	—	—
	38	—	1·58	—
	term	25·5	—	—
	beyond term	30·4	2·07	—

Both creatinine and uric acid show a steady rise throughout pregnancy though urea stays steady or falls slightly in mid-pregnancy and then rises as the foetus matures. Among the first to suggest a clinical application for liquor creatinine levels were Pitken & Zwirek (1967). They noted that liquor creatinine level remained constant or increased very gradually till approximately 34 weeks when a more abrupt increase occurred; after 37 weeks the level was 2 mg% or greater in 94% of patients.

The use of liquor creatinine level in assessing foetal maturity has been confirmed by Droegemueller *et al* (1969) and Doran *et al* (1970) but rejected by Mandelbaum & Evans (1969) who considered the test unreliable as many mature foetuses had levels below 1·8 mg%.

The value of creatinine levels in abnormal pregnancies is not clear. Pitken & Zwirek (1967) found no difference between normal pregnancies and cases of pre-eclamptic toxaemia, chronic hypertension or diabetes mellitus while Roopnarinesingh (1970) found that liquor creatinine levels in cases of pre-eclamptic toxaemia, chronic hypertension or diabetes mellitus were higher than normal for each stage of pregnancy. Mandelbaum & Evans (1969) also reported higher levels of creatinine in liquor from cases of diabetes mellitus.

The increasing concentrations of the urea, uric acid and creatinine is thought to represent maturing foetal renal function or reflect increasing foetal muscle mass (Droegemueller *et al* 1969). Although Doran *et al* (1970) and Roopnarinesingh (1970) found a correlation between infant body weight and liquor levels, Droegemueller *et al* (1969) showed no correlation at all.

While urea, uric acid and creatinine are found in foetal urine (see above) the liquor levels are the end result of removal by foetal swallowing, exchange across membranes and possibly from other sites. It is not known whether there is any control mechanism for the liquor levels of these substances as their concentrations appear to rise at a steady rate even when the liquor volume is diminishing as happens after 38 weeks. In fact it is possible that despite a rising concentration in the liquor, the total amount of substance in the amniotic sac may fall after 38 weeks as the liquor volumes fall by a greater percentage than the concentration rises.

The membranes may play a part in influencing the levels of urea and creatinine. Moore & Ward (1970) have shown that the chorio-amnion *in vitro* is significantly less permeable to creatinine than to urea. This may mean that creatinine and urea are removed from the liquor at different rates and thus provide a partial explanation of liquor concentrations.

Apart from foetal urine, other routes may be relevant. Pitken (1969) suggests that raised maternal plasma creatinine levels can influence the liquor levels by perhaps raising foetal plasma levels. The creatinine could also enter the liquor via the chorio-amnion, the foetal surface of the placenta or via the umbilical cord.

Substances usually excreted by the kidney may enter liquor by other routes. Reynolds *et al* (1969) injected Diodrast-I^{131} into the foetal circulation of normal and nephrectomized foetal monkeys. They found that the concentration of Diodrast in the liquor was approximately the same whether the foetus was intact or nephrectomized and suggested that the substance entered the liquor via the respiratory tract, umbilical cord or embryonic membranes. It is quite possible that urea, uric acid and creatinine have similar multiple sites of entry into the liquor.

The clinical value of creatinine measurements for estimation of maturity remains uncertain.

BILIRUBIN

Liquor analysis is used to assess the severity of haemolytic disease. The amount of bilirubin and bilirubin-like pigments is measured by chemical or spectrophotometric means and thus the degree of the haemolytic disease in the fetus can be predicted. Morris et al (1967) considered that more accurate predictions of the degree of severity of the disease (and thus more correct timing of induction or decision on intra-uterine transfusion) were made when the predictions were based on the ratio of bile pigment to protein in liquor.

Bilirubin has been found in normal liquor from 10 weeks of gestation (Berk & Sussman 1970). The concentration rises to 18 weeks, levels off from then till 25 weeks and falls from 26 to 36 weeks (Walker 1970). Levels are so high in early and mid-pregnancy that Berk and Sussman pointed out that if they had been obtained after 25 weeks, the foetus would be considered to be severely affected by rhesus disease.

The disappearance of bilirubin after 36 weeks has been used by Mandelbaum et al (1967) to confirm foetal maturity but Andrews (1970) noted that in 16% of a series of cases near term the liquor had an optical density greater than 0. He considered this to be a major disadvantage in the use of liquor bilirubin as an indicator of foetal maturity. The analysis of the pigment found in the liquor and measured by optical density methods has proved difficult. It is now generally agreed that the major pigment is unconjugated bilirubin and/or related hydrophobic compounds (Heirwegh et al 1969); these authors suggest that some conjugated bilirubin may be present.

Liquor from cases of diabetes mellitus (Andrews, 1970) anencephaly with and without hydramnios (Lee & Wei 1970) small bowel atresia with polyhydramnios and maternal ss haemoglobinopathy, hepatitis and chronic cirrhosis (Grimes & Cassidy 1970) has been shown to have a high optical density using spectrophotometry. It has been assumed by some of these authors that because the optical density is raised at 450 mμ the pigment is necessarily bilirubin and is the same pigment as is present in liquor from cases of rhesus isoimmunization. This may be an unwarranted assumption.

The route of entry of the bile pigments into the liquor is unknown. It is unlikely to be foetal urine. Mandelbaum & Evans (1969) examined the urine from three severely erythroblastic foetuses and could not detect bilirubin. Reynolds *et al* (1969) injected bilirubin C^{14} into a rhesus monkey *in utero*. While radioactivity promptly appeared in the liquor none was found in foetal urine. Free (unconjugated) bilirubin is not excreted by the kidney in the human. Goodlin & Lloyd (1968) found unconjugated bilirubin in the tracheal fluid of foetal rabbits and of three human foetuses kept alive in a hyperbaric oxygen chamber. Both they and Grimes and Cassidy suggested a pulmonary origin for the pigments but in view of the physiology described for the respiratory tract, this is considered to be an unlikely source.

There is a high correlation between the concentration of bile pigments and protein in normal liquor (Murray *et al* 1970) and in rhesus-affected liquor (Dunstan 1968). Cherry *et al* (1970) showed that the bilirubin is strongly bonded to the liquor albumin and found that when they injected albumin into the liquor, the amount of bilirubin in the liquor rose. They further suggest that bilirubin could cross the foetal surface of the placenta, umbilical cord or foetal skin to enter the liquor.

PROTEINS

Queenan *et al* (1970) measured total protein concentration in 115 amniotic fluids from 80 normal patients of gestational age 11 to 43 weeks. Using an ultraviolet technique and the biuret method to estimate the proteins, they obtained fair correlation with the two methods. From 11 to 15 weeks gestation, the range is wide with values as wide apart as 0·1 g/100 ml and 0·6 g/100 ml. Values in the gestational age 16–24 weeks appear to show a slight rise with concentrations varying between 0·247 to 0·765 g/100 ml. The mean for 16–20 weeks is 0·536 g/100 ml and for 21–24 weeks 0·636g/100 ml. From 24 weeks till term there is a definite decreasing trend with the mean for 37–40 weeks being 0·279 g/100 ml.

These authors also investigated the liquor protein from rhesus sensitized pregnancies. Samples from cases of mildly and moderately affected infants showed the same decreasing trend of values with increasing foetal maturity. When the foetus was severely affected the protein levels were spread over a wide range without the pronounced downward trend with increasing foetal age seen normally, while in fluid from hydrops foetalis the protein levels rose, with some very high readings (>1 g/100 ml) being recorded.

Dunstan (1968) measured the liquor proteins by the biuret method in 53 liquors from 11 rhesus sensitized women but gave no details on how seriously the infants were affected. He found that there was a gradual fall in the protein concentration with advancing pregnancy but that the variation in values was so wide that the apparent trend was not significant.

Cherry et al (1965) suggested a direct correlation between protein concentration and liquor volume but Dunstan (1968) found no correlation. Queenan et al (1970) considered that in general, the larger the infant's birth weight, the lower the liquor protein value before delivery. They considered the role of foetal swallowing but were uncertain of its importance for foetal growth. In contrast, however, Belo'ussova (1969) found that the level of liquor protein was directly related to the infant's weight; the larger the infant, the higher the total protein level of the liquor.

There appears to be a large range of proteins present in liquor and as more sensitive techniques are developed more proteins are found in the liquor. Mendenhall (1970) using the technique of single radial immunodiffusion, examined liquor from normal term pregnancies and found albumin levels of 204 mg% (90–800 mg%), IgG, 91 mg% (66–120 mg%) and small amounts of IgA, IgM, ceruloplasmin, transferrin α-2-macroglobulin and α-1-antitrypsin.

It should be remembered, as Heron (1966) points out, that the values obtained for albumin and globulin concentrations in liquor vary with the technique employed in their assessment.

There is evidence that maternal plasma, foetal plasma and amnion all contribute to the protein pool of liquor. Abbas & Tovey (1959) concluded that the protein pattern of the liquor resembled a dialysate of maternal serum while Derrington & Soothill (1961) favoured the suggestion that the liquor protein was an ultrafiltrate of maternal rather than foetal serum. Ruoslahti et al (1966) and Seppala et al (1966) provided genetic evidence for the maternal origin of liquor proteins when they measured Gc protein and transferrin, both of which they examined for the genetic type of protein. In all cases where the maternal and foetal type were different, the liquor protein was of the maternal type.

Further evidence for maternal origin was provided by Dancis *et al* (1960) when they injected I^{131}-albumin and labelled globulin into the mother before termination of pregnancy. Their experiments showed a small but significant transfer into liquor but it is difficult to draw any conclusions concerning the pathways involved in the transfer of the proteins.

Brzezinski *et al* (1964) argued for a foetal origin of protein when, using moving boundary electrophoresis, they found a greater similarity between foetal plasma and liquor proteins than between maternal plasma and liquor. They also described one case in which there was bis-albuminaemia in the foetal serum and liquor but not in the maternal serum.

Bevis (1968) noted that the membranes are capable of producing protein when he cultured amnion from various sites. An electron-microscopic examination of the amnion cell confirms this capability to produce protein by showing the presence of the necessary intracellular organelles.

The evidence presented suggests that liquor proteins have multiple origins, entering liquor from maternal plasma, foetal plasma as a dialysate or ultrafiltrate and probably also being produced by the amnion.

HUMAN CHORIONIC GONADOTROPHIN (HCG) AND SOMATOMAMMOTROPHIN (HCS)

Berle (1969) found that chorionic gonadotrophin was present in liquor, the umbilical vein and the umbilical artery in the ratio of 3:1·5:1 at term. From the 9th week of gestation there appears to be a striking rise in concentration in the liquor, reaching a peak around the 13th week. This peak is followed by a substantial decrease reaching relatively constant low values, sustained till term. In severe rhesus isoimmunization and diabetes mellitus, there are increased concentrations of chorionic gonadotrophin in liquor.

Crosignani & Polvani (1969) measured HCG in liquor from third trimester pregnancies using radioimmunoassay and found values of 0·39 I.U./ml. They found slightly higher values in liquor from patients in labour (0·45 I.U./ml).

Teoh (personal communication) measured the levels of human chorionic somatomammotrophin in liquor at term, finding only small amounts (0·2–1·0 μg/ml).

GLUCOSE, PYRUVATE AND LACTATE CONCENTRATIONS

Wood et al (1963) measured glucose levels in liquor at term, finding that the mean was 13 mg/100 ml whilst when pregnancy was prolonged the figure fell to 8 mg/100 ml. They felt that the liquor glucose levels reflect fasting maternal blood glucose levels. Furthermore when the liquor glucose was less than 10 mg/100 ml, there was often evidence of foetal distress or delay in the onset of regular respiration in the newborn. Liquor lactic acid levels at term or past term were found to average 38·7 mg/100 ml and when the level was greater than 40 mg/100 ml, delay in the onset of newborn respiration was more common.

Schreiner & Schmid (1969) quoted the levels of liquor glucose in early pregnancy (10–11 weeks) as 53·7 mg/100 ml, falling to 40·2 mg/100 ml (14–16 weeks) and to 22·2 mg/100 ml at term. When pregnancy was prolonged (291 days or more) the liquor level was 18·1 mg/100 ml but if the foetus was also clinically postmature, the fluid level was always below 6 mg/100 ml.

Lactate and pyruvate levels in liquor fell from 92·1 and 1·64 mg/100 ml respectively in the first half of pregnancy to 64 and 0·79 mg/100 ml during the last month. There was no change with postmaturity.

Cohen (1970) measured (among other substances) glucose, lactate and pyruvate levels in liquor collected at induction. He found no correlation between the levels of these three substances in the liquor with the occurrence of foetal distress in labour or with the state of the infant at birth,

These conflicting reports suggest that the value of measuring these products of carbohydrate metabolism is uncertain at the present moment.

ACID-BASE AND GAS TENSION OF THE LIQUOR

The possibility that foetal acidosis or hypoxia might be accompanied by characteristic changes in liquor prompted a number of investigations. Cassady & Barnett (1969) reviewed the early papers which suggested that diminishing liquor Po_2 and rising liquor Pco_2 accompanied foetal compromise. However, when this work was followed up, there appeared to be no correlation between the condition of the foetus and the pH and gas levels in the liquor.

Corson & Bolognese (1968) measured the pH of liquor obtained from patients in labour. They found values ranging from 6·8 to 7·3 and there was no correlation with foetal asphyxia at delivery. These authors felt that liquor was a moderately good acid buffer which might be of physiologic value in supporting foetal homeostatic mechanisms. Quilligan (1966) measured liquor Po_2 and Pco_2 in cases near term. The oxygen tension was 12·6 ± 4·1 mm Hg (range 4–25·6 mm) but it was of interest that where foetal death *in utero* had occurred there was no difference in the liquor oxygen tension compared with that of liquor from normal pregnancies. Carbon dioxide tension was found to average 47·2 mm Hg in liquor from patients in labour. Quilligan also found no correlation between foetal arterial gas tension and the liquor gas tensions. The oxygen in the liquor was thought to come from many sources; foetal skin, some from foetal urine and perhaps from the chorion amnion.

Seeds & Helligers (1968) determined the acid-base status of the liquor throughout pregnancy. They showed that liquor becomes more acid as pregnancy progresses. From 10–23 weeks the mean pH was 7·227 while at term there was a significant difference, the level being 7·105. Carbon dioxide tension rose from 41·01 mm Hg in early pregnancy (10–23 weeks gestation) to 50·9 mm Hg at term while the bicarbonate concentration fell from 16·57 mm/l in early pregnancy to 14·82 mm/l at term. With both the gas tension and bicarbonate concentrations the differences are significant.

The reasons advanced by Seeds and Helligers for these changes are:

(1) The increase in liquor carbon dioxide tension may be associated with the slight elevation of foetal plasma Pco_2 as maternal uterine venous Pco_2 rises during later stages of gestation.

(2) The decrease in the bicarbonate concentration parallels a decrease in the liquor total solute concentration in the latter half of pregnancy.

(3) The decrease in the liquor bicarbonate could be secondary to increased renal excretion of fixed acids by the foetus later in pregnancy.

The changes in liquor pH and Pco_2 have been confirmed by Cassady & Barnett (1969). There is a wide range of values at each stage of pregnancy when the results obtained by Helligers and Seeds and Cassady and Barnett are combined, making it very doubtful whether this information is of any value in the diagnosis of foetal age. Cassady and Barnett also found that there was no correlation in their series between pH,

P_{CO_2}, P_{O_2} and either foetal distress or maternal complications (toxaemia, diabetes mellitus). They investigated one case of foetal death *in utero* in which the liquor acid-base balance and oxygen tension were within the normal limits.

All these results indicate that liquor pH and gas tensions have no value in the diagnosis of foetal distress but as Cassady & Barnett (1969) pointed out, foetal capillary or arterial values should be measured in parallel with the liquor values before abandoning all interest in these measurements.

Notwithstanding the lack of correlation in the human between foetal state and liquor values, Seeds *et al* (1967) found that in the rhesus monkey liquor P_{CO_2} and pH paralleled acute changes in the foetus during acidosis associated with raised foetal carbon dioxide tension. Perhaps species differences account for these results.

STEROIDS

The realization that the content or pattern of steroids in liquor might reflect foetal and foeto-placental metabolism has stimulated much interest in this field though the differentiation of maternal as opposed to foetal or foeto-placental contributions is proving difficult. The following groups of steroids will be discussed.

(1) 17-oxosteroids.
(2) cortisol and 17-hydroxycorticosteroids.
(3) progesterone and metabolites.
(4) Δ5-3β hydroxy compounds (including C-16 hydroxylated steroids).
(5) oestrogens.
(6) pregnanetriol.

17-OXOSTEROIDS

These compounds, reflecting androgen metabolism in part, are found in liquor from early pregnancy. 1·4 to 2·7 μg were found in the amniotic sac of foetuses aged 11–12 weeks, the concentrations ranging from 2·2–6 μg/100 ml. The amounts in the liquor rose linearly as the crown-rump length of the foetus increased so that at 18 weeks 7–14 μg of steroid were present. The concentration of 17-oxosteroid appeared to drop at 18 weeks (range 2·6–3·2 μg/100 ml) but this appeared to be a reflection of the greater increase in liquor volume per week in later pregnancies (Abramovich & Wade 1969).

In pregnancies of 26–36 weeks gestation the mean concentration was 7·4 µg/100 ml with 14% of the steroids present as the free (unconjugated) form 48% as glucosiduronates and 38% as sulphates. At term the concentration had risen to 12 µg/100 ml (range 4–20 µg/100 ml) with a similar pattern of free and conjugated steroids. There was no correlation between the weight or sex of the foetus at term with the steroid concentration nor between the concentration and amount of steroid in early pregnancy with the foetal sex (Wade & Abramovich 1967, Abramovich & Wade 1969b).

The concentration of steroid in liquor from anencephalic pregnancies is approximately one-quarter of the average value at term (2·9 µg/100 ml). As in most anencephalic foetuses the adrenal gland is atrophied, the lowered steroid concentration is to be expected. However, when the polyhydramnios seen in most of these cases is taken into account, the amount of steroid in the amniotic sac can equal or exceed that found in normal pregnancies (Wade & Abramovich 1967). It is not known whether some of the anencephalic foetuses can produce normal amounts of steroid or whether the state of affairs which produces the polyhydramnios also influences the steroid levels by slowing their exit from the amniotic cavity.

Testosterone, though not a 17-oxosteroid, is included conveniently in this section. Its levels in liquor have been measured by Gandy (1971) who found a mean concentration of 0·06 µg/100 ml. Dehydroepiandrosterone (DHA—this steroid can also be classified as a Δ^5-3β-hydroxy compound, section 5) was found by Schindler and Siiteri in pooled term liquor in a concentration of 0·8 µg/100 ml, while Gandy reported levels of 0·05 µg/100 ml for DHA and 0·54 µg/100 ml of dehydroepiandrosterone sulphate (DHAS).

It is difficult to distinguish maternal from foetal contributions to the liquor oxosteroids. The foetus certainly contributes to the liquor levels as concentrations within the normal range were found at 38 weeks in liquor from an adrenalectomized patient who was being treated with 5 mg prednisone thrice daily (Abramovich & Wade 1969b). No quantitative estimate of foetal production could be obtained as the liquor volume was not measured.

Passage of oxosteroid from the maternal blood stream across the placenta and possibly the membranes also occurs. Abramovich (1970a) infused the mother with labelled DHA intravenously and showed that while DHA does not enter liquor from mother, other oxosteroids do, as radioactivity was found in the liquor fraction which was known to contain these steroids.

Foetal urine is certainly one pathway by which these steroids enter liquor. When urine was collected from mid-trimester foetuses whose mothers had been infused with tritiated DHA as noted above, radioactivity was found mainly in the conjugated fraction (67–92% of total activity was conjugated), and in three out of four specimens more radioactivity was conjugated as a sulphate than as a glucuronide.

CORTISOL AND 17-HYDROXYCORTICOSTEROIDS

17-hydroxycorticosteroids (17-OHCS) are present in liquor from at least 11 weeks gestation. The amount rises linearly with increasing foetal crown-rump length from 2·8 µg at 11 weeks gestation (c.r.l. 5·5 cm) to 19–20 µg with foetal c.r.l. 17 cm. The concentration in the liquor in early pregnancy ranged from 2·7 µg to 8 µg/100 ml.

In pregnancies of 26–36 weeks, the mean concentration was 14·5 µg/100 ml while at term it had risen to 24 µg/100 ml (6·6–47 µg/100 ml). 17% of the total steroid was present in the unconjugated or free form, while the majority was conjugated, 47% of the total steroid being present as the glucosiduronate and 36% as the sulphate (Wade & Abramovich 1967, Abramovich & Wade 1969b).

Individual steroids have been measured in the liquor. Baird & Bush (1960) found that the concentration of cortisol was 2·6 µg/100 ml and of cortisone 1·3 µg/100 ml while Lambert & Pennington (1963, 1964) found 6β-hydroxycortisol and then the 20β-hydroxy derivatives of both 6β-hydroxycortisol and 6β-hydroxycortisone to be present.

As with oxosteroids the maternal and foetal or foeto-placental contributions to the corticosteroid levels are difficult to differentiate. In liquor from the adrenalectomized patient mentioned above, the concentration of corticoids was 17·4 µg/100 ml (Abramovich & Wade 1969b). Though the metabolites of prednisone would be included in the measurement in the liquor by the chemical method used it was not thought that these metabolites made a significant contribution to the liquor levels.

Some of the corticoids in liquor probably arise from the maternal circulation. Cortisol crosses the placenta and has been found in the foetal circulation (Migeon *et al* 1961) and following maternal infusion of tritiated cortisol, radioactive metabolites were found in the liquor.

While foetal urine is one pathway by which both free and conjugated corticosteroids enter the amniotic sac, the steroids can enter the liquor through the reflected membranes, the foetal surface of the placenta and possibly the umbilical cord (Abramovich & Wade 1969a). It is not known whether there are any enzymes present in the membranes which influence the passage of the conjugated oxo- and corticosteroids as is seen with oestriol glucosiduronate and sulphate.

PROGESTERONE AND ITS METABOLITES

Progesterone has been found in liquor in concentrations ranging from 8·3 to 16·7 mg/100 ml (Wiest 1967) though Schindler & Siiteri (1968) found it in smaller quantities (4 μg/100 ml). It is not known in what state the progesterone exists in the liquor, whether unconjugated, as an enolic glucosiduronate or bound to a lipoprotein (Klopper 1970). The question of conjugation is important as it may help identify the route of entry into liquor; progesterone is usually unconjugated and so insoluble in water, thus making it unlikely that foetal urine is the major route of entry.

Pregnanediol is usually present as a conjugate. Klopper (1970) found that the levels in later normal pregnancy were 40 μg/100 ml with a substantial proportion of the pregnanediol present as the sulphate. On the other hand, Schindler & Siiteri (1968) noted a concentration of 14·5 μg/100 ml present mainly as pregnanediol glucosiduronate.

There does not appear to be any evidence at the moment linking the liquor levels of progesterone and pregnanediol with the clinical state of the foetus or neonate.

Δ^5-3β HYDROXYCOMPOUNDS (INCLUDING C-16 HYDROXYLATED STEROIDS)

The levels of major Δ^5 compounds are shown in Table 6.9. Care has to be taken in interpreting the concentrations quoted by various authors as the chemical methods used have varied and these steroids have not yet been fully classified into free and conjugated compounds. It seems, for example, that 16 hydroxydehydroepiandrosterone and 16-ketoandrostenediol are present in liquor mainly conjugated with glucuronic acid or as a double conjugate of the sulphate glucuronide type (Luukkainen et al 1970).

The sulphates of these steroids can be further divided into mono- and disulphates. Luukkainen et al (1970) identified 16α OHDHA in both the monosulphate fraction (3·9 µg/100 ml) and in the disulphate fraction (4·8 µg/100 ml, see table). The levels of monosulphates appear to be low in liquor yet they are high in foetal blood (Luukkainen et al 1970) but the biological significance of this is unexplained.

Also unexplained is how the presence of a keto group ($=$ o) or a hydroxyl ($-$OH) group in the 16 position seems to facilitate the entry of this substance into the liquor from the foetal blood stream. Schindler

TABLE 6.9. The concentration of Δ^5-3β-hydroxysteroids (including sixteen oxygenated or hydroxylated compounds)

Steroid	Authors	Concentration (µg/100 ml)
dehydroepiandrosterone	Schindler & Siiteri (1968)	0·8
16α-hydroxydehydroepiandrosterone	Schindler & Siiteri (1968)	79·8
(16α DHA)	Siegel et al (1969)	102·0
16α DHA SO$_4$ as the disulphate	Luukkainen et al (1970)	4·8
16β DHA SO$_4$	Luukkainen et al (1970)	8·8
16-ketoandrostenediol	Schindler & Siiteri (1968)	57·5
16-ketoandrostenediol	Siegel et al (1969)	110·0
16-ketoandrostenediol sulphate	Luukkainen et al (1970)	0·6
3β, 17α-androstenediol	Luukkainen et al (1970)	22·4
3β, 17β-androstenediol	Luukkainen et al (1970)	4·2
androstenetriol	Schindler & Siiteri (1968)	4·8
16α-hydroxypregnenolone	Schindler & Siiteri (1968)	11·9

& Siiteri (1968) found 160 µg/100 ml of dehydroepiandrosterone and 100 µg/100 ml of 16-hydroxydehydroepiandrosterone in foetal blood while in liquor the concentration of the 16-hydroxylated compound is nearly 1000-fold greater than the parent compound. It is not known whether this is an excretion mechanism, whether the hydroxylation facilitates urinary excretion or passage through membranes or cord or whether there is increased protein binding of 16-hydroxylated steroids.

The concentrations of the main neutral steroids in the sulphate fraction of early and midterm liquor have been measured by Huhtaniemi & Vihko (1970) who found that while the qualitative composition closely resembled that of term liquor the total concentration was only about 10% of that at term.

The high concentration of the main Δ^5 compounds in liquor compared with the oxosteroids and hydroxycorticoids is explained by the higher levels of Δ^5 steroids in foetal plasma. The mode of entry into liquor is presumed to be foetal urine as in newborn urine Δ^5-3β hydroxy compounds predominate over the Δ^5-3-ketones both in the androstene (C19) and pregnene (C21) series (Klopper 1970). There is little knowledge available on other routes of entry or on exit routes.

Schindler & Ratanasopa (1968) presented an interesting report on the levels of DHA, pregnanediol, 16 keto-androstenediol, 16αOHDHA and oestriol in liquor from rhesus sensitized pregnancies. They found that while DHA and pregnanediol levels were within normal limits the concentrations of 16-oxygenated steroids (16-keto-androstenediol, 16αOHDHA and oestriol) were low. It is also known that maternal urinary oestriol excretion in rhesus cases is unaffected (Klopper & Stevenson 1966). Schindler & Ratanasopa (1968) suggested that there was a decreased production of the oestriol precursors to blood levels just below foetal renal threshold while there was increased conversion of those precursors to oestriol in the placenta. More information is needed before this suggestion can be confirmed.

OESTROGENS

The liquor levels of the classical oestrogens were described by Diczfalusy & Magnusson (1958). Both oestrone and oestradiol are present in liquor from the 12th week, oestrone levels rising from 0·25 to 0·44 μg/100 ml at term and oestradiol from 0·12 to 0·31 μg/100 ml.

The presence of other oestrogens has been noted by Schindler & Siiteri (1968) and Siegel et al (1969) who found 2-methoxy-oestrone, 16α and β-hydroxy-oestrone, 16-keto-oestradiol, epioestriol and the 15 hydroxy derivatives of both oestrone and oestriol (the latter compound being an oestetrol). It may be that oestetrol originates solely from the foetus in which case its concentration could then give an accurate reflection of foetal well-being.

More interest has been shown in the levels of liquor oestriol as it has been felt that these values may have a clinical application. The values found by various workers in liquor from term pregnancies are listed in Table 6.10.

The levels of oestriol throughout pregnancy were measured by Michie & Livingstone (1969) who noted a rise from 2 μg/100 ml at 10–20 weeks to 19·8 μg/100 ml at 30–38 weeks while Aleem *et al* (1969) found the concentration to be 28·5 μg/100 ml at 15–20 weeks and 57·1 μg/100 ml at 33–36 weeks. Both sets of figures show a sharp rise after 36 weeks, paralleling the increase in maternal oestriol output.

A clinical application of liquor oestriol levels has been suggested by the finding of a correlation between liquor oestriol concentration and maternal urinary oestriol (Berman *et al* 1968, Klopper & Biggs 1970). However, Michie & Livingstone (1968) found no correlation between the levels in the two fluids.

TABLE 6.10. Average oestriol values in liquor amnii in term pregnancies

Author	Oestriol concentration (μg/100 ml)
Diczfalusy & Magnusson (1958)	79·4
Schindler & Herrmann (1966)	97·9
Schindler & Ratanasopa (1968)	67·4
Berman *et al* (1968)	97·1
Aleem *et al* (1969)	135·2
Michie & Livingstone (1969)	91·1
Klopper & Biggs (1970)	101·0

All the figures mentioned above have been for the total oestriol content of liquor, but Troen *et al* (1961) measured the various fractions of oestriol; 2·5% is free and 97·5% is conjugated with 85% of the conjugated material being glucosiduronate and 15% sulphate.

The great preponderance of oestriol glucosiduronate when compared with the sulphate has been explained by the differential rate of both entry and exit for the two compounds. The placental membranes have been shown to contain a sulphatase which hydrolyses the conjugate, thus liberating the free compound which more easily crosses the membranes and thus leaves the amniotic cavity (Levitz 1966). There is no β-glucuronidase in placenta and membranes and any oestriol glucosiduronate leaving the liquor via the membranes must cross as the conjugate which is only slowly transported.

The data of Troen *et al* (1961) and Diczfalusy *et al* (1964) suggest that foetal urine is the main source of the glucosiduronates in liquor. Their work showed that 98·5% of urinary oestriol was conjugated and that

89% of the conjugated material was oestriol glucosiduronate and 11% sulphate. However, when urine from mid-pregnancy foetuses was examined no oestriol sulphate was found (Goebelsmann *et al* 1966). The explanation may be that as the foetus matures the pattern of excretion changes, or that changes in conjugation in the urine occur on standing or more likely, that it is incorrect to equate urine passed by the neonate from birth to 48 hours with that passed immediately at delivery. It is suggested that only urine passed at delivery is representative of urine voided by the foetus *in utero* and only this urine should be used for steroid estimations.

The route of entry of oestriol sulphate into the liquor is unclear but Klopper (1970) has shown that this substance can pass from the foetal circulation via the umbilical cord into the liquor. Thus present evidence suggests that the main route of glucosiduronate entry into the liquor is by foetal urine and that of the sulphates by the umbilical cord, but more work is needed to confirm this statement. Klopper (1970) has also pointed out that oestriol sulphate is readily hydrolysed to free oestriol in the liquor and that for physiological purposes free oestriol and oestriol sulphate may be considered as a single entity.

The only situation in which there has been a correlation between liquor oestriol and the clinical state of the foetus has been rhesus isoimmunization. Schindler *et al* (1967) showed that while plasma and urinary oestriol were unaffected by severe rhesus isoimmunization, liquor oestriol levels reflected the foetal condition and were particularly useful following intra-uterine transfusion. Aleem *et al* (1969) studied 67 patients, finding no difference in the mean oestriol concentration in mildly affected cases and normals while in moderately affected cases, the oestriol concentration was lowered only in the period 37–40 weeks. In severe rhesus cases, the oestriol concentration was significantly lowered from as early as 26 weeks. This lowering of oestriol concentration may be related to the lowering of the oestriol precursors in the liquor.

Maternal urinary oestriol levels provide a guide to foetal health but as measurements on liquor are closer to the foetus than those on maternal urine, liquor investigations may prove to be a valuable clinical tool (Klopper 1970). However, basic information on levels of the oestriol fractions and their correlation with foetal well-being is still lacking. Until they are available, the potential value of liquor measurements is uncertain.

PREGNANETRIOL

One of the early indications that liquor might reflect foetal steroid metabolism was the diagnosis of congenital adrenal hyperplasia *in utero* by Jeffcoate *et al* (1965) who found a raised level of pregnanetriol (10·6 μg/100 ml) in liquor compared with their normal value of 2·3 μg/100 ml. Nichols & Gibson (1970) quoted two other cases of this syndrome similarly diagnosed by finding raised liquor levels of pregnanetriol. However, Merkatz *et al* (1969) were unable to confirm these findings as they examined liquor pregnanetriol levels in normal infants and those which were diagnosed as adrenal hyperplasia at birth and found no difference. It is not known whether differing biochemical techniques could explain these different results.

No firm conclusions can yet be drawn on the value of measuring steroids in liquor. It has proved difficult to differentiate between maternal and foetal contributions and there is very little definite information on the pathways of entry and exit. They may reflect foetal hypothalamic-pituitary-adrenal axis activity and thus help to define the foetal role in the onset of labour. In early pregnancy any difference between male and female steroid patterns may be highlighted thus helping to understand the differentiation of the reproductive tract. Later in pregnancy foetal endocrine abnormalities may be diagnosed thus enabling treatment *in utero* while lastly foetal well-being and growth may be more accurately charted. These aims are far from being fulfilled at the moment.

CELLS OF THE LIQUOR AMNII

Various authors have described the cells found in the liquor amnii. They used different staining methods and differing modes of cell classification thus making any comparison of their findings difficult.

Hoyes (1968a) using light and electron microscopy examined liquor cells in gestations up to 21 weeks. There were few cells present in smears of fluid from foetuses younger than 16 weeks, a fact confirmed by Wachtel *et al* (1969). In specimens from fetuses of 16–21 weeks menstrual age, cells were much more numerous but only a proportion contained nuclei, the number of nucleated cells falling from 50–75% in the younger specimens to 30% in the oldest (21 weeks).

Hoyes described two types of cell. Type I were relatively large, often polygonal and anucleate. Surface microvilli were easily seen by electron microscopy and the cells contained much glycogen. As they matured (i.e. from 16 to 21 weeks) the glycogen disappeared and the nuclei became pyknotic. This cell type originated from the foetal epidermis.

The second cell type was about half the size of the first type with a prominent nucleus and rather dense, sometimes vacuolated cytoplasm. Electron microscopy showed a well-developed Golgi apparatus and a number of membrane-found bodies containing electron dense material in the cytoplasm. These cells often showed direct evidence of phagocytic activity. Their origin was thought to be the amniotic epithelium.

Huisjes (1970) examined the cells in liquor from pregnancies of 25 to 43 weeks, staining them by the Harris-Shorr method. He found the following groups of cells:

(1) Nucleated and anucleate eosinophilic cells. These had red to orange cytoplasm and came from the foetal mouth.

(2) Nucleated and anucleate cyanophilic cells. The cytoplasm is light blue and they originate from the vulva and oral cavity.

(3) Small nucleated and anucleate cyanophilic cells. These cells are more or less round and tend to form clusters. It is believed they are urinary in origin.

(4) Polygonal cells. The cytoplasm is transparent orange. They tended to cluster, were rare before the 38th week and originated from foetal epidermis.

Two other less regularly occurring cell forms probably originating from amnion and umbilical cord were described by Huisjes who also considered that the anucleate forms of cells described under 1 and 2 above were degenerative forms of the nucleated cells.

Examination of the liquor cells became important clinically when it was suggested that foetal maturity could be estimated by counting the percentage of cells which stained orange after the liquor was mixed with Nile blue sulphate. Gordon & Brosens (1967) showed that from 34 to 38 weeks $1-10\%$ of cells stained orange while after 38 weeks the percentage rose, the figure being $10-50\%$. They considered that the cells originated from the foetal sebaceous glands.

The origin of the orange-staining cells was discussed by Sharp (1968) who pointed out that sebaceous glands are holocrine glands and that it would be unlikely for them to excrete intact cells in large numbers. Huisjes (1970) believed that the orange cells seen in increased numbers in the liquor after the 38th week were the same as the polygonal cells found in preparations stained by the Harris-Shorr procedure. He felt that the orange droplets and patches of fat are not situated within the cell but on its surface.

Floyd et al (1969) used the Papanicolaou stain and nomenclature to classify the cells found in liquor. They, too disagreed with the belief that the orange-stained cells were sebaceous in origin believing that neutral lipids adhered to the exfoliated cells in increasing amounts as gestation proceeded. Not all authors believe that the Nile blue staining of cells indicated foetal maturity accurately. Chan et al (1969) found that more than 10% of cells stained orange only after the 38th week but that this percentage was recorded in only 60% of cases of gestational age 38–42 weeks. However, Sharma & Trussell (1970) in a large series found the technique valuable. They confirmed the findings of Bishop & Corson (1968) that the proportion of orange-staining cells is directly related to foetal maturity and not to birth weight. Sharma and Trussell did find a low cell count in a number of cases (4%) but in all of them there were large particles of vernix caseosa clearly visible and orange-staining droplets of free lipid were seen microscopically. These authors believe that these three factors should be taken into account in assessing foetal maturity.

The factors are:

(1) percentage of orange-staining cells,
(2) presence of free particles of vernix caseosa, and
(3) presence of orange-staining droplets seen microscopically.

Whether these factors can explain the conflicting reports over the value of Nile blue staining of cells in assessing foetal maturity is unknown. Until further information is available, this test, like liquor creatinine and bilirubin levels, must be used only as a guide, the clinical assessment of the patient remaining paramount.

Important advances have been made in the field of intra-uterine paediatrics by examining the cells of the liquor amnii. The information obtained in this way may be listed as follows:

(1) Direct examination of cells; (a) sex-chromatin studies, (b) blood group determination, and (c) enzymes.

(2) Examination of cultured cells, (a) karyotyping, (b) enzymes, (c) autoradiography, (d) histochemistry.

The sex of the foetus can be determined by examination of the liquor cells for Barr bodies. The presence of sex-chromatin positive cells in 5% or more of cells counted is considered by Abbo & Zellweger (1970) to indicate two X-chromosomes. Pre-natal sex determination should be performed when the mother is a true or probable carrier of a mutant gene of X-linked diseases such as Duchenne and Becker type of muscular dystrophy, the X-linked variant of mucopolysaccharidoses, haemophilia, Lesch-Nyhan hyperuricaemia and Lowe's syndrome (Abbo & Zellweger, 1970). It is probable that foetal sex can be determined accurately before the 20th week (Emery, 1970).

Foetal ABO blood group can be determined from foetal squames in liquor but not the rhesus blood group (Scott *et al* 1969). There does not as yet appear to be any practical application of this knowledge.

It is possible to detect some inborn errors of metabolism before birth by enzyme measurements in cells from the liquor. Many enzymes have been demonstrated in cultured cells obtained from liquor as early as 10 weeks gestation (Nadler 1969).

A large number of congenital metabolic diseases have been diagnosed from study of amniotic cells. Emery (1970) has sounded a note of caution in pointing out that the normal values of these enzymes in cultured cells from all stages of pregnancy are unknown and it can thus be dangerous to diagnose an enzyme deficiency when normal values are unknown. Nadler & Gerbie (1969) demonstrated a wide range of enzymes in uncultured liquor cells.

Chromosome analysis of liquor cells is now possible but there are many technical difficulties. The cells must first be cultured successfully and then not all are found to be satisfactory for karyotyping. The analysis is used where one of the parents, usually the mother, is known to be a carrier of a translocation such as a D/G translocation. If the translocated chromosome as well as the normal chromosome -21 is passed on to the foetus then it is effectively trisomic for chromosome -21, though still having forty-six chromosomes (chapter 1).

Autoradiography, an extension of biochemical studies on cultured liquor cells, has been used to diagnose the Lesch-Nyhan syndrome (uric aciduria) which is inherited as an X-linked recessive trait. Sufferers of this disease lack the enzyme hypoxanthine-guanine phosphoribosyltransferase, the presence of which can be shown by autoradiography.

In a number of hereditary disorders metachromatic granules can be demonstrated by staining cultured skin fibroblasts with toluidine blue. The disorders thus diagnosed include the mucopolysaccharides, fibrocystic disease, Gaucher's disease, amaurotic idiocy and Marfan's syndrome (Emery 1970).

This technique has been used to diagnose Hunter and Hurler's syndrome *in utero* (Fratantoni *et al* 1969). The disadvantage of this technique is that it may take many weeks for the metachromasia to develop in the cultured liquor cells.

There is no doubt that examination of liquor cells is an important advance in preventative medicine. The techniques are specialized, much has yet to be learned about normal values and ranges and as yet there are difficulties in the way of early and quick diagnosis. The rate of advance in this field is expected to be rapid.

AMINO ACIDS

Levels of amino acids in liquor throughout pregnancy (Emery *et al* 1970) and in foetal fluids from 15–20 weeks (Cockburn *et al* 1970) have recently been documented. Emery and co-workers measured the levels of twenty-seven amino acids and noted three patterns of change at various stages of pregnancy. The concentration of some amino acids remained more or less the same throughout pregnancy (cysteic acid phosphoethanolamine, ethanolamine and proline). In the second group of compounds (serine, glycine, phenylalanine, lysine and arginine) the concentration decreased progressively from the first trimester to term while the remaining substances appeared to have a lower concentration in the first trimester than in the period 13–16 weeks. The levels then fell till 36 weeks after which the concentration rose till term (Emery *et al* 1970).

Cockburn *et al* (1970) measured 30 amino acids in maternal and foetal plasma, foetal urine and liquor in mid-pregnancy (15–20 weeks) finding that foetal plasma concentration was higher than liquor. The amino acids were present in foetal urine which obviously forms one pathway of entry of these compounds into the liquor. However, in view of the figures quoted earlier for liquor turnover and foetal urine output at this stage of pregnancy, it is likely that amino acids enter liquor by other routes as well as foetal urine. The figures of Cockburn *et al* (1970) are interesting as they showed that there were a high number of significant positive correlations between the amino-acid concentration in foetal urine and liquor suggesting that close relationship existed between the two fluids. Thus these workers suggested that foetal urine makes a significant contribution to the liquor levels or that the chorio-amnion transfers amino acid in a pattern comparable to that maintained by the foetal kidney tubule. The true position is unknown at present.

The clinical importance of these findings lies in the fact that ante-natal diagnosis of some inborn errors of amino-acid metabolism (e.g. cystinuria, glycinuria, Hartnup's disease) may now be possible.

LIPIDS

Lipids are present in liquor in steady concentrations throughout pregnancy with phospholipids being the only group that increase in concentration as term approaches, the level rising from 3·15 mg/100 ml at 27–33 weeks to 5·19 mg/100 ml at 34–40 weeks (Biezinski et al 1968). These workers found that the average values (mg/100 ml) for other lipid classes were: monoglycerides 0·23, diglycerides 0·9, triglycerides 1·4, free fatty acids 2·11, free cholesterol 1·41, cholesterol ester 1·97.

Nelson (1969) also measured the lipids in liquor; the concentration of total lipids was 60 mg/100 ml and of phospholipids 15 mg/100 ml. The most abundant phospholipid was lecithin but lysolecithin, sphingomyelin, phosphatidyl serine and phosphatidyl ethanolamine were also identified. In liquor from premature infants suffering from respiratory distress syndrome and in anencephaly with polyhydramnios Nelson found low total lipids and phospholipids and a decreased percentage of lecithin. It is unknown whether there are any clinical implications in this association of lowered liquor lipid values and the respiratory distress syndrome. The origin of these substances is unknown but they may come from the foetal lung as lecithin has been identified as a major constituent of alveolar surfactant (Nelson 1969).

PROSTAGLANDINS

The discovery of prostaglandins E_1, E_2, $F_1\alpha$ and $F_2\alpha$ in liquor is of great interest as it has been shown that both E_2 and $F_2\alpha$ cause contraction of strips of pregnant myometrium *in vitro* while intravenous infusions stimulate the uterus to contract in the first and second trimesters of pregnancy (Karim & Hillier 1970).

In early and mid-pregnancy only E_1 is present with concentrations of 0·06 to 3·1 mg/ml while term liquor, collected at artificial rupture of the membranes where labour had not begun, contained E_1 (0·11–4·1 mg/ml) and in three out of eleven cases E_2 (Karim & Devlin 1967). Karim & Hillier (1970) then examined the prostaglandin content of liquor obtained from the amniotic sac following spontaneous abortion at 13–23 weeks. They found high concentrations of E_2 (2·6–18 mg/ml) and F_2 (3·9–36·0 mg/ml) while in a matched series of control liquor from hysterotomies they found no E_2 or $F_2\alpha$. Likewise in liquor from normal patients at term in labour, $F_2\alpha$ in high concentrations (5·5–57·0 mg/ml) was always present (Karim & Devlin, 1967). E_2 was also present in the majority of cases though in lower concentrations (0·04–2·3 mg/ml).

These results suggest that the prostaglandins might play a role in the aetiology of spontaneous abortion and perhaps in the onset of normal labour though the evidence is still circumstantial. Karim & Devlin (1970) suggested that the decidual cells form the prostaglandins which then pass to both liquor and myometrium (Chapter 4).

RENIN

Brown *et al* (1964) found high concentrations of renin (or a renin-like substance) in liquor at 11 weeks gestation (1060 units/l) and at term (37–1560 units/l). The enzyme concentration exceeded that in the umbilical vein plasma and in the maternal peripheral venous blood. Skinner *et al* (1968) found that the chorion contained the highest concentration of renin when compared with liquor, amnion, deciduum, myometrium and maternal and foetal plasma. They concluded that the chorion was the source of liquor renin and that the decidua effectively limited outward diffusion, promoting intra-amniotic accumulation.

The function of the large amounts of renin in liquor is unknown. It might influence sodium homeostasis *in utero* by affecting passage of the ion through the membranes or it may contribute to maternal circulating renin thus influencing maternal sodium homeostasis.

CONCLUSIONS

Physiological conditions can change at each stage of pregnancy and it is therefore necessary to investigate the various aspects of liquor in pregnancies of similar age or physiological development.

While water makes up the bulk of liquor and examination of the sites of exchange and methods of control of water exchange may help to decide if there is any single factor which controls the volume, each individual group of constituents needs to be examined for the route of entry, exit and mode of control.

It is thought that an understanding of the biophysical forces which control flow of water across membranes is necessary before any mechanisms controlling liquor volume can be postulated. At present the general picture is far from clear, as neither clinical nor experimental evidence has been correlated. There is scanty knowledge of what happens *in utero*. Foetal swallowing, voiding and turnover of water, the importance of diffusion, osmotic and hydrostatic pressures, and the effect that the foeto-placental circulation may have on these forces, is unknown.

There is increasing diagnostic use of liquor and a proper understanding of the basic physiology of liquor could extend the clinical benefits available from its examination.

REFERENCES

ABBAS T.M. & TOVEY J.E. (1960) Proteins of the liquor amnii. *Brit. med. J.* **1,** 476.
ABBO G. & ZELLWEGER H. (1970) Prenatal determination of fetal sex and chromosomal complement. *Lancet* **1,** 216.
ABRAMOVICH D.R. (1968) The volume of amniotic fluid in early pregnancy. *J. Obstet. Gynaec. Brit. Cwlth.* **75,** 728.
ABRAMOVICH D.R. (1970a) *Some Aspects of Liquor Amnii.* Ph.D. thesis, University of Liverpool.
ABRAMOVICH D.R. (1970b) Fetal factors influencing the volume and composition of liquor amnii. *J. Obstet. Gynaec. Brit. Cwlth.* **77,** 865.
ABRAMOVICH D.R. & WADE A.P. (1969a) Transplacental passage of steroids. *J. Obstet. Gynaec. Brit. Cwlth.* **76,** 610.
ABRAMOVICH D.R. & WADE A.P. (1969b) Levels and significance of 17-oxosteroids and 17-hydroxycorticosteroids in amniotic fluid throughout pregnancy. *J. Obstet. Gynaec. Brit. Cwlth.* **76,** 893.
ADAMS F.H., DESILETS D.T. & TOWERS B. (1967) Control of flow of fetal lung fluid at the laryngeal outlet. *Resp. Physiol.* **2,** 302.
AHLFELD F. (1888) quoted by S.Farber & L.K.Sweet. (1931) Amniotic sac contents in the lungs of infants. *Amer. J. Dis. Child.* **42,** 1372.
ALEEM F.A., NEILL D.W. & PINKERTON J.H. (1968) A method for oestriol estimation in amniotic fluid and its use in the study of normal and abnormal pregnancy. *Steroids* **12,** 651.
ALEEM F.A., PINKERTON J.H. & NEILL D.W. (1969) Clinical significance of the amniotic fluid oestriol level. *J. Obstet. Gynaec. Brit. Cwlth.* **76,** 200.
ALEXANDER D.P., NIXON D.A., WIDDAS W.F. & WOHLZOGEN F.X. (1958) Renal function in the sheep foetus. *J. Physiol.* **140,** 14.
ANDREWS B.F. (1970) Amniotic studies to determine maturity. *Ped. Clin. Nth. America* **17,** 49.
ARMSTRONG W.D., WILT J.C. & PRITCHARD E.T. (1968) Vacuolation in the human

amnion cell studied by time-lapse photography and electron microscopy. *Amer. J. Obstet. Gynec.* **102,** 932.

BAIN A.D. & SCOTT J.S. (1960) Renal agenesis and severe urinary tract dysplasia. *Brit. med. J.* **1,** 841.

BAIRD C.W. & BUSH I.E. (1960) Cortisol and cortisone content of amniotic fluid from diabetic and non-diabetic women. *Acta. endocr. (Kgh.)* **34,** 97.

BATTAGLIA F.C. & HELLEGERS A.E. (1964) Permeability to carbohydrates of human chorion laeve *in vitro. Amer. J. Obstet. Gynec.* **89,** 771.

BECKER R.F., WINDLE W.F., BARTH E.E. & SCHULZ M.D. (1940) Fetal swallowing, gastro-intestinal activity and defecation *in utero. Surg. Gynec. Obstet.* **70,** 603.

BEHRMAN R.E., PARER J.T. & DE LANNOY C.W. (1967) Placental growth and the formation of amniotic fluid. *Nature* **214,** 678.

BEISCHER N.A., BROWN J.B. & TOWNSEND C. (1969) Studies in prolonged pregnancy III. *Amer. J. Obstet. Gynec.* **103,** 496.

BELO'USSOVA V.I. (1969) Protein composition of the serum of the newborn and of the amniotic fluid in normal pregnancy. *Voprosy Okhrany Materinstva i Detstva* **14,** 90.

BENIRSCHKE K. & MCKAY D.G. (1953) The anti-diuretic hormone in the fetus and infant. *Obstet. Gynec.* **1,** 638.

BERK H. & SUSSMAN L. (1970) Spectrophotometric analysis of amniotic fluid during early pregnancy. *Obstet. Gynec.* **35,** 170.

BERLE P. (1969) Der Gehalt an Chorialen Gonadotropin im Fruchtwasser wahrend normaler und pathologischer Schwangerschaft. *Acta. Endocr. (Kbh.)* **61,** 369.

BERMAN A.M., KALCHMAN G.G., CHATTORAJ S.C. & SCOMMEGNA A. (1968) Relationship of amniotic fluid oestriol to maternal urinary oestriol. *Amer. J. Obstet. Gynec.* **100,** 15.

BEVIS D.C.A. (1968) Errors arising from amniocentesis and the origin of the liquor amnii. *J. Obstet. Gynaec. Brit. Cwlth.* **75,** 1214.

BIEZINSKI J.J., POMERANCE W. & GOODMAN J. (1968) Studies on the origin of amniotic fluid lipids. *Amer. J. Obstet. Gynec.* **102,** 853.

BISHOP E.H. & CARSON S. (1968) Estimation of fetal maturity by cytological examination of amniotic fluid. *Amer. J. Obstet. Gynec.* **102,** 654.

BOURNE G. (1962) *The Human Amnion and Chorion.* London, Lloyd-Luke Ltd.

BROWN J.J., DAVIES D.L., DOAK P.B., LEVER A.F., ROBERTSON J.I.S. & TREE M. (1964) The presence of renin in human amniotic fluid. *Lancet* **2,** 64.

BRZEZINSKI A., SADOVSKY E. & SHAFRIR E. (1961) Electrophoretic distribution of proteins in amniotic fluid and in maternal and fetal serum. *Amer. J. Obstet. Gynec.* **82,** 800.

BUFE W. (1936) quoted by E.L.Potter & G.P.Bohlender (1941) Intrauterine respiration in relation to development of the fetal lung. *Amer. J. Obstet. Gynec.* **42,** 14.

CAMERON G. & CHAMBERS R. (1938) Direct evidence of function in kidney of an early human fetus. *Amer. J. Physiol.* **123,** 482.

CARTER C.O. (1960) *Congenital Malformations. Ciba Foundation Symposium,* p. 266. Churchill, London.

CASSADY G. & BARNETT R. (1969) Acid-base and gas tension studies of the amniotic fluid in human gestation. *Biol. Neonat.* **14,** 251.

CHAN W.H., WILLIS J. & WOODS J. (1969) The value of the Nile blue sulphate stain in the cytology of the liquor amnii. *J. Obstet. Gynaec. Brit. Cwlth.* **76,** 193.

CHARLES D. & JACOBY H.E. (1966) Preliminary data on the use of sodium aminohippurate to determine amniotic fluid volumes. *Amer. J. Obstet. Gynec.* **95,** 266.

CHARLES D., JACOBY H.E. & BURGESS F. (1965) Amniotic fluid volumes in the second half of pregnancy. *Amer. J. Obstet. Gynec.* **93,** 1042.

CHERRY S.H., KOCHWA S. & ROSENFIELD R.E. (1965) Bilirubin-protein ratio in amniotic fluid as an index of the severity of erythroblastosis fetalis. *Obstet. Gynec.* **26,** 826.

CHERRY S.H., ROSENFIELD R.E. & KOCHWA S. (1970) Mechanism of accumulation of amniotic fluid pigment in erythroblastosis fetalis. *Amer. J. Obstet. Gynec.* **106,** 297.

CHEZ R.A., SMITH F.C. & HUTCHINSON D.L. (1964) Renal function in the intrauterine primate fetus. *Amer. J. Obstet. Gynec.* **90,** 128.

COCKBURN F., ROBINS S.P. & FORFAR J.O. (1970) Free amino-acid concentrations in fetal fluids. *Brit. med. J.* **3,** 747.

COHEN B.M. (1970) The relationship between the fetal condition and the biochemical analysis of amniotic fluid at induction of labour. *J. Obstet. Gynaec. Brit. Cwlth.* **77,** 496.

CORSON S. & BOLOGNESE R.J. (1968) Amniotic fluid pH as an indicator of fetal asphyxia. *Obstet. Gynec.* **31,** 397.

CROSIGNANI P.G. & POLVANI F. (1969) Protein and human chorionic gonadotrophin in amniotic fluid in the third trimester of normal pregnancy. *J. Obstet. Gynaec. Brit. Cwlth.* **76,** 424.

DANCIS J., LIND J. & VARA P. (1960) In C.A.Villee (ed.) *Transfer of Labelled Proteins Across Human Placenta from the Placenta and Fetal Membranes*, p. 185. New York, Williams & Wilkins.

DAVIDSON W.M. & ROSS G.I.M. (1954) Bilateral absence of the kidneys and related congenital anomalies. *J. Path. Bact.* **68,** 459.

DAVIS M.E. & POTTER E.L. (1946) Intra-uterine respiration of the fetus. *J. Amer. med. ass.* **131,** 1194.

DAWES G.S. (1954) In discussion in *Water Metabolism*, R.A.McCance & E.M. Widdowson. *Cold Spring Harb. Symp. quant. Biol.* **19,** 164.

DAWES G.S., FOX H.E., LEDUC B.M., LIGGINS G.C. & RICHARDS R.T. (1970) Respiratory movements and paradoxical sheep in the foetal lamb. *J. Physiol.* **210,** 47P.

DERRINGTON M.M. & SOOTHILL J.F. (1961) An immunochemical study of the proteins of amniotic fluid and of maternal and foetal serum. *J. Obstet. Gynaec. Brit. Cwlth.* **68,** 755.

DICZFALUSY E. (1961) In discussion in *Transplacental Passage of Steroids*. C.J. Migeon, J.Bertrand & C.A.Gemzell (1968) *Rec. Progr. Horm. Res.* **17,** 246.

DICZFALUSY E., BARR M. & LIND J. (1964) Oestriol metabolism in an anencephalic monster. *Acta. endocr. (Kbh.)* **46,** 511.

DICZFALUSY E. & MAGNUSSON A.M. (1958) Tissue concentration of oestrone, oestradiol-17β and oestriol in the human fetus. *Acta endocr. (Kbh.)* **28**, 169.
DORAN T.A., BJERRE S. & PORTER C.J. (1970) Creatinine, uric acid and electrolytes in amniotic fluid. *Amer. J. Obstet. Gynec.* **106**, 325.
DROEGEMUELLER W., JACKSON C., MAKOWSKI E.L. & BATTAGLIA F.C. (1969) Amniotic fluid estimation as an aid in the assessment of gestational age. *Amer. J. Obstet. Gynec.* **104**, 424.
DUNSTAN M.K. (1968) Amniotic fluid volume and protein concentration in rhesus sensitized women. *J. Obstet. Brit. Cwlth.* **75**, 732.
EHRHARDT K. (1937) Der trinkende Fotus. *Munch. Med. Wschr.* **84**, 1699.
ELLIOTT P.M. & INMAN W.H.W. (1961) Volume of liquor amnii in normal and abnormal pregnancy. *Lancet* **2**, 835.
EMERY A.E.H. (1970) In A.E.H.Emery (ed.) *Modern Trends in Human Genetics*, p. 267. London, Butterworth.
EMERY A.E.H., BURT D., NELSON M.M. & SCRIMGEOUR J.B. (1970) Antenatal diagnosis and aminoacid composition of amniotic fluid, *Lancet* **1**, 1307.
FELIKS M. (1969) Permeability of the vascular vessels in the umbilical cord. *Pol. Med. J.* **7**, 520.
FLOYD W.S., GOODMAN P.A. & WILSON A. (1969) Amniotic fluid filtration and cytology. *Obstet. Gynec.* **34**, 583.
FRATANTONI J.C., NEUFELD E.F., UHLENDORF B.W. & JACOBSON C. (1969) Intrauterine diagnosis of the Hurler and Hunter syndromes. *New Eng. J. Med.* **280**, 686.
FRIEDBERG V. (1955) Untersuchungen über die fetale Urinbildung. *Gynecologia* **140**, 34.
GADD R.L. (1966) The volume of liquor amnii in normal and abnormal pregnancies. *J. Obstet. Gynaec. Brit. Cwlth.* **73**, 11.
GADD R.L. (1970) In E.E.Philipp, Josephine Barnes & M.Newton (eds.) *Scientific Foundations of Obstetrics and Gynaecology*, p. 254. London, Heinemann.
GANDY H.M. (1971) In F.Fuchs & A.Klopper (eds.) *Endocrinology of Pregnancy*, p. 124. New York, Harper & Row.
GARBY L. (1957) Studies on transfer of matter across membranes with special reference to the isolated human amniotic membrane and the exchange of amniotic fluid. *Acta Physiol. Scand.* **40**, Suppl. 137.
GERSH I. (1937) The correlation of structure and function in the developing mesonephros. *Carnegie Cont. to Embryol.* **26**, 33.
GILLIBRAND P.N. (1969a) The rate of water transfer from the amniotic sac with advancing pregnancy. *J. Obstet. Gynec. Brit. Cwlth.* **76**, 530.
GILLIBRAND P.N. (1969b) Changes in the electrolytes, urea and osmolality of the amniotic fluid with advancing pregnancy. *J. Obstet. Gynaec. Brit. Cwlth.* **76**, 893.
GOEBELSMANN U., ERIKSSON G., DICZFALUSY E., LEVITZ M. & CONDON G.P. (1966) Fate of oestriol-3-sulphate and oestriol-16-glucosiduronate in the intact foetoplacental unit at midpregnancy. *Acta. endrocr. (Kbh.)* **53**, 391.
GOODLIN R. & LLOYD D. (1968) Fetal tracheal excretion of bilirubin. *Biol. Neonat.* **12**, 1.

GORDEN H. & BROSENS I. (1967) A new test for fetal maturity. *Obstet. Gynec.* **30**, 652.
GOWAR F.J.S. (1935) Anephrogenesis. *J. Obstet. Gynaec. Brit. Emp.* **42**, 871.
GRIMES L.D. & CASSADY G. (1970) Fetal gastrointestinal obstruction. *Amer. J. Obstet. Gynec.* **106**, 1196.
GUTHMAN H. & MAY W. (1930) Gibt es eine intrauterine Nierensekretion? *Arch. Gynäk.* **141**, 450.
HEIRWEGH K.P.M., MEUWISSEN J.A.T.P. & JANSEN F.H. (1969) On the quantitation and analysis of bile pigments in amniotic fluid. *Biol. Neonat.* **14**, 74.
HELLMAN L.M., FLEXNER D.B., WILDE W.S., VOSBURGH G.J. & PROCTOR N.K. (1948) The permeability of the human placenta to water and the supply of water to the human fetus as determined with deuterium oxide. *Amer. J. Obstet. Gynec* **56**, 861.
HERON H.J. (1966) The electrophoresis of proteins of amniotic fluid. *J. Obstet. Gynaec. Brit. Cwlth.* **73**, 91.
HEWER E.F. (1924) Secretion by the human foetal kidney. *Quart. J. Exp. Physiol.* **14**, 49.
HIBBARD B.M. (1959) Hydrops foetalis in one of uniovular twins. *J. Obstet. Gynaec. Brit. Emp.* **66**, 649.
HIBBARD B.M. (1962) Polyhydramnios and oligohydramnios. *Clin. Obstet. Gynec.* **5**, 1044.
HOYES A.D. (1968a) Ultrastructure of the cells of the amniotic fluid. *J. Obstet. Gynaec. Brit. Cwlth.* **75**, 164.
HOYES A.D. (1968b) Fine structure of human amniotic epithelium in early pregnancy. *J. Obstet. Gynaec. Brit. Cwlth.* **75**, 949.
HOYES A.D. (1968c) Electron microscopy of the surface layer (periderm) of the human foetal skin. *J. Anat.* **103**, 321.
HOYES A.D. (1969d) Ultrastructure of the epithelium of the human umbilical cord. *J. Anat.* **105**, 149.
HUHTANIEMI I. & VIHKO R. (1970) Quantitation of neutral steroid sulphates in the amniotic fluid of early and mid-pregnancy. *Ann. Med. Exp. Fenn.* **48**, 188.
HUISJES H.J. (1970) Origin of the cells in the liquor amnii. *Amer. J. Obstet. Gynec.* **106**, 1222.
HUNTINGFORD P.J. & MARSDEN D. (1966) Osmotische Beziehungen zwischen Mutter, Fotus und Fruchtwasser. *Geburtsh. Frauenheilk.* **5**, 867.
HUTCHINSON D.L., HUNTER C.B., NESLEN E.D. & PLENTL A.A. (1955) The exchange of water and electrolytes in the mechanism of amniotic fluid formation and the relationship to hydramnios. *Surg. Gynec. Obstet.* **100**, 391.
HUTCHINSON D.L., GRAY M.J., PLENTL A.A., ALVAREZ H., CALDEYRO-BARCIA R., KAPLAN B. & LIND J. (1959) The role of the fetus in the water exchange of the amniotic fluid of normal and hydramniotic patients. *J. clin. Invest.* **38**, 971.
JAMES L.S. (1961) *Somatic Stability in the Newly Born. Ciba Foundation Symposium*, p. 295. London, Churchill.
JEFFCOATE T.N.A. (1931) Dystocia due to dilatation of the urinary tract in a foetus. *J. Obstet. Gynaec. Brit. Emp.* **38**, 814.
JEFFCOATE T.N.A. & SCOTT J.S. (1959) Polyhydramnios and oligohydramnios. *Canad. med. Ass. J.* **80**, 77.

JEFFCOATE T.N.A., FLIEGNER J.R., RUSSELL S.H., DAVIS J.C. & WADE A.P. (1965) Diagnosis of the adrenogenital syndrome before birth. *Lancet* **2**, 553.
JOST A. (1954) in discussion in *Water Metabolism*, R.A.McCance & E.M.Widdowson. *Cold Spring Harb. Symp., quant. Biol.* **19**, 164.
KARCHMER S., ONTIVEROS E., SHOR V. & ALMAREZ R. (1969) Estudio de la deglucion por el feto anencefalo in utero. *Ginec. Obstet. Mex.* **25**, 489.
KARIM S.M.M. & DEVLIN J. (1967) Prostaglandin content of amniotic fluid during pregnancy and labour. *J. Obstet. Gynaec. Brit. Cwlth.* **74**, 230.
KARIM S.M.M. & HILLIER K. (1970) Prostaglandins and spontaneous abortion. *J. Obstet. Gynaec. Brit. Cwlth.* **77**, 837.
KJELLBERG S.R. & RUDHE V. (1949) The fetal renal secretion and its significance in congenital deformities of the uterus and urethra. *Acta. Radiol.* **31**, 243.
KLOPPER S. & STEVENSON R. (1966) The excretion of oestriol and of pregnanediol in pregnancy complicated by Rh immunization. *J. Obstet. Gynaec. Brit. Cwlth.* **73**, 282.
KLOPPER A. (1970) Steroids in amniotic fluid. *Annals of Clinical Research* **2**, 289.
KLOPPER A. & BIGGS J. (1970) The correlation between urinary oestriol excretion and the oestriol concentration in liquor amnii. *J. Endocr.* **48**, 310.
LAMBERT M. & PENNINGTON G.W. (1963) 6β-hydroxy-cortisol in liquor amnii. *Nature* **197**, 391.
LAMBERT M. & PENNINGTON G.W. (1964) Isolation and identification of the 20β-hydroxy derivatives of 6β-hydroxy-cortisol and 6β-hydroxy-cortisone in liquor amnii. *Nature* **203**, 656.
LEE T.Y. & WEI P.Y. (1970) Spectrophotometric analysis of amniotic fluid in anencephalic pregnancies. *Amer. J. Obstet. Gynec.* **107**, 917.
LEVITZ M. (1966) Conjugation and transfer of feto-placental steroid hormones. *J. clin. Endocr. & Metab.* **26**, 773.
LILEY A.W. (1963) In H.M.Carey (ed.) *Modern Trends in Reproductive Physiology*, p. 227. London, Butterworth.
LIND T., PARKIN F.M. & CHEYNE G.A. (1969) Biochemical and cytological changes in liquor amnii with advancing gestation. *J. Obstet. Gynaec. Brit. Cwlth.* **76**, 673.
LLOYD J.R. & CLATWORTHY H.W. (1958) Hydramnios as an aid to the early diagnosis of congenital obstruction of the alimentary tract. *Pediatrics*, **21**, 903.
LUUKKAINEN T., SIEGEL A. & VIHKO R. (1970) Neutral steroid sulphates in amniotic fluid. *J. Endocr.* **46**, 391.
MCCANCE R.A. & WIDDOWSON E.M. (1953) Renal function before birth. *Proc. roy. Soc. B.* **141**, 488.
MAKEPEACE A.W., FREMONT-SMITH F., DAILEY, M.E. & CARROLL M.P. (1931) The nature of the amniotic fluid. *Surg. Gyn. Obstet.* **53**, 635.
MANDELBAUM B., LA CROIX G.C. & ROBINSON A.R. (1967) Determination of fetal maturity by spectrometric analysis of amniotic fluid. *Obstet. Gynec.* **29**, 471.
MANDELBAUM B. & EVANS T.N. (1969) Life in the amniotic fluid. *Amer. J. Obstet. Gynec.* **104**, 365.
MENDENHALL H.W. (1970) Serum protein concentrations in pregnancy. *Amer. J. Obstet. Gynec.* **106**, 581.

MENEES T.O., MILLER, J.D. & HOLLY L.E. (1930) Amniography. *Am. J. Roentg.* **24**, 363.
MERKATZ I.R., NEW M.I., PETERSON R.E. & SEAMAN M.P. (1969) Prenatal diagnosis of adrenogenital syndrome by amniocentesis. *J. Pediat.* **75**, 977.
MICHIE E.A. & LIVINGSTONE J.R. (1969) Oestriol concentration in amniotic fluid. *Acta. endocr. (Kbh.)* **61**, 329.
MIGEON C.J., BERTRAND J. & GEMZELL C.A. (1961) Transplacental passage of steroids. *Rec. Progr. Horm. Res.* **17**, 207.
MILES P.A. & PEARSON J.W. (1969) Amniotic fluid osmolality in assessing fetal maturity. *Obstet. Gynec.* **34**, 701.
MOORE W.M.O., HELLEGERS A.E. & BATTAGLIA F.C. (1966) *In vitro* permeability of different layers of human placenta to carbohydrates and urea. *Amer. J. Obstet. Gynec.* **96**, 951.
MOORE W.M.O. & WARD B.S. (1970) Placental membrane permeability to creatinine and urea. *Amer. J. Obstet. Gynec.* **108**, 635.
MORRIS E.D., MURRAY J. & RUTHVEN C.R.J. (1967) Liquor bilirubin levels in normal pregnancy: a basis for accurate prediction of haemolytic disease. *Brit. med. J.* **2**, 352.
MURRAY J., NORRIE D.L. & RUTHVEN C.R.J. (1970) Liquor bilirubin levels in normal pregnancy. *Brit. med. J.* **4**, 387.
NADLER H.L. (1969) Prenatal detection of genetic defects. *J. Pediat.* **74**, 132.
NADLER H.L. & GERBIE A.B. (1969) Enzymes in noncultured amniotic fluid cells. *Amer. J. Obstet. Gynec.* **103**, 710.
NELSON G.H. (1969) Amniotic fluid phospholipid patterns in normal and abnormal pregnancies. *Amer. J. Obstet. Gynec.* **105**, 1072.
NICHOLS J. & GIBSON G.G. (1970) Antenatal diagnosis of the adrenogenital syndrome. *Lancet* **2**, 1068.
NICHOLS J. & SCHREPFER R. (1966) Polyhydramnios in anencephaly. *J. Amer. med. ass.* **197** (2), 549.
O'LEARY J.A. & FELDMAN M. (1970) Amniotic fluid osmolality in the determination of fetal age and welfare. *Obstet. Gynec.* **36**, 525.
PARMELY T.H. & SEEDS A.E. (1970) Fetal skin permeability to isotopic water in early pregnancy. *Amer. J. Obstet. Gynec.* **108**, 128.
PARRY E.W. & ABRAMOVICH D.R. (1970) Some observations on the surface layer of full-term human umbilical cord epithelium. *J. Obstet. Gynec.* **77**, 878.
PEDERSEN J. (1954) Glucose content of the amniotic fluid in diabetic pregnancies. *Acta. endocr. (Kbh.)* **15**, 342.
PEDERSEN J. & JORGENSEN G. (1954) Hydramnios in diabetes. *Acta. endocr. (Kbh.)* **15**, 333.
PITKIN R.M. (1969) In discussion in role of chemical and cytologic analysis of amniotic fluid in determination of fetal maturity. C.A.White, D.E.Doorenbos & J.T.Bradbury. *Amer. J. Obstet. Gynec.* (1969), **104**, 670.
PLENTL A.A. (1961) Transfer of water across perfused umbilical cord. *Proc. Soc. Exp. Biol. Med.* **107**, 622.
POTTER E.L. & BOHLENDER G.P. (1941) Intrauterine respiration in relation to development of the fetal lung. *Amer. J. Obstet. Gynec.* **42**, 14.

POTTER E.L. (1961) *The Pathology of the Fetus and Newborn*, 2nd ed. Chicago, Year Book Publishers.
PRITCHARD J.A. (1965) Deglutition by normal and anencephalic fetuses. *Obstet. Gynec.* **25**, 289.
PRITCHARD J.A. (1966) Fetal swallowing and amniotic fluid volume. *Obstet. Gynec.* **28**, 606.
QUEENAN J.T. & GADOW E.C. (1970) Polyhydramnios. *Amer. J. Obstet. Gynec.* **108**, 349.
QUEENAN J.T., GADOW E.C. BACHNER P. & KUBARYCH S.F. (1970) Amniotic fluid proteins in normal and Rh sensitized pregnancies. *Amer. J. Obstet. Gynec.* **108**, 406.
QUILLIGAN E.J. (1966) Amniotic fluid gas tensions. *Amer. J. Obstet. Gynec.* **84**, 20.
REIFFERSCHEIDT K. (1911) quoted by S.Farber & L.K.Sweet (1931) Amniotic sac contents in the lungs of infants. *Amer. J. Dis. Child.* **42**, 1372.
REIFFERSCHEIDT W. & SCHMIEMANN R. (1939) quoted by E.L.Potter & G.P.Bohlender, Intrauterine respiration in relation to development of the fetal lung. *Amer. J. Obstet. Gynec.* **42**, 14.
REYNOLDS S.R.M. (1953) A source of amniotic fluid in the lamb: the nasopharyngeal and buccal cavities. *Nature* **172**, 307.
REYNOLDS W.A., PITKIN R.M. & HODARI A.A. (1969) Transfer of iodine into amniotic fluid by the normal and the nephrectomized subhuman primate fetus. *Amer. J. Obstet. Gynec.* **104**, 633.
RHODES P. (1966) The volume of liquor in early pregnancy. *J. Obstet. Gynaec. Brit. Cwlth.* **73**, 23.
ROOPNARINESINGH S. (1970) Amniotic fluid creatinine in normal and abnormal pregnancies. *J. Obstet. Gynaec. Brit. Cwlth.* **77**, 785.
ROSA P. (1951) Etude de la circulation du liquide amniotique humain. *Gynec. Obstet.* **50**, 463.
RUOSLAHTI E., THALLBERG TH. & SEPPALA M. (1966) Origin of proteins in amniotic fluid. *Nature* **212**, 841.
SCHILLER W. & TOLL R.M. (1927) An inquiry into the cause of oligohydramnios. *Amer. J. Obstet. Gynec.* **12**, 689.
SCHINDLER A.E. & HERRMANN W.L. (1966) Estriol in pregnancy urine and amniotic fluid. *Amer. J. Obstet. Gynec.* **95**, 301.
SCHINDLER A.E. & RATANASOPA V. (1968) Profile of steroids in amniotic fluid of normal and complicated pregnancies. *Acta endocr. (Kbh.)* **59**, 239.
SCHINDLER A.E., RATANASOPA V., LEE T.Y. & HERRMANN W.L. (1967) Estriol and Rh isoimmunization. *Obstet. Gynec.* **29**, 625.
SCHINDLER A.E. & SIITERI P.K. (1968) Isolation and quantitation of steroids from normal human amniotic fluid. *J. clin. Endocr. & Metab.* **28**, 1189.
SCHREINER W.E. & SCHMID J. (1969) The clinical significance of biochemical tests on the amniotic fluid in the early detection of fetal hypoxia, in P.J.Huntingford, K.A.Huter & E.Saling (eds.) *Perinatal Medicine*, p. 20. New York, Academic Press.
SCOGGIN W.A., HARBERT G.M., ANSLOW W.P., VAN'T RIET B. & MCGAUGHEY H.S. (1964) Fetomaternal exchange of water at term. *Amer. J. Obstet. Gynec.* **90**, 7.

SCOTT J.S. & WILSON J.K. (1957) Hydramnios as an early sign of oesphageal atresia. *Lancet* **2**, 569.

SCOTT J.S., COULSON A. & GOULDEN R. (1969) Antenatal determination of fetal blood groups. *J. Obstet. Gynaec. Brit. Cwlth.* **76**, 330.

SEEDS A.E. (1967) Water transfer across the human amnion in response to osmotic gradients. *Amer. J. Obstet. Gynec.* **98**, 568.

SEEDS A.E. (1968) In C.A.Barnes (ed.) *Amniotic Fluid and Fetal Water Metabolism. Intrauterine Development*, p. 129. Philadelphia, Lea & Febiger.

SEEDS A.E. (1970) Osmosis across term human placental membranes. *Amer. J. Physiol.* **219**, 551.

SEEDS A.E. & HELLEGERS A.E. (1968) Acid-base determinations in human amniotic fluid throughout pregnancy. *Amer. J. Obstet. Gynec.* **101**, 257.

SEEDS A.E., KOCK H.C., MYERS R.E., STOLTE L.A.M. & HELLEGERS A.E. (1967) Changes in rhesus monkey amniotic fluid pH, pCO_2 and bicarbonate following maternal and fetal hypercarbia and fetal death *in utero*. *Amer. J. Obstet. Gynec.* **97**, 67.

SEPPALA M., RUOSLAHTI & THALBERG T. (1966) Genetical evidence for maternal origin of amniotic fluid proteins. *Ann. Med. Exp. Fenn.* **44**, 6.

SHARMA S.D. & TRUSSELL R.R. (1970) The value of amniotic fluid exfoliative cytology. *J. Obstet. Gynaec. Brit. Cwlth.* **75**, 812.

SHARP F. (1968) Estimation of fetal maturity by amniotic fluid exfoliative cytology. *J. Obstet. Gynec. Brit. Cwlth.* **75**, 812.

SIEGEL A.L., ADLERCREUTZ H. & LUUKKAINEN T. (1969) Gas chromatographic and mass spectrometric identification of neutral and phenolic steroids in amniotic fluid. *Ann. Med. Exp. Fenn.* **47**, 22.

SIDDAL R.S. (1924) Chorioangiofibroma. *Amer. J. Obstet. Gynec.* **8**, 430.

SKINNER S.L., LUMBERS E.R. & SYMONDS E.M. (1968) Renin concentration in human fetal and maternal tissues. *Amer. J. Obstet. Gynec.* **101**, 529.

SUZUKI K. & PLENTL A.A. (1969) Chronic implantation of instruments in the neck of the primate fetus for physiological studies and production of hydramnios. *Amer. J. Obstet. Gynec.* **103**, 272.

SYLVESTER P.E. & HUGHES D.R. (1954) Congenital absence of both kidneys. *Brit. med. J.* **1**, 77.

TAHTI E. (1966) Kidney function in early foetal life. *Brit. J. Radiol.* **39**, 226.

TAUSCH M. (1936) Der Fetalharn. *Arch. Gynäk.* **162**, 217.

TAUSSIG F.J. (1927) The amniotic fluid and its quantitative variability. *Amer. J. Obstet. Gynec.* **14**, 505.

TERVILA L. (1964) Transfer of water from maternal blood to amniotic fluid of live and dead fetuses in health and in some pathological conditions of the mother. *Ann. Chir. et Gyn. Fenn.* Suppl. **53**, 131.

THOMAS C.E. (1965) The ultrastructure of human amnion epithelium. *J. Ultrastruct. Res.* **13**, 65.

TOWERS B. (1968) In N.S.Assali (ed.) *Biology of Gestation*, vol. 2, p. 203. New York, Academic Press.

TROEN P., NILSSON B., WIQVIST N. & DICZFALUSY E. (1961) Pattern of oestriol

conjugates in human cord blood, amniotic fluid and urine of newborns. *Acta Endocr. (Kbh)* **38,** 361.
VERNIER R.L. & SMITH R.G. (1968) In N.S.Assali (ed.) *Biology of Gestation*, vol. 2, p. 248. New York, Academic Press.
VILLEE C.A. (1954) In discussion in *Water Metabolism*, R.A.McCance & E.M. Widdowson. *Cold Spring Harb. Symp. quant. Biol.* **19,** 165.
VOSBURGH G.J., FLEXNER L.B., COWIE D.B., HELLMAN L.M., PROCTOR N.K. & WILDE W.S. (1948) The rate of renewal in women of the water and sodium of the amniotic fluid as determined by tracer techniques. *Amer. J. Obstet. Gynec.* **56,** 1156.
WACHTEL E., GORDON H. & OLSEN E. (1969) Cytology of amniotic fluid. *J. Obstet. Gynaec. Brit. Cwlth.* **76,** 596.
WADE A.P. & ABRAMOVICH D.R. (1967) The distribution of 17-oxosteroids and 17-hydroxycorticoids in amniotic fluid. *Steroids* **10,** 669.
WAGNER G. & FUCHS F. (1962) The volume of amniotic fluid in the first half of human pregnancy. *J. Obstet. Gynaec. Brit. Cwlth.* **69,** 131.
WALKER W. (1970) The role of liquor examination. *Brit. med. J.* **2,** 220.
WIENER M. (1883) quoted by R.E. Shaw & H.J.Marriott (1949) The origin of amniotic fluid and the bearing on this problem of foetal urethral atresia. *J. Obstet. Gynaec. Brit. Emp.* **56,** 1004.
WIEST W.G. (1967) Estimation of progesterone in biological tissues and fluids from pregnant women by double isotopic derivative assay. *Steroids* **10,** 279.
WINDLE W.F., BECKER R.F., BARTH E.E. & SCHULZ M.D. (1939) Aspiration of amniotic fluid by the fetus. *Surg. Gynec. Obst.* **69,** 705.
WOOD C., ACHARYA P.T., CORNWELL E. & PINKERTON J.H.M. (1963) The significance of glucose and lactic acid concentration in the amniotic fluid. *J. Obstet. Gynec. Brit. Cwlth.* **70,** 274.
ZANGEMEISTER W. & MEISSL T. (1903) Untersuchungen über mutterliches und kindliches Blut und Fruchtwasser nebst Bemerkungen über die foetale Harnsekretion. *Münch. Med. Wchschr.* **50,** 673.

7
Myometrial and Tubal Physiology

CARL WOOD

MYOMETRIAL FUNCTION

CONTRACTILITY

Techniques

Since 1889 obstetricians and gynaecologists have endeavoured to investigate the non-pregnant human uterus, by inserting into it a variety of recording devices. Amongst the techniques employed were rubber balloons filled with fluid or air, radio-opaque contrast material, carbon granules forming a transducer, open-ended catheters, and sponges. In evaluating the relative merits of these techniques special attention must be paid to the physical characteristics of the pressure 'sensors' and the procedures employed for recording intra-uterine pressure (Csapo 1970).

A number of factors introduce errors into the measurement of intra-uterine pressure in the non-pregnant state.

(1) Size of uterine cavity. Pressure represents force per unit area and area will change with the size of the uterine cavity. It is negligible during the menstrual cycle and is considerable at the time of menstruation. Furthermore balloons have been placed in the uterus and filled to volumes varying from 0·5 to 15 ml—pressure-volume studies show the uterus is overdistended at 2·0 ml.

(2) State of cervical canal. Whether the uterine cavity is open or not is in doubt. At the time of menstruation it is clearly open, but during the rest of the cycle the cervix may be occluded to some extent by the viscosity of the cervical mucus. During a myometrial contraction intra-uterine pressure will be lost through an open cervix, the loss being

determined by the dilatation of the cervical canal and the method of measurement.

The two commonest methods of recording uterine contractility are the open-ended catheter (Hendricks & Brenner 1964, Caldeyro-Barcia & Pose 1958) and the closed microballoon (Csapo 1964). The closed microballoon measures maximum intra-uterine pressure as the myometrium contracts isometrically in a closed system. The optimum volume is determined at which maximum intra-uterine pressure develops. However, the isometric system may not represent the physiological state. The open-ended cannula or sponge must be syringed from time to time to maintain patency of the system. This may introduce an error due to acute and small changes in volume stretching the uterine muscle and producing reflex contractions. The open-ended system has the additional disadvantage that its placement is uncertain. Significantly higher pressures are recorded if the tip abuts on the endometrial wall rather than lying free in the cavity (pressure measured being a function of myometrial tensions rather than intra-uterine pressure). A further problem with this technique is that leakage from the cervical canal may be intermittent with consequent variations in pressure measured.

Terminology

According to classic concepts muscle function is graded at the organ rather than at the cellular level. Overall activity is controlled by recruitment of different numbers of cells, and is determined by differences in the 'threshold of excitability' for each cell. All the cells are activated in a tetanic contraction. Physiological activity is usually submaximal, however, when activity is limited to one area of muscle, muscle function being local or non-propagating (Csapo 1970). When activity spreads to distant regions it is called propagating. If several pacemakers activate regions locally, muscle function is said to be asynchronic. Activity is synchronic when a single pacemaker activates the whole of a propagating muscle. Between the two extreme conditions, of local and propagating activity, various intermediates exist. Electrophysiological studies have demonstrated the presence of both local and propagating activity in the myometrium. Local or non-propagating, asynchronic activity is characterized by high frequency and low active pressure; the contraction cycle is irregular and the increment of pressure is linear. In contrast, propagating synchronic activity is characterized by low frequency and high active pressure; the contraction cycle is regular and the increment of pressure is quadratic. Electrophysiological terminology may now replace arbitrary terms such as A and B waves, which have no physiological meaning.

During early attempts to quantitate intra-uterine pressure the significance of amplitude (A) and frequency (F) of pressure (P) was well recognized. More recently, however, the values of these two parameters have been multiplied. This product of A × F has become known as the 'Montevideo Unit' (MU). It may be a meaningful unit when only A, or F, change, but becomes meaningless if both change in the opposite direction (Csapo 1970). Such is the case during the normal menstrual cycle when a decrease in A coincides with an increase in F and vice versa. Thus while the separate terms A and F clearly define the cyclic changes in uterine activity, the produce of A × F (MU) completely obscures the 'Myometrial Cycle'. Several investigators have measured the area beneath the contraction cycles. The same shortcomings are inherent in this concept.

In this chapter the terms Resting Pressure (RP), Active Pressure (AP), Rate of Rise of Pressure (AP/T) and Frequency (F) will be employed in accordance with the terminology of classic muscle physiology. The shape of the recording of the rise of pressure will provide information about local and propagating activity.

Non-pregnant myometrial cycle (*Fig.* 7.8)

The activity of the non-pregnant myometrium is cyclical and incorporates both an active and quiescent phase. The quiescent phase in the early part of the cycle favours cyclical endometrial development, compatible with the possibility of conception and implantation of the blastocyst, whilst the active phase subsequently enables the expulsion of blood and endometrial debris at the time of menstruation.

Muscle activity at the time of menstruation is propagating in nature. This is indicated by the rapid and quadratic rises in pressure, the sensitivity to oxytocin and the sequential changes in pressure from the fundus to cervix (Csapo & Sauvage 1968, Behrman et al 1969). After menstruation the activity becomes non-propagating with slow and linear rates of pressure rise, insensitivity to oxytocin and asynchronic changes in pressure in different parts of the uterus. The frequency of contractions is high when the activity is non-propagating. The activity

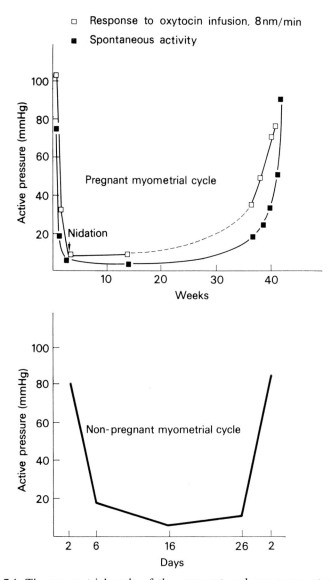

FIG. 7.1. The myometrial cycle of the pregnant and non-pregnant uterus. The active pressure was recorded by the microballoon technique. The average active pressures at different times during the menstrual cycle and pregnancy are plotted from studies carried out by Csapo & Sauvage (1968) and Csapo *et al* (1966). The figures have been adapted from their work.

during menses is similar to that during labour except that the pressure rises more rapidly in the non-pregnant than pregnant uterus.

At the time of menstruation the contractions may actually contribute to the process of endometrial degeneration. Pressures up to 80 mm Hg will produce venous stasis, elevate intra-capillary pressure and favour endometrial oedema and degeneration. It is also possible that the increase in intramyometrial pressure during menstruation would reduce arterial flow, as occurs during parturition (Borell *et al* 1965). Menstruation may be an autocatalytic process, contractions favouring endometrial degeneration, and degenerative products of the endometrium favouring myometrial contractility. Both serotonin and prostaglandins have been found in menstrual fluid.

Pregnant myometrial cycle (*Fig.* 7.1)

Like the non-pregnant, the pregnant uterus has an active and quiescent phase. A 3–4000-fold increase in uterine capacity and a large increase in uterine blood supply occur during the inactive myometrial phase, while expulsion of the foeto-placental unit occurs during the active phase. The major difference between the non-pregnant and pregnant cycle is the more prolonged inactive phase and more gradual evolution of activity at the end of pregnancy.

At the 7th week of pregnancy uterine activity is similar to day 16 of the normal menstrual cycle (Csapo & Sauvage 1968). The pressure of the contraction cycle is low, frequency is high and oxytocin response, at a rate of 128 MU/min, is poor. At 14 weeks uterine activity is still virtually absent but oxytocin response has increased—at a rate of 128 MU/min, there is a tendency to contracture (in obstetric terminology a tetanic contraction).

At 36 weeks uterine activity is apparent, the average active pressure being 18 mm Hg. Thereafter uterine activity increases week by week, as shown by re-examination of the same patient or by the group average of a number of patients. The average active pressure at 40 weeks is 32 mm Hg. Evolution of uterine activity in the last 4 weeks of pregnancy is usually gradual and incremental but there is considerable variation between individual patients. For example, two types of contraction cycles are seen, one with high frequency and low amplitude and another of low frequency and high amplitude. The latter are the Braxton-Hicks contractions, and when palpated resemble the uterine contractions of labour. Examination of the pressure recording of Braxton-Hicks contractions reveal differences from the contractions of established labour. The pressures are comparable in magnitude, but their rate of increase differs. During labour pressure increases rapidly and in a quadratic manner while during Braxton-Hicks contractions pressure rises more slowly and in a linear manner. In electrophysiological terms the Braxton-Hicks contraction is formed by the summation of a number of partially propagating activities of muscle while the labour contraction represents propagating muscle activity. Thus the nature of a contraction cycle may be expressed not only by the magnitude of the pressure but also by the shape of the contraction wave. Progressive labour occurs when propagating muscle activity results in fast and quadratic pressure increments (Csapo 1964).

Uterine activity increases at the onset of clinical labour. The change is not abrupt, however, and there is considerable overlap between values of patients who are clinically in labour and those who are not. At a similar gestation period the average values of active pressure for patients in prelabour is 32 mm Hg and those patients in early labour is 45 mm Hg (Csapo & Sauvage 1968). While the averages are distinct, considerable overlap between pre-labour and labour values prevents diagnostic or prognostic use of the uterine tracing. Furthermore, the onset of *clinical* labour is physiologically undramatic. Consideration of the evolution of uterine contractility suggests that *physiological* labour commences gradually during mid-late pregnancy.

A more marked increase in uterine activity occurs during the first stage of labour (Csapo & Sauvage 1968, Caldeyro-Barcia 1964, Hendricks et al 1962). The average active pressure at this time increases from 45 to 85 mm Hg whilst the rate of pressure rise increases nearly three-fold.

There is a gradual increase in oxytocin sensitivity over the last 4 weeks of pregnancy, the pattern of change being similar to the increase of active pressure of spontaneous uterine contractions.

At 36 weeks a 1-hour infusion of 8 MU/min produces an average active pressure of 37 mm Hg and at 40 weeks, 65 mm Hg. The increment of pressure follows a different pattern of change over the last 4 weeks, not altering significantly until after the onset of labour when it increases markedly. As the rate of pressure rise reflects the propagating state of the muscle the most dramatic change in clinical labour is the increase in propagation of contractile impulses through the myometrium.

CONTROL OF MYOMETRIAL CONTRACTILITY

What factors control the cyclic activity of the non-pregnant and pregnant uterus? The pregnant uterus has aroused more interest than the non-pregnant. Hippocrates considered that at the appointed hour the foetus put its feet against the fundus and pushed. The foetus was, in fact, considered to be responsible for its own birth until well into the sixteenth century. At that time Fabricius ab Aquapendento, an Italian anatomist, put forward the idea, amidst great opposition, that the chief agent concerned in parturition was the muscular action of the uterus itself. This debate has continued into the twentieth century. It is undeniable that the maintenance of pregnancy and expulsion of the foeto-placental unit ultimately depends on myometrial function. More precise information as to how the myometrium achieves these objectives is still required however. On the other hand it would be remarkable if the foeto-placental unit did not influence the myometrium in some way. Evidence is available which indicates that this is the case.

A reasonable working hypothesis is that myometrial function is controlled by the corpus luteum in the non-pregnant and the foeto-placental unit in the pregnant. Other factors probably modify these basic controlling systems (Fig. 7.2).

Foeto-placental unit

The foeto-placental unit may control myometrial function in a variety of ways.

Volume. Uterine volume is dependent upon growth of the foeto-placental unit and may affect myometrial function by increasing acto-myocin concentration (Michael & Schofield 1969), muscle excitability (Kuriyama 1961, Csapo *et al* 1963a and b) or by altering length-tension relationships (Csapo 1955, Schofield & Wood 1964). The significance of increased uterine volume as a factor regulating uterine activity has been neglected, most probably because of the misconception that membrane rupture facilitates uterine activity by decreasing uterine volume.

It has been established that the stimulating effect on the uterus of membrane rupture is influenced by the stage of gestation. When patients less than 36 weeks pregnant are induced by rupturing membranes, delay in the clinical onset of labour is observed four times more frequently than in patients more than 36 weeks pregnant (Lebherz, Boyce & Huston 1961). That the effect of membrane rupture is not due to volume

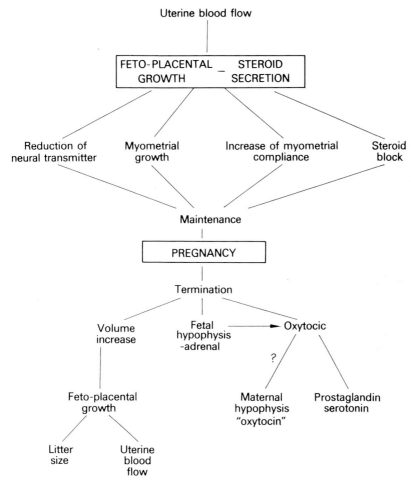

FIG. 7.2. Diagram presenting factors controlling uterine contractility.

loss has been suggested by the lack of correlation between the volume of the amniotic fluid removed and the interval between the induction and onset of labour (Garrett 1960). Clinical trials have been made which support this contention.

When 500–1000 ml of amniotic fluid was withdrawn from patients at term by trans-abdominal amniocentesis (Csapo et al 1963a) the intrauterine pressure remained unaltered or was reduced. This volume reduction did not precipitate clinical labour. It would appear, therefore, that rupturing of the membranes increases uterine activity by a mechanism other than that of decreasing uterine volume.

In contrast to the uncertainties about the mechanism of the effect of membrane rupture, there is clinical and experimental evidence that excessive uterine volume precipitates labour. Where there is associated hydramnios, 44% of the patients commence labour before the 37th week of gestation (Yordan & d'Esopo 1955). Twin pregnancies have a mean gestation period 22 days shorter than singleton, and 32% of twins are born prematurely (Guttmacher 1939). The length of gestation becomes progressively shorter in triplets and quadruplets.

Increasing the uterine volume by trans-abdominal instillation of 250–500 ml of isotonic saline solution increased uterine activity in normal patients at term and precipitated labour in the majority (Csapo et al 1963a). The same effect was observed (Wood, Elstein & Pinkerton 1963) when albumin or dextran, rather than isotonic saline, was used to increase uterine volume near term. In mid-pregnancy increase of uterine volume results in transitory and small increments of activity (Pulkinnen 1969, Anderson & Turnbull 1968). This evidence is in accord with the results from experimental animal research which have indicated that increased uterine volume affects both the excitability and the contractility of the uterus *in situ* (Csapo & Lloyd-Jacob 1963). This stretch effect may, however, be modified by other factors which influence myometrial activity (compare different response of mid and term uterus).

Since the human uterus undergoes such a large change in its volume during pregnancy, the length-tension properties of this muscle also may be important in understanding normal uterine function. At a certain length, skeletal and smooth muscle contracts maximally, whereas above or below this length, the tension developed during the contraction diminishes (Csapo & Goodall 1954). By comparing the optimal length of the same strip before it was cut from the uterus, Wood (1964a) and Schofield (1968) concluded that during pregnancy the myometrium is modified so that contraction strength is better maintained with alteration in length. This minimizes the effect that sudden alteration of uterine volume would otherwise have upon myometrial activity during pregnancy, labour and the puerperium. Furthermore compensatory changes occur in the myometrium which prevent it from becoming functionally overstretched. In the first half of pregnancy the myometrium is understretched (70–90% of the optimal length for contraction), whereas in late pregnancy it is close to its optimal length (90–100%). Thus despite the huge expansion of uterine capacity in pregnancy, the myometrium adjusts to a length which is optimal for the expulsion of the foetus.

Hormones. Both the placenta and foetus may secrete hormones which influence myometrial function. The importance of the placenta was recognized by De Snoo (1919), who suggested that it produced an inhibitory effect on uterine contractility. Different evidence is available in the human to suggest the importance of the placenta in parturition (Csapo & Wood 1968).

In the rat the placenta influences the myometrium particularly at its site of implantation, blocking myometrial contractility. This functional asymmetry has been demonstrated in a variety of ways but technical difficulties prevent its display in the human. Nevertheless evidence does exist that the placental site in the human may impose a functional asymmetry. A functional asymmetry between the non-pregnant and pregnant horns of a bicornuate human uterus has been shown (Wood & Pinkerton 1963). In the laboratory animal electrical activity is more evident at the anti-placental than the placental site. If this were so in the human, the presence of a fundal placenta would determine that myometrial activity was maximal in the lower part of the uterus and this might have a deleterious effect on the nature of labour. Early study of the influence of placental site on the duration of labour showed no such effect (Wood *et al* 1962, Little 1964). By standardizing the time of onset of labour with oxytocin infusion, and other conditions of the experiment, and by continuously monitoring uterine activity, Csapo *et al* (1968) were able to show a significant effect of the site of placental attachment on the length of labour. The early phase of labour is markedly prolonged in patients with high or fundal placentas, 173 ± 14 minutes, $P < 0.01$. As the magnitude of uterine activity was unrelated to placental location it is most likely that placental site influences the direction of propagation of the contractile wave and this, when unfavourable, results in slow progress of labour. It was also found that failure to induce labour was more frequently associated with high or fundal placentas. These findings support the concept of a placental-dependent functional asymmetry of the parturient human uterus.

Circumstantial evidence has accumulated that progesterone may be important in the maintenance of human pregnancy (Csapo 1969). First, plasma progesterone levels, measured repeatedly in the same patient, have been shown to increase steadily during pregnancy, but plateau before the onset of labour, and decline in labour (Csapo et al 1970). Secondly plasma progesterone levels decrease after the injection of hypertonic saline to induce mid-trimester abortion. This fall roughly parallels the development of uterine activity (Wiest et al 1970, Pulkkinen 1969). Finally plasma progesterone levels have been shown to be directly related to the delay which follows induction of labour at term (Johanssen 1968). The failure to influence markedly myometrial contractility by parenteral progesterone therapy may be due to limitations imposed by the method of administration, or to metabolism of the drug before it can act on the myometrial cell.

The concentration of progesterone is higher in the myometrium than in the systemic circulation. The hormone may pass from the placenta to the myometrium by a local transport system or travel systemically. Ultimately it may be bound by a receptor in the myometrial cell. Such possibilities limit the biological interpretation of plasma hormone levels. The role of progesterone is further complicated by the finding that various metabolites are biologically active. For example Gyermek (1968), using isolated strips of rat uteri showed that pregnanolone was a more potent inhibitor of oxytocin induced contractions than progesterone.

Oestrogen has been shown to induce myometrial hypertrophy, actomyosin formation, storage of high energy phosphates, the development of membrane potentials in the myometrial cell and also myometrial excitability (Reynolds 1949, Csapo 1962). Michael (1970) has shown that actomyosin concentration becomes progressively higher when measured in post-menopausal, pre-menopausal and pregnant human uteri. There is a further increase between early and late pregnancy the level reaching maximum by about the 32nd week (Fig. 7.3). The increase of actomyosin depends upon the effect of oestrogen and increase of uterine volume (Csapo 1962, Michael 1970). In the human, oestrogen levels may be extremely low in cases of missed abortion yet labour can be induced by increasing the volume of the uterus or by oxytocin (Csapo et al 1963b). In rats oestrogen deficiency results in failure of excitability and pharmacological activity and in the human a significant negative correlation has been found between the level of oestriol in the urine and the length of gestation (Turnbull et al 1967, Klopper & Billewicz 1963). However, the latter finding was not confirmed by Beischer et al (1968). The role of oestrogen in human pregnancy requires further elucidation. Like progesterone, blood levels of the hormone may not accurately reflect biological activity as specific protein receptors for oestradiol have been found in nuclei of the uterus (see Chapter 2).

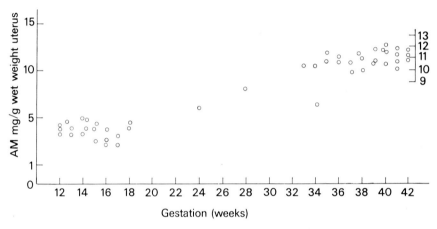

FIG. 7.3. Actomyosin concentration in the human uterus (work of C.A. Michael, 1970, to be published, Department of Obstetrics and Gynaecology, University of Western Australia).

Ryan (1968) has emphasized the importance of a comparative approach in elucidating the role of various factors in the initiation of labour. There is good evidence that absolute and relative amounts of oestrogen and progesterone are critical for pregnancy maintenance in animals where the endogenous source of hormones has been removed by oophorectomy (Catchpole 1959, Deanesley 1966). While the evidence in the human is not clear-cut further study may elucidate oestrogen–progesterone relationships and their role in the human.

Evidence that the foetal endocrine system may influence the onset of labour was first produced by Malpas in 1933. He noticed that anencephaly predisposed to prolonged pregnancy except when hydramnios was present. In a well-documented series, premature labour was recorded in all of the twenty-nine patients who had both anencephaly and hydramnios whilst prolongation of pregnancy occurred in nine out of fifteen cases of anencephaly without hydramnios. The average gestation of the nine was 327 days. Malpas suggested that the absence of the foetal pituitary and adrenals in the anencephalic was responsible for the prolongation of pregnancy.

Observations in the ewe (Binna, James & Shupe 1964) and in Guernsey cattle (Holm 1963) also suggest that the foetus can influence the duration of gestation. In the ewe Liggins (1969a) has shown that foetal infusion of ACTH or glucocorticoids initiates premature labour, whereas foetal hypophysectomy or bilateral adrenalectomy delays parturition. He suggested that parturition normally occurred in the ewe because of maturation of the hypothalamic response, which stimulated foetal pituitary ACTH and, as a result, cortisol secretion by the foetal adrenal gland. He further suggested that the glucocorticoid activity of the cortisol may diminish progesterone secretion by the trophoblast, thereby removing the progesterone block. This particular foetal mechanism may not act in the human being, since injection of cortisone into the foetus does not alter uterine contractility (Liggins 1969b). Experiments attempting to exclude a role for the foetus in the initiation of labour are not conclusive. In monkeys and mice all foetuses have been surgically removed leaving the placentas, which have been delivered at a time characteristic of the species (Newton 1938, Van Wagenen & Newton 1943). However, the large reduction of uterine volume after removal of the foetuses in these experiments precludes valid comparison with the situation in normal pregnancy.

In the human foetus adrenal size also has been related to the onset of premature labour. The adrenals of infants delivered as a result of the 'unexplained' onset of labour before 36 weeks were larger than normal (Turnbull & Anderson 1969) and this increase of adrenal size could not be accounted for by an overall increase in foetal weight. This finding is of interest as it provides further support of the importance of the foetal adrenal (Malpas 1937, Holm 1963, Liggins 1969 and see Chapter 4).

Serotonin may act as a physiologic oxytocic. It is excreted by the foeto-placental unit and accumulates in myometrium (Koren & Romney 1970). At the beginning of labour levels in myometrium are increased. Serotonin causes contraction not only of myometrium but also of vascular smooth muscle so that if it has a physiological role in the initiation of labour, a mechanism must exist to prevent harmful effects on the uterine or foetal circulation. Such a mechanism may exist for human myometrium since it becomes more sensitive to serotonin as pregnancy progresses (Contractor et al 1968). This may be due to declining levels of monoamine oxidase in the myometrium. This enzyme is the principal factor capable of inactivating serotonin in the myometrium.

Prostaglandins have been found in the human umbilical cord, amniotic fluid, decidua and maternal and foetal blood. Prostaglandin $F_2\alpha$ stimulates uterine activity. It first appears in maternal blood during labour, and its secretion from the foeto-placental unit may have a physiological role in increasing uterine activity during clinical labour (Karim 1968). This possibility is enhanced by Karim's finding that the level rises in maternal vein blood before each uterine contraction. The increase of uterine contractility during the latter part of pregnancy cannot be explained by prostaglandin, however, unless the present assay is insufficiently sensitive to detect very small amounts present at that time.

Chard et al (1970) have found high levels of oxytocin in arterial cord blood and conclude that oxytocin is produced in the baby's pituitary in labour. This may be another mechanism whereby the foetus contributes to its own expulsion.

Maternal

Ovary. Bilateral ovariectomy, carried out during the first trimester, provides conclusive evidence that the ovaries are dispensable, as early as the 2nd month of pregnancy (Asdell 1928, Calatroni et al 1954). However, if corpora lutea are removed at or before the 7th week abortion usually occurs while removal of cysts from the ovary opposite to that containing the corpus luteum does not (Wilson 1937). In Hungary the corpus luteum has been removed from women awaiting therapeutic abortion. At the 7th week abortion occurs within 3–7 days, while at the 10th week it may be delayed beyond 14 days or not occur at all (Csapo 1969). In the human the corpus luteum may be important for pregnancy maintenance before the 7th week. Fifty milligrammes of parenteral progesterone delays the onset of uterine activity after the corpus luteum is removed at the 7th week. This suggests a supportive role for progesterone in early pregnancy.

The non-pregnant myometrial cycle is dependent upon ovarian

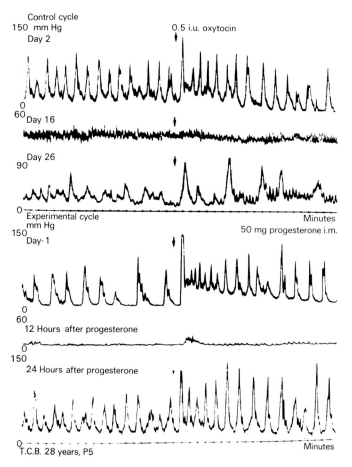

FIG. 7.4. Effect of progesterone on non-pregnant uterus. Control and experimental cycle show similar uterine contractility on day 2 and 1, respectively. In the experimental cycle progesterone (50 mg) blocks activity after 12 hours and this continues to 24 hours. The contrast between contractility on day 2 of the experimental and control cycle is seen. From T.D.Kerenyi et al. (1969) in G.E.W.Wolstenholme & Julie Knight (eds.) *Progesterone: its Regulatory Effect on the Myometrium*. London, Churchill.

function. Maximum activity occurs after the cessation of oestrogen and progesterone secretion. The quiescent phase occurs during the time of fluctuating oestrogen secretion and maximum progesterone secretion. Progesterone, in doses of 50 mg intramuscularly, has been shown to block myometrial activity in the non-pregnant patient (Kerenyi 1969,

Fig. 7.4). A physiological role for progesterone in the control of myometrial contractility in the non-pregnant appears likely. The role of oestrogen has still to be clarified.

Pituitary. The blood levels of oxytocin and oxytocic substances increase during labour, particularly during the second stage (Hawker *et al* 1963, Caldeyro-Barcia 1964, Fitzpatrick 1966). This suggests that oxytocin has an important role in the completion of labour (Csapo & Wood 1968) and the prevention of post-partum haemorrhage (Wood 1964a). This view has been challenged by Chard *et al* (1970). Using a specific radioimmunoassay they found no oxytocin at all in maternal blood samples.

Hypophysectomized patients, and also those suffering from diabetes insipidus, commence labour spontaneously near term (Kaplan 1961, Little *et al* 1958, Stan 1961). The significance of these findings has been contested on the grounds that patients with either diabetes insipidus or previous hypophysectomy may continue to produce normal amounts of oxytocin from the hypothalamus (Sloper 1960). However, patients with diabetes insipidus require completion of delivery by forceps or caesarean section in 68% of cases (Stan, Balaceano & Peterson 1961).

The limited oxytocin response during the mid-trimester implies that either oxytocin is destroyed more rapidly at this time or that the myometrium is insensitive. The fact that high oxytocin dosages in mid-pregnancy stimulate the uterus, but tend to produce contracture (tetanic contraction) rather than normal contractility (Csapo & Sauvage 1968) suggests that the explanation for the altered oxytocin response lies in the myometrium itself rather than in humoral factors breaking down oxytocin (Csapo & Wood 1968). This view is supported by the findings that myometrial strips taken from the mid-pregnant uterus and studied *in vitro* have less spontaneous activity and less oxytocin sensitivity than strips taken from term uteri (Fuchs & Fuchs 1963).

Autonomic nervous system. When infused during pregnancy, noradrenaline increases and adrenaline decreases uterine contractility (Wansbrough *et al* 1967, 1968). Evidence has accumulated that autonomic influences upon the human uterus may have functional significance (Wood 1969).

(1) Noradrenergic nerves supplying smooth muscle fibres in the myometrium have been demonstrated by fluorescent histochemistry (Nakanishi *et al* 1969, Silva 1969) (Fig. 7.5).

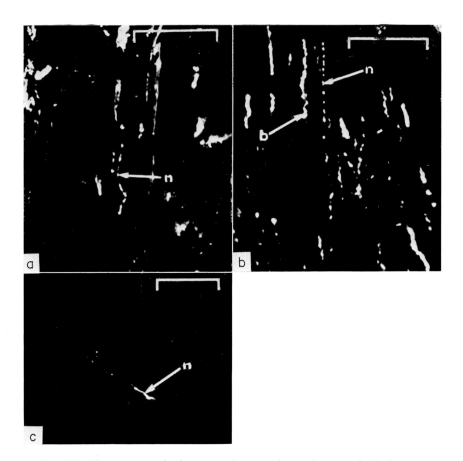

FIG. 7.5. Fluorescent stain demonstrating noradrenergic nerves in the human uterus: incubated in formaldehyde vapour for one hour, calibration 100 μm. (a) Non-pregnant uterus showing sparse innervation of the myometrium of the upper two-thirds of the body. Isolated varicose nerves (n) run within the smooth muscle. Large areas are devoid of fluorescent nerves. (b) Non-pregnant uterus showing numerous fluorescent nerves within the myometrium of the lower part of the body and isthmus: note both fine varicose nerves (n) and small nerve bundles (b). (c) Pregnant uterus showing isolated nerve fibre (n) in the lower uterine segment; this contrasts with the abundant nerve fibres in the lower part of the non-pregnant uterus.

(2) Nerves in the myometrium have been stimulated *in vitro* (Nakanishi *et al* 1969). The response is characteristic of a sympathetic nerve supply; optimal frequency was at 30 cycles/second, and guanethidine and α-receptor blocking drugs reduced the response (Fig. 7.6).

(3) Alpha and beta receptors have been demonstrated in the human uterus. Receptor blocking drugs influence myometrial contractility, both *in vitro* and *in vivo* (Wansbrough *et al* 1968).

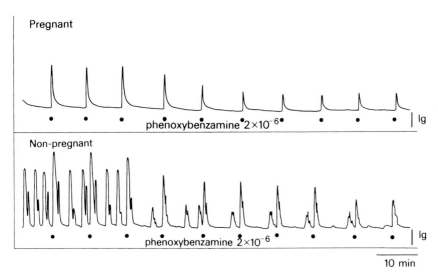

FIG. 7.6. Evidence of a functional noradrenergic innervation of the human uterus. Blocking the adrenergic α-receptors with phenoxybenzamine diminishes the contractile response to intramural nerve stimulation both in pregnant and non-pregnant uterus, *in vitro* study.

(4) Stimulation of the hypogastric nerve increases uterine (myometrial) contractility in early pregnancy and in the luteal phase of the menstrual cycle (Alvarez *et al* 1965).

(5) The presence of cholinergic nerves is less certain. The demonstration of cholinesterase activity in the myometrium does not necessarily indicate their presence. Nevertheless, atropine decreases and physostigmine increases the contractile response to transmural nerve stimulation (Nakanishi *et al* 1968).

The influence of the autonomic nervous system upon the uterus changes in pregnancy. Noradrenergic nerves supplying smooth muscle become sparser, so that the stimulating effect of neural noradrenaline is reduced (Nakanishi *et al* 1969). These finding are consistent with those of Vasicka & Kretchmer (1961) and Alvarez *et al* (1959), who showed that spinal anaesthesia as high as C.6 to T.2 did not alter uterine contractility during labour. During pregnancy the proportion of circulating adrenaline to noradrenaline increases approximately four-fold; furthermore, when adrenaline is infused into pregnant patients it results in uterine inhibition (Garrett 1954, Wansbrough *et al* 1967). Thus in pregnancy the increased humoral adrenaline and the decreased neural noradrenaline would favour uterine inhibition. These findings support the hypothesis that the autonomic influence upon the human uterus changes in pregnancy to favour its maintenance.

Uterine blood supply. The possible role of uterine blood flow in controlling myometrial function has received scant attention. Acute changes of flow may influence uterine contractility (Whiteside *et al* 1967) but are not relevant to the understanding of physiological mechanisms. The haemodynamic hypothesis has been discussed by Dawes (1968) as a factor influencing foetal and placental growth. This receives support from experiments in which the arterial supply to selected rat foetuses was reduced (Wigglesworth 1964). As the foeto-placental unit may be an important factor controlling the myometrium (see above) there may be interdependence between uterine blood flow, foeto-placental growth and myometrial contractility. Although data are limited in the human and other species the increase in uterine blood flow during pregnancy is in general proportional to the foeto-placental weight (Makowski *et al* 1968, Lewis 1969). However, repeated measurements of uterine blood flow on the same animal (close to the end of pregnancy and using methods of high precision) are not yet available. Furthermore, the interrelationship between uterine blood flow, foeto-placental function and length of gestation may depend upon the absolute level of flow throughout pregnancy rather than a change of flow at the end of pregnancy.

There may also be a direct relationship between uterine contractility and blood flow. An interesting result from Makowski *et al* (1968) using radioactive microspheres in the ewe, is that while flow to the placenta increases during the last third of pregnancy, that to the myometrium and endometrium remains constant. Increasing myometrial expansion in the latter part of pregnancy in the presence of an unchanging myometrial flow may alter the topographical relationship between capillaries and smooth muscle cells. This may change access of blood-borne substances, such as progesterone or oxytocin, to the smooth muscle cell.

Alteration of blood flow at the onset of menstruation or labour may also influence contractility. Each uterine contraction is associated with reduction of venous and arterial flow (Borell *et al* 1965). This may affect foeto-placental function unfavourably. The rapid and fairly predictable completion of parturition in various species may depend upon a feedback process involving a sequence of changes dependent upon increasing strength of uterine contractions and decreasing uterine blood flow. Brotanek *et al* (1969) found that uterine blood flow was reduced both before and during contractions. They suggested that a decrease in blood flow may actually stimulate uterine contractions. While this is an attractive hypothesis their method of study is subject to some limitations. The single sensor was placed in the cervix not the myometrium and heat loss from the thermistor is influenced by its movement in relation to adjacent blood vessels and to changes in the temperature of the tissue which may be quite unrelated to blood flow.

ACCOMMODATION

The non-pregnant and pregnant myometrium undergo changes which accommodate the growing endometrial mass or foeto-placental unit. When the volume of the non-pregnant human uterus is increased experimentally from 0·5 to 1·0 ml a difference of about 10–20 mm Hg in the increment of resting pressure is observed between days 2 and 17 of the cycle (Csapo *et al* 1966). The uterus at day 17 shows greater extensibility in spite of the fact that its cavity is already expanded by endometrium. Progesterone may be responsible for the cyclic change in the compliance of the non-pregnant uterus. Compliance has been artificially increased in the non-pregnant patient using a retroprogesterone, duphaston (Sharp & Wood 1966).

Maintenance of pregnancy can take place only if the uterus expands to accommodate the growing products of conception. This massive expansion involves changes in both muscle and connective tissue and involves three separate adaptive mechanisms.

The uterus undergoes tremendous growth which affects muscle, connective tissue, and blood vessels. At parturition the average net weight of the uterus is 880 g compared with a normal resting weight of 75 g (Woessner, Ferguson & Brewer 1961). Study of ultrastructure shows that growth is largely due to increase in the size of the individual muscle fibres (Laguens & Lagrutta 1964; Silva 1969). This probably results from the influence of increased oestrogen secretion as well as from mechanical distension (Reynolds 1969, Csapo *et al* 1965). The increase in muscle mass was thought to occur mainly in the first half of pregnancy (Reynolds 1949, Gillespie 1950) but Hytten & Cheyne (1969) found that the average uterine weight at term (1100 g) was greater than at mid-pregnancy (300–400 g). The nitrogen proportion remained constant throughout pregnancy.

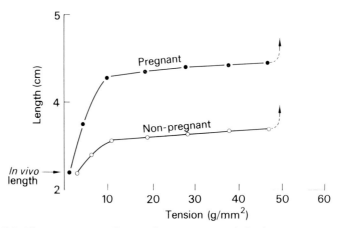

FIG. 7.7. The greater compliance of pregnant muscle is shown. At any given increase of tension following stretch, the pregnant muscle is longer than the non-pregnant. The initial lengths of both strips were similar *in vivo*. From C. Wood (1969) in R.J.Kellar (ed.) *Modern Trends in Obstetrics*, 4. London, Butterworth.

Pregnant myometrium is more compliant than non-pregnant (Wood 1964b, Conrad *et al* 1966) (Fig. 7.7). This greater compliance is due to an increase in the plastic or viscous property of the muscle but not its elasticity. Harkness & Harkness (1960) have shown that connective tissue plays a more important role in determining expansile behaviour than contractile elements, as collagen is relatively inextensible. The microfibrils of collagen run in bundles which make a three-dimensional network in the uterus. The first effect of stretching muscle is merely to pull the collagen microfibrils straight, and this occurs without significant increase in tension. Subsequent stretch is by slip between microfibrils, which allows slow extension but is accompanied by a marked rise in tension. An increase in the expansile properties of connective tissue could be attained either by expansion of the collagenous framework or by softening of the 'cement' between the fibrils of collagen (Harkness & Harkness 1960). The increase in compliance of the pregnant myometrium may result from production of oestrogen and progesterone by the placenta (Cullen & Harkness 1960, Wood 1964b). On the other hand pregnancy has been regarded as a relentless series of 'injurious' stimuli to the connective tissue in the uterus (Bhussry & Rao 1968). These authors regard the changes in connective tissue as similar to those which take place in healing dermal wounds—collagen bundles split and collagen degeneration occurs.

Finally expansion of the uterus is also aided by the rearrangement of spiral bundles of muscle fibres as in the uncoiling of a spring (Goerrtler 1920).

RESISTANCE TO UTERINE EXPULSIVE FORCES

The factors influencing the rate of progress of labour include not only the uterine expulsive forces but the resistance to these forces. This resistance is composed of the lower uterine segment, the cervix and the friction between the foetal presenting part and uterine wall.

By studying measured myometrial strips *in vitro*, Wood (1965) has shown that the resting tension of the myometrium rises in late pregnancy. During labour the myometrium of the lower segment becomes overstretched, resting tension being four times greater than that of the upper segment. The birth canal becomes progressively harder to dilate as the part of the lower segment which lies below the maximum diameter of the presenting part becomes overstretched (Wood 1965). On the other hand, this overstretching decreases the contractile response of the lower segment, leaving the upper segment dominant. As the lower segment is stretched during labour, the uterus must work against the increasing resistance. This is why, in some circumstances, the cervix can stop dilating although uterine contractility has not decreased. High resistance of the lower segment will occur when it is over-distended by a large presenting part.

In multiparae the amount of uterine work required to dilate the cervix is less than in the primigravida, although there is no difference in the amount of uterine work required to efface the cervix (Alvarez, Cibils & Gonzalez-Panizza 1961). The cervix of the multigravida is usually 1–2 cm more dilated than that of the primigravida at term, and it is the lesser dilatation required of the former rather than its greater dilatability which determines the more rapid course to delivery.

By measuring the pressure at different sites between the head and uterine wall, Lindgren & Holmlund (1969) have been able to calculate the coefficient of friction. This was low, 0·20, presumably due to the presence of vernix or other lubricating substance. However, in several patients in whom labour was delayed the force hampering descent of the head was greater than the expulsive force. They attributed this to spasm of the lower uterine segment but it is more likely overstretch of the muscle, due to either a large presenting part or the lower compliance of muscle as expressed in the passive length-tension property of muscle.

The quantitative importance of the resistance to uterine expulsive forces is as yet unknown. The rapid progression to uterine rupture in the presence of a mechanical factor obstructing labour, e.g. transverse lie, suggests that uterine contractility is more important. However, in some situations, e.g., in determining the length of labour in the primigravida, the resistance to expulsive forces may be of practical importance.

APPLIED PHYSIOLOGY

CONTRACTILITY

Emotions and sexual arousal

Both pain and fear are known to increase uterine activity in the non-pregnant patient (Robertson 1939, Alvarez & Caldeyro-Barcia 1954). Kelly (1962) found that fear and the rapid injection of adrenaline both resulted in an almost identical increase in activity and he considered that fear produced this effect through the release of adrenaline.

Sexual arousal from visual stimuli and sexual excitation and orgasm from physical stimulation have been shown to increase uterine contractility in the non-pregnant (Masters & Johnson 1966, Bardwick & Behrman 1967). However, the technique of recording in these studies leaves their conclusions open to doubt.

Garrett (1960) produced some evidence that fear may alter uterine contractility during labour, but only as a temporary phenomenon.

Patients receiving ante-natal preparation (Thomas, Raphael & Isbister 1966) or ante-natal hypnosis (Perchard, 1961) do not have significantly shorter labour than other patients. In a prospective study Engstrom *et al* (1964), were able to show a greater frequency of inert labours in women with psychosocial problems. These included poor emotional or sexual relationships with their husband, fear of pregnancy or delivery, or a predominantly negative attitude to the pregnancy. Whether there is a cause and effect relationship between psychosocial problems and inertia is unknown.

Contraceptives

By mounting strain gauges on a Lippes loop Beyer and Behrman (1970) have studied the effect of different oestrogen–progestogen mixtures on uterine activity. Oral contraceptives with higher doses of oestrogen increased uterine activity while those with higher progestogen doses inhibited activity. While the findings are consistent with known uterine physiology the possible effect of the Lippes loop on the uterus limits the significance of the results.

It was previously thought that the intra-uterine device acted as a contraceptive by increasing tubal or uterine motility. However, studies in the monkey show it does not cause premature expulsion of the egg, effect ovulation, sperm transport, fertilization or early embryonic development (Marston *et al* 1969). Evidence favours the view that fertilized eggs undergo premature degeneration.

Clinical assessment

The clinician is concerned with several factors which influence the assessment of the strength of uterine contractions during labour.

(1) Patients feel pain during uterine contractions at pressures which vary from 15 to 60 mm Hg. Some never feel pain. The severity of the pain felt by a patient is a poor index of the strength of contractions.

(2) The strength of a contraction can be roughly gauged by abdominal palpation. At greater than 30 mm Hg it is difficult to indent the uterus with one finger, although the thickness of the abdominal wall affects the accuracy of this sign.

(3) The measurement of the duration of contraction by hand is not accurate. The true duration of the contractions is about twice as long

as the duration felt by the hand. Individual observers detect contractions at different intra-amniotic pressures. Usually a pressure of 20 mm Hg is easily felt with the hand, but it may vary from 10 to 25 mm Hg depending upon the observer and the patient.

Posture

When the patient changes from the dorsal to the lateral position during labour, the intensity of contractions usually increases and their frequency decreases (Caldeyro-Barcia *et al* 1960). The average change in forty-two patients was an increase of 7·6 mm Hg in the amplitude of contraction and a decrease in frequency of 0·7 contractions per 10 minutes. Most patients assume the position which is most comfortable during labour. The lateral position should be encouraged, however, when contractions are excessively frequent (e.g. late labour, oxytocin excess). It is possible that high amplitude, low frequency contractions are more effective in dilating the cervix.

Cervical dilatation

In pregnant women near term, dilatation of the cervix by means of intermittent traction on the cervix with a balloon of 8 cm diameter has been shown to increase uterine activity (Fisch, Sala & Schwarcz 1964, 1965). Labour was induced effectively with the technique in seven out of nine patients. Intramammary pressure, which was also measured, did not change in seven of the nine patients. This suggests that the increase in uterine activity which resulted from cervical dilatation was not due to the reflex release of oxytocin. Nevertheless an oxytocin-dependent reflex has been previously described by Ferguson in the rabbit.

Many obstetricians stretch the cervix at the time of amniotomy in the belief that induction of labour is more effective. On the other hand intra-amniotic pressure tracings before and after manual cervical dilation of short duration usually show no change or only a temporary increase of uterine activity. Using the small balloon of a Foley catheter Embrey has successfully induced labour in patients with an unripe cervix. However, the risk of infection may prevent its widespread use.

When the foetus is delivered using the vacuum extractor before the cervix is fully dilated, both the frequency and the duration of contractions are increased (Footit 1967).

Effect upon foetus

Studies in the animal (Dawes 1968) have suggested there is normally a fall in the percentage of oxygen saturation of foetal arterial blood towards the end of each contraction. However, some caution in accepting these results is warranted because of the difficulty in establishing physiological conditions even in chronic animal experiments. In the human, uterine blood flow is reduced during contractions (Borell *et al* 1965) both venous and arterial compression occurring in the myometrium. In addition compression of the aorta and inferior vena cava may occur in some patients, particularly in the supine position (Poseiro *et al* 1964). Investigation of the effect of uterine contractions has been limited in the human to the first stage of labour (Renou *et al* 1968, Walker *et al* 1970). Foetal scalp blood pH, P_{CO_2} and P_{O_2} were measured during and between contractions in twenty-eight patients while foetal heart rate and intra-amniotic pressure were monitored continuously. In seventeen other cases foetal scalp tissue P_{O_2} was measured continuously with foetal heart rate and intra-amniotic pressure. In the majority of cases the foetal heart rate did not change during the uterine contraction and in this situation there was no significant change in foetal blood pH, P_{CO_2} or P_{O_2} or foetal scalp tissue P_{O_2}. On the other hand, when the foetal heart rate slowed during the contraction by more than 15 beats per minute there was a small but significant fall in foetal scalp blood P_{O_2} and in five out of eight patients a significant reduction in foetal scalp tissue P_{O_2}. In the first stage of labour it would seem that uterine contractions do not usually produce hypoxaemia of foetal scalp blood or tissue. However, in some circumstances hypoxaemia may result and this is usually associated with a slowing of the foetal heart rate. In one patient scalp tissue P_{O_2} fell when the heart rate increased with contractions.

It is possible that the present methods of studying the human foetus are incapable of detecting very small changes in foetal scalp blood or tissue P_{O_2}. It is also possible that in the presence of mild asphyxia during contractions, circulatory shunts develop and these help to maintain normal gas tensions in blood perfusing the brain and scalp. This would mask to some extent the effect of the uterine contraction on the foetus when studies are restricted to the foetal scalp. Again the strength of uterine contractions could influence the possible production of foetal asphyxia—in our own investigation the intra-amniotic pressure varied between 15 and 50 mm Hg.

INDUCTION OF LABOUR

Oxytocin

This has become established as a safe and effective method for inducing labour (Theobald 1963). There is considerable variation in the dosage and the route of administration which are employed in different centres. The principle of employing the minimum dose to produce adequate uterine response is best. Theobald advocates the use of 2 MU/minute for intravenous infusion purposes. In many centres 20 MU/minute are used initially, and it is claimed that results are good and perinatal mortality is low. Because of the wide variation in the response of the uterus to various concentrations of oxytocin at term, it is safer to start at low dosage, 2 MU/min., and increase this until adequate uterine activity is attained.

The safety and efficiency of oxytocin induction can be increased by using two pieces of equipment; an infusion pump which accurately controls the rate of oxytocin infusion, and an extra-amniotic balloon connected to a tocograph in order to monitor intra-uterine pressure continuously. Doses have been doubled every 12·5 minutes up to 128 MU/min. in order to shorten the induction delivery interval (Francis *et al* 1970, Pawson & Simmons 1970). This reduces the hazard of infection without apparent ill effect on the foetus so that study of larger samples in a controlled trial with neonatal follow-up are indicated.

The hazards of oxytocin infusion include the following

(1) The resting pressure may increase before adequate contractions are produced; this change may not be appreciated by the patient or her attendant, and if prolonged may impair foetal oxygenation. Even pressure recording after amniotomy may be misleading as measurement of resting pressure in an open amniotic sac may be inaccurate. Caldeyro-Barcia & Poseiro (1958) emphasized the danger of excessive uterine contractility diminishing placental blood flow, and demonstrated a marked reduction of intervillous clearance of ^{131}I in a patient with a resting amniotic pressure close to 33 mm Hg. From their experience of recording uterine contractility, they state that infusion rates greater than 16 MU/min. will produce excessive contractility in most patients beyond the 36 weeks of pregnancy (this infusion rate is equivalent to running a drip at 30 drops/min with 8 units of oxytocin per litre of fluid). While uterine tetany will not usually kill a foetus unless it is unduly prolonged or severe, or occurs in the presence of some other obstetric complication,

it might produce sufficient hypoxia to cause brain damage in a surviving infant. Until adequate trials of various oxytocin infusion regimens are made and the babies are followed after birth, it would be wise to use the minimal effective dose and not to use high rates of infusion unless resting pressure is measured continuously.

(2) The sensitivity of the uterus may change during the course of the infusion and, particularly with higher doses, a tetanic contraction may result. The dose of oxytocin usually can be reduced as labour progresses.

(3) The rapid i.v. injection of 0·2 unit of oxytocin (Syntocinon) is sufficient to decrease blood pressure (Caldeyro-Barcia & Poseiro 1958). However, oxytocin infusion is unlikely to cause hypotension since the drip would have to be run at the rate of 200 MU/min., which is the equivalent of a solution containing 50 units oxytocin per litre. Such a rate of infusion has been used in cases of missed abortion.

(4) Water intoxication may result from high rates of fluid infusion over a prolonged period of time, particularly when renal function is impaired.

Buccal oxytocin. This has been successfully employed to induce labour over the past 5 years. An analysis of 550 inductions showed the success rate to be 68%, but treatment beyond 2 days was not usually successful. However, a number of cases of uterine rupture have been recorded following the administration of buccal oxytocin; while in some of these cases excessive doses were given, these results led Chalmers & Prakash (1967) to explore the use of small doses. This has resulted in a similar success rate, but without obvious uterine hypertonicity. Except in patients with foetal death or missed abortion, these authors have not used intravenous oxytocin over the last 4 years. Their previous experience suggested that when buccal oxytocin failed, so also did intravenous oxytocin.

More extensive trials with buccal oxytocin are required to show whether the incidence of uterine rupture and perinatal death and morbidity is comparable to that of patients induced with intravenous oxytocin.

Prostaglandin

Both labour and abortion have been successfully induced by prostaglandin (PG) (Karim *et al* 1968, Embrey 1970, Karim & Filshie 1970).

The rapid onset and physiological pattern of uterine contractility and the lack of side effects of PGF 2α in late pregnancy suggests it may be a useful oxytocic and larger trials are now being carried out. The absence of risk of water intoxication when using PGF 2α in inducing labour may be of particular value when maternal pre-eclampsia or renal disease is present (Roberts et al 1970). The prostaglandins effect a variety of smooth muscle containing organs and maternal side effects of vomiting, diarrhoea and vasomotor disturbance may result. Follow-up studies of infants delivered subsequent to prostaglandin are also necessary. A small series of abortions have been induced by prostaglandin F2α. Some failures were recorded and a few patients suffered from drug-induced side effects. Embrey (1970) considers PGE to be equal or more potent than PGF 2α. This may be related to the potentiating action of PGE on some other oxytocics as well as its direct effect on uterine muscle (Clegg 1966).

SUPPRESSION OF LABOUR

This may be indicated in premature labour, uterine hypercontractility (e.g. when ergometrine is given before the birth of an undiagnosed second twin) or when abnormal FHR patterns are associated with uterine contractions and caesarean section is planned (e.g. pressure on umbilical cord). Most drugs used for this purpose have been either relatively ineffective or had side effects limiting their use at therapeutic dose levels, e.g. valium, progesterone, isoxsuprine, methanesulphonaamide. More recently alcohol has been used by Fuchs (1965) and even oral alcohol, 20 g has been shown to reduce uterine activity temporarily in normal labour (Luukainen et al 1967).

The only drug whose effectiveness has been proven in a multi-centre-controlled trial is ritodrine, DU 21220, 4-hydroxy-x-(1-2-4-hydroxyphenyl ethylamino ethyl) benzylalcohol-ACI Phillips-Duphar (Wesselius de Casparis 1970) (Fig. 7.8). This is a β-mimetic drug which blocks myometrial contractility at dose levels between 100–700 μg/min. without producing maternal hypotension. The absence of this side effect is important as blood flow to the uterus is closely dependent upon systemic blood pressure. Tachycardia sometimes occurs at therapeutic levels but is usually insufficient to disturb the patient and may be blocked by a specific cardiac beta receptor blocking agent, practolol, ICI (Barret et al 1967) without affecting the blocking action of the beta mimetic drug on the uterus (Beveridge & Wood 1970). Another β-mimetic drug, alupent, has a similar action to ritodrine, being a myometrial blocking agent whose only apparent side effect is tachycardia (Eskes *et al* 1969). Whether successful treatment of premature labour will reduce perinatal mortality and morbidity remains to be determined.

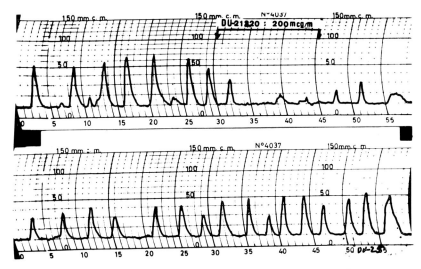

FIG. 7.8. Effect of DU-21220 (ritodrine) a β-mimetic drug on uterine contractility in labour. At infusion rate of 200 μg/min activity is blocked in the absence of change in maternal pulse or blood pressure (work of Professor V.Conhill-Serra, by permission of Phillips-Duphar Limited).

Addendum. The biochemical basis of smooth muscle contraction has only been dealt with when this has specific relevance to control of uterine activity. Detailed review of biochemical aspects of smooth muscle and myometrial contractility have been made by Needham & Shoenberg (1967) and Pulkkinen (1969).

REFERENCES

ALVAREZ H. & CALDEYRO-BARCIA R. (1954) Fisiopatologia de la contraccion uterina y sus applicaciones en la clinica obstetrica. *Relato Oficial presentado al Segundo Congreso Latino-americano de Obstetricia y Ginecologia,* p. 13, Sao Paulo, Brazil.

ALVAREZ H., POSEIRO J.J., POSE S.V. & SICA-BLANCO Y. (1959) Effects of the anesthetic blockage of the spinal cord on the contractility of the pregnant human uterus, in *Communs of 21st Int. Physiol. Sci.* p. 14, Buenos Aires.

ALVAREZ H., CIBILS L.A. & GONZALEZ-PANIZZA V.H. (1961) Cervical dilatation and uterine 'work' in labour induced by oxytocin infusion, in R.Caldeyro-Barcia & H.Heller (eds.) *Oxytocin*, pp. 203–211. Oxford, Pergamon Press.

ALVAREZ H., BLANCO Y.S., PANIZZA V.G., ROZADA H. & LUCAS O. (1965) Effects of the electrical stimulation of the presacral nerve on contractility of the human pregnant uterus. *Amer. J. Obstet. Gynec.* **93**, 131–135.

ANDERSON A.B.M. & TURNBULL A.C. (1968) Spontaneous contractility and oxytocin sensitivity of the human uterus in mid-pregnancy. *J. Obstet. Gynaec. Brit. Cwlth.* **75**, 271–277.

BARDWICK J.M. & BEHRMAN S.J. (1967) Investigation into the effects of anxiety, sexual arousal, and menstrual cycle phase on uterine contractions. *Psychosomatic Medicine* **29**, 468–482.

BARRETT A.M., CROWTHER A.F., DUNLOP D., SHANKS R.G. & SMITH L.H. (1968) Cardio-selective B-blockade. *Arch. Pharmakol. Exp. Path.* **259**, 152–153.

BEISCHER N.A., BHARGAVA V.L., BROWN J.B. & SMITH M.A. (1968) Incidence and significance of low oestriol excretion in an obstetric population. *J. Obstet. Gynaec. Brit. Cwlth.* **75**, 1024–1033.

BEVERIDGE H. & WOOD C. (1970) DU21220—A new mimetic drug; a pilot trial to determine its use as a uterine relaxant in conjunction with the B-blocking drug practolol (in press).

BEYER G. & BEHRMAN S.J. (1970) Myometrial activity and the I.U.C.D.: III. Effect of contraceptive pills. *Amer. J. Obstet. Gynec.* **106**, 81–92.

BINNS W., JAMES L.F., SHUPE J.L. & EVERETT G. (1963) Congenital cyclopian-type malformation in lambs induced by maternal ingestion of a range plant, *Veratrum Californicum*. *Amer. J. Vet. Res.* **24**, No. 103, 1164–1174.

BOOTH R.T., WOOD C., BEARD R.W., GIBSON J.R. & PINKERTON J.H. (1962) Significance of site of placental attachment in uterus. *Brit. med. J.* **1**, 1732–1734.

BORELL U., FERNSTRÖM I., OHLSON L. & WIQVIST N. (1965) Influence of uterine contractions on the utero placental blood flow at term. *Amer. J. Obstet. Gynec.* **93**, 44–67.

BROTANEK V., HENDRICKS C.H. & YOSHIDA T. (1969) Changes in uterine blood flow during uterine contraction. *Amer. J. Obstet. Gynec.* **103**, 1108–1116.

BHUSSRY B.R. & RAO S. (1968) Histochemical response to experimental skin injury in rats. *Biochem. Pharmacol.* **17** (Suppl.) 51–62.

CALATRONI C.J., RUIZ V. & DI PAOLA G. (1954) La castracion en el embarazo. *Obstet. Ginec. Lat.-am.* **3**, 145–151.

CALDEYRO-BARCIA R., NORIEGA-GUERRA L., CIBILS L.A., ALVAREZ H., POSEIRO J.J., POSE S.V., SICA-BLANCO Y., MENDEZ-BAUER C., FIELITZ C. & GONZALEZ-PANIZZA V.H. (1960) Effect of position changes on the intensity and frequency of uterine contractions during labour. *Amer. J. Obstet. Gynec.* **80**, 284–290.

CALDEYRO-BARCIA R. (1964) Regulation of myometrial activity in pregnancy, in W.M.Paul, E.E.Daniel, C.M.Kay & C.Monckton (eds.) *Proceedings of the*

Symposium held at the Faculty of Medicine, University of Alberta, pp. 317–347. Oxford, Pergamon Press.

CATCHPOLE H.R. (1959) Endocrine mechanisms during pregnancy, in H.H.Cole & P.T.Cupps (eds.) *Reproduction in Domestic Animals*, vol. 1, p. 469. New York, Academic Press.

CHALMERS J.A. & PRAKASH A. (1967) Buccal oxytocin induction using reduced dosage, in *Fifth World Congress of Gynaecology and Obstetrics*, p. 901. Sydney, Butterworth.

CHARD T., BOYD N.R.H. & HUDSON C.N. (1970) Immunoassay of oxytocin in maternal and foetal blood. *Int. J. Obstet. Gynaec.* (in press).

CLEGG P.C. (1966) Antagonism by prostaglandins of the responses of various smooth muscle preparations to sympathomimetics. *Nature* **209**, 1137–1139.

CONRAD J.T., JOHNSON W.L., KUHN & HUNTER C.A. (1966) Passive stretch relationships in human uterine muscle. *Amer. J. Obstet. Gynec.* **96**, 1055–1059.

CONTRACTOR S.F., JONES J.J. & ROUTLEDGE A. (1968) Response of human myometrial strips to 5-hydroxytryptamine and oxytocin and its monoamine oxidase activity at various stages of gestation. *J. Obstet. Gynaec. Brit. Cwlth.* **75**, 1113–1116.

COUTHINHO E.M. (1964) Hormone induced ionic regulation of labour, in *2nd International Congress of Endocrinology*. International Congress Series, 83. London.

CSAPO A. (1955) The mechanism of myometrial function and its disorders, in K.Bowes (ed.) *Modern Trends in Obstetrics and Gynaecology*, 2nd series, pp. 20–49. London, Butterworth.

CSAPO A.I. & GOODALL M. (1954) Excitability, length tension relation and kinetics of uterine muscle contraction in relation to hormonal status. *J. Physiol. (Lond.)* **126**, 384–395.

CSAPO A.I. (1964) Extraovular pressure—its diagnostic value. *Amer. J. Obstet. Gynec.* **90**, 493–504.

CSAPO A.I. & SAUVAGE J. (1968) The evolution of uterine activity during human pregnancy. *Acta obstet. gynec. scand.* **47**, 181–212.

CSAPO A.I., KNOBIL E., VAN DER MOLEN H.J. & WIEST W.G. Plasma progesterone concentrations during pregnancy, before the onset of and during labour (to be published).

CSAPO A., PINTO-DANTAS C.A., KERENYI T.D., SOUZA O. DE, DARSE E., LINDBLADE J. & KAO S.M. (1966) Progesterone and myometrial activity, in B.Westin & N.Wiqvist (eds.), *Fertility and Sterility*, pp. 429–435, Proceedings of the Fifth World Congress, Excerpta Medica Foundation, International Congress Series, 133. Stockholm.

CSAPO A.I., JAFFIN H., KERENYI T., LIPMAN J.I. & WOOD C. (1963a) Volume and activity of the pregnant human uterus. *Amer. J. Obstet. Gynec.* **85**, 819–835.

CSAPO A.I., JAFFIN H., KERENYI T., MATTOS C.E.R. DE & SOUSA FILHO M.B. DE (1963b) Fetal death *in utero. Amer. J. Obstet. Gynec.* **87**, 892–905.

CSAPO A.I. (1962) Smooth muscle as a contractile unit. *Physiol. Rev.* **42**, Suppl. 5, 7–33.

M*

CSAPO A.I. & LLOYD-JACOB M.A. (1963) Effect of uterine volume on parturition. *Amer. J. Obstet. Gynec.* **85**, 806–812.
CSAPO A. (1969) The four direct regulatory factors of myometrial function, in G.E.W.Wolstenholme and J. Knight (eds.) *Progesterone: its Regulatory Effect on the Myometrium*, pp. 13–55. J. & A. Churchill, London.
CSAPO A.I. & WOOD C. (1968) In V.H.T.James (ed.) *Recent Advances in Endocrinology*, pp. 207–239. London, Churchill.
CSAPO A., ERDOS T., MATTOS C.R. DE, GRAMSS E. & MOSCOWITZ C. (1965) Stretch-induced uterine growth, protein synthesis and function. *Nature (Lond.)* **207**, 1378–1379.
CSAPO A. (1970) The diagnostic significance of the intra-uterine pressure. *Obstet. & Gynae. Surv.* **52**, 403–435.
CULLEN B.M. & HARKNESS R.D. (1960) The effect of hormones on the physical properties and collagen content of the rat's uterine cervix. *J. Physiol. (Lond.)* **152**, 419–436.
DAWES G.S. (1968) *Foetal and Neonatal Physiology*, p. 42. Chicago, Year Book Medical Publications.
DEANESLY R. (1966) The endocrinology of pregnancy and foetal life, in A.S.Parkes (ed.) *Marshall's Physiology of Reproduction*, 3rd ed., vol. III, pp. 891–1063. London, Longmans, Green & Co. Ltd.
DE SNOO, quoted by Kloosterman G.J. (1963) The placenta, the duration of pregnancy and perinatal mortality, Margaret Oxford Lecturer for 1963. *Transactions of the College of Phys., Surg. & Gynaec. of South Africa* **7**, 18.
ENGSTROM L., GEIJERSTAM G. AF, HOLMBERG N.G. & UHRUS K. (1964) A prospective study of the relationship between psychosocial factors and course of pregnancy and delivery. *J. Psychoms. Res.* **8**, 151–155.
EMBREY M. (1970) 6th World Congress of Obstetrics and Gynaecology, New York (in press).
ESKES T., SEELEN J. & GENT L. VAN (1966) The effect of β-mimetic adrenergic drugs on the activity of the pregnant human uterus, tested with intra-uterine pressure method. *Arzneimittle-Forschung*, 16. Jahrgang, pp. 762–766.
FISCH L., SALA N.L. & SCHWARCZ R.L. (1964) Effect of cervical dilatation upon uterine contractility in pregnant women and its relation of oxytocin secretion. *Amer. J. Obstet. Gynec.* **90**, 108–114.
FITZPATRICK R.J. (1966) The posterior pituitary gland and the female reproductive tract, in G.W.Harris & B.T.Donovan (eds.) *The Pituitary Gland*, vol. III, pp. 453–504. London, Butterworth.
FITZPATRICK R.J. & WALMSLEY C.F. (1965) In J.H.M.Pinkerton (ed.) *Advances in Oxytocin Research*, pp. 51–73. Oxford, Pergamon Press.
FOOTIT G. (1967) The effect of vacuum extraction on uterine action. Obstetrical commentary in book submitted for M.R.C.O.G. examination.
FRANCIS J.G., TURNBULL A.C. & THOMAS F.F. (1970) Automatic oxytocin infusion equipment for induction of labour. *J. Obstet. Gynaec. Brit. Cwlth.* **77**, 594–602.
FUCHS A.R. & FUCHS F. (1963) Spontaeous motility and oxytocin response of the pregnant and nonpregnant human uterine muscle *in vitro*. *J. Obstet. Gynaec. Brit. Cwlth.* **70**, 4, 658–664.

FUCHS A.R. (1965) Role of oxytocin in the initiation of labour. *2nd International Congress of Endocrinology*, pp. 753–758. Excerpta Medica Congress Series, 83. London.
GARRETT W.J. (1960) Inefficient uterine action in labour. *Med. J. Aust.* **2**, 482–489.
GARRETT W.J. (1954) The effects of adrenaline and noradrenaline on the intact human uterus in late pregnancy and labour. *J. Obstet. Gynaec. Brit. Emp.* 586–589.
GARRETT W.J. (1960) Prognostic signs in surgical induction of labour. *Med. J. Aust.* **1**, 929–931.
GILLESPIE E.C. (1950) Principles of uterine growth in pregnancy. *Amer. J. Obstet. Gynec.* **59**, 949–959.
GOERTTLER F. (1930) Die Architektur dur Muskelwand des menschilichen Uterus und ihre funktionelle Bedeutung. *Morph. Jb.* **65**, 45–128.
GUTTMACHER A.F. (1939) Analysis of 573 cases of twin pregnancy. II. Hazards of pregnancy itself. *Amer. J. Obstet. Gynec.* **38**, 277–288.
GYERMEK L. (1968) Effect of pregnanolone and progesterone. *Lancet* **2**, 1195.
HENDRICKS C.H. & BRENNER W.E. (1964) Patterns of increasing uterine activity in late pregnancy and the development of uterine responsiveness to oxytocin. *Amer. J. Obstet. Gynec.* **90**, 485–492.
HENDRICKS C.H., ESKES T.K.A.B. & SAAMELI K. (1962) Uterine contractility at delivery and in the puerperium. *Amer. J. Obstet. Gynec.* **83**, 890–906.
HOLM L.W. (1967) Prolonged pregnancy. *Adv. vet. Sci.* **11**, 159–205.
HYTTEN F.E. & CHEYNE G.A. (1969) Size and composition of the human pregnant uterus. *J. Obstet. Gynaec. Brit. Cwlth.* **76**, 400–403.
JOHANSSON E.D.B. (1968) Progesterone level and response to oxytocin at term. *Lancet* **2**, 570.
KAPLAN N.M. (1961) Successful pregnancy following hypophysectomy during the twelfth week of gestation. *J. Clin. Endocr. Metab.* **21**, 1139–1145.
KARIM S.M.M. (1968) Appearance of prostaglandin F2α in human blood during labour. *Brit. med. J.* **4**, 618–621.
KARIM S.M.M. (1970) Proc. of VI World Congress of Obstetrics and Gynaecology. *Int. J. Gynae. Obstet.* (in press).
KARIM S.M.M., TRUSSELL R.R., PATEL R.C. & HILLIER K. (1968) Response of pregnant human uterus to prostaglandin F2α-induction of labour. *Brit. med. J.* **4**, 621–623.
KARIM S.M.M. & FILSHIE G.M. (1970) Therapeutic abortion using prostaglandin F2α. *Lancet* **1**, 157–158.
KELLY J.V. (1962) Effect of fear upon uterine motility. *Amer. J. Obstet. Gynec.* **83**, 576–581.
KERENYI T.D., PINTO-DANTAS C.A., SOUSA O. DE & DARZE E. (1969) The effect of progesterone on the nonpregnant and early pregnant human uterus, in *Progesterone: its Regulatory Effect on the Myometrium*, pp. 120–132. J. & A. Churchill, London.
KLOPPER A. & BILLEWICZ W. (1963) Urinary excretion of oestriol and pregnanediol during pregnancy. *J. Obstet. Gynaec. Brit. Cwlth.* **70**, 1024–1033.

KOREN Z. & ROMNEY S.L. (1970) Fetoplacental metabolism of C-5-HT hydroxytryptamine (serotonin) in pregnant rats. *Proceedings of VIth World Congress of Obstetrics and Gynaecology* (in press).

HARKNESS M.L.R. & HARKNESS R.D. (1960) Physical properties of the reproductive tract in relation to pregnancy, in A.C.Copley & G.Stainsby (eds.) *Flow Properties of Blood and Other Biological Systems*, pp. 207–222. Oxford, Pergamon Press.

HAWKER R.W., GOODRICKE B., KLEMM G.H., PEARSON I., CHRISTIE W., DEVIETTI A. COCKBURN K.G. (1963) Uterine and oxytocic activity in women in labour. *Aust. N.Z. J. Obstet. Gynaec.* **3,** 84–89.

KURIYAMA H. (1961) In G.E.W.Wolstenholme & Margaret P.Cameron (eds.) *Progesterone and the Defence Mechanism of Pregnancy*, pp. 51–70. Ciba Federation Study Group No. 9. London, Churchill.

LAGUENS R. & LAGRUTTA J. (1964) Fine structure of human uterine muscle in pregnancy. *Amer. J. Obstet. Gynec.* **89,** 1040–1047.

LEBHERZ T.B., BOYCE, C.R. & HUSTON J.W. (1961) Premature rupture of the membranes. *Amer. J. Obstet. Gynec.* **81,** 658–665.

LEWIS B.V. (1969) Uterine blood flow. *Obstet. & Gynaec. Surv.* **24,** 1211–1233.

LIGGINS G.C. (1969a) The foetal role in the initiation of parturition in the ewe. *Ciba Foundation Symposium on Foetal Autonomy*, pp. 218–231. London, Churchill.

LIGGINS G.C. (1969b) Personal communication.

LINDGREN L. & HOLMLUND D. (1969) Friction between the fetal head and uterine wall during normal labour and lower uterine spasm. *Amer. J. Obstet. Gynec.* **103,** 939–941.

LITTLE B., SMITH O.W., JESSINAU A.G., SELANKOW II.A., VANT HOFF W., EGLIN J.M. & MOORE F.D. (1958) Hypophysectomy during pregnancy in a patient with cancer of the breast. *J. clin. Endocr. & Metab.* **18,** 425–443.

LITTLE W.A. (1964) Significance of placental position *in utero*. *Amer. J. Obstet. Gynec.* **90,** 328–333.

LUUKKAINEN T., VAISTO L. & LARVINEN P.A. (1967) The effect of oral intake of ethyl alcohol on the activity of the pregnant human uterus. *Acta Obstet. et Gynecologica, Scand.* **46,** 486–493.

MAKOWSKI E.L., MESCHIA G., DROEGEMUELLER W. & BATTAGLIA F.C. (1968) Distribution of uterine blood flow in the pregnant sheep. *Amer. J. Obstet. Gynec.* **101,** 409–412.

MALPAS P. (1933) Postmaturity ewe malformations of foetus. *J. Obstet. Gynaec. Brit. Emp.* **40,** 1046–1053.

MARSTON J.H., KELLY W.A. & ECKSTEIN P. (1969) The effect of an I.U.D. on uterine motility in women and rhesus monkey. *Post-grad. Med. J.* **45,** 75–77.

MICHAEL C.A. & SCHOFIELD B.M. (1969) The influence of the ovarian hormones on the actomyosin content and the development of tension in uterine muscle. *J. Endocrin.* **44,** 501–511.

NAKANISHI H., McLEAN J., WOOD C. & BURNSTOCK G. (1969) The role of sympathetic nerves in control of the nonpregnant and pregnant human uterus. *J. Rep. Med.* **2,** 20–33.

NEEDHAM D.H. & SHOENBERG C.F. (1967) The biochemistry of the myometrium, in R.M.Wynn (ed.) *Cellular Biology of the Uterus*, pp. 291-352, New York, Appleton, Century-Crofts.

NEWTON W.H. (1938) Hormones and placenta. *Physiol. Rev.* **18**, 419-446.

PAWSON M.E. & SIMMONS S.C. (1970) Routine induction of labour by amniotomy and simultaneous syntocinon infusion (synthetic oxytocin). *Brit. med. J.* **3**, 191-193.

PERCHARD S. (1961) Trial of hypnosis in ante-natal preparation. Personal communication.

POSEIRO J.J. (1958) Fetal and maternal dangers due to misuse of oxytocin. *Int. Congr. Gynaec. Obstet., Montreal*, p. 3.

POSEIRO J.J., MASSI C.B., BIENIAZ J., CROTHOGINI J.J., CURUCHET E. & CALDEYRO-BARCIA R. (1964) In R.Caldeyro-Barcia, Mendez-Bauer & C.S.Dawes (eds.) *Effects of Labor on the Fetus and Newborn Symposium, Montevideo*. Oxford, Pergamon Press.

POSE S.V. (1958) Measurements of uterine response to oxytocin at different gestational ages in normal and abnormal conditions. *Int. Congr. Gynaec. Obstet., Montreal*, p. 1.

PULKKINEN M.O. (1969) The significance of progesterone in myometrial regulation during mid-trimester human pregnancy, in G.E.W.Wolstenholme & J.Knight (eds.) *Progesterone: its Regulatory Effect on the Myometrium*, pp. 133-158. London, J. & A. Churchill.

PULKKINEN M.O. (1970) Regulation of uterine contractility. *Acta obstet. gynec. scand.* **49**, Suppl. 1, 23-41.

RENOU P., NEWMAN W., LUMLEY J. & WOOD C. (1968) Fetal scalp blood changes in relation to uterine contractions. *J. Gynaec. Brit. Cwlth.* **75**, 629-635.

REYNOLDS S.R.M. (1949) *Physiology of the Uterus*, 2nd ed. New York, Hoeber.

ROBERTS G., ANDERSON A., MCGARRY J. & TURNBULL A.C. (1970) Absence of antidiuresis during administration of prostaglandin. *Brit. med. J.* **2**, 152-154.

ROBERTSON E.M. (1939) Effects of emotional stress on contractions of human uterus: preliminary report. *J. Obstet. Gynaec. Brit. Emp.* **46**, 741-747.

RYAN K.J. (1968) Theoretical basis for endocrine control of gestation—a comparative approach, in A.Pecile (ed.) *Proceedings of the International Symposium on Foeto-placental unit*, pp. 120-131. Excerpta Medica International Congress Series, 183.

SCHWARCZ R.L., FISCH L. & SALA N.L. (1965) Effect of cervical dilatation upon milk ejection in humans and its relation to oxytocin secretion. *Amer. J. Obstet. Gynec.* **91**, 1090-1094.

SCHOFIELD B.M. (1968) In A.McLaren (ed.) *Advances in Reproductive Physiology*. London, Logos Press.

SCHOFIELD B.M. & WOOD C. (1964) Length-tension relation in rabbit and human myometrium. *J. Physiol. (Lond.)* **175**, 125-133.

SHARP A.H. & WOOD C. (1966) The effect of duphaston (dydrogesterone) upon uterine and pressure-volume relationships in the nonpregnant human uterus. *Aust. N.Z. J. Obstet. Gynaec.* **6**, 321-326.

SILVA D. (1967) Electron microscopy of human myometrium—muscle and neural components, in R.J.Kellou (ed.) *Modern Trends in Obstetrics*, vol. 4, p. 58. London, Butterworth.

STAN M., BALACEANO M. & PETERSEN S. (1961) Cercetari in domen il formelor functionale de diabet insipid. *Studii Cerc. Endocr.* **14**, 685.

THEOBALD G.W. (1963) The induction of labour and of premature labour, in A. Claye (ed.) *British Obstetric Practice*, pp. 1055–1088, 3rd ed. vol. 1, London, Heinemann.

THOMAS J., RAPHAEL M. & ISBISTER C. (1966) Evaluation of a preparation-for-childbirth programme. *Med. J. Aust.* **1**, 776–783.

TURNBULL A.C., ANDERSON A.B. & WILSON G.R. (1967) Maternal urinary oestrogen excretion as evidence of a foetal role in determining gestation at labour. *Lancet* **ii**, 627–629.

TURNBULL A.C. & ANDERSON A.B.M. (1969) The influence of the foetus on myometrial contractility, in G.E.W.Wolstenholme & J. Knight (eds.) *Progesterone: its Regulatory Effect on the Myometrium*, pp. 106–119. London, J. & A. Churchill.

VAN WAGENEN G. & NEWTON W.H. (1943) Pregnancy in monkey after removal of fetus. *Surgery, Gynec. Obstet.* **77**, 539–543.

VASICKA A. & KRETCHMER H. (1961) Effect of conduction and inhalation anesthesia on uterine contractions. *Amer. J. Obstet. Gynec.* **82**, 600–611.

WALKER A., MADDERN L., DAY E., RENOU P., TALBOT J. & WOOD C. (1970) Foetal scalp tissue PO_2: measurements in relation to maternal dermal PO_2 and foetal heart rate. *J. Obstet. Gynaec. Brit. Cwlth.* (in press).

WANSBROUGH H., NAKANISHI H. & WOOD C. (1967) Effect of epinephrine on human uterine activity *in vitro* and *in vivo*. *Obstet. & Gynec.* **30**, 779–780.

WANSBROUGH H., NAKANISHI H. & WOOD C. (1968) The effect of adrenergic receptor blocking drugs on the human uterus. *J. Obstet. Gynaec. Brit. Cwlth.* **75**, 189–198.

WESSELIUS DE CASPARIS A. (1970) Results of a double blind multicentre study with DU-21220 in premature labour. *6th World Congress of Obstetrics and Gynecology*. New York (in press).

WHITESIDE J. H., BRAME R.G. & MCGAUGHER H.S. Jr. (1967) Myometrial response to vascular pressure changes in primates. *Surgical Forum* **18**, 422–424.

WIEST G.W., PULKKINEN M.O., SAUVAGE J. & CSAPO A.I. (1970) Plasma progesterone levels during saline-induced abortion. *J. clin. Endocr. & Metab.* **30**, 6, pp. 774–777.

WIGGLESWORTH J.S. (1964) Experimental growth retardation in the foetal rat. *J. Path. & Bact.* **88**, 1–13.

WILSON K.M. (1937) Pregnancy complicated by ovarian and parovarian tumors. *Amer. J. Obstet. Gynec.* **34**, 977–986.

WOESSNER J.F., FERGUSON J.H. & BREWER T.W. (1961) Post-partum involution of the uterus. A.M.A. exhibit, New York.

WOOD C. (1969) Uterine activity in late pregnancy, in G.E.W.Wolstenholme & J.Knight (eds.) *Progesterone: its Regulatory Effect on the Myometrium*, pp. 159–184. London, J. & A. Churchill.

WOOD C. (1965) Resting tension in the human uterus. *Aust. N.Z. J. Obstet. Gynaec.* **5**, 219–221.

WOOD C. (1964a) Physiology of uterine contractions. *J. Obstet. Gynaec. Brit. Cwlth.* **7**, 360–373.

WOOD C. (1964b) The expansile behaviour of the human uterus. *J. Obstet. Gynaec. Brit. Cwlth.* **71**, 615–620.
WOOD C., ELSTEIN M. & PINKERTON J.H. (1963) Uterine volume and myometrial function. *J. Obstet. Gynaec. Brit. Cwlth.* **70**, 396–401.
YORDAN E. & D'ESOPO D.A. (1955) Hydramnios: review of 204 cases at Sloane Hospital for Woman. *Amer. J. Obstet. Gynec.* **70**, 266–273.

TUBAL PHYSIOLOGY

Studies in the human have been limited by the paucity of methods for studying tubal function *in vivo* and the difficulty in identifying the relative importance of neuromuscular, endocrine and circulatory influences from *in vitro* experiments. Human tubal physiology will be described but studies of other species will be presented when human data are lacking.

Fertilization depends upon the properly prepared ovum and spermatozoon meeting at an appropriate time in the ampulla of the tube. Normal development of the fertilized ovum into a blastocyst depends upon the tubal and uterine environment. Its passage from the tube into the uterus occurs at the time when implantation into the primed endometrium is most favoured.

SPERM TRANSPORT

Investigations in several species have shown that the number of sperm reaching the site of fertilization is considerably less than the number which enter the cervix and uterus (Chang & Pincus 1951, Doak *et al* 1957, Hartman 1962, Marcus 1965, Rigby 1966). Although hundreds of millions of sperms are deposited in the vagina, only thousands enter the oviducts, and fewer still reach the site of fertilization. The principal sites where sperm numbers diminish are the cervix, utero-tubal junction and the isthmus (Sobfero 1963). In a number of species the cervix acts as a major sperm reservoir (Mattner 1963a, 1966; Quinlivan & Robinson 1969). There is very slight dilatation of the cervix after sexual excitation in the human female (Masters & Johnson 1966) but it is not known whether the human cervix acts as a reservoir for spermatozoa. In the sow a high concentration of sperm persists at the utero-tubal junction, maintaining a reservoir to supply the ampulla of the tube. It is evident from animal experiments and clinical findings in the human that the minimum number of sperms necessary to achieve fertilization is much less than the number present in the average ejaculate (Chang & Pincus 1951). Witschi refers to the 'squandering' of male germ cells at 'incomprehensible rates' (Witschi, 1968). The squandering may be explained largely in terms of the restricted anatomical apertures which intervene between the spacious cavities of the vagina, uterus and tubal ampulla. Making a number of anatomical assumptions and assuming that no biologic factors except for motility affect sperm transport, it has been calculated that the average human ejaculate results in less than one hundred sperms entering the tube (Handsjuk 1970).

The explanation for this apparently incomprehensible rate of sperm loss referred to by Witschi (1968) may be:

(1) A large loss ensures that only biologically fit sperm survive to fertilize the ovum, for which there is no evidence.

(2) A factor is present in the ejaculate or the vagina which hinders fertilization or ovum development. This would necessitate separation of the site of fertilization from the site of ejaculation and entail loss of sperm numbers. In the rabbit there is suggestive evidence that seminal plasma does have a deleterious effect upon development of fertilized ova (Chang 1950, Hadek 1959) but this is not so in sheep (Killeen & Moore 1970).

Separation of the site of conception and site of coition may also provide mechanical and bacteriological protection of the ovum and diminish the possibility of loss of the single ovum (compared to uterine fertilization).

After penetration of the lumen of the cervix a complex system involving muscular contractions, fluid currents, ciliary action, anatomical barriers and sperm motility may influence the rate at which sperms ascend to the site of fertilization (Noyes 1968). It is also possible that chemical and electrical energy might influence sperm transport.

Studies in a number of species have shown that tubal transport of particles is dependent upon their size and on the current endocrine status (Woodruff 1969). Small particles are transported through the tubes in both directions at all times in the cycle. Noyes (1968) considered that active sperm motility was an important factor.

However, sperm transport through the uterus and into the tube cannot be accounted for by sperm motility alone (Chang & Pincus 1951, Olds & Van Demark 1957). In the human for example, sperms have been recovered from the fallopian tube 30 minutes after being placed on the cervix (Rubenstein 1951).

Walton et al (1960) have proposed a theory to account for the small proportion of sperm which progresses up the genital tract when the main flow of mucus is downwards. Until the volume and direction of currents in the genital tract are known it would seem purposeless to pursue this theory. Nevertheless the possibility of wave dynamics and 'surf' game theory applying to sperm dynamics is attractive—wipe-out from a wave descending down the genital tract may be a mechanism of sperm survival!

SPERM CAPACITATION

In 1951, evidence was produced that spermatozoa undergo some physiological change within the female reproductive tract whereby the capacity to penetrate the zona pellucida of the ovum is acquired. This phenomenon was called 'capacitation' (Austin 1952) and has been shown to be necessary in the rabbit, rat, hamster, mouse and ewe (Chang 1951, Yanagimachi 1966, Braden & Austin 1954, Mattner 1963b). Using spermatozoa recovered from the rabbit uterus after normal mating fertilization has been achieved *in vitro*. Following transfer of the fertilized eggs to suitably prepared recipient rabbits, live young have been obtained (Bavister 1969). The success of these experiments seemed to depend upon use of uterine spermatozoa. It was surprising therefore that capacitation of spermatozoa and *in vitro* fertilization was achieved in the golden hamster using epididymal spermatozoa. It is unknown which of the constituents of the genital tract normally facilitate capacitation. Follicular, tubal or uterine fluid may be responsible. In the human it is not even clear whether capacitation occurs at all because sperm penetration of the ovum has been achieved *in vitro* using washed ejaculate and oocytes recovered from ovarian tissue (Menkin & Rock 1948, Edwards et al 1969).

FERTILIZATION

Normally the sperm fertilizes the ovum in the ampulla of the tube. The process can only occur within a limited period in each cycle of the limited life span of the spermatozoon and ovum. The fertilizing life span of spermatozoa is about half the motile life span (Chang & Pincus 1951,

Hartman 1962, Pincus 1965). In the human motile sperms have been found in peritoneal washings 24 hours after coitus (Horne & Thibault 1962). An upper limit of 24 hours has been estimated as being the survival time of human ova (Hartman 1962). Tubal fluid may play a role in normal fertilization. The fact that early stages in the fertilization of the human oocytes have been achieved *in vitro* (Edwards *et al* 1969) and that ovarian-uterine translocations (Estes operation) are occasionally successful, suggests that it is not vital however. More recently Seitz *et al* (1970) have been able to fertilize rabbit oocytes recovered from the ovary.

Fertilization of the ovum by sperm poses two questions: (a) How does the sperm find the ovum? Is it by chance, or by physical or chemical direction? (b) What ensures the monogamous relationship between sperm, and ovum, because it is known that penetration of the ovum by more than one sperm is incompatible with normal development. In sheep Moore (1970) has shown that mechanical penetration of the zona pellucida does not prevent entry of other spermatozoa. It is probable that a chemical reaction prevents polyspermy.

CLEAVAGE

In the human and nearly all mammalian species, the fertilized ovum remains in the oviduct for 3 days prior to entering the uterus. Whilst tubal fluid may not be an absolute prerequisite for the maintenance of the fertilized ovum, evidence suggests that this is normally the case. It is certain that the cytoplasm of the mammalian ovum does not contain sufficient nutrients to maintain the zygote. The nutritional role of tubal fluid has been demonstrated by the passage of radioactive sulphur from tubal lining to the ovum and blastomere of the rabbit. Studies using culture techniques have shown that the mouse embryo needs both exogenous energy and exogenous amino-nitrogen sources (Brinster 1969). The energy source can be pyruvate or oxaloacetate between ovulation and the two-cell stage, and pyruvate, oxaloacetate, lactate, or phosphoenolpyruvate between the two- and eight-cell stage. The ineffectiveness of glucose is remarkable since it serves both *in vitro* and *in vivo* as a major source of energy for most tissues. After the eight-cell stage, a variety of single compounds, including glucose and certain amino acids, may serve as the sole source of energy. Eight-celled mouse embryos develop into blastocysts during a period of 48 hours, with only a single compound to act as a carbon source. An exogenous amino-nitrogen source is not required. Using U-14C-glucose, Quinn & Wales (1970) found a rapid increase in glucose uptake as the rabbit embryo developed from a zygote to blastocyst, with incorporation of the label in protein fractions also. Tubal fluid has been shown to contain lactate, pyruvate, glucose and amino acids. These constituents may vary with the endocrine status of the animal and thereby provide a suitable milieu to support segmentation and blastulation.

OVUM TRANSPORT

The ovum reaches the tube either by, (a) direct transport when the fimbria of the tube and ovary converge (Decker 1951, Westman 1959) or (b) migration across the peritoneal cavity (Doyle 1951, Berlind 1960). Suction is not necessary for ovum capture (Cleive & Mastroianni 1958).

The fertilized ovum rapidly traverses the oviduct to the ampullary isthmic junction where it resides for one or more days before passing to the uterine cavity (Anderson 1927, Kirton & Hafs 1965, Lutwak-Mann 1962, Holst 1970). The time spent by ova in the tubes is most important as the uterus provides a hostile environment for underdeveloped eggs. Two questions arise. What transports the ovum and what controls the delay in ovum transport? It is not particularly remarkable that an ovum of 130 μ passes through a tube of 5–8 cm length when the tube is known to undergo peristalsis and allow movement of particles in either direction. On the other hand the precise timing of the ovum's movements is quite remarkable. This movement may result from ciliary action, tubal muscular action, or fluid flow in the tubal lumen.

Ciliary action

Passage from the end of the tube to the ampulla may depend partly on ciliary action. In the rabbit the cilia act on the cumulus oophorus covering the ovum (Blandau, quoted by Woodruff, 1969). Denuded ova spin, but are not transported into the ampulla.

Tubal muscular action

Tubal movement has been observed by culdoscopy and peritoneoscopy at the time of ovulation. Three movements have been described: (1) peristalsis passing from the ampulla to isthmus, (2) segmental to and

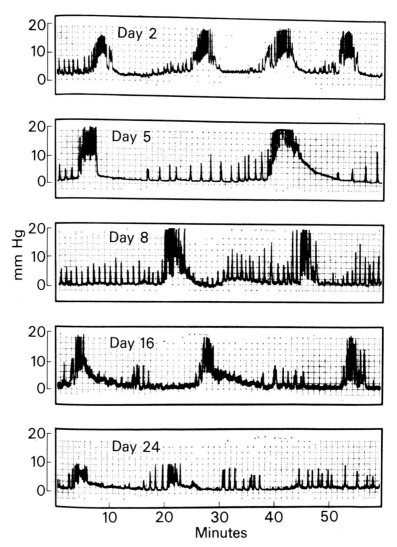

Fig. 7.9. Patterns of tubal motility of a 26-year-old woman during the menstrual cycle. Burst of activity most frequent during menstruation. From H.Maia & E.M.Coutinho (1968) *Amer. J. Obstet. & Gynec.* **102**, 1043.

fro movements at the junction of the ampulla and isthmus and (3) peristalsis from the isthmus to uterine cornu (Doyle, 1954).

By inserting polyvinyl catheters into the fallopian tubes of patients and fixing the tube to the uterine fundus by catgut, Maia & Coutinho (1970) have recorded pressure changes in different sections of the tube during various stages of the menstrual cycle (Fig. 7.9). This is the best *in vivo* demonstration of tubal function although the recorded changes in isometric pressure may differ from the predominantly isotonic tubal muscular activity. They found peristalsis and anti-peristalsis in both the proliferative and luteal phase. The peristaltic wave was slowest in the luteal phase. This may be responsible for the delay in passage of the ovum, which occurs at this time. Peristalsis was strongest during menstruation when there was no anti-peristalsis. Retrograde tubal menstruation may thus be prevented. In addition uterine and tubal contractility have been found to be asynchronic (Coutinho & Maia 1969) so that the tube is not the functional pacemaker of the uterus. Details of tubal movement as deduced from analysis of Rubin's insufflation procedure or hysterosalpingographic screening will not be considered. These techniques are unphysiological.

Control of tubal contractility

Endocrine control of tubal transport has been subject to a large number of studies in a variety of species. The results frequently differ and occasionally conflict (Woodruff & Pauerstein 1969). In a number of mammalian species, particles approximating the size of ova are transported very rapidly from the peritoneal ostium of the tube to the ampullary-isthmic junction, at which point there is considerable delay. This delay is critical to the arrival of the fertilized ova in the uterus at an appropriate time in the ovarian cycle. Tubal transport of ova has been shown to vary with the ratio of oestrogen and progesterone in the genital tract and may normally depend upon an appropriate ratio of these hormones. This ratio is specific for any particular species.

The delay in ovum transport occurs in the isthmical region. It is not known, however, whether this occurs at the utero-tubal or ampullary isthmic junction. As there is no clear-cut evidence in the human that an anatomical block exists at either of these places, a physiological block involving the muscle of the isthmus has been postulated. Evidence for this is as follows:

(1) A difference in pressure on either side of the ampullary-isthmic junction has been demonstrated in the human oviduct *in vitro* (Seithchik et al 1968).

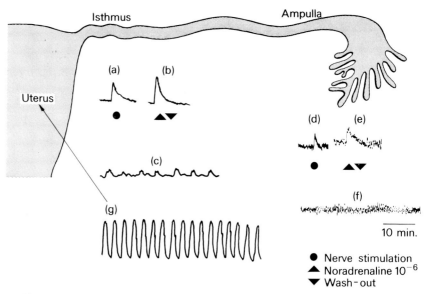

FIG. 7.10. Nerve supply of human tube. (a) Contractile response of isthmic region to perivascular nerve stimulation 930 pulses/sec frequency, 1 msec pulse width, submaximal intensity, 10 sec duration. (b) Contractile response of isthmic region to noradrenaline (10^{-6}). (c) Spontaneous movement of isthmic muscle. (d) Contractile response of ampullary region to perivascular nerve stimulation (30 pulses/sec frequency, 1 msec pulse width, submaximal intensity, 10 sec duration). (e) Contractile response of ampullary region to noradrenaline (10^{-6}). (f) Spontaneous movement of ampullary muscle. Note the higher sensitivity of isthmic muscle to nerve stimulation and noradrenaline. The spontaneous contractions of the isthmus have a lower frequency and higher amplitude compared to the ampulla. (g) Spontaneous movement of muscle from uterine body. From H.Nakanishi & C.Wood (1968) *J. Reprod. Fert.* **16**, 24.

(2) In the human tube the isthmus has a dense noradrenergic innervation of its circular muscle while the ampulla has a sparse innervation (Owman et al 1967).

(3) Using a perivascular-nerve preparation of the human tube the isthmus has been shown to be more sensitive than the ampulla both to

the noradrenergic transmitter, noradrenaline, and to nerve stimulation, suggesting that it may act as a sphincter (Nakanishi & Wood 1968) (Fig. 7.10).

Control of the physiologic sphincter may be neural or hormonal. The nerve supply to the tube has been shown to be post synaptic, noradrenergic in type and to have both alpha and beta adrenergic receptors in the muscle cells (Nakanishi et al 1967, Rosenblum & Stein 1966, Seitchik 1968). The neural transmitter noradrenaline, normally produces contraction of the tube following nerve stimulation. However, its effect on the tube consists of the summation of a strong excitatory and a weak inhibitory influence, due to stimulation of the alpha and beta adrenergic receptors respectively (Nakanishi & Wood 1968a). Blockage of the alpha receptor by a blocking agent such as phenoxybenzamine reveals a weak inhibition following nerve stimulation. That this weak inhibitory effect results from stimulation of the β-receptor can be shown by also blocking the β-receptor with a β-receptor blocking drug such as propranolol. *In vivo* the excitatory effect of noradrenaline released after neural stimulation may be modified. If the sensitivity of the alpha (excitatory) or beta (inhibitory) receptors are changed by hormones, as suggested by Miller & Marshall (1965) for the uterus, or if either the alpha or beta receptors are preferentially occupied by humoral catecholamines the release of neural noradrenaline may produce little or no tubal contraction or even its inhibition. Control of the physiologic sphincter certainly alters in the pregnant patient. The fallopian tube at this time is less sensitive to noradrenaline and nerve stimulation than in the non-pregnant state and the response of the isthmus is similar in magnitude to that of the ampulla (Nakanishi & Wood 1968). Under these circumstances the isthmus can no longer act as a physiologic sphincter. Furthermore, in post-menopausal women, nerve stimulation produces only slight excitation and noradrenaline inhibition. The change of neural response in both pregnant and post-menopausal women may be related to endocrine status.

A wide variety of pharmacologic agents have been tested on the tube (Woodruff, 1969) but the physiologic significance of the findings is difficult to gauge. Prostaglandins F_1 and F_2 have a stimulatory effect and the alpha fraction of F_1, which occurs naturally, increases tonus mostly in the proximal segment. Zetler et al (1969) studied the potency of various agents in their ability to stimulate the fallopian tube *in vitro*. They demonstrated a decreasing response from the tachykinin peptides, prostaglandin $F_2\alpha$, 5 hydroxytryptamine, and noradrenaline to oxytocin. Their findings, however, have not been substantiated by other workers who have found greater tubal sensitivity to noradrenaline.

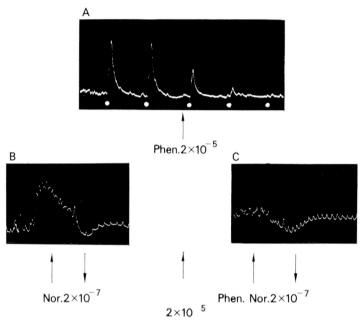

FIG. 7.11. The excitatory effect of noradrenaline (neural transmitter) can be changed to inhibition in the presence of an α-receptor blocking agent, e.g. phentolamine, phenoxybenzamine or prostaglandin. Acting in this manner prostaglandin released at coitus may cause isthmic relaxation and aid sperm passage. (A) complete inhibition of nerve stimulated response by phentolamine, 2×10^{-5}. (B) excitatory response to noradrenaline, 2×10^{-7}. (C) inhibitory response to noradrenaline, 2×10^{-7}, in the presence of phentolamine, 2×10^{-5}. Stimulation 30 pulses/sec, 1 msec, 15 v/cm, for 10 sec. From H.Nakanishi, H.Wansbrough & C.Wood (1967) *Amer. J. Phys.* **213**, 615.

Because spermatozoa have been found in the isthmus of the tube soon after coitus (Hartman & Ball 1930), it has been postulated that relaxation of the isthmus occurs after coitus. It has been suggested that oxytocin release at the time of orgasm may do this (Fox & Knaggs 1969). This is unlikely as very high concentrations of oxytocin are needed to relax the isthmus *in vitro* (Nakanishi & Wood 1968c). However, prostaglandin in semen may dilate the isthmus. It has a marked adrenergic α-blocking action (Eliasson & Pose 1965, Hedqvist 1970). We have shown that the contractile response to nerve stimulation of the isthmus changes to relaxation in the presence of adrenergic α-blocking agents, (Nakanishi & Wood 1968) (Fig. 7.11a). Unlike oxytocin release the action of prostaglandin is not dependent upon the occurrence of orgasm.

In summary tubal transport of the fertilized ovum is probably controlled by a physiological sphincter, the isthmus, whose contractile state may be related to the inter-relationship between neural and steroid influences. A possible mechanism for the interaction of steroid and neural influences is suggested by the study of Nakanishi & Wood (1968b), whereby alterations of magnesium and calcium ionic concentration changed the tubal response to both neural stimulation and noradrenaline. Steroids have been shown to influence calcium and magnesium binding in smooth muscle cells of the genital tract.

Fluid movement

Another factor which may influence ovum transport is fluid flow through the tube. In the ewe it has been shown that the flow is increased by oestrogen and decreased by progesterone (McDonald & Bellve 1969). A reduction in flow resulting from progesterone secretion in the luteal phase may retard ovum transport and aid sphincter action of the isthmus. The *Macaca mulatta* monkey which has a similar menstrual cycle to the human, increases tubal secretion 2 days before ovulation, after which it decreases rapidly (Mastroianni *et al* 1961).

The possibility of replacing tubal function in patients with infertility resulting from tubal disease depends on an understanding of the role of the tube in conception. Possible mechanisms have been discussed by Rioux *et al* (1969). Bilateral ileocornual anastomosis in eight dogs resulted in one pregnancy (Wingate *et al* 1970) while appendix, uterine muscle, veins and arteries have been used unsuccessfully to replace tubes in the human. More recently the technical feasibility of a system of artificial transport of the ovum from the ovary to the uterus has been demonstrated in the human (Wood *et al* 1970). The biological requirements determining the success of such a system may be stringent but may well be elucidated by further research. The first step of an alternative system, artificial transport of human ovum from the ovary and early *in vitro* fertilization, has been achieved (Edwards *et al* 1969). *In vitro* blastocyst development and replacement in the human uterus may follow.

ACKNOWLEDGMENTS

I wish to thank Dr Peter Paterson and Mrs Philippa Gillard for help in the preparation of the manuscript. I also wish to thank Dr Neil Moore, University of Sydney, McGaughey Memorial Institute, Jerilderie, for his advice in preparing the section on tubal physiology.

REFERENCES

ANDERSON D. (1927) The rate of passage of the mammalian ovum through various portions of the fallopian tube. *Amer. J. Physiol.* **82**, 557–569.

AUSTIN C.R. (1951) Observation on the penetration of the sperm into the mammalian egg. *Aust. J. Sci. Res.* **B 4**, 581–589.

AUSTIN C.R. (1952) The 'capacitation' of the mammalian sperm. *Nature* **170**, 326.

BAVISTER B.D. (1969) Fertilization of mammalian eggs *in vitro*. *Research in Reproduction* **1(2)**, 1–2.

BERLIND M. (1960) The contralateral corpus luteum—an important factor in ectopic pregnancies. *Obstet. Gynec.* **16**, 51–52.

BRADEN A.W.H. & AUSTIN C.R. (1954) Number of sperms about the eggs in mammals and the significance for normal fertilization. *Aust. J. Biol. Sci.* **7**, 543–551.

BRINSTER R.L. (1969) Metabolism of pre-implantation mammalian embryos. *Research in Reproduction* **1(1)**, 2–3.

CHANG M.C. (1950) Effect of seminal plasma on fertilized rabbit ova. *Proc. nat. Acad. Sci., U.S.A.* **36**, 188–191.

CHANG M.C. (1951) Fertilizing capacity of spermatozoa deposited into fallopian tubes. *Nature* **168**, 697–698.

CHANG M.C. & PINCUS G. (1951) Physiology of fertilization in mammals. *Physiol. Rev.* **31**, 1–26.

CHANG M.C. & YANAGIMACHI R. (1963) Fertilization of ferret ova by deposition of epididymal sperm into the ovarian capsule with special reference to the fertilizable life of ova and the capacitation of sperm. *J. Exp. Zool.* **154**, 175–183.

CLEIVE P.H. & MASTROIANNI L. Jr. (1958) Mechanisms of ovum pickup: Functional capacity of rabbit oviducts ligated near the fimbria. *Fertil. & Steril.* **9**, 13–17.

COUTINHO E.M. & DA SILVA MAIA H. (1969) Asynchronism between tubal and uterine activity in woman. *J. Reprod. Fert.* **19**, 591–593.

DECKER A. (1951) Culdoscopic observations on tubo-ovarian mechanism of ovum reception. *Fertil. & Steril.* **2**, 253–259.

DOYLE J.B. (1951) Exploratory culdotomy for observation of tubo-ovarian physiology at ovulation time. *Fertil. & Steril.* **2**, 475–586.

EDWARDS R.G., BAVISTER B.D. & STEPTOE P.G. (1969) Early stages of fertilization *in vitro* of human oocytes matured *in vitro*. *Nature (Lond.)* **221**, 632–635.

ELIASSON R. & POSSE N. (1965) Rubin's test before and after intravaginal application of prostaglandin. *Int. J. Fert.* **10**, 373–377.
FOX C.A. & KNAGGS G.S. (1969) Milk ejection activity (oxytocin) in peripheral venous blood in man during lactation and in association with coitus. *J. Endocrinol.* **45**, 145–146.
HADEK R. (1959) Study of the sperm capacitation factor in the genital tract of the female rabbit. *Amer. J. Vet. Res.* **20**, 753–755.
HANDSJUK L. (1970) Medical Student Elective. Monash University, Melbourne. 'Theoretical Calculation of Sperm Loss in the Human Genital Tract.'
HARTMAN C.G. (1962) *Science and the Safe Period*. Baltimore, Williams & Wilkins.
HARTMAN C.G. & BALL J. (1930) On the almost instantaneous transport of spermatozoa through the cervix and the uterus in the rat. *Proc. Soc. Exp. Biol. Med.* **28**, 312–319.
HEDQVIST P. (1970) Control by prostaglandin F_2 of sympathetic neurotransmission in the spleen. *Life Science* **2**, part 1, 269–278.
HOLST P.J. (1970) Ovum transport in the ewe, in J.R.Cuming & I.A.Cuming (eds.) *Proc. Aust. Soc. Rep. Biol.* Melbourne, University Press.
HORNE H.W. & THIBAULT J.P. (1962) Sperm migration through the human female reproductive tract. *Fertil. & Steril.* **13**, 444–447.
KILLEEN I.D. & MOORE N.W. (1971) The morphological appearance and development of sheep ova fertilized by surgical insemination. *J. Reprod. Fert.* **24**, 63–70.
KIRTON K.T. & HAFS H.D. (1965) Sperm capacitation by uterine fluid or beta-amylase *in vitro*. *Science* **150**, 618–619.
LUTWAK-MANN C. (1962) Glucose, lactic acid and bicarbonate in rabbit blastocysts fluid. *Nature (Lond.)* **193**, 653–654.
MCDONALD M.F. & BELLVE A.R. (1969) Influence of oestrogen and progesterone on flow of fluid from the fallopian tube in the ovariectomized ewe. *J. Reprod. Fert.* **20**, 51–61.
MAIA H. & COUTINHO E.M. (1968) A new technique for recording human tubal activity *in vivo*. *Amer. J. Obstet. Gynec.* **102**, 1043.
MARCUS S.L. (1965) The passage of rat and foreign spermatozoa through the uterotubal junction of the rat. *Amer. J. Obstet. Gynec.* **91**, 985–989.
MASTERS W.H. & JOHNSTON V.E. (1966) *Human Sexual Response*. London, Churchill.
MATTNER P.E. (1963a) Spermatozoa in the genital tract of the ewe. Distribution after coitus. *Aust. J. Biol. Sci.* **16**, 688–694.
MATTNER P.E. (1963b) Capacitation of ram spermatozoa and penetration of the ovine egg. *Nature* **199**, 772–773.
MATTNER P.E. (1966) Formation and retention of the spermatozoa reservoir in the cervix of the ruminant. *Nature* **212**, 1479–1480.
MENKIN M.F. & ROCK J. (1948) *In vitro* fertilization and cleavage of human ovarian eggs. *Amer. J. Obstet. Gynec.* **55**, 440–452.
MILLER M.D. & MARSHALL J.M. (1965) Uterine response to nerve stimulation; relation to hormonal status and catecholamines. *Amer. J. Physiol.* **209**, 859–865.
MOORE N.W. (1970) Insemination of sheep ova, in J.R.Goding & I.A.Cuming (eds.) *Proc. of Aust. Soc. Rep. Biol.* Suppl. Melbourne University Press.

NAKANISHI H., WANSBROUGH H. & WOOD C. (1967) Postganglionic sympathetic nerve innervating human fallopian tube. *Amer. J. Physiol.* **213**, 613–619.

NAKANISHI K. & WOOD C. (1968c) Biphasic effects of oxytocin on human uterine and tube motility. *Aust. N.Z. J. Obstet. Gynaec.* **8**, 181–188.

NAKANISHI H. & WOOD C. (1968a) Effects of adrenergic blocking agents on human fallopian tube motility *in vitro*. *J. Reprod. Fert.* **16**, 21–28.

NAKANISHI H. & WOOD C. (1968b) Effects of calcium and magnesium on sympathetic transmission in human fallopian tube. *Arch. int. Pharmacodyn.* **174**, 469–480.

NOYES R.W. (1968) Sperm transport, in S.J.Behrman & R.W.Kistmer (eds.) *Progress in Infertility*, pp. 181–194. Boston, Little Brown.

OLDS D. & VAN DEMARK N.L. (1957) Physiological aspects of fluids in female genitalia with special reference to cattle. *Amer. J. Vet. Res.* **18**, 587–602.

OWMAN C., ROSENGREN, E. & SJOBERG N. (1967). Adrenergic innervation of the human female reproductive organs: a histochemical and chemical investigation. *Obstet. Gynec.* **30**, 763–773.

PINCUS G. (1965) *The Control of Fertility*. New York, Academic Press.

QUINN P. & WALES R.J. (1970) The *in vitro* metabolism of U-14C-glucose by the preimplantation rabbit embryo, in J.R.Goding & I.A.Cuming (eds.) *Proc. Aust. Soc. Rep. Biol.* p. 4. Melbourne University Press.

QUINLIVAN T.D. & ROBINSON T.J. (1969) Numbers of spermatozoa in the genital tract after artificial insemination of progestagen-treated ewes. *J. Reprod. Fertil.* **19**, 73–86.

RIGBY J.P. (1966) The persistence of spermatozoa at the uterotubal junction of the sow. *J. Reprod. Fertil.* **11**, 153–155.

RIOUX J., TREMBLAY A., BASTIDE A.F., & BRASSARD A. (1969). Artificial transport of the ovum from the ovary to the uterus. *J. of Reproductive Medicine: Lying-in* **2**, 168–175.

ROSENBLUM I. & STEIN A.A. (1966) Autonomic response of the circular muscles of the isolated human fallopian tube. *Amer. J. Physiol.* **210**, 1127–1129.

RUBENSTEIN B.B., STRAUSS H., LAZARUS M.L. & HANKIN H. (1951) Mobile sperm in fundus and tubes of surgical cases. Sperm survival in women. *Fertil. & Steril.* **2**, 15–19.

SEITCHIK J., GOLDBERG E., GOLDSMITH J.P. & PAUERSTEIN C.J. (1968) Pharmacodynamic studies of the human fallopian tube *in vitro*. *Amer. J. Obstet. Gynec.* **102**, 727–735.

SEITZ H.M., BRACKETT B.G. & MASTROIANNI L. (1970) *In vitro* fertilization of ovulated rabbit ova recovered from the ovary. *Biol. Reprod.* **2**, 262–267.

SOBRERO A.J. (1963) Sperm migration in the female genital tract, in Conference on Physiological Mechanisms concerned with Conception, *Mechanics Concerned with Conception*, C.G.Hartman (ed.). New York, Macmillan, 1959.

WALTON A., BISHOP M.W.H. & PARKES A.S. (1960) In F.H.A.Marshall, *Physiology of Reproduction*. **1**, Part 2. Boston, Little Brown.

WINGATE M.B., WINGATE L. & LACHLAN S. (1970) Pregnancy following bilateral ileocornual anastomosis in the dog. *Obstet. Gynec.* **35**, 63–68.

WITSCHI E. (1968) Natural control of fertility. *Fertil. & Steril.* **19**, 1–4.

WOOD C., LEETON J. & TAYLOR R. (1971) A preliminary design and trial of an artificial human tube. *Fertil. & Steril.* **22**, 446–450.

WOODRUFF J.D. & PAUERSTEIN C.J. (1969) *The Fallopian Tube*, p. 70. Baltimore, Williams & Wilkins Co.

YANAGIMACHI R. (1966) Time and process of sperm penetration into hamster ova *in vivo* and *in vitro*. *J. Reprod. Fert.* **11**, 359–370.

ZETLER G., MÖNKEMEIR D. & WIECHELL H. (1969) Stimulation of fallopian tubes by prostaglandin F2α, biogenic amines and peptides. *J. Reprod. Fert.* **18**, 147–149.

8
Foetal Cardio-Respiratory Physiology

E.D.BURNARD

The physiology of the foetus has been a challenge since Harvey's time. More recent awareness that respiration and metabolism were carried out in an environment very low in oxygen compared with the adult did nothing to simplify understanding, but some of the problems have been clarified in the past two decades. In part this is the result of new discoveries of general biological importance. Also, in the human case, interest has steadily grown and new techniques have been introduced. But the main reason is the general acceptance of the animal model, the contributions of the physiologist working on this basis, and the resulting productive exchange between experimentalist and clinician. Some of the more confusing hypotheses have been abandoned or qualified and the way cleared for further study.

The reader may note that many of the references given below are to the monograph by G.S.Dawes (1968a) rather than to the original papers, as a convenient source of more information on some main topics.

PLACENTA AND GAS EXCHANGE

ANATOMICAL CONSIDERATIONS

Theories of gas exchange and consequent inferences about the foetal oxygen environment have inevitably been influenced by prevailing views on anatomy and on the manner of circulation envisaged within the placenta. This has usually been considered rather sluggish on the maternal side but a different picture emerges from studies in recent years. Inferences drawn from carefully injected and dissected preparations have been related to the appearances *in vivo* obtained by contrast injection and radiography, both in primates and the human subject, and a model arrived at which reasonably satisfies the haemodynamic requirements for exchange of fluid as well as those for gaseous diffusion.

Recognition of the unitary character of the foetal cotyledon (Boe 1953) was perhaps the first step to the modern position. He suggested also that the maternal spiral arteries entered cotyledons at their bases. Following writers agreed, with some dissentients, and others considered the possible existence of a central space in the cotyledon (see Freese 1968). In the meantime radio-opaque injections *in vivo* showed that maternal blood entered through the basal plate in spurts (Borell *et al* 1958, Ramsey *et al* 1963) at pressures presumably near arterial. Bartels *et al* (1962) emphasized the capillary-like character of the intervillous space though it was recognized that a good part was of greater than capillary dimension (Boyd & Hamilton 1967). A satisfactory synthesis has been achieved by Freese (1966, 1968) whose injected preparations show the spiral artery entering the base of a cotyledon and an injection mass in a central space within the villi, while his cineradiographs show a spurt spreading to form a ring exactly as would be expected from the static preparation. Other injection and radiographic studies are in agreement (Ramsey *et al* 1963; Wigglesworth 1966, 1969; Panigel 1969).

Thus the functional unit of the haemochorial placenta is the cotyledon and its related spiral artery. Blood enters and distends the intracotyledonary space at high pressure (Fig. 8.1). Some reaches the chorionic plate and drains to the subchorionic space or adjacent cotyledons. Most of it flows outwards between the foetal villi, a 3-dimensional region approaching capillary dimensions (Fig. 8.2), where there is likely therefore to be a substantial pressure drop. From here it drains to decidual veins.

The suggestion of such a relationship harmonizes also with ideas about placental growth. Reynolds (1966) considered that the cotyledons must have become organized and develop around the spiral artery, and Wigglesworth (1969) points out that preferential growth around the incoming arterial stream would determine an intervillous space of capillary size there (compare the tight packing of villi next to the central space, Fig. 8.2) whereas further away, with less active growth, dimensions become larger and suited for venous drainage.

Considering gas exchange in relation to circulatory pattern, counter-current flow would in principle be the most efficient. The first histological observations appeared to support this view for the rabbit and the lamb (Mossman 1926, Barcroft 1946, cited by Dawes 1968a). The idea depended on the demonstration of opposed length of capillaries, but later interpretation of histological appearances did not support it, and reflecting on other complexities as well, Dawes states 'we should think

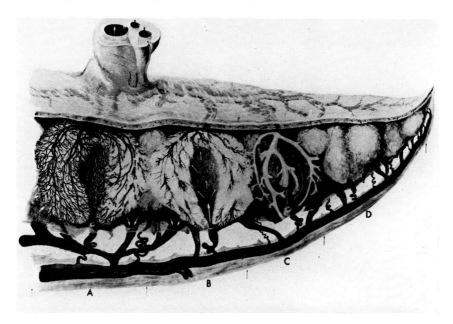

FIG. 8.1. Diagram of placenta haemochorialis. Note relationship between central space of cotyledon and spiral artery. Arrows indicate direction of flow. From U.E.Freese (1968) *Amer. J. Obstet. Gynec.* **101**, 8.

in terms of gas transfer between the two contiguous surface areas through which capillaries are running in every direction'. Calculations of diffusing capacity (Bartels 1970) have to take into account the postulated type of flow, and it is doubtful if for the present they tell us more than do general comparisons (e.g. Table 8.1). The pattern of flow now thought most likely is cross-current (Dawes 1968a) a term synonymous with 'multivillous' (Bartels *et al* 1962); this would fit very well with the anatomical characteristics just described.

The total surface area of the villi of the human placenta at term is about 11 m^2 and that part of it where the capillaries are near enough to the surface for gas exchange estimated as 1·8 m² (Aherne & Dunhill 1966). Compared with conditions after birth, on the basis of foetal and newborn oxygen consumption, the placenta performs only a little less well than do the lungs in resting conditions (Table 8.1).

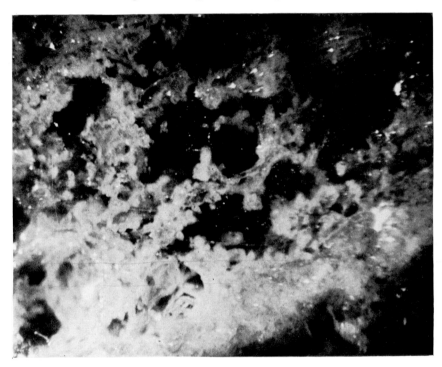

FIG. 8.2. Tightly packed terminal villi adjacent to central space. From U.E. Freese (1968) *Amer. J. Obstet. Gynec.* **101**, 8.

The modern description is quite at variance with Spanner's view of the placental circulation, on which textbooks have drawn heavily. However, Spanner had been under strong criticism for a long time (for review see Ramsey 1960) and the unreliability of classic morphological methods in studying the placenta has been emphasized by Freese (1966).

FOETAL BLOOD GASES AND pH

Difficulties in the way of establishing normal values have always been freely acknowledged, both from limitations imposed in animal experiment by the interference involved and in man from the extreme variability in conditions at birth. Findings for the exteriorized lamb are set out in Table 8.2. They are in quite close agreement with chronic catheter preparations in the same animal (Meschia et al 1965). In monkeys aortic Po_2 was somewhat higher, 38–42 mm Hg (Adamsons et al 1965). In man, umbilical venous saturation is reckoned to be about

TABLE 8.1. Comparison of O_2 transfer across the placenta at term and in the lung of newborn and adult man (Dawes 1968a)

	Placenta at term	Infant (basal)	Adult (basal)
Diffusion area (M^2)	1·8	2·8	75
VO_2 (ml/min)	16*	16*	240
VO_2/M^2	8·9	5·7	3·4
Minimum length of diffusion path (μ)	3·5	2·5	0·5

* Calculated as 4·6 ml/kg/min in a 3·5 kg infant.

80% (James 1964) and pH was 7·35 by scalp blood sampling early in labour, before the likelihood of interference with placental circulation (Kubli & Berg 1965). The question of foetal acidosis as a permanent feature of intra-uterine life, suggesting some unusual metabolic adapta-

TABLE 8.2. Mean values of blood gas tensions and pH in nine mature foetal lambs under chloralose anaesthesia (Dawes 1968a)

	Po_2 (mm Hg)	Pco_2 (mm Hg)	pH
Maternal artery	82	36	7·44
Maternal placental cotyledonary vein	43	40	7·40
Foetal placental cotyledonary vein	32	43	7·36
Foetal femoral artery*	24	45	7·33

* This is equivalent to an umbilical arterial sample since both are branches of the descending aorta.

tion, has been raised from time to time, but it is now clear that H-ion and CO_2 differences between foetus and mother are quite small, comparable to the usual order of difference between arterial and venous blood.

PLACENTAL OXYGEN CONSUMPTION

The possibility that the metabolic activity of the placenta might influence the oxygen gradient between mother and foetus was examined systematically by Campbell *et al* (1966). They showed that the placental oxygen consumption amounted on average to rather more than a third of the foeto-placental complex, a remarkably high proportion. However, any effect on foetal oxygen supply would depend on the relation of incoming arterial blood in the maternal vessels to the site of this placental activity. If such tissue were quite separate from the path to the foetus, i.e. in parallel with it, then the foetal villi would still be exposed to blood at the maternal partial pressure of oxygen, but if the two were in series a very different situation would obtain. While the real position is certain to be complex, not excluding the possibility that the foetus itself supplies some of the placental oxygen uptake, Dawes (1968a) produces convincing arguments for the very important role of placental oxygen consumption in determining foetal oxygen environment. For a critique, see Bartels (1970).

Another concept which may prove to be important deserves brief mention. Just as departures from the 'ideal' ventilation-perfusion relationship in the lung are mainly responsible for variation in the postnatal alveolar–arterial oxygen gradient in both health and illness, so variations in perfusion on the two sides of the placental membrane might affect the maternal–foetal gradient. Investigation so far has been inconclusive (Dawes 1968a) but more is likely to be heard of this approach (Ross 1967).

HAEMATOLOGICAL AND PHYSICO-CHEMICAL CONSIDERATIONS

The oxygen capacity of foetal bood is high. Haemoglobin concentration rises with gestation and in most species at term is usually greater than the adult level (Table 8.3). The difference has been regarded, generally speaking, as an adaptation to the low foetal oxygen level, and in particular has the increase with gestational age been thought evidence that the growth of the foetus was outstripping the ability of the placenta to maintain oxygen supply. However, many errors were likely in the data and experimental work on which these ideas were based (for review, see Dawes 1968a). When a fresh method of study was applied to the lamb, using a well-controlled chronic preparation, no increase in the oxygen carrying capacity was found beyond 100 days gestation. Thus the relative elevation of Hb cannot be taken as evidence that hypoxia of the foetus, judged by adult standards, is an abnormal state which has to be dealt with by some form of adaptation.

Foetal haemoglobin (HbF) comprises 70% of the content at term. It differs chemically from HbA in certain amino-acid constituents and configurations, and is only slowly denatured by strong alkali, a fact which sometimes provides a useful way of quickly identifying blood loss from the foetus in the labour room. The greater affinity of foetal blood

TABLE 8.3. Haemoglobin concentrations in newborns and adults of different species (Dawes 1968a)

Species	Hb (gm/100 ml whole blood)	
	Newborn at term	Adult
Man	18	15
Pig	11·8	13·8
Sheep	15	10
Cat	12	12
Rabbit	14·5	12·8–14·7
Rat	7·6–10·3	12·8–14·9
Goat	11	13

for oxygen (Fig. 8.3), which influences foetal oxygen uptake advantageously at the placenta, has always been something of a puzzle, since solutions of the two haemoglobins have similar oxygen dissociation curves (Allen et al 1963). Hence a cellular factor was suspected, but this remained unidentified until the work of Tyuma & Shimizu (1970) and Oski et al (1970). They showed that red cell organic phosphates tend to keep HbA in the reduced form, whereas this action is much less pronounced for HbF.

An increase in H-ion concentration shifts the haemoglobin dissociation curve to the right—the Bohr effect— with a fall in oxygen affinity (Fig. 8.3). Maternal blood in the placenta, by virtue of a fall in pH after taking up CO_2 from the foetus, will thus release more oxygen at a given partial pressure, and conversely foetal blood as it loses CO_2 will have an increased affinity for oxygen.

The actual extent of the gain in oxygenation through these devices is uncertain. With certain assumptions about pH differences, Bartels *et al* (1962) estimated that the Bohr effect might enhance oxygen transfer by 30% and that, together with the greater oxygen affinity of foetal cells,

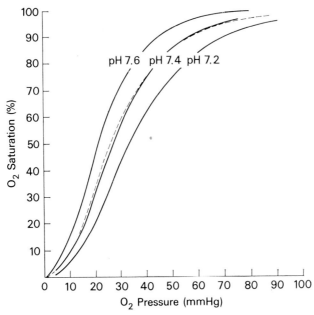

FIG. 8.3. Haemoglobin-oxygen dissociation curves for adult haemoglobin at three different pH values. The curve for foetal haemoglobin (HbF) at pH 7·4 is the same as for adult haemoglobin at pH 7·6 (Handbook of Respiration, 1958, *Fed. Amer. Soc. Exp. Biol.*).

the foetal blood would be 13% more saturated than if the two mechanisms did not exist. However, the pH differences which they assumed were greater than those now generally accepted.

The CO_2 dissociation curves of foetal and maternal blood are closely similar. In principle the Haldane effect, by which is meant the reduction occurring in CO_2 content when oxygen is taken up, might favour CO_2 transfer from foetus to mother (like the Bohr effect for oxygen in the opposite direction). However this is unlikely to play any part in CO_2 transfer since the gas is so diffusible.

Energy metabolism

Carbohydrate is the main fuel for the foetus. Fat can be metabolized as shown by tissue slice experiments. At birth there is a sharp rise in plasma free fatty acids, probably as a result of enhanced sympathetic activity (see below). Blood glucose is higher in the newborn than the foetus, and an explanation may very well be the inhibitory effect of fatty acids on insulin (van Duyne *et al* 1965 and see Chapter 11).

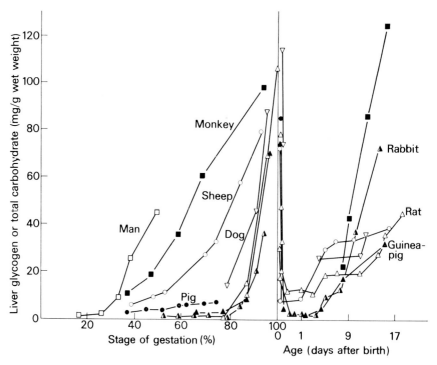

FIG. 8.4. Changes in liver glycogen in different species before and after birth. Modified from Heather J.Shelley (1961) *Brit. med. Bull.* **17**, 138.

Glycogen accumulation in foetal tissues, and in particular its behaviour in liver and heart, is of main interest. In the liver it rises steadily during gestation, and at delivery there is a precipitous fall (Fig. 8.4). In the heart the amount of glycogen at delivery has a direct relationship to the length of time that the newborn of different species can survive asphyxia. The purpose served by the cardiac store in this regard was nicely demonstrated experimentally (Fig. 8.5). The glycogen which steadily accumulates in skeletal muscle is not dissipated in asphyxia. Early in gestation the cardiac glycogen is higher than at term, for reasons that are not known; glycogen also accumulates in the lung at an early stage, and falls with maturity.

Foetal oxygen consumption, calculated from umbilical blood-flow measurements and arteriovenous oxygen differences, is 4–5 ml/kg/min in the lamb near term (it is in the same range in the animal after birth, and also in the newborn infant in basal conditions). If conditions

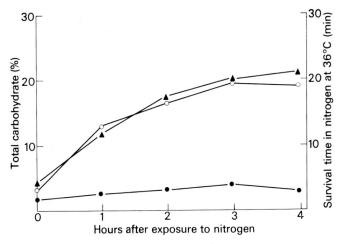

FIG. 8.5. Cardiac and liver carbohydrate, and survival in asphyxia. Litters of newborn rats were exposed to nitrogen for periods just short of lethal. Concentration of carbohydrate was then measured during recovery in heart (▲) and liver (●), and subsequently the survival times of littermates on re-exposure to nitrogen (○). From G.S.Dawes (1968a), Year Book Medical Publishers Inc.

become abnormally hypoxic the foetus can maintain oxygen consumption to the extent that redistribution of the circulation is achieved (see below) but then it *must* fall, by contrast with the adult in whom it is maintained or increased by the stimulation of respiratory and muscular activity to the time of collapse (which is much earlier than in the foetus). This economy in the face of limitation of supply is useful up to a point, and Po_2 is greater than it would otherwise be, but the cost is the developing metabolic acidosis which characterizes incomplete carbohydrate breakdown with lactic acid production when tissues become hypoxic.

Foetal temperature is 0·5°C above maternal, contrary to the more usual assumption that they are identical (Adamsons & Towell 1965, Adamsons 1966). On first principles there would have to be a gradient to allow transfer of metabolic heat. The foetus, however, has no power to regulate its own temperature. The danger of maternal fever to the baby is common experience. While a serious outcome is usually attributed to concomitant infection, it is now clear that the capacity of foetal metabolism to deal with fever is very different from the adult, and this alone may explain the rapid damage.

FOETAL CIRCULATION

COURSE

The course taken by oxygenated blood returning to the foetus in the umbilical vein is characterized by mixing with far more reduced blood from regions within the foetus, and by several shunts. Speculation on the method and purpose of these has sometimes made understanding difficult as, for example, in the controversy which existed for a long time on the 'crossing of streams' in the right atrium, favouring the brain and heart with better oxygenated blood (Sabbatier hypothesis).

The main facts about the course of the foetal circulation are well established (Fig. 8.6). The first shunt encountered by oxygenated blood returning in the umbilical vein is at the ductus venosus. In the lamb about 60% of umbilical venous blood is diverted by this route (Rudolph & Heyman 1967, Rudolph 1969). The biological role of this hepatic by-pass is not very clear, however, since it is closed before term in the pig and foal. Having reached the inferior vena cava, blood mixes with the small but very desaturated contribution from the lower extremities, and with that from the liver. Moving on to the right atrium, most of the flow is directed to the left atrium through the foramen ovale. The foramen opens directly off the inferior vena cava in the foetus and the semblance of its being interposed between the two atria is an autopsy artefact (Dawes 1968a). Thus incomplete mixing of blood from inferior and superior vena caval streams in the right atrium is explained. The consequent net difference for ascending aortic and carotid oxygen saturation is 4% for the sheep, or about 0·5 ml O_2/100 ml blood, confirming the

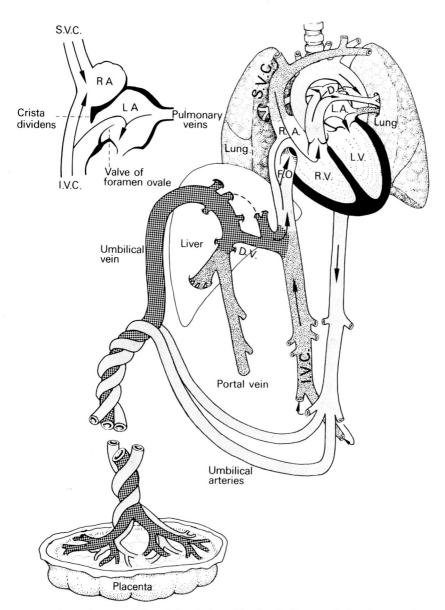

FIG. 8.6. Diagram of foetal circulation. Shading indicates relative degrees of oxygen desaturation, arrows the direction of flow. The liver is pulled up to expose its dorsal aspect. After G.S.Dawes (1968) in *Textbook of Physiology and Biochemistry*, Edinburgh, Livingstone.

Sabbatier hypothesis and indicating its role in foetal oxygen economy.

Inferior vena caval blood which is not diverted through the foramen mixes with very desaturated blood returning from the head and upper extremities, and is ejected by the pulmonary artery and ductus arteriosus to join the aortic stream after the diversion of a certain amount to the lungs. The abdominal viscera are supplied and then the umbilical arteries carry the descending aortic flow back to the placenta from the internal iliacs. The remaining aortic flow carries on to supply the lower extremities, in parallel with the return to the placenta.

CONTROL AND DISTRIBUTION; RESPONSE TO STRESS

Blood pressure rises steadily during gestation. Chemoreceptors are anatomically well developed relatively early (Purves & Biscoe 1968) and baroreceptor activity is also demonstrable. Similarly, adrenal medullary activity is demonstrable in the relatively young foetus (see below). All this evidence suggests that the capacity for autonomic regulation of the foetus is well developed by term, and later work has indeed shown that the circulation is under tonic nervous control in the later part of gestation (Dawes et al 1969).

The two ventricles of the foetal heart work in parallel, ejecting blood to the systemic circulation via the aorta and the pulmonary artery plus ductus arteriosus. Their output does not have to be the same (Table 8.4). Foetal cardiac output is large. For the lamb it is 310–360 ml/kg/min as against 110–120 for the grown animal (Dawes 1968a); later estimates are even higher (Rudolph 1969, and Fig. 8.7). Cardiac output is distributed in roughly equal proportions between foetus and placenta (Fig. 8.7, Table 8.4). Changes with gestational age are also illustrated in Figs. 8.7 & 8.8, and more details of distribution within the foetus at term in Table 8.4. Some points deserve further comment. Pulmonary blood flow is very low (Table 8.4) but rises with age (Fig. 8.8) perhaps as an indication of increasing local metabolism relating to formation of surfactant (see below). The large and increasing blood supply to the heart (Fig. 8.8) reflects the burden it carries. The proportion of flow to the placenta is the same throughout gestation although foetal growth greatly outstrips that of the placenta in the latter third. This development requires some change in haemodynamic relationships between the two, and is explicable by a fall in vascular resistance of the placenta in later gestation (Dawes 1962). Finally, it might be expected that in the human, the amount of blood going to the brain would be greater than shown in the diagram.

The foetal pulmonary circulation is exceedingly reactive. Pronounced sensitivity to changes in alveolar and arterial Po_2, Pco_2 and H-ion concentration has been demonstrated in numerous experiments. Quite early in gestation catecholamines and other agents produce effects in

TABLE 8.4. Blood flow through the major vessels of the foetal lamb

	% of combined output of both ventricles	Flow (ml/kg/min)
R + L ventricle	100	315
R ventricle	45	142
L ventricle	55	174
Thoracic inferior cena cava	76	240
IVC flow to R heart	29	92
IVC flow through foramen ovale	46	145
Superior cava + coronary veins	15	47
Lungs	10	31
Ductus arteriosus	35	110
Aortic isthmus	38	120
Hinderpart of body	19	60
Umbilical flow	57	180

Note: The sum of blood flow through the foramen ovale (46%) and ductus arteriosus (35%) is less than the combined output of both ventricles by a figure which is *double* pulmonary blood flow (10%) because this blood alone passes through both ventricles in series. From G.S.Dawes (1968a). Courtesy of Year Book Medical Publishers.

concentrations far less than for the adult. Reflex control of the pulmonary vascular bed is demonstrable from 2/3 gestation in the lamb. A low pulmonary blood flow is of course a necessary condition of intrauterine life. Conversely at birth the large change in gaseous environment, particularly for oxygen, releases vasoconstrictor tone as a main component in establishing the postnatal circulation and extrauterine respiration.

The umbilical circulation (i.e. from umbilical arteries to placenta and back) is a low resistance area. Furthermore, it is virtually unreactive to pharmacological and chemical stimuli. Umbilical arteries are innervated in their intra-abdominal course, and there is a scanty sympathetic innervation of the amnion, presumably in relation to vessels (Jacobson 1967, Fox & Jacobson 1969). In hypoxia, a small rise in resistance in the placenta can be demonstrated, probably from released catecholamines rather than any effect of blood gas levels directly on

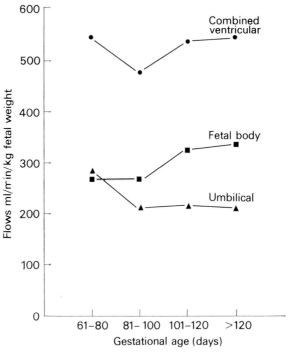

FIG. 8.7. Gestational changes in combined ventricular output, umbilical blood flow and blood flow to the foetal body, related to foetal body weight. From A.M.Rudolph (1969) in *Foetal Autonomy*. London, Churchill.

vessels or from reflex sympathetic activity. However, placental blood flow is on balance maintained (Rudolph 1969) or rises (Dawes 1968a) in response to hypoxia as a result of substantial readjustment within the foetus, analogous to the centralization of the circulation which occurs in the adult in emergency. The main components in the foetal reaction are a fall by half in pulmonary blood flow, and similar reduction in flow to the lower extremities, with a consequent rise in blood pressure (Dawes 1968a). Rudolph (1969), using rather different methods, found no change in arterial pressure and the cardiac output remained at its already very high level (Fig. 8.7). These are minor disagreements, the lesson from experimental work being the capacity of the foetus to redistribute its circulation in emergency by reflex and endocrine means and to maintain the placental circulation.

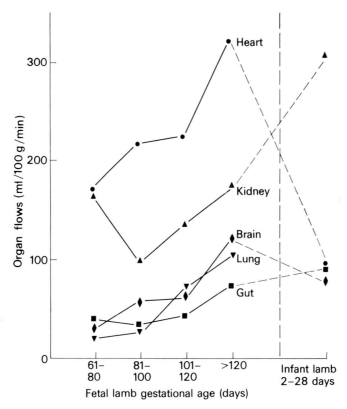

FIG. 8.8. Changes in organ blood flow with gestational age, and after birth. From A.M.Rudolph (1969) in *Foetal Autonomy*. London, Churchill.

The absence of reactivity in the umbilical circulation when tested in physiological preparations is in contrast with the prompt constriction of umbilical arteries at birth, but the two situations are hardly comparable. Observations on the perfused human placenta suggested that the arteries might react directly to gas tensions (Panigel 1962), but the experiments have been criticized and no inferences can be safely drawn from them about *in vivo* responses (Dawes 1968a).

LUNG DEVELOPMENT

In the development of the lung in the later part of gestation two points are of particular importance. The first is the appearance of surfactant, the phospholipid-protein complex produced by the alveolar cell. Activity can be demonstrated in the human from about 24 weeks. In other species it appears relatively later. It increases steadily to term. After birth this material, by its influence in reducing surface tension at the air-liquid interface, stabilizes the alveolus at low volume and permits the existence of the normal functional residual capacity of air at low pressure (Fig. 8.9). The responsible cells have high metabolic activity and hence supply of the material is vulnerable to asphyxial insult.

The second concerns the lung liquid. At term the lung is expanded by a volume of fluid which is about the same as its functional residual capacity for gas after the start of breathing. This liquid differs to a marked degree from amniotic fluid in most of its constituents, notably urea, bicarbonate, chloride and H-ion. It is produced in the alveoli, as an ultra-filtrate of plasma. The fluid is swallowed and little, if any, enters the amniotic fluid. The history of these observations is interesting. Tracheal occlusion in the foetus was shown some time ago to lead to accumulation of fluid in the lung, both animal and human (Dawes 1968a, citing Jost & Policard 1946), but since the hypothesis of a tidal circulation of amniotic fluid to the alveoli was then dominant the significance of the observation was uncertain. Workers using the exteriorized foetus were then impressed by the steady flow of liquid from the trachea. Some, for a time, even felt it might make a significant contribution to amniotic fluid. A balance of opinion has now been reached. There is no 'tidal' circulation of amniotic fluid, but the rate of production of lung liquid is quite low (1–2 ml/kg/hr).

Barcroft in 1946 described rhythmic respiratory movements in the young foetus from tactile stimulation, and the disappearance of this response before mid-gestation. Misconceptions which arose from similar observations in later pregnancy, leading to inferences about amniotic fluid entering and leaving the lungs as a physiological event (Davis & Potter 1946) were clarified with the realization that the movements in later pregnancy were caused by asphyxia during the experiments. Barcroft's statement that 'the stage is set for the first breath though only half the gestation period has elapsed' points none the less to the early development of this basic postnatal function. Cessation of the normal

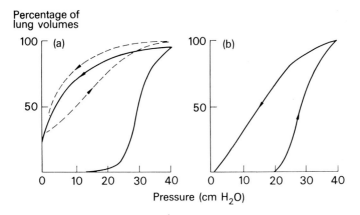
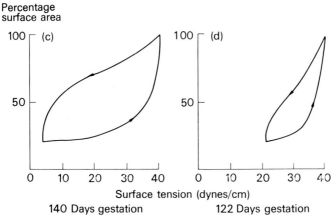

140 Days gestation   122 Days gestation

FIG. 8.9. Pressure–volume relationships in lungs, and surface tension of lung extracts, at two different gestational ages. Observations on a mature foetal lamb (140 days gestation), (a) to show the pressure volume curves of the first (solid line) and subsequent (broken line) pulmonary inflations. In an immature lamb (122 days gestation), (b) the effects of subsequent inflations were no different from the first. The lower figures (c, d) show the corresponding surface tension-area diagrams of lung extracts spread as a film on the surface of liquid in a trough. During compression of the film, the surface tension falls much more in the extract from the mature lung, indicating the presence of surface-active material. From G.S.Dawes (1968a) Year Book Medical Publishers Inc.

movements was presumably related to the maturing nervous system, which led to some theorizing about apnoea after birth—if prolonged, might this be evidence of damage to higher brain centres which prevented their arousal to permit release of the inhibitory influence which had taken over at lower levels? More recent work, described below, has removed the need for speculation of this sort.

LABOUR AND DELIVERY

Birth is normally an asphyxiating process in some degree, with hypoxia and resulting lactic acid production (metabolic acidosis) and hypercapnia (respiratory acidosis). The evidence comes from human data. The biochemical results of asphyxia were demonstrated in the cord blood findings of virtually all babies at birth (James *et al* 1958); within an hour the acidosis had recovered in those who had been vigorous with high Apgar scores. Oxygen content varied widely at the moment of birth, and had no relationship to the clinical state of the babies, but metabolic acidosis with rapid recovery was presumptive evidence of preceding hypoxia. This partly inferential evidence for the usual presence of some degree of asphyxiation during birth was soon confirmed, and its influence has been far reaching. Direct evidence for progressive falls in oxygen and pH in the course of normal labour came later by the method of sampling blood from the foetal scalp (Dawes 1968a).

There are probably two main reasons for regular interference with the foetal blood supply during labour. The first is reduction in placental flow during uterine contractions, which has been visualized in the primate (Ramsey *et al* 1963). Fall in oxygen content during the contraction and recovery afterwards have also been measured (Dawes 1968a). The second is interference with the cord from chance entanglement with foetal parts. Proof that this has occurred depends on showing a wide arteriovenous difference, for Po_2 and acid-base values, in a segment of the cord, clamped immediately after birth, and in fact such wide orders of difference were found by James *et al* (1958) in 30% of apparently normal, vigorous babies. Other possibilities are partial separation of the placenta, postural effects of the gravid uterus on major blood vessels, and maternal hypotension. These would ordinarily be classified as abnormal complications of labour. Nevertheless it is uncertain how often lesser degrees may go unrecognized in the course of an apparently normal labour.

Foetal asphyxia leads to release of catecholamines from the adrenal medulla. The response is well marked at the end of gestation (Comline & Silver 1966) and less so earlier. Release in the foetus is mediated directly through asphyxia (mainly the element of hypoxia, rather than the hypercarbia) acting on the medullary cells, by comparison with the asphyxiated adult when splanchnic nerve stimulation via the central nervous system is predominant. The proportions of noradrenaline and adrenaline vary in the medulla of different species, but the usual view that preponderance of noradrenaline indicates relative immaturity may not be correct (Comline & Silver 1966). The large release of catecholamines at birth has important effects on metabolism and circulation, referred to elsewhere.

RESISTANCE TO ASPHYXIA

Reasons for the newborn's capacity to sustain respiratory efforts longer than the adult, both in the intact animal and for the isolated head, have been extensively sought and the experimental evidence is complex (Dawes 1968a). Broadly speaking, the conclusions are that resistance does not depend on any unique metabolic property of neural tissue. The most important limiting factor to tissue survival in the brain is probably pH.

Survival is longer early in gestation and between species it is greater in those who are relatively immature at birth. It is closely correlated with hepatic and cardiac glycogen content. On present evidence newborn resistance to asphyxia depends on maintenance of the circulation and on the availability of energy from hepatic stores. Anaerobic glycolysis is less than a tenth as efficient as aerobic, so the stores though large (see above) are soon dissipated; furthermore, this is at the expense of the accumulation of lactic acid and a rise in H-ion concentration.

ASPHYXIA AND FOETAL HEART RATE

There are at least three interrelated mechanisms, first tachycardia from catecholamines, secondly reflex bradycardia from rise in blood pressure and thirdly bradycardia from myocardial anoxia. Still another cause of bradycardia is compression of the head during contractions; the rate rarely drops below 100 for this reason.

Two distinctive patterns are recognized by foetal electrocardiography in relation to uterine contractions, first a U-shape, with a prompt drop and immediate recovery, which has no statistical relation to clinical foetal distress or abnormal outcome in the foetus and secondly bradycardia which recovers far more slowly (Caldeyro-Barcia's type 2 dips); this has a definite relationship to foetal morbidity and mortality. The first is commonly attributed to compression of the head whereas in the second myocardial hypoxia is at work. However, it is also important to remember that bradycardia with prompt recovery can be a sign of asphyxia in its early stages, particularly from cord compression (Hon 1963). Should such incidents continue the more dangerous pattern appears in due course. The relationships between heart rate patterns and foetal condition after birth are shown in Fig. 8.10. Perhaps the important point is that although there are two causes for the U-type dip, one of them entirely innocent, a large fall in rate is likely to signify hypoxia even though return to the base line is prompt.

OXYGEN TO THE MOTHER

Oxygen breathing causes a large rise in maternal oxygen partial pressure though the actual quantity added to the blood is small, less than 2 ml/100 ml. Elevation of foetal Po_2 is much less, the lag being mainly attributable to the placental oxygen consumption. Even so, a small change in partial pressure when it is low is associated with significant change in saturation and content (Fig. 8.3). The value of this measure for the relief of foetal hypoxia from complications of labour has been thought dubious (Towell 1966) both because of theoretical considerations and some clinical results. However, since administration of oxygen to the mother is capable of abolishing abnormal electrocardiographic patterns (Dawes 1968a, citing Caldeyro-Barcia *et al* 1966) the method has some place. Perhaps one should add that administration should be efficient, i.e. by mask, at sufficient flow rate (Caterall *et al* 1967).

HYPERVENTILATION

Hyperventilation of the ewe leads to a fall in the Po_2 and also to metabolic acidosis of its foetus (Motoyama *et al* 1966). A fall in uterine blood flow from the vasoconstriction which follows hypocapnia is the likely explanation. A counterpart to this observation in the anaesthetized human is depression and acidosis of the foetus delivered by caesarean section after excessive maternal hyperventilation, to the point of pronounced maternal alkalosis (Moya *et al* 1965); moderate hyperventilation in the same circumstances slightly ameliorated foetal condition, presumably from better oxygenation without hypocapnia during the

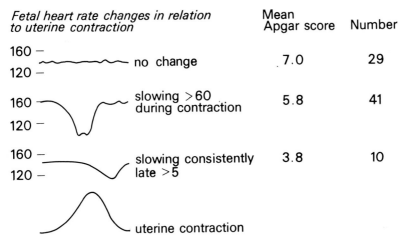

FIG. 8.10. Foetal heart rate and Apgar Score. The Apgar Score of babies born when the foetal heart rate slowed by more than 60 beats per min. during a contraction, or by more than 5 beats per min. consistently late in relation to a contraction, was significantly lower than the normal group (P < ·05 and < ·001 respectively). From C.Wood (1969) Spastics Int. Med. Pub.

operation. Whether conscious hyperventilation in the human subject can produce untoward effects on the foetus is uncertain. Wood (1969) found no significant changes in six patients.

ONSET OF BREATHING

Chemical, thermal and tactile stimuli combine to initiate breathing in the baby. Much experimental work has been undertaken to disentangle their several roles in two distinct functions, the initiation of breathing on the one hand and its rhythmic continuation on the other. While the distinction may seem academic, it is important to keep in mind for its bearing on possible situations in the human. For example, lambs kept warm and supplied by the maternal circulation could be stimulated to gasp by painful stimuli but spontaneous breathing was not sustained. Tying of the umbilical cord, with resulting asphyxia, stimulated both the first gasps and the continuation of breathing. Extreme cold also initiated breathing, and would maintain it in the absence of other stimuli (for further discussion see Dawes 1968a).

At the risk of over-simplification, gasping according to current views is attributable both to the chemoreceptor response to asphyxia and to multiple physical stimuli, while continued rhythmic breathing depends on chemoreceptor activity. Arousal of chemoreceptors after birth remains to be explained—it seems remarkable that they should be dormant *in utero*, at low levels of blood Po_2, and yet tonically active within a short time of birth when Po_2 in the blood is much higher. However, there is good evidence that the 'setting' of these highly vascular organs may be changed rapidly. Their signals are triggered by oxygen supply, not Po_2 as such, and the actual supply is effectively lower after birth from reflex and sympathetic vasoconstrictor activity, as compared with the foetal state.

Fluid is expressed from the lung during delivery, to the extent of a third or more of the total volume (Karlberg 1960). That which is re-aspirated with the start of breathing is absorbed through the circulation and, notably, also through the lymphatics (Strang 1967). Absence of the 'big squeeze' might conceivably hamper the prompt establishment of normal lung function and this has often been considered as contributing to the morbidity of caesarean section. There is little evidence one way or another. Additional factors which complicate caesarean delivery, such as time lapse with the possibility of compromising foetal blood supply (Lumley *et al* 1970) should be remembered in considering morbidity.

The 'first breath' is usually a succession of substantial gasps, achieving transpulmonary pressures of 40 to 60 cm H_2O, expanding the fluid-filled lung. The surfactant has little part in this initial process, its role being to maintain volume and allow the continuation of rhythmic breathing with small expense of energy. Should its activity be reduced or absent continued breathing becomes in effect a succession of first breaths (Fig. 8.9), and the dyspnoea which overwhelms affected babies is readily understood.

APNOEA AFTER BIRTH

The analysis of breathing patterns in the experimentally suffocated newborn animal has greatly improved the understanding of apnoea at birth (Davis 1961, Dawes 1968a). A brief phase of gasping, analogous to the hyperpnoea of the hypoxic adult, is followed by *primary* apnoea, the animal becoming suddenly inert and toneless, with an abrupt fall in heart rate (Fig. 8.11). After an interval, gasping movements begin. Blood pressure, which had already risen in primary apnoea, continues to do so, and the heart remains slow. Chemical asphyxiation is steadily increasing (see measurements, Fig. 8.11). If air or oxygen now enters the lungs spontaneous breathing and recovery soon follow. With continuing suffocation, however, the circulation begins to deteriorate with falling blood pressure and then the animal ceases to gasp, entering the phase of *secondary* or *pre-terminal* apnoea.

In primary apnoea any stimulus will initiate gasping. On the other hand, sedative drugs prolong this phase, to the point where it may become very difficult to distinguish from secondary apnoea. Artificial ventilation is necessary for recovery in secondary apnoea, with the addition of other steps depending on its duration (see below).

APGAR SCORE (Table 8.5)

The items of the score are a measure of neurological arousal and of respiratory and circulatory performance. It was devised two decades ago as an improvement on the criteria then in use for prediction of survival and also to compare the results of resuscitative methods and experiences in different hospitals (Apgar 1966). The approach was validated subsequently with the finding of a reasonably close relationship between biochemical status at birth and the score at one minute of age (James et al 1958). Later still the predictive value for neurological status in the infant at follow-up was found to correlate better with the score at 5 minutes rather than at one (Abramson 1966). It was not, of course, intended to be a sole guide to management, or to delay whatever steps in resuscitation might be required immediately after birth. The subjective element in scoring introduces a degree of imprecision which can be improved up to a point (see Apgar 1966). Other methods of assessment have been devised for particular reasons (e.g. Tizard 1964) but the score

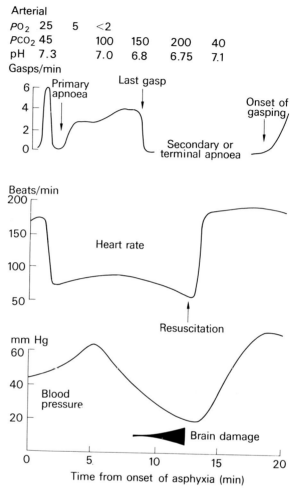

FIG. 8.11. Schematic diagram of changes in rhesus monkeys during asphyxia and on resuscitation by positive pressure ventilation. Brain damage was assessed by histological examination weeks or months later. From G.S. Dawes (1968a) Year Book Medical Publishers Inc.

as originally set out remains a most useful clinical tool, as attested by its world-wide adoption.

RESUSCITATION

The present-day approach is based on considerations outlined above. A few babies begin to breathe before delivery is completed, and continue without interruption. The majority are born in primary apnoea. Anaesthesia and drugs play a part here, to an extent which is very difficult to ascertain in the individual case. The main question which arises in this connection is the administration of morphine antagonists (there are no other specific drug antidotes). Morphine and its analogue meperidine readily cross the placenta. The limits both for time and dosage beyond which reasonable excretion from the foetus might be expected are unknown. Furthermore, the clinical effect in a particular baby is unpredictable since the drug action is to potentiate existing

TABLE 8.5. Apgar Score. After J.S.Drage & H.W. Berendes (1966) *Pediat. Clin. N. Amer.* **13**, 635.

		Age at time of scoring (min.)	1	2	Etc.
Heart rate	0 absent	1 below 100	2 over 100		
Respiratory effort	0 absent	1 weak cry hypoventilation	2 crying lustily		
Muscle tone	0 flaccid	1 some flexion	2 well flexed		
Reflex irritability	0 no responce	1 some motion	2 cry		
Colour	0 blue/pale	1 blue hands and feet	2 entirely pink		
		TOTAL			

asphyxia, the degree of which itself is unknown (except in the circumstance of scalp blood sampling). On empirical grounds a working rule is to give the antagonist nalorphine if the mother has received 100 mg of meperidine within the preceding 4 hours, or if multiple doses have been given earlier. The slightly depressant effect of the antidote by itself is beside the point when its congener is already present.

In secondary apnoea the object is to restore the internal environment as quickly as possible—monkeys resuscitated more than 4 minutes after the 'last gasp' had brain damage (James 1966). Ventilation of the lungs is the first measure and if a high oxygen concentration is also supplied it will promote relaxation of pulmonary vessels, improving oxygen uptake. Correction of acidosis further relaxes pulmonary vessels and improves the situation for tissues generally. Dextrose is given because of the likelihood of depletion of carbohydrate stores—in fact blood glucose may be immeasurably low in seriously asphyxiated infants. Cardiac massage is indicated. This is because although cardiac action may still be present the blood pressure is very low and even a brief period of artificial elevation and improved blood supply to vital tissues may contribute significantly to quicker reversal of the consequences of asphyxia. Effective systolic ejection can be readily produced by squeezing the infant's thorax between thumb and fingers (Moya *et al* 1962).

The value of these measures has been clearly demonstrated in animals. Monkeys resuscitated soon after the 'last gasp' were intact; if several minutes elapsed, those revived by ventilation alone were brain-damaged, but this was absent if alkali and glucose had also been infused. Application in the labour room requires appreciation of priorities, preparedness, and familiarity with procedures like umbilical venous catheterization. Ventilation of the lungs has to be performed after endotracheal intubation, since a pressure of 30 to 40 cm H_2O is needed for the first inflation. Risks have been exaggerated. When pneumothorax, for example, has developed in resuscitated babies in this hospital serial radiographs have commonly revealed it as a later event, presumably from check-valve obstruction of airways by inhaled material as at other ages, rather than an immediate consequence of the procedure. Chemical correction should be controlled by appropriate measurements, but when facilities are absent guidance on empirical ground is available, e.g. 5 mEq/kg of bicarbonate given slowly is an effective dose in most circumstances. Tris buffer (THAM) may have some advantages, but it is not yet in general clinical use.

The problem remains of identifying the baby in need of active resuscitation and the point at which to proceed with vigorous measures. From observation alone it is not always possible to distinguish primary from secondary apnoea. Hence in the interest of those babies at most serious risk, some who would breathe spontaneously are likely to be intubated. This consideration and its attendant dangers were perhaps of more consequence a decade ago when labour room staff was less likely to be trained in resuscitative procedures than today.

There are, of course, many other methods which have been tried in resuscitation. Deficiencies in those which still have a certain advocacy are revealed when they are examined in the light of newer knowledge and especially when tested in animals. For example, analeptic drugs have no place. Their administration will certainly promote breathing in primary apnoea, but in secondary apnoea they have no stimulant effect and furthermore increase the rate at which blood pressure falls (Cross 1968). Thus they are dangerous for the babies in most need of resuscitation. Hypothermia, which does prolong gasping time in experimental asphyxia, did not lead to spontaneous breathing in monkeys to whom cooling was applied after the last gasp, i.e. in secondary apnoea. When cooling was applied earlier the results showed no advantage over controls in respect of brain damage, whereas other controls treated with alkali as well showed no histological change later in the brain (Daniel *et al* 1966). Hyperbaric oxygen, for which there had been some clinical trials in humans with equivocal results, proved valueless when tested in animals (Campbell *et al* 1966). Barbiturates, like cold, prolong time to the last gasp; although this fact has been of interest to investigators, they have warned against incorporating the idea in any policy for resuscitation.

CEREBRAL BIRTH INJURY, 'SHOCK', AND APNOEA

When breathing has been delayed there may later be signs of damage to the brain. Thus in a loose fashion 'cerebral birth trauma' (by implication mechanical in origin) is often in the mind of the accoucheur when a baby remains apnoeic after delivery. The notion that apnoea might be related to unspecified mechanical trauma from the forces of delivery is not one that has been tested experimentally, nor probably could it be, although an attempt was made (Schwartz 1961). Possibly the idea arose by analogy with the brief apnoea of concussion in the adult, but the comparison is implausible (Burnard 1962). It is interesting that Victorian observers assigned a large part of the risks of complicated birth to asphyxia, although on terms admittedly different from today's (Snow 1841, Little 1862). Nevertheless the fact may be set beside recent pathological studies, reporting asphyxia and gross trauma in the proportion of 5:1 as a cause of neonatal death (Clyne 1964, Courville 1963).

Further historical considerations, together with the knowledge gained from experimental work, put the perspective right. Holland (1922) showed 50 years ago that massive subdural bleeding was a common intracranial lesion associated with profound depression at birth, and he related it to obstetric practices of the era. Methods were changed as a result of these and other contemporary studies, and the lesion is now rare. In the meantime, however, notions such as 'obstetric shock' or 'vasomotor paralysis' were invoked to explain a clinical picture attributed to intracranial bleeding of any kind, including the common petechial haemorrhages in brain substance, or larger haemorrhage in ventricles and subarachnoid space. The opinion of pathologists is now quite clear that these forms of bleeding are asphyxial in origin (for review see Burnard 1962) and this would be entirely in agreement with inferences from experimental studies. Signs of shock ('asphyxia pallida') may indeed be present but they are explicable in ordinary physiological terms rather than hypothetical 'damage to vital centres'. Brain damage at higher levels may indeed be a sequel, but it is an effect of the asphyxiation, not a cause of the apnoea.

RECOVERY FROM ASPHYXIA

The circulation improves first and breathing begins later. The time for the return of breathing in the animal resuscitated after asphyxia beyond the last gasp has a close relationship to the interval before starting ventilation. There is also a relationship to subsequent brain damage depending on the supplementary measures used (see above). While all this emphasizes the need for speed in resuscitation, time relationships in the monkey and the lamb, the two animals most studied, should be applied with reservation to the human, for whom survival capacity may be longer (Dawes 1968a). Furthermore, the obstetric situation can only rarely be compared directly with the experimental, in which total asphyxiation was produced. Shoulder dystocia at delivery is perhaps the nearest example, But as a rule asphyxia of the human foetus has been intermittent in the course of many different complications of labour, with opportunity for partial recovery.

It would be rare nowadays for resuscitation not to be instituted on a baby so long as it has a heart beat. The above points are raised, however, for their bearing on the difficult question of when to cease resuscitative attempts. Judgement is bound to differ on this, and only two general guide lines might be suggested. The first is that when the response to resuscitation is not proceeding as well as might be expected, a congenital anomaly like diaphragmatic hernia or pulmonary hypoplasia should be considered. The second concerns the diagnosis of irreversible intracranial damage. This is often suspected when spontaneous breathing is not resumed despite improvement in the circulation after ventilation and administration of alkali and glucose. However, in the writer's experience management in such circumstances has often turned out to be less than ideal and the situation has often improved after review and consultation. Resuscitative efforts in the labour room should not be abandoned without much consideration.

In the hours after birth the baby who has experienced undue asphyxia is first depressed, then irritable and often both by turn. Desmond *et al* (1959) showed that the time scale of these phases may vary greatly, and related this to the severity of the insult. Their observations may be used to clarify a curious misconception which arises sometimes among nursing and resident staff to the effect that nalorphine has stimulant properties. It is prompted by the marked irritability after birth in some babies who, having received the drug following maternally administered pethidine, display at an early stage the irritability of recovering asphyxia unmasked from narcotic depression.

CHANGES IN CIRCULATION AFTER BIRTH

The foetal pattern of ejection from the two ventricles in parallel promptly changes to the adult on expansion of the lungs. The sudden increase in pulmonary blood flow and hence in the volume of blood returning to the left atrium causes a rise in left atrial pressure and at the same time inferior vena caval flow to the right atrium falls with ligation of the cord. Pressure relationships between the atria are thus reversed and the valve of the foramen ovale held closed. Systemic blood pressure rises and exceeds that in the pulmonary artery, which itself has fallen somewhat with the reduction in pulmonary vascular resistance; flow through the ductus arteriosus becomes left-to-right or bi-directional. The ductus closes by muscular constriction, its wall being comprised of smooth muscle which contracts in response to rising Po_2 (it reacts also to catecholamines). Functional closure is usually complete within a few hours. Anatomical obliteration does not begin till later in infancy.

Cardiac output is high in the hours after birth (Gessner *et al* 1965, Burnard *et al* 1966). Possible reasons are the existence still of foetal shunts, metabolic demands of the newborn period and sympathetic activity. Pulmonary arterial pressure continues to decline over 2 or 3 weeks. Renal blood flow is relatively low at first and shortly rises. The ductus venosus is closed within an hour or two of birth, although the mechanism is not clear.

In the infant after birth the circulation is in a transitional phase between foetal and adult state. The foetal shunts are potentially open. The pulmonary arterioles still have the capacity for powerful vasoconstriction in response to hypoxia and acidosis. Thus, if pulmonary complications occur in the newborn period—and this is the commonest neonatal abnormality after birth—the circulation is liable to revert to the foetal condition, with the greater part of the cardiac output bypassing the lungs. Profound systemic hypoxia is the result. The position is unique and makes comparisons difficult between respiratory illness in the newborn and later ages. It does, however, explain why, given adequate supportive treatment while the pulmonary lesions resolve, the neonate may survive illnesses which by both clinical and biochemical criteria would be lethal at any other age.

PLACENTAL TRANSFUSION

A good deal of heat is removed from the long-standing controversy about the proper time for cord ligation after birth with the realisation that a quarter of the placental blood available to the infant reaches it within 15 seconds and a half within a minute (Usher *et al* 1963). Thus it may be inferred that a modicum of transfer is normal and this would agree with the difference in blood volume between babies after immediate ligation (78 ml/kg) and the normal 90 ml/kg at 3 days of age (Usher *et al* 1963). There remains the question of whether a maximal transfer is desirable. Comparisons have usually been made between the effects of clamping within 15 seconds and at 5 minutes. In the latter case, there is a marked temporary rise in blood volume, to 124 ml/kg in the series of Usher *et al* (1963), falling to 89 ml/kg later, with concomitant rise in haematocrit. Systemic venous and arterial pressures are elevated after delayed ligation, as is pulmonary arterial pressure (Burnard & James 1963, Buckels & Usher 1963, Arcilla *et al* 1966). Bilirubin is higher in premature infants (Taylor *et al* 1963). Most observers have detected only trivial clinical differences in haemodynamic and pulmonary status which might be important at the clinical level after delayed ligation, although a study in premature babies shows a worse record for respiratory distress (Yao *et al* 1969). The whole question is one about which speculation has far exceeded reliable facts (for review see Moss & Monset-Couchard 1967). For example, it is suggested from time to time that ligation should be delayed till after the first breath (Brown 1959,

Redmond et al 1965). This would indeed allow larger transfer to the baby for several reasons, if that is thought desirable, but a shift of blood from placenta to lung is no part of the normal neonatal adaptation, since the increment of pulmonary blood volume from relaxation of resistance vessels must be minute; the few measurements which have been made support this opinion (Usher et al 1963).

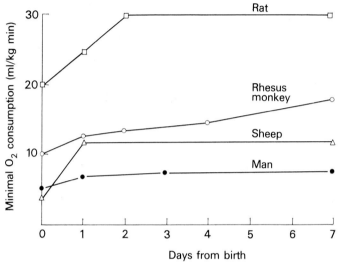

FIG. 8.12. Changes in minimal O_2 consumption, in a neutral thermal environment, in different species after birth. From G.S.Dawes (1968a), Year Book Medical Publishers Inc.

One may conclude that prompt ligation is sensible in the premature, when in any case the volume available for transfer is large because of the relatively bigger placenta. In babies at term the margin of safety is wide. Ill effects from immediate ligation are unlikely, as is borne out in common experience when the cord has to be clamped for entanglement around the neck. Prolonged delay in ligation is also likely to be well borne, but in principle the resulting hypervolaemia and later haemoconcentration do constitute something of a hazard.

FOETAL HAEMORRHAGE

On animal evidence the newborn circulation is as efficient as the adult in withstanding blood loss (Mott 1966). Observations in infants suggested that the human may not be so well placed (Young & Cottom 1966, Wallgren et al 1967) and in practice acute blood loss from the baby during delivery is a very dangerous hazard, needing prompt treatment. The subject has been reviewed by Raye et al (1970).

RESPIRATION AND METABOLISM IN THE NEWBORN

Evaporative heat loss is very large in the wet baby after delivery. When there is delay in the establishment of breathing heat production is negligible and body temperature rapidly moves to that of the surroundings (Hey & Kelly 1968). The fall in body temperature considerably aggravates any existing metabolic acidosis (Gandy et al 1964), thus compounding problems of resuscitation. Methods for reducing heat loss in this emergency are difficult to apply, but its very real risks should always be kept in mind.

Basal oxygen consumption (metabolic rate) increases after birth, though to a smaller extent in man than in some other animals (Fig. 8.12). The change is probably explained by increase in tissue and organ activity. In response to cold the baby increases heat production and oxygen uptake by increased muscular activity, by shivering (which contributes much less than in the adult and does not occur in premature babies), and notably by increased metabolism in the brown fat deposits (Dawes 1968a), mediated by sympathetic stimulation. Oxygen consumption is almost trebled by these means and then body temperature begins to fall. Rectal temperature is thus a relatively crude guide to management when it is desired to minimize expenditure of energy, as in the case of sick babies. Term infants sweat freely in the heat, but not premature babies.

The 'neutral range' of environmental temperature is that within which metabolism is minimal. It is surprisingly narrow for the naked baby at first (32–34°C), even more so for the premature and it widens in the following days. These are important considerations in current practices of incubator nursing. Furthermore, intrinsic limitations of the apparatus itself may not always be realized. Air temperature registered by the incubator thermometer, for instance, is an imperfect indication of the effective temperature at the baby's skin because of losses from radiation to incubator wall which will vary with the room temperature. Radiation losses can be minimized and a more stable condition produced by any device to provide a double wall, such as a plastic cover over the infant within the incubator. Clothing and blankets improve the tolerance of low temperatures, but increasing the number of blankets adds surprisingly little benefit and a room which is comfortable for an adult may remain too cold for a baby. These questions are well reviewed by Hey & Katz (1970) and Hey & O'Connell (1970).

ACKNOWLEDGMENT

I am indebted to Dr Uwe E. Freese for help in preparing the section on placental circulation.

REFERENCES

ABRAMSON H. (1966) The search for new and precise knowledge, in H.Abramson (ed.) *Resuscitation of the Newborn Infant*, p. 350. St Louis, C.V. Mosby.
ADAMSONS K. Jr. (1966) The role of thermal factors in fetal and neonatal life. *Pediat. Clin. N. America* **13**, 599–619.
ADAMSONS K. Jr. & TOWELL M.E. (1965) Thermal homeostasis of the fetus and newborn. *Anesthesiology* **26**, 531–548.
ADAMSONS K. Jr., JAMES L.S., TOWELL M.E. & LUCEY J.F. (1965) Physiologic observations during induced anemia *in utero* in the rhesus monkey. *J. Pediat.* **67**, 1042–1048.
ALLEN D.W., WYMAN J. & SMITH C.A. (1953) The oxygen equilibrium of fetal and adult hemoglobin. *J. Biol. Chem.* **203**, 81–87.
AHERNE W. & DUNNILL M.S. (1966) Morphometry of the human placenta. *Brit. med. Bull.* **22**, 5–8.
APGAR V. (1966) The newborn (Apgar) scoring system: reflections and advice. *Pediat. Clin. N. America* **13**, 645–650.
ARCILLA R.A., OH W., LIND J. & GESSNER I.H. (1966) Pulmonary arterial pressures of newborn infants born with early and late clamping of the cord. *Acta Paediat. Scand.* **55**, 305–315.
BARCROFT J. (1946) *Researches on Prenatal Life*. Oxford, Blackwell Scientific Publications, cited by Dawes (1968a).
BARTELS H. (1970) *Prenatal Respiration*, pp. 115ff. Amsterdam, North-Holland Publishing Company.
BARTELS H., MOLL W. & METCALFE J. (1962) Physiology of gas exchange in the human placenta. *Amer. J. Obstet. Gynec.* **84**, 1714–1730.
BØE F. (1953) Studies on vascularization of human placental relationships. *Acta obst. gynec. scand.* **32** (Suppl. 5), 1–92.

BORELL U., FEMSTRÖM I., WESTMAN A. (1958) Eine arteriographische Studie des Plazentarkreislafs. *Geburtsch. u. Frauenh.* **18**, 1–9.

BOYD J.D. & HAMILTON W.J. (1967) Development and structure of the human placenta from the end of the 3rd month of gestation. *J. Obstet. Gynaec. Brit. Cwlth.* **74**, 161–226.

BROWN R.J.K. (1959) Respiratory difficulties at birth. *Brit. med. J.* **1**, 404–408.

BUCKELS L.J. & USHER R. (1965) Cardiopulmonary effects of placental transfusion. *J. Pediat.* **67**, 239–247.

BURNARD E.D. (1962) The relative dangers of asphyxia and mechanical trauma at birth. *Med. J. Aust.* **2**, 487–992.

BURNARD E.D. & JAMES L.S. (1963) Atrial pressures and cardiac size in the newborn infant. *J. Pediat.* **62**, 815–826.

BURNARD E.D., GRAUAUG A. & GRAY R.E. (1966) Cardiac output in the newborn infant. *Clin. Sci.* **31**, 121–133.

CALDEYRO-BARCIA R., MÉNDEZ-BAUER C., POSEIRO J.J., ESCARCENA L.A., POSÉ S.V., BIENIERZ J., ARNT I., GULIN L. & ALTHABE O. (1966) Control of human fetal heart rate during labour, in D.E.Cassels (ed.) *The Heart and Circulation in the Newborn and Infant*, pp. 7–36. New York, Grune & Stratton, Inc., cited by Dawes (1968a).

CAMPBELL A.G.M., CROSS K.W., DAWES G.S. & HYMAN A.I. (1966) A comparison of air and O_2, in a hyperbaric chamber or by positive pressure ventilation, in the resuscitation of newborn rabbits. *J. Pediat.* **68**, 153–163.

CAMPBELL A.G.M., DAWES G.S., FISHMAN A.P., HYMAN A.I. & JAMES G.B. (1966) The oxygen consumption of the placenta and foetal membranes in the sheep. *J. Physiol.* (*Lond.*) **182**, 439–464.

CATERALL M., KAZANTZIS G. & HODGES M. (1967) The performance of nasal catheters and a face mask in oxygen therapy. *Lancet* **i**, 415–417.

CLYNE D.G.W. (1964) Traumatic *versus* anoxic damage to the foetal brain. *Develop. Med. Child. Neurol.* **6**, 455–457.

COMLINE R.S. & SILVER M. (1966) Development of activity in the adrenal medulla of the foetus and newborn animal. *Brit. med. Bull.* **22**, 16–20.

COURVILLE C.B. (1963) Birth and brain damage: traumatic versus anoxic damage to the fetal brain. *Bull. Los Angeles Neurol. Soc.* **28**, 209–215.

CROSS K.W. (1968) Aspects of applied physiology of neonatal circulation. *Brit. Heart J.* **30**, 483–492.

DANIEL S.S., DAWES G.S., JAMES L.S., ROSS B.B. & WINDLE W.F. (1966) Hypothermia and the resuscitation of infant monkeys. *J. Pediat.* **68**, 45–53.

DAVIS M.E. & POTTER F.L. (1946) Intrauterine respiration of the human fetus. *J. Amer. med. Ass.* **131**, 1194–1201.

DAVIS J.A. (1961) The effect of anoxia in newborn rabbits. *J. Physiol.* (*Lond.*) **155**, 56P.

DAWES G.S. (1962) The umbilical circulation. *Amer. J. Obstet. Gynec.* **84**, 1634–1648.

DAWES G.S. (1968) In G.H.Bell, J.N.Davidson & H.Scarborough (eds.) *Textbook of Physiology and Biochemistry*, p. 1168. Edinburgh, E. & S. Livingstone Ltd.

DAWES G.S. (1968a) *Foetal and Neonatal Physiology*. Chicago, Year Book Medical Publishers, Inc.

DAWES G.S. (1969) Foetal blood gas homeostasis, in G.E.W.Wolstenholme & M.O'Connor (eds.) *Foetal Autonomy*, pp. 162–172, London, J. & A. Churchill Ltd.

DAWES G.S., DUNCAN S.L.B., LEWIS B.V., MERLET C.L., OWEN-THOMAS J.B. & REEVES J.T. (1969) Hypoxaemia and aortic chemoreceptor function in foetal lambs. *J. Physiol. (Lond.)* **201**, 105–115.

DESMOND M.M., KAY L.J. & MEGARITY A.L. (1959) The phases of 'transitional distress' occurring in neonates associated with prolonged umbilical cord pulsations. *J. Pediat.* **55**, 131–151.

DRAGE J.S. & BERENDES H. (1966) Apgar scores and outcome of the newborn. *Pediat. Clin. N. America* **13**, 635–643.

FOX H. & JACOBSON H.N. (1969) Innervation of the human umbilical cord and umbilical vessels. *Amer. J. Obstet. Gynec.* 384–389.

FREESE U.E. (1966) The fetal–maternal circulation of the placenta (I). *Amer. J. Obstet. Gynec.* **94**, 354–360.

FREESE U.E., RANNIGER K. & KAFLAN H. (1966) The fetal–maternal circulation of the placenta (II). *Amer. J. Obstet. Gynec.* **94**, 361–365.

FREESE U.E. (1968) The uteroplacental vascular relationship in the human. *Amer. J. Obstet. Gynec.* **101**, 8–16.

GANDY G.M., ADAMSONS K. Jr., CUNNINGHAM N., SILVERMAN W.A. & JAMES L.S. (1964) Thermal environment and acid-base homeostasis in human infants during the first few hours of life. *J. clin. Invest.* **43**, 751–758.

GESSNER I.H., KROVETZ L.J., BENSON R.W., PRYSTOWSKY H., STENGER V. & EITZMAN D.V. (1965) Hemodynamic adaptations in the newborn infant. *Pediatrics* **36**, 752–762.

HEY E.N. & KELLY J. (1968) Gaseous exchange during endotracheal ventilation for asphyxia at birth. *J. Obstet. Gynaec. Brit. Cwlth.* **75**, 414–424.

HEY E.N. & KATZ G. (1970) The optimum thermal environment for naked babies. *Arch. Dis. Childh.* **45**, 328–334.

HEY E.N. & O'CONNELL B. (1970) Oxygen consumption and heat balance in the cot-nursed baby. *Arch. Dis. Childh.* **45**, 335–343.

HOLLAND E. (1922) Cranial stress in the foetus during labour and the effects of excessive stress on the intracranial contents; with an analysis of eighty-one cases of torn tentorium cerebelli and subdural haemorrhage. *J. Obstet. Gynaec. Brit. Emp.* **29**, 549–571.

HON, E.H. (1963) The classification of fetal heart rate. *Obstet. & Gynec.* **22**, 137–146.

JACOBSON H.N. (1967) Intrinsic innervation of the human placenta. *Nature* **214**, 102–104.

JAMES L.S. (1966) Onset of breathing and resuscitation. *Pediat. Clin. N. America* **13**, 621–634.

JAMES L.S., WEISBROT I.M., PRINCE C.E., HOLADAY D.A. & APGAR V. (1958) The acid-base status of human infants in relation to birth asphyxia and the onset of respiration. *J. Pediat.* **52**, 379–394.

JAMES L.S. & ADAMSONS K. Jr. (1964) Respiratory physiology of the fetus and newborn. *New Eng. J. Med.* **271**, 1352–1360, 1405–1409.

JOST A. & POLICARD A. (1946) Contribution experimentale à l'étude du developpement prénatal du poumon chez le lapin. *Arch. Anat. micr.* **37**, 323–332, cited by Dawes (1968a).

KARLBERG P. (1960) The adaptive changes in the immediate postnatal period, with particular reference to respiration. *J. Pediat.* **56**, 585–604.

KUBLI F. & BERG D. (1965) The early diagnosis of foetal distress. *J. Obstet. Gynaec. Brit. Cwlth.* **72**, 507–513.

LEMTIS H.G. (1969) New insights into the maternal circulatory system of the human placenta, in A.Pecile & C.Finci (eds.) *The Foeto-placental Unit*, pp. 23–33. Amsterdam, Excerpta Medica Foundation.

LITTLE W.J. (1862) On the influence of abnormal parturition, difficult labour, premature birth, and asphyxia neonatorum on the mental and physical condition of the child, especially in relation to deformities. *Trans. Lond. Obstet. Soc.* **3**, 293–344.

LUMLEY J., WALKER A., MARUM J. & WOOD C. (1970) Time: an important variable at caesarean section. *J. Obstet. Gynaec. Brit. Cwlth.* **77**, 10–23.

MOSS A.J. & MONSET-COUCHARD M. (1967) Placental transfusion: early *versus* late clamping of the umbilical cord. *Pediatrics* **40**, 109–126.

MOSSMAN H.W. (1926) The rabbit placenta and the problem of placental transmission. *Amer. J. Anat.* **37**, 433–497, cited by Dawes (1968a).

MOTOYAMA E.K., RIVARD G., ACHESON F. & COOK C.D. (1966) Adverse effect of maternal hyperventilation on the foetus. *Lancet* **i**, 286–288.

MOTT J.C. (1966) Cardiovascular function in newborn mammals. *Brit. med. Bull.* **22**, 66–69.

MOYA J., JAMES L.S., BURNARD E.D. & HANKS E.C. (1962) Cardiac massage in the newborn infant through the intact chest. *Amer. J. Obstet. Gynec.* **84**, 798–803.

MOYA F., MORISHIMA H.O., SHNIDER S.M. & JAMES L.S. (1965) Influence of maternal hyperventilation on the newborn infant. *Amer. J. Obstet. Gynec.* **91**, 76–84.

OSKI F.A., GOTTLIEB A.J., MILLER W.W. & DELIVORIA-PAPADOPOULOS M. (1970) The effects of deoxygenation of adult and fetal hemoglobin on synthesis of red cell 2,3-diphosphoglycerate and its *in vivo* consequences. *J. clin. Invest.* **49**, 400–407.

PANIGEL, M. (1962) Placental perfusion experiments. *Amer. J. Obstet. Gynec.* **84**, 1664–1683.

PANIGEL M. (1969) Comparative anatomical, physiological and pharmacological aspects of placental permeability and haemodynamics in the non-human primate placenta and in the isolated perfused human placenta, A.Pecile & C.Finzi (eds.) *The Foeto-placental Unit*, pp. 279–295. Amsterdam, Excerpta Medica Foundation.

PURVES M.J. & BISCOE T.J. (1966) Development of chemo-receptor activity. *Brit. med. Bull.* **22**, 56–60.

RAMSEY R.M. (1960) The placental circulation, in C.A.Villee (ed.) *The Placenta and Fetal Membranes*, pp. 36–62. Baltimore, Williams & Wilkins Company.

RAMSEY E.M., CORNER G.W. & DONNER M.W. (1963) Serial and cineradiographic visualization of maternal circulation in primate (hemochorial) placenta. *Amer. J. Obstet. Gynec.* **77**, 213–225.

RAYE J.R., GUTBERLET R.L. & STAHLMAN M. (1970) Symptomatic posthemorrhagic anemia in the newborn. *Pediat. Clin. N. America* **17**, 401–413.

REDMOND A., ISANA S. & INGALL D. (1965) Relation of onset of respiration to placental transfusion. *Lancet* **i**, 283–285.

REYNOLDS S.R.M. (1966) Formation of fetal cotyledons in the hemochorial placenta. *Amer. J. Obstet. Gynec.* **94**, 425–439.

ROSS B.B. (1967) Comparative properties of the lungs and the placenta; a graphical analysis of placental gas exchange, in A.V.S. de Reuck & R.Porter (eds.) *Development of the Lung*, pp. 238–254. London, J. & A. Churchill Ltd.

RUDOLPH A.M. (1969) The course and distribution of the foetal circulation, in G.E.W.Wolstenholme and M.O'Connor (eds.) *Foetal Autonomy*, pp. 147–156. London, J. & A. Churchill Ltd.

RUDOLPH A.M. & HEYMAN M.A. (1967) The circulation of the fetus *in utero*; methods for studying distribution of blood flow, cardiac output and organ blood flow. *Circulation Res.* **21**, 163–184.

SCHWARTZ P. (1961) *Birth Injuries of the Newborn*, pp. 29ff. New York, Hafner.

SHELLEY H.J. (1961) Glycogen reserves and their changes at birth. *Brit. med. Bull.* **17**, 137–143.

SNOW J. (1841) On asphyxia and on the resuscitation of still-born children. *Westminster med. Gaz.* **29**, 222–227.

STRANG L.B. (1967) Uptake of liquid from the lungs at the start of breathing, in A.V.S. de Reuck & Ruth Porter (eds.) *Development of the Lung*, pp. 348–361. London, J. & A. Churchill Ltd.

TAYLOR P.M., BRIGHT N.H., BIRCHARD E.L., DERINOZ M.N. & WATSON D.W. (1963) The effects of race, weight loss and time of clamping the umbilical cord on neonatal bilirubinemia. *Biol. Neonat.* **5**, 299–318.

TIZARD J.P.M. (1964) Indications for oxygen therapy in the newborn. *Pediatrics* **34**, 771–786.

TOWELL M.E. (1966) The influence of labor on the fetus and newborn. *Pediat. Clin. N. America* **13**, 575–598.

TYUMA I. & SHIMIZU K. (1970) Effect of organic phosphates on the difference in oxygen affinity between fetal and adult hemoglobin. *Federation Proc.* **29**, 1112–1114.

USHER R., SHEPHARD M. & LIND J. (1963) The blood volume of the newborn infant and placental transfusion. *Acta Paediat. (Stockh.)* **52**, 497–512.

VAN DUYNE C.M., PARKER H.R. & HOLM L.W. (1965) Metabolism of free fatty acids during perinatal life of lambs. *Amer. J. Obstet. Gynec.* **91**, 277–285.

WALLGREN G., HANSON J.S. & LIND J. (1967) Quantitative studies of the human neonatal circulation. III. Observations on the newborn infant's central circulatory response to moderate hypovolemia. *Acta paediat. Scand.* Suppl. 179, 43–54.

WIGGLESWORTH J.S. (1967) Vascular organization of the human placenta. *Nature (Lond.)* **216**, 1120–1121.

WIGGLESWORTH J.S. (1969) The vascular organization of the human placenta and its significance for placental pathology, in A.Pecile & C.Finci (eds.) *The Foetoplacental Unit*, pp. 34–40. Amsterdam, Excerpta Medica Foundation.

WOOD C. (1969) Use of fetal blood sampling and fetal heart rate monitoring in K.Adamsons (ed.) *Diagnosis and Treatment of Fetal Disorders*, pp. 163–174. New York, Springer Verlag Inc.

YAO A.C., LIND J., TIISALA R. & MICHELSSON K. (1969) Placental transfusion in the premature infant with observation on clinical course and outcome. *Acta Paediat. Scand.* **58,** 561–566.

YOUNG M. & COTTOM D. (1966) Arterial and venous blood pressure response during a reduction in blood volume and hypoxia and hypercapnia in infants during the first two days of life. *Pediatrics* **37,** 733–742.

ID# 9

The Adrenal Cortex

A.W.STEINBECK

INTRODUCTION

The crucial role of the adrenal cortex and action of its secretions, whether in a 'regulatory' or a 'permissive' manner, is undoubted.

The relevance of the adrenal cortex to pregnancy remains a problem and the nature of its function, 'regulatory', 'permissive' or 'protective', can be debated. The designation of pregnancy as a 'physiological Cushing's syndrome' was more virtual than real although there is increased cortisol activity.

NORMAL FUNCTIONS OF THE ADRENAL CORTEX

Embryology

The adrenal cortex is of mesodermal origin and derived from the intermediate cell mass, the origin of the primitive kidney and gonads as well as the foetal cortex. The adrenal glands are large in foetal life, due to the foetal cortex which regresses in infancy. The glands grow progressively in weight from infancy to adolescence, with a final 'pubertal' stimulation.

Anatomy

Single adrenal glands weigh 3·5–4 g in the female. Cortical tissue is identifiable by its 11β-hydroxylase activity and more is in the body of the gland than elsewhere.

There is a rich blood supply to the gland and two types of circulation occur in the gland, a cortico-medullary 'portal' system in the head and parts of the body and a single circuit elsewhere. Also, emissary veins protect the gland from infarction with central vein occlusion. Nodular hyperplasia is probably a response to focal ischaemic loss of cortical substance with ageing.

Histology

The zona glomerulosa is not prominent in the human gland, focal in distribution and its cells are the only source of aldosterone. The zona fasciculata has an outer and inner zone, with different cell types and the interface cells may be a target for ACTH. The zona reticularis has similarities with this area. On the basis of functional zonation and histological features some describe the cortex as a double gland.

Functional zonation

There is a functional zonation of steroid hormones in the cortex (Dobbie *et al* 1968). The zona glomerulosa produces only aldosterone, and the zona reticularis and fasciculata function as a single unit. Excepting aldosterone, steroids are derived from blood precursors under ordinary conditions. The zona fasciculata is a reserve for steroid precursors in response to increased ACTH activity and stress. The zona reticularis is wider when plasma ACTH levels are raised, as in disease, and following ACTH administration. When ACTH levels fall after hypophysectomy, the zona fasciculata and reticularis atrophy which corresponds with decreased cortisol production but the zona glomerulosa and aldosterone secretion are initially unimpaired, although atrophy and decreased secretion can finally occur.

ACTH effects

ACTH stimulation has an immediate and a slower effect in man. The immediate effect is to increase blood flow and 10 units natural ACTH intravenously doubles or trebles the rate of blood flow, increases the cortisol secretion rate and the cortisol/corticosterone ratio in cannulation studies before adrenalectomy (Grant *et al* 1957, 1968). Prolonged ACTH stimulation increases ribonucleic acid, steroid 11β-hydroxylation, cortical width and alters cell morphology, provided vascular perfusion is normal. Natural ACTH has thirty-nine amino acids in a single polypeptide chain, with the common residues between species 1–24 and 33–39, and human ACTH has a more prolonged action in man than the others. Synthetic polypeptides have maximum physiological activity if they contain twenty-four amino-acid residues, and substitution alters activity. Structural similaries with melanophore stimulating hormones explain the pigmentary effects of ACTH and synthetic polypeptides, including tetracosactrin (β^{1-24}ACTH) which produces significant pigmentation, with melanin spots, after protracted use.

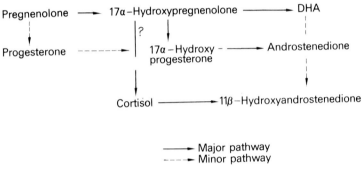

FIG. 9.1. The major and minor biosynthetic pathways of androgen and cortisol in the human adrenal gland as studied by perfusion *in situ* (Deshpande *et al* 1970). The conclusions were based upon incorporation studies and the pool sizes.

Biosynthesis of steroid hormones

Biosynthesis has been discussed generally in Appendix A but, in this chapter, adrenocortical aspects will be emphasized. During ACTH stimulation, depot cholesterol ester is used up and further cholesterol is derived from the blood stream. In the 'classical' pathway for biosynthesis, pregnenolone is converted to progesterone and, following hydroxylations steps at C-17, C-21 and C-11, cortisol or corticosterone is formed. However 17α-hydroxypregnenolone is converted in substantial quantity to cortisol and is a significant (Cameron 1969) and probably the major pathway for cortisol synthesis (Deshpande *et al* 1970). It is common to cortisol and dehydroepiandrosterone and a summary, from studies with the human gland *in situ*, is shown in Fig. 9.1. Studies with adenomas suggest the possibility of 17α, 21-dihydroxypregnenolone as an intermediate to dehydroepiandrosterone or 11-deoxycortisol synthesis (Cameron 1969), as shown in Fig. 9.2.

Hydroxylations are important and their inborn quantitative defects underlie the common forms of congenital adrenal hyperplasia. A defect in the 21-hydroxylase enzyme system impairs cortisol synthesis, results in compensatory ACTH hypersecretion and the development of adrenal hyperplasia. Quantitation determines when it becomes evident, at birth, during infancy or even after puberty. The 11- and 17-hydroxylations may be involved, likewise the conversion of cholesterol to pregnenolone, and pregnenolone to progesterone. Metyrapone, an 11β-hydroxylase inhibitor sometimes used in the study of hypothalamic-pituitary-adrenal function, reproduces some features of the inborn enzymatic defects. As a defect at any stage of biosynthesis blocks forward progression, precursor compounds accumulate and their metabolites appear in blood and urine. When the enzymatic defect does not involve androgen pathways, virilization occurs. Both the synthetic and clinical abnormalities can be considerable.

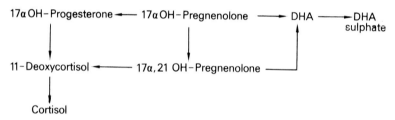

FIG. 9.2. Biosynthetic routes from 17α-hydroxypregnenolone to cortisol and dehydroepiandrosterone sulphate (Pasqualini et al 1964; Cameron 1969).

The principal pathway for aldosterone synthesis is by way of corticosterone and 18-hydroxycorticosterone. In man, sodium deprivation and depletion increases corticosterone production, not unexpectedly with a constant ratio for the secretion rates of aldosterone and corticosterone (Biglieri et al 1969). In the sheep, angiotensin can increase aldosterone synthesis but not corticosterone conversion to aldosterone suggesting that, like ACTH with cortisol, its effect is on early stages of synthesis.

The conversion of pregnenolone to progesterone by Δ^5-3β-hydroxysteroid dehydrogenase is irreversible, introduces the Δ^4-3-ketone group essential for biological activity and the enzyme activity limits the rate of biosynthesis. As androgenic substances inhibit cortisol production at this early stage, a mild defect in synthesis could be aggravated by androgen increase and inhibition of steroid dehydrogenase activity.

Gonadal androgenic and oestrogenic substances modify adrenocortical secretion at puberty when ACTH stimulates androgen production in addition to cortisol. Also, ovarian abnormality makes that of the adrenal obvious.

Dehydroepiandrosterone sulphate is a significant secretory steroid and principal androgen (C-19 form) of the human gland, is without activity in metabolic studies but converts to other C-19 steroids including testosterone (see also Chapter 3). 11β-hydroxyandrostenedione is present in adrenal venous blood, quantitatively exceeding androstenedione (Weinheimer et al 1966) and is the major product of this series, with cortisol a major and androstenedione a minor precursor. Progesterone is synthetized in small amounts and, like 17α-hydroxyprogesterone, is identifiable in adrenal venous blood. These syntheses are shown in Fig. 9.1, et seq.

Catabolism of steroids

Cortisol catabolism is rapid by reduction to ring A saturated steroids, dihydro proceeding to tetrahydro forms. Also, cortisol relates to cortisone in a reversible oxidation-reduction which is one determinant for the circulating level of cortisol, differently oriented in the foetus to the adult. Tetrahydro compounds are conjugated with glucuronic acid at C-3, and the water-soluble conjugates are excreted in urine. The plasma half-life for cortisol is 90–115 min and cortisone 30 min. These are one measure of the distribution of the steroid—the time taken for the 'initial' plasma level with intravenous injection to be halved. This is a measure of the biological half-life (Cope 1965) but is different from the body half-life, a measure of metabolic fate, which is 4 hours for cortisol. As shown in Fig. 9.3, initial enzymatic reduction of ring A gives two stereo-isomers, leading to *allo*-tetrahydro (5α), and tetrahydro (5β) forms. Occurring predominantly in the liver, this step is irreversible producing biologically inactive derivatives. The tetrahydro forms undergo side chain reduction at C-20, producing cortols from cortisol, cortolones from cortisone. Quantitatively, these are second in importance to the tetrahydro derivatives as excretory metabolites. The derivatives are made water-soluble by conjugation with formation mainly of 'glucuronides' (correctly glucosiduronates).

Cortisol and cortisone are interchangeable although some tissues may not make the conversion to cortisol (Bailey & West 1969). Cortisone

o*

and its analogues have biological activity when the 11-oxo group is reduced to an 11β-hydroxyl and conversions and metabolism are related to the plasma half-life. Values for this measurement with natural and synthetic steroids are shown in Table 9.1. Cortisone forms account for two-thirds of cortisol metabolites and, normally, tetrahydrocortisone is the major and 'unique' metabolite but loses this position in abnormal conditions (Kowarski *et al* 1969, New *et al* 1969).

TABLE 9.1. Plasma half-life (biological half-life) in minutes of natural and synthetic glucocorticoids in adults,* after Cope (1965)

Steroid	Half-life (min)
Cortisol	80–115
Cortisone	30
Corticosterone	60–80
Aldosterone	50
Prednisone	60
Prednisolone	200
Methylprednisolone	210
Fluorocortisol	90
Dexamethasone	200

* In the newborn the half-life of cortisol is considerably prolonged (Reynolds *et al* 1962).

Removal of the C-17 side chain, with production of 17-oxosteroids, gives C-19 steroids with an oxygen function at C-11, either 11β-hydroxy- or 11-oxosteroids, but cortisol conversion to 17-oxosteroids is less than 10%.

A small percentage of cortisol as the free, relatively insoluble steroid is excreted unchanged in urine, its amount increasing with the plasma levels of the unbound or free steroid and correlating with plasma levels of cortisol. Its estimation forms the basis of certain tests but values below the age of 10 years, expressed as μg/kg body weight, are less than the adult (Minick 1966). Small amounts of cortisol are also excreted as sulphate and glucuronide, and metabolites as sulphates. Conjugation

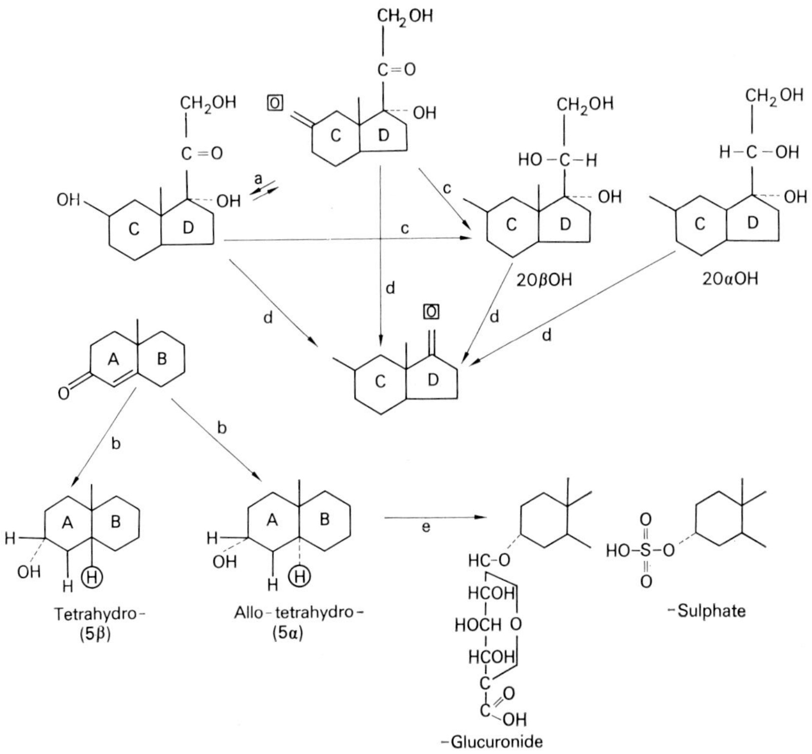

FIG. 9.3. Enzymatic actions in the metabolism of adrenocortical steroids (Visser 1966).

a, oxidation at C-11 (cortisol→cortisone, corticosterone→11-dehydrocorticosterone, etc).

b, reduction of ring A to give the tetrahydro metabolites, as tetrahydrocortisol or allo-tetrahydrocortisol, etc.

c, reduction at C-20 with formation of cortols, cortolones.

d, removal of C-17 side chain to give 17-oxosteroids (17-ketosteroids) of the 11-oxy-17-oxo series.

e, conjugation at C-3 with formation of glucosiduronates (glucuronides) and sulphates.

patterns suggest that corticosteroid sulphates are important in the newborn (Drayer & Giroud 1965). Finally, hydroxylation of cortisol also occurs at C-6, 6β-hydroxycortisol is formed, and being relatively water-soluble, is excreted without conjugation. Although a minor catabolic product in adults, relatively more occurs in normal infants (Ulstrom et al 1960); it is increased in pregnancy and toxaemia and may indication functional changes in the cortex.

Corticosterone catabolism is basically similar to that of cortisol and tetrahydro metabolites of 11-dehydrocorticosterone (the 11-oxo form) are produced but the side chain remains. Total cortisol and corticosterone metabolite excretions are in proportion to their secretion rates. Corticosterone sulphate excretion in urine is a measure of its production (Kielmann et al 1966) except in pregnancy (Jänne & Vihko 1970).

According to the 5α- or 5β-configuration in ring A saturation, 5β/5α ratios are sometimes used for 17-oxosteroids (Slaunwhite et al 1964): cortisol gives rise mainly to 5β-forms; 11-deoxy-C-19 steroids such as androstenedione and testosterone give a 5β-predominance with a ratio of 1·5–2; 11-hydroxy-C-19 steroids give a 5α-predominance and for 11β-hydroxyandrostenedione the ratio is 0·12:0·32. 11β-hydroxyandrosterone is 60% of the metabolic product, or a measure of adrenal 'androgen' production. Childhood steroid patterns, with minimal gonadal hormone production are different from adults and 11-oxygenated forms exceed 11-deoxy forms (Paulsen et al 1966). The 11-deoxy forms approximate the 11-oxygenated forms shortly before puberty, and exceed them afterwards. These changes represent the relative contribution of corticosteroids. The actual 17-oxosteroid values are also low from shortly after birth until puberty. Dehydroepiandrosterone, a major adult 11-deoxy-17-oxosteroid does not appear before 7 years and its appearance correlates with biological maturation.

Metabolism of the adrenal 'androgens' is summarized in Fig. 9.4. In normal puberty, increased excretion of 11β-hydroxyandrosterone is matched by increased 11β-hydroxyaetiocholanolone, mainly from cortisol metabolism. The activity of the reductase system for ring A saturation is influenced conversely by androgens and cortisol, and shows changes with puberty and abnormal conditions.

Children excrete relatively more tetrahydrocortisone than adults, more cortols and cortolones, more 6β-hydroxycortisol but the metabolism of cortisol and other corticosteroids is similar for both ages. Although there is an increased production coinciding with rapid growth, sexual maturation and biological maturity, the values expressed in terms of surface area are not significantly different for the different ages, except that cortisol and corticosterone production may be higher in infants.

Aldosterone metabolism is similar to other Δ^4-3-ketone steroids and its rate disposes most of the hormone in urine within 24–48 hr after an injection. The kidneys extract aldosterone from plasma, produce half of the urinary 18-glucuronide and account for up to one-fifth of plasma clearance. A small fraction appears in the urine as the free steroid, about 10% as the 18-glucuronide and the remainder as other metabolites including tetrahydroaldosterone (5β-form).

The metabolism of cortisol, corticosterone, aldosterone is summarized in Fig. 9.5.

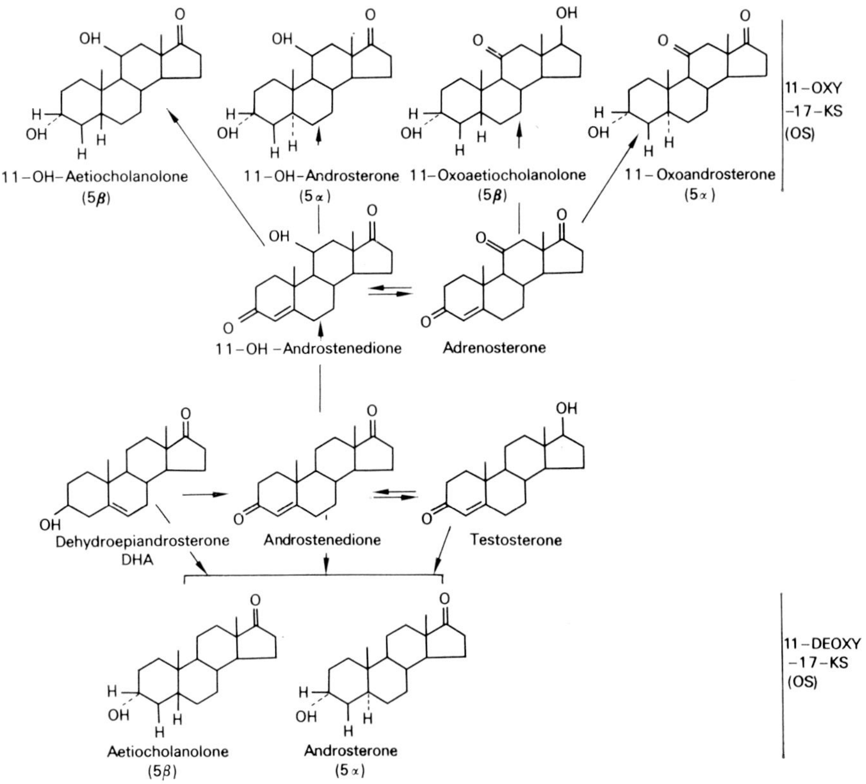

FIG. 9.4. Major metabolites of adrenal androgens (Visser 1966). As in Fig. 9.3, 5β- and 5α- forms are shown; -OH- refers to a hydroxyl function, -oxo- to a ketone function and -oxy- to presence of oxygen at the carbon position. 17-KS = 17-ketosteroids (17-oxosteroids).

NEONATAL ADRENOCORTICAL FUNCTION

The foetal zone regresses rapidly after birth and the definitive cortex increases in size. Removal of precursor placental steroids and loss of circulating progesterone stimulates development of Δ^5-3β-hydroxysteroid dehydrogenase. Rapid changes in biosynthesis and catabolism of cortisol/corticosterone, progesterone and oestrogens reveal a transitional, rapidly developing functional capacity of the cortex that prevents

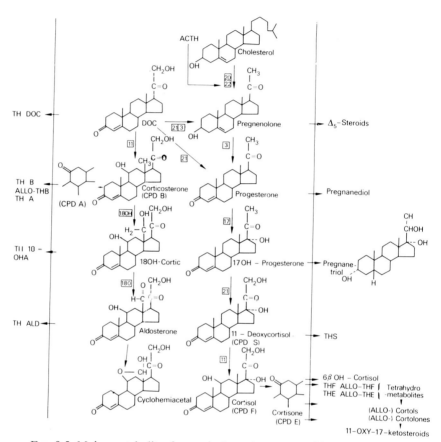

FIG. 9.5. Major metabolites for cortisol, corticosterone, aldosterone and their precursors (Visser 1966). DOC = 11-deoxycorticosterone; cortic = corticosterone; TH = tetrahydro; other abbreviations as earlier. Numbers in boxes refer to enzymatic hydroxylation and dehydrogenation steps. Urinary metabolites are shown outside the vertical lines.

relative hypofunction, and cortisol has a prolonged biological life at this time. Cord levels of cortisol are similar to those of non-pregnant women and, in the first five months of life (Hillman & Giroud 1965, Ulstrom *et al* 1961) plasma levels do not fall, although a low cortisol/

cortisone ratio persists for two weeks after birth.* Cord levels of cortisol and cortisone are low with maternal adrenocortical insufficiency. Inadequately developed, liver enzyme systems of the newborn explain some differences between the newborn and adult; dehydroepiandrosterone is not found in newborn urine but 16β-hydroxydehydroepiandrosterone is identifiable, and in pregnancy urine. Intra-uterine existence requires steroid hydroxylations at C-6, C-15 and C-16 (Luukkainen *et al* 1970) which persist in infancy with formation of water-soluble metabolites, at a time when liver glucuronide conjugation is underdeveloped and inadequate. This accounts for patterns of steroid excretion in infancy, unchanged compounds being excreted because of their water solubility with significant amounts as sulphates. The glucuronides rise as the liver matures, tetrahydrocortisone glucuronide corresponding to less than $\frac{1}{2}\%$ radio-labelled cortisol in the first few days of life but close to 5% after 2 weeks. Sulphate conjugation of Δ^4-3-ketone steroids is predominant over glucuronides early in life (a persistence of the foetal pattern) and has to be considered when studying excretions, possibly also disulphates. In the 1st week of life, more corticosterone conjugates are found than cortisol and infants secrete proportionately more corticosterone than cortisol (about one-seventh in the 3rd week) than the adult or older child.

Foetal adrenal function

Foetal adrenal function has been discussed in chapter 4 but some aspects of the maternal excretion of steroids are relevant. Adrenal hypoplasia is associated with low oestriol excretion in the third trimester and low cord levels of cortisol. Lipid hyperplasia is associated with low pregnanetriol and oestriol excretions but high pregnanediol giving abnormal oestrogen ratios; with the common type of adrenal hyperplasia pregnanetriol is sometimes raised. Suggestively, the urinary steroid excretions in pregnancy may be useful diagnostically for enzymatic abnormalities before delivery, safer and, perhaps, more convincing than amniotic fluid analysis (Luukkainen *et al* 1970).

* Where the information is available, specific reference will be made to cord artery or cord vein levels. Where this information is not clear in the original text the level will refer to 'cord blood'.

STEROID SECRETION RATES
(see Appendix C)

Usually, the secretion rates for normal subjects are in the range 9·0–31·0 mg/24 hr for cortisol and 0·9–4·4 mg/24 hr for corticosterone. Lower values can be found normally, with a corticosterone/cortisol ratio about one-tenth. Increasing age lowers secretion rates but childhood values are comparable with the adult and obesity may induce a higher secretion rate than normal. A representative mean value for cortisol secretion in woman is 16·1 ±0·8 mg/24 hr (Layne *et al* 1962).

Published aldosterone secretion rates have a wide range. In some of these studies sodium intake was uncontrolled, or not comparable with others; in some, subjects were active and upright, factors that also affect aldosterone production. Rates determined from tetrahydroaldosterone excretion rather than the 18-glucuronide are higher, except in hypertensive patients. Secretion rates derived from tetrahydroaldosterone excretion are in the range 50–275 μg/24 hr and from the 18-glucuronide 14–195 μg/24 hr. There is a progressive increase in secretion throughout the menstrual cycle, in one study from 138 to 263 μg/24 hr, and luteal phase values are higher than follicular (Gray *et al* 1968). These changes are discernible in excretion studies, as shown in Table 9.2. The increase in aldosterone corresponds with that of angiotensin (Sundsfjord & Aakvaag 1970).

For non-stressed men and women, the plasma levels of testosterone and oestradiol in the adrenal vein are only slightly above peripheral levels, suggesting low adrenal secretions (Baird *et al* 1969). Also, castration in men lowers plasma testosterone levels to approximately zero and in women similarly lowers oestradiol levels. Thus, these two 17-hydroxylated steroids are produced in the main by gonadal tissue. Plasma levels for androstenedione of 10–40 μg/100 ml in the adrenal vein suggest a secretion rate of about 2 mg/day in men; in women, the blood production rate, or that amount delivered into the blood from precursor sources, is greater in the follicular phase suggesting that only 50% derives from the adrenal cortex. 11β-hydroxyandrostenedione is a more important secretion than androstenedione and, provided side chain cleavage of cortisol to form 11β-hydroxyandrostenedione does not occur, a rate of 4·5 mg/24 hr is possible (Goldzieher & Beering 1969); otherwise, production rates only can be derived. The secretion rate of dehydroepiandrosterone sulphate approximates that of cortisol.

Plasma levels of cortisone in the adrenal vein of normal subjects suggest a secretion of 10 μg/24 hr.

Steroid blood levels

Plasma steroids are bound to, or associated with, carrying proteins so that bound and free (or unbound) steroid fractions coexist in plasma. Likewise, steroid conjugates bind to protein. For non-conjugated steroids, the free fraction is regarded as the tissue active form. Arguing

TABLE 9.2. The excretion of aldosterone 18-glucuronide during the menstrual cycle in three normal women. Aldosterone (Aldo) excretion is μg/24 hr and 17-hydroxycorticosteroids (Norymberski chromogens; 17-OHCS) mg/24 hr

Age in years	Day of cycle	Aldo	17-OHCS
23	19	23·6	13·0
	26	44·0	10·6
	9	15·5	7·6
	16	21·7	7·0
31	9	11·5	8·0
	20	29·9	7·9
	9	12·6	13·1
	16	13·0	12·1
	26	21·9	9·8
25	6	4·9	6·7
	10	7·9	7·1
	26	11·0	7·1

from the low cortisol levels in cerebrospinal fluid and filtrates, the steroid-protein complex does not cross the vascular barrier, except in the liver. Although transcortin exists outside the vascular system, the free moiety would seem to be the only active fraction. Most methods of steroid estimation, because of an initial extraction step, estimate total plasma levels of the free plus bound steroid, or active plus inactive forms. These levels correlate with physiological variables, and free levels relate to them so that there is little error in most situations; the plasma 'cortisol'* response to stimulation and suppression procedures

* This is usually measured as total levels by non-specific methods.

in tests of the hypothalamic-pituitary-adrenal axis is a good example. However, free cortisol correlates better with clinical conditions than the total levels, so that some reservation is appropriate, and better with the rate of cortisol reduction in liver incubations. The free cortisol excreted in urine over a 24-hour period is derived from the free or unbound cortisol of plasma, that undergoes filtration at the glomerulus, less that lost by passive back diffusion in the tubules. It correlates with the amount of free cortisol available for diffusion into tissues over 24 hours and can be related to the excretion of 17-hydroxycorticosteroids (Norymberski chromogens), which is a measure of the degradation of adrenocortical steroids, without correlation except in the broadest sense.

Cortisol is bound to a high affinity, low capacity α_1-globulin, transcortin or corticosteroid-binding-globulin (CBG), leaving 6–8% free at 37°C, that is approximately 10% at body temperature. It is also bound to the higher capacity, lower affinity albumin. Corticosterone, 11-deoxycortisol, 17-hydroxyprogesterone and progesterone, amongst others, are highly competitive with cortisol for transcortin binding, cortisone only moderately so. Some therapeutic steroids, such as prednisolone, are highly competitive but others, like dexamethasone and prednisone, are less. Transcortin may be virtually absent in some families, rarely idiopathically elevated, and 'stress' decreases binding sites, increasing free cortisol levels. Transcortin levels are elevated with oestrogen therapy and in pregnancy, without significant evidence of hypercortisolism. If free cortisol levels increased in pregnancy, the extent and duration would be insufficient probably to establish features of Cushing's syndrome. Transcortin increases to reach a peak in the last trimester, double that of normal, but is already elevated in the first (Rosenthal et al 1969). Cortisol is 80% bound, with 13% bound to albumin, in the last trimester; the unbound fraction decreases as pregnancy advances but the percentage of free cortisol remains constant, as with oestrogen therapy. However, free cortisol does not vary significantly through the menstrual cycle with the oestrogen fluctuations. The effect of oestrogen treatment is summarized in Table 9.3.

Transcortin normally has a plasma half-life of 6 days, shorter with oestrogen treatment from increased hormone synthesis. There is no significant day–night variation in transcortin concentrations.

Aldosterone is loosely bound to transcortin and albumin and its transcortin binding increases like cortisol. It is easily displaced by cortisol and progesterone which has a high affinity with transcortin and albumin and explains failure of the binding to increase in pregnancy. Three different plasma proteins have an affinity for progesterone including transcortin and albumin but the third, orosomucoid, binds little under physiological conditions. Progesterone competition with cortisol for transcortin binding could lead to higher levels of free cortisol and explain the higher levels of free cortisol found in pregnancy than with oestrogen treatment, likewise the low levels of transcortin in venous cord blood and subsequently in the newborn.

TABLE 9.3. Concentration of cortisol, μg/100 ml plasma, for various plasma fractions before and after oestrogen treatment in the same subject, as calculated by Tait & Burstein (1964). A = Albumin and T = Transcortin. The miscible pool of cortisol in the body was increased, the metabolic clearance and secretion rates were decreased, and the metabolic turnover reduced. Total plasma levels of cortisol were increased but the free levels were low and unchanged.

	Cortisol secretion rate mg/24 hr	Un-bound	Cortisol concentration μg/100 ml plasma A-bound	T-bound	Total	Cortisol metabolic clearance rate litres/24 hr
Normal	28·5	1·5	2·0	12·5	16·0	178
After oestrogen	15·6	1·5	2·0	36·5	40·0	30

Sex-steroid binding plasma protein (SBP) binds testosterone and oestradiol and is discussed elsewhere (chapter 3). Albumin exhibits non-specific binding in relation to water-solubility of the steroids, and their sulphates.

The role of the binding proteins is regarded as a reservoir or a buffer system for the free steroid. They prevent increased levels of free steroid during temporary increases in secretion and ensure steroid availability with metabolism and delivery to tissues.

Variations in plasma steroid levels throughout the day and day–night variations

Highest plasma cortisol levels are found at 6–10 a.m., the lowest close to midnight to 4 a.m., but actual timing varies and may relate to location. The rhythm is individual, repeated at near daily intervals and more correctly termed a circadian rhythm than a diurnal variation, the often used term. It is reflected in the urinary excretion of steroids, either cortisol or its related metabolites, and an 8-hour time-shift in the sleep–wake schedule is followed by a corresponding time-shift in the timing of the plasma variations. The timing of the cortisol rhythm is dependent upon the sleep–wake activity of the individual and its length and timing is correspondingly determined by the length and timing of his habitual sleep–wake pattern (Orth et al 1967).

The usual description, of high morning levels falling throughout the day to become low at night or in the early morning hours with an increase close to waking, is an inadequate description if smooth curves are visualized. A smooth curve obtained by frequent sampling is shown in Fig. 9.6, but 'stress' can involve the individual and lead to episodic increases of cortisol secretion and plasma levels throughout the day, and other episodic fluctuations in levels of both ACTH and cortisol occur. Episodic cortisol secretion is evident when repeated blood sampling is possible, occurring both during the early morning hours and the waking day (Hellman et al 1970). For this reason, isolated, single blood levels of cortisol give little information about the activity of the pituitary-adrenal system: there is not a steady state and the amount of cortisol available for or involved in biological activity at any time is unknown; also fluctuations in ACTH activity itself and the dynamic nature of the system emphasize that plasma levels alone do not describe its overall activity.

The nocturnal surge of ACTH secretion, and cortisol secretion, is closely related to rapid eye movement (REM) sleep, and an anticipatory rise occurs during sleep with ordinary sleep cycles. The rhythm can be re-staged, after a delay period of adjustment, by altering sleep patterns and a new circadian rhythm of plasma cortisol is established in relation to the new sleep cycle. The rhythm is not evident in some blind subjects, fails in some cerebral diseases and its programming can be altered by drugs and destroyed by prolonged glucocorticoid therapy. Then, following steroid withdrawal, there can be a long period before reappearance of a cycle and return of normal hypothalamic-pituitary-adrenal function (Graber et al 1965). At first, plasma cortisol levels will be low with narrow limits of variation; in normal subjects, low values show wider variations and one low value at 8–9 a.m. can indicate a high earlier level. Sampling times may not correspond with peak or trough levels. Also, a pattern is not evident in some with repeated 2-hourly sampling;

obesity can flatten the curve, likewise myxoedema. Oral contraceptive therapy increases plasma cortisol levels with maintained day–night variation and higher night levels than expected normally, similar to morning levels for some. Notwithstanding, free cortisol excretion in urine remains low. Possibly, a photo-period is involved in the timing of the cortisol rhythm for there is a delay in the 'anticipatory' rise of plasma cortisol (decrease in the sleep-related peak) when darkness is

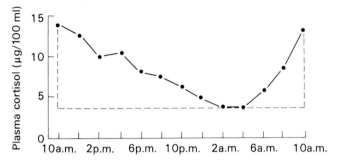

FIG. 9.6. Example of a nycthemeral rhythm (diurnal rhythm) for plasma cortisol of a smooth type (Doar et al 1970). Repetition of the pattern is the circadian cycle. The plasma cortisol level is obtained by the Mattingly (1962) method (SAFA value). The authors used the area enclosed by the curve, obtained from 2-hourly samplings, and the dashed lines at the extremes, as an index of the ability of the hypothalamic-pituitary-adrenal system to alter cortisol levels throughout the day. This 'diurnal plasma cortisol area' was useful in their investigation of hypothalamic-pituitary-adrenal function in pituitary disease, combined with repeated resting cortisol levels, ACTH stimulation and insulin-induced hypoglycaemia tests.

extended after awakening and there is close timing of the peak in conjunction with illumination and a failure of the cycle with total ocular blindness, features which suggest that light perception is concerned with the timing.

Dehydroepiandrosterone sulphate is the major plasma 17-ketosteroid sulphate and its production rate and that of dehydroepiandrosterone respond to ACTH stimulation but are not totally suppressed in castrates by dexamethasone. Secretion of dehydroepiandrosterone sulphate rises after 7 years and does not appear to respond to ACTH in adrenal vein studies although plasma levels suggest a response; the free steroid shows a response and plasma levels of both alter with dexamethasone. The free steroid has high morning levels, and progressively lower values during the day, but the sulphate has low levels between midnight and

8 a.m., highest between mid-day and 4 p.m. The rhythms are not established as for cortisol, there is interconversion between the two steroids and their significance is not evident.

Diurnal sensitivity of adrenal cortex

The adrenocortical response in ACTH stimulation tests is greater in the morning than the evening, reflecting 'diurnal sensitivity'. As the variation in response is also found in pyrogen administration, lysine-vasopressin stimulation and metyrapone inhibition tests, this indicates morning priming of the cortex by ACTH in its circadian secretory rhythm.

ACTH AND THE ADRENAL CORTEX

ACTH is primarily a stimulating and trophic hormone for the adrenal cortex, concerned with steroidogenesis and the regulation of adrenocortical growth. Its effect upon cortisol secretion takes some minutes to develop and near maximum secretory responses are achieved with 3 mU/100 ml or greater, most clinical tests using supra-maximal stimulation. The half-life of ACTH is short and its action ceases shortly after completion of an infusion. Corticosteroid secretion then decreases but some cortisol returns to the plasma from extravascular spaces and the levels do not fall as quickly.

The first change following stimulation is due to cyclic AMP, later RNA increases and enzyme activity—with cell hypertrophy; with continued stimulation, DNA and protein increases indicate hyperplasia. These changes are dose- and time-dependent and, for steroidogenesis, the same dose is more effective over 24 hours than over 4–6 hours. A prolonged infusion produces hypertrophy more rapidly than intermittent short infusions. The capacity of the cortex to respond to ACTH is dependent upon its previous trophic history: an atrophic cortex responds more slowly, takes longer to attain its maximal but limited capacity, than a normal; cortisol levels in blood rise slowly during an 8-hour infusion to reach normal or lower than normal levels and the corresponding urinary excretion is diminished. This is the feature of secondary adrenocortical failure with post-partum hypopituitarism, but repeated stimulation with exogenous ACTH increases the capacity to respond. However, continued suppression of endogenous ACTH secretion, as by dexamethasone, reduces the capacity to respond over some days and, finally, induces adrenocortical failure, measured by the capacity to respond to ACTH stimulation.

If an injection or infusion of ACTH is given to a normal or pregnant subject in the morning, plasma levels of cortisol increase, for $1\frac{1}{2}$–2 hours after the injection, during the infusion and for $1\frac{1}{2}$–2 hours afterward, but return to normal. On the subsequent morning, plasma levels of cortisol increase following the nocturnal surge of ACTH secretion. A hypothalamic abnormality is suggested when this does not occur, notwithstanding low midnight levels, and failure of an adequate insulin-induced hypoglycaemia to increase plasma cortisol confirms the hypothalamic failure. This occurs with some pituitary tumours and post-encephalitic states.

Plasma ACTH levels have a circadian rhythm: 6 a.m. values are 0·1–0·5 mU/100 ml and corresponding 6 p.m. figures are below this range (<0·1 mU). High ACTH levels precede the high morning levels of cortisol, and are sufficiently out of phase to be the secretory stimulus. 11β-hydroxylase inhibition by metyrapone increases ACTH secretion, dexamethasone suppression reduces ACTH secretion and conditions where cortisol secretion is reduced (congenital adrenal hyperplasia and Addison's disease) are associated with increases in plasma ACTH. Plasma ACTH increases during laparotomy, an increase not inhibited by large doses of cortisol or synthetic steroids given 2 hours before surgery (Estep *et al* 1963). With this 'stress', ACTH secretion is not affected by hypothalamic-pituitary inhibition; 'stress' hypersecretion is determined at another level. With cortisol deficiency, a circadian rhythm for ACTH secretion persists, although suppressible. 'Stress' hypersecretion of ACTH is relevant to clinical situations: a young woman aged 23 years with regular ovulation but obese, hirsute with considerable acne and reactive hypertension had consistently elevated levels of 'cortisol' and, over some months, a progressive increase occurred in the absence of a circadian rhythm, and the cortex was not suppressible by dexamethasone. Tested by her own colleagues, when she did not feel besieged by clinicians committed to an adrenalectomy, she had normal control levels and responses.

Insulin-induced hypoglycaemia

Adequate hypoglycaemia, measured by a 40% fall in the fasting value within 20–30 minutes after injection (both the rate of fall and its extent are significant), with features of adrenaline secretion (including pallor, sweating and tachycardia) induces ACTH secretion. Some obese subjects, most with Cushing's syndrome, those with psychiatric disorders and 'stressed' fail to respond in the test, although an adequately functioning cortex is present; subjects with adrenocortical insufficiency (with Addison's disease or hypopituitarism) or suffering starvation are unduly sensitive to adrenaline effects. The test will decide whether previous glucocorticoid therapy has impaired the hypothalamic-pituitary-adrenal response, and the need for cortisone coverage of operations.

ACTH and cortisol secretion

The quantity of cortisol produced is related to available ACTH and modifications in cortisol secretion, plasma production or delivery to the blood, are results of alterations in ACTH secretion. If metabolism is constant, ACTH secretion and cortisol production are also. Normally, a circadian secretory cycle for ACTH is evident and a 'stress' secretion can occur. With pelvic laparotomy, ACTH levels rise more than threefold above the normal range, with increases in plasma and urinary levels of Porter-Silber chromogens, although the secretory response is not maximal when compared with an ACTH infusion of more than 1 U/hr. Minor operations are less stressful.

Control of ACTH secretion

Regulation of ACTH secretion starts with the limbic system—hypothalamus, although generally described as transmission of hypothalamic information via corticotrophin releasing hormone (Schally *et al* 1970), to the adenohypophysis. There is feedback of information to the hypothalamic-pituitary area and cortisol modifies several activities of the brain. Feedback regulation is broadly evident, in terms of plasma levels of cortisol, but 'direct' feedback is unlikely even for the pituitary, and areas beyond the hypothalamus are involved. 'Indirect' control mechanisms are well established.

Corticosteroids act at a locus near the median eminence to reduce adrenocortical secretion, which can induce adrenocortical atrophy. Concentrations of corticotrophin releasing hormone increase in the same area after adrenalectomy, simultaneously with plasma ACTH. The inhibitory effect of cortisol, its negative feedback bias, is largely mediated through the hypothalamic mechanism, but there may be more direct effects upon the pituitary in some animals (Berthold *et al* 1970). In rats, but not in man, exogenous ACTH inhibits endogenous ACTH secretion; there is a positive feedback bias as oestrogen implanted in the median eminence increases ACTH secretion. There appear to be both corticoid-sensitive and corticoid-resistant components to the response; 'short' feedback mechanisms seem likely as ACTH plus dexamethasone together block stress responses that dexamethasone fails to. In animals, gonadal hormones modify adrenocortical function directly and also indirectly by influencing centres outside the hypothalamus. In man, the factors that establish a circadian rhythm for ACTH are not overcome by cortisol in Addison's disease, and a circadian rhythm persists in the presence of continued cortisol deficiency, although stress mostly interferes with it.

In animals, chlorpromazine, reserpine and endotoxins activate pituitary-adrenocortical activity or ACTH secretion. In man, endotoxin increases cortisol levels, chlorpromazine and other phenothiazines have variable effects, also reserpine and other centrally active drugs. Amphetamines increase cortisol levels in some but not all and the response as for other centrally active drugs may be individual. The pyrogen response assesses pituitary-adrenal function, at times unsatisfactorily, and has a pituitary locus of action. Morphine blocks release of corticotropin releasing hormone, but not pituitary activity, and suppresses the vasopressin response. Lysine-vasopressin tests are frequently unsatisfactory as a test for release of corticotrophin releasing hormone as discomfort with the test makes it a 'stress' test rather than a specific measure. Diffuse areas of the brain are sensitive to corticosteroids and the EEG rhythm alters with cortisol excess. The mid-brain is involved in regulatory mechanisms with the limbic system intimately concerned and 'internal' or 'short' feedback mechanisms and resistant components may be significant. Computer models have been programmed to test theoretical modes (Yates & Brennan 1969).

Recovery of pituitary-adrenal function after chronic corticosteroid suppression

There is evidence that 'short' feedback mechanisms occur, because ACTH plus dexamethasone block a 'stress' response that dexamethasone alone fails to and pre-treatment with ACTH reduces metyrapone responses (Sussman *et al* 1965); such mechanisms would explain suppression of ACTH response by administration of ACTH or cortisol.

Prolonged glucocorticoid therapy depresses hypothalamic-pituitary-adrenal function, demonstrated by insulin-hypoglycaemia tests (Amatruda *et al* 1960, Graber *et al* 1965), which may have untoward effects, but the incidence is unpredictable. Following cortisol excess in glucocorticoid treatment, also with adrenal adenomas, function may not return. When treatment is stopped, ACTH secretion is at first low, then increases over 2 to 5 months to normal levels, and finally exceeds them; cortisol production and levels are also low, adrenocortical responsiveness is impaired but the rising ACTH levels develop a circadian rhythm and cortisol levels return to normal, usually between the 5th and the 9th month after prolonged treatment.

Extra-adrenal action of ACTH

Extra-adrenal actions are described for ACTH: in adrenocortical insufficiency it interferes with the half-life measurement of cortisol (Kusama *et al* 1970), and accelerates cortisol transference into the tissues, increasing its distribution volume and inhibiting its re-entry into the vasculature, slowing metabolism.

PHYSIOLOGICAL CONTROL OF ALDOSTERONE SECRETION

Six types of stimuli influence aldosterone secretion: hypotonicity, hyponatraemia with normal tonicity, potassium ions, ACTH, angiotensin and renin-activity, and sodium deprivation (which does not operate entirely through the other mechanisms).

Effect of ACTH upon aldosterone

Prolonged hypopituitarism lowers aldosterone levels, from atrophy of the glands with loss of the trophic effect of ACTH. An 8-hour infusion of ACTH increases aldosterone levels but prolonged infusion depresses aldosterone secretion, although 11-deoxycorticosterone and corticosterone outputs increase, without relation to sodium retention or restriction (Biglieri *et al* 1969). Repeated ACTH stimulations have the same effect and aldosterone levels are also decreased in Cushing's syndrome.

ACTH stimulation has a variable effect upon the quantitative aldosterone response and depends upon conditions of study, being greater in sodium depletion (Tucci et al 1967) which induces a response in insulin-hypoglycaemia (James et al 1968). Potassium ions sensitize the adrenal to respond to ACTH stimulation of aldosterone secretion (Williams et al 1970). 'Stress' acts in part through the ACTH control system.

Although a possibility of neural influences has long been contemplated, the pineal, sub-commissural zone and near-by areas of the brain do not play important roles in aldosterone secretion. The brain-stem affects aldosterone secretion to an extent because it affects ACTH secretion, although its role is not physiologically important.

The renin-angiotensin system is important to aldosterone control and increased angiotensin is a consistent stimulus to aldosterone production; angiotensin also stimulates cortisol and corticosterone output, but to a smaller extent than ACTH. Acute, severe renal-artery constriction induces hyperaldosteronism with frank symptoms and biochemical changes. Stimuli to renin production include a low sodium diet, renal hypertension and pregnancy.

Relationships between aldosterone and other hormones

Human growth hormone does not increase aldosterone secretion although it increases adrenal size. The increased aldosterone secretion of diabetes inspidus is the result of dehydration with hypovolaemia and changes in the renin-angiotensin system. Aldosterone metabolism is reduced in hypothyroidism, with normal blood levels, but its metabolism is increased in thyrotoxicosis, which follows the pattern of other steroid degradations.

Aldosterone secretion rises during the day, in good relationship with the upright position of daytime activity, both renin and aldosterone showing the same variations during the day and increasing with the upright position. A day–night variation occurs but other factors, including anxiety, increase both aldosterone and cortisol output; the problem of discriminating between a selective stimulation of aldosterone secretion and that associated with increased glucocorticoid output occurs with surgery (Hume et al 1962), air-encephalography and pregnancy.

Direct stimulatory influences upon aldosterone secretion include high potassium intake, and low sodium intake. With sodium deprivation of various types, both renin and angiotensin increase; other influences are mediated through cardiovascular changes and the renin-angiotensin system including renal artery constriction, standing, heat, among others. Plasma electrolyte changes appear to be a 'back-up' mechanism to increase aldosterone output. Stimuli that decrease aldosterone secretion include sodium retention; deoxycorticosterone administration suppresses aldosterone production without evidence of direct feedback control of aldosterone secretion within the gland. Clinically, sodium deprivation effectively stimulates aldosterone secretion, no matter how deprivation is established. In this regard, electrolyte balance needs to be established before investigating aldosterone production and body deficit of potassium hinders aldosterone output studies.

Hyperaldosteronism is only one factor, perhaps 'permissive', in the sodium retention of oedematous patients. It occurs in some women troubled by intermittent or idiopathic oedema, an exaggerated orthostatic response and where 'escape' may fail, as in most instances of oedema, except in hypopituitarism.

Metabolic clearance rate

The virtual volume of blood that is irreversibly and completely cleared of a substance in unit time measures the metabolic clearance rate for that substance. The metabolic clearance rate for aldosterone in the non-pregnant subject is close to 2200 l whole blood/24 hr, which approximates hepatic bood flow. Other values are used, with appropriate care, to describe certain aspects of steroid metabolism and transport and include volumes of distribution, plasma half-life and body half-life. The total volume of distribution for aldosterone is 40 litres, about 12 and 20 litres for cortisol and corticosterone, respectively. Metabolic clearance rates are decreased in hepatitis and cirrhosis, where the liver is structurally affected, and in hypothyroidism; they are increased in thyrotoxicosis. The clearance rate is determined by the enzymatic activity reducing Δ^4-3-ketone groupings, and will be affected where this is affected by metabolic changes (e.g. thyroid disease), by vascular perfusion or by structural changes in the liver, among other factors.

With cortisol excess as in Cushing's syndrome or with glucocorticoid therapy, metabolic clearance rates of cortisol increase, and rates appear related to previous steroid levels, a factor to be considered in emergency use of steroids. Oestrogen therapy reduces cortisol clearance rates. A reduction of the metabolic clearance rate for aldosterone occurs in pregnancy (Tait *et al* 1962), with metabolic features similar to oestrogen therapy and cirrhosis.

PREGNANCY: HORMONOLOGY OF PREGNANCY

Plasma corticosteroids

Pregnancy is associated with a progressive increase in plasma cortisol levels throughout its course, measured specifically, as Porter-Silber chromogens or fluorogenic steroids ('cortisol'). Table 9.4 illustrates

TABLE 9.4. µg Porter-Silber chromogens/100 ml plasma during the course of pregnancy

Non-pregnant women	1st trimester	2nd trimester	3rd trimester	Reference
9·5(3–16)	10 ± 1·9	13 ± 1·1 (3rd mth) to 20 ± 1·7 (6th mth)	21 ± 1·8 (7th mth) to 22 ± 1·3 (9th mth)	Bayliss et al (1955)
6·6 ± 0·95	11 ± 1·53 (8th wk) 9·8 ± 1·49 (12th wk)	13·3 ± 1·45 (16th wk) 12·7 ± 1·91 (24th wk)	17·9 ± 2·68 (28th wk) 21·5 ± 2·08 (32nd wk)	Birke et al (1958)
9·9 ± 2·5	9·9 ± 2	11·3 ± 3·4	15·1 ± 3·4	Kawakara (1958)

changes for Porter-Silber chromogens during the course of normal pregnancies. The plasma level at term can be considerably increased (three to five times normal in some women), or only moderately so, but all women show some alteration and Figs. 9.7 and 9.8 demonstrate the changes diagrammatically. However, the conjugated metabolites, tetrahydrocortisone and tetrahydrocortisol, maintain normal levels and the circadian rhythm for cortisol persists during pregnancy. Levels are much higher at delivery, whether vaginally or by caesarean section, from associated 'stress'. Corticosterone levels follow the trend for cortisol and cortisol/corticosterone ratios are higher in pregnant women than in the non-pregnant. Plasma cortisol levels increase to a greater extent than normal with ACTH stimulation (Steinbeck & Theile 1961). Urinary excretions increase with ACTH stimulation and metyrapone suppression, and decrease with dexamethasone suppression (Maeyama & Nakagawa 1970), but the change with ACTH suggests a plasma-urine dissociation.

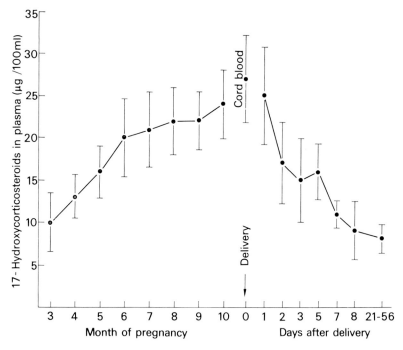

FIG. 9.7. The mean level (and standard deviation) of plasma Porter-Silber chromogens at monthly intervals throughout pregnancy and during the puerperium (Bayliss *et al* 1955). Thirty normal pregnancies were followed to term; fifteen cord bloods were obtained; follow-up during the puerperium was incomplete.

In general, less cortisol is present in cord blood than maternal (see Table 9.5); the average ratio is 1:3 and it is unchanged by ACTH, or becomes smaller. In anencephaly, the ratio is the same as in normal children (Nichols *et al* 1958); as the foetal cortex is atrophic the cord cortisol is of maternal origin. Cortisol and corticosterone transfer from the maternal to the foetal circulation, and isotope-labelled cortisol transfers from the amniotic fluid to the maternal circulation. After maternal infusion of cortisol the gradient of concentrations was 203:111:79 μg/100 ml for maternal: cord venous: foetal heart plasma (Leyssac 1961). Corticosterone levels are relatively increased in the cord blood and in the newborn compared with the adult levels, and proportionately to cortisol. Although maternal blood levels of 'cortisol' are from 2 to 5 times higher (average 3 times) than the cord blood (Aarskog

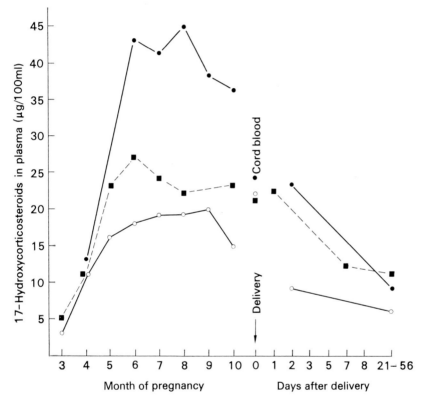

FIG. 9.8. Three examples of individual variation of plasma Porter-Silber chromogens (derived as for Fig. 9.7) at monthly intervals throughout pregnancy, with associated cord blood levels.

1965), a specific method produced average maternal cortisol levels of 47·4 and 52·4 μg/100 ml corresponding to 9·2 and 7·8 μg/100 ml, respectively, in cord plasma (Bro-Rasmussen *et al* 1962, Hillman & Giroud 1965) indicating a greater disparity. Cortisol levels on either side of the placenta (although placental sinus blood is similar to, but no higher than, maternal) reflect the difference in transcortin levels in the two circulations; in the maternal circulation, transcortin is elevated but it is low in foetal plasma, and albumin is the more important binding protein. Significantly, although total levels of cortisol are different in the two circulations, the free or diffusible moiety has the same concentration (Aarskog 1965), suggesting that the levels of active hormone are

TABLE 9.5. Plasma levels of steroids at delivery and to first week of life

Steroid	Sample source	Plasma concentration μg/100 ml	Author(s)
A Cortisol	Cord	38[a]	
	Maternal	48[a]	Bayliss et al[1] (1965)
	Cord	13–25[b]	Aarskog (1965)[2]
		12·9[b]	Hughes et al (1962)
	Cord	7·8[c]	
	Maternal	52·4[c]	Hillman & Giroud (1965)
	Cord	9·2[c]	
	Maternal	47·4[c]	Bro-Rasmussen et al (1962)
	0–7 days	0–18·8[a]	Bertrand et al (1963)
		5–30[b]	Franks (1967)
		1·1–7·4[c]	Hillman & Giroud (1965)
		2·9–8·1[a]	Hughes et al (1962)
B Cortisone	Cord	13·5[c]	Bro-Rasmussen et al (1962), Hillman
	Maternal	5[c]	& Giroud (1965)
C Cortico-sterone	Cord	5·8[b]	
	0–7 days	6·3[b]	Hughes et al (1962)

[1] Specimens obtained at caesarean section.
[2] Natural delivery.
[a] Porter-Silber chromogens. [b] fluorogenic steroids. [c] Specific estimates (isotopic dilutional method).

at equilibrium on either side of the placenta, because of the relative binding capacity of the two transcortins. Plasma levels of cortisone in the maternal circulation average close to 4 μg/100 ml at term but cord levels of cortisone exceed those of cortisol and are higher than maternal levels; there is a different metabolism of cortisol in the foetus and infant compared with later life. The ratio of cortisol to cortisone is 0·7:1 in mixed (arterial and venous) cord blood and is the reverse of the maternal ratio (approximately 11:1) and this could result from 'milking' of venous blood at delivery. In cord plasmas, cortisol levels are the same (7 µg/100 ml) in both artery and vein; but cortisone levels are lower in the artery (10·5 µg/100 ml) than the vein (14 µg/100 ml), as expected from a placental source. This shift in cortisol:cortisone equilibrium results from placental 11β-hydroxy dehydrogenase activity; the placenta simply 'supplies' the foetus with inactive cortisone until cortisol activation occurs.

Urinary corticosteroids and metabolites

The urinary excretions of 17-hydroxycorticosteroids (Norymberski chromogens) increase with advancing pregnancy: many individual excretions are in the normal range but progressively increase from the 28th week. 21-deoxyketols (Norymberski chromogens) and 17-ketosteroids also increase. 17-hydroxycorticosteroid excretions can decrease in some pregnancies during the last months. Excretions of Porter-Silber chromogens decrease during the first 6 months and return to non-pregnant levels at term (Migeon *et al* 1968). The excretions of both free cortisol and cortisone increase throughout pregnancy, but their tetrahydro metabolites do not, and more free steroid is excreted after ACTH (Cohen *et al* 1958). In the third trimester, metabolite increase after ACTH is less than in the non-pregnant (indicating some plasma-urine dissociation).

The ratio for 17-hydroxycorticosteroids to 17-deoxycorticosteroids (Normymberski chromogens) also decreases in the late months of pregnancy, corresponding with increased excretions of corticosterone and its tetrahydro metabolites in late pregnancy, compared with cortisol and its metabolites; the cortisol/corticosterone metabolite ratio at term is about 3:1, compared with 5:1 in the non-pregnant. In this regard, corticosterone sulphate passes the placenta without hydrolysis and with neutral disulphates is excreted in urine. 6β-hydroxycortisol occurs in increased amount in pregnancy urine (3–5 times normal, or 13–22% of the metabolite fraction) and the excretions of free and more polar (water-soluble) forms of Porter-Silber chromogens increase. Excretions of 11-deoxycortisol (Compound S) and its tetrahydro metabolite increase during pregnancy, and tetrahydro-S decreases with intra-uterine death (Dässler 1970).

P

Foetal metabolism of cortisol is different from the maternal using the newborn metabolism for comparison. Steroid conjugation is defective so that more free forms are excreted; polar (water-soluble) forms, and especially 6β-hydroxycortisol, are increased and large amounts of 20-dihydro metabolites occur due to deficient reduction of Δ^4-3-ketone groupings. Corticosterone and 11-dehydrocorticosterone levels are proportionately high. In the foetus, enzymatic steroid hydroxylations, which are appropriate for foetal existence, occur at C-6, C-11, C-15 and C-21, persist to term and into the newborn period; likewise, enzymatic sulphations persist, affecting particularly corticosterone and Δ^5-steroid metabolism. These latter steroids are found in high levels in cord plasma, in neonatal urine, indicating a quantitative deficiency at birth of the dehydrogenase needed for Δ^5-steroid conversion to Δ^4-forms. It is quantitative, not qualitative, as otherwise the steroids essential for life could not be synthesized and the newborn could not withstand birth stresses. The foetal adrenal cortex synthesizes corticosteroids from placental progesterone, with provision for synthesis from acetate before progesterone is withdrawn, and contains enzymes appropriate for these syntheses. The extent of actual interchange between foetal and maternal corticosteroids is unknown; a placental barrier exists to the transfer of dehydroepiandrosterone sulphate but not to corticosterone sulphate, although free steroids of the several series transfer unhindered.

The plasma half-life of cortisol is increased in pregnancy and becomes normal after delivery (Cohen *et al* 1958) but its urine clearance is unaltered (Thomas & Flynn 1964). Estimation of cortisol secretion rates in pregnancy has inherent difficulties for the body compartments are not identical with those of the non-pregnant state, and specific activities of marker steroids may be different. Also, there is a significant foetal output. Production rates have not appeared elevated above the non-pregnant in general and secretion rates have not appeared altered.

Cortisol production rates decrease in mid-pregnancy, increase with labour and correspond with Porter-Silber chromogen values. These relate well to the increased plasma concentrations associated with the prolonged plasma half-life and altered metabolism; a greater fraction of cortisol passes unconjugated into the urine, less as glucuronides, and the total metabolite excretions are decreased or normal at most. The lowered production rate and decreased metabolite excretion is similar to changes found with oestrogens, and lowered production occurs with norethinodrel (Layne *et al* 1962).

In pregnancy, cortisol disposition into tetrahydrocortisone is reduced to less than half the non-pregnant amount. Plasma cortisol is retained in the circulation by the increased transcortin and its metabolism is slowed and altered, as increases in certain metabolites show. Radioactivity appearance in urinary steroids during studies suggests slight over-production of cortisol occurs (Cope & Black 1959, Jayle & Pasqualini 1965) but production rates in general do not confirm this (Migeon et al 1968). Cushing's hyperplasia is not found in pregnancy and the increase in zona fasciculata and zona reticularis does not suggest ACTH stimulation, although there seems no definitive study of corticotrophin levels in pregnancy.

Controversy exists concerning the actual amount of free cortisol present in pregnancy. Oestrogen treatment is not an equivalent of pregnancy but is followed by the expected increases in transcortin, the concentrations of free cortisol remaining normal; likewise, norethinodrel is associated with transcortin increases (Doe et al 1969). However, some techniques of estimation suggest free cortisol increases and, in this way, pregnancy seemingly has either normal or elevated levels of free cortisol (Doe et al 1969, O'Connell & Walsh 1969). The discrepancy is a methodological one, normal levels being found if albumin-binding is not allowed for. In pregnancy, the circadian rhythm for cortisol persists and free cortisol levels are increased both at 9 a.m. and 9 p.m., although the morning level is not significantly increased compared with oestrogen therapy in women but the night level is. Significantly, the elevation of free cortisol correlates with a cortisol-type amino-aciduria that occurs in pregnancy and changes in steroid concentration found in parotid fluid. These changes are best demonstrated in pooled pregnancy plasma: gradual increases occur in pregnancy so that in the third trimester free cortisol levels are three times normal (Rosenthal et al 1969) but the percentages of transcortin and albumin bound cortisol and free cortisol do not alter, remaining similar to normal.

In diabetics receiving cortisol, oestrogen therapy increases glucosuria and leucocytosis, measures of cortisol activity that suggest increases in free cortisol. Oestrogens increase transcortin levels and progesterone competitively displaces cortisol but the changes are not sufficient to explain the increases in free cortisol levels; in pregnancy, the progesterone-transcortin complex may equal one-half cortisol-transcortin. Elevation of free cortisol levels suggests resetting of the hypothalamic-pituitary-adrenal axis for pregnancy and cortisol secretion is normal or reduced. At the same time, cortisol inactivation by the liver is lowered and differences between intravascular and extravascular transcortin could be such that only normal amounts of free cortisol are presented to the tissues, making cortisol effects improbable in pregnancy. Much transcortin is extravascular and oestrogens do not increase the extravascular cortisol compartment or its diffusion volume, although ACTH does. Thus, to consider all cortisol changes to be transcortin effects, all transcortin changes to be oestrogen effects seems unsound particularly as other parameters may be altered by oestrogens and progesterone, and other hormones.

Plasma cortisol levels increase with oestrogenic substances, singly or in combination, in parallel with increases in transcortin (and its half-life remains constant). With deliberate use of oestrogens plasma cortisol levels can be made to increase greatly and corticosterone increases in parallel, but to a lesser extent; urinary cortisol and Porter-Silber chromogens decrease; ACTH produces a relatively greater increase in plasma 'cortisol' levels; the plasma half-life of cortisol and corticosterone are prolonged, and that of the dihydro- and tetrahydro metabolites of cortisol, but that of cortisone is unchanged. Total corticosterone levels increase in pregnancy with transcortin binding but the cortisol to corticosterone ratio is unaltered from the non-pregnant. In general, the oestrogen effect appears contradictory for it augments total plasma cortisol levels and seemingly reduces the free fraction, although indirect indices of cortisol activity suggest an increase; the miscible pool for cortisol is increased, doubled or more, and the rate of cortisol renewal is decreased; the cortisol secretion rate is much reduced. Other descriptions of the effect, using the metabolism rate constant or metabolic clearance rate, confirm that turnover is less.

Pregnancy was once termed 'a physiological Cushing's syndrome', an issue that continues to interest some: the similarities between the two states, one physiological and the other pathological, are resistance to insulin, lymphopenia and eosinopenia, amino aciduria and glucosuria, amelioration of rheumatoid arthritis and pulmonary sarcoidosis, among others. Dissimilarities are that minor increases only are detectable in free cortisol levels in pregnancy, that cortisol levels maintain a circadian rhythm and that a resistance to the metabolic activity of cortisol may be evident in pregnancy. Nonetheless, in some pregnancies clinical features suggest the possibility of adrenocortical hyperfunction; 'it has been very difficult to give an answer without implications' (Jayle & Pasqualini 1965).

ALDOSTERONE

Urinary aldosterone excretions increase during normal pregnancy and decrease in the toxaemias of pregnancy, although in some patients with toxaemia there may be an increase. Excretions of free aldosterone and its 18-glucuronide conjugate increase throughout pregnancy, some values remaining within normal limits and occasioning doubts about electrolyte intakes, but free aldosterone does not show this change and becomes less. Aldosterone 'secretion rate' studies give variable results: in some pregnancies they increase in the late stages but, allowing for much variability, considerable increases are evident by the 15th week and maintained to the last month when further increases occur. Even with a constant, high sodium intake, secretion rates increase in the last trimester but are most variable in the same patient.

In pregnancy, with allowance for those features that alter aldosterone levels—posture, respiratory movements, renal perfusion, oedema, among others—aldosterone levels should have considerable variability and its absence would be unusual. However, with a constant, high sodium diet, although progressive increases in secretion rate occur especially after the 32nd week, a small number of women fail to show increases; this cannot be correlated with other factors. The secretion rates show an expected increase with lowered sodium intake and decrease with augmented intake. In mild pre-eclampsia, secretion rates are mostly normal for pregnancy or slightly reduced; in severe pre-eclampsia, the secretion rates are reduced and there is a slight reduction in sodium excretion (Sims et al 1964), but there is no correlation between sodium excretion and abnormal aldosterone secretion in toxaemia.

The metabolic clearance rate for aldosterone is similar to the non-pregnant value; although a minor alteration in the pattern of metabolism occurs from an oestrogen effect upon the liver, the liver 'handles' aldosterone normally. In oestrogen therapy, the metabolism rate constant and metabolic clearance rate decrease while the secretion rate and plasma levels increase. In pregnancy, there is no evidence of increased aldosterone activity although sodium retention and oedema occur more easily than in the non-pregnant. The relationship between aldosterone and oedema is 'permissive'; pregnancy is associated with some resistance to the effect of aldosterone, even with adrenocortical insufficiency. Increased aldosterone production is a homoeostatic mechanism, incidental to sodium metabolism, and foetal and gestational demands. However, the natriuretic effect of progesterone may be factor, through direct antagonism or indirectly by a renal tubular effect; the increase could be compensatory for the increased delivery of sodium to the tubules as a result of the increased glomerular filtration of pregnancy. Progesterone increases aldosterone secretion and excretion in both pregnancy and the non-pregnant, and increases sodium excretion.

Toxaemia of pregnancy

Urinary corticosteroid excretions differ little from normal pregnancy in toxaemia, assessed by 17-hydroxycorticosteroid (Norymberski chromogens) or Porter-Silber chromogen levels, except that in severe toxaemias the excretions fall. Plasma cortisol levels are little different from normal pregnancy; cortisol excretions are normal in mild toxaemias but much decreased in severe toxaemia, and not from renal factors. In severe toxaemias, aldosterone secretion is reduced, but the excretion of its 18-glucuronide is unaltered, and its levels are very variable; renal factors may be involved in the 'improved' sodium retention and consequent decreased aldosterone secretion and explain the reduction of urinary 18-glucuronide in some instances. Although biological assays and Na/K ratios indirectly suggested an increase of free aldosterone in toxaemia, aldosterone antagonists do not affect sodium excretion.

17-KETOSTEROIDS IN PREGNANCY

Plasma 17-ketosteroids

Plasma levels of 17-ketosteroids fall during pregnancy and the levels at term are much lower than those of the non-pregnant and less than in cord plasma. High values in early pregnancy fall to low levels before term, increase at delivery then fall to very low levels in the post-partum period, subsequently rising gradually to non-pregnant levels over some months. These features are due mainly to dehydroepiandrosterone and its sulphate, dehydroepiandrosterone and 16α-hydroxydehydroepiandrosterone levels being lower in maternal than cord plasma; arterial cord plasma values are higher than the venous, with a greater difference for the 16α-hydroxylated form. The increased maternal plasma levels of 17-ketosteroids found at delivery are due to transfer from the foetal compartment. Production rates of dehydroepiandrosterone and its sulphate into the maternal compartment increase during pregnancy:

as this is a measure of the total amount of the steroids entering the blood, irrespective of secretion or peripheral synthesis, they could derive from the foetal compartment, which could also utilize steroids. Foetal dehydroepiandrosterone was transferred to the maternal circulation of an adrenalectomized woman after the 17th week (Siiteri & MacDonald 1967). Plasma levels of the sulphate show considerable individual variation but decrease consistently from early pregnancy (by the 10–14th week) to term without consistent increases at any time and the decrease corresponds with its utilization by the placenta for conversion to oestrogens. Androsterone sulphate levels decrease more than those of dehydroepiandrosterone (Hankin 1967), correlating with increased production of dehydroepiandrosterone sulphate.

Urinary 17-ketosteroids

In pregnancy, minor increases occur in the urinary excretion of total 17-ketosteroids, although scarcely evident in many patients. Methodological differences exist but 20-ketosteroids cause some of the apparent increase through colorimetric interference. Androsterone and aetiocholanolone excretions decrease, dehydroepiandrosterone excretion is unaltered except for increases in early pregnancy, and at this stage excretions increase considerably with ACTH. Dexamethasone suppresses 17-ketosteroid excretions to low levels as expected with adrenal origins and ACTH stimulation and metyrapone inhibition have expected effects.

ABNORMAL ADRENOCORTICAL STATES AND PREGNANCY

Adrenocortical insufficiency

Pregnancy may occur successfully in Addison's disease without any treatment, but the morbidity is high: with full replacement therapy, or cortisone and fludrocortisone as required, successful pregnancies without morbidity occur in Addison's disease and following adrenalectomy. Such experiences emphasize that the adrenal cortex is essential for the normal growth of the placenta and foetus, in contra-distinction to earlier views, and emphasize the need to appropriately alter treatment for delivery. Pregnant Addisonian patients lack cortisol and aldosterone: urinary aldosterone is absent, as it is after adrenalectomy, but some patients with the partial form excrete the 18-glucuronide, and tetrahydro metabolites are recognizable in some instances. This situation is similar to that for cortisol, which is absent, but tetrahydro metabolites are recognizable. In unrecognized Addison's disease restriction of sodium intake near term has caused deterioration.

Adrenocortical hyperfunction

Cushing's syndrome produces usually anovulation and infertility; if conception occurs abortion, foetal death and premature labour are common, particularly with large cortisol increases and androgen excess. Pregnancy may occur successfully, although often with toxaemia but whether causally or fortuitously is unknown. Pregnancy may occur normally in a woman whose adrenal androgen production is excessive, with clitoral enlargement in the child. In general, hyperfunction is associated with pregnancy complications, but mostly with infertility or sterility, as with high glucocorticoid dosage.

TRANSPLACENTAL TRANSFER OF STEROIDS

Cortisol crosses the placenta in either direction with an equilibrium for the free fraction. A concentration gradient exists across the placenta for total cortisol levels in response to the transcortin levels in the two circulations. Foetal binding of cortisol is predominantly a function of albumin and this situation of a high affinity maternal binding protein opposed to low affinity foetal binding protein enables the mother to 'capture' cortisol from the foetus. The significant difference between cortisol concentrations in the circulations is one aspect of this, like the equilibrium for the free fractions across the placenta. Cortisol probably does not spill-over into the foetal circulation in significant amounts until delivery. Foetal and maternal distribution of plasma glucocorticoids is shown in Tables 9.5 and 9.6.

Cortisol by crossing the placenta probably suppresses the foetal adrenal (Dickey & Thompson 1969) but foetal cortisol secretion cannot be greatly different from the newborn, and foetal cortex preparations respond to ACTH stimulation. Some mothers treated with glucocorticoids have given birth to infants with hypoplastic adrenals (Lanman 1961), although mostly cortisol secretion is unimpaired. Similarly, glucocorticoids suppress cord levels of dehydroepiandrosterone sulphate and the 16α-hydroxylated form, with depression of urinary oestrogen excretions; as ACTH is detectable at 16 weeks or later, possible depression could follow. Corticosterone and 11-deoxycorticosterone sulphates occur in higher amount in the foetal plasma, like cortisol levels, and corticosterone may be hindered in placental transfer; free corticosterone is more significant than the sulphate in maternal plasma. The disparity between the concentrations in the two circulations is obvious in high cord to maternal plasma ratios. Cortisone levels are probably a function of placental 11-hydroxysteroid dehydrogenase during transfers. Inactivation of cortisol could be protective in the same way that sulphation prevents 11-deoxycorticosterone exerting mineralocorticoid activity.

TABLE 9.6. Plasma concentration of cortisol and cortisone in cord* (C) and maternal (M) plasma as μg/100 ml plasma (Schweitzer et al 1969)

	Cortisol		Cortisone	
	C	M	C	M
Mean	3·6	34·4	9·0	6·7
Standard deviation	1·5	4·3	1·4	1·6

The ratios for C/M were 0·05–0·23 for cortisol and 1·03–1·80 for cortisone, respectively.

* Samples obtained from distal end of excised cord.

METABOLIC ACTIONS OF CORTICOSTEROIDS

Glucocorticoids

The actions of the hormones may be pharmacological or physiological. 'Pharmacological' actions refer to effects observed when steroids are used in amounts that are greater than exist under natural or physiological conditions. 'Physiological' actions refer to natural actions or, in clinical usage, replacement dosage effects. There appears no single hormonal effect that can satisfactorily explain the responses to cortisol-like steroids. Two concepts, that of the hormone-receptor and that of an action upon single enzyme systems, seem inadequate to explain the hormone actions. The concept of a glucocorticoid action comprehends diverse effects of the hormones and 'physiological' action may be conceived as 'permissive' or 'regulatory', or both.

Molecular structure of metabolically active glucocorticoids suggests that potent pharmacological agents have the following structural features, e.g. Δ^4-3-ketone grouping, a 17β-side chain, and an 11β-hydroxyl group. Modifications enhancing potency are well known. Relationship of structure to biological activity (Bush 1962) suggested the importance of the 11β-hydroxy group, and cortisol is the biologically active form, not cortisone. Retention of biological activity requires an α,β-unsaturated ketone and cortical side chain, the 20-keto-21-hydroxy grouping. Although structure and its modifications may allow the steroids to relate to hypothetical receptor sites, knowledge of the relationship between structure and function is incomplete.

General effects

The observation that nitrogen excretion decreased in fasting rats after adrenalectomy and returned to normal with adrenocortical extracts, established the relevance of the adrenal cortex to protein metabolism. The extracts also increased nitrogen excretion in normal fasting rats and evidence of protein catabolic effects included loss of carcass weight, thinning of integument, fall in antibody titre or lack of immune response and involution of the thymus. The potency of a cortisol analogue can be assessed by a protein catabolic action and it has been assumed that catabolic activity correlates with anti-inflammatory action.

Action upon enzymes

Glucocorticoids may affect regulation of gluconeogenesis by at least two mechanisms: a rapidly acting 'fine control' involving feedback or substrate effects and a different, slower 'coarse control', arising from alteration in the rate of enzyme synthesis. The anti-inflammatory effects of the glucocorticoids may involve some effect on lysosomal enzyme systems but the relationship between the anti-inflammatory effect and enzyme systems is not sure. Glucocorticoid administration results in altered activity of enzymes (Knox *et al* 1956), an increased synthesis of enzyme protein in some circumstances and evidence of effect in more diffuse ways. Activity of some enzymes involved in carbohydrate metabolism (glucose-6-phosphatase, fructose-1,6-diphosphatase) appears to increase in pharmacological experiments. An alteration of enzyme activity in amino-acid metabolism appears to occur but reviewers comment upon difficulties in the data. In acute experiments, glucocorticoids increase hepatic production of urea. Uptake of amino-acids may precede the alteration in urea production suggesting an increased peripheral catabolism of proteins with transport of amino acids to the liver and their rapid metabolism there. A rapid entry of amino acids into the liver may account for some metabolic effects of the steroids: in the liver, amino acids not only contribute to urea but also glucose, and increased glucose production appears to precede changes in nitrogen metabolism. Corticosteroids increase liver glycogen consistently and sufficiently sensitively to test the potency of new synthetic forms. A primary action of cortisol upon the liver has been suggested with the increased peripheral breakdown of protein a consequence of an increased uptake of amino acids. However, in eviscerate rats, plasma amino-acid levels increase after steroid administration.

Protein metabolism

Although protein depletion occurs in most body tissues, nitrogen content increases in some. In rats, cortisone administration increases total liver protein, that of the gastrointestinal tract and urogenital organs, and total plasma albumin but depletes the remainder of the carcass. There is a gain in liver weight due to both carbohydrate and protein, with a cytoplasmic gain. There is an increased amino-acid incorporation into protein of the cells and an increased protein synthesis due to microsomal changes. Earlier workers considered the probability of a translocation of protein and, with its peripheral catabolism by steroids, relevant amino acids could be transported and taken up by other tissues. Cortisone increases liver ribonucleic acid and protein, and changes in the former precede the latter by several hours in acute experiments. There is enzyme induction and/or synthesis but the effects of corticoids on blood glucose and liver glycogen precede the apparent changes in enzyme activity and synthesis of ribonucleic acid.

Immunological experiments suggest that the hormones stimulate specific enzyme systems, effects mediated by nucleic acids. In rats, an increase in formation of specific messenger ribonucleic acid followed 1–2 hours after the appropriate transferase enzyme showed increased activity, some 30 minutes after cortisol injections, and there was a subsequent increase in protein synthesis. This suggested that the hormone action was upon the nucleus to activate specific ribonucleic acid

(Lang & Sekeris 1964), with formation of enzyme proteins. The appropriate types of liver ribonucleic acid (precursor, messenger and transfer) show increased rates of formation with the hormones. Except at the periphery, where catabolic effect is exerted with inhibition of protein synthesis, the corticoid action is probably to regulate the synthesis of messenger ribonucleic acid. Induction of tryptophane pyrrolase has been shown in human liver but a simple interpretation of enzyme control may not apply to human tissues, such as liver and adipose tissue. Also, the possibility that the equilibrium for cortisol/cortisone interconversion differs with different tissues (Bailey & West 1969) requires evaluation. That adrenalectomy impairs amino-acid incorporation into protein emphasizes the role of corticoids in optimal protein synthesis.

Small local application of corticoids *in vivo* produce cytological changes (Bush 1962) that suggest cellular sensitivity to their actions. Many clinical features are terminal or gross from protracted or intense corticoid therapy. The effect of corticoids on protein metabolism is debated as to its relationship with non-nitrogenous metabolism, whether it is protein catabolism, inhibition of anabolism or both, and whether it is peripheral or hepatic. ACTH will produce negative nitrogen balance in man, likewise cortisone or cortisol, and this may be diminished by a high calorie, high protein, high carbohydrate diet. Glucocorticoid amino-aciduria occurs, serum concentrations of globulin decrease, although that of albumin increases proportionately with corticoid dosage. In Cushing's syndrome, the total exchangeable albumin pool is lower than normal, degradation rates of albumin are normal and synthesis increased, suggesting an increased turnover (Sterling 1960, Hoffenberg & Black 1963), but results are not consistent.

Carbohydrate metabolism

Corticoid administration in the adrenalectomized rat increases blood glucose levels and conversely experimental diabetes is ameliorated by adrenalectomy. Although blood glucose increases from decreased utilization or increased liver production, cortisone increases production markedly in rats. Liver slices from animals, previously chronically treated with corticoids show increased glucose production from pyruvate. Addition *in vitro* is not so successful although triamcinolone and dexamethasone (both relatively resistant to liver inactivation) increase glucose synthesis from pyruvate and alanine. Increased glucose release probably contributes to an abnormal glucose tolerance in man; also, normal glucosuria increases proportionately with steroid dosage, likewise renal glucosuria. In Cushing's syndrome, this is not accounted for by protein breakdown and urinary nitrogen loss. In fasting, adrenalectomized rats, liver glycogen increases after corticoid administration and the increased non-protein nitrogen excretion more than corresponds to the carbohydrate formed. The steroid effect occurs within a few hours and much of the labelled hormone is excreted before its effect is evident. The mechanism whereby corticoids promote liver glycogen deposition is not settled and an interaction of factors appears likely. The reduction of corticoid effect in pyridoxine deficiency, the reduction with puromycin and actinomycin have been interpreted as either supporting or negating involvement of enzyme synthesis. None the less, cortisol may decrease degradation of liver glycogen.

Increased glucose production takes place in the rat kidney after cortisone treatment. In man, there is an increased glomerular filtration, which explains part of the increased glucosuria, but tubular reabsorptive capacity is unaltered.

There is no clear-cut evidence that corticoids inhibit glucose utilization but some to suggest that glucose utilization is enhanced in man, as in the rat. Glucose utilization may not alter with cortisol administration and may be increased in Cushing's syndrome (Pupo et al 1966). Here, plasma insulin levels are elevated and there may be increased utilization on this basis although insulin output is inadequate for the glucose levels. The diabetic state secondary to steroids can return to normal despite their continued use and there is a relative resistance or antagonism to insulin in 'steroid diabetes' and Cushing's syndrome. Glucose phosphorylation in heart muscle is only partly regulated by corticoids but glucose utilization by adipose tissue is reduced by corticoid treatment and addition *in vitro*, although some early effects can be overcome by insulin. Corticoids also decrease the rate of re-esterification in adipose tissue.

Large amounts of glucocorticoids have long been known to reduce glucose tolerance in animals and man and ACTH stimulation has the same effect. Strangely, large intravenous doses of cortisol in ill Addisonian patients do not increase plasma glucose levels. In normal patients, prednisone and cortisone decrease glucose tolerance in the intravenous test within 2 hours and the oral, cortisone or prednisone tolerance tests are based upon similar data. They are used to detect impairment in glucose tolerance or the increase in glucosuria after steroid administration and the detection of early diabetes mellitus. Whether corticoids impair glucose tolerance or produce relative insulin deficiency in certain individuals and thereby reduce glucose utilization, remains debatable.

Lipid metabolism

Hepatic lipogenesis is depressed in cortisone treated animals. Adrenalectomy reverts lipogenesis and liver synthesis of long-chain fatty acids towards normal in diabetic animals but cortisol inhibits fatty acid synthesis in the adrenalectomized preparation. This inhibition occurs several hours after its administration and at the time of pyruvate-carbon incorporation into glucose. These effects are not found in isolated adipose tissue where corticoids affect fatty acid release, decrease the rate of re-esterification, and inhibit glucose-carbon incorporation into fatty acids and depress glucose uptake and glucose-carbon incorporation into carbon dioxide. High corticoid concentrations are lipolytic, increasing glycerol and free fatty acid release. Fatty acid synthesis in adipose tissue is related to glucose utilization and insulin addition without glucose does not stimulate pyruvate-carbon incorporation into fatty acids, although glucose increases lipogenesis. Corticoids are involved in the mobilization of fatty acids from adipose tissue. Adrenaline does not mobilize the expected amount of fatty acids in an adrenalectomized animal, but cortisone restores the response. Glucocorticoids may alter glucose utilization through release of fatty acids, according to the glucose-fatty acid cycle. Liver production of aceto-acetate depends upon an intact adrenal cortex and the presence of insulin. Glucocorticoid has its effects at the periphery with delivery of fatty acids to the liver, rather than a direct effect.

Glucocorticoid effects upon adipose tissue are evident from the observation of patients with Cushing's syndrome. Glucocorticoids are anti-ketogenic, as indicated, but ACTH is ketogenic, through extra-adrenal adipokinetic effects (Lebovitz & Engel 1964), so that Cushing's syndrome may not be a pure example of glucocorticoid effect. Serum cholesterol is raised in many patients with Cushing's syndrome and with significant corticoid dosage.

Diabetes mellitus in Cushing's syndrome may differ from the primary form, and potassium levels can modify 'steroid diabetes'. Also, inhibition of pyruvate utilization increases its availability and the response to risk, medico-legal connotations cannot be disregarded when using glucocorticoids if parents are doubtful about deformity. Protein-binding of cortisol and analogues probably gives some protection to the foetus, and the loss of corticoid in the liver as it is absorbed reduces the apparent dosage, or delivery to the circulation. In this regard, 20–40 mg prednisone and 15 mg triamcinolone daily have been trouble-free and children have had normal secretion rates. In one review (Bongiovanni & McPadden 1960), only one instance of presumptive perinatal failure occurred in 260 pregnancies, although there are other instances, some with adrenal pathology. Hypoplasic adrenals are a possibility (Lanman 1961) but maternal oestriol excretions could check this prospectively.

REFERENCES*

AARSKOG D. (1965) Cortisol in the newborn infant. *Acta paediat. Scand.* Suppl. 158.
AMATRUDA T.T., HOLLINGSWORTH D.R., D'ESPOSO N.D., UPTON G.V. & BONDY P.K. (1960) A study of the mechanism of the steroid withdrawal syndrome. Evidence for integrity of the hypothalamic-pituitary-adrenal system. *J. clin. Endocr.* **20**, 339–354.
BAILEY E. & WEST H.F. (1969) The secretion, interconversion and catabolism of cortisol, cortisone and some of their metabolites in man. *Acta endocr. (Kbh.)* **62**, 339–359.
BAIRD D.T., HORTON R., LONGCOPE C. & TAIT J.F. (1969) Steroid dynamics under steady-state conditions. *Rec. Progr. Horm. Res.* **25**, 611–656.
BARTTER F.C. (1963) Regulation of the volume and composition of extracellular and intracellular fluid. *Ann. New York Acad. Sci.* **110**, 682–703.
BAYLISS R.I.S., BROWNE J.C.McC., ROUND B.P. & STEINBECK A.W. (1955) Plasma 17-hydroxycorticosteroids in pregnancy. *Lancet* **i**, 62–64.
BEAS F., ZURBRUGG R.P., CARA J. & GARDNER L.I. (1962) Urinary C_{19} steroids in normal children and adults. *J. clin. Endocr.* **22**, 1090–1094.
BERTHOLD K., ARIMURA A. & SCHALLY A.V. (1970) Comparison of effects of 6-dehydro-16-methylene-hydrocortisone (STC407) and dexamethasone on the suppression of the hypothalamo-pituitary-adrenal system. *Acta endocr. (Kbh.)* **63**, 423–430.
BERTRAND J., LORAS B., GILLY R. & GAUTENET B. (1963) Contribution à l'étude de la sécrétion et du métabolisme du cortisol chez le nouveau-né et le nourrison de moins de trois mois. *Path. Biol., Paris* **11**, 997–1022.
BIGLIERI E.G., SHAMBELAN M. & SLATON P.E. (1969) Effect of adrenocorticotrophin on desoxycorticosterone, corticosterone and aldosterone excretion. *J. clin. Endocr.* **29**, 1090–1101.

* Included are general references and those specifically cited in the text.

BIRD C.E., WIQUIST N., DICZFALUSY E. & SOLOMON S. (1966) Metabolism of progesterone by the perfused previable human fetus. *J. clin. Endocr.* **26**, 1144–1154.
BIRKE G., GEMZELL C.A., PLANTIN L.O. & ROBBE A. (1958) Plasma levels of 17-hydroxycorticosteroids and urinary excretion pattern of ketosteroids in normal pregnancy. *Acta endocr. (Kbh.)* **27**, 389.
BOJESEN E. (1964) Aldosterone in peripheral plasma of normal man, in E.E.Baulieu & P. Robel (eds.) *Aldosterone*, pp. 163–167. Oxford, Blackwell.
BONGIOVANNI A.M. & MCPADDEN A.J. (1960) Steroids during pregnancy and possible fetal consequences. *Fertil. & Steril.* **11**, 181–186.
BRODIE A. & TAIT J.F. (1968) Assay of aldosterone and related compounds, in R.I.Dorfman (ed.) *Methods of Hormone Research*, 2nd ed., pp. 323–382. New York, Academic Press.
BRO-RASMUSSEN F., BUUS O., LUNDWALL F. & TROLLE D. (1962) Variations in plasma cortisol during pregnancy, delivery and the puerperium. *Acta endocr. (Kbh.)* **40**, 571–578.
BRO-RASMUSSEN F., BUUS O. & TROLLE D. (1962) Ratio cortisone/cortisol in mother and infant at birth. *Acta endocr. (Kbh.)* **40**, 579–583.
BROWN J.J., FRASER R., LEVER A.F. & ROBERTSON J.I.S. (1968) Renin and angiotensin in the control of water and electrolyte balance: relation to aldosterone, in V.H.T. James (ed.) *Recent Advances in Endocrinology*, 8th ed., pp. 271–292. London, Churchill.
BUSH I.E. (1962) Actions of adrenal steroids at the cellular level. *Brit. med. Bull.* **18**, 141–147.
BUSH I.E. (1962) Chemical and biological factors in the activity of adrenocortical steroids. *Pharmacol. Rev.* **14**, 317–445.
CAMERON E.H.D. (1969) Steroid biosynthesis in the human adrenal cortex, in K. Griffiths & E.H.D. Cameron (eds.) *The Human Adrenal Gland and its Relation to Breast Cancer*, pp. 47–55. Cardiff, Alpha omega alpha.
CATHRO D.M. (1969) Adrenal cortex and medulla, in D.Hubble (ed.) *Paediatric Endocrinology*, pp. 187–327. Oxford, Blackwell.
CHRISTY N.P. (1966) Adrenal cortical steroids in various types of hypertension in W.M. Manger (ed.) *Hormones and Hypertension*, pp. 169–187. Springfield (Ill.), Thomas.
CLAYTON B.E. (1968) Assessment of pituitary-adrenal function in children. *Mem. Soc. Endocr.* **17**, 237–246.
COHEN M., STIEFEL M. REDDY W.J. & LAIDLAW J.C. (1958) The secretion and disposition of cortisol during pregnancy. *J. clin. Endocr.* **18**, 1076–1092.
COOKE A.R., PRESHAW R.M. & GROSSMAN M.I. (1966) Effect of adrenalectomy and glucocorticoids on the secretion and absorption of hydrogen ions. *Gastroenterology* **50**, 761–767.
COPE C.L. (1965) *Adrenal Steroids and Disease*. London, Pitman.
COPE C.L. & BLACK E.G. (1959) The hydrocortisone production in late pregnancy. *J. Obstet. Gynaec.* **66**, 404–408.
DALLMAN M.F. & YATES F.E. (1968) Anatomical and functional mapping of central neural input and feedback pathways of the adrenocortical system. *Mem. Soc. Endocr.* **17**, 39–71.
DÄSSLER C-G. (1970) Die Ausscheidung von 11-Desoxycortisol und von Tetrahydro-

11-desoxycortisol in der normalen Schwangerschaft und bei intrauterinem Fruchtod. *Acta endocr. (Kbh.)* **63**, 437–440.

DAUGHADAY W.H. (1967) The binding of corticosteroids by plasma protein, in A.B.Eisenstein (ed.) *The Adrenal Cortex*, pp. 385–403. Boston, Little Brown.

DESHPANDE N., JENSEN V., CARSON P., BULBROOK R.D. & DOOUSS T.W. (1970) Adrenal function in breast cancer: biogenesis of androgens and cortisol by the human adrenal gland *in vivo*. *J. Endocr.* **47**, 231–242.

DICKEY R.P. & THOMPSON J.P. (1969) Effect of ACTH and metyrapone on estriol, 17-hydroxycorticosteroid, 17-ketosteroid, pregnanediol and pregnanetriol late in pregnancy. *J. clin. Endocr.* **29**, 701–706.

DOBBIE J.W., MACKAY A.M. & SYMINGTON T. (1968) The structure and functional zonation of the human adrenal cortex. *Mem. Soc. Endocr.* **17**, 103–111.

DOE R.P., DICKINSON P., ZINNEMAN H.H. & SEAL U.S. (1969) Elevated nonproteinbound cortisol (NPC) in pregnancy, during estrogen administration and in carcinoma of the prostate. *J. clin. Endocr.* **29**, 757–766.

DRAYER H.M. & GIROUD C.J.P. (1965) Corticosteroid sulfates in the urine of the human neonate. *Steroids* **5**, 289–317.

EBERLEIN W.R. (1965) Steroids and sterols in umbilical cord blood. *J. clin. Endocr.* **25**, 1101–1118.

EGDAHL R.H. (1968) Excitation and inhibition of ACTH secretion. *Mem. Soc. Endocr.* **17**, 29–37.

EISENBERG E. & GORDAN G.S. (1961) Skeletal dynamics in man measured by nonradioactive strontium. *J. clin. Invest.* **40**, 1809–1825.

EISENSTEIN A.B. (1967) Nutritional factors and the adrenal cortex, in A.B.Eisenstein (ed.) *The Adrenal Cortex*, pp. 315–344. Boston, Little Brown.

ESTEP H.L., ISLAND D.P., NEY R.L. & LIDDLE G.W. (1963) Pituitary-adrenal dynamics during surgical stress. *J. clin. Endocr.* **23**, 419–425.

FALUDE G., MILLS L.C. & CHAYES Z.W. (1964) Effect of steroids on muscle. *Acta endocr. (Kbh.)* **45**, 68–78.

FRANKS R.C. (1967) Diurnal variation of plasma 17-hydroxycorticosteroids in children. *J. clin. Endocr.* **27**, 75–78.

FRANTZ A.G. & RABKIN M.T. (1964) Human growth hormone. Clinical measurement, response to hypoglycaemia and suppression by corticosteroids. *New Engl. J. Med.* **271**, 1375–1381.

GANDY H. & PETERSON R.E. (1968) Measurement of testosterone and 17-ketosteroids in plasma by the double isotope dilution derivative technique. *J. clin. Endocr.* **28**, 949–977.

GARDNER L.I. (1956) Plasma neutral 17-ketosteroids in pregnancy, in L.I. Gardner (ed.) *Adrenal Function in Infants and Children*, pp. 205–209. New York, Grune & Stratton.

GOLDZIEHER J.W. & BEERING S.C. (1969) Metabolism of 11β-hydroxyandrostenedione, adrenosterone and hydrocortisone to urinary 11-oxy-17-ketosteroids. *J. clin. Endocr.* **29**, 171–178.

GRABER A.L., NEY R.L., NICHOLSON W.E., ISLAND D.P. & LIDDLE G.W. (1965) Natural history of pituitary-adrenal recovery following long-term suppression with corticosteroids. *J. clin. Endocr.* **25**, 11–16.

GRANT J.K., FOREST A.P.M. & SYMINGTON T. (1957) The secretion of cortisol and corticosterone by the human adrenal cortex. *Acta endocr. (Kbh.)* **26**, 195–203.
GRANT J.K., GRIFFITHS K. & LOWE M. (1968) Biochemical investigations of the actions of corticotrophins on the adrenal cortex. *Mem. Soc. Endocr.* **17**, 113–123.
GRAY M.J., STRAUSFELD K.S., WATANABE M., SIMS E.A.H. & SOLOMON S. (1968) Aldosterone secretory rates in the normal menstrual cycle. *J. clin. Endocr.* **28**, 1269–1275.
HANKIN M.E. (1967) Hormones in pregnancy with particular reference to the plasma 17-oxosteroids. PhD. Thesis, University of Adelaide, Australia.
HECHTER O. & HALKERSTON I.D.K. (1964) On the action of mammalian hormones, in G.Pincus, K.V.Thimann & E.B.Ashwood (eds.) *The Hormones*, vol. v, pp. 697–825. New York, Academic Press.
HELLMAN L., NAKADA F., CURTI J., WEITZMAN E.D., KREAM J., ROFFWARG H., ELLMAN S., FUKUSHIMA D.K. & GALLAGHER T.F. (1970) Cortisol is secreted episodically by normal man. *J. clin. Endocr.* **30**, 411–422.
HILLMAN D.A. & GIROUD C.J.P. (1965) Plasma cortisone and cortisol levels at birth and during the neonatal period. *J. clin. Endocr.* **25**, 243–248.
HOFFENBERG R. & BLACK E.C. (1963) Some observations on the metabolism of ^{131}I-albumin in Cushing's syndrome. *S. Afric. Med. J.* **27**, 114–117.
HUGHES E.R., SEELY J.R., KELLEY V.C. & ELY R.S. (1962) Corticosteroid levels before and after corticotropin. *Amer. J. Dis. Childh.* **104**, 168–173.
HUIS IN'T VELD L. (1954) L'excrétion des 17-cétosteroides au cours de la grossesse. *Gynéc et Obstét.* **53**, 42–56.
HUME D.M., BELL C.C. & BARTTER F.C. (1962) Direct measurement of adrenal secretion during operative trauma and convalescence. *Surgery* **52**, 174–187.
HUNT A.B. & MACONAHEY W.M. (1953) Pregnancy associated with diseases of the adrenal glands. *Amer. J. Obstet. Gynec.* **66**, 970–987.
JAMES V.H.T., LANDON J. & FRASER R. (1968) Some observations on the control of corticosteroid secretion in man. *Mem. Soc. Endocr.* **17**, 141–156.
JÄNNE O. & VIHKO R. (1970) Neutral steroids in urine during pregnancy. *Acta endocr. (Kbh.)* **65**, 50–68.
JAYLE M.F. & PASQUALINI J.R. (1965) Metabolism of adrenocortical hormones in the course of human pregnancy. *Europ. Rev. Endocrin.* Suppl. 1, pp. 127–153.
JEANRENAUD B. & RENOLD A.E. (1969) Metabolic effects of corticosteroids, in E.F.Pfeiffer *Handbuch des Diabetes Mellitus*, pp. 591–613. Munich, Lehmanns.
JENKINS J.S. (1968) The pituitary-adrenal response to pyrogen. *Mem. Soc. Endocr.* **17**, 205–212.
KAWAHARA H. (1958) Plasma levels of 17-hydroxycorticosteroids in umbilical cord blood, with special reference to variations in level between *A. umbilicalis* and *V. umbilicalis. J. clin. Endocr.* **18**, 325–327.
KENDALL J.W., LIDDLE G.W., FEDERSPIEL C.F. & CORNFIELD J. (1964) Dissociation of corticotropin-suppressing activity from the eosinopenic and hyperglycemic activities of corticosteroid analogues. *J. clin. Invest.* **42**, 396–403.
KIELMANN N., STACHENKO J. & GIROUD C.J.P. (1966) Production rate of corticosterone sulfate in normal human adults. *Steroids* **8**, 993–1025.

KNOX W.E., AUERBACH V.H. & LIN E.C.E. (1956) Enzymatic and metabolic adaptations in animals. *Physiol. Rev.* **49**, 164–254.

KONO T., YOSHIMI T. & MIYAKE T. (1966) Metabolic clearance of aldosterone, cortisol and corticosterone in various clinical conditions, in G.Pincus, T.Nakao & J.F.Tait (eds.) *Steroid Dynamics*, pp. 429–461. New York, Academic Press.

KOWARSKI A., LAWRENCE B., HUNG W. & MIGEON C.J. (1969) Interconversion of cortisol and cortisone in man and its effect on the measurement of cortisol secretion rate. *J. clin. Endocr.* **29**, 377–381.

KUMAR D., FELTHAM L.A.W. & GORNALL A.G. (1959) Aldosterone excretion and tissue electrolytes in normal pregnancy and pre-eclampsia. *Lancet i*, 541–545.

KUSAMA K., ABE O., SAKAUCHI N., TAKATANI O., MAYAMA T., DEMURA R. & KUMAOKA S. (1970) Extra-adrenal action of adrenocorticotrophin on cortisol metabolism. *J. clin. Endocr.* **30**, 778–784.

LANG N. & SEKERIS C.E. (1964) Zum Wirkung-mechanismus der Hormone, III: Einfluss von Cortisol auf den Ribonucleinsäure und Proteinstoff-wechsel in Rattenleber. *Z. physiol. Chem.* **339**, 238–248.

LANMAN J.T. (1961) The adrenal gland in the human fetus. *Pediatrics* **27**, 140–158.

LAYNE D.S., MEYER C.J., VAISHWANAR P.S. & PINCUS G. (1962) The secretion and metabolism of cortisol and aldosterone in normal and in steroid-treated women. *J. clin. Endocr.* **22**, 107–118.

LEBOVITZ H.E. & ENGEL F.L. (1964) Relationships between the structure and biological activities of corticotrophin and related peptides. *Metabolism* **13**, 1230–1246.

LEYSSAC P. (1961) Hydrocortisone in foetal plasma following intravenous administration of hydrocortisone to the mother. Part I, with special reference to the binding of hydrocortisone by plasma proteins. *Acta obstet. gynec. scand.* **40**, 174–180. Part II, with special reference to foetal elimination of hydrocortisone. *Acta obstet. gynec. scand.* **40**, 181–186.

LIDDLE G.W. (1966) An analysis of circadian rhythms in human adrenocortical secretary activity. *Arch. intern. Med.* **117**, 739–744.

LIDDLE G.W. (1969) Physiological review of adrenocortical function, in E.B.Stear & A.H.Kadish (eds.) *Hormonal Control Systems*, pp. 1–19. New York, Elsevier.

LIDDLE G.W., ISLAND D.P. & MEADOR C. (1962) Normal and abnormal regulation of corticotropin secretion in man. *Rec. Progr. Horm. Res.* **18**, 125–166.

LUETSCHER J.A., DOWDY A.J., ARNSTEIN A.R., LUCAS C.P. & MURRAY C.L. (1964) Idiopathic oedema and increased aldosterone excretion, in E.E.Baulieu & P. Robel (eds.) *Aldosterone*, pp. 487–558. Oxford, Blackwell.

LUUKKAINEN T., SIEGEL A. & VIHKO R. (1970) Neutral steroid sulphates in amniotic fluid. *J. Endocr.* **46**, 391–399.

MAEYAMA M. & NAKAGAWA T. (1970) Effects of ACTH, metopirone and dexamethasone on urinary steroid excretion in late pregnancy. *Steroids* **15**, 267–274.

MARTIN J.D. & MILLS I.H. (1956) Aldosterone excretion in normal and toxaemic pregnancies. *Brit. med. J.* **ii**, 571–573.

MESS B. & MARTINI L. (1968) The central nervous system and the secretion of anterior pituitary trophic hormones, in V.H.T.James (ed.) *Recent Advances in Endocrinology*, 2nd ed., pp. 1–49. London, Churchill.

MIGEON C.J. (1956) Dehydroepiandrosterone and androsterone levels in maternal and cord plasma, in L.I.Gardner (ed.) *Adrenal Function in Infants and Children*, pp. 21–30. New York, Grune & Stratton.

MIGEON C.J., BERTRAND J., GEMZELL C.A. (1961) The transplacental passage of various steroid hormones in mid-pregnancy. *Rec. Progr. Horm. Res.* **17**, 207–248.

MIGEON C.J., KENNY F.M. & TAYLOR F.H. (1968) Cortisol production rate. VIII. Pregnancy. *J. clin. Endocr.* **28**, 661–666.

MINICK M.C. (1966) Cortisol and cortisone excretion from infancy to adult life. *Metabolism* **15**, 359–363.

NEW M.I., SEAMAN M.P. & PETERSON R.E. (1969) A method for the simultaneous determination of the secretion rates of cortisol, 11-desoxycortisol, corticosterone, 11-desoxycorticosterone and aldosterone. *J. clin. Endocr.* **29**, 514–522.

NEY R.L., ORTH D.N. & LIDDLE G.W. (1968) Evaluation of pituitary adrenal function in man. *Mem. Soc. Endocr.* **17**, 285–300.

NICHOLS J., LESCURE O.L. & MIGEON C.J. (1958) Levels of 17-hydroxysteroids and 17-ketosteroids in maternal and cord plasma in term anencephaly. *J. clin. Endocr.* **18**, 444–452.

O'CONNELL M. & WALSH G.W. (1969) Unbound plasma cortisol in pregnant and Enovid-E treated women as determined by ultrafiltration. *J. clin. Endocr.* **29**, 563–568.

ORTH D.N. & ISLAND D.P. (1969) Light synchronization of the circadian rhythm in plasma cortisol (17-OHCS) concentration in man. *J. clin. Endocr.* **29**, 479–486.

ORTH D.N., ISLAND D.P. & LIDDLE G.W. (1967) Experimental alteration of the circadian rhythm in plasma cortisol (17-OHCS) concentration in man. *J. clin. Endocr.* **27**, 549–555.

PAULSEN E.P., SOBEL E.M. & SHAFRAN M.S. (1966) Urinary metabolites in children 1. Individual 17-ketosteroids in children with normal sexual development. *J. clin. Endocr.* **26**, 329–339.

PEKKARINEN A., RAURAMO L. & THOMASSON B. (1960) Studies on the content of free and conjugated 17-hydroxycorticosteroids and its diurnal rhythm in the plasma during normal and toxemic pregnancy. *Acta endocr. (Kbh.)* Suppl. 51.

PETERSON R.E. (1964) Determination of peripheral plasma aldosterone, in E.E. Baulieu & P.Robel (eds.) *Aldosterone*, pp. 145–161. Oxford, Blackwell.

PETERSON R.E. (1959) The miscible pool and turnover rate of adrenocortical steroids in man. *Rec. Progr. Horm. Res.* **15**, 231–274.

PUPO A.A., WAJCHENBERG B.L., & SCHNAIDER J. (1966) Carbohydrate metabolism in hyperadrenocorticism. *Diabetes* **15**, 24–29.

REYNOLDS J.W., COLLE E. & ULSTROM R.A. (1962) Adrenocortical steroid metabolism in newborn infants. V. Physiologic disposition of exogenous cortisol loads in the early neonatal period. *J. clin. Endocr.* **22**, 245–253.

ROSENTHAL H.E., SLAUNWHITE W.R. & SANDBERG A.A. (1969) Transcortin: a corticosteroid binding protein of plasma. X. Cortisol and progesterone interplay and unbound levels of these steroids in pregnancy. *J. clin. Endocr.* **29**, 352–367.

ROSS E.J. (1959) *Aldosterone in Clinical and Experimental Medicine*. Oxford, Blackwell.

Sabeh G., Alley R.A., Robbins T.J., Narduzzi J.V., Kenny F.M. & Danowski T.S. (1969) Adrenocortical indices during fasting in obesity. *J. clin. Endocr.* **29**, 373–376.

Sambhi M.P., Weil M.H. & Udhoji V.I. (1965) Acute pharmacodynamic effects of glucocorticoids. Cardiac output and related hemadynamic changes in normal subjects and patients in shock. *Circulation* **21**, 523–530.

Sandberg A.A., Rosenthal H., Schneider S.L. & Slaunwhite W.R. (1966) Protein-steroid interactions and their role in the transport and metabolism of steroids, in G.Pincus, T.Nakao & J.Tait (eds.) *Steroid Dynamics*, pp. 1–61. New York, Academic Press.

Schally A.V., Arimura A., Bowers C.Y., Wakabayashi I., Kastin A.J., Redding T.W., Mittler J.C., Nair R.M.G., Pizzolato P. & Segal A.J. (1970) Purification of hypothalamic releasing hormones of human origin. *J. clin. Endocr.* **31**, 291–300.

Schweitzer M., Branchaud C. & Giroud C.J.P. (1969) Maternal and umbilical cord plasma concentrations of steroids of the pregn-4-ene C-21-yl sulfate series at term. *Steroids* **14**, 519.

Siiteri P.K. & MacDonald P. (1967) The origin of placental estrogen precursors during human pregnancy, in L.Martini, F.Faschini & M.Motta (eds.) Excerpta Medica International Congress Series 132, 726.

Sims E.A.H., Meeker C.I., Gray M.J., Watanabe M. & Solomon S. (1964) The secretion of aldosterone in normal pregnancy and pre-eclampsia, in E.E. Baulieu & P.Robel (eds.) *Aldosterone*, pp. 499–508. Oxford, Blackwell.

Slaunwhite W.R., Neely L. & Sandberg A.A. (1964) The metabolism of 11-oxyandrogens in human subjects. *Steroids* **3**, 391–415.

Stark G. (1962) Zum Abbau und der Bildung des Aldosterons bei Nichtschwangeren und Schwangeren. *Arch. Gynäk.* **197**, 484–493.

Steinbeck A.W. & Theile H. (1961) Urinary steroids in pregnant women before and after corticotrophin. *Acta endocr. (Kbh.)* **36**, 479–484.

Steinbeck A.W. & Theile H. (1962) Urinary steroids during pregnancy. *Acta endocr. (Kbh.)* **40**, 123–132.

Sterling K. (1960) The effect of Cushing's syndrome upon serum albumin metabolism. *J. clin. Invest.* **39**, 1900–1908.

Sundsfjord J.A. & Aakvaag A. (1970) Plasma angiotensin 11 and aldosterone excretion during the menstrual cycle. *Acta endocr. (Kbh.)* **64**, 452–458.

Sussman L., Lubrik L. & Clayton G.W. (1965) Effect of prior ACTH administration on ACTH release in man. *Metabolism* **14**, 583–589.

Symington T. (1960) The morphology of the adrenal cortex, in F.Clark & J.K. Grant (eds.) *The Biosynthesis and Secretion of Adrenocortical Steroids*, pp. 40–49. Cambridge, University Press.

Tait J.F. & Burstein S. (1964) *In vivo* studies of steroid dynamics in man, in G. Pincus, K.V.Thimann & E.B.Astwood (eds.) *The Hormones*, vol. V, pp. 441–557. New York, Academic Press.

Tait J.F., Little B., Tait S.A.S. & Flood C. (1962) The metabolic clearance rate of aldosterone in pregnant and nonpregnant subjects estimated by both single-injection and constant-infusion methods. *J. clin. Invest.* **41**, 2093–2100.

Thomas J.P. & Flynn T.G. (1964) Adrenal function in normal pregnancy and toxaemia. *Clin. Sci.* **26**, 69–79.
Thrasher K., Werk E.E., Choi Y. Sholiton L.J., Meyer W. & Olinger C. (1969) The measurement, excretion and source of urinary 6-hydroxycortisol in humans. *Steroids*, **14**, 455–468.
Tucci J.R., Espiner E.A., Jagger P.I., Pauk G.L. & Lauler D.P. (1967) ACTH stimulation of aldosterone secretion in normal subjects and in patients with chronic adrenocortical insufficiency. *J. clin. Endocr.* **27**, 568–575.
Ulstrom R.A., Colle E., Burley J. & Gunville R. (1960) Adrenocortical steroid metabolism in newborn infants. II. Urinary excretion of 6β-hydroxycortisol and other polar metabolites. *J. clin. Endocr.* **20**, 1080–1094.
Ulstrom R.A., Colle E., Reynolds J.E. & Burley J. (1961) Adrenocortical steroid metabolism in newborn infants. IV. Plasma concentrations of cortisol in the early meonatal period. *J. clin. Endocr.* **21**, 414–425.
Venning E.H., Dyrenfurth I., Lowenstein L. & Beck J. (1959) Metabolic studies in pregnancy and the puerperium. *J. clin. Invest.* **19**, 403–424.
Venning E.H., Primrose T., Caligaris L.C.S. & Dyrenfurth I. (1959) Aldosterone excretion in pregnancy. *J. clin. Endocr.* **17**, 473–482.
Visser H.K.A. (1966) The adrenal cortex in childhood. Part 1. Physiological aspects. Part 2. Pathological aspects. *Arch. Dis. Childh.* **41**, 2–16 and 113–136.
Watanabe M., Dominguez O.V., Meeker C.I., Sims E.A.H., Gray M.J. & Solomon S. (1962) Aldosterone and progesterone secretion in pregnancy. *J. clin. Invest.* **41**, 1408–1409.
Watanabe M., Meeker C.I., Gray M.J., Sims E.A.H. & Solomon S. (1963) Secretion rate of aldosterone in normal pregnancy. *J. clin. Invest.* **42**, 1619–1631.
Weinheimer B., Oertel G.W., Leppla W., Blaise H. & Bette L. (1966) Plasma steroid concentrations of adrenal venous blood from women with and without hirsutism, in A.Vermeulen & D.Exley (eds.) *Androgens in Normal and Pathological Conditions*, p. 36. New York and London, Excerpta Medica Foundation.
Weissmann G. & Thomas L. (1964) The effects of corticosteroids upon connective tissue and lysosomes. *Rec. Progr. Horm. Res.* **20**, 215–245.
Werthamer S. & Amaral L. (1970) The effect of cortisol on the incorporation of RNA and protein precursors into the nucleus and cytoplasma of human neutrophils cultured *in vitro*. *Acta endocr.* (*Kbh*.) **65**, 497–501.
Wilhelm D.L. (1962) The mediation of increased vascular permeability in inflammation. *Pharmacol. Rev.* **14**, 251–280.
Wilkins L. (1965) *The Diagnosis and Treatment of Endocrine Disorders in Childhood and Adolescence*. Springfield (Ill.), Thomas.
Williams G.H., Dluhy R.G. & Underwood R.H. (1970) The relationship of dietary potassium intake to the aldosterone stimulating properties of ACTH. *Clin. Sci.* **39**, 489–496.
Wolff H.P., Lommer D., Jahnecke & Torbica M. (1964) Hyperaldosteronism in oedema, in E.E.Baulieu & P.Robel (eds.) *Aldosterone*, pp. 471–486. Oxford, Blackwell.

WYNN V. (1968) The assessment of hypothalamic-pituitary-adrenocortical function in man. *Mem. Soc. Endocr.* **17,** 213–234.

YATES F.E. & BRENNAN R.D. (1969) Study of the mammalian adrenal glucocorticoid system by computer stimulation, in E.B. Stear & A.H. Kadish (eds.) *Hormonal Control Systems*, pp. 20–65. New York, Elsevier.

10

Thyroid Function

H.K.IBBERTSON

The thyroid gland is notable amongst the endocrine hierarchy for its ready accessibility, its unique requirement for iodine as a substrate, its ability to store its synthetic product thyroid hormone, and finally its recently discovered role in calcium homeostasis. These properties and the availability of radioiodine have allowed precise documentation of thyroid pathophysiology in childhood and adult life. During pregnancy the remote foetal thyroid has been less subject to scrutiny. The natural reluctance to use radioiodine and lack of precise information on hormone levels has resulted in an incomplete knowledge of foeto-maternal relationships. Much of the information derived from animal studies is not directly applicable to the human and precise knowledge must await the application of newly developed techniques of hormone measurement.

REQUIREMENT AND UTILIZATION OF IODINE
(Fig. 10.1)

The normal daily intake of iodine is around 70–200 µg and this is easily achieved by the regular use of iodized salt. Absorption from the small gut is rapid and the plasma inorganic iodide (PII) whose concentration is normally less than 1µg/100 ml varies with dietary intake, thyroid utilization and renal loss. Since the renal clearance of iodide is normally fixed, homeostatic adjustments to maintain normal iodine economy are mediated by changes in thyroid clearance rate. There is a normal concentration gradient of iodide between thyroid and serum (T/S 20:1) which depends on the activity of pituitary thyrotrophic hormone and the gland iodide content. It can increase several hundredfold in iodine deficiency. The trapping mechanism by which this gradient is achieved is oxygen dependent and probably enzymatic. Although usually rate limiting it can be by-passed by iodide excess and the ion can enter the gland by simple diffusion at high levels. The process is blocked by thiocyanate, perchlorate and nitrate ions which have the same charge

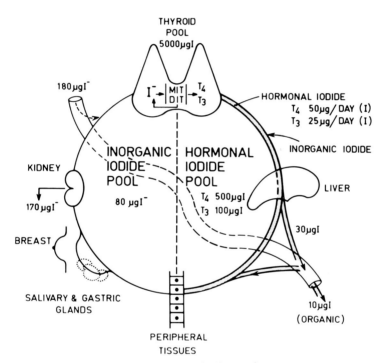

FIG. 10.1. The iodine cycle.

and ionic volume as iodide but which are not all selectively concentrated in the gland. Most extrathyroidal cell membranes are relatively impermeable to iodide though concentration occurs in the red cell, salivary glands and gastric mucosa, mammary tissue, choroid plexus of the brain, sebaceous glands and hair follicles. The iodide in these tissues is in equilibrium with that in the plasma. During the child-bearing period a number of factors mitigate against efficient iodide conservation. In addition to foetal demands there is a unique increase in renal iodide clearance which allows iodide excretion at low levels of PII (Aboul-Khair et al 1964). During lactation selective concentration of iodide and organic binding (short of actual thyronine formation) may result in a loss of 20–50μg of iodide in the milk per day; a loss which of course does not affect the total iodide balance of mother and child. Menstrual loss of iodide is slight and certainly insufficient to account for the usual female: male goitre incidence of 4:1, a predilection for thyroid enlargement which is seen even in the nulliparous woman.

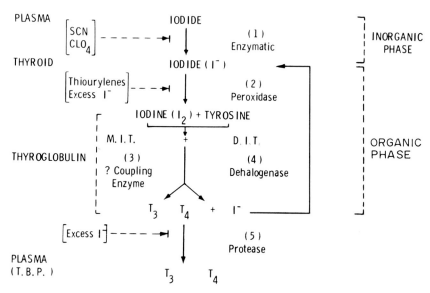

Fig. 10.2. Biosynthesis of thyroid hormones. The figures indicate points at which enzymatic activity is involved.

BIOSYNTHESIS OF THYROID HORMONES
(Fig. 10.2)

Iodide enters the thyroid gland as a chemically inert, negatively charged ion (I^-). Conversion to active molecular iodine (I_2) by a peroxidase is a prerequisite to the subsequent iodination of tyrosyl molecules of the polpyeptides which form thyroglobulin. The iodination products are 3-monoiodotyrosine (MIT), 3,5-diiodotyrosine (DIT), and their various oxidative condensation products, mostly thyroxine (T_4) and triiodothyronine (T_3). Since the thyroid cell potential is 50 mv below that of the extracellular fluid and the follicular colloid, iodide transport requires energy. The process is ATP dependent and is blocked by cardiac glycosides such as oubain which inhibit potassium and sodium transport (Wolff 1964). Both the oxidation and coupling reactions are blocked by antithyroid drugs such as thiouracil and carbimazole, some sulphonamides and also by excess iodide. These successive synthetic steps occur at the cell-colloid interface within the preformed thyroglobulin molecule and are not inhibited by treatment with puromycin, a drug which blocks protein synthesis at the translation stage.

Thyroglobulin is a protein with a molecular weight of 660,000 and a sedimentation constant of 19S. Ten per cent of the molecule is carbohydrate. Synthesis of its constituent amino acids occurs within the endoplasmic reticulum by a process which is independent of RNA formation (Fig. 10.3). After 'packaging' and formation into secretion granules in the Golgi apparatus where carbohydrate is added, the protein is transported into the follicular lumen. Degradation of thyroglobulin occurs when pseudopod-like extensions of the cell engulf colloid to form vacuoles (pinocytosis). These are digested in lysosomes by proteolytic enzymes after prior oxidation of SH bonds. Some thyroglobulin escapes hydrolysis and reaches the circulation (via the lymphatics) where it can be measured by radioimmunoassay in a concentration of 30–100 ng/ml (Roitt & Torrigiani 1967). The iodotyrosines DIT and MIT are deiodinated by a dehalogenase enzyme and the released iodide re-utilized. The iodothyronines T_3 and T_4 are transferred serially from the intrathyroidal *reservoir* (thyro-)globulin to serum *binding* proteins and thence to intracellular *receptor* proteins which determine hormonal localization and facilitate metabolic action (Jorgensen et al 1962).

Genetic abnormalities affecting each stage of thyroid hormone biosynthesis have been described (Stanbury & McGirr 1957). The clinical picture in each case is similar, with variable thyroid enlargement and clinical evidence of hypothyroidism. A defect of the trapping mechanism is characterized by a failure of iodine concentration in both thyroid and salivary glands. Thyroxine production can be restored to normal by the administration of large doses of iodine which reaches the intact intrathyroidal synthetic mechanism by passive diffusion (Stanbury & Chapman 1960). The peroxidase defect in which there is a failure of organic binding of iodine is most commonly seen in Pendred's syndrome in which there is an associated gene defect leading to deafness. The presence of unbound iodine in the gland can be demonstrated by the administration of potassium perchlorate following a dose of radio-iodine. Discharge of most of the radioactivity indicates a defect in organic binding. Coupling defects and abnormalities of thyroglobulin synthesis have both been described, the latter being associated with the presence of an abnormal prealbumin iodoprotein only 10% of which has the properties of thyroglobulin.

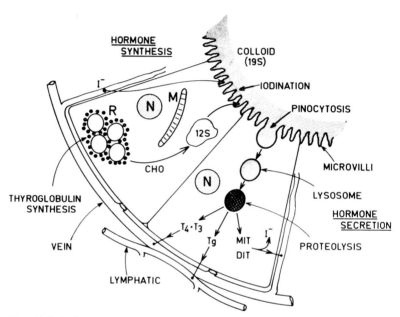

FIG. 10.3. A diagrammatic representation of the synthesis and degradation of thyroglobulin in the follicular cell. The synthesis of globulin and the iodination steps proceed independently. N, nucleus; R, ribosomes; M, mitochondria.

Transport and Metabolic activity

Free T_3 and T_4 circulate in reversible equilibrium with protein bound hormone (TBP) in plasma.

$$\text{Free } T_4 + TBP \rightleftharpoons T_4 \cdot TBP$$

Biological activity is a function of the free hormone though there is some evidence that in the liver, the bound moiety may participate in metabolic activity. The three major binding proteins are thyroxine binding globulin (TBG), thyroxine binding prealbumin (TBPA) and albumin. Their thyroxine binding affinity is inversely proportional to their capacity* being strongest for TBG, intermediate for TBPA and least for albumin. By serving as a reservoir for thyroxine, they regulate its distribution, degradation and metabolic action. TBG is normally absent in the bird and also occasionally in the human as a rare genetic defect. In this situation the peripheral disposal of thyroxine is markedly increased. In pregnancy there is an *increase* in TBG binding capacity which results in a fall in the proportion of free T_4 and a retardation of the fractional degradation rate of the hormone. These changes are countered by an increase in thyroxine production (via the servo mechanism), a restoration to normal of the free thyroxine level and absolute rate of turnover, with consequent expansion in the total hormonal pool due primarily to an increased amount of bound hormone. In contrast to T_4, T_3 is only weakly protein bound and circulates with TBG and albumin alone. Alterations in protein binding thus have little influence on either the serum level of T_3 or its peripheral metabolism.

Metabolic contribution of T_3 and T_4

Although the total serum concentration of T_3 is only 3% of that of T_4 its metabolic clearance is more rapid and the daily turnover of 50–60 µg† is almost as high as that of T_4 (80 µg). Since the biological activity of T_3 is at least three times that of T_4 it can be calculated that approximately two-thirds of the peripheral hormone effect is normally attributable to T_3 (Sterling 1971). A disproportionate increase in T_3 secretion will result in an even greater metabolic contribution. Such a situation is seen in iodine deficiency, following radioiodine therapy and in certain patients with untreated thyrotoxicosis (T_3-thyrotoxicosis). Here euthyroidism may be associated with *low*, and thyrotoxicosis with *normal* serum PBI levels. The demonstration of conversion of T_4 to T_3 in athyreotic human subjects (Braverman *et al* 1970) has revived an old theory that thyroxine is a 'pro-hormone' exerting its effect only after conversion to T_3 in target tissues.

* Binding Power = Binding Affinity × Capacity.
† Expressed in terms of *absolute* hormonal turnover.

EFFECTS OF THYROID HORMONE

The earliest described physiological action of thyroid hormone was its ability to increase oxygen consumption ('metabolism') in both hypothyroid and euthyroid subjects (Magnus Levy 1895). This action is now thought to be a secondary phenomenon related to protein synthesis and regulation of the Na^+-pump (*vide infra*). It can be demonstrated *in vitro* in skeletal muscle, heart, liver and kidney tissue from pre-treated animals but not in gonads and genital tract, nervous tissue, lung, smooth muscle and lymphatic tissues (Barker 1964). The latter organs share with the tadpole a dependence on iodothyronines for protein synthesis and other metabolic actions unaccompanied by any change in oxidative phosphorylation. Clinical observation of patients with iodothyronine deficiency or excess have given insight into the many actions of thyroid hormone. It has a general stimulatory effect on growth. In the cretin the total cell count as judged by DNA content per gramme of tissue is normal but the cell size is reduced reflecting defective protein synthesis (Cheek 1968). The abnormality can be reversed with thyroxine. In juvenile hypothyroidism the cells are if anything, larger than normal and shrink with treatment. These changes are consistent with observations made on brain tissue, that protein synthesis is more severely impaired by thyroxine lack in the young than in the adult animal. Thyroid hormone appears to be exclusively involved in the maturation of certain specific tissues, particularly brain, bone and skin. Epiphyseal maturation is markedly retarded in the cretin at birth, but body length is normal and stunting comes later. The first centres of ossification *in utero* appear at about the 18th week at a time when there is an acceleration in brain cell multiplication (Winick & Rosso 1969). Such changes are believed to be partly dependent on the surge of thyroid activity which occurs at about this time; the degree of intellectual retardation in cretinism correlates closely with the immaturity of epiphyseal development. The general effects of thyroid hormone on growth are *permissive* in nature. In children with hypothalamic lesions, growth may be impaired through deficiency of both thyroid and growth hormone. Replacement of growth hormone without thyroxine leads to sub-optimal growth. On the other hand growth hormone secretion may be impaired in primary hypothyroidism. The growth response which follows the administration of thyroxine in this condition is in part due to the reappearance of growth hormone in plasma. The circulatory effects of thyroid hormone are obvious in thyrotoxicosis. Altered sensitivity to circulating catecholamines has previously suggested an indirect action on the myocardium. More recent work has failed to confirm these observations though in thyrotoxicosis there appears to be an increase in free catecholamines at myocardial receptor sites. The demonstration that thyroid hormone itself has direct inotrophic and chronotrophic action on the myocardium explains the failure of beta blocking agents such as propranolol to abolish completely the augmentation of heart rate and contractility observed in patients with thyrotoxicosis (Levey 1971).

Reproductive function

Thyroid hormone has been shown in amphibians and fish to be concerned with sex differentiation and gonadal maturation (Lynn 1969). In the mouse testicular and ovarian maturation are retarded by thyroxine deficiency (Maqsood 1952) and in the human cretin gonadal maturation is similarly delayed though reproductive capacity may be normal. Although a popular scapegoat, hypothyroidism is an uncommon cause of infertility (less than 2%) either in the human or the mouse (Bruce & Sloviter 1957) yet anovulatory cycles are common and frequently accompanied by irregular or heavy periods (Goldsmith *et al* 1952). Following thyroidectomy in the baboon Gillman & Gilbert (1953) found infrequent ovulation and amenorrhoea with irregular menstrual bleeding from a proliferative endometrium. As in the human, treatment with thyroxine promptly restored the normal cyclical menstrual pattern. The lack of ovarian shrinkage in this study suggested continuing FSH secretion and this together with inadequate LH secretion may also be the basis for the abnormality in the human. The application of radio-immunoassay techniques for FSH and LH should resolve this problem though cross-reaction with the high levels of TSH in hypothyroid sera demands an FSH antiserum of exceptional specificity. A reduction in menstrual flow was a feature in 60% of thyrotoxic patients in the author's series. Cycles are usually regular and ovulatory though ovulation may fail in severe cases with amenorrhoea. Restoration of the normal pattern usually follows improvement in thyroid function and often precedes full clinical remission.

In addition to the postulated (though not proven) effects on pituitary gonadotrophins, thyroid dysfunction alters the metabolism of both oestrogens and androgens. In hypothyroidism the urinary androsterone-aetiocholanolone ratio is reduced and the hydroxylation of oestrone at the two position is diminished with the formation of more oestriol. In hyperthyroidism the changes are opposite (Fischmann *et al* 1965 and Gallagher *et al* 1960). When one adds to these changes the complexities of thyroid hormone action on cellular metabolism and the influence of nutritional and emotional disturbance which frequently complicate clinical thyroid dysfunction it is not surprising that menstrual disturbances are diverse in their manifestations.

MECHANISM OF ACTION OF THYROID HORMONE

Although the diverse effects of thyroid hormone on cellular metabolism are well recognized, the mechanism of its action is ill understood. Major debate has concerned its influence on oxidative phosphorylation and the primacy of alterations in mitochondrial function and protein synthesis. Mitochondria are the major site of oxidative phosphorylation, a process in which oxygen utilization is coupled to energy production by the formation of energy rich phosphate bonds ($\sim$P). The ability of thyroxine to uncouple this process suggested that this was its primary locus of action. Other uncoupling agents such as dinitrophenol, salicylates and chlorpromazine do not however mimic other effects of thyroid hormones and the profound tissue changes of metamorphosis in the tadpole is affected by thyroxine without change in mitochondrial oxidative function. It now seems likely that this action may be important only at the toxic level (as in thyrotoxicosis) and that there may be other mitochondrial effects not necessarily related to uncoupling. A characteristic feature of thyroxine action is the latent period between administration and metabolic effect. Such delay suggests an obligatory metabolic transformation before biological action though at least part of it is due to protein binding of hormone in the circulation. Peripheral deiodination of thyroid hormone is a constant accompaniment of its metabolic action (Galton 1968). It has even been suggested that T_4 merely carries iodine to its intracellular site of action though it is difficult to reconcile this concept with the finding that several T_4 analogues possess thyromimetic activity yet contain no iodine (Benua *et al* 1960). Of more relevance is the possibility of *partial* deiodination of T_4 to T_3. Although previously discounted, this theory has gained recent support from the observation of peripheral transformation of T_4 to T_3. In a kinetic study in normal subjects, the average daily rate of extrathyroidal conversion of T_4 to T_3 was 33% of total T_4 production. This means a daily peripheral generation of about 20 µg of T_3 (Pittman *et al* 1971). Thyroidectomized rats respond metabolically to 3' but not 5'-substituted 3,5-iodothyronine (Barker 1964). The primary step in the hormonal action of T_4 could involve the formation of 'active T_3' by the binding of

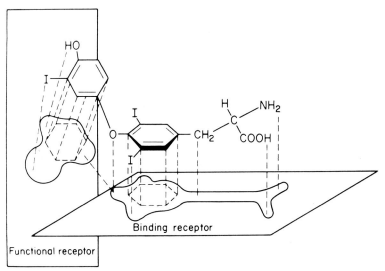

FIG. 10.4. Theoretical representation of triodothyronine interacting with a tissue receptor. The 3'-iodo and 4'-hydroxyphenyloxy part of this molecule is active and the 3,5-diiodotyrosyl residue bound to a protein receptor. (From E.C.Jorgenson, P.A.Lehman, C.Greenberg & N.Zenher, 1962). Thyroxine Analogues VIII, Antigastrogenic, Calorigenic and Hypocholestraemic 'Activities of Some Oliphatic, Acyclic and Aromatic Ethers of 3,5-Diiodotyrosines in the Rat'. *J. biol. Chem.* **237**, 3832.

the 3,5,3'-isomer with a tissue 'receptor protein' (Fig. 10.4), the 3,5,5'-isomer which is not bound being metabolized by non-specific deiodinating mechanisms. Such prior alteration of T_4 partially explains the 'lag period' though the profound changes in protein metabolism which are involved in the metabolic response to thyroid hormone may contribute. Thyroxine is known to stimulate protein synthesis in the tadpole without change in oxidative metabolism, and the feedback inhibition of TSH secretion from human pituitary thyrotropes is mediated by a thyroxine induced inhibitory protein (*vide infra*). Tata (1967) has demonstrated that rapid synthesis of nuclear RNA precedes the incorporation of labelled amino acids into mitochondrial and microsomal proteins. Inability to demonstrate a specific increase in messenger-RNA (mRNA) following T_4 administration suggests that its action is primarily concerned with the process of *translocation* in which coded information carried in mRNA, programmes the ribosomes for synthesis of specific protein molecules. This co-ordinated sequence of changes in protein synthesizing mechanisms occurs before the biological effects of thyroid hormone becomes apparent. Such work relegates the mitochondria to a secondary role. Sokoloff & Kaufman (1961) on the other hand have shown that L-T_4 stimulates protein synthesis in cell-free systems only when mitochondria are present, and selective localization of the hormone has been demonstrated in sub-mitochondrial particles (Arbogast & Hoch 1968). It is obvious that mitochondria play an important part in thyroxine induced energy transfer but the sequence of their involvement is still debated.

The membrane bound enzyme adenyl cyclase mediates the effects of a variety of hormones on their target tissues. Thyroid stimulating hormone (TSH) acting as a 'first messenger' activates adenyl cyclase by interaction with a receptor protein on the thyroid cell membrane. The enzyme in turn catalyses the formation of cyclic 3',5',-AMP (cyclic-AMP) from ATP. The cyclic-AMP acting as a 'second messenger' amplifies the signal through a variety of enzyme pathways and initiates the main physiological response—thyroid hormone production. L-thyroxine itself has been shown to activate adenyl cyclase in particulate preparations of cat myocardium (Levey & Epstein 1969) but T_4 concentrations in this and other experiments exceed by many times the estimated normal tissue levels of 10^{-7}M to 10^{-8}M. It seems unlikely that this mechanism is involved in the peripheral action of T_4 at physiological levels but may be important in thyrotoxicosis. Here adenyl cyclase synthesis is increased but ATP (and cyclic-AMP) is reduced with a resultant defect in energy production.

It has recently been suggested that thyroxine may be specifically concerned in the utilization of energy by the Na^+ transport mechanism in the cell membrane (Strauss 1971). Twenty to forty-five per cent of the resting oxygen consumption of mammalian tissues is expended in the transmembrane extrusion of Na^+ against an electrochemical gradient (Whittam 1964). The energy for this process is derived from conversion ATP→ADP. Ouabain, a digitalis glycoside blocks the Na^+ pump by specific inhibition of $(Na^+ + K^+)$-activated ATP-ase. This drug has been used in *in vitro* studies on rat liver and skeletal muscle obtained from euthyroid animals previously treated with T_3 or T_4. In both tissues ouabain inhibited more than 90% of the increase in oxygen consumption produced by the injection of thyroid hormone. The drug was without effect when the tissue slices were suspended in Na^+-free media. These observations suggest a selective effect of thyroid hormones on $(Na^+ + K^+)$-activated ATP-ase, though other membrane bound enzymes may be involved. Thus much of the thyroxine-dependent oxygen consumption of mammalian cells is utilized to provide energy for Na^+ extrusion. It seems possible that this basic mechanism may be involved in many of the cellular effects of thyroid hormone.

CONTROL SYSTEMS (Fig. 10.5)

The secretion rate of thyroid hormones (iodothyronines) is dependent on the level of pituitary thyrotrophin (TSH) stimulation. TSH responsiveness and biosynthetic activity is in turn controlled by an intrinsic process of autoregulation dependent on the size of the intrathyroidal organic iodine pool. The secretion of TSH is regulated by two interacting systems; feedback control by thyroid hormones at the pituitary level and neural control by the hypothalamus. The 'set point' of the negative feedback system is the normal resting level of free thyroid hormone in plasma. Minor deviations stimulate appropriate inverse changes in TSH secretion which by varying thyroid hormone production maintains the serum level within a narrow normal range. The level of the set point around which these feedback responses are elicited is determined by a tonic influence from the hypothalamus mediated by a TSH releasing hormone (TRH). This hormone which has been characterized and synthesized (Bolar *et al* 1969) is a tripeptide containing the amino acids glutamic acid (glu), histidine (his) and proline (pro) which is active in the form L-pyroglutamyl-L-histidyl-L-proline-amide. It is normally synthesized in the median eminence of the hypothalamus and reaches the pituitary thyrotrope via the portal system investing the pituitary stalk. TRH stimulation of TSH secretion involves movement of calcium ions into the thyrotrope, a process common to a number of secretory mechanisms and known as 'stimulus secretion coupling'. Free T_3 and T_4 antagonize TRH and thus inhibit TSH production by a process which involves the synthesis of an inhibitory protein. Such a

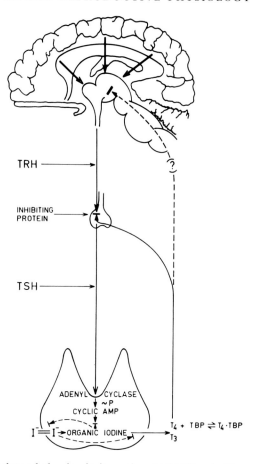

FIG. 10.5. The hypothalamic-pituitary-thyroid (HPT) axis. The tonic influence of TRH on the anterior pituitary thyrotroph is blocked by an inhibitory protein whose synthesis is stimulated by the level of free iodothyronine in blood perfusing the anterior pituitary gland. The action of TSH on the thyroid gland is mediated by cyclic AMP and modified by the intrathyroidal iodine level.

mechanism accords with the view that a primary action of thyroxine in all tissues involves RNA activation and new protein synthesis. It is supported by the observations that thyroxine inhibition of TSH secretion can be blocked by actinomycin (Vale *et al* 1968) and is not instantaneous but delayed. The blocking effect of a single dose of T_3 or T_4 may be apparent for several days suggesting persistence of the inhibitory protein at receptor sites.

Thus both the servo mechanism which operates at a pituitary level and the basic hypothalamic influences are nicely integrated. Certain stimuli may 'break through' the feedback control. Emotional stress has been implicated though the evidence is tenuous; the role of ambient temperature is, however, proven. A reduction in body temperature causes TSH release and thyroid activation in animals and also in the neonatal infant (Fisher & Odell 1969) but not in adults, suggesting a relatively primitive mechanism which is preserved in the immediate post-natal period becoming redundant in adult life.

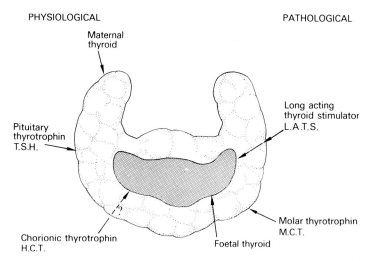

FIG. 10.6. Thyroid stimulators. Maternal LATS may influence the foetal thyroid gland by direct transplacental passage. Maternal and foetal TSH are confined to their respective circulations. There is no evidence that maternal HCT and HMT cross the placental barrier.

Thyroid stimulators

In addition to pituitary TSH the thyroid gland in the pregnant woman may be influenced by a number of other stimulators (Fig. 10.6) which include chorionic thyrotrophin (HCT) (Hennen 1965), molar thyrotrophin (Hershman & Higgins 1970) or the long acting thyroid stimulator (LATS) of Grave's disease (Adams & Purves 1956). The last two may cause over-production of thyroid hormone and clinical thyrotoxicosis. Increased thyroid function without convincing clinical hyperthyroidism occurs in about 10% of patients with malignant trophoblastic disease and molar thyrotrophin has been demonstrated in plasma and tumour extract from a patient with frank thyrotoxicosis (Hershman & Higgins 1971).

The long acting thyroid stimulator is not unique to the pregnant patient but is found in serum concentrates in up to 80% of all patients with Grave's Disease. This IgG gamma globulin is an antibody of lymphoid origin which is believed to stimulate secretion of thyroid hormone by interaction with thyroid microsomes. In contrast to TSH, these stimulators do not observe the niceties of servo control and are uninfluenced by the circulating level of free hormone. Their presence can be inferred when relative or absolute autonomy of thyroid function is demonstrated in the T_3-suppression test in which there is a failure of administered triiodothyronine to suppress radioiodine uptake or serum thyroxine.

THYROCALCITONIN

The discovery of a calcium lowering hormone—calcitonin, by Copp *et al* (1962), has added yet another dimension to thyroid physiology. The original experiments in which perfusion of the thyroid–parathyroid gland complex of dogs with high calcium blood caused a *fall* in the systemic serum calcium suggested a parathyroid origin for this new hormone. Subsequent work (Hirsch *et al* 1963) in the rat and pig showed that a very potent hypocalcaemic substance could be extracted from the thyroid gland (thyrocalcitonin) and this appears to be the main source in mammals. The cell responsible for synthesis of the hormone, the parafollicular (or C cell) has a different embryological origin from the follicular cells of the thyroid gland. In non-mammalian vertebrates the sixth branchial pouch remains distinct as the ultimo branchial gland. Large quantities of calcitonin have been extracted from the ultimo branchials (but not the thyroid) of the domestic fowl and dog fish. This structure has a much earlier phylogenetic origin than the parathyroid glands which appear first in the amphibians. In mammals ultimo branchial cells migrate into the thyroid gland during foetal development and can be identified by specific staining techniques in an interacinar position. These same cells have subsequently been demonstrated in the human parathyroid and thymus (Galante *et al* 1968) though the thyroid is the main source of calcitonin in man.

Within 6 years of the discovery of this hitherto unknown polypeptide its chemical structure was deduced, its synthesis accomplished and specific bioassay and radioimmunoassay techniques developed to measure the hormone in plasma (Hirsch & Munson 1969). Thyrocalcitonin is a single chain 32 amino-acid polypeptide with a unique amino-acid

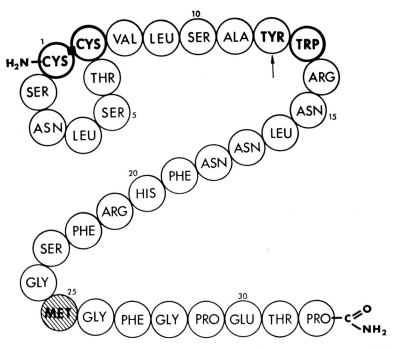

FIG. 10.7. Porcine calcitonin molecule. Porcine, human and salmon calcitonin all have 32 amino acids but these differ in the three species. They resemble oxytocin and vasopressin in having an intra-chain disulphide ring at the N terminus and an amide at the C terminus. From T.T.Potts Jnr. *et al* (1968) in *Parathyroid Hormone and Thyrocalcitonin (Calcitonin)* pp. 54–68. R.V.Tomage & L.F.Belanger (eds.) Amsterdam, Excerpta Medica Foundation.

sequence and a molecular weight of approximately 3500 (Fig. 10.7). In contrast to other polypeptide hormones such as ACTH the entire molecule is necessary for biological activity. There are, however, striking species differences and the porcine hormone differs from the human material by 18 amino acids. These differences reflect marked variation in biological potency. Salmon calcitonin was first extracted and isolated from 200 pounds of ultimo branchial tissue from approximately 5000 tons of salmon (Copp 1970). This material has a potency of 5000 MRC units/mg compared with 120 units/mg for human thyrocalcitonin. It appears to be the most potent natural biological product yet isolated and has obvious therapeutic potential.

The principal action of thyrocalcitonin at least in mammalian species is inhibition of bone resorption. Its hypocalcaemic effect is directly proportional to the rate of this process, being greater in the young than in adults. Its effect can still be demonstrated following nephrectomy, evisceration or parathyroidectomy though under normal conditions it acts as a physiological antagonist to parathyroid hormone. In bone tissue culture, thyrocalcitonin blocks the parathyroid hormone-stimulated incorporation of uridine into RNA (Raisz & Niemann 1967) but is not itself inhibited by actinomycin D, suggesting the protein synthesis is not an essential feature of its action. The action of parathyroid hormone in causing bone resorption is mediated by an increase in cyclic AMP. Theophylline facilitates this action by inhibition of phosphodiesterase, an enzyme which normally increased cyclic AMP degradation. The action of thyrocalcitonin is blocked by theophylline suggesting that this hormone normally acts by potentiation of phosphodiesterase. However, thyrocalcitonin does not inhibit the rise in cyclic AMP in bone after parathyroid hormone treatment (Aurbach & Chase 1970) suggesting that it acts by preventing the normal action of cyclic AMP in stimulating bone resorption. Both parathormone and calcitonin act on most body cells. Parathormone stimulates passive movement of calcium into the cell and calcitonin stimulates calcium efflux (Rasmussen 1971). These changes in intracellular calcium presumably regulate the activity of the osteoclast, a cell specifically concerned with the resorption of calcium from bone. Thyrocalcitonin also has an action on the kidney. The deceased urinary excretion of calcium magnesium and hydroxyproline following its administration appears to be secondary to its effect on bone. The characteristic phosphaturia can be demonstrated in parathyroidectomized animals and represents a direct action on renal tubular resorption though the doses used have generally been unphysiological. More recently the hormone has been shown to have a powerful naturetic and diuretic effect on young rats and in human subjects (Copp 1970). Its potency appears to be some 5000 times as great as that of frusemide, a powerful diuretic agent.

It seems likely that other extra-osseous actions of thyrocalcitonin will be discovered in the future. The high concentration of the hormone present in the ultimo branchial glands of the dogfish which has a cartilagenous skeleton suggests an entirely different role at this level of development. The relative importance of thyrocalcitonin in the regulation of the normal serum calcium is still debated. Thyroid deficient patients even when receiving thyroxine fail to restore the serum calcium normally following an intravenous calcium infusion (Ibbertson *et al* 1967). Such an infusion leads to a rise in serum thyrocalcitonin and an inhibition of parathyroid hormone secretion. These changes lead to a reduction in the rate of bone resorption and a consequent fall in the serum calcium level. The slight effect of injected calcitonin on the serum calcium level in normal adults, however, throws doubt on the importance of bone resorption in normal serum calcium homeostasis. It is possible that the thyrocalcitonin is mainly concerned with bone remodelling and mineral turnover and that serum calcium changes are incidental. Nevertheless Munson & Gray (1970) recently reported that fasting thyroidectomized rats but not controls showed persisting hypercalcaemia after eating food containing a normal concentration of calcium (1%). Such observations require confirmation in man but do suggest a physiological role for thyrocalcitonin in protecting the normal animal against hypercalcaemia. It remains to be shown whether failure of this mechanism is attended by more obvious sequelae such as nephrocalcinosis or renal stone formation.

Thyrocalcitonin in pregnancy

There is little information available concerning thyrocalcitonin in human pregnancy. Subcutaneous injection of porcine thyrocalcitonin into the rat foetus decreases foetal calcium levels, maternal calcium being unaffected. The mild foetal hypocalcaemia which follows maternal injection of thyrocalcitonin is attributable to the fall in the maternal serum calcium and not to transplacental passage of the hormone. Experiments in which ^{125}I-thyrocalcitonin was injected into foetus or mother have failed to demonstrate placental transfer of the hormone (Milhaud *et al* 1970).

An interesting case study has been reported (Verdy *et al* 1971) in which four of nine children of a 32-year-old woman with medullary carcinoma of the thyroid (*vide infra*) were shown to have osteopetrosis. Plasma calcitonin levels were high in the mother but normal in the two children who were studied. Since the mother also had osteopetrosis the bone condition was considered to be unrelated to the calcitonin levels and probably represented dominant transmission of this genetic disorder.

Thyrocalcitonin in disease

Doubts about the physiological function of thyrocalcitonin have not prevented the recognition of certain disease states in animals and man in which alterations in thyrocalcitonin secretion may be of considerable importance. In dairy cows the condition of parturient paresis characterized by hypocalcaemic tetany will lead to paralysis and death unless treated by intravenous calcium infusion. In this condition there is a depletion of C cells in the thyroids of diseased animals and the thyrocalcitonin content of the gland is reduced by about 75% (Kaplan & Young 1967). These findings suggest that the acute hypocalcaemia is precipitated by a massive discharge of calcitonin. However injection of calcitonin causes a fall in serum calcium in young cows but not in animals over 3 years old in whom the syndrome usually occurs. This would appear to rule out calcitonin as an aetiological factor. It remains possible, however, that the absence of response reflects the low rate of bone resorption which has been demonstrated, and that this in turn is due to suppression by endogenous calcitonin. Under these circumstances the maintenance of a normal serum calcium level is more dependent on the dietary intake of calcium which is often reduced at this time (Mayer 1970). Actual measurement of serum thyrocalcitonin should clarify this problem.

In man increased thyrocalcitonin has been demonstrated in the serum and tumour tissue of patients with medullary carcinoma of the thyroid. This unusual tumour is derived from the C cells within the thyroid gland. Although the tumour is poorly differentiated the prognosis is relatively good and amyloid tissue is often present in the stroma. The occurrence of multiple mucosal neuromas and occasional bilateral phaeochromocytomas in a familial form of the disorder has led to the suggestion that the C cells derive ultimately from neural tissues (Pearse 1970) and merely 'travel' with ultimo branchial tissues during embryological development. These cells normally convert 5 hydroxytryptophan to 5 hydroxytryptamine and the secretion of serotonin and kinins are said to be responsible for the flushing attacks which occur in this syndrome. Intractable diarrhoea from the secretion of prostaglandins and occasional tetany completes this bizarre syndrome. The absence of hypocalcaemia in many patients despite high levels of thyrocalcitonin may indicate parathyroid compensation or more likely the relative unimportance of bone resorption in the maintenance of the normal serum calcium.

TESTS OF THYROID FUNCTION

A knowledge of the methods currently available for the measurement of thyroid function is essential to a proper understanding of thyroid patho-physiology. In pregnancy because of the shortcomings of basal metabolic rate and *in vivo* radioiodine measurements, major emphasis is placed on plasma levels of circulating hormones.

Free thyroid hormone

Approximately 0.05% of T_4 and 0.5% of T_3 are present in the serum in a free form. Since these levels reflect a balance between thyroid hormone secretion and peripheral degradation, their measurement should in theory provide the most precise single test of thyroid function. Most methods require preliminary dialysis of serum to which has been added ^{125}I-labelled hormone ($*T_4$). Measurement of radioactivity in the dialysate ($F*T_4$) and of bound absolute (BT_4) and bound radioactive ($B*T_4$) hormone in serum allows a calculation of the absolute free hormone level according to the formula:

$$\text{Serum free thyroxine (FT}_4\text{)} = \frac{\text{Serum BT}_4 \times F*T_4}{\text{Serum B}*T_4}$$

Rigid laboratory control is required to minimize interassay variation though the method is being increasingly used as a final arbiter. Its value in this role is slightly diminished by the finding of high levels in non-thyroidal illness without obvious explanation (Liewendahl & Lamberg 1969). The normal free hormone level varies with the method used but is in the range of 2–4 ng/100 ml for T_4 and 0·75–1·5 ng/100 ml for T_3. Both are normal in pregnancy though one report suggests increased maternal FT_4 levels in the first 11–18 weeks (Fisher *et al* 1970).

Since free and protein bound hormone are in reversible equilibrium and because of the difficulties inherent in the above technique, measurement of protein bound hormone is more often used to assess thyroid function in clinical practice.

Protein bound thyroid hormone

The affinity of TBP (thyroid binding protein) for triiodothyronine is low and most 'bound hormone' in the circulation is thyroxine. The binding capacity of TBG in 100 ml of human serum is approximately 20 μg of T_4 though the protein is normally only 20% saturated. When the TBG concentration is normal the level of bound T_4 (or PBI) reflects the free thyroxine concentration and is in stable equilibrium with it. Methods measure either hormonal iodide or the hormone itself.

Hormonal iodide measurement

The normal serum protein bound iodine of 4·0–8·0 μg/100 ml measures mainly the iodine content of thyroxine (T_4) bound to protein, T_3 contributing only about 0·2 μg/100 ml. Chronically administered iodine may elevate the PBI through non-specific binding to albumin. Organic iodine compounds are similarly measured and PBI levels may be elevated for 10 to 14 days after an intravenous pyelogram, several months after cholecystogram and years after myelography. Transplacental passage accounts for high PBI levels in infants of mothers subjected to these procedures years previously. In one report (Hung 1966) an elevated PBI in an 8-year-old child was ascribed to the radio contrast compound iophenoxic acid (Teridex) given to the mother 4 and 5 years before conception. It was calculated that the PBI would not be expected to reach normal levels for a further 33 years! The PBI may also give misleading information in patients with subacute thyroiditis and with certian intrathyroidal enzyme defects in whom the iodotyrosines, MIT and DIT, circulate in the plasma. A more certain measure of thyronine iodine is obtained by the acid butanol extraction of serum (BEI) or by the T_4 'by column' method (Pillegi *et al* 1964). Both give normal levels about 1 μg below the normal range of PBI. The column method offers some advantage in being less influenced by organic iodine compounds which are largely butanol soluble.

In all these procedures iodine is measured by a catalytic method involving the conversion of ceric to cerous ion in the presence of arsenious acid and iodine. The reaction is inhibited by mercury which may be absorbed through the skin from such antiseptics as methiolate. Although usually expressed in terms of iodine the actual thyroxine concentration can be simply calculated since iodine represents 65·3% of the total weight of the thyroxine molecule.*

Direct hormone measurement

Direct measurement of total serum thyroxine by competitive protein binding (Murphy et al 1966) overcomes the problem of iodide contamination inherent in the above methods. Preliminary extraction of serum

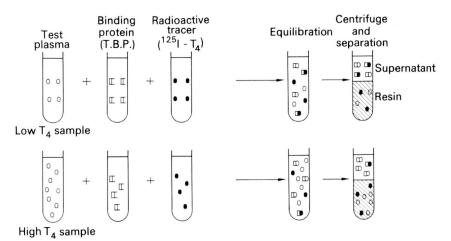

FIG. 10.8. Measurement of plasma total thyroxine (TT_4) by competitive protein binding. The proportion of radioactive T_4 in the final supernatant is dependent on the absolute T_4 level in the original plasma.

with alkaline ethanol separates thyroxine from its binding protein. Incubation of this extract with a fixed amount of TBG and a trace of $^{125}I-T_4$ establishes a system in which labelled T_4 is progressively displaced from binding sites on TBG as the T_4 content of the extract increases (Fig. 10.8). By shaking with an anionic exchange resin the free T_4 is removed and the bound hormone remains in the supernatant after centrifuging. Construction of a standard curve using known concentrations of T_4 allows estimation of the hormone in an unknown plasma. The normal range (Mean ±2 SD) in the author's laboratory is 9·3 ±4·8

* PBI (μg/100 ml) × 100/65·3 = Serum thyroxine (μg/100 ml).

µg/100 ml. This is equivalent to 6.0 ± 3.1 µg/100 ml expressed as hormonal iodide.

This technique has also been applied to the measurement of T_3 though extensive concentration and chromatography is necessary (Nauman et al 1967). A more promising approach is the production in rabbits of an antibody to T_3 following its preliminary coupling to albumin (Gharib et al 1970). The antibody can then be used in an immunoassay system to detect total T_3 at physiological levels. The normal serum total T_3 level is about 150–250 ng/100 ml.

Influence of protein binding

Implicit in the interpretation of all these protein bound hormone methods is the assumption that the measurement reflects the free thyroxine level. Such is the case, however, only when thyroxine binding protein (TBP) levels are normal. The interaction between T_4 and TBP can be designated—

$$\text{Free } T_4 + TBP_u \rightleftharpoons T_4.TBP$$

where TBP_u represents unoccupied binding sites and $T_4.TBP$ is the bound T_4. Since at equilibrium 99.95% of total serum T_4 is bound to TBP, the PBI is a measure of $T_4.TBP$ and free T_4 is a tiny fraction of the total. The equilibrium is expressed:

$$K = \frac{T_4.TBP}{(FT_4)(TBP_u)} \propto \frac{PBI}{(FT_4)(TBP_u)}$$

$$\text{i.e. Free thyroxine } (FT_4) \propto \frac{PBI}{TBP_u}$$

Thus concentration of free T_4 is proportional to the serum PBI and inversely proportional to the unoccupied binding capacity (TBP_u). An increase in TBP occurs in pregnancy, in women taking oral contraceptives and in certain liver diseases. Low levels are seen in nephrosis, portal cirrhosis, during androgen administration and occasionally as a genetic defect. The total hormone concentration may then be high or low and yet the free thyroxine level is quite normal. Another source of error is the low protein bound thyroxine level caused by displacement of the hormone by certain drugs such as the hydantoins (phenytoin), salicylates and quinidine.

THYROID FUNCTION

The frequency with which thyroid disease occurs in woman of childbearing age on 'the pill' imposes a serious diagnostic limitation on protein bound hormone measurements and the TBP abnormality persists for about 6 weeks after the contraceptive is stopped. The difference between the thyrotoxic and euthyroid subject on 'the pill' or pregnant, all with a high PBI, lies in the different proportions of unoccupied (or reserve) thyroxine binding capacity (Fig. 10.9). In thyrotoxicosis there is less and in the pregnant patient more.

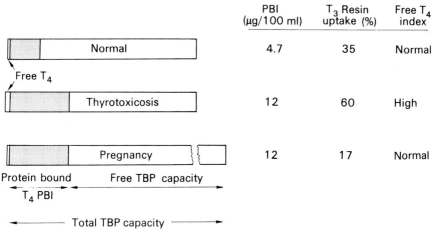

	PBI (µg/100 ml)	T_3 Resin uptake (%)	Free T_4 index
Normal	4.7	35	Normal
Thyrotoxicosis	12	60	High
Pregnancy	12	17	Normal

FIG. 10.9. TBP capacity and the serum PBT. In thyrotoxicosis both the serum FT_4 and PBI are high. In pregnancy the serum PBI is high because of an increase in TBP capacity. Serum FT_4 is normal. The free thyroxine index corrects for TBP alterations and is directly proportional to the absolute free thyroxine level.

T_3 resin uptake test

Measurement of the unoccupied TBP (TBP_u) can be achieved by the relatively simple triiodothyronine resin uptake test (T_3R). This test differs in principle from the thyroxine method outlined above only in that the *whole* serum is incubated with ^{125}I-T_3 before being shaken with resin. The resin uptake of residual ^{125}I-T_3 is inversely proportional to the binding sites on TBP left unoccupied by the T_4 in the sample. In thyrotoxicosis when TBP is relatively saturated with T_4 the resin uptake of ^{125}I-T_3 is high. Serum from a pregnant patient will contain a large amount of unoccupied TBP available for T_3 binding and the resin uptake is low. Thus both the PBI and T_3 resin uptake are high in thyrotoxicosis while in the pregnant patient the T_3 resin uptake is low despite a high PBI. This discordance of PBI and T_3 resin uptake in patients with altered levels of TBP is utilized in the free thyroxine 'index' (FT_4I). Since the free thyroxine level is proportional to PBI/TBP_u and since T_3R is inversely proportional to TBP_u it follows that—

$$FT_4 \propto PBI \times T_3R = \text{Free thyroxine index (FT}_4\text{I)}$$

The absolute value of the index will depend on the method used for measurement of thyroxine and resin uptake but is normal in pregnancy and in women taking an oral contraceptive.

Measurement of the T_3 resin uptake thus adds precision to bound hormone measurements and is a necessary refinement to allow their proper interpretation in the pregnant patient. Since neither the total thyroxine measurement by competitive binding nor the T_3 resin uptake is influenced by iodine, the free thyroxine index is best calculated from these two measurements.

The emphasis placed on the serum thyroxine level disregards the important (two-thirds) metabolic contribution of T_3. Although the concentration of these two hormones usually runs parallel there is increasing evidence of preferential T_3 secretion in a number of clinical conditions. It seems likely in the future that measurement of both T_4 and T_3 levels will be considered mandatory whenever a full assessment of thyroid function is required.

Other thyroid function tests

Several developments promise to simplify and add precision to tests of pituitary-thyroid integrity. The availability of radioimmunoassay capable of measuring TSH in the serum in physiological concentrations will obviate many of the problems (Utiger 1971). The normal serum level of TSH is 0–3·1 $\mu U/ml$ (Patel *et al* 1971). Assay sensitivity does not at present always allow distinction between low and normal levels though this is likely to improve. The distinction between primary and secondary hypothyroidism can often be made on clinical grounds, though careful testing is sometimes necessary. Primary thyroid pathology can be inferred when radio-iodine uptake or serum thyroxine levels have failed to rise following injection of TSH (TSH Stimulation Test). In thyrotoxicosis, thyroid autonomy can be demonstrated by the T_3-suppression test in which there is failure of triidothyronine to suppress radioiodine uptake or serum thyroxine when given to the patient in a dose of 120 μg daily for 7 days. Such autonomy may arise from nodules within the gland or from stimulation by LATS. In both instances suppression of TSH is inferred. Such stimulation and suppression tests are time consuming, require several patient visits and are potentially dangerous. Measurements of serum TSH, however, may eventually eliminate the need for these manœuvres. High levels are almost invariably present in primary hypothyroidism but often also in apparently euthyroid patients following thyroidectomy or with Hashimoto's disease. Such findings may indicate that TSH levels are a more sensitive index of iodothyronine deficiency than are the conventional tests of peripheral hormone action. In secondary hypothyroidism serum TSH is low. An abnormality of release can be confirmed by an intravenous or oral dose of the newly available synthetic thyrotrophin releasing hormone (TRH). Following an oral dose of 40 mg, serum TSH and PBI normally rise to a maximum at 20 minutes and 2 hours respectively (Ormston et al 1971). A normal response in a patient with secondary hypothyroidism suggests a hypothalamic lesion while a failure of TSH release indicates that the anterior pituitary is at fault. In mild thyrotoxicosis the demonstration of low (suppressed) levels of TSH will in the future allow confirmation of the diagnosis when there is only a borderline elevation of serum thyroid hormone.

Conventional tests of peripheral hormone action lack specificity and are often unhelpful. The BMR is misleading in pregnancy and even the half relaxation time of the ankle reflex which is often a valuable index of hypothyroidism in the non-pregnant patient may be falsely prolonged. The electrocardiogram tends to be neglected and often gives useful evidence of hypothyroidism when other indices are borderline. Confirmation may be obtained by finding a change following a 1 month course in triiodothyronine 20 μg b.i.d. (two-thirds normal replacement). A recent newcomer is the estimation of the sodium content of the red cells (Goolden et al 1971). In thyrotoxicosis there is a breakdown in normal energy transfer which maintains the 'Na^+ pump' with a rise in intracellular Na^+. This has proved at least as good as the serum PBI in confirming thyrotoxicosis in our own clinic though there is considerable overlap in hypothyroidism. The normal mean (± 2 SD) for red cell sodium in the author's laboratory is $6\cdot85 \pm 2\cdot24$ mM/litre.

In both the pregnant and non-pregnant patient there is an outstanding need for a precise test of thyroid hormone deficiency at the tissue level. Although the free thyroxine index corrects for disturbances of thyroxine binding protein, the T_3 resin uptake is relatively insensitive at low levels. Direct measurement of free thyroxine is the ideal but not often available in routine practice. A veritable battery of tests may be necessary to confirm the diagnosis on occasions (Winikoff 1971). It remains to be seen whether measurement of TSH will bridge the diagnostic gap.

THYROID FUNCTION IN PREGNANCY

THE MATERNAL THYROID

The physiological changes of pregnancy may closely simulate the features of early thyrotoxicosis. Mild anxiety, heat intolerance, urinary frequency and amenorrhoea, together with findings of thyroid enlargement, bruits in the neck, palmar erythema and warm extremities are all suggestive. Further persuasion is provided by an increased thyroid radioiodine uptake (Halnan 1958), a high serum PBI and an increased metabolic rate (Freedberg *et al* 1957). Despite the convincing façade other explanations can be found for this constellation of symptoms and signs.

Thyroid enlargement

This has been recognized since antiquity when a cheap though presumably inaccurate pregnancy test was practised by the ancient Egyptians. The breaking of a reed tied round the neck of a maid 'at risk' was said to herald the onset of pregnancy. The incidence of thyroid enlargement in pregnancy varies with the local iodine intake. In Aberdeen, Crookes *et al* (1967) showed it occurs twice as frequently in pregnant as in non-pregnant women. There was no difference in a similar group examined in Iceland. (Table 10.1). The pregnancy goitre incidence in these two populations correlated with the serum inorganic iodide which was significantly higher in the Icelandic population where the diet of fish is high in iodine. Pregnant patients are, moreover, more sensitive to variations in dietary iodine. The requirements of the developing foetus and a unique increase in renal iodide clearance by the maternal kidney combine to render the pregnant woman more liable to iodine deficiency. Renal clearance of iodine doubles within the first trimester and remains elevated throughout pregnancy falling to normal by the sixth post-partum week (Fig. 10.10) (Aboul-Khair et al 1964). The mechanism of this phenomenon is unknown. It is unrelated to the increased glomerular filtration rate of pregnancy which reaches only 50% of its maximum by the 15th week and does not appear to be due to oestrogenic influences which, if anything, reduce iodide clearance. Its consequence is a low plasma inorganic iodide (PII) which is apparent by the 16th week on-

TABLE 10.1. Incidence of goitre in pregnancy. Incidence of visible and palpable thyroid glands in pregnant and non-pregnant women in Aberdeen and Reykjavik by the same observers. From Crooks J., Tulloch M.I., Turnbull A.C., Davidsson D., Skulason T. & Snaedal, G. (1967) Comparative incidence of goitre in pregnancy in Iceland and Scotland. *Lancet* ii, 625–627.

	No. of patients	Visible and palpable thyroids
ABERDEEN		
Pregnant	184	70·6%
Non-pregnant	116	37·9%
REYKJAVIK		
Pregnant	227	23·0%
Non-pregnant	108	19·0%

wards. In the above series the mean ($\pm$S.E.) was $0·10\pm0·1$ μg/100 ml in the pregnant subject and $0·20\pm0·025$ μg/100 ml in the non-pregnant controls.

The possible contribution of chorionic thyrotrophin to the thyroid enlargement of early pregnancy is discussed below.

The basal metabolic rate

There is a rise in the BMR to $+20-+30\%$ by the third trimester. About three-quarters of this is due to the uterus and its contents; the remainder resulting from increased maternal cardiac and respiratory work (Freedberg et al 1957).

Thyroid iodide turnover

Radioiodine studies are usually avoided in pregnancy though ^{132}I with a half-life of 2·3 hours is probably safe. The thyroid uptake and clearance of ^{132}I is increased from about the 12th week of pregnancy to the 6th post-partum week. This increase reflects the reduction in plasma inorganic iodide (PII), the stable iodide transfer or absolute iodine uptake (AIU) being normal or only slightly increased.

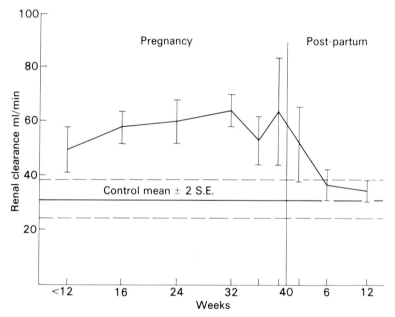

FIG. 10.10. Renal clearance of iodine in pregnancy. From S.A.Aboul-Khair, J.Crooks, A.C.Turnbull & F.E.Hytten (1964). The physiological changes in thyroid function during pregnancy. *Clin. Sci.* **27**, 195–207.

$$\begin{array}{ccc} \text{PII} & \times \text{Thyroid } ^{132}\text{I clearance} = & \text{AIU} \\ (\mu g/100 \text{ ml}) & (\text{ml/hr}) & (\mu g/\text{hr}) \\ \textit{Low} & \textit{High} & \textit{Normal} \end{array}$$

The actual secretion rate of thyroid hormone may be measured by following the disappearance from the plasma of ^{131}I-T_4 or ^{131}I-T_3 following their intravenous injection. Analysis of plasma curves over a period of days provides values for the half-life of the injected material in the circulation and the fractional degradation rate constant (k) as well as the volume of distribution of T_4 or T_3 (TDS). If, in addition, the absolute plasma concentration of T_4 and T_3 are measured, the combined data allows calculation of the extrathyroidal pool of T_4 and T_3 (ETT) and the rate at which each is degraded, and hence in the steady state, secreted. In pregnancy (Table 10.2) the volume of distribution of T_4 is not significantly different from that in non-pregnant women. The slightly slower fractional rate of degradation (k) is offset by the increased total extrathyroidal content of T_4 (ETT). The net hormonal turnover is thus increased in pregnancy but when expressed in terms of surface area, the daily turnover rate (D) is normal. Thus hormonal requirements remain proportional to body size (Dowling *et al* 1967).

TABLE 10.2. Thyroxine turnover in human pregnancy. Data from Dowling J.T., Appleton W.G., & Nicoloff J.T. (1967) *J. Clin. Endocr.* **27**, 1749–1750.

	No.	PBI (μg/100 ml)	k	TDS (L)	ETT (μg I$^-$)	D (μg I$^-$/Day)	D/m^2 (μg I$^-$/Day)
Non-pregnant	7	5·1	0·118	10·2	470	56·0	35·0
Pregnant	8	7·4	0·096	9·1	663	60·6	35·0

The serum protein bound iodine

The serum PBI and total thyroxine rise in the first trimester and remain elevated throughout pregnancy falling to normal within 6 weeks of delivery. The normal PBI range in pregnancy is 7·5–13·5 μg/100 ml. The change results from an oestrogen stimulated increase in TBG synthesis by the liver. The resulting increase in total thyroxine (ETT) reflects the increase in bound hormone. The serum-free thyroxine is in fact increased in the first trimester and falls progressively throughout pregnancy though the levels are still higher than normal at term (Fisher *et al* 1970b). Early studies indicated that when the serum PBI failed to rise during pregnancy there was a high risk of abortion (Mann *et al* 1951). Treatment with thyroxine raised the PBI level temporarily but failed to prevent this complication. It seems likely that these changes are in no way causal but merely secondary to the reduced oestriol production characteristic of the failing foeto-placental unit.

Control systems

The increased thyroidal clearance of ^{132}I in iodine deficiency is mediated by the normal servo increase of TSH and is suppressible by the administration of a pharmacological dose of T_3. In pregnancy however,

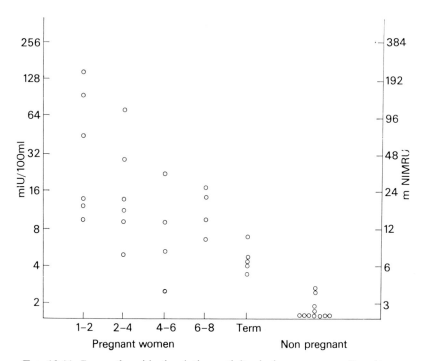

FIG. 10.11. Serum thyroid-stimulating activity during pregnancy. Results are expressed in International Units (IU) and National Institute for Medical Research NIMR—Thyrotrophin Research Standard A. Levels are maximum in the first 2 months but are still higher than normal at term. From G. Hennen, J.G.Pierce & O.Freychet (1969). Human chorionic thyrotrophin: further characterization and study of its secretion during pregnancy. *J. clin. Endocr.* **29**, 581–594.

the clearance is increased before the PII falls and suppression with T_3 is less than complete. Werner (1958) administered doses of 75–100 μg of T_3 daily to pregnant normal females for 7 days and found only minimal suppression in some. Raiti *et al* (1967) gave larger doses of 150–300 μg/day and found that the plasma T_4 'by column' was suppressed normally in less than half and smaller doses were ineffective. Preferential placental detoxification of T_3 could account for this failure of suppression but actual measurement of serum T_3 following loading doses has shown adequate serum levels in pregnant patients (Dussault et al 1969). Observations such as this suggest relative autonomy of the hypothalamic pituitary axis in pregnancy though immunoassay has failed to demonstrate an increase in TSH in pregnant subjects (Hershman & Pittman 1971).

The demonstration of increased thyroid stimulating activity in human pregnancy serum and chorionic tissue (Hershman & Starnes 1969) suggests an alternative explanation. The serum activity is highest early in pregnancy though still elevated at term (Fig. 10.11). The levels of stimulation measured in the serum by a mouse bioassay are of the same order as those resulting from elevated levels of pituitary TSH in human myxoedema. Despite a similar molecular weight (28,000) this human chorionic thyrotrophin (HCT) is immunologically distinct from human TSH being inhibited in the assay system by antisera to *bovine* but not *human* TSH. Another thyroid stimulating substance first isolated in molar pregnancy (Hershman et al 1970) is present in the urine of apparently normal pregnant woman. Molar thyrotrophin (HMT) has a much higher molecular weight (60,000–70,000) than HCT and is again immunologically distinct. The specific molecular relationship between these two thyrotrophins remains to be defined. The early appearance of HCT and its persistence in normal pregnancy plasma could provide an explanation for thyroid enlargement, increased ^{132}I clearance and free thyroxine levels (Fisher et al 1970b), which may appear before overt iodine deficiency develops, and for the relative thyroid autonomy which persists throughout.

THE FOETAL THYROID

Within the primitive foetal thyroid, follicular differentiation is apparent by the 11th day in the developing chick, 19th day in the rat, 22nd day in the rabbit and 12th week in man. This change is relatively rapid with the appearance of colloid droplets in syncytial cell masses and later formation of micro follicles (Gorbman 1955). The process is apparently independent of TSH and proceeds after foetal hypophysectomy (Jost 1953) though TSH has been demonstrated by radioimmunoassay in the human foetal pituitary at the 12th week (Fukuchi et al 1970). Once

TABLE 10.3. Thyroxine and TSH levels in maternal and foetal serum during pregnancy. Summary of maternal and foetal serum thyroxine (T_4), free thyroxine (FT_4) and TSH concentration data between 11-week gestation and term. (Numbers in parentheses indicate number of samples.) From Fisher D.A., Calvin J., Hobel M.D., Garza R. & Pierce C. (1970) Thyroid function in the preterm foetus. *Paediatrics* **46**, 208–216.

Period of gestation (weeks)	T_4 (μg/100 ml)	Maternal FT_4 (mμg/100 m$^\cdot$)	TSH (μU/ml)	T_4 (μg/100 ml)	Foetal FT_4 (mμg/100 ml)	TSH (μU/ml)
11–18	12·9 ± 1·1 (10)	2·97 ± 0·27 (10)	4·2 ± 0·68 (14)	2·6 ± 0·24 (9)	1·85 ± 0·17 (8)	2·4 ± 0·14 (16)
22–34	12·2 ± 0·51 (16)	2·82 ± 0·12 (16)	3·8 ± 0·38 (21)	7·2 ± 0·61 (16)	2·49 ± 0·17 (16)	9·6 ± 0·93 (22)
38–40	11·5 ± 0·56 (17)	2·30 ± 0·13 (17)	4·3 ± 0·40 (16)	11·2 ± 0·43 (17)	2·90 ± 0·10 (17)	8·9 ± 0·93 (16)

initiated, differentiation is accelerated by TSH to which the newly formed follicles are sensitive. This sensitivity has been demonstrated in the rabbit where thiouracil will cause hyperplasia of the foetal thyroid as soon as microfollicles appear (Logothetopoulous & Scott 1956).

At about the 13th week following follicular differentiation the foetal thyroid begins to concentrate radioiodine, synthesize MIT and DIT and later the metabolically active hormones T_3 and T_4. These iodothyronines are bound in the serum to foetal TBP whose synthesis begins at about the 15th week. Studies by Fisher et al (1970b) indicate a low level of thyroid function in the human foetus between the 11th and 18th week (Table 10.3). The low serum total T_4 ($2 \cdot 6 \pm 0 \cdot 24$ μg/100 ml) reflects the low maximal binding capacity of foetal TBP (Greenberg et al 1970) and despite a high dialysable T_4 the absolute free T_4 concentration is low ($1 \cdot 85 \pm 0 \cdot 17$ m μg/100 ml). The thyroxine is presumably of foetal origin since T_4 does not cross the placenta early in pregnancy (Myant 1958). The TSH which is detectable by radioimmunoassay in the foetal serum at this time is also likely to be foetal in origin since studies in a number of animal species have failed to show placental transfer of thyrotrophin (Sobel et al 1960). Maternal serum contains a relatively high concentration of chorionic thyrotrophin (HCT) early in gestation (Hennen et al 1969). This might conceivably reach the foetal circulation by direct transfer or secretion. There is some cross-reactivity of this material in the thyrotrophin immunoassay though probably insufficient to account for the TSH levels measured. It seems unlikely to be a significant influence in the foetus though more data are needed.

Synthesis of thyroid hormone by the foetal thyroid is known at least in animals to be TSH dependent (Jost 1966) and the above data suggest that in the human the foetal pituitary exerts an influence as early as 12 weeks gestation. The control system is however immature, since serum TSH levels are low despite low free and total thyroxine concentrations. In the human a goitre has in fact been reported in the aborted 20-week foetus of a thiouracil treated mother (Davis & Forbes 1945) indicating negative feedback control of TSH at least as early as mid-gestation. As pregnancy proceeds there is a fairly abrupt change in foetal thyroid function between the 18th and 22nd week (Fisher et al 1970b, Fig. 10.12). Foetal serum TSH rises from a mean of $2 \cdot 4 \pm 0 \cdot 14$ μU/ml to $9 \cdot 6 \pm 0 \cdot 93$ μU/ml between the 22nd and 34th week and is maintained at this level until term. There is an increase in pituitary TSH content at about the same time and a parallel rise in thyroid radioiodine uptake

(Evans *et al* 1967) indicating a normal thyroid response. There is a progressive increase in total serum thyroxine to near maternal levels between 12 and 40 weeks which is in part due to a rise in TBG but an independent increase in free thyroxine concentration indicates increasing T_4 secretion. At term in the sheep the mean foetal serum free thyroxine is also higher than the maternal level and both T_4 and T_3

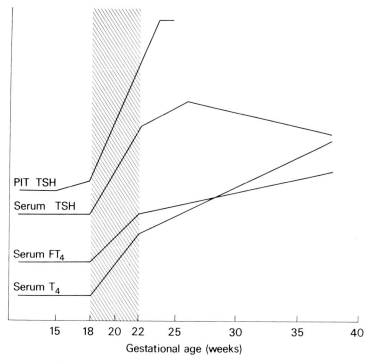

FIG. 10.12. Changes in foetal thyroid during pregnancy. By courtesy of D.A. Fisher. Unpublished figure. Data from *Pediatrics* (1970) **46**, 208.

turnover in the foetus exceeds the maternal rate by eight and three times respectively (Fisher 1971). In the human, measurement of serum thyroxine levels in the foetal scalp vein before delivery (and exposure to the external environment) indicates levels of free and bound thyroxine clearly higher than those in the maternal circulation. Although clinical and experimental evidence favours limited materno-foetal iodothyronine transfer, this foetal–maternal gradient of free thyroxine favours movement in the opposite direction.

Foetal control systems

The available data thus suggest a maturation at about 20 weeks of the hypothalamic transducer system which controls pituitary TSH synthesis by TRH release; the increase in foetal pituitary FSH and LH content which occurs at about the same time (Levina 1968) being consistent. This intrauterine hypothalamic maturation has been likened to amphibian metamorphosis (Fisher *et al* 1970b) involving as it does histological and functional maturation of hypothalamus and median eminence and a progressive increase in thyroxine secretion rate. In the first trimester the feedback system is immature and the low levels of free iodothyronine elicit little TSH response. By mid-gestation both TSH and free thyroxine levels are increasing and negative feedback operates. By term there is evidence of a relative autonomy of foetal hypothalamic function with TSH and free thyroxine levels higher than those observed in the mother.

This form of relative autonomy which involves a shift upwards of the 'set point' of the servo mechanism results in a state of 'physiological hyperthyroidism' geared perhaps to meet the high metabolic requirements of the foetus and the needs of the extra-uterine environment. Oestradiol benzoate has been shown to augment TSH secretion in newborn infants (Fisher & Oddie 1963). The rising oestrogen level of pregnancy could be implicated though there is no parallel rise in maternal thyroid function at this time. Despite this alteration in regulation, the occurrence of foetal goitre at term following maternal antithyroid drug treatment indicates that the servo mechanism is intact.

Developmental implications

The extent to which the developing foetus requires thyromimetic hormones for tissue growth and differentiation is little understood. The evident retardation of skeletal maturation and intellectual development in the athyreotic cretin indicates that the definitive development of these two tissues is iodothyronine dependent. It is perhaps significant that the thyroid begins to function in rats and in man shortly after the appearance of the first centres of ossification (Osorio & Myant 1960) and that the degree of residual intellectual defect in the cretin correlates less with the time of onset of therapy than with the degree of skeletal maturation.

The fact that women with apparently complete thyroid failure may give birth to perfectly normal children and the evidence of absent ^{131}I-T$_4$ transfer from mother to child in early pregnancy indicates that the foetus neither receives nor requires maternal T$_4$ for its early development. Although foetal serum free T$_4$ is low before the 20th week the levels are significant and could have a permissive role, particularly if tissue sensitivity were increased. It remains possible that the metabolic contribution of T$_4$ is supplemented by foetal T$_3$ production at this time though no evidence on this point is as yet available. The clinical evidence is against significant early T$_3$ transfer from the mother though this needs testing. Since foetal thyroglobulin synthesis has not been shown to occur before the 12th week it must be concluded at present that cel-

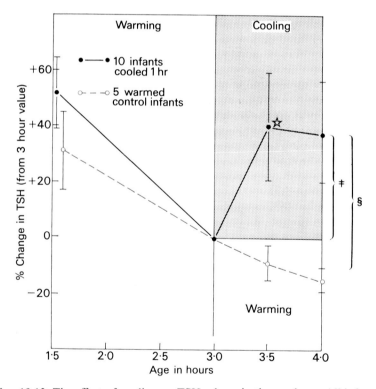

FIG. 10.13. The effect of cooling on TSH release in the newborn. All infants were warmed during the first 3 hours after birth. Warming did not prevent a serum TSH peak at 30 minutes and cooling of ten infants between 3 and 4 hours caused a secondary rise. From D.A.Fisher & W.D.Odell (1969). Acute release of thyrotrophin in the newborn. *J. clin. Invest.* **48**, 1670–1677.

lular growth is independent of thyroid hormone before this time. Later in pregnancy the situation is clear and foetal iodothyronine turnover has been estimated at about 35 μg/day for an average 7·5 lb infant (Fisher 1967); this secretion rate approaches that of the adult (50–60 μg/day) and is in accord with the high serum free T_4 and disposal rates at that time.

The neonate

Within 1 hour of birth there is an abrupt increment in TSH (Hershman & Pittman 1971) which is at least partly caused by the fall in body temperature at that time (Fig. 10.13). Warming of the infant during the first 3 hours of life does not prevent this early acute release of thyrotrophin but cooling between 3 and 4 hours has been shown to cause a significant increase in serum TSH concentration (Fisher & Odell 1969). The increased TSH secretion is reflected by an increase in the serum PBI to a peak level of about 11 μg/100 ml at 24 hours. Thereafter both TSH and PBI fall though normal levels are not reached until about the 6th week. Since the serum T_3 probably rises concordantly, the total caloric contribution of the two hormones is considerable and gives rise to a state of hypermetabolism which is reflected in the high rate of thyroxine utilization in the newborn (Cottino *et al* 1961). Despite this hypersecretion, the neonatal servo-control system like that in the foetus, is operating and is sensitive to T_4 administered between the 3rd and 5th day of post-natal life (Fisher & Oddie 1964). The current data do not, however, indicate whether the 'set point' of the negative feedback mechanisms is still elevated.

FOETO-MATERNAL RELATIONSHIPS (Fig. 10.14)

Thyroid hormones

The degree to which maternal and foetal thyroid function are interrelated is still incompletely known. Current evidence suggests that the thyroid economy of both is largely independent. The mothers of sporadic athyreotic cretins are usually euthyroid, and hypothyroid mothers produce euthyroid infants (Myant 1964). Administration of physiological dosage of T_4 or T_3 to laboratory animals will prevent the goitrogenic action of antithyroid drugs yet foetal goitre is either unaffected (D'Angelo 1967) or has regressed only at comparatively high dosage of administered thyroid hormones (Knobil & Josimovich 1958). Such protection by pharmacological dosage has been claimed in the human. Carr et al (1959) showed that the administration of desiccated thyroid to mothers during pregnancy in doses of 1200 mg/day prevented cretinism in two infants subsequently shown to be athyreotic. In retrospect it may be relevant that this amount contains about 200 μg of T_3. Newborn BEI levels although significantly elevated were considerably below maternal levels in another study in which thyroxine was given in single doses as high as 8000 μg some hours before delivery (Fisher et al 1964). Studies have confirmed slow and incomplete transfer of ^{131}I-T_4 from mother to foetus at term (Grumbach & Werner 1956) and a virtual absence in early pregnancy (Osorio & Myant 1960). Thus for most of its intra-uterine life the foetus determines its own thyroidal environment. The observation that the athyreotic cretin often shows little evidence of thyroxine lack for some days after delivery suggests a limited maternal contribution at term, though the situation differs from the normal where both maternal and foetal TBP are normally saturated (see below). Any form of transplacental equilibrium implies a corresponding foeto-maternal transfer of thyroid hormone. Evidence for this in the human is slight though the increased foeto-maternal gradient of free thyroxine (Fisher et al 1970a) would favour it. A number of factors may be concerned in limiting transplacental passage of thyroid hormone. The ratio of the respective binding powers of foetal and maternal TBP doubtless determines the concentration of *total* thyroxine on either side of the placenta (Osorio & Myant 1960). It has been suggested that in maternal hypothyroidism, the increased TBP capacity of the maternal serum would favour passage of thyroxine from foetus to mother, a situation perpetuated by the use of triiodothyronine in treatment since this hormone maintains euthyroidism without significant TBP binding (Van Wyk 1967). Such a sequence would be expected to lower foetal serum T_4 with a resultant increase in production of foetal TSH and goitre; this phenomenon has not been reported, suggesting that even in pathological states, placental 'permeability' is of primary importance.

Placental permeability and T_3

Placental blood flow, tissue adsorption and metabolism may all play a part in determining the permeability of the placental membranes to thyroid hormone. Molecular size is unlikely to be involved. Active placental metabolism is suggested by the obervation that ^{131}I-T$_3$ given to pregnant monkeys appears in the foetus either as iodide or as an organic compound which differs chromatographically from T$_3$ (Shultz et al 1965). Placental permeability may be particularly significant for triiodothyronine whose protein binding is minimal. Following injection of ^{131}I-T$_4$ and ^{131}I-T$_3$ into pregnant women 2–8 hours before delivery, cord blood was shown to contain 5% of the T$_4$ and 25% of T$_3$ (Kearns & Hutson 1963). The administration of unlabelled T$_3$ to eight women in doses of 150–300 μg daily during the last 4 to 6 weeks of pregnancy was shown by Raiti et al (1967) to cause a reduction in foetal serum T$_4$ 'by column' in only four despite adequate suppression of maternal serum T$_4$. Actual measurement of total and free T$_3$ levels in foetal and maternal serum in a similar study (Dussault 1969) showed a significant increase in serum T$_3$. The values, however, did not approach those of the maternal serum and the reduction in T$_4$/T$_3$ ratio was less for the foetus than the mother. There is thus significant foeto-maternal transfer of T$_3$ when this hormone is given to the mother in a pharmacological dose. The relatively small change in the T$_4$/T$_3$ ratio in the foetus may at least be partly explained by a relative autonomy of foetal TSH release. Two mothers in this study received only 50 μg T$_3$ daily over a period of 3 weeks. The cord serum T$_3$ values were 0·52 and 0·58 μg/100 ml significantly increased over the mean level of the control group of 0·32 ± 0·13. Such dosage is within the physiological range and if confirmed, this observation could have important biological and therapeutic implications.

Iodide and thyrotrophin

In contrast to the considerable doubt about placental passage of thyroid hormone the evidence for iodide and thyrotrophin is fairly clear. In the guinea pig placenta there exists a mechanism for the active concentration of iodide on the foetal side of the placenta. This can be blocked by thiocyanate and is analogous to the thyroid trapping mechanism (London et al 1964). There is no evidence for this in the human and injected ^{131}I rapidly equilibrates in equal concentration on both sides of the placenta (Kearns & Hutson 1963). High doses of iodide may thus cross the placenta and reach the foetus where by blocking intrathyroidal organic binding of iodine (Wolff-Chaikoff effect) it may cause hypothyroidism and goitre. There is no evidence that thyrotrophic hormone crosses the placenta in any of the animal species so far studied (Myant

1964). In contrast the long-acting thyroid stimulator (LATS) crosses freely, the situation being analogous to the transient passive immunity of the newborn dependent on the passage of other maternal IgG globulin antibodies. LATS titres tend to be higher in the foetus than the mother and its estimated half-life of 6–14 days is shorter than that of other gamma

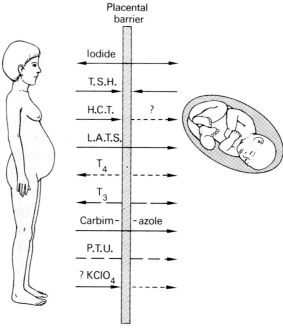

FIG. 10.14. A summary of foetomaternal relationships. No attempt is made to indicate the direction of gradients. The information on the antithyroid drugs is yet to be confirmed in the human. Modified from Furth E.D. (1969) *Proc. Fifth Midwest Conference on the Thyroid.* Eds. Senhauser, D.A. & Anderson R.R., p. 79.

globulins (16–28 days) measured in the neonate, a difference attributed to selective utilization in the target organ (thyroid) and increased turnover in the presence of hyperthyroidism. The observed duration of the clinical symptoms of thyrotoxicosis in the neonate of 3 weeks to 2 months is consistent. There is no information available concerning the presence or otherwise of human chorionic thyrotrophin (HCT) and molar thyrotrophin (HMT) in the foetal circulation.

Antithyroid drugs

Of particular interest to the clinician is the behaviour of antithyroid drugs at the placental barrier. Indirect evidence of transfer is based on the observation of hypertrophy or decreased radioiodine concentration in the foetal thyroid (Peterson & Young 1952) when such drugs are given to the mother. Rapid placental transfer of ^{35}S thiourea (Sheppard 1963) and ^{14}C thiouracil (Hayashi & Gilling 1967) has been demonstrated in rats. More recently placental transfer of ^{35}S methimazole (and carbimazole) and ^{35}S propylthiouracil PTU has been demonstrated (Brownlie et al 1971). The rat studies showed foeto-maternal serum ratios of 1:1 at 20 minutes for methimazole and 0·71:1 at 60 minutes for PTU after intravenous injections of these labelled materials. Similar results were obtained in humans (Brownlie, personal communication). The delayed transfer of PTU may be related to the protein binding of this drug. Methimazole which is more soluble is not protein bound in the circulation. These studies suggest that propylthiouracil may have a theoretical advantage in pregnancy though further information concerning absolute drug levels is required. Preliminary animal data indicate that perchlorate transfer is even less than the other antithyroid drugs though the occurrence of aplastic anaemia with this compound has restricted its use in humans.

REFERENCES

ABOUL-KHAIR S.A., CROOKS J., TURNBULL A.C. & HYTTEN F.E. (1964) The physiological changes in thyroid function during pregnancy. *Clin. Sci.* **27**, 195–207.

ADAMS D.D. & PURVES H.D. (1956) Abnormal responses in the assay of thyrotrophin. *Proc. Univ. Otago Med. School* **34**, 11–12.

ARBOGAST B. & HOCH F.L. (1968) Iodine content of submitochondrial particles prepared from rat liver by drastic sonification. *F.E.B.S. Letters* **1**, 315–376.

AUBACH G.D. & CHASE L.R. (1970) Cyclic 3',5'-adenylic acid in bone and the mechanism of action of parathyroid hormone. *Federation Proc.* **29**, 1179–1182.

BARKER S.B. (1964) In *Physiological Activity of Thyroid Hormones and Analogues in the Thyroid Gland*, R.V.H. Pitt-Rivers & W.G.Trotter (eds.), vol. 1, pp. 199–236. London, Butterworth.

BENUA R.S., LEEPER R.D., KUMAOKA S. & RAWSON R.W. (1960) Metabolic effects of thyroxine analogues in human myxoedema. *Ann. New York Acad. Sci.* **86**, 563–581.

BOLER J., ENZMANN F., FOLKERS K., BOWERS C.Y. & SCHALLY A.V. (1969) The identity of chemical and hormonal properties of the thyrotrophin releasing hormone and pyroglutamyl-histidyl-proline amide. *Biochemical and Biophysical Research Communications*, **37**, 705–710.

BRAVERMAN L.E., INGBAR S.H. & STERLING K. (1970) Conversion of thyroxine (T_4) to triiodothyronine (T^3) in athyreotic human subjects. *J. clin. Invest.* **49**, 855–864.

BROWNLIE B.E.W. (1971) Personal communication.
BROWNLIE B.E.W., MARCHANT B. & ALEXANDER W.D. (1971) Placental transfer of 35S-propylthiouracil and 35S-methimazole in the rat, in R.Hofer (ed.) *Further Progress in Thyroid Research.* Proceedings of the Sixth International Thyroid Congress.
BRUCE H.M. & SLOVITER H.A. (1957) Effect of destruction of thyroid tissue by radioactive iodine on reproduction in mice. *J. Endocr.* **15**, 72–82.
CARR E.A., BEIERWALTES W.H., RAMAND G., DODSON V.N., TANTON J., BETZ J.S. & STANBAUGH R.A. (1959) The effect of maternal thyroid function on foetal thyroid function and development. *J. clin. Endocr.* **19**, 1–5.
CHEEK D.B. (1968) Cellular growth hormones, nutrition and time. *Paediatrics* **41**, 30–46.
COPP D.H., CAMERON E.C., CHENEY B.A., DAVIDSON B.G.H. & HENZE K.G. (1962) Evidence for calcitonin—a new hormone from the parathyroid that lowers blood calcium. *Endocrinology* **70**, 638–649.
COPP D.J. (1970) Calcitonin and calcium metabolism. *Canad. Med. Ass. J.* **103**, 821–824.
COTTINO F., COLOMBO G., FERRARA G.C. & COSTA A. (1961) Investigations on the metabolism of the thyroid hormone in children by means of radiothyroxine *Panminerva Medica*, **3**, 471–474.
CROOKS J., TULLOCH M.I., TURNBULL A.C., DAVIDSSON D., SKULASON T. & SNAEDAL G. (1967) Comparative incidence of goitre in pregnancy in Iceland and Scotland. *Lancet* **ii**, 625–627.
D'ANGELO S.A., (1967) Pituitary thyroid inter-relations in maternal foetal and neonatal guineapigs. *Endocrinology* **81**, 132–138.
DAVIS L. & FORBES W. (1945) Thiouracil in pregnancy. Effect on foetal thyroid. *Lancet* **ii**, 740–742.
DOWLING J.T., APPLETON W.G. & NICOLOFF J.T. (1967) Thyroxine turnover during pregnancy. *J. clin. Endocr.* **27**, 1749–1750.
DUSSAULT J., ROW V.V., LICKRISH G. & VOLPÉ R. (1969) Studies of serum tri-iodothyronine concentration in maternal and cord blood: transfer of triidothyronine across the human placenta. *J. clin. Endocr.* **29**, 595–603.
EVANS T.C., KRETZSCHMAR R.M., HODGES R.E. & SONG C.W. (1967) Radioiodine uptake studies of the human foetal thyroid. *J. Nuclear Med.* **8**, 157–165.
FISCHMAN J., HELLMAN L., ZUMOFF B. & GALLAGHER T.F. (1965) Effects of thyroid on hydroxylation of oestrogen in man. *J. clin. Endocr.* **25**, 365–368.
FISHER D.A. & ODDIE T.H. (1963) Thyroxine secretion rate during infancy: the effect of oestrogen. *J. clin. Endocr.* **23**, 811–819.
FISHER D.A., LEHMAN H., & LACKEY C. (1964) Placental transfer of thyroxine. *J. clin. Endocr.* **24**, 393–400.
FISHER D.A. & ODDIE T.H., (1964) Neonatal thyroidal hyperactivity. *Amer. J. Dis. Child.* **107**, 574–581.
FISHER D.A. (1967) Hyperthyroidism in the pregnant woman and the neonate. Panel discussion. Moderator S.C.Werner. *J. clin. Endocr.* **27**, 1637–1654.
FISHER D.A. & ODELL W.D. (1969) Acute release of thyrotrophin in the newborn. *J. clin. Invest.* **48**, 1670–1677.

Fisher D.A., Odell W.D., Hobel C.J. & Garza, B.S. (1970) Thyroid function in the term foetus. *Paediatrics* **44**, 526–535.

Fisher D.A., Hobel C.J., Garza R. & Pearce C.A. (1970) Thyroid function in the preterm foetus. *Paediatrics* **46**, 208.

Fisher D.A. (1971). Personal communication.

Freedberg I.M., Hamolsky M.W. & Freedberg A.S. (1957) Thyroid gland in pregnancy. *New Eng. J. Med.* **256**, 505–510, 551–555.

Fukuchi M., Inoue T., Abe H. & Kumahara Y. (1970) Thyrotropin in human foetal pituitaries. *J. clin. Endocr.* **31**, 565–569.

Galante L., Gudmundsson T., Matthews E.T.A., Williams E., Woodhouse N. & McIntyre I. (1968) Thymic and parathyroid origin of calcitonin in man. *Lancet* **ii**, 537–538.

Gallagher T.F., Hellman L., Bradlow H.L., Zumoff B. & Fukushima D.K. (1960). Effects of thyroid hormones on metabolism of steroids. *Ann. New York Acad. Sci.* **86**, 605–611.

Galton V.A. (1968) Thyroid hormone metabolism, in V.H.T.James (ed.) *Recent Advances in Endocrinology*, pp. 181–206. London, Churchill.

Gharib H., Mayberry W.E. & Ryan R.J. (1970) Radioimmunoassay for triiodothyronine: a preliminary report. *J. clin. Endocr.* **31**, 709–712.

Gillman J. & Gilbert C. (1953) Thyroid gland and its relations to the menstrual cycle of the baboon. *J. Obstet. Gynaec. Brit. Emp.* **60**, 445–475.

Goolden A.W.G., Bateman D. & Torr, S. (1971) Red cell sodium in hyperthyroidism. *Brit med. J.* **2**, 552–554.

Gorbman, A. (1955) The thyroid. Report of Symposium, June 9–11, 1954, p. 3. New York, Brookhaven National Laboratory.

Greenberg A.H., Czernichow P., Reba R.C., Tyson J. & Blizzard R.M. (1970) Observations on the maturation of the thyroid function in early foetal life. *J. clin. Invest.* **49**, 790–803.

Grumbach M.M. & Werner S.C. (1956) Transfer of thyroid hormone across human placenta at term. *J. clin. Endocr.* **16**, 1392–1395.

Halnan K.E. (1958) The radioiodine uptake of the human thyroid in pregnancy. *Clin. Sci.* **17**, 281–290.

Hayashi T.T. & Gilling B. (1967) Placental transfer of thiouracil. *Obstet. & Gynaec.* **30**, 736–740.

Hennen G. (1965) Detection and study of a human-chorionic-thyroid-stimulating factor. *Arch. Int. Physiol.* **73**, 689–695.

Hennen G., Pierce J.G. & Freychet P. (1969) Human chorionic thyrotropin, further characterization and study of its secretion during pregnancy. *J. clin. Endocr.* **29**, 581–594.

Hershman J.M. & Higgins H.P. (1971) Hydatidiform mole—a cause of clinical hyperthyroidism. *New Engl. J. Med.* **284**, 573–577.

Hershman J.M., Higgins H.P. & Starnes W.R. (1970) Differences between thyroid stimulator in hydatidiform moles and human chorionic thyrotrophin. *Metabolism* **19**, 735–744.

Hershman J.M. & Pittman J.A. (1971) Utility of the radioimmunoassay of serum thyrotrophin in man. *Ann. Int. Med.* **74**, 481–490.

HERSHMAN J.M. & STARNES W.R. (1969) Extraction and characterization of a thyrotropic material from the human placenta. *J. clin. Invest.* **48**, 923–929.

HIRSCH P.F. & MUNSON P.L. (1969) Thyrocalcitonin. *Physiol Rev.* **49**, 548–622.

HIRSCH P.F., GAUTHIER G.F. & MUNSON P.L. (1963) Thyroid hypocalcaemic principle and recurrent laryngeal nerve injury as factors affecting the response to parathyroidectomy in rats. *Endocrinology* **73**, 244–252.

HUNG W. (1966) Elevation of protein-bound iodine in an eight year old due to transplacental passage of an organic iodine dye. *Paediatrics* **37**, 677–680.

IBBERTSON H.K., ROCHE A.H.G. & PYBUS J. (1967) The thyroid and calcium homeostasis in man: evaluation by calcium infusion. *Australasian Annals of Med.* **16**, 121–124.

JORGENSEN E.C., LEHMAN P.A., GREENBERG C. & ZENKER N. (1962) Thyroxine analogues VII antigastrogenic, calorigenic and hypocholesteraemic activities of some oliphatic, acyclic and aromatic ethers of 3, 5-diiodotyrosine in the rat. *J. biol. Chem.* **237**, 3832.

JOST A. (1953) Sur le développement de la thyroide chez le foetus de lapin. *Arch. Anat. Microsc. Morphol. Exper.* **42**, 168–183.

JOST A. Anterior pituitary function in foetal life, in G.W.Harris & V.T.Donovan (eds.) *The Pituitary Gland*, vol. II, pp. 299–323. London, Butterworth.

KAPLAN C.C. & YOUNG D.M. (1967) Thyrocalcitonin: evidence for release in a spontaneous hypocalcaemic disorder. *Science* **157**, 205–206.

KEARNS J.E. & HUTSON W. (1963) Tagged isomers and analogues of thyroxine (their transmission across human placenta and other studies). *J. Nuclear Med.* **4**, 453–461.

KNOBIL E. & JOSIMOVICH J.B. (1958) Placental transfer of thyrotropic hormone—thyroxine, triiodothyronine and insulin and rats. *Ann. New York Acad. Sci.* **75**, 895–904.

LEVEY G.S. (1971) Catecholamine sensitivity, thyroid hormone and the heart. A reevaluation. *Amer. J. Med.* **50**, 413–420.

LEVEY G.S. & EPSTEIN S.E. (1969) Activation of adenyl cyclase by glucagon in cat and human heart. *Circl. Res.* **24**, 151.

LEVINA S.E. (1968) Endocrine features in development of human hypothalamus hypophysis and placenta. *Gen. Comp. Endocr.* **11**, 151–154.

LIEWENDAHL K. & LAMBERG B.A. (1969) Free thyroxine in serum and its use in clinical diagnosis. *Acta Endocrinol.* **61**, 343–349.

LOGOTHETOPOULOUS J. & SCOTT R.F. (1956) Histology and function of the developing foetal thyroid in normal and goitrous guinea pigs. *J. Endocr.* **14**, 217–227.

LONDON W.T., MONEY W.L. & RAWSON R.W. (1964) Placental transfer of ^{131}I-labelled iodide in the guinea pig. *J. Endocr.* **28**, 247–252.

LYNN W.G. (1969) The thyroid gland and reproduction in cold blooded vertebrates. *Proceedings of the Fifth Mid-West Conference on Thyroid*, pp. 17–30.

MAGNUS-LEVY A. (1895) Über den respiratorischen Gaswechsel unter den Einfluss der Thyroidea sowie unter verschiedenden pathologischen Zustanden. *Berlin Klin. Wschr.* **32**, 650–652.

MANN L.I., LUTZ M. & SCHULMAN H. (1967) Hydatidiform mole with hyperthyroidism. *Amer. J. Obstet. Gynec.* **98**, 1151–1152.

MAQSOOD M. (1952) Thyroid function in relation to reproduction of mammals and birds. *Biol. Rev.* **27**, 281–319.

MAYER G.P. (1970) Possible relation of calcitonin to the development of hypocalcaemia in parturient cows, in S.Taylor & G.Foster (eds.) *Calcitonin 1969*, pp. 386–391. Proceedings of the Second International Symposium. London, Heinemann.

MILHAUD G., MOUKHTAR M.S., PERAULT-STAUB A.M., COURTIS G., ARDAILLOU R., BLOCH-MICHEL H., GAREL J.M., KLINGER E., TUBIANA M. & SIZONENKO P.C. (1970) Studies in thyrocalcitonin, in S.Taylor & G.Foster (eds.) *Calcitonin 1969*, pp. 182–193. Proceedings of the Second International Symposium. London, Heinemann.

MUNSON P.L. & GREY T.K. (1970) Function of thyrocalcitonin in normal physiology. *Federation Proc.* **29**, 1206–1208.

MURPHY B.E.P., PATTEE C.J. & GOLD A. (1966) Clinical evaluation of a new method for the determination of serum thyroxine. *J. clin. Endocr.* **26**, 247–256.

MYANT N.B. (1958) Passage of thyroxine and triiodothryonine from mother to foetus in pregnant women. *Clin. Sci.* **17**, 75–79.

MYANT N.B. (1964) The thyroid and reproduction in mammals, in R.V.H.Pitt-Rivers & W.R.Foster (eds.) *The Thyroid Gland*, vol. I, pp. 283–302. London, Butterworth.

NAUMAN J.A., NAUMAN A., WERNER S.C. (1967) Total and free triiodothyronine in human serum. *J. clin. Invest.* **46**, 1346–1355.

ORMSTON B.J., KILBORN J.R., GARRY R., AMOS J. & HALL R. (1971) Further observations on the effect of synthetic thyrotrophin-releasing hormone in man. *Brit. med. J.* **ii**, 199–202.

OSORIO C. & MYANT N.B. (1960) Thyroid hormone in pregnancy. *Brit. med. Bull.* **16**, 159–163.

PATEL Y. (1971) Personal communication.

PEARSE A.G.E. (1970) The characteristics of the C cell and their significance in relation to those of other endocrine polypeptide cells and to the synthesis storage and secretion of calcitonin, in S.Taylor & G.Foster (eds.) *Calcitonin 1969*, pp. 125–140. Proceedings of the Second International Symposium, London, Heinemann.

PETERSON R.R. & YOUNG W.C. (1952) The problem of placental permeability for thyrotrophin, propylthiouracil and thyroxine in the guinea pig. *Endocrinology* **50**, 218–225.

PILEGGI V.J., SEGAL H.A. & GOLUB O.J. (1964) Determination of organic iodine compounds in serum III iodotyrosines in normal human serum. *J. clin. Endocr.* **24**, 273–280.

PITTMAN C.S., CHAMBERS J.B. JR. & READ V.H. (1971) The extrathyroidal conversion rate of thyroxine to triiodothyronine in normal man. *J. clin. Invest.* **50**, 1187–1196.

RASMUSSEN H. (1971) Ionic and hormonal control of calcium homeostasis. *Amer. J. Med.* **50**, 567–588.

RAISZ L.G. & NIEMANN I. (1967) Early effects of parathyroid hormone and thyrocalcitonin on bone in organ culture. *Nature* **214**, 486–487.

RAITI S., HOLZMAN G.B., SCOTT R.L. & BLIZZARD R.M. (1967) Evidence for the placental transfer of triiodothyronine in human beings. *New Eng. J. Med.* **277**, 456–459.

ROITT I.M. & TORRIGIANI G. (1967) Symposium on the thyroid gland. *Suppl. J. clin. Path.* **20**, 391–394.

SCHULTZ M.A., FORSANDER J.B., CHEZ R.A. & HUTCHINSON D.L. (1965) The bi-directional placental transfer of ^{131}I 3:5:3:triiodothyronine in the rhesus monkey. *Paediatrics* **35**, 743–752.

SHEPARD T.H. (1963) Metabolism of thiourea S-35 by the foetal thyroid of the rat. *Endocrinology* **72**, 223–230.

SOBEL E.H., HAMBURN M. & KOBLIN R. (1960) Development of the foetal thyroid in rats—evidence for placental transfer of thyroxine. *Amer. J. Dis. Child.* **100**, 709–710.

SOKOLOFF L. & KAUFMAN S. (1961) Thyroxine stimulation of aminoacid incorporation into protein. *J. biol. Chem.* **236**, 795–803.

STANBURY J.B. & CHAPMAN E.M. (1960) Congenital hypothyroidism with goitre. Absence of iodide-concentrating mechanism. *Lancet* **ii**, 1162–1165.

STANBURY J.B. & MCGIRR E.M. (1957) Sporadic or non-endemic familial cretinism with goitre. *Amer. J. Med.* **22**, 712–723.

STERLING K. (1971) The importance of circulating triiodothyronine. Editorial. *New Eng. J. Med.* **284**, 271–272.

STRAUSS M.B. (1971) Mechanism of thyroid stimulation of metabolism. Editorial. *Ann. Int. Med.* **74**, 793–794.

TATA J.R. (1967) The formation and distribution of ribosomes during hormone induced growth and development. *Biochem. J.* **104**, 1–16.

UTIGER R.D. (1971) Thyrotrophin, radioimmunoassay: another test of thyroid function. *Ann. Int. Med.* **74**, 627–629.

VALE W., BURGUS R. & GUILLEMIN R. (1968) Mechanism of action of TRF: Effects of cycloheximide and actinomycin on the release of TSH stimulated *in vitro* by TRF and its inhibition by thyroxine. *Neuroendocrinology* **3**, 34–36.

VAN WYK J.J. (1967) Hyperthyroidism in the pregnant woman and the neonate. Panel Discussion. Moderator S.C.Werner. *J. clin. Endocr.* **27**, 1637–1654.

VERDY M., BEAULIEU R., DEMERS L., STURTRIDGE W.C., THOMAS P. & KUMAR M.A. (1971) Plasma calcitonin activity in a patient with thyroid medullary carcinoma and her children with osteopetrosis. *J. clin. Endocr.* **32**, 216–221.

WERNER S.C. (1958) Effect of triiodothyronine administration on elevated protein-bound iodine level in human pregnancy. *Amer. J. Obstet. Gynec.* **75**, 1193–1196.

WHITTAM R. (1964) In J.F.Hoffman (ed.) *The Interdependence of Metabolism and Active Transport in the Cellular Functions of Membrane Transport*, p. 139. New York, Prentice-Hall Inc.

WINICK M. & ROSSO P. (1969) The effects of severe early malnutrition on cellular growth of human brain. *Paediatric Res.* **3**, 181–184.

WINIKOFF D. (1971) Oral contraceptives and thyroid function tests—the diagnosis of thyroid disease. *Med. J. Aust.* **1**, 1059–1063.

WOLFF J. (1964) Transport of iodide and other anions in the thyroid gland. *Physiol. Rev.* **44**, 45–90.

11

Carbohydrate and Lipid Metabolism During Pregnancy

JOHN R. TURTLE

INTRODUCTION

During normal pregnancy there are significant changes in carbohydrate and lipid metabolism. These are most obvious in patients suffering from diabetes mellitus, in whom pregnancy imposes a major stress and increases the clinical and biochemical manifestations of diabetes. Frequently the clinical signs of diabetes appear initially during pregnancy only to subside at least temporarily, when the pregnancy is over. During pregnancy in the non-diabetic individual the requirements for insulin secretion are markedly increased. If pregnancy occurs in a patient who has relative or absolute insulin deficiency due to diabetes, then the insulin requirements increase but the pancreas cannot respond to the added load. As a result the patient is more susceptible to hyperglycaemia, ketosis and the other complications of pregnancy.

The profound alterations in carbohydrate, lipid and insulin metabolism during pregnancy depend to a large extent on maternal hormone changes induced by the foeto-placental unit. Before considering in detail these metabolic alterations, it is necessary to briefly review normal carbohydrate and lipid metabolism in the non-pregnant individual.

CARBOHYDRATE METABOLISM

Glucose must enter the cell before it can be utilized. Within the cell it is promptly phosphorylated by hexokinase to glucose 6-phosphate. From this point onwards, glucose can be metabolized by two important pathways, either via anaerobic glycolysis or the pentose shunt; the products of these pathways are subsequently oxidized in the tricarboxylic acid cycle. Alternatively, glycogen synthesis or carbohydrate storage can occur. It is convenient to consider these pathways separately, although it is not implied that any metabolic process exists in isolation from the others, rather all pathways are proceeding simultaneously.

Anaerobic glycolysis

The glycolytic sequence is shown in Fig. 11.1. It commences with glucose and terminates with pyruvate or lactate. There is no utilization of oxygen during glycolysis. Although ATP is consumed in the generation of glucose 6-phosphate and fructose diphosphate, it is recovered at the diphosphoglycerate and phosphoenolpyruvate steps. As each glucose molecule is split into two trioses at the triose phosphate step, anaerobic glycolysis results in a net yield of 2 moles of ATP per mole of glucose utilized. Nicotinamide-adenine dinucleotide (NAD) is reduced to NADH in the synthesis of diphosphoglycerate, but the NADH is oxidized to NAD again in the formation of lactate from pyruvate. In skeletal muscle, particularly under aerobic conditions, it is this process of glucose utilization terminating with lactic acid that produces most of the energy for contraction. In the presence of oxygen, pyruvate is oxidized via the tricarboxylic acid cycle, as will be discussed in detail shortly.

Pentose shunt

An alternative to the glycolytic pathway is active in many tissues e.g. liver, adipose tissue, pancreatic islets. In the pentose shunt glucose 6-phosphate is oxidized to 6-phosphogluconate which in turn is decarboxylated to ribulose 5-phosphate (Fig. 11.2). A different nucleotide is involved in the oxidative steps of the pentose shunt. Nicotinamide-adenine nucleotide phosphate (NADP) is reduced to NADPH, an essential cofactor required for the synthesis of fatty acids from acetoacetyl CoA. The pentose shunt is favoured under conditions of increased fatty acid synthesis in liver and adipose tissue. The shunt may be relatively inactive in diabetes mellitus with ketosis when the synthesis of fatty acids is suppressed.

CARBOHYDRATE AND LIPID METABOLISM IN PREGNANCY

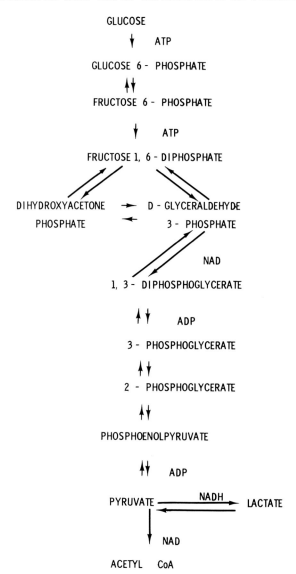

FIG. 11.1. A summary of the metabolic sequence of anaerobic glycosis.

The pentose shunt differs from anaerobic glycolysis in that carbon dioxide is produced and oxygen is consumed. ATP is not required for the oxidation of glucose 6-phosphate. The pathway provides a source for a number of sugars other than hexoses which are generated by oxidation of 6-phosphogluconate. Among these are the pentose derivatives, particularly ribose, an essential component for ribonucleic acid synthesis.

Tricarboxylic acid cycle

Pyruvate may be decarboxylated to acetyl CoA or may be enzymatically combined with CO_2 to form oxaloacetate. Thus pyruvate oxidation provides the two major sources of fuel (acetyl CoA and oxaloacetate) for the

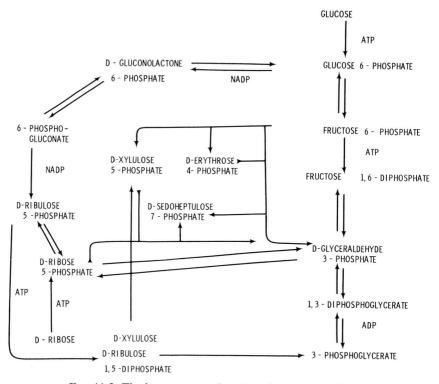

FIG. 11.2. The hexose monophosphate (or pentose) shunt.

tricarboxylic acid cycle. This cycle involves oxidation of substrates at several points (Fig. 11.3) and is a major source of CO_2 and ATP. Each turn of the cycle produces 12 molecules of ATP. This is accomplished by the oxidation of 8 H atoms from the components of the cycle. Oxidation of α-ketoglutarate provides 4 ATP but succinate oxidation produces only 2 molecules of ATP per 2 H atoms oxidized. In addition pyruvate, isocitrate and malate oxidation yields either NADH or NADPH and each mole of NADH oxidized by mitochondria can produce a maximum of 3 moles of ATP. Thus complete oxidation of pyruvate yields 15 moles of ATP. As each mole of glucose gives 2 moles of pyruvate with the production of 8 moles of ATP (calculating direct ATP production and oxidation of NADH), then the complete oxidation of glucose via glycolytic and tricarboxylic acid pathways might be expected to yield a total of 38 moles of ATP.

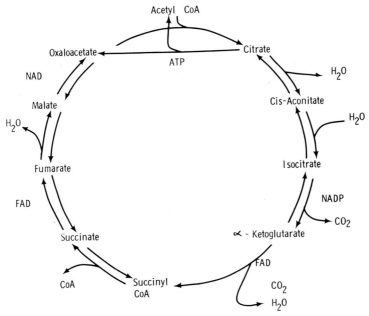

FIG. 11.3. The tricarboxylic acid cycle.

Glycogen synthesis and glycogenolysis

Following ingestion of carbohydrate, complex sugars are split to simple hexoses in the small intestine where they are absorbed and travel to the liver via the portal vein. As feeding is intermittent in the human, some form of storage is required to provide fuel for oxidation during periods of food deprivation. The glycogen concentration in liver is quite high indicating that the liver has a central role in energy storage. Glycogen is a complex branched polysaccharide comprised of multiple glucose units in 1–4 linkage with branches at 1–6 positions. Glucose must be phosphorylated then converted to glucose 1-phosphate by phosphoglucomutase. Subsequently uridine triphosphate is oxidized to UDP-glucose in the presence of glucose 1-phosphate. UDP-glucose then acts as a transfer system increasing the size of the glycogen molecule by adding a series of single glucose units. This whole system is sensitive to insulin and glycogen synthesis is accelerated by insulin. In the presence of cyclic 3'5' adenosine monophosphate (cyclic AMP), inactive phosphorylase b is converted to phosphorylase a, and the process of glycogenolysis, or glycogen breakdown to glucose, is initiated. A number of hormones, specifically epinephrine and glucagon, increase the intracellular concentration of cyclic AMP in the liver by activating adenyl cyclase, the membrane enzyme responsible for the synthesis of cyclic AMP from ATP. An increase in the intracellular concentration of cyclic AMP activates phosphorylase and glycogen is broken down to release glucose. During fasting, circulating insulin levels are low; thus glycogen synthesis is suppressed. Epinephrine and glucagon then mobilize glycogen from the liver and make glucose available for oxidation in essential tissue, e.g. brain.

LIPID METABOLISM

It has been recognized for many years that force-feeding of carbohydrate caused animals to accumulate fat greatly in excess of that present in the diet (Devel & Morehouse 1964, Rapport 1930). Conversion of glucose to fat has been demonstrated in the intact animal (Masoro *et al* 1949) and in liver slices (Chernick *et al* 1950). Under normal conditions lipogenesis from ingested carbohydrate is a quantitatively important process and it has been estimated that up to 90% of the glucose carbons stored in the body are retained as fat (Stetten & Boxer, 1944).

Lipogenesis and cholesterol synthesis

Although fatty acid synthesis can occur in the liver, the conversion of glucose to fatty acids takes place primarily in adipose tissue (Masoro *et al* 1949). The major biochemical steps in fatty acid synthesis are shown in Fig. 11.4. With minor exceptions, all fatty acids are derived from the common precursor acetyl CoA. Two molecules of acetyl CoA are condensed to form acetoacetyl CoA which is then reduced by NADH to yield β-hydroxybutyryl CoA. Subsequently this is hydrated and reduced to form butyryl CoA or the first step in sequential 2-carbon fatty acid synthesis. In the reduction step leading to the formation of butyryl CoA, NADPH acts as the H-donor and this may be a limiting step in fatty acid synthesis. Successive condensations with acetyl CoA and reductions by NADH and NADPH lead to the synthesis of long chain fatty acids, of which palmitic acid appears to be physiologically the most important.

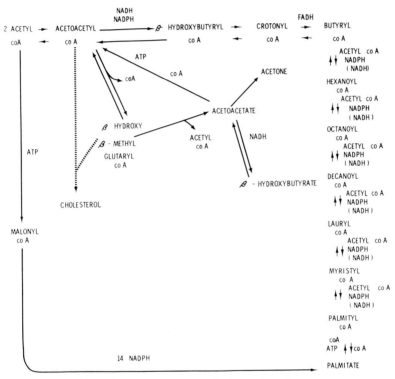

FIG. 11.4. Fatty acid and cholesterol synthesis.

Acetyl CoA is the principal substrate for the synthesis of cholesterol proceeding via acetoacetyl CoA and mevalonic acid (Fig. 11.4). In the formation of mevalonic acid, 2 molecules of NADPH are required. A series of ATP-consuming condensation steps follow to produce squalene, which then undergoes cyclization to form lanosterol and cholesterol with further oxidation of NADPH.

It should be noted that there is an extensive requirement for reduced NADPH in both fatty acid and cholesterol synthesis, This NADPH is usually provided by oxidation of glucose via the pentose shunt. Many of the abnormalities in lipid synthesis in diabetes may be explained by a reduced availability of NADPH normally generated in the pentose phosphate pathway.

Fatty acid oxidation

Triglycerides, the storage form of fatty acids condensed in a ratio of 3:1 with alpha glycerophosphate, are hydrolysed by a variety of tissue lipases to release free fatty acids and glycerol. Rapid re-esterification of fatty acids to triglyceride occurs providing there is a continuous source of alpha glycerophosphate from glycolysis in adipose tissue. As there is no glycerokinase in adipose tissue, glycerol itself cannot be re-utilized for fatty acid re-esterification. Before fatty acid oxidation can provide energy for tissues such as skeletal muscle during fasting, lipolysis must occur to release fatty acids from triglycerides. ATP and coenzyme A activate the long chain fatty acids to form the fatty acid COA derivative. This undergoes rapid stepwise oxidation and a molecule of NADH and FADH are produced at each step with the release of acetyl CoA. The 8 molecules of acetyl CoA produced by the oxidation of palmitic acid then can enter the tricarboxylic acid cycle to be further oxidized to CO_2, each yielding 12 molecules of ATP. The NADH and FADH can also lead to ATP formation through subsequent oxidation.

In diabetes mellitus, fatty acids are oxidized at an increased rate and this can be reversed by the administration of insulin. When carbohydrate is withheld from the diet, tissue triglycerides are converted to free fatty acids and glycerol, increasing circulating level of both. Fatty acids then provide the major source of energy during oxidation for skeletal muscle, leaving glucose available for essential tissues, such as brain. During starvation the changes in fatty acid oxidation and glucose utilization are similar to those produced by insulin deficiency in diabetes. It becomes difficult to explain the increased cholesterol synthesis seen in mild diabetes in terms of the availability of NADPH. A possible explanation is that fatty acid oxidation increases the availability of cholesterol precursors when glucose is limited so that a net increase in cholesterol synthesis results. Although this theory has not been proved it is supported by the fact that cholesterol synthesis in diabetes is closely related to the rate of ketone body production which is typical of the disease. Although ketone body oxidation is not significantly depressed in diabetes, the synthesis of acetoacetic acid, β-hydroxybutyric acid and acetone is markedly increased due to the accumulation of acetyl CoA during fatty acid degradation. Acetyl CoA inhibits citrate synthetase preventing further oxidation of acetyl CoA in the tricarboxylic acid cycle. If acetyl CoA does not accumulate, the tricarboxylic acid cycle is not inhibited, complete substrate oxidation occurs and ketosis does not develop.

PLACENTAL HORMONES AND THEIR ACTIONS ON CARBOHYDRATE AND LIPID METABOLISM

Pregnancy poses a number of unusual metabolic problems both for the mother and the developing foetus. Theoretically the mother should conserve exogenous nutrients in anticipation of the demands of the foetus which continue during periods of maternal food deprivation. Thus, enhanced anabolism should occur during feeding and enhanced catabolism during fasting in order that a constant nutrient can be maintained for the foetus (Freinkel 1965). During pregnancy net maternal weight gain is greater than would be expected from foetal growth alone (Scow et al 1964, Kumaresan & Turner 1968). The liver increases in size (Poo et al 1939) and net nitrogen retention occurs (Zuspan & Goodrich 1968). However catabolism is also accelerated with enhanced fat mobilization and frequently ketosis (Scow et al 1964, Bergman & Sellers, 1960).

These complex metabolic changes are induced by the development of the 'foeto-placental unit' and to a large extent they depend on gradual changes in the maternal endocrine environment due to direct secretion of placental hormones into the maternal circulation. Only those hormones which may be responsible for the changes in carbohydrate and lipid metabolism during pregnancy will be considered here.

Steroid hormones

The placenta is the source of massive amounts of progesterone produced during pregnancy. The enzymes necessary for the synthesis of cholesterol and subsequently progesterone from acetate are all present in the placenta (Ryan et al 1966) although the subsequent metabolism of progesterone by the placenta is limited (Sobrevilla et al 1964, Villee 1969). There is considerable evidence that oral contraceptive preparations may alter carbohydrate tolerance (Gershberg *et al* 1964, Kalkhoff *et al* 1968, Beck & Wells 1960) but the actions of progesterone itself on carbohydrate metabolism are less clear. Beck (1969) has shown that long-term treatment of monkeys with progesterone is associated with an exaggerated insulin response to intravenous glucose. Also in man, progesterone has catabolic effects on protein metabolism (Landau & Lugibihl 1961, 1967).

The enormous amounts of oestriol produced during pregnancy are dependent upon certain precursors provided to the placenta by the foetus (Chapter 4). Synthetic oestrogens such as mestranol produce carbohydrate intolerance post-partum in women who have demonstrated subclinical diabetes during pregnancy (Beck & Wells 1969). It is possible that oestrogens act similarly to glucocorticoids and increase the requirements for pancreatic insulin secretion (Welt *et al* 1952, Perley & Kipnis 1966); however it appears more likely that oestrogens induce end-organ resistance to glucose transport in peripheral tissue (Riddick *et al* 1962, Beck & Wells 1969), leading to decreased sensitivity of the intact organism to the hypoglycaemic action of insulin.

In addition to oestriol and progesterone, small amounts of corticosteroids and aldosterone have been found in placenta (Neher & Stark 1961) although it is unlikely that the production of any one of these would be sufficiently high to change any aspect of the metabolism of carbohydrates or lipids during pregnancy. Although the precise quantitative contribution of each of these hormones to the metabolic changes of pregnancy has not been established, it seems likely that their continuous interaction with human placental lactogen (HPL) may be implicated in lipolysis, the increase in plasma free fatty acids, pancreatic islet cell hyperplasia and insulin insensitivity during pregnancy (Herrera *et al* 1969).

Human placental lactogen

A peptide produced by the normal placenta has been found to be distinct from chorionic gonadotrophin (HCG). This peptide has been referred to by various investigators as human placental lactogen (HPL) chorionic growth substance, purified placental protein and more recently, human chorionic somatomammotrophin (Josimovich & McLaren, 1962, Kaplan & Grumbach 1964, Florini *et al* 1966). The peptide is present in human plasma only during pregnancy and in the serum of patients with trophoblastic tumours (Beck et al 1965). It is probably responsible for the lipolytic activity of pregnancy plasma (Fleicher et al 1964). HPL first appears in pregnancy plasma as early as 3 weeks gestation, reaches maximum levels at term and rapidly disappears after parturition. It shares with growth hormone certain physicochemical properties such as size, amino-acid composition, behaviour on electrophophoresis and immunological activity (Catt et al 1967, Josimovich et al 1963). The biological properties of HPL are also similar to those of growth hormone although its potency is much less. Friesen (1965) demonstrated that HPL increased the body weight of hypophysectomized rats and Florini et al (1966) showed that HPL potentiated the effect of a small dose of growth hormone on weight gain. Grumbach et al (1966) reported that HPL increased the plasma concentration of free fatty acids in children with hypopituitarism although Beck & Daughaday (1967) were unable to show any change in plasma free fatty acids during intravenous infusion of HPL into normal male adults. In vitro, Turtle & Kipnis (1967) showed that HPL accelerated lipolysis in adipose tissue and that HPL potentiated the lipopolytic action of a small amount of growth hormone. In animal experiments using either monkeys or rabbits, Riggi et al (1966) were able to show consistently that HPL increased plasma free fatty acids in similar fashion to growth hormone, although the potency of HPL was only about 0.1% that of growth hormone.

The concentration of HPL in pregnancy plasma reaches a maximum level of 2–5 $\mu g/ml$ at term (Beck & Daughaday, 1967). Assuming a half-life of approximately 20 minutes, the placenta must produce 2–3 g of HPL daily. Such vast quantities of a peptide with the lipolytic properties of HPL could account for the progressive increase in plasma-free fatty acids during pregnancy. Although HPL circulates in maternal plasma it cannot be detected in foetal blood; thus any changes in foetal metabolism which are induced by HPL must be mediated via the mother.

In addition to activating lipolysis, HPL has other effects on carbohydrate and protein metabolism. It inhibits glucose uptake (Turtle & Kipnis 1967) and maternal gluconeogenesis, sparing both glucose and protein (Grumbach et al 1968). The insulinogenic action of HPL increases circulating insulin levels which in turn favour protein synthesis and ensure a mobilizable source of maternal amino acids for transport to the foetus. Interpreting these metabolic actions of HPL,

Grumbach postulates that HPL is the 'growth hormone' of the latter half of pregnancy and that it has its metabolic actions entirely in the mother, to provide fuel for the foetus during its period of rapid growth.

Whatever the biological role of HPL might be, it certainly increases insulin secretion from the pancreas and induces morphologic changes in the beta cells similar to those seen during pregnancy (Malaisse et al 1969). Furthermore, it induces diabetic glucose tolerance curves in subclinical diabetic subjects, together with a delayed response in serum insulin (Kalkhoff et al 1969). Alone, it could account for many of the metabolic changes during pregnancy; however it is circulating together with increased concentration of other hormones. Although it is difficult to determine the precise contributions of each of these hormones, it seems likely that their continuous integrated action may be implicated in (i) the lipolytic action of pregnancy plasma, (ii) the increase plasma free fatty acids, (iii) the insulin resistance and (iv) the hyperinsulinaemia and pancreatic islet hyperplasia.

CARBOHYDRATE METABOLISM IN PREGNANCY

Carbohydrate tolerance

Many factors may alter glucose tolerance. Pregnancy is associated with changes in gastrointestinal absorption (Hytten & Leitch 1964) and renal function (Welsh & Sims 1960). These physiological variables make it difficult to interpret glucose tolerance in pregnancy and to determine the presence and significance of minor deviations from normal.

Jackson (1952) found that the glucose tolerance curve in late pregnancy was flat, but if this was compared with the same person in the non-pregnant state, it was abnormally high for that individual. In an attempt to overcome the difficulties in interpretation of blood glucose levels during an oral glucose tolerance test, many investigators have defined arbitrary limits, above which glucose tolerance is regarded as 'abnormal' (Hurwitz & Jensen 1946, Welsh 1960, Wilkerson & O'Sullivan 1963). In a careful study of unselected pregnant patients, O'Sullivan and Mahan (1964) have defined upper limits for a 100 gramme oral glucose tolerance test, using venous blood as follows:

Fasting: 90 mg/100 ml
1 Hour: 165 mg/100 ml

2 Hours: 145 mg/100 ml
3 Hours: 125 mg/100 ml

When these criteria were used, the subsequent rate of development of clinical diabetes in the 'abnormal' group was similar to that found by Conn & Fajans (1961) using the cortisone glucose tolerance test. It is interesting to compare O'Sullivan's and Mahan's figures with the upper limits established by the American Diabetes Association, i.e. fasting less than 100 mg/100 ml, not above 160 mg/100 ml at any point, less than 120 mg/100 ml at 1½ hours and less than 110 mg/100 ml at 2 hours. These figures for non-pregnant individuals are lower at each point after the glucose load than those of O'Sullivan and Mahan, providing further evidence that normal pregnancy provides a diabetogenic stress to the pancrease.

The intravenous glucose tolerance test using 25 g of glucose is highly reproducible and is preferred to the oral test by many investigators (Duncan 1956, McIntyre *et al* 1964). Using this test, Silverstone *et al* (1961) have shown that carbohydrate tolerance improves transiently in the first trimester, coinciding with an increase in the distribution volume of glucose. At this stage there is no significant change in insulin sensitivity. As soon as HPL reaches significant levels, insulin resistance appears and carbohydrate tolerance progressively deteriorates. Although Silverstone suggested that the early increase in carbohydrate tolerance could could be due to an increased secretion of insulin, there is no evidence in rats that either pancreatic insulin content or insulin secretion increases until pregnancy is beyond the 10th day in animals having a total gestation period of 22 days (Malaisse *et al* 1969). This would be analogous to the second trimester in humans and could not explain the transient improvement in carbohydrate tolerance during the first few weeks of pregnancy.

Although Silverstone regards the intravenous test as being capable of identifying the potential or pre-diabetic accurately, Benjamin & Casper (1967) found that the oral 100 g glucose tolerance test was more reliable. Comparisons of both tests on the same patients are limited, but in general, the intravenous test yields a much lower percentage of abnormals than the oral test (Ocampo *et al* 1964, Benjamin & Casper 1967). In an attempt to overcome these difficulties, the steroid modified glucose tolerance test has been used (Jackson 1961, Kyle *et al* 1964, Joplin *et al* 1961). The value of this test is doubtful as glucocorticoids merely increase insulin resistance by decreasing the effectiveness of insulin on peripheral tissues. Thus steroids non-specifically stress the pancreas in a similar way to pregnancy. It is not known how accurately the glucose tolerance test, performed in any of the two ways described above, can detect the potential or pre-diabetic. Herein the difficulty lies. Glucose intolerance cannot be regarded as a satisfactory marker for the presence or absence of diabetes. Therefore variable and unreliable results must occur if this crude test is used in an attempt to discriminate and to determine the presence of borderline changes as a consequence of an underlying cellular biochemical disorder.

Plasma insulin

The major endocrine defect in diabetes mellitus is an absolute or relative deficiency of insulin. When we recognize the role of insulin in the pathogenesis of diabetes and therapeutically in the correction of the gross metabolic derangement, then it becomes important to consider insulin metabolism in the normal individual, the derangement that occurs in diabetes and the effects of pregnancy (Fig. 11.5). Normally insulin release is prompt after the ingestion or intravenous injection of glucose, the glucose and insulin curves are in phase and the concentration of circulating plasma free fatty acids falls. The maturity onset diabetic has a delayed plasma insulin response to glucose, but the insulin concentration continues to rise to a level higher than in the non-diabetic (Yalow & Berson 1968). In contrast, the juvenile diabetics have no insulin response to glucose (Berson & Yalow 1962). An essential feature in both types of diabetes mellitus is the decreased insulin response in terms of the absolute blood glucose concentration. Although the maturity onset diabetic secretes more insulin than the normal person, his insulin response is less than would be expected from the glucose levels (Ricketts *et al* 1966). In addition to this relative insulin deficiency Cerasi & Luft (1967) have noted a characteristically sluggish insulin response to glucose. After an intravenous glucose infusion a normal person responds promptly with an increase in circulating insulin whereas a genetic diabetic responds only after a delay of 10 to 60 minutes.

In the normal non-diabetic woman, pregnancy causes significant peripheral antagonism to insulin. Although there may be a minimal change in blood glucose, the plasma insulin response to glucose and tolbutamide is considerably greater during pregnancy (Kalkhoff *et al*

1964). This increase in circulating insulin appears early in the second trimester and progressively rises towards parturition, coinciding with a progressive resistance to the effects of endogenous and exogenous insulin (Spellacy & Goetz 1963, Kalkhoff et al 1964, Freinkel 1964). Pregnancy imposes a major stress on the normal pancreas and frequently precipitates carbohydrate intolerance in the genetic diabetic. In a patient with a clinical diabetes, pregnancy further decreases glucose

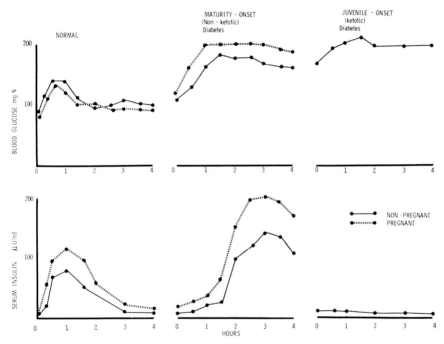

FIG. 11.5. Blood glucose and serum insulin response patterns following a glucose tolerance test in normal, maturity onset and juvenile onset diabetics. The effects of pregnancy are shown on both normal and diabetic patients.

tolerance and increases the exogenous insulin requirement. Following parturition, as HPL rapidly disappears from the maternal circulation, the normal plasma insulin response pattern reappears within a few days. In the normal pregnant woman HPL does not produce any carbohydrate intolerance; however, in subclinical diabetics during pregnancy or parturition, HPL infusion induces deterioration of carbohydrate tolerance as well as a delayed abnormal plasma insulin response (Kalkhoff et al 1969).

S

The most accurate marker for diabetes is the presence of a delayed plasma insulin response to a glucose load. In many patients this abnormal delay may be present although glucose levels are quite normal during a glucose tolerance test. In the presence of relative insulin deficiency, the additional stress of pregnancy increases insulin resistance and the insulin deficiency becomes more pronounced. When the pancreatic insulin output fails to maintain euglycaemia, carbohydrate intolerance will appear. Although it appears likely that HPL is responsible for the insulin deficiency, the dynamic interplay between HPL, oestrogens and progesterone is probably more important than any single hormone in the development of insulin resistance and the changes in carbohydrate metabolism which are characteristic of pregnancy.

Gluconeogenesis and ketogenesis

Circulating insulin is increased in normal pregnancy to levels which would be expected to inhibit lipolysis, hepatic gluconeogenesis, ketogenesis and amino-acid release (Cahill *et al* 1968). Although it is possible that some of this circulating insulin is less potent yet immunologically reactive, current evidence suggests that this is not so and full biological activity is retained. This hyperinsulinism may be fully biologically active in the fed pregnant woman, but during fasting, insulin levels fall to those seen in the non-pregnant state. Under these circumstances all the adaptations to starvation (i.e. mobilization of fat, activation of gluconeogenesis ketogenesis, muscle breakdown, etc.) are triggered more rapidly and more severely in the pregnant woman (Herrera *et al* 1969). Thus during food deprivation in pregnancy the 'normal' levels of insulin are insufficient to prevent gluconeogenesis and ketogenesis. As blood glucose falls, additional factors, e.g. catecholamines, are called into play to provide further catabolic fuels in the maternal circulation. What is the physiological reason for this response? It appears that this process of accelerated maternal catabolism during starvation is protective for the mother as well as the foetus. With an intermittently eating mother and a foetus that requires continuous nutrition, some form of rapid and efficient transport is essential. The increasing placental elaboration of a lipolytic gluconeogenic peptide such as HPL would activate maternal catabolism when insulin levels were limiting. High insulin levels during feeding would ensure the maximum conservation of maternal glucose and gluconeogenic precursors until they were required by both foetus and maternal brain. As insulin levels fell during starvation, gluconeogenesis would be promptly initiated thus maintaining an ideal temporal relationship at all times for complete nutrition of mother and foetus.

LIPID METABOLISM IN PREGNANCY

During pregnancy, the circulation of every general lipid fraction increases with the exception of the phospholipid lysolecithin (Robertson & Sprecher 1968). Although the hyperlipidaemia in the mother decreases in the first few days post-partum, normal levels are not reached for 2 to 5 days for free fatty acids and 2 to 6 months for the other lipid fractions. This slow decline suggests that a simple mechanism of foetoplacental transfer to the maternal circulation cannot be responsible for the lipid change seen during pregnancy.

Free fatty acids

Although maternal plasma free fatty acid levels rise progressively during pregnancy it is difficult to draw conclusions about the placental transport of free fatty acids, due to the complexity of changes in maternal and foetal circulation in the perinatal period. When maternal free fatty acids are elevated, there is no change in foetal umbilical cord free fatty acid levels, suggesting that free fatty acid transport does not occur. However, as glucose transport across the placenta is so rapid, relative maternal hyperglycaemia may permit sufficient glucose to enter the foetal circulation to esterify fatty acids rapidly and prevent any increase in umbilical vein blood.

Several investigators have shown that the umbilical cord blood free fatty acid pattern at birth does not resemble the mother's circulating free fatty acids (Renkonen 1966, Chen et al 1965, Robertson et al 1968). It appears likely that, *in utero*, the foetal free fatty acids are derived both from foetal sources and by selective placental transport from the mother, certainly as far as linoleic acid is concerned (King et al 1967). This fatty acid appears in the foetal circulation, it is of maternal origin and it rapidly falls in foetal blood following parturition. Although no definitive experiments have been performed in the human to demonstrate placental free fatty acid transport, both monkeys and rabbits transport radioactive free fatty acids from maternal to foetal circulation.

We can only infer that humans behave similarly and that placental transport of free fatty acid does occur. Providing this is so, a definitive role can be proposed for HPL as a lipolytic hormone providing constant fuel for transport to the foetus during maternal starvation.

Triglycerides

There are no data available which suggest direct placental transport of intact triglycerides from maternal to foetal circulation. It seems probable that maternal triglycerides could be hydrolysed by lipoprotein lipase to produce fatty acids and glycerol which could cross the placental barrier, although the intact triglyceride could not. The triglycerides of the foetal circulation probably arise from the foetal hepatic synthesis of new triglycerides (Shorland *et al* 1966).

Cholesterol

Although there is good evidence that maternal cholesterol crosses the placenta and enters the foetal circulation, it is not known whether cholesterol esters are transported (Goldwater & Stetten 1947, Chevallier 1964, Connor & Lin 1967). In rats, rabbits and guinea pigs, the percentage of foetal cholesterol which is derived from the maternal circulation falls from 70% towards the end of the first half of the pregnancy to 15 to 20% near term. Cholesterol esters are not hydrolysed by placental tissue, but can be formed in both maternal and foetal serum by the transfer of a fatty acid from lecithin to cholesterol.

Phospholipids

After the injection of labelled phospholipids into the pregnant animal, small amounts of radioactivity are found in foetal phospholipids (Nielson 1941). Although maternal phospholipids enter the placenta from the maternal circulation they probably do not pass directly into the foetal circulation. Selective transport, as phospholipid esters, may occur. However most foetal phospholipid is probably synthesized by foetal tissues.

Foetal lipids

Umbilical cord levels of the various lipid fractions are consistently lower than the corresponding maternal values (Renkonen 1966), with the exception of lysolecithin. Following delivery infant plasma-free fatty acids rise to a maximum within the next 12 hours (Novak *et al* 1964), coinciding with the peak of foetal plasma triglycerides (Zee 1967). Total cholesterol does not change within the first 12 hours. There is no acute decline in any foetal lipid component following delivery which would suggest that the newborn child had not been dependent on maternal lipids or had rapidly developed autonomous control of his own lipid metabolism immediately after birth.

PREGNANCY AND THE SUBCLINICAL DIABETIC

Pregnancy as a stress factor

It is probable that primarily HPL and, secondarily, a combination of oestrogens and progesterone are responsible for the diabetogenic stress of pregnancy. This stess is manifest mainly towards the end of pregnancy in the third trimester in all women, but the clinical manifestations of diabetes and hyperglycaemia occur in relatively few. If HPL is given to a normal woman during pregnancy, there is no alteration in glucose tolerance, but an HPL infusion in a pregnant subclinical diabetic woman consistently impairs glucose tolerance (Kalkhoff *et al* 1969). These same patients respond to other stress factors e.g. corticosteroids, by a similar breakdown in glucose tolerance. Patients with subclinical diabetes mellitus have a characteristic delay in insulin release from the pancreas during stress, whether this is a glucose load, tolbutamide or a combination of, for example, corticosteroid and glucose. If the insulinogenic index is calculated (an expression for the insulin response in terms of the corresponding glucose response), then subclinical and clinical diabetes have relative insulin deficiency. In other words, the pancreas cannot respond normally to the stresses imposed. In this clinical situation a further stress, such as pregnancy, increases the pancreatic insulin secretory load and the degree of relative insulin deficiency becomes more pronounced. Should the insulin deficiency become marked, then clinical hyperglycaemia and glycosuria will appear. At this stage a severe metabolic abnormality is already present and a crude measurement of blood glucose at random intervals throughout the day cannot indicate the degree of metabolic disorder in the mother, or the extent to which the foetus is exposed to a constant hyperglycaemic stimulus.

Other stress factors during pregnancy

Any other factors which increase the requirement for pancreatic insulin secretion will cause a further deterioration in the relative insulin deficiency of the subclinical or clinical diabetic. Although these factors all operate in the non-pregnant state, their effects become manifest more readily during pregnancy.

Obesity. In obesity insulin resistance increases and there is marked hyperinsulinism during a glucose tolerance test (Perley & Kipnis 1965), frequently in the absence of impaired carbohydrate tolerance. Excessive body weight imposes a marked stress on the pancreas to compensate for the peripheral antagonism; this aggravates the relative insulin deficiency in an individual who is genetically predisposed to diabetes.

Glucocorticoids. These hormones decrease glucose utilization in muscle and adipose tissue, accelerate gluconeogenesis and increase circulating free fatty acids. Perley & Kipnis (1966) showed that glucocorticoids cause a marked increase in the plasma insulin response to glucose and tolbutamide in both normal and diabetic individuals. As in the obese person, hyperinsulinism is secondary to steroid induced insulin resistance (Berger *et al* 1966).

Drugs. The pregnant woman is likely to be given a variety of drugs during pregnancy, particularly if she develops toxaemia of pregnancy. Of these, several may alter the pancreatic insulin secretory mechanism. The most important are dizaoxide (Seltzer & Allen 1965, Graber *et al* 1966) which inhibits insulin release from the pancreas similarly to epinephrine, and the benzothiadiazine diuretics which cause potassium loss, deplete the pancreatic islet potassium and impair insulin secretion (Fajans *et al* 1964).

Thyroxine. The thyrotoxic patient has marked insulin resistance. In the pregnant diabetic woman the coincidental development of thyrotoxicosis further increases insulin resistance and aggravates the metabolic consequence of diabetes.

Many factors may be responsible for alterations in the pancreatic insulin secretory mechanism of the development of insulin resistance. These should always be considered in the management of the pregnant subclinical or clinical diabetic. The elimination of any or all of these stress factors will decrease the pancreatic load and reduce the difficulties in control of the metabolic disorder in diabetes.

DIABETES AND THE FOETUS

Size

Although it was thought that the excessive size of many infants born to diabetic mothers was due to water retention, many studies have shown that these infants have less extracellular water and total body water than normal infants (Cheek *et al* 1961, Clapp *et al* 1962, Cokingitin *et al* 1963). They are fat, but the amount of adipose tissue is consistent with the increase in size (Osler 1960). Visceromegaly contributes to the excessive weight, but the weights of organs are proportional to the size of the babies (Jackson 1967).

An increased deposition of glycogen may occur in infants of diabetic mothers as well as a decrease in total body water.

Blood glucose and free fatty acids

Infants of diabetic women, whether or not they are treated with insulin, tend to become hypoglycaemic during the first hours of life (McCann *et al* 1966, Farquhar & Isles 1967). These infants dispose of glucose load faster than normal. It is now generally recognized that this neonatal hypoglycaemia may cause irreversible cerebral damage in these infants whose brains are no less susceptible to hypoglycaemia than adult brains (Cornblath *et al* 1959, Brown & Wallis 1963, Anderson *et al* 1966).

Immediately after birth the infant of a diabetic mother has lower plasma ketones and free fatty acids than a normal child (Melichar & Wolf 1967, Joassin *et al* 1967). In general the findings in the newborn infant of a diabetic mother are similar to those observed in carbohydrate fed adult animals subjected to sudden caloric deprivation. It appears likely that these infants are exposed to persistent hyperglycaemia, irrespective of the degree of apparent maternal diabetic control. The metabolic disorder in the newborn is a consequence of the sudden withdrawal of the hyperglycaemic stimulus at parturition. Although control of hyperglycaemia in diabetes is regarded as good when a random blood glucose may be less than 120 mg %, this bears little relationship to the physiological control of blood glucose when variation over a 24-hour period may be 20 to 30 mg % below this level. Further support for the importance of hyperglycaemia in the foetal disorder is the negative correlation between maternal blood glucose and foetal birth weight (O'Sullivan *et al* 1966) and the fact that foetal insulin rises after an infusion of glucose into the mother (Milner & Hales 1965). Furthermore, rigid control of hyperglycaemia lowers not only the foetal mortality, but also the foetal birth weight (Oakley 1961).

The most important argument against the importance of hyperglycaemia and foetal size has been that women who develop diabetes at a later time tend to have had children of high birth weights. Many of these have had hyperglycaemia during pregnancy, but this cannot always be confirmed. It remains unknown whether other factors may be responsible in part for the foetal weight, although the importance of these is likely to be much less significant than relative maternal hyperglycaemia.

The pancreas and insulin

A characteristic morphological change in infants of diabetic mothers is pancreatic islet beta cell hypertrophy (Woolf & Jackson 1957, Driscoll *et al* 1960). Baird & Farquhar (1962) have shown that these infants have a greater than normal plasma insulin response to glucose injected intravenously within a few minutes of birth. It appears to be most unlikely that maternal insulin passes the placenta to the foetal circulation in significant amounts (Goodner & Freinkel 1961, Cohen *et al* 1968) and it seems more reasonable to suppose that the characteristic hyperinsulinism shown by the infant of the diabetic mother immediately after birth is the result of abnormal stimulation of the foetal pancreas *in utero* by transfer of maternal glucose across the placenta to the foetal circulation.

Growth hormone

Since growth hormone does not cross the placenta (Gitlin *et al* 1965), any effect of maternal growth hormone would have to be indirectly mediated. Although early immunoassays showed cross-reaction between growth hormone and HPL (Josimovich & McLaren 1962), later assays have been able to distinguish these two peptides and true growth hormone levels are not elevated during pregnancy in normal or diabetic women (Laron et al 1967).

As would be expected from the suppression of growth hormone by hyperglycaemia, foetal growth hormone levels from diabetic pregnancies are lower than those obtained at birth from normal pregnancies. The excessive size of the infant of a diabetic mother cannot be explained on this basis (Cornblath et al 1965).

Perinatal mortality

The foetal loss rate in diabetic pregnancies does not fall below about 5% even if control is 'good' (Farquhar 1962). 'Good' control still exposes the foetus to hyperglycaemia. Although it has been difficult to prove a direct relationship between effectiveness of diabetic control and foetal mortality, nevertheless it is certain that the mortality can be reduced if hyperglycaemia and especially ketosis are avoided (Knowles et al 1965, Larsson et al 1962, Pederson & Brandstrup 1956). The presence of ketosis is frequently fatal to the foetus and the mere avoidance of this will significantly improve perinatal mortality. However it is only with the most stringent control of maternal hyperglycaemia that perinatal mortality will be reduced to a minimum.

SUMMARY

The present state of knowledge of the changes in carbohydrate and lipid metabolism during pregnancy have been reviewed. Biochemically, many of the metabolic changes in pregnancy are a result of the complex synthesis and secretion of human placental lactogen, oestrogens and progesterone by the placenta. These hormones are responsible for (a) the lipolytic activity of pregnancy plasma (b) the increase in plasma free fatty acids (c) the insulin resistance (d) the pancreatic islet beta cell hyperplasia and (e) the hyperinsulinism, which are all characteristics of normal pregnancy.

Pregnancy imposes a severe diabetogenic stress in the normal individual. In a woman who has the relative insulin deficiency of subclinical or clinical diabetes, the additional stress of pregnancy will consistently impair carbohydrate tolerance and may be responsible for the first appearance of the gross metabolic abnormalities of diabetes.

The foetus of a diabetic mother is exposed to the hazards of persistent hyperglycaemia from the maternal circulation, causing high birth weight, a tendency to neonatal hypoglycaemia and a high perinatal mortality. Superadded risks of maternal ketosis increase the perinatal mortality even further. Stringent avoidance of maternal hyperglycaemia and ketosis will decrease the metabolic abnormalities in mother and foetus and reduce the foetal loss rate.

REFERENCES

ANDERSON J.M., MILNER R.D.G. & STRICH S.J. (1966) Pathological changes in the nervous system in severe neonatal hypoglycaemia. *Lancet* ii, 372–375.

BAIRD J.D. & FARQUHAR J.W. (1962) Insulin secreting capacity in newborn infants of normal and diabetic women. *Lancet* 1, 71–74.

BECK P. (1969) Progestin enhancement of the plasma insulin response to glucose in rhesus monkeys, *Diabetes* 18, 146–152.

BECK P. & DAUGHADAY W.H. (1967) Human placental lactogen. Studies of its acute metabolic effects and disposition in normal man. *J. clin. Invest.* 46, 103–110.

BECK P., PARKER M.L. & DAUGHADAY W.H. (1965) Radioimmunologic measurement of human placental lactogen in plasma by a double antibody method during normal and diabetic pregnancies. *J. clin. Endocr. & Metab.* 25, 1457–1462.

BECK P. & WELLS S.A. (1969) Comparison of the mechanisms underlying carbohydrate intolerance in subclinical diabetic women during pregnancy and during post partum oral contraceptive steroid treatment. *J. clin. Endocr.* 29, 807–818.

BENJAMIN F. & CASPER D.J. (1967) Comparative validity of oral and intravenous glucose tolerance tests in pregnancy. *Amer. J. Obstet. Gynec.* 97, 488–492.

BERGER S., DOWNEY J.L., TRAISMAN H.S. & METZ R. (1966) Mechanism of the cortisone modified glucose tolerance test. *New Eng. J. Med.* 274, 1460–1464.

BERGMAN E.N. & SELLERS A.F. (1960) Comparison of fasting ketosis in pregnant and nonpregnant pigs. *Amer. J. Physiol.* 198, 1083–1086.

BERSON S.A. & YALOW R.S. (1962) Immunoassay of plasma insulin. *Ciba Found. Colloq. Endocrinol.* 14, 182.

BLEICHER S., MOLDAU J., SCHERRER J. & GOLDNER M.G. (1964) A lipid mobilizing substance in the serum of pregnant women, of probable placental origin. *Metabolism* 13, 583–586.

BROWN R.J.K. & WALLIS P.G. (1963) Hypoglycaemia in the newborn infant. *Lancet* 1, 1278–1282.

CAHILL G.F., OWEN O.E. & FELIG P. (1968) Insulin and fuel homeostasis. *Physiologist* 11, 97–102.

CATT K.J. MOFFAT B. & NIALL H.D. (1967) Human growth hormone and placental lactogen: structural similarity. *Science* 157, 3–321.

CERASI E. & LUFT R. (1967) What is inherited—What is added. Hypothesis for the pathogenesis of diabetes mellitus. *Diabetes* 16, 615–627.

CHEEK D.B., MADDISON T.G., MALINEK M. & COLDBECK J.H. (1961) Further observations on the corrected bromide space of the neonate and investigations of water and electrolyte status in infants born of diabetic mothers. *Pediatrics, Springfield* **28**, 861–869.

CHEN C.H., ADAM P.A.J., LASKOWSKI D.E., MCCANN M.L. & SCHWARTZ R. (1965) The plasma free fatty acid composition and blood glucose of normal and diabetic pregnant women and of their newborns. *Pediatrics* **36**, 843–855.

CHERNICK S.S., MASORO E.J. & CHAIKOFF I.L. (1950) The *in vitro* conversion of ^{14}C labelled glucose to fatty acids. *Proc. Soc. Exp. Biol. Med.* **73**, 348–352.

CHEVALLIER F. (1964) Transferts et synthèse du cholestérol chez le rat au cours de sa croissance. *Biochem. Biophys.* **84**, 316–339.

CLAPP W.M., BUTTERFIELD J. & O'BRIEN D. (1962) Body water compartments in the premature infant, with special reference to the effects of the respiratory distress syndrome and of maternal diabetes and toxaemia. *Pediatrics, Springfield* **29**, 883–906.

COHEN W.M., ALEXANDER D.P., BRITTON H.G., NIXON D.A. & PARKER R.A. (1968) Plasma insulin and the pancreas in foetal sheep. British Diabetic Association Meeting, Oxford, April 1968.

COKINGITIN L., HANNA F.M. & JACKSON R.L. (1963) Body composition in infants and normal newborn infants. *J. Pediat.* **63**, 482–483.

CONN J.W. & FAJANS S.S. (1961) The prediabetic state. A concept of dynamic resistance to a generic diabetogenic influence. *Amer. J. Med.* **31**, 839–850.

CONNOR W.E. & LIN D.S. (1967) Placental transfer of cholesterol-4-^{14}C into rabbit and guinea pig foetus. *J. Lip. Res.* **8**, 558–564.

CORNBLATH M., ODELL G.B. & LEVIN E.Y. (1959) Symptomatic neonatal hypoglycaemia associated with toxaemia of pregnancy *J. Pediat.* **55**, 454–462.

CORNBLATH M., PARKER M.L., REISNER S.H., FORBES A.E. & DAUGHADAY W.H. (1965) Secretion and metabolism of growth hormone in premature and full term infants. *J. clin. Endocr. & Metab.* **25**, 209–218.

DRISCOLL S.G., BENIRSCHKE K. & CURTIS G.W. (1960) Neonatal deaths among infants of diabetic mothers: post-mortem findings in ninety-five infants. *Amer. J. Dis. Child.* **100**, 818–835.

DEUEL H.J. & MOREHOUSE M.G. (1946) The interrelation of carbohydrate and fat metabolism, in *Advances in Carbohydrate Chemistry*, p. 119. New York, Academic Press Inc.

DUNCAN L.J.P. (1956) Intravenous glucose tolerance test. *Quant. J. Exp. Physiol.* **41**, 85–96.

FAJANS S.S., FLOYD J.C., KNOPF R.F., RULL J., GUNTSCHE E.M. & CONN J.W. (1964) Benzothiadiozine suppression of insulin release from normal and abnormal islet tissue in man. *J. clin. Invest.* **45**, 481.

FARQUHAR J.W. (1962) Birthweight and the survival of babies of diabetic women. *Arch. Dis. Childh.* **37**, 321–329.

FARQUHAR J.W. & ISLES T.E. (1967) Glycemics et comportement clinique chez les nouveau nés de mères normales et diabétiques. *Journées Annuelles de diabétologie de l'Hôtel Dieu*, pp. 99–115.

FLORINI, J.R., TONELLI, G., BREVER, C.B., COPPOLA, J., RINGLER, I. & BELL, P.H. (1966) Characterization and biological effects of purified placental protein (human). *Endocrinology* **79**, 692–708.

FRIESEN H. (1965) Purification of a placental factor with immunological and chemical similarity to human growth hormone. *Endocrinology* **76**, 369–381.

FREINKEL N. (1964) The effect of pregnancy on insulin homeostasis. *Diabetes* **13**, 260–267.

FREINKEL N. (1965) Effects of the conceptors on maternal metabolism during pregnancy, in B.S.Leibel & G.A.Wrenshall (eds.) *On the Nature and Treatment of Diabetes*, p. 679. Excerpta Medica Foundation, New York.

GERSHBERG H., JAVIER Z. & HULSE M. (1964) Glucose tolerance in women receiving an ovulatory suppressant. *Diabetes* **13**, 378–382.

GITLIN D., KUMATE J. & MORALES C. (1965) Metabolism and maternofoetal transfer of human growth hormone in the pregnant woman at term. *J. clin. Endocr. & Metab.* **35**, 1599–1608.

GOLDWATER W.H. & STETTEN D. (1947) Studies in foetal metabolism. *J. biol. Chem.* **169**, 723–738.

GOODNER C.J. & FRIENKEL N. (1961) Carbohydrate metabolism in pregnancy. IV studies on the permeability of the rat placenta I to I^{131} insulin. *Diabetes* **10**, 383–392.

GRABER A.L., PORTE D. & WILLIAMS R.H. (1966) Clinical use of diazoxide and mechanism for its hyperglycaemic effects. *Diabetes* **15**, 143–148.

GRUMBACH M.M., KAPLAN S.L., ABRAMS C.L., BELL J.J. & CONTE E.A. (1966) Plasma free fatty acid response to the administration of chronic 'growth hormone-prolactin'. *J. clin. Endocr.* **26**, 478–482.

GRUMBACH M.M., KAPLAN S.L., SCIARRA J.J. & BURR I.M. (1968) Chronic growth hormone prolactin (CGP): secretion, disposition, biologic activity in man, and postulated function as 'growth hormone' of second half of pregnancy. *Ann. New York Acad. Sci.* **148**, 501–531.

HERRERA E., KNOPP R.W. & FREINKEL N. (1969) Carbohydrate metabolism in pregnancy. VI. Plasma fuels, insulin, liver composition, gluconeogenesis and nitrogen metabolism during late gestation in the fed and fasted rat. *J. clin. Invest.* **48**, 2260–2272.

HURWITZ D. & JENSEN D. (1946) Carbohydrate metabolism in normal pregnancy. *New Eng. J. Med.* **234**, 327–329.

HYTTEN F. & LEITCH I. (1964) *The Physiology of Human Pregnancy*, pp. 128–137. Oxford, Blackwell Scientific Publications.

JACKSON W.P.U. (1952) Studies in pre-diabetes. *Brit. med. J.* **2**, 690–696.

JACKSON W.P.U. (1961) The cortisone-glucose tolerance test with special reference to the prediction of diabetes. Diagnosis of pre-diabetes. *Diabetes* **10**, 33–48.

JACKSON W.P.U. (1967) Diabetes and pregnancy. *Acta Diabetologica Latina* **4**, 1–36.

JOASSIN G., PARKER M.L., PILDES R.S. & CORNBLATH M. (1967) Infants of diabetic mothers. *Diabetes* **16**, 306–312.

JOPLIN G.F., FRASER T.R. & KEELEY K.J. (1961) Prednisone-glycosuria test for prediabetes. *Lancet* **2**, 67–70.

JOSIMOVICH J.B., ATWOOD B.L. & GOSS D.A. (1963) Luteotrophic, immunologic

and electrophoretic properties of human placental lactogen. *Endocrinology* **73**, 410–420.
JOSIMOVICH J.B. & MCLAREN J.A. (1962) Presence in placenta and term serum of a highly lactogenic substance immunologically related to pituitary growth hormone. *Endocrinology* **71**, 209–220.
KALKHOFF R.K., KIM H. & STODDARD F.J. (1968) Acquired subclinical diabetes mellitus in women receiving oral contraceptive agents. Programme of the 28th Annual Meeting of the American Diabetes Association, San Francisco, California, June 16.
KALKHOFF R.K., RICHARDSON B.L. & BECK P. (1969) Relative effects of pregnancy, human placental lactogen and prednisolone in carbohydrate tolerance in normal and subclinical diabetic subjects. *Diabetes* **18**, 153–175.
KALKHOFF R., SCHALCH D.S., WALKER J.L., BECK P., KIPNIS D.M. & DAUGHADAY W.H. (1964) Diabetogenic factors associated with pregnancy. *Trans. Ass. Amer. Phys.* **77**, 270–280.
KAPLAN S.L. & GRUMBACH M.M. (1964) Studies of a human and simian placenta hormone with growth hormone-like and prolactin-like activities. *J. clin. Endocr. & Metab.* **24**, 80–100.
KING K., ADAM P., LASKOWSKI D.E. & SCHWARTZ R. (1967) Sources of plasma free fatty acids in the new born. American Pediatric Society, 77th Meeting, Atlantic City, April 1967, p. 40.
KNOWLES H.C., GUEST G.M., LAMPE J., KESSLER M. & SKILLMAN T.C. (1965) The course of juvenile diabetes treated with unmeasured diet. *Diabetes* **14**, 239–274.
KYLE G.C., YALCIN S., DREWYER R. & CARRUTHERS B. (1964) The prednisolone glucose tolerance test in pregnancy. *Diabetes* **13**, 572–579.
LANDAU R.L. & LUGIBIHL K. (1961) The catabolic and natriuretic effects of progesterone in man. *Rec. Prg. Horm. Res.* **17**, 249–292.
LANDAU R.L. & LUGIBIHL K. (1967) The effect of progesterone on the concentration of plasma amino acids in man. *Metabolism* **16**, 1114–1122.
LARON Z., MANNHEIMER S., NITZEM M. & GOLDMANN J. (1967) Growth hormone, glucose, and free fatty acid levels in mother and infant in normal, diabetic and toxaemic pregnancies. *Arch. Dis. Childh.* **42**, 24–28.
LARSSON Y., STERKY G. & CHRISTIANSSON G. (1962) Long term prognosis in juvenile diabetes mellitus. *Acta paediat. Stockh.* **51**, 1–76.
MCCANN M.C., ADAM P.A.J., LIKLY B.F. & SCHWARTZ R. (1966) Prevention of hypoglycaemia by fructose in infants of diabetic mothers. *New Eng. J. Med.* **275**, 8–12.
MCINTYRE N., HOLDSWORTH C.D. & TURNER D.S. (1964) New interpretation of oral glucose tolerance. *Lancet* **2**, 20–21.
MALAISSE W.J., MALAISSE LAGAE F., PICARD C. & FLAMENT DURAND J. (1969) Effects of pregnancy and chorionic growth hormone upon insulin secretion. *Endocrinology* **84**, 41–44.
MASORO E.J., CHAIKOFF I.L. & DAUBEN W.G. (1949) Lipogenesis from glucose in the normal and liverless animal as studied with C^{14}-labelled glucose. *J. biol. Chem.* **179**, 1117–1125.
MELICHAR V. & WOLF H. (1967) Postnatal changes in the blood serum content of

glycerol and free fatty acids in premature infants. Influence of hypothermia and of respiratory distress. *Biol. Neonat.* **11**, 50–60.

MILNER R.D.G. & HALES C.N. (1965) Effect of intravenous glucose on concentration of insulin in maternal and umbilical-cord plasma. *Brit. med. J.* **1**, 284–286.

NEHER R. & STARK G. (1961) Nachweis von Corticosteroiden in menschlicher Placenta und Isolierung von 16α Hydroxytestosterone. *Experimentia* **17**, 510–512.

NIELSON P.E. (1941) A study with radioactive phosphorus of the permeability of the rat placenta to phospholipid. *Amer. J. Physiol.* **135**, 670–675.

NOVAK M., MELICHAR V. & HAHN P. (1964) Postnatal changes in the blood serum content of glycerol and fatty acids in human infants. *Biol. Neonat.* **7**, 179–184.

OAKLEY W.G. (1961) Symposium on diabetes during gestation. *Proceedings of the Fourth Congress of the International Diabetes Federation* **1**, 380–382.

OCAMPO P.T., COSERIN V.G. & QUILLIGAN E.J. (1964) Comparison of standard oral glucose tolerance test and rapid intravenous glucose tolerance test in normal pregnancy. *Obstet. Gynec. New York* **24**, 580–583.

OSLER M. (1960) Body fat of newborn infants of diabetic mothers. *Acta. Endocr. Copenh.* **34**, 277–286.

O'SULLIVAN J.B., GELLIS S.S., TENNEY B.O. & MAHAN C.M. (1965) Aspects of birth weight and its influencing variables. *Amer. J. Obstet. Gynec.* **92**, 1023–1029.

O'SULLIVAN J.B. & MAHAN C.M. (1964) Criteria for the oral glucose tolerance test in pregnancy. *Diabetes* **13**, 278–286.

PEDERSEN J. & BRANDSTRUP E. (1956) Foetal mortality in pregnant diabetics. *Lancet* **2**, 607–610.

PERLEY M. & KIPNIS D.M. (1965) Insulin secretion and biological effectiveness of endogenous hormone in normal, obese-diabetic and non-obese diabetic subjects. *Clin. Res.* **13**, 331.

PERLEY M. & KIPNIS D.M. (1966) Effect of glucocorticoids in plasma insulin. *New Eng. J. Med.* **274**, 1237–1241.

PERLEY M.J. & KIPNIS D.M. (1967) Plasma insulin responses to oral and intravenous glucose, studies in normal and diabetic subjects. *J. clin. Invest.* **46**, 1954–1962.

POO L.J., LEW W. & ADDIS T. (1939) Protein anabolism of organs and tissues during pregnancy and lactation. *J. biol. Chem.* **128**, 69–77.

RAPPORT D. (1930) The interconversion of the major foodstuffs. *Physiol. Rev.* **10**, 349–472.

RENKONEN O.V. (1966) Serum lipids of labouring mothers and newborn babies. *Amer. Med. Exp. Fenn.* **44**, 1–48.

RICKETTS H.T., CHERRY R.A. & KIRSTEINS L. (1966) Biochemical studies of 'prediabetes'. *Diabetes* **15**, 880–888.

RIDDICK F.A., REISLER D.M. & KIPNIS D.M. (1962) The sugar transport system in striated muscle. Effect of growth hormone hydrocortisone and alloxan diabetes. *Diabetes* **11**, 171–178.

ROBERTSON A. & SPRECHER H. (1966) Human placental lipid metabolism. 1. Synthesis of phosphatidylcholine from lysophosphatidylcholine. *Pediatrics* **38**, 1028–1033.

ROBERTSON A., SPRECHER H. & WILCOX J. (1968) Free fatty acid patterns of human maternal plasma, perfused placenta, and umbilical cord plasma. *Nature* **217**, 378–379.

RYAN K.J., MEIGS R. & PETRO Z. (1966) Formation of progesterone by human placenta. *Amer. J. Obstet. Gynec.* **96**, 676–686.
SCOW R.O., CHERNICK S.S. & BRINLEY M.S. (1964) Hyperlipemia and ketosis in the pregnant rat. *Amer. J. Physiol.* **206**, 796–804.
SELTZER H.S. & ALLEN E.W. (1965) Inhibition of insulin secretion in diazoxide-diabetes. *Diabetes* **14**, 439.
SHORLAND F.B., BODY, D.R. & GASS J.P. (1966) The foetal and maternal lipids of Romney sheep. II. The fatty acid composition of the lipids from the total tissues. *Biochem. Biophys. Acta* **126**, 217–225.
SILVERSTONE F.A., SOLOMONS E. & RUBRICIUS J. (1961) The rapid intravenous tolerance test in pregnancy. *J. clin. Invest.* **40**, 2180–2189.
SOBREVILLA L., HAGERMAN D. & VILLEE C. (1964) Metabolism of pregnenolone and 17-α-hydroxyprogesterone by homogenates of human term placentas. *Biochem. Bioyhus. Acta* **93**, 665–657.
SPELLACY W.N. & GOETZ F.C. (1963) Plasma insulin in normal late pregnancy. *New Eng. J. Med.* **268**, 988–991.
STETTEN D. & BOXER G.E. (1944) Studies in carbohydrate metabolism. 1. The rate of turnover of liver and carcass glycogen studied with the aid of deuterium. *J. biol. Chem.* **155**, 231–236.
TURTLE J.R. & KIPNIS D.M. (1967) The lipolytic action of human placental lactogen on isolated fat cells. *Biochem. Biophys. Acta* **144**, 583–593.
VILLEE D.B. (1969) Development of endocrine function in the human placenta and foetus. *New Eng. J. Med.* **281**, 473–484.
WELSH G.H. (1960) Studies of abnormal glucose metabolism in pregnancy. *Diabetes* **9**, 466–471.
WELSH G.W. & SIMS A.H. (1960) The mechanisms of renal glycosuria in pregnancy. *Diabetes* **9**, 363–369.
WELT I.D., STETTEN D., INGLE D.J. & MORLEY E.H. (1952) Effect of cortisone upon rates of glucose production and oxidation in rats. *J. biol. Chem.* **197**, 57–66.
WILKERSON H. & O'SULLIVAN J.B. (1963) A study of glucose tolerance and screening criteria in 752 unselected pregnancies. *Diabetes* **12**, 313–319.
WOOLF N. & JACKSON W.P.U. (1957) Maternal prediabetes and the foetal pancreas. *J. Path. Bact.* **74**, 223–226.
YALOW R.S. & BERSON S.A. (1960) Immunoassay of endogenous plasma insulin in man. *J. clin. Invest.* **39**, 1157–1175.
ZEE P. (1967) Lipid metabolism in the first 12 hours of life. Society of Pediatric Research 37th Annual Meeting, Atlantic City, April 1967, p. 152.
ZUSPAN F.P. & GOODRICH S. (1968) Metabolic studies in normal pregnancy. 1. Nitrogen metabolism. *Amer. J. Obstet. Gynec.* **100**, 7–14.

12

Cardiovascular Function in Pregnancy

WILLIAM A.W. WALTERS

Excluding the reproductive organs, the maternal cardiovascular system demonstrates *par excellence* physiological adaptation to pregnancy. The resulting modifications of the normal circulatory physiology in the non-pregnant state are described in the following text.

With the development of the placental vascular bed in the maternal circulation it would seem reasonable to expect that more blood would be required to fill the additional vascular space and more cardiac work would be required to circulate the blood contained therein, thus accounting for the haemodynamic changes occurring in pregnancy. However, a closer examination of these changes reveals some unusual features which are not in accord with so facile an explanation. Obvious haemodynamic modifications have occurred long before the placenta has developed as a definite entity and therefore must have their origin in some other factor or factors. It is more likely that the very early appearance of some of the changes observed is due to a chemical agent produced at the onset of pregnancy and having widespread systemic effects. One or more of the sex steroid hormones would come into this category. Oestrogens and/or gestagens, which are secreted in progressively increasing amounts throughout pregnancy must be suspected first as aetiological agents. Research is proceeding in our laboratory and in other centres to gain more information in this field.

BLOOD VOLUME

Total blood volume is measured most accurately by the simultaneous use of independent methods for measuring both plasma and red cell volumes. Leucocytes are present in such small numbers that white cell volume is not taken into account.

The fundamental principle upon which modern techniques for measuring plasma volume or red cell volume depend is the dilution of an indicator after its introduction into the blood stream. It is essential that the indicator mixes completely with the blood and remains entirely within the vascular system for an adequate period of time. Knowing the exact quantity of indicator used, the degree to which it is diluted in the blood is a measure of the blood volume.

Because of technical difficulties plasma and red cell volumes have seldom been measured simultaneously. Usually one component of the total blood volume has been measured along with the haematocrit (packed cell volume) from which the other component is derived.

PLASMA VOLUME

Plasma volume is measured by employing a dye substance or a radioactive isotope as the indicator. A dye such as Evans blue (T-1824), which becomes attached to albumin, is preferable for use during pregnancy, especially if repeated measurements of plasma volume are required in the same subject, as radioactive isotopes might have adverse effects on the developing foetus. The indicators used for plasma volume measurement do not cross the placental barrier, nor is there any evidence for their sequestration in the choriodecidual space or absorption by the trophoblast. Therefore, they measure only maternal plasma volume.

If measurements are to be comparable between subjects and within the same subjects, plasma volume must be measured under standard conditions, viz., with the subject fasting after a restful night, recumbent for at least 30 minutes in a warm bed. Plasma volume may be influenced by a host of factors; the erect posture and a bout of muscular exercise reduce it, while emotional excitement and high environmental temperature increase it (Best & Taylor 1966).

Some of the variation in published results of plasma volume in pregnancy is probably due to the heterogeneity of the groups of women investigated. Very few studies have been reported where any attempt has been made to select for age, parity, stature, socio-economic status and ethnic grouping, all of which may influence plasma volume. For example, the mean increase in plasma volume during pregnancy is consistently higher in multigravidae than in primigravidae (Hytten & Paintin 1963).

There is almost universal agreement that plasma volume increases progressively throughout pregnancy reaching a maximum in the 9th calendar month and decreasing slightly in the 10th calendar month (Thomson et al 1938, Roscoe & Donaldson 1946, McLennan & Thouin 1948, Tysoe & Lowenstein 1950, Hytten & Paintin 1963). In a careful study reported by Hytten & Paintin (1963) the maximum increase in plasma volume in thirty-nine normal, healthy, primigravidae ranged from 630–1940 ml (representing 28–80% of the non-pregnant value) with a mean of 1230 ml. A similar value was obtained by other workers (Thomson et al 1938, Tysoe & Lowenstein 1950, Caton et al 1951). However, all investigations have shown wide individual variation in determinations of plasma volume, partly for the reasons mentioned earlier.

Prolonged pregnancies are associated with higher plasma volumes than pregnancies of normal duration and in such patients there is a relationship between increasing baby and placental weights and plasma volume (Cope 1958). Even in pregnancies of normal duration, Hytten & Paintin (1963) found that the maximum plasma volume was more closely related to the birth weight of the baby than to the weight or height of the mother. However, other workers have found that blood volume during pregnancy is clearly related to maternal weight (Statzer 1959, Lund & Donovan 1967).

In an attempt to compare results in different subjects, volumetric data are often recorded in relation to weight, height, both weight and height or body surface area. The validity of using body surface area in pregnancy is in doubt, since the original Du Bois formula was constructed from data obtained from non-pregnant subjects. Indeed, the relationship between basal metabolism and body surface area is probably fortuitous and body weight may be a better reference factor (Kleiber 1947).

The pattern of plasma volume increase during pregnancy is well demonstrated by the findings of Hytten & Paintin (1963), shown in Fig. 12.1. At approximately 10 weeks gestation the mean plasma volume was 2768 ml, at 33–36 weeks it was 4000 ml, after which there was slight fall of about 300 ml before term. According to more recent studies this decrease may not be significant (Rovinsky & Jaffin 1965, Lund & Donovan 1967). The peak increment in plasma volume probably occurs at about 34 weeks gestation although individual variation is wide. Six to eight weeks after delivery the mean plasma volume (2700 ml) was close to that found in early pregnancy (Hytten & Paintin 1963). It is interesting to note that women show a distinct tendency to repeat the same plasma volume pattern in successive pregnancies (Lund & Donovan 1967).

The time at which the plasma volume begins to increase in early pregnancy is not known, but Thomson et al (1938) reported a 5% increase in total blood volume 3 weeks after conception in one patient and Lund & Donovan (1967) reported an increase as early as the 6th

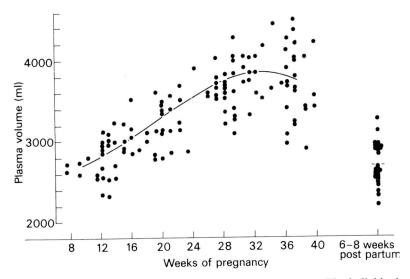

FIG. 12.1. The increase in plasma volume during pregnancy. The individual points have been adjusted to the mean height and weight of the group. From F.E.Hytten & D.B.Paintin (1963) *J. Obstet. Gynaec. Brit. Cwlth.* **70**, 402.

week of pregnancy. A complete picture of plasma volume changes during pregnancy will only be obtained when the same subjects are studied repeatedly before, during and after pregnancy.

Many workers in this field have reported plasma volume changes in pregnancy as a percentage of the non-pregnant value. In this regard, the findings of Rovinsky & Jaffin (1965) are typical; plasma volume in normal singleton pregnancy was 34% above non-pregnant levels at 21–24 weeks, after which it rose to a maximum of 49% above non-pregnant levels at 33–36 weeks, remaining essentially at this level until term. In twenty patients with twin pregnancies the same workers found the maximum increase in plasma volume (67% above normal non-pregnant levels) at 37–40 weeks gestation. This increment in twin pregnancy was 43% greater than that observed in singleton pregnancy. Earlier workers had also noticed larger plasma volumes in occasional subjects in their series with multiple pregnancy (Thomson et al 1938). However, for comparative purposes, absolute measurements of plasma volume are more reliable, since there is little or no correlation between non-pregnant plasma volume and the increment gained during pregnancy (Hytten & Paintin 1963).

Plasma volume decreases at delivery by approximately 500–600 ml and probably returns to normal non-pregnant levels by 6–8 weeks post-partum (McLennan & Thouin 1948, Tysoe & Lowenstein 1950, Hytten & Paintin 1963, Lund & Donovan 1970). The major factor contributing to this decrease is blood loss at delivery (Quinlivan & Brock 1970). It has also been suggested that a shift of fluid from the vascular to extra-vascular compartments also contributes to the decreased plasma volume (Statzer 1959, Duhring 1962). Two to three days after delivery there is a gradual increase in plasma volume, suggesting a return of fluid from the extravascular space (Duhring 1962, Landesman & Miller 1963).

RED CELL VOLUME

A direct method for measuring red cell volume consists of the intravascular injection of red cells labelled with a radioactive isotope such as ^{55}Fe, ^{59}Fe, ^{57}Cr or ^{32}P which mixes with the recipient's red cells. The degree to which the 'tagged' cells are diluted by the recipient's cells is a measure of total red cell volume. Alternatively, red cell volume can be calculated indirectly by measuring plasma volume and the haematocrit in a sample of blood obtained from the opposite vein. In using this method, a correction must be made to allow for uneven distribution of red cells throughout the circulation. For example, in capillaries, the proportion of red cells to plasma is usually less than in the large vessels because of axial streaming of red cells in the small vessels. Thus, in pregnancy, the whole body haematocrit, i.e. the ratio of red cell volume to total blood volume is not the same as the venous haematocrit but is lower by a factor 0·875 (Davis 1962, Hytten & Paintin 1963, Pritchard & Rowland 1964). There is evidence to suggest that pregnancy does not alter the whole body–large vessel haematocrit ratio (Paintin 1963, supine to lateral recumbent or sitting positions resulted in a mean increase of 7 mm Hg in blood pressure.

Many surveys of blood pressure changes in pregnancy have been reported but most have failed to take into consideration standardization of technique and observer bias. In one study by MacGillivray et al (1969) these factors were taken into consideration; a sphygmomanometer specially designed to prevent observer bias was used to measure blood pressure in 226 primigravidae. The lowest mean pressures (systolic and diastolic) were found between 16–20 weeks gestation. Subsequently, both pressures rose gradually throughout pregnancy, the increases in systolic pressure being substantially less than those in diastolic pressure, and the highest readings were found 6 weeks after delivery. Assuming that the blood pressure 6 weeks after delivery is representative of that before pregnancy, there must have been a fall in blood pressure early in the first trimester. These findings of a slight fall in systolic pressure and a relatively greater fall in diastolic pressure during early and mid-pregnancy have confirmed those of earlier workers (Landt & Benjamin 1936, Burwell et al 1938, Walters et al 1966).

Systolic blood pressure taken with the subject recumbent at all stages of pregnancy is 6–12 mm Hg higher than when taken in the sitting position, whereas in the case of diastolic pressure, readings are similar in both positions (MacGillivray et al 1969). Fig. 12.2 shows the trend in blood pressure throughout pregnancy. The mean systolic pressures taken in the sitting position were 102·8 (S.D. $\pm 9·8$) mm Hg before 16 weeks, 101·7 (S.D. $\pm 9·5$) mm Hg between 16–20 weeks rising gradually to 108·9 (S.D. $\pm 11·1$) mm Hg at 39–40 weeks. Corresponding diastolic pressures were 54·4 (S.D. $\pm 7·7$) mm Hg, 53·7 (S.D. $\pm 8·6$) mm Hg and 67·4 (S.D. $\pm 11·2$) mm Hg. Six weeks post-partum the mean systolic pressure was 110·4 (S.D. $\pm 12·4$) mm Hg and the mean diastolic pressure 70·3 (S.D. $\pm 11·2$) mm Hg (MacGillivray et al 1969).

Robinson & Brucer (1939) regarded the normal range of blood pressure in healthy non-pregnant women as being within the range of 90–120 mm Hg systolic and 60–80 mm Hg diastolic pressure. Although Browne (1947) recommended adopting a blood pressure of 120/80 mm Hg as the upper limit of normal in early pregnancy, most obstetricians now regard 140/90 mm Hg as the more realistic upper limit. This is in agreement with the recommendations of the American Committee on Maternal Welfare (1952) and the findings of an epidemiological survey in Aberdeen (MacGillivray 1961). McClure Browne (1961) substantiated that a blood pressure greater than 148/90 mm Hg in women between 15–30 years of age in early pregnancy must be regarded as abnormal, although he stressed that it was difficult to draw a sharp, dividing line between normal and abnormal blood pressure in pregnancy.

Summarizing present knowledge in this field, systolic blood pressure falls slightly in pregnancy below that in the non-pregnant state, while diastolic blood pressure is significantly lower during early and mid-pregnancy than in the non-pregnant state and tends to return to normal levels in late pregnancy.

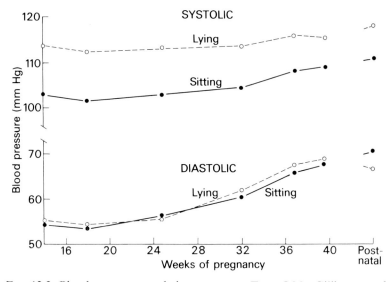

FIG. 12.2. Blood pressure trends in pregnancy. From I.MacGillivray *et al.* (1969). *Clin. Sci.* **37**, 395.

VENOUS PRESSURE

During pregnancy venous pressure is greatly dependent upon posture and the period of gestation. Pressure in veins of the lower limbs in standing subjects increases as pregnancy advances in contrast with venous pressure in the upper limb veins which remains within normal limits (McLennan 1943). The rise of venous pressure in the lower limbs may be due partly to increasing obstruction to venous return imposed by the enlarging uterus and partly to increased venous distensibility related to the increased amounts of circulating progesterone.

Pressure in the lower inferior vena cava is also increased in late pregnancy especially when the subject is lying supine. It has been clearly demonstrated that this finding is due to caval compression just below the level of the diaphragm by the gravid uterus (Scott & Kerr 1963).

Confusion exists concerning pressure changes in the right atrium, right ventricle and pulmonary artery in pregnant women, some workers finding that these pressures were increased (Palmer & Walker 1949, Rose *et al* 1956), some finding them decreased (Werkö *et al* 1948), and others finding no change (Hamilton 1949, Angelino *et al* 1954, Bader *et al* 1955). It is reasonable to conclude from the data available that if there are any changes in these parameters, they are likely to be small deviations from readings obtained in normal non-pregnant subjects.

Increased venous distensibility has been reported in the finger-tips in pregnancy (McCausland *et al* 1961) and in the calf and forearm (Goodrich & Wood 1964). However, in a detailed serial study of forearm venous tone measured by a direct method in the same subjects before and after parturition, Duncan (1968) was unable to demonstrate any significant increase in venous distensibility during pregnancy.

CARDIAC OUTPUT

The output of the heart is probably the most important of the circulatory changes occurring during pregnancy because it provides a measure of any alteration in the circulatory burden.

The principal function of the heart is to convert chemical energy into mechanical energy or work so that blood is propelled through the circulatory system. Cardiac output may be defined as the volume of blood expelled by either ventricle of the heart in a unit of time, usually expressed in litres per minute. In the normal heart under steady state conditions the output of each ventricle is identical. The work of the heart is the expenditure of energy in ejecting a certain volume of blood into an already distended aorta against the resistance offered by the blood pressure and in imparting velocity to the blood ejected. Thus, the energy of myocardial contraction is converted partly into potential energy in the form of distension of the arterial wall and partly into kinetic energy represented by the momentum of the moving column of blood (Davson & Eggleton 1968). The external work done by the heart is measured by the quantity of blood ejected and the pressure developed during each ventricular contraction. For clinical purposes cardiac work is the product of cardiac output and mean arterial pressure. Cardiac output, in turn, is determined by the product of heart rate and stroke volume.

Comprehensive surveys of methods used for cardiac output measurement have been made by Cournand (1945), McMichael (1945) and Hamilton (1953). Of all the methods devised, there are only two which are reliable and thoroughly tested for determining cardiac output in man viz., the direct Fick and Stewart-Hamilton indicator dilution methods. The former depends upon respiratory gas exchange in the lungs which, in turn, depends upon the volume of blood circulating through the lungs and hence the amount of blood ejected by the right ventricle. Thus cardiac output in litres per minute is equal to the exchange of either oxygen or carbon dioxide per minute divided by the arteriovenous difference of the gas in millilitres per litre of blood. A major problem with the application of this method to man was the difficulty in obtaining a truly representative sample of mixed venous blood which is best obtained as close as possible to its entrance to the pulmonary vascular bed. However, since Forssmann (1929) and Klein (1930) first described cardiac catheterization, which was subsequently developed and applied to cardiac output determination in man by Cournand & Ranges (1941), it is now possible to sample mixed venous blood from the right heart and pulmonary artery by a catheter, usually inserted into a peripheral vein in the cubital fossa at the elbow and advanced through part of the peripheral and central venous systems to reach its destination in the pulmonary artery. Obvious disadvantages of the direct Fick method are its component procedures of arterial puncture and cardiac catheterization, both of which are uncomfortable for the subject even when local anaesthesia is used, and which may lead to apprehension and anxiety thereby causing a spurious rise in cardiac output. Moreover, ethical considerations preclude the use of these techniques in normal, healthy subjects especially when serial studies are required.

Before development of the direct Fick method, indirect applications of the Fick principle were used in man, utilizing foreign gases such as nitrous oxide or acetylene to measure blood flow through the lungs. Unfortunately, the indirect Fick methods are fraught with inaccuracies and have now become obsolete.

The second reliable method for measuring cardiac output in man utilizes the Stewart-Hamilton dye-dilution principle. A known amount of dye is injected into the venous circulation and after mixing with blood in the heart its concentration is sampled in the arterial circulation. The extent to which the dye has been diluted in a given time provides a measure of the volume of blood passing through the heart. This method has now surpassed all others in popularity because of its simplicity and the ease with which it can be applied to clinical problems. Furthermore, it is much less formidable and relatively nontraumatic when compared with the direct Fick method. It permits observation of transient changes in cardiac output allowing repeated measurements to be made in the same subject within a short period of time, whereas the direct Fick method is accurate only when the heart output and pulmonary gas exchange remain stable during the period of measurement.

Results obtained by both direct Fick and dye-dilution methods have been shown to agree closely (Taylor & Shillingford 1959, Phinney et al 1963). The average resting cardiac output in a subject of average size (70 kg) is of the order of 5–6 litres per minute (Guyton 1963).

To compare cardiac output values between individuals of varying size an adjustment is made according to body surface area. The result obtained is known as the cardiac index, expressed in litres per minute per square metre. Although it has been widely accepted that cardiac output is closely related to body size, Kleiber (1947) questioned the validity of this relationship. However, Jegier et al (1963) have substantiated the earlier assumption by showing a close correlation between cardiac output and body surface area. Whether such a relationship holds good for pregnant women is not known. In non-pregnant, normal, healthy subjects, the cardiac index has a normal range of 2·5–4·5 litres/min/M^2 (Marshall & Shepherd 1968). It is calculated by dividing cardiac output in litres per minute by body surface area in square metres.

The early literature on maternal circulation in pregnancy contains suggestions, based upon clinical observation, that the work of the heart is probably increased in pregnancy. Since the latter part of the nineteenth century many workers have been fascinated with cardiac output measurement in pregnancy and successive investigators have utilized new and improved techniques thereby improving upon the results of their predecessors.

Since Lindhard (1915) first measured cardiac output in human pregnancy by the indirect Fick method, numerous investigations of this parameter have been made employing various techniques. This is not the appropriate place to review all of the work in the field, but it is worth while considering more recent studies in which the direct Fick and dye-dilution methods were used.

Palmer & Walker (1949) studied the cardiac output successfully in eighty-four pregnant women in normal health, using the technique of cardiac catheterization and the direct Fick principle. A single estimation only was made in each subject. The mean cardiac output of all the pregnant subjects was 5·8 litres/minute, significantly higher than the mean cardiac output of the non-pregnant control series which was 4·6 litres/minute. With the exception of a low value at the 7th lunar month, the cardiac output was found to be uniformly high from the 3rd lunar month of pregnancy until term. The low value in the 7th month was explained as a postural effect. Using the same technique Hamilton (1949) performed sixty-eight successful determinations of cardiac output in sixty-eight normal pregnant subjects and showed a rise in cardiac output beginning at the 10th week of pregnancy, reaching a maximum between 26–29 weeks and thereafter decreasing to approach normal levels between 38–40 weeks. Very few observations were made in early pregnancy.

Bader et al (1955) improved the accuracy of the direct Fick method by using radiology to control localization of the cardiac catheter, thereby ensuring collection of adequately mixed venous blood. An additional improvement was the collection of arterial blood by arterial puncture so that arterial oxygen saturation could be measured and not just assumed to be 95% as in earlier studies. Bader and associates studied forty-six normal pregnant women between 14 weeks gestation and term, and found that the maximum cardiac output occurred at 25–27 weeks, when it was approximately 40% greater than normal. Subsequently, it declined gradually to approach normal levels in the last 4 weeks of pregnancy. Again, only a single determination of cardiac output was made in each subject, and no observations were made before the 14th week of pregnancy.

Adams (1954) was the first to report results of a study of cardiac output in pregnancy using the dye-dilution principle with Evans' blue (T-1824) as the indicator. Arterial puncture was used to collect blood samples for analysis of dye content in order to construct the dye-dilution curve. A mean cardiac output graph for the thirty-one normal pregnant subjects studied, showed that the maximum increase of 32%

occurred at 28 weeks gestation, thereafter declining and approaching normal levels near term. Immediately after delivery the cardiac output increased by 29% and remained at this level for several days before returning to normal levels. A total of 94 determinations of cardiac output were made in the 31 subjects, an average of 3 determinations per subject. Hence, the serial nature of this study was much restricted.

Within the last decade another five studies of cardiac output at rest during pregnancy have been reported. All have utilized the dye-dilution principle.

Using Evans' blue and arterial puncture to monitor arterial dye concentration, de Schwarcz et al (1964) made 48 determinations of cardiac output in 33 pregnant and 5 non-pregnant women. By the 8th week of pregnancy the output had increased by 58%. The maximum increase of 64% above normal was found at 18 weeks, after which there was a gradual decline although it was still 15% above normal levels at term. The increase in cardiac output was due mainly to an increase in stroke volume with little change in heart rate. Few subjects in this series had more than one determination of cardiac output.

Roy et al (1966) reported results of 82 determinations of cardiac output in 29 normal, pregnant Indian women. Dye-dilution curves were recorded by an arterial cuvette-densitometer using indigo-carmine dye. An increase in cardiac output was not apparent until after the 21st week of pregnancy, and the maximum increase of 39% above postpartum values occurred at 28–34 weeks. Again the increase in output was thought to be due mainly to an increase in stroke volume associated with a reduction in total peripheral resistance.

A comprehensive serial study of cardiac output in pregnancy was carried out by Walters et al (1966). Thirty normal young women had their cardiac outputs measured at monthly intervals throughout pregnancy after peripheral venous injections of Coomassie blue (sodium anoxynaphthonate) monitored continuously in the arterial circulation by a photo-electric earpiece. Cardiac output was found to be elevated in the first trimester reaching a maximum of 6·7 litres per minute at 24–32 weeks gestation, thereafter decreasing to normal levels near term. An increase in stroke volume was confirmed as being responsible for the increase in cardiac output (Fig. 12.3).

Rovinsky & Jaffin (1966) used the dye-dilution technique to determine cardiac output in 34 women with single pregnancies, 20 women with twin pregnancies, 2 women with triplet pregnancies and 6 non-pregnant controls. In the women with single pregnancies, the mean cardiac index was maximal (44·2% above normal) at 25–28 weeks, thereafter declining to normal levels. By contrast, in women with twin pregnancies, the cardiac index remained at the same level (48% above normal) between 21–32 weeks and subsequently decreased to 15·1% above control values at term. Changes in cardiac output and left ventricular work were found to be independent of changes in blood volume during pregnancy.

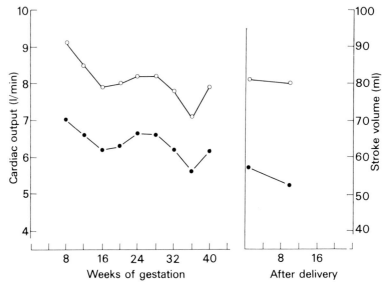

FIG. 12.3. The mean levels of cardiac output (●) and stroke volume (○) in thirty subjects in whom serial investigations were made at intervals of 4 weeks before and after parturition. Cardiac output was measured with all subjects in the supine position. From W.A.W.Walters *et al* (1966) *Clin. Sci.* **30**, 1.

Another variation in the dye-dilution method was employed by Lees *et al* (1967) who used a venous catheter advanced into an intrathoracic vein through which indocyanine green was injected. Another catheter was passed into the root of the aorta where blood was obtained for passage through a cuvette-densitometer to record continuously the arterial dye concentration. Three serial determinations of cardiac output were obtained in 5 pregnant subjects, and single determinations were made in 11, 8 and 9 women at 11–13, 24–27, and 34–41 weeks gestation respectively. The resting cardiac output increased during pregnancy by 1·2–3·1 litres/minute (30–40%). Most of this increase had occurred by the end of the first trimester and was maintained at this level until term. For the first time, it was demonstrated that the decrease in cardiac output in late pregnancy, reported by earlier workers, is due to the supine position and can be prevented by placing the subject in a lateral position (Fig. 12.4). As there was no consistent change in heart rate, an increase in stroke volume was postulated as the cause for the increase in cardiac output during pregnancy.

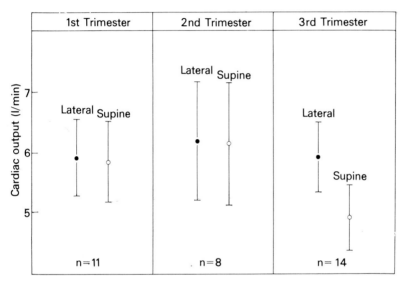

FIG. 12.4. Average mean values of cardiac output in the supine and lateral positions in twenty subjects, to illustrate the effect of the supine position in lowering cardiac output in the third trimester. The vertical lines represent ± one standard deviation. From M.M.Lees *et al.* (1967) *J. Obstet. Gynaec. Brit. Cwlth.* **74**, 319.

In summary, there is no doubt that cardiac output is increased by 30–50% during pregnancy. According to most workers this increase develops early in pregnancy and reaches a maximum at 24–28 weeks, thereafter returning to near normal levels just before term. This decline in cardiac output in the last few weeks of pregnancy is almost certainly due to the supine position, which promotes occlusion of the inferior vena cava by retrodisplacement of the pregnant uterus (Lees *et al* 1967). Hence, venous return is impaired and cardiac output falls. The magnitude and times of changes in cardiac output during pregnancy are still not well established. In this regard, the lack of uniformity in results reported by different workers may be due to one or more of a number of factors viz., variation in techniques, different environmental conditions, individual variation in response to the techniques employed, difference in number, sizes and ages of subjects and intrinsic experimental errors.

STROKE VOLUME

Stroke volume or the amount of blood expelled by each beat of the ventricle is technically difficult to measure directly. However, it can be calculated indirectly from cardiac output and heart rate. Thus, Adams (1954) found that stroke volume increased during pregnancy from a mean level of 84 ml in the non-pregnant state to 98 ml at 28 weeks gestation followed by a decrease to 67 ml at 40 weeks. Other workers have confirmed that stroke volume increases in pregnancy (Rovinsky & Jaffin 1966, Roy et al 1966, de Schwarcz et al 1966, Walters et al 1966). In contrast Bader et al (1955) and Werkö et al (1950) reported that stroke volume in early pregnancy was not significantly different from that in the non-pregnant subject, while it was below normal levels in late pregnancy. The discrepancy between results obtained in these two studies and those mentioned above can probably be explained on technical grounds. The weight of evidence favours an increase in stroke volume during pregnancy with a further transient increase immediately after delivery (Adams 1954, Cunningham 1966).

LEFT VENTRICULAR WORK

The energy output of the left ventricle appearing as work is a function of cardiac output and arterial blood pressure. As might be expected from a knowledge of the trends of the last two parameters during pregnancy, left ventricular work is increased at all stages of pregnancy (Bader et al 1955, Lees et al 1967).

CENTRAL BLOOD VOLUME

Central blood volume refers to the amount of blood occupying an ill-defined vascular space, usually incorporating the heart, lungs, great vessels and sometimes part of the major peripheral vasculature. Because of difficulties in measuring this parameter quantitatively, changes in central blood volume during pregnancy are poorly understood and are usually derived from measurements of cardiac output and mean circulation time. Studies utilizing this method for calculation of central blood volume have shown that it increases by 19% above normal levels in mid-pregnancy (Adams 1954, Rovinsky 1966).

REGIONAL BLOOD FLOW

UTERINE BLOOD FLOW

Our knowledge of human uterine blood flow is extremely limited because of the great technical and ethical problems involved in its investigation. Even if it were possible to measure total blood flow through the human uterus, the next problem would be to find some means of completely differentiating between placental and myometrial blood flow.

Human uterine blood flow at term has been estimated to be about 500–700 ml/min, using the nitrous oxide diffusion technique based upon the Fick principle (Assali et al 1953, Metcalfe et al 1955). However, because of the length of time taken for nitrous oxide to equilibrate between blood and tissues, the validity of this method has been questioned.

Browne & Veall (1953) were the first to use radio-isotopes to measure maternal blood flow through the intervillous space of the placenta by measuring the rate of clearance of ^{24}Na from the placental site after its injection directly into the space. This technique is only feasible when the placenta is located on the anterior uterine wall, in which case the intervillous space can be penetrated by a needle passed through the anterior abdominal wall and myometrium. Besides the difficulty in knowing the exact position of the point of the needle, the volume of the space cannot be measured but is assumed to contain 250 ml blood at term. Under these circumstances normal human choriodecidual blood flow was found to be approximately 500 ml/min.

With the advent of the electromagnetic flowmeter for measuring blood flow directly, Assali et al (1960) applied cuff electromagnetic flowmeters to the uterine arteries in twelve pregnant women who were undergoing therapeutic abortion by abdominal hysterotomy between 10–28 weeks gestation. Total uterine blood flow was found to increase from a mean of 51·7 ml/min at the 10th week to 185 ml/min at the 28th week of gestation. These measurements were performed under conditions which were far from the physiological norm and therefore must be interpreted with caution.

A cineangiographic technique was developed by Borell et al (1965), who studied uterine blood flow at term in three women bearing malformed foetuses. Marked retardation of flow occurred during uterine contractions, which presumably compressed the vessels.

BLOOD FLOW IN THE EXTREMITIES

The most reliable method for measurement of blood flow in the limbs is venous occlusion plethysmography. It is based on the principle that when the veins are briefly occluded, the congestive swelling that results is due to arterial inflow, and the rate of swelling is, for the first few seconds, a precise measure of the rate of arterial blood flow into the part.

At the beginning of the present century it was thought that peripheral vasoconstriction must occur in pregnancy to allow the increasing demands of the growing uterus for blood to be met (Huchard, quoted by Abramson 1943). However, Abramson et al (1943) observed an increase in hand blood flow during pregnancy but no change in forearm or leg blood flow. The measurements were made in a higher room temperature (27°C) than normal, which might have influenced their results by causing a persistent vasodilatation. In contrast Burt (1950) found slight increases in both hand and forearm blood flow in the second half of pregnancy. An initial vasoconstriction found in the hands during the first trimester was replaced by vasodilatation as pregnancy progressed. Herbert et al (1958) made 5660 measurements of forearm and leg blood flow in fifty-eight normal pregnant women and 3750 measurements in twenty-five non-pregnant women, demonstrating an increase in blood flow in both forearm and leg only in the latter half of pregnancy. Spetz (1964) and Spetz & Jansson (1969) confirmed that the increase in resting forearm blood flow was only significant in the second half of pregnancy and attributed this to a marked fall in peripheral resistance.

Ginsburg & Duncan (1967) observed that mean blood flow through the hands increased from 2·7 ml/100 ml tissue/min in early pregnancy to 18·1 ml/100 ml tissue/min at term with subjects resting quietly in the supine position. By the 6th week of the puerperium, the hand blood flow had decreased to 9·1 ml/100 ml tissue/min. A similar but smaller increase in blood flow through the foot was recorded during pregnancy. Unlike results of previous studies, no change in blood flow through the forearm or calf was observed at any time.

In an attempt to establish whether multiple pregnancy evoked a greater response in the peripheral circulation than singleton pregnancy, Duncan & Ginsburg (1968) compared blood flow to the hand, foot, forearm and calf in singleton, twin and triplet pregnancies. There was no significant increase in mean blood flow to these four limb segments in twin pregnancies when compared with singleton pregnancies, but a significant increase was observed in triplet pregnancies.

Digital electrocapacitance plethysmography has revealed that various external stimuli such as cold, pain, sound and mental stress cause a predominance of vasodilator responses in early pregnancy, but as pregnancy advances vasoconstrictor responses become more prominent until at term they outnumber vasodilator responses. In men and non-pregnant women the same stimuli produce vasoconstriction. It was postulated that changes in blood levels of chorionic gonadotrophin associated with changes in the central nervous system might be responsible for the altered vascular responses during pregnancy (Doležal & Figar 1965).

Vasodilatation is apparent in the skin during pregnancy as indicated by palmar erythema and a rise in skin temperature in the peripheral parts of the limbs (Herbert *et al* 1958). Further evidence for vasodilatation of skin vessels is provided by the development of 'spider' naevi (Bean *et al* 1947) and capillary dilatation in the nail-folds on microscopy (Melbard 1938).

Thus, there is good evidence, both qualitative and quantitative, for vasodilatation and an increase in peripheral blood flow in the extremities during pregnancy. Furthermore, most of this increase in blood flow appears to be destined for the skin rather than for muscle. In this way extra heat generated by the conceptus could be dissipated more effectively, thereby preventing maternal and foetal hyperthermia.

RENAL BLOOD FLOW (see also Chapter 13)

In the normal non-pregnant state renal blood flow accounts for approximately 25% of the total output of the heart. Therefore in the presence of an increased cardiac output in pregnancy a proportional increase in blood distribution to the kidneys might be expected.

Usually, renal blood flow is calculated from renal plasma flow and venous haematocrit, the former being estimated by the clearance of a known amount of para-amino-hippurate (PAH) from the blood stream by the kidneys. As it is possible that not all blood circulating is cleared of PAH, clearance is often taken to indicate 'effective' renal plasma flow. Total renal plasma flow can be determined if a sample of renal vein blood is obtained and its PAH concentration measured. The arteriovenous PAH difference divided by the arterial concentration of PAH gives the extraction of PAH from blood. When 'effective' renal plasma flow is divided by the last value, total renal plasma flow is obtained. In normal man, extraction seldom differs much from 0·9, indicating that 'effective' renal plasma flow is a good approximation of total renal plasma flow (Wesson 1969). Mean 'effective' renal plasma flow in normal nonpregnant young women in the basal state, is about 600 ml/min/1·73 M^2, total renal plasma flow 660 ml/min/1·73 M^2 and assuming an average haematocrit of 40%, total renal blood flow is approximately 1100 ml/min/1·73 M^2. Glomerular filtration rate under the same basal conditions averages 120 ml/min/1·73 M^2 between 20–40 years of age (Wesson 1969).

In a careful serial study of renal function in 9 non-pregnant control subjects and another 12 subjects between 15–40 weeks gestation and post-partum, Sims & Krantz (1958) found that renal plasma flow measured by PAH clearance increased by one-third in early and mid-pregnancy but decreased to normal levels in the third trimester. A further decrease occurred in the puerperium with levels below normal. They also found that glomerular filtration rate measured by inulin clearance was increased by 50% until the last week of pregnancy when it returned to normal levels.

Dignam et al (1958) also found a progressive increase in both PAH and inulin clearance during pregnancy but failed to show any decrease near term. A slightly different pattern was described by Chesley (1960) who reported an increase of 25–50% in PAH clearance from early pregnancy until the last trimester when it decreased to approach normal levels. However, inulin clearance increased by 50% from early pregnancy until the last week when it decreased to approach the normal level, the decline being less marked than that of PAH clearance. Mahran et al (1968) also reported an increase in glomerular filtration rate in the first and second trimesters of normal pregnancy, a decrease in the third trimester, with an increase again after parturition.

In contrast, de Alvarez (1958) found that both PAH and inulin clearances increased in the first trimester but decreased to approach normal levels in the second trimester, and both were below normal levels in the third trimester.

It is important to remember that the posture of the subject may influence renal function as demonstrated by Pritchard *et al* (1955), who found that glomerular filtration rate and renal blood flow were lower when the pregnant subject was supine than when in the lateral recumbent position.

Summarizing the data available, renal blood flow and glomerular filtration rate are increased by approximately 50% above normal levels in early pregnancy, thereafter gradually decreasing to pre-pregnancy levels in the 4 weeks before term. The precise time of onset of these changes in early pregnancy is still obscure. It is difficult to offer any explanation for these findings since renal blood flow is apparently lowest when one would expect the load on the kidneys to be greatest.

HEPATIC BLOOD FLOW

Using hepatic vein catheterization and bromsulphthalein as the indicator, Bradley *et al* (1945) estimated blood flow through the liver by application of the Fick principle. The liver probably contains most of the blood in the splanchnic circulation which is estimated to hold 20% of the total blood volume. Hence blood flow through the liver is likely to indicate blood flow through the entire splanchnic area. In the normal adult subject mean liver blood flow is 1530 (S.D. ±300) ml/min (Clearfield 1965).

Subsequently, Munnell & Taylor (1947) used the same method to investigate hepatic blood flow in pregnant women. They found no difference between pregnant and non-pregnant subjects, the mean hepatic blood flow being 1·5 litres/min/1·73 M^2.

CEREBRAL BLOOD FLOW

Very few studies of cerebral blood flow in pregnancy have been reported. McCall (1949) measured cerebral blood flow by the nitrous oxide method of Kety & Schmidt (1945) in nine normal young women between 34–40 weeks gestation. The results, when compared with those obtained by the same method in fourteen normal young men, showed no significant differences.

MAMMARY BLOOD FLOW

The visible dilatation of veins in the overlying skin and engorgement of the breasts during pregnancy suggest that mammary blood flow increases. However, at the time of writing, no quantitative changes in human mammary blood flow in pregnancy have been reported.

TOTAL PERIPHERAL RESISTANCE

Total peripheral resistance is a measure of all factors affecting peripheral blood flow and is largely dependent upon the calibre of the arterioles and capillaries and the viscosity of the blood.

As the systemic arterial pressure is a product of cardiac output and total peripheral resistance, and since there is little change in arterial blood pressure during pregnancy, it follows that the increase in cardiac output in pregnancy must be accompanied by a decrease in total peripheral resistance. This has been confirmed by several investigators most of whom have found the decrease to be greatest in mid-pregnancy (Adams 1954, Rose *et al* 1956, Rovinsky & Jaffin 1966, de Schwarcz *et al* 1966).

THE EFFECTS OF POSTURE ON CARDIOVASCULAR DYNAMICS DURING PREGNANCY

When the non-pregnant subject assumes the vertical position from recumbency, regardless of whether this postural change has been effected by active or passive means, heart rate increases (Drischel *et al* 1963, Pyörälä 1966, Tuckman & Shillingford 1966), systolic blood pressure remains the same or falls slightly, while diastolic blood pressure also remains the same or rises slightly (Pyörälä 1966, Tuckman & Shillingford 1966). Simultaneously, cardiac output decreases by 20–30% (Pyörälä 1966, Tuckman & Shillingford 1966, Wang *et al* 1960), heart volume decreases (Kjellberg 1953, Pyörälä 1966), stroke volume decreases by 20–50% (Pyörälä 1966, Stevens 1966, Tuckman & Shillingford 1966, Wang *et al* 1960), total peripheral resistance increases (Pyörälä 1966) and peripheral blood flow diminishes in the upper limbs (Brigden *et al* 1950).

In most studies of cardiovascular function in pregnancy subjects have been investigated in the supine position. Few workers have observed the effects of different posture, and therefore no firm conclusions can be reached concerning the haemodynamic consequences of body position in pregnancy. For convenience, the meagre information available on the influence of posture on haemodynamics in pregnancy will be considered in early (<24 weeks) and late (>32 weeks) pregnancy.

EARLY PREGNANCY

In early pregnancy a change from recumbency to the upright position induces a variable increase (Dahlström & Ihrman 1960, Widlund 1945) or no significant alteration in heart rate, systolic blood pressure falls (Pyörälä 1966) diastolic blood pressure rises slightly (Dahlström & Ihrman 1960) or remains the same, heart volume and cardiac output decrease while stroke volume and total peripheral resistance show no significant change (Pyörälä 1966).

The adoption of a lateral recumbent position after being in the supine position results in a fall in both systolic and diastolic blood pressures as recorded by brachial sphygmomanometry (Trower & Walters 1968) or no significant change when recorded directly (Lees *et al* 1967). Heart rate, cardiac output and stroke volume show no significant difference in supine or lateral recumbent positions.

LATE PREGNANCY

When assuming the upright position from recumbency, subjects in late pregnancy show an increase (Dahlström & Ihrman 1960, Widlund 1945) or no significant change in heart rate, systolic blood pressure remains the same (Pyörälä 1966) or falls slightly (Dahlström & Ihrman 1960), diastolic blood pressure rises slightly (Dahlström & Ihrman 1960) or remains the same, while heart volume, cardiac output, stroke volume and peripheral resistance are maintained at the same levels in both positions (Pyörälä 1966), indicating an enhanced compensatory response to postural change.

In late pregnancy, when subjects change from supine to lateral recumbent positions, heart rate decreases slightly (Ueland & Hansen 1969) or remains unaltered (Lees *et al* 1967) while systolic and diastolic arterial pressures fall (Trower & Walters 1968, Tůmová 1964) or remain the same (Lees et al 1967, de Rezende 1950), In another study 11% of pregnant women at term had a significant fall in systolic pressure when turned from lateral recumbent to supine positions (Howard et al 1953). However, a similar investigation showed no significant alteration in blood pressure (Quilligan & Tyler 1959). The cardiac output tends to be higher in the lateral recumbent than in the supine position (Lees et al 1967, Vorys et al 1963).

AORTO-CAVAL OCCLUSION AND SUPINE HYPOTENSION

The haemodynamic changes observed in late pregnancy when subjects adopt the supine position is thought to be due largely to an abrupt reduction in venous return from the lower part of the body consequent upon varying degrees of compression of the inferior vena cava by the gravid uterus.

This phenomenon is believed to cause the clinical picture known as the supine hypotensive syndrome, characterized by bradycardia, hypotension, pallor, sweating and syncope. Relief is obtained when any position other than the supine one is assumed. Estimates of the incidence of the syndrome in late pregnancy vary between 4–11% (Howard et al 1953, Holmes 1953).

Further confirmation of the syndrome was obtained by Kerr (1965) who reported a sudden fall in cardiac output in three out of six subjects in advanced pregnancy when turned into the supine position. At the same time a reflex increase in total peripheral resistance occurred thus maintaining systolic blood pressure.

Complete occlusion of the inferior vena cava has been demonstrated in some supine patients during caesarean section but in the majority venous return occurred satisfactorily via an alternative collateral venous system involving the azygos and vertebral veins. Supine hypotension only developed in subjects in whom collateral channels were inadequate (Kerr et al 1964, Mundow 1967, Scott & Kerr 1963).

Other investigations have shown that a similar compression of the abdominal aorta occurs in normotensive women lying supine in late pregnancy (Bieniarz et al 1963, Bieniarz et al 1969). Consequently, despite normal brachial arterial pressure, blood pressure in the arterial system distal to the compression may be reduced causing some impairment of uterine blood flow (Bieniarz et al 1968).

CARDIOVASCULAR DYNAMICS DURING LABOUR AND PARTURITION

Labour and parturition are periods of great physiological instability and hence the little information that has been collected at this time must be interpreted in the light of the unfavourable conditions existing.

Heart rate increases progressively during labour along with increasing frequency and severity of uterine contractions and is at a maximum immediately after parturition. Pain and anxiety probably contribute appreciably to this tachycardia since in subjects labouring under epidural analgesia, no change in heart rate occurred (Lees *et al* 1970). Winner & Romney (1966) recorded an increase of 6–29 beats/min during uterine contractions in the first stage of labour, whereas during the second stage the increase ranged from 10–52 beats/min. A marked reduction in heart rate occurred during the first hour after delivery.

A reciprocal relationship exists between heart rate and arterial blood pressure during a uterine contraction cycle. Heart rate increases at the commencement of a contraction, begins to decrease at its height and decreases still further as the contraction subsides. By contrast, arterial blood pressure shows little change during the initial half of the contraction cycle but rises appreciably in the second half (Hendricks 1958).

Between uterine contractions blood pressure remains unchanged throughout labour. However, during contractions systolic and diastolic pressures rise in the first and second stages of labour. The increase in systolic pressure (15–35 mm Hg) is usually greater than that of the diastolic pressure (5–25 mm Hg). Furthermore, the increases in both systolic and diastolic pressures during the second stage are quantitatively greater than in the first stage of labour (Cunningham 1966, Lees *et al* 1970, Ueland & Hansen 1969, Winner & Romney 1966).

Uterine contractions may increase cardiac output by increasing venous return or by producing pain, they may stimulate the sympathetic nervous system and hence the output of the heart.

Most workers have found a variable increase (20–30%) in cardiac output during effective uterine contractions in the first stage of labour (Adams & Alexander 1958, Cunningham 1966, Hendricks & Quilligan 1956, Lees *et al* 1970). Although Cunningham (1966) found a progressive rise in cardiac output during the first stage of labour and a more dramatic rise during contractions in the second stage, Winner &

Romney (1966) found that cardiac output was within the normal range in all stages of labour.

There is some doubt about what happens to cardiac output immediately after delivery. Cunningham (1966) described a transient fall, while Lees *et al* (1970) described a rise of 20% above levels found during uterine contractions. Nevertheless in both of these studies stroke volume was found to be increased after parturition.

It is important to remember that pain, posture, analgesic drugs, oxytocics, the Valsalva manœuvre and the mode of delivery may all have an effect on the haemodynamic status of the patient during labour and parturition. Therefore, to isolate the effects of uterine contractions *per se* may be extremely difficult.

The major component of the cumulative rise (40%) in cardiac output during labour is probably due to pain and apprehension, as it can be prevented by continuous caudal analgesia (Lees *et al* 1970, Ueland & Hansen 1969).

AETIOLOGY OF THE HAEMODYNAMIC CHANGES OF PREGNANCY

The cause of the altered haemodynamics of pregnancy is obscure and may be multifactorial. Consideration will be given to the four major factors, viz., hormonal, volumetric, haemodynamic and nutritional, that have been postulated as causes for the marked circulatory changes observed.

Hormonal

Possibly sex steroid hormones are bound to receptors in muscle cells of the heart and blood vessels, where they might initiate production of increased amounts of contractile protein. A corollary of this might be an increase in myocardial contractility resulting in an increased stroke volume and cardiac output or an increase in vascular tone resulting in an increase in peripheral resistance with a secondary increase in cardiac output. An alternative mode of action of sex steroids might be a direct effect on cardiac and vasomotor centres in the medulla with subsequent changes in cardiac output or vasomotor tone (Fig. 12.5).

The interplay between the female sex hormones and the renin-angiotensin system might also be highly relevant to altered cardiovascular function in pregnancy. Plasma renin, renin activity, renin substrate, aldosterone and blood angiotensin levels are significantly higher in pregnancy than in the non-pregnant state (Brown *et al* 1965, Chesley *et al* 1965, Fasciolo *et al* 1963, Helmer & Judson 1967). Furthermore,

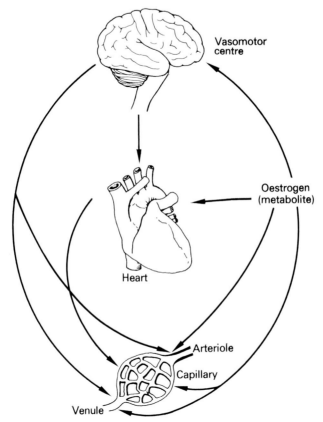

FIG. 12.5. A diagrammatic representation of the possible sites of action of sex steroids (in this case, oestrogen) in the cardiovascular system.

oestrogen administration to men increases plasma renin substrate (Helmer & Griffith 1952) and women taking combined oestrogen-progesterone oral contraceptives show an increase in plasma renin substrate leading to an increase in renin activity associated with suppression of renin secretion by the kidney (Skinner *et al* 1969). In addition blood angiotensin II levels have been shown to increase three-fold in women taking oral contraceptives (Walters & Lim 1970).

Plasma volume in non-pregnant women increases after oestrogen therapy of short duration (Witten & Bradbury 1951), while in non-pregnant ewes oestrogen infusion increases heart rate and cardiac output and decreases mean blood pressure and peripheral resistance. Similar changes are seen in normal ovine pregnancy (Ueland & Parer 1966). These investigators suggested that a combination of selective peripheral vasodilatation and a digitalis-like effect of oestrogens on the heart were responsible.

While it has been demonstrated that oestrogens cause vasodilatation and an increase in blood flow in the uterine circulation in animals (Greiss & Marston 1965, Holden 1939) and humans (Borell et al 1965), there is little knowledge of the effects of oestrogens on regional blood flow elsewhere in the body. During human pregnancy, vasodilatation with an increase in blood flow occurs in the hand and foot with minimal changes in the forearm and calf circulations (Ginsburg & Duncan 1967). The suggestion that female sex hormones might be responsible is provided by the work of Goodrich & Wood (1964) who found an increase in forearm blood flow in women taking oral contraceptives.

Gestagen production is also increased in pregnancy as indicated by maternal blood levels of progesterone (Zelenik 1965). Because of their smooth-muscle-relaxing properties and increased concentration in peripheral blood during pregnancy, vascular tone might be reduced. Some support for this theory is provided by the work of Brehm & Kindling (1955) who found a fall in peripheral resistance and a simultaneous rise in cardiac output during the luteal phase of the normal menstrual cycle. It is apparent that we know even less about the cardiovascular effects of gestagens than of oestrogens.

Other hormones, such as aldosterone, cortisol, oxytocin and thyroxine may also influence the cardiovascular system in pregnancy, directly or indirectly. It is conjectural whether human chorionic gonadotrophin, human placental lactogen or the pituitary gonadotrophins have any action on the cardiovascular system as there is no information available.

Volumetric

Hypervolaemia may contribute to the circulatory changes of pregnancy. Plasma volume increases, probably as a result of retention of sodium and water (MacGillivray & Buchanan 1958, Seitchik 1967). This increment in plasma volume may lead to an increase in cardiac output and the other hyperkinetic circulatory features seen in pregnancy. Witten & Bradbury (1951) attributed sodium retention to the increased production of oestrogen and progesterone by the placenta, while Venning et al (1959) attributed it to increased production of adrenal corticosteroids during pregnancy, especially aldosterone.

It is unlikely that hypervolaemia alone is responsible for all the haemodynamic changes of pregnancy, since cardiac output is already elevated in early pregnancy before blood volume has increased sufficiently to account for this phenomenon.

Haemodynamic

At first sight, the circulatory changes in pregnancy appear analogous to those occurring in a patient with an arteriovenous fistula. Features common to both include an increase in heart rate, pulse pressure, cardiac output, central venous oxygen saturation, central arteriovenous oxygen difference and blood volume. Clinically, cardiac murmurs and vasodilatation are evident in both conditions.

Burwell et al (1938) first advanced the theory of the intervillous circulation behaving as an arteriovenous fistula and attributed the decline in cardiac output in the last trimester to the closing down of the shunt as the placenta gradually became senescent. However, there is little evidence for placental senescence (Crawford 1962). Moreover, comparison of the peripheral circulation at 28 weeks gestation when cardiac output is at its maximum with that of an arteriovenous shunt reveals that peripheral resistance is decreased in pregnancy but increased in case of the arteriovenous shunt, while total blood flow is at a maximum at 28 weeks gestation but at a minimum in the presence of an arteriovenous shunt. In addition cardiac output and pulse rate increase most rapidly in early pregnancy, when changes in uterine blood flow are trivial (Winner 1965).

Nutritional

Increasing requirements of the mother and foetus for food and oxygen during pregnancy would be most efficiently served by a hyperkinetic circulation facilitating transport and exchange mechanisms. However, the maximum increase in maternal oxygen consumption (10% above the normal level in the non-pregnant) is relatively much less than the maximum increase in cardiac output (40% above normal levels in the non-pregnant) during pregnancy (Bader *et al* 1955, Lees *et al* 1967). This discrepancy together with a steadily increasing oxygen consumption until term in the presence of a static or perhaps declining cardiac output suggests that nutritional demand is not a major factor in causing the haemodynamic changes of pregnancy.

CONCLUSION

Having considered the four major factors which have been postulated as causes of the cardiovascular changes in pregnancy, it would seem both reasonable and logical to suppose that hormonal influence is largely responsible for these changes. The dramatic changes in hormone production in very early pregnancy could readily explain the early appearance of circulatory changes long before the uteroplacental circulation is of any haemodynamic significance. The mode of action of hormones at cellular level is only just being elucidated and until our knowledge increases in this sphere, it is unlikely that the hormonal theory will be proven.

REFERENCES

ABRAMSON D.I., FLACHS K. & FIERST S.M. (1943) Peripheral blood flow during gestation. *Amer. J. Obstet. Gynec.* **45,** 666.

ADAMS J.Q. (1954) Cardiovascular physiology in normal pregnancy: studies with the dye-dilution technique. *Amer. J. Obstet. Gynec.* **67,** 741.

ADAMS J.Q. & ALEXANDER A. (1968) Alteration in cardiovascular physiology during labour. *Obstet. Gynec.* **12,** 542.

DE ALVAREZ R.R. (1958) Renal glomerular mechanisms during normal pregnancy. *Amer. J. Obstet. Gynec.* **75,** 931.

American Committee on Maternal Welfare (1952)—EASTMAN J.J. *et al, The Mother,* p. 10. Chicago.

American Heart Association—KIRKENDALL W.M., BURTON A.C., EPSTEIN F.H. & FREIS E.D. (1967) Recommendation for human blood pressure determinations by sphygmomanometers. *Circulation* **36,** 980.

ANGELINO P.F., ACTIS-DATO A., LEVI V., SILIQUINI P.N. & REVELLI E. (1954) Nuovi concetti diemodinamica in gravidanza da indagini con cateterismo angiocardiaco. *Minerva Ginec.* **6,** 517.

ASSALI N.S., DOUGLASS R.A., BAIRD W.W., NICHOLSON D.B. & SUYEMOTO R. (1953) Measurement of uterine blood flow and uterine metabolism. *Amer. J. Obstet. Gynec.* **66,** 248.

BADER M.E., BADER R.A., ROSE D.J. & BRAUNWALD E. (1955) Hemodynamics at rest and during exercise in normal pregnancy as studied by cardiac catheterization. *J. clin. Invest.* **34**, 1524.
BEAN W.B., DEXTER M.W. & COGSWELL R.C. (1947) Vascular spiders and palmar erythema in pregnancy. *J. clin. Invest.* **26**, 1173.
BEST C.H. & TAYLOR N.B. (1966) *The Physiological Basis of Medical Practice*, 8th ed., p. 805 (stroke vol.) p. 500 (blood vol.). Baltimore, Williams & Wilkins Co.
BIENIARZ J., MAQUEDA E. & CALDEYRO-BARCIA R. (1966) Compression of the aorta by the uterus in late human pregnancy. *Amer. J. Obstet. Gynec.* **95**, 795.
BIENARZ J., BRANDA L.A., MAQUEDA E., MOROZOVSKY, J. & CALDEYRO-BARCIA R. (1968) Aorto-caval compression by the uterus in late pregnancy. III. Unreliability of the sphygmomanometric method in estimating uterine artery pressure. *Amer. J. Obstet. Gynec.* **102**, 1106.
BIENIARZ J., YOSHIDA T., ROMERO-SALINAS G., CURUCHET E., CALDEYRO-BARCIA R. & CROTTOGINI J.J. (1969) Aortocaval compression by the uterus in late human pregnancy. IV. Circulatory homeostasis by preferential perfusion of the placenta. *Amer. J. Obstet. Gynec.* **103**, 19.
BERLIN N.I. GOETSCH C., HYDE G.M. & PARSON R.J. (1953) The blood volume in pregnancy as determined by P^{32}-labelled red blood cells. *Surg. Gynec. Obstet.* **97**, 173.
BORELL V., FERNSTRÖM I., OHLSON L. & WIQVIST N. (1965) Influence of uterine contractions on the uteroplacental circulation at term. *Amer. J. Obstet. Gynec.* **93**, 44.
BRADLEY S.E., INGELFINGER F.J., BRADLEY G.P. & CURRY J.J. (1945) Estimation of hepatic blood flow in man. *J. clin. Invest.* **24**, 890.
BREHM H. & KINDLING E. (1955) Der Kreislauf während Schwangerschaft und Wochenbett. *Arch. Gynäk.* **185**, 696.
BRIGDEN W., HOWARTH S. & SHARPEY-SCHAFER E.P. (1950) Postural changes in the peripheral blood flow of normal subjects with observation on vasovagal fainting reactions as a result of tilting, the lordotic posture, pregnancy and spinal anaesthesia. *Clin. Sci.* **9**, 79.
BROWNE F.J. (1947) Chronic hypertension and pregnancy. *Brit. med. J.* **2**, 283.
BROWN J.J., DAVIES D.L., DOAK P.B., LEVER A.F., ROBERTSON J.I.S. & TRUST P. (1965) Plasma-renin concentration in hypertensive disease of pregnancy. *Lancet* **ii**, 1219.
BROWNE J.C.McL. & VEALL N. (1953) The maternal placental blood flow in normotensive and hypertensive women. *J. Obstet. Gynaec. Brit. Emp.* **60**, 141.
BURT C.C. (1950) Symposium on haemodynamics in pregnancy. IV. Peripheral circulation in pregnancy. *Edin. Med. J.* **57**, 18.
BURWELL C.S., STRAYHORN W.D., FLICKINGER D., CORLETTE M.B., BOWERMAN E.P. & KENNEDY J.A. (1938) Circulation during pregnancy. *Arch. intern. Med.* **62**, 979.
CATON W.L., ROBY C.C., DUNCAN E., CASWELL R., MALESKOS C.J., FLUHARTY R.G. & GIBSON J.G. (1951) The circulatory red cell volume and body hematocrit in normal pregnancy and the puerperium. *Amer. J. Obstet. Gynec.* **61**, 1207.

CHESLEY L.C. (1960) Renal functional changes in normal pregnancy. *Clin. Obstet. Gynec.* **3**, 349.

CHESLEY L.C., TALLEDO E., BOHLER C.S. & ZUSPAN F.P. (1965) Vascular reactivity to angiotensin II and norepinephrine in pregnant and nonpregnant women. *Amer. J. Obstet. Gynec.* **91**, 837.

CLEARFIELD H.R. (1965) In H.L. Bockus (ed.) *Gastroenterology*, 2nd ed., vol. 3, p. 94. Philadelphia and London, W.B. Saunders Company.

COHEN M.E. & THOMSON K.J. (1936) Studies on the circulation in pregnancy. I. The velocity of blood flow and related aspects of the circulation in normal pregnant women. *J. clin. Invest.* **15**, 607.

COPE I. (1958) Plasma and blood volume changes in late and prolonged pregnancy *J. Obstet. Gynaec. Brit. Emp.* **65**, 877.

COURNARD A.F. (1945) Symposium on cardiac output. *Federation Proc.* **4**, 183.

COURNARD A. & RANGES H.A. (1941) Catheterization of the right auricle in man. *Proc. Soc. Exp. Biol. Med.* **46**, 462.

CRAWFORD, J.M. (1962) Vascular anatomy of the human placenta. *Amer. J. Obstet. Gynec.* **84**, 1543.

CUNNINGHAM I. Cardiovascular physiology of labour and delivery. *J. Obstet. Gynaec. Brit. Cwlth.* **73**, 500.

DAHLSTRÖM H. & IHRMAN K. (1960) A clinical and physiological study of pregnancy in a material from Northern Sweden. I. Observations with special regard to cardiopulmonary function during the first trimester of pregnancy. *Acta Soc. Med. Upsalien* **65**, 117.

DAVIS H.A. (1962) *Blood Volume Dynamics*, p. 106. Springfield, Illinois, Charles C. Thomas.

DAVSON H. & EGGLETON M.G. (1968) *Principles of Human Physiology*, 14th ed., p. 161. London, J.&A.Churchill Ltd.

DIGNAM W.J., TITUS P. & ASSALI N.S. (1958) Renal function in human pregnancy. I. Changes in glomerular filtration rate and renal plasma flow. *Proc. Soc. Exp. Biol. Med.* **97**, 512.

DOLEŽAL A. & FIGAR S. (1965) The phenomenon of reactive vasodilatation in pregnancy. *Amer. J. Obstet. Gynec.* **93**, 1137.

DONOVAN J.C., LUND C.J. & HICKS E.L. (1965) Effect of lactation on blood volume in the human female. *Amer. J. Obstet. Gynec.* **93**, 588.

DRISCHEL H., FANTER H., GÜRTLER H., LABITZKE H. & PRIEGNITZ F. (1963) Das Verhalten des Herzfrequenz gesunder Menschen beim Übergang vom Liegen zum Stehen. *Arch. Kreisl.-Forsch.* **40**, 135.

DUNCAN S.L.B. & GINSBURG J. (1968) Peripheral circulation in multiple pregnancy. *Amer. J. Obstet. Gynec.* **101**, 223.

DUHRING J.L. (1962) Blood volume in pregnancy. *Amer. J. Med. Sci.* **243**, 808.

FASCIOLO J.C., DE VITO E., ROMERO J.C. & CUCCHI J.N. (1963) The renin content of the blood of humans and dogs under several conditions. *Canad. Med. Ass. J.* **90**, 206.

FORSSMANN W. (1929) Die Sondierung des rechten Herzens. *Klin. Wschr.* **8**, 2085.

GINSBURG J. & DUNCAN S.L.B. (1967) Peripheral blood flow in normal pregnancy. *Cardiov. Res.* **1**, 132.

GINSBURG J. & DUNCAN S. (1969) Direct and indirect blood pressure measurement in pregnancy. *J. Obstet. Gynaec. Brit. Cwlth.* **76**, 705.

GOODRICH S.M. & WOOD J.E. (1964) Peripheral venous distensibility and velocity of venous blood flow during pregnancy or oral contraceptive therapy. *Amer. J. Obstet. Gynec.* **90**, 740.

GREENSTEIN N.M. & CLAHR J. (1937) Circulation time studies in pregnant women. *Amer. J. Obstet. Gynec.* **33**, 414.

GUYTON A.C. (1963) *Circulatory Physiology: Cardiac Output and its Regulation*, p. 10. Philadelphia and London, W.B.Saunders Co.

HAMILTON H.F.H. (1950) Blood viscosity in pregnancy. *J. Obstet. Gynaec. Brit. Emp.* **57**, 530.

HAMILTON H.F.H. (1949) The cardiac output in normal pregnancy as determined by the Cournand right heart catheterization technique. *J. Obstet. Gynaec. Brit. Emp.* **56**, 548.

HAMILTON W.F. (1953) The physiology of the cardiac output. *Circulation* **8**, 527.

HELMER O.M. & GRIFFITH R.S. (1952) The effect of the administration of estrogens on the renin substrates. *Endocrinology* **51**, 421.

HELMER O.M. & JUDSON W.E. (1967) Influence of high renin substrate level on the renin angiotensin system in pregnancy. *Amer. J. Obstet. Gynec.* **99**, 9.

HENDRICKS C.H. (1958) The hemodynamics of a uterine contraction. *Amer. J. Obstet. Gynec.* **76**, 969.

HERBERT C.M., BANNER E.A. & WAKIM K.G. (1958) Variation in the peripheral circulation during pregnancy. *Amer. J. Obstet. Gynec.* **76**, 742.

HOLDEN R.B. (1939) Vascular reactions of the uterus of the immature rat. *Endocrinology* **25**, 593.

HOLMES F. (1953) Incidence of the supine hypotensive syndrome in late pregnancy. *J. Obstet. Gynaec.* **1**, 371.

HOWARD B.K., GOODSON J.H. & MENGERT W.F. (1953) Supine hypotensive syndrome in late pregnancy. *Obstet. Gynec.* **1**, 371.

HUCHARD H. (1943) *Maladies du cœur artériosclerose*. J.B.Baillière et Fils, Paris. Quoted by Abramson *et al.*

HYTTEN F.E. & LEITCH I. (1964) *The Physiology of Human Pregnancy*. Oxford, Blackwell Scientific Publications.

HYTTEN F.E. & PAINTIN D.B. (1963) Increase in plasma volume during normal pregnancy. *J. Obstet. Gynaec. Brit. Cwlth.*, **70**, 402.

JEGIER W., SEKELJ P., AULD P.A.M., SIMPSON R. & MCGREGOR M. (1963) The relation between cardiac output and body size. *Brit. Heart J.* **25**, 425.

KERR M.G. (1965) The mechanical effects of the gravid uterus in late pregnancy. *J. Obstet. Gynaec. Brit. Cwlth.* **72**, 513.

KERR M.G., SCOTT D.B. & SAMUEL E. (1964) Studies of the inferior vena cava in late pregnancy. *Brit. med. J.* **1**, 532.

KETY S.S. & SCHMIDT C.F. (1954) *Amer. J. Physiol.* **143**, 53.

KJELLBERG S.R. (1953) The roentgenologic determination of heart volume. *Acta Med. Scand.* **145**, Suppl. 277, 25.

KLEE F. (1924) Die Strömungsteschwindigkeit des Blutes in Schwangerschaft. *Z. Geburtsh. Gynäk.* **88**, 303.

KLEIBER M. (1947) Body size and metabolic rate. *Physiol Rev.* **27**, 511.

KLEIN O. (1930) Zur Bestimmung des Zirkulatorischen Minutenvolumens beim Menschen nach dem Fickschen Prinzip. (Gewinnung des gemischten venosen Blutes mittels Herzsondierung.) *München Med. Wschr.* **77**, 1311.

KOTTE J.H., IGLAUER A. & MCGUIRE J. (1944) Measurements of arterial blood pressure in the arm and leg: comparison of sphygmomanometric and direct intra-arterial pressures, with special attention to their relationship in aortic regurgitation. *Amer. Heart J.* **28**, 476.

LANDESMAN R. & MILLER M.M. (1963) Blood volume changes during the immediate postpartum period. *Obstet. & Gynec.* **21**, 40.

LANDT H. & BENJAMIN J.E. (1936) Cardiodynamic and electrocardiographic changes in normal pregnancy. *Amer. Heart J.* **12**, 592.

LEES M.M., TAYLOR S.H., SCOTT D.B. & KERR M.G. (1967) A study of cardiac output at rest throughout pregnancy. *J. Obstet. Gynaec. Brit. Cwlth.* **74**, 319.

LEES M.M, SCOTT D.B., KERR M.G. & TAYLOR S.H. (1967) The circulatory effects of recumbent postural change in late pregnancy. *Clin. Sci.* **32**, 453.

LEES M.M., SCOTT D.B. & KERR M.G. (1970) Haemodynamic changes associated with labour. *J. Obstet. Gynaec. Brit. Cwlth.* **77**, 29.

LINDHARD J. (1915) Über das Minutenvolumens des Herzens bei Ruhe und bei Muskelarbeit. *Pflüg. Arch. ges. Physiol.* **161**, 233.

LUND C.J. & DONOVAN J.C. (1967) Blood volume during pregnancy. Significance of plasma and red cell volume. *Amer. J. Obstet. Gynec.* **98**, 393.

MCCLURE BROWNE J.C. (1961) Survey of eclampsia—clinical aspects. *Path. Microbiol.* **24**, 542.

MACGILLIVRAY I. & BUCHANAN T.J. (1958) Total exchangeable sodium and potassium in non-pregnant women and in normal and pre-eclamptic pregnancy. *Lancet* ii, 1090.

MACGILLIVRAY I. (1961) Hypertension in pregnancy and its consequences. *J. Obstet. Gynaec. Brit. Cwlth.* **68**, 557.

MACGILLIVRAY I., ROSE G.A. & ROWE B. (1969) Blood pressure survey in pregnancy. *Clin. Sci.* **37**, 395.

MCCALL M.L. (1949) Cerebral blood flow and metabolism in toxemias of pregnancy. *Surg. Gynec. Obstet.* **89**, 715.

MCCAUSLAND A.M., HYMAN C., WINSOR T. & TROTTER A.D. (1961) Venous distensibility during pregnancy. *Amer. J. Obstet. Gynec.* **81**, 472.

MCLENNAN C.E. (1943) Antecubital and femoral venous pressures in normal and toxemic pregnancy. *Amer. J. Obstet. Gynec.* **45**, 568.

MCLENNAN C.E. & THOUIN L.G. (1948) Blood volume in pregnancy. *Amer. J. Obstet. Gynec.* **55**, 189.

MCMICHAEL J. (1945) Notes on cardiac output methods. *Federation Proc.* **4**, 212.

MAHRAN M., SABOUR M.S., FADEL H.E., EL-MAHALLAWI M.E. & ELDIN S. (1968) Glomerular filtration rate in normal pregnancy and the early postpartum period. *Obstet. & Gynec.* **31**, 621.

MANCHESTER B. & LOUBE S.D. (1946) The velocity of blood flow in normal pregnant women. *Amer. Heart J.* **32**, 215.

MARSHALL R.J. & SHEPHERD, J.T. (1968) *Cardiac Function in Health and Disease*, p. 11. Philadelphia, London, Toronto, W.B.Saunders Co.

MELBRAND S.M. (1938) Valeur diagnostique de la capillaroscopic dans la grossesse et dans la sepsie puerpérale. *Gynéc. et Obstét.* **37**, 200.

METCALFE J. ROMNEY S.L., RAMSEY L.H., REID D.E. & BURWELL C.S. (1955) Estimation of uterine blood flow in normal human pregnancies at term. *J. clin. Invest.* **34**, 1632.

MOSS A.J. & ADAMS F.H. (1963) Index of indirect estimation of diastolic blood pressure. *Amer. J. Dis. Child.* **106**, 364.

MUNDOW L.S. (1967) The supine hypotensive syndrome of pregnancy. *J. Irish Med. Ass.* **40**, 194.

MUNNELL E.W. & TAYLOR H. JR. (1947) Liver blood flow in pregnancy—hepatic vein catheterization. *J. clin. Invest.* **26**, 952.

OSOFSKY H.J. & WILLIAMS J.A. (1964) Changes in blood volume during parturition and early postpartum period. *Amer. J. Obstet. Gynec.* **88**, 396.

PAINTIN D.B. (1962) The size of the total red cell volume in pregnancy. *J. Obstet. Gynaec. Brit. Cwlth.* **69**, 719.

PALMER A.J. & WALKER A.H.C. (1949) The maternal circulation in normal pregnancy. *J. Obstet. Gynaec. Brit. Emp.* **56**, 537.

PEILLISSIER P. DE (1912) De la viscosité du sang totale chez la femme enceinte et accouchée et de ses rapports avec la tension artérielle. *Archiv. mens d'obst. et gynéc.* Par. II, 306.

PHINNEY A.O. JR., CLASON W.P.C., STOUGHTON P.V. & MCLEAN C.E. (1963) Measurement of cardiac output using the photoelectric earpiece: a comparison with simultaneous Fick measurements. *Circ. Res.* **13**, 80.

PRITCHARD J.A., BALDWIN R.M., DICKEY J.C. & WIGGINS K.M. (1962) Blood volume changes in pregnancy and the puerperium. II. Red blood cell loss and changes in apparent blood volume during and following vaginal delivery, cesarean section, and cesarean section plus hysterectomy. *Amer. J. Obstet. Gynec.* **84**, 1271.

PRITCHARD J.A., BARNES A.C. & BRIGHT R.H. (1955) The effects of the supine position on renal function in the near term pregnant woman. *J. clin. Invest.* **34**, 777.

PRITCHARD J.A. & ROWLAND R.C. (1964) Blood volume changes in pregnancy and the puerperium. III. Whole body and large vessel hematocrits in pregnant and nonpregnant women. *Amer. J. Obstet. Gynec.* **88**, 391.

PRITCHARD J.A. & ADAMS R.H. (1960) Erythrocyte production and destruction during pregnancy. *Amer. J. Obstet. Gynec.* **79**, 750.

PYÖRÄLA T. (1966) Cardiovascular response to the upright position during pregnancy. *Acta Obstet. Gynaec.* **45**, Suppl. 5.

QUILLIGAN E.J. & TYLER C. (1959) Postural effects on the cardiovascular status in pregnancy. A comparison of the lateral and supine postures. *Amer. J. Obstet. Gynec.* **78**, 465.

QUINLIVAN W.L.G. & BROCK J.A. (1970) Blood volume changes and blood loss associated with labour. I. Correlation of changes in blood volume measured by I^{131}-albumin and Evans blue dye, with measured blood loss. *Amer. J. Obstet. Gynec.* **106**, 843.

DE REZENDE J.M. (1950) Pressão arterial e pulso no final da gestação; hipotensão clinostática e colapso vasomotor condicionados pelo decubito dorsal. *Arch. Bras. Med.* **40**, 375.

ROBERTS L.N., SMILEY R.A. & MANNING G.W. (1953) A comparison of direct and indirect blood pressure determinations. *Circulation* **8**, 232.

ROBINSON S.C. & BRUCER M. (1939) Range of normal blood pressure: statistical and clinical study of 11,383 persons. *Arch. Int. Med.* **64**, 409.

ROMAN J., HENRY J.P. & MEEHAN J.P. (1965) Validity of flight blood pressure data. *Aerospace Med.* **36**, 436.

ROSCOE M.H. & DONALDSON G.M.M. (1946) The blood in pregnancy. II. The blood volume and haemoglobin mass. *J. Obstet. Gynaec. Brit. Emp.* **53**, 527.

ROSE D.J., BADER M.E., BADER R.A. & BRAUNWALD E. (1956) Catheterization studies of cardiac hemodynamics in normal pregnant women with reference to left ventricular work. *Amer. J. Obstet. Gynec.* **72**, 233.

ROVINSKY J.J. (1966) Cardiovascular hemodynamics in pregnancy. III. Cardiac rate, stroke volume, total peripheral resistance and central blood volume in pregnancy. *Amer. J. Obstet. Gynec.* **95**, 787.

ROVINSKY J.J. & JAFFIN H. (1965) Cardiovascular hemodynamics in pregnancy. I. Blood and plasma volumes in multiple pregnancy. *Amer. J. Obstet. Gynec.* **93**, 1.

ROVINSKY J.J. & JAFFIN H. (1966) Cardiovascular hemodynamics in pregnancy. II. Cardiac output and left ventricular work in multiple pregnancy. *Amer. J. Obstet. Gynec.* **95**, 781.

ROY S.B., MALKANI P.K., VIRIK R. & BHATIA M.L. (1966) Circulatory effects of pregnancy. *Amer. J. Obstet. Gynec.* **96**, 221.

DE SCHWARCZ S.B., ARAMENDIA P. & TAQUINI A.C. (1964) Variaciones Hemodinamicas en el embarazo normal. *Medicina (B. Air.)* **24**, 113.

SCOTT D.B. & KERR M.G. (1963) Inferior vena caval pressure in late pregnancy. *J. Obstet. Gynaec. Brit. Cwlth.* **70**, 1044.

SEITCHIK J. (1967) Total body water and total body density of pregnant women. *Obstet. & Gynec.* **29**, 155.

SIMS E.A. & KRANTZ K.E. (1958) Serial studies of renal function during pregnancy and the puerperium in normal women. *J. clin. Invest.* **37**, 1764.

SKINNER S.L., LUMBERS E.R. & SYMONDS E.M. (1969) Alteration by oral contraceptives of normal menstrual changes in plasma renin activity, concentration and substrate. *Clin. Sci.* **35**, 67.

SPETZ S. (1964) Peripheral circulation during normal pregnancy. *Acta obstet. gynec. scand.* **43**, 309.

SPETZ S. & JANSSON I. (1969) Forearm blood flow during normal pregnancy studied by venous occlusion plethysmography and ^{133}Xenon muscle clearance. *Acta obstet. gynec. scnad.* **48**, 285.

SPITZER W. (1933) Die Blutströmungsgeschwindigkeit in normaler und gestörter Schwangerschaft. Beitrag zur funktionsprüfung des Herzens in der Schwangerschaft und vor der Geburt. *Arch. Gynäk.* **154**, 449.

STATZER D.E. (1959) Blood volume in pregnancy with iodinated human serum albumin. *Obstet. & Gynec.* **14**, 37.

STEELE J.M. (1941) Comparison of simultaneous indirect (auscultatory) and direct intra-arterial) measurements of arterial pressure in man. *J. Mount Sinai Hosp.* **8**, 1042.

STEVENS P.M. (1966) Cardiovascular dynamics during orthostasis and the influence of intravascular intrumentation. *Amer. J. Cardiol.* **71**, 211.

TAYLOR S.H. & SHILLINGFORD J.P. (1959) Clinical applications of Coomassie blue. *Brit. Heart J.* **21**, 497.

THOMSON K.J., HIRSHEIMER A., GIBSON J.G. 2nd & EVANS W.A. JR (1938) Studies on the circulation in pregnancy. III. Blood volume changes in normal pregnant women. *Amer. J. Obstet. Gynec.* **36**, 48.

TROWER R. & WALTERS W.A.W. (1968) Brachial arterial blood pressure in the lateral recumbent position during pregnancy. *Aust. New Zeal. J. Obstet. Gynaec.* **8**, 146.

TUCKMAN J. & SHILLINGFORD J. (1966) Effect of different degrees on tilt on cardiac output, heart rate and blood pressure in normal man. *Brit. Heart J.* **28**, 32.

TŮMOVÁ Z. (1964) Posturální vliv na krevní tlak těhotných. Poloha na zádech a poloha na boku. *Čs. Gynek.* **29**, 10.

TYSOE F.W. & LOWENSTEIN L. (1950) Blood volume and hematologic studies in pregnancy and the puerperium. *Amer. J. Obstet. Gynec.* **60**, 1187.

UELAND K. & HANSEN J.M. (1969) Maternal cardiovascular dynamics. II. Posture and uterine contractions. *Amer. J. Obstet. Gynec.* **103**, 1.

UELAND K. & PARER J.T. (1966) Effects of estrogens on the cardiovascular system of the ewe. *Amer. J. Obstet. Gynec.* **96**, 400.

VENNING E.H., DYRENFURTH I., LOWENSTEIN L. & BECK J. (1959) Metabolic studies in pregnancy and the puerperium. *J. Clin. Endocr.* **19**, 403.

VEREL D., BURY J.D. & HOPE A. (1956) Blood volume changes in pregnancy and the puerperium. *Clin. Sci.* **15**, 1.

VORYS N., ULLERY J.C. & HANUSEK G. (1963) The cardiac output changes in various positions in pregnancy. *Surg. Obstet. Gynec.* **116**, 511.

WALTERS W.A.W., MCGREGOR W.G. & HILLS M. (1966) Cardiac output at rest during pregnancy and the puerperium. *Clin. Sci.* **30**, 1.

WALTERS W.A.W. & LIM Y.L. (1970) Haemodynamic changes in women taking oral contraceptives. *J. Obstet. Gynaec. Brit. Cwlth.* **77**, 1007.

WANG J., MARSHALL R.J. & SHEPHERD J.T. (1960) The effect of changes in posture and of graded exercise on stroke volume in man. *J. clin. Invest.* **39**, 1051.

WESSON L.G. (1969) *Physiology of the Human Kidney*, p. 96. New York and London, Grune & Stratton.

WERKÖ I., LAGERLÖF H., BUCHT H. & HOLMGREN A. (1948) Cirkulationen vid graviditet. *Nord. Med.* **40**, 1868.

WIDLUND G. (1945) The cardio-pulmonary function during pregnancy. *Acta obstet. gynec. scand.* **25**, Suppl. 1.

WINNER W. & ROMNEY S.L. (1966) Cardiovascular responses to labour and delivery. *Amer. J. Obstet. Gynec.* **95**, 1104.

WINNER W. (1965) The role of the placenta in the systemic circulation. *Obstet. Gynec. Survey* **20**, 545.

WITTEN C.L. & BRADBURY J.T. (1951) Hemodilution as a result of estrogen therapy. Estrogenic effects in the human female. *Proc. Soc. Exp. Biol. Med.* **78**, 626.

ZELENIK J.S. (1965) Endocrine physiology of pregnancy. *Clin. Obstet. Gynec.* **8**, 528.

13
Normal Renal Physiology and Changes in Pregnancy*

RANJIT SINGH NANRA† &
PRISCILLA KINCAID-SMITH

FUNCTIONAL ANATOMY

The nephron is the basic functional unit of the kidney, and is composed of the glomerulus, the proximal convoluted tubule, the loop of Henle and the distal convoluted tubule which opens into the renal pelvis through collecting tubules and larger ducts of Bellini (Fig. 13.1). Each human kidney has approximately 1 million nephrons and may be divided into an outer part, the cortex, and an inner part, the medulla. The bulk of the cortex is composed of glomeruli, proximal and distal convoluted tubules and blood vessels. The majority of glomeruli are found in the outer cortex, but some are situated in the juxta-medullary region. The medulla consists of loops of Henle, vasa recta and collecting tubules only, and in man is divided into twelve to fourteen medullary pyramids per kidney which project into the renal pelvis as papillae. Only one out of seven nephrons have long loops of Henle which go down deep into the papilla where they form hairpin bends, and return to the cortex to join the distal convoluted tubules. The long loops of Henle belong mainly to the juxta-medullary glomeruli which are particularly adapted to salt conservation as opposed to the nephrons supplied by outer cortical glomeruli which are better adapted to salt excretion (Barger 1966).

---

* This work was supported by the National Health and Medical Research Council of Australia and by a personal grant to one of us (P.K.-S.) from the Wellcome Trust.

† Present address: Renal Physician, Royal Newcastle Hospital, Newcastle, New South Wales.

The glomeruli are supplied from the interlobular arteries by afferent arterioles, which break up into twenty to forty intraglomerular capillary loops to form a vast filtration network. The glomerular filtrate has to pass through the wall of the capillary loop to enter into Bowman's

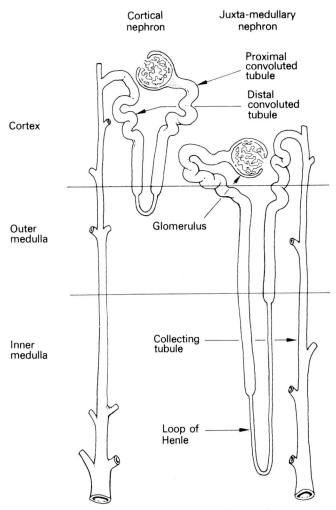

FIG. 13.1. Cortical and juxta-medullary nephrons.

capsular space and from there down the nephron. Each capillary loop is lined on the luminal surface by endothelial cells and on the outer or capsular surface by epithelial cells and between them is the glomerular basement membrane which in the adult is up to 3000 Å units thick. The endothelial cells are separated by fenestrations with an average diameter of 600 Å. Although no pores are visible in the glomerular basement membrane even on electron microscopy, it behaves as if it had pores of 75 to 100 Å in diameter. The epithelial cell is only in contact with the basement membrane by foot processes or podocytes, which are separated by slit pores with a mean width of 400 Å. The capillary loops join up to form the efferent arteriole, which breaks up into a diffuse capillary meshwork surrounding the tubules in the cortex and the outer medulla. The efferent arterioles of the juxta-medullary glomeruli enter the renal medulla, and divide to form compact bundles of vasa recta. The longest of these arterial vasa recta descend to the tip of the renal papilla, where individual vessels unite through short capillaries with venous vasa recta, which drain into the arcuate veins at the cortico-medullary junction.

This unique hairpin configuration of the loops of Henle and their close proximity to similarly arranged vasa recta are responsible for the concentrating ability of the kidney by means of the counter-current multiplier and exchanger mechanisms. The collecting duct traverses the whole length of the renal medulla and papilla and is thereby subjected to the medullary concentration gradient generated by the counter-current mechanisms. Under the influence of the antidiuretic hormone (ADH) which determines its permeability to water, the collecting duct is responsible for the final volume of urine produced.

In the wall of the afferent arteriole at the point of its entry into the glomerular tuft is the juxta-glomerular apparatus, a series of granular secretory cells which are the source of the enzyme, renin (Peart 1965). At this point the distal convoluted tubule comes into contact with the vascular pole of the glomerulus and the nuclei of the epithelial cells of the tubule become crowded, and appear darker staining. This prominent group of cells is called the macula densa and is responsible for sampling the Na^+ concentration in the distal tubule (Fig. 13.2). The arrangement of these structures appears to be ideally suited for the control of sodium excretion, intra-renal blood flow and systemic blood pressure through the renin-angiotensin-aldosterone system.

The luminal surface of the proximal tubular cells has a characteristic brush border, which consists of numerous microvilli, about 150 per square micron. The microvilli increase enormously the absorbing surface area of the proximal convoluted tubule. The large number of mitochondria and ribosomes in the cells of the proximal tubule reflect the tremendous metabolic activity of this part of the nephron. A similar concentration of intracellular organelles is also found in the cells of the thick ascending portion of the loop of Henle.

Both the afferent vessels which supply the glomeruli and the efferent vessels which drain the glomeruli have characteristics of arterioles. They have prominent smooth muscle fibres which are richly supplied with sympathetic nerve endings and exert a sphincter-like action to control intraglomerular pressure and blood flow. The smooth muscle cells of the efferent arterioles of the glomeruli continue on as far as the arterial vasa recta (Moffat 1967). The demonstration of adrenergic

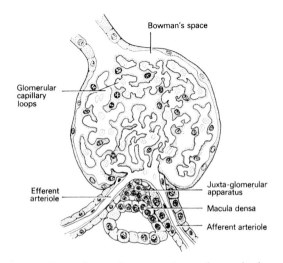

FIG. 13.2. Juxta-glomerular apparatus and macula densa.

innervation of arterial blood vessels, including the vasa recta (with the exception of efferent arterioles) appear to give a role to the sympathetic nervous systems and catecholamines in the control of renal blood flow (McKenna & Angelakos 1968).

NORMAL RENAL PHYSIOLOGY

The kidneys are vital organs of the body and maintain the composition of the 'internal environment' within very narrow limits. To achieve this important function, 20 to 25% of the total cardiac output passes through the kidneys and in 24 hours 180 litres of glomerular filtrate are produced. The volume and composition of the glomerular filtrate active tubular transport of the substance up to a point at which there is a saturation of the transport mechanism. This maximum rate of tubular reabsorption of a substance is referred to as the tubular maximum (Tm) of the substance. Similarly there is a Tm for tubular secretion. Passive tubular transport depends on diffusion down osmotic or electrochemical gradients.

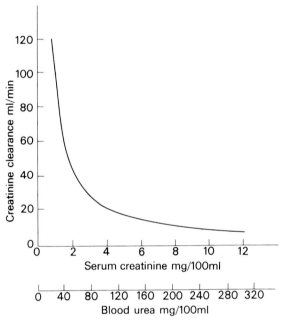

FIG. 13.3. Relationship between serum creatinine and blood urea concentrations and GFR.

Tubular reabsorption

The substances which are actively reabsorbed and have a mechanism which is limited by the Tm for that substance are glucose, phosphate, sulphate, amino acids, certain organic acids of the Krebs cycle, uric acid and proteins. Other substances are either dependent upon a gradient-time limited transport system, e.g. sodium, chloride and bicarbonate or passively reabsorbed, e.g. water, chloride and urea. The tubular handling of these substances is separately described below.

Glucose. Glucose is actively reabsorbed in the proximal tubule by a mechanism common to xylose, fructose and galactose. Competitive for the creation of a more significant medullary concentration gradient where the tissue osmolality is maximal at the tip of the papilla (1200 mOsm/l) and lowest at the cortico-medullary junction (300 mOsm/l). This process is termed the '*counter-current multiplication of concentration*' and is the main function of the loop of Henle. The Na^+, together with other solutes, and water which enters the medullary interstitium, are removed from the medulla by the blood flowing through the vasa recta loops. However, the apposition of the afferent and efferent flow in the hairpin vasa recta loops combined with the sluggish flow of blood in these vessels tend to reduce the excessive loss of osmotically active solutes from the medulla and therefore help to conserve the medullary concentration gradient (Fig. 13.4). The vasa recta thus form the '*counter-current exchanger system*'. The factors which therefore determine the maximal osmolar concentration at the papillary tip, hence the concentrating ability of the kidney, are the Na^+ pump in the ascending limb of the loop of Henle, the delivery of Na^+ to the loop, the rate of blood flow through the vasa recta loops and the level of circulating ADH.

During hydropoenia the high circulating levels of ADH render the epithelium of the distal and collecting tubules freely permeable to water. The action of ADH is mediated through intracellular adenosine-3'5'-phosphate (cyclic-AMP) (Orloff & Handler 1967). The tubular fluid which is hypotonic (100 mOsm/l) when delivered to the distal convoluted tubule becomes iso-osmotic with the cortical interstitium (300 mOsm/l). In the distal tubule Na^+ removal from tubular fluid comes under the influence of aldosterone. The concomitant passive diffusion of water with Na^+ transport further reduces the volume of tubular fluid to a small percentage of the original glomerular filtrate. As the fluid moves down the collecting tubules, water moves out freely into the hypertonic medullary interstitium. The volume of urine falls progressively while its osmolar content progressively increases. Finally a small volume of highly concentrated urine is passed into the renal pelvis. During a water diuresis when the level of circulating ADH is low, the epithelium of the distal and collecting tubules is impermeable to water, and the hypotonicity of the tubular fluid leaving the loop of Henle is maintained throughout the rest of the nephron. Therefore, the final urine is dilute and the volume large. By these means the osmolar content of the urine may be varied from 37 mOsm/l to 1400 mOsm/l. Aldosterone may contribute to the production of the final volume and composition of urine for it has been shown to influence the Na^+ pump in the loop of Henle and to decrease the permeability of the distal nephron to water (Hierholzer et al 1965, Wiederholt & Wiederholt 1968).

The amount of water in urine in excess of what is needed to make the urine iso-osmolar with plasma is called 'free water'. During a water diuresis when the osmolality of urine is less than that of plasma, free water clearance (C_{H_2O}) is said to be positive and may be as high as 15 ml/min. C_{H_2O} is not a true clearance and may be mathematically expressed as:

$$C_{H_2O} = V - \frac{U_{osm} V}{P_{osm}}$$

where C_{H_2O} = free water clearance in ml/min
V = urine volume/min
U_{osm} = urine osmolality in mEq/l
P_{osm} = plasma osmolality in mEq/l

When the urine is concentrated and has an osmolality greater than that of plasma, C_{H_2O} is said to be negative. This increase in urine osmolality represents medullary reabsorption of free water, T^cH_2O. The mathematical formula may now be written as follows:

$$T^c_{H_2O} = \frac{U_{osm} V}{P_{osm}} - V$$

During an osmotic diuresis with mannitol with simultaneous intravenous administration of ADH, the medullary reabsorption of water approaches a maximal rate, $Tm^c_{H_2O}$. $Tm^c_{H_2O}$ is approximately 5 ml/min and is an index of the functioning mass of distal tubules and loops of Henle.

Sodium

The conservation of Na^+ by the kidney is one of the major functions of the nephron and bears a direct relationship to renal oxygen consumption. The bulk of Na^+ transport is an active process and in the proximal tubule is dependent upon a Na^+-K^+-activated ATP-ase enzyme system (Katz & Epstein 1967). However, some Na^+ transport may occur passively by a physical phenomenon called solvent drag (Ullrich 1967).

Seventy-five to eighty per cent of the filtered Na^+ is actively reabsorbed in the proximal tubule where both physiological and anatomical evidence suggests that the intracellular channels may play an important role. Passive reabsorption of Cl^- occurs with Na^+.

An increased filtered load of Na^+ is accompanied by an increase in its proximal tubular reabsorption, the fractional fluid reabsorption remaining constant. This is a reflection of glomerulo-tubular balance and is believed to be not only a function of GFR (Rector et al 1966), but also of tubular volume (Gertz et al 1965) and intratubular pressure (Leyssac 1966).

In 1961 de Wardener and his colleagues demonstrated that expansion of the intravascular volume is accompanied by an augmented urinary excretion of Na^+ and other studies have substantiated the fact that this is independent of aldosterone, ADH and GFR (Lindheimer et al 1967, Rector et al 1967). This mechanism of natriuresis has been attributed to the 'third factor', the first and second major factors being GFR and aldosterone. Micropuncture studies have confirmed that the increase in the urinary excretion of Na^+ following expansion of intravascular volume results from diminished proximal tubular reabsorption of Na^+ (Dirks et al 1965). The nature of the third factor is still not known. A number of physical theories have been postulated to explain the natriuresis which accompanies volume expansion. Among these are increased medullary blood flow with resultant washout of the medullary concentration gradient (Earley & Friedler 1965, Horster & Thurau 1968), changes in the geometry of the proximal tubule (Gertz et al 1965, Rector et al 1966, Widerholt et al 1967) and alterations in the oncotic pressure in peritubular capillaries (Earley 1964, Martino & Earley 1967, Windhager et al 1969). A natriuretic hormone has been postulated by de Wardener (1961 and 1969) and McDonald et al (1967), but micropuncture studies have not substantiated the presence of a hormonal factor (Rector & Seldin 1969). Mills and his colleagues have postulated an intrarenal mechanism dependent upon a pressure-sensitive area in the region of the juxta-glomerular cells (Wilson et al 1970). Part of the natriuresis may be due to depression of Na^+ transport in the loop of Henle by prostaglandin-E_2, a vasodepressor substance that has been extracted from the renal medulla (Daniels et al 1967, Mills 1970). The adrenergic innervation of the kidney has also been incriminated because the natriuretic response to volume expansion may be blocked by catecholamine depletion with reserpine (Schrier et al 1967).

Ten per cent of the filtered Na^+ is reabsorbed in the loop of Henle by the active Na^+ pump in the thick ascending portion under the influence of ADH. Micropuncture studies have revealed that the thin part of the ascending portion of the loop also partakes in the Na^+ pump and therefore contributes to counter-current multiplication (Jamison et al 1967, Jamison 1969). During volume expansion, the depression of proximal tubular Na^+ reabsorption results in an increased delivery of Na^+ to the loop of Henle and consequently increased Na^+ reabsorption by the loop (Cortney et al 1965).

In the distal and collecting tubules Na^+ reabsorption occurs under the influence of aldosterone. Here the Na^+ transport is an active ionic exchange and is coupled with the excretion of H^+ and K^+. This is a very efficient mechanism, for it is capable of making the urine almost free of Na^+ during states of Na^+ depletion. There is some evidence to suggest that aldosterone may also affect Na^+ transport in the proximal tubule and loop of Henle (Hierholzer et al 1965, 1966).

Potassium

Almost all the filtered load of K^+ is actively reabsorbed in the proximal tubule under the influence of Na^+-K^+-activated ATP-ase (Malnic et al 1966a, 1966b). The urinary excretion of K^+ is mainly dependent upon the active ionic exchange between Na^+, and K^+ and H^+ in the distal nephron under the influence of aldosterone (Giebisch et al 1967). The capacity of the kidney to conserve K^+ is not as efficient as that for Na^+.

Acidification

The kidney regulates the acid-base balance of the body by maintaining the serum bicarbonate (HCO_3^-) concentration between 26 and 28 mEq/l while the respiratory system contributes to this regulation by stabilizing the plasma carbonic acid at a level of 1·3 to 1·4 mEq/l. Renal stabilization of the serum HCO_3^- involves the excretion of hydrogen ion (H^+) and the complete reabsorption of HCO_3^- in the proximal tubule. When the serum HCO_3^- exceeds 28 mEq/l HCO_3^- is excreted in the urine, and in stituations where HCO_3^- reserves are depleted, compensation occurs by excretion of titrable acid and ammonia.

H^+ is generated in renal tubular cells from carbonic acid, under the influence of carbonic anhydrase.

$$H_2O + CO_2 \xrightarrow{\text{Carbonic Anhydrase}} H_2CO_3 \rightleftharpoons H^+ + HCO_3^-$$

It is actively secreted along the entire length of the nephron in exchange for Na^+ which enters tubular cells passively and is then actively extruded into peritubular capillaries. The basic ionic exchange mechanisms involved in the excretion of H^+ are:

Reabsorption of HCO_3^-. HCO_3^- is the major urinary buffer in the proximal tubule. The HCO_3^- in tubular fluid combines with H^+ to form carbonic acid and ultimately water and carbon dioxide. Carbon dioxide freely diffuses back into the tubular epithelial cell where it is rapidly converted to HCO_3^- under the influence of carbonic anhydrase (Fig. 13.5). Negligible amounts of HCO_3^- are found in the distal tubules. At urinary pH below 6, the amount of HCO_3^- is so small that it can be ignored for practical purposes.

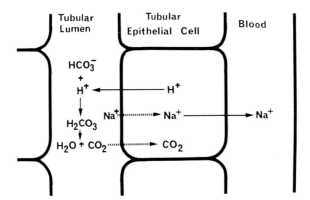

FIG. 13.5. Reabsorption of HCO_3. The interrupted arrow represents passive transport.

Excretion of titrable acid. Titrable acid is defined as the amount of alkalis required in mEq to titrate urine to a pH of 7·4 and refers to the amount of free H^+, and H^+ buffered with phosphate and creatinine. Phosphate and creatinine form the main buffers in the distal tubules. At a pH of 7·4 the ratio of HPO_4^{--} to $H_2PO_4^-$ is 4:1 (Fig. 13.6).

Excretion of ammonia (NH_3). NH_3 is formed in kidney cells from glutamine under the influence of glutaminase, and by the diamination of amino acids, glycine, alanine, aspartic acid and leucine. In its non-ionic form, NH_3 diffuses freely through the outer lipid layer of tubular epithelial cells into the tubular lumen; this is called *non-ionic diffusion.*

NH_3 combines with the free H^+ to form NH_4^+ which is excreted in urine. Because of its ionized state NH_4^+ cannot diffuse back through the lipid layer into tubular epithelial cells (Fig. 13.7).

The total amount of H^+ excreted daily may therefore be mathematically expressed as follows:

$$H^+ = NH_4^+ + T.A. - HCO_3^-$$

where

H^+ = Total amount of H^+ excreted in mEq/day or μEq/min
NH_4^+ = Ammonium in mEq/day
T.A. = Titrable acid in mEq/day

and

HCO_3^- = Bicarbonate in mMol/day.

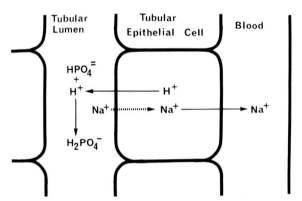

FIG. 13.6. Excretion of titrable acid. The interrupted arrow represents passive transport.

Urea

Urea is the major urinary solute. Forty to fifty per cent of the filtered urea is passively reabsorbed in the proximal tubule (Clapp 1965). It is preferentially trapped in the renal medulla where it is actively transported from the distal and collecting tubules into the interstitium under the influence of ADH (Truniger & Schmidt-Nielsen 1964). Urea is an important solute in the renal medulla and contributes to the creation of the medullary osmotic gradient and hence the renal concentrating ability. A low protein diet causes a fall in urinary urea concentration and a reduction of the maximum urinary osmolality and medullary reabsorption of free water ($Tm^C_{H_2O}$).

RENAL BLOOD FLOW (RBF)

In the resting state, normal human kidneys are perfused by 1200 ml of blood per minute and this is 20 to 25% of the normal cardiac output. Washout studies with ^{85}Kr and ^{133}Xe (Thorburn et al 1963, Ladefoged & Pedersen 1967) suggest that 90% of the blood flows through the cortex, 8–10% to the medulla and 1–2% to the papilla. Over a range of arterial pressures between 80 to 180 mmHg the cortical blood flow and hence the GFR remains constant. This phenomenon is called *auto regulation of renal blood flow* and is independent of renal nerves and extrinsic hormones. In contrast medullary blood flow is not controlled by auto-regulation. Both plasma skimming and variations in intra-renal

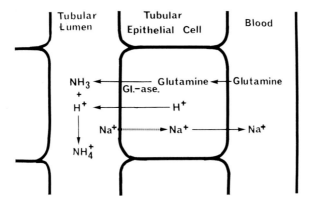

FIG. 13.7. Excretion of NH$_3$. The interrupted arrow represents passive transport.

pressure have been postulated as mechanisms explaining renal auto-regulation (Pappenheimer & Kinter 1956, Hinshaw et al 1959a, 1959b). However, the myogenic theory, the contractile response of smooth muscles in the afferent arteriole, and the intra-renal feedback mechanism of the renin-angiotensin system appear to be the more favoured explanations (Thurau 1964, 1967).

RBF may be modified by a number of extra- and intra-renal mechanisms. Renal sympathetic nerves have a vasoconstrictor role in the kidney. During emergency situations, e.g. anaesthesia, or even changes in posture, when the sympathetic drive is increased, RBF may fall without detectable changes in GFR. The action of sympathetic nerves is mediated largely through the release of catecholamines which when administered in larger doses will not only reduce RBF but also GFR (Balint & Chatel 1967). Hydropoenia has been reported to compromise medullary blood flow (Moffat 1968) while an osmotic or water diuresis has a reverse effect (Thurau 1964). Ureteral obstruction causes a fall in capillary perfusion in spite of unchanged RBF suggesting that arterio-venous shunting may occur (Harsing et al 1967).

RENIN-ANGIOTENSIN SYSTEM

The renin-angiotensin system serves as a sensitive and complex hormonal feedback loop for regulating blood volume and blood pressure. Renin, a proteolytic enzyme, has been shown by microdissection, ultracentrifugation and immunohistological techniques to be produced by the secretory granular cells of the juxta-glomerular apparatus (Peart 1965). Renin acts on a plasma renin substrate, an α-globulin called angiotensinogen, to produce an active decapeptide, angiotensin I by cleavage of the leucine-leucine bond in angiotensinogen. Angiotensin I is converted to its active octapeptide principle, angiotensin II by a 'converting enzyme' in the pulmonary circulation. The liver is the major site of renin degradation (Heacox et al 1967) while angiotensin II is inactivated by angiotensinases probably in the peripheral capillary bed. Angiotensin II is a powerful pressor agent whose major function is to stimulate the release of aldosterone thereby promoting renal Na^+ conservation. Angiotensin itself depicts a paradoxical dose related effect on renal handling of salt and water. In small doses it produces a marked antidiuresis and antinatriuresis while with higher rates of infusion it causes a diuresis and natriuresis (Louis & Doyle 1965).

The extra- and intra-renal factors responsible for release of renin have been reviewed extensively (Vander 1967, Mills 1970). The extra-renal factors are first, the drive of the sympathetic nervous system and secondly, angiotensin which acts as a negative feedback. The intra-renal mechanisms involve pressure sensitive areas in the wall of the afferent arteriole in the region of the juxta-glomerular cells, and the macula densa which is a Na^+ sensing device (Vander 1967, Thurau 1967) and is strategically situated at a point where the osmolality of tubular fluid is lowest (Windhager 1968). Catecholamines may have a direct effect on juxta-glomerular cells in the release of renin (Assaykeen et al 1969). These mechanisms are compatible with the distribution of renin in glomeruli and the variations in GFR and renal blood flow of individual nephrons. Renin is mainly found in the superficial cortical glomeruli and not in the juxta-medullary glomeruli (Brown et al 1966) while micropuncture studies suggest that the GFR and consequently renal blood flow is highest in juxta-medullary glomeruli (Horster & Thurau 1968).

RENAL PHYSIOLOGY IN PREGNANCY

During pregnancy there are profound physiological changes involving the cardiovascular system and renal function. The basic changes in the kidneys are a hypertrophy of function, leakage of some substances, altered haemodynamics, and exaggerated responses to posture.

ANATOMICAL CHANGES IN PREGNANCY

During pregnancy both the pelvi-calyceal system and the ureters undergo dilatation, the changes being more marked on the right than the left (Crabtree 1942, Fainstat 1963). The dilatation of the ureters commonly occurs as far as the pelvic brim, the ureters being of normal calibre below this point. There is an associated decrease in amplitude of the peristaltic waves of the ureter starting in the 3rd month of pregnancy and becoming maximal during the 7th and 8th months (Traut & McLane 1936). Normal peristaltic waves return in the last month. The dilated ureters are said to be hypotonic, oedematous and hypertrophied and may hold up to 200 ml of urine. This increases the ureteric dead space two-fold and reaches a peak between 20 and 35 weeks of gestation (Longo & Assali 1960). These changes have been attributed largely to the action of hormones, mainly progesterone (Van Wagenam & Jenkins 1939). The mechanical compression of the ureters by the gravid uterus (Bieniarz et al 1966) is unlikely to be an important factor.

GLOMERULAR FILTRATION RATE AND RENAL BLOOD FLOW

The hypertrophy of function seen in pregnancy has been well documented both in clinical and experimental studies. This is characterized by an increase in GFR and in RBF by about 25 to 50% (Bonsnes & Lange 1950, Bucht 1951, Page et al 1954, Sohar et al 1956, Dignam et al 1958, Sims & Krantz 1958, Chesley & Sloan 1964, Robb et al 1970).

Before 1950, it had been firmly believed that normal pregnancy had little or no effect on GFR and RBF (reviewed by Chesley 1951). The discrepancies in results were partly related to the difficulties in accurate measurements close to term, when there is a decline in function and when posture appears to have a maximal effect in altering renal function. The many pitfalls that may be encountered in studies of renal function during pregnancy have been reviewed by Chesley (1960) and Sims (1963).

TABLE 13.1. *Serial changes in GFR and RBF in pregnancy*

		First trimester	Second trimester	Third trimester		
				Early	Mid	Late
Bonsnes & Lange (1950)	GFR	—	↑54%	—	—	—
	RBF	—	—	—	—	—
Bucht (1951)	GFR	↑34%	↑43%	↑23%	↑57%	↑25%
	RBF	↑38%	↑37%	↑44%	↑18%	↑8%
Sohar et al (1956)	GFR	—	↑50%	No change		
	RBF	—	↑50%	Fall in RBF		
Sims & Krantz (1958)	GFR	↑49%	↑53%	↑44%	↑58%	↑42%
	RBF	↑42%	↑36%	↑11%	↑21%	0%
Assali et al (1959)	GFR	↑25%	↑45%	↑55%	—	—
	RBF	↑12%	↑33%	↑40%	—	—

Serial studies of C_{IN} and C_{PAH} in pre-pregnancy, during the three trimesters of pregnancy and immediately post-partum reveal that the GFR which is elevated by 23 to 58% throughout pregnancy tends to decline in the last weeks prior to delivery and becomes normal in early puerperium (Bonsnes & Lange 1950, Bucht 1951, Sohar et al 1956, Sims & Krantz 1958, Assali et al 1958). Part of the latter reduction in GFR has been explained on the basis of posture (Chesley & Sloan, 1964). The reduction of C_{PAH}, however, tends to occur earlier than that of C_{IN} so that there is a rise in the filtration fraction. The serial changes in GFR and RBF during the various stages of pregnancy found in the different studies are shown in Table 13.1. In contrast to the above studies the reports of de Alvarez (1958) showed a 40 to 60% rise in both C_{IN} and C_{PAH} in the first trimester and a subsequent progressive fall throughout the second and third trimesters. However, the technical flaws in the physiological methods used in this study have been criticized (Chesley 1960).

During pregnancy C_{CR} increases proportionately to C_{IN} and therefore may be used as a valid and routine clinical assessment of GFR and renal function (Bucht 1951, Evans 1955, Hayashi 1957, Sims & Krantz 1958).

The augmented filtration rate during pregnancy results in the reduction below normal of serum creatinine and blood urea concentrations. Sims & Krantz (1958) found the mean serum creatinine and blood urea concentrations to be 0.46 ± 0.14 mg per 100 ml and 16.34 ± 3.0 mg per 100 ml respectively compared to non-pregnant values of 0.67 ± 0.14 mg per 100 ml and 26.0 ± 6.0 mg per 100 ml. Similar values were reported by Pollak & Nettles (1960). 'Normal values' of serum creatinine, blood urea and C_{CR} actually indicate a reduction in GFR and may lull an unsuspecting physician into a clinical trap. The other major physiological consequence of an increase in GFR is the increased filtered load of all the constituents of glomerular filtrate, including electrolytes, glucose, amino acids and protein. This is partly responsible for the development of glycosuria and amino-aciduria (Sims 1968).

The increase in the GFR and RBF in pregnancy are unique physiological phenomena; there is no other known way to increase the filtration rate by 50% or better for prolonged periods (Smith 1956). In animals this augmented state of renal function is not accompanied by anatomic hypertrophy of the kidney (Sims 1963). It has been postulated that an increase in growth hormone during pregnancy may be a factor in the hypertrophy of function (Corvilain & Abramow 1962, Sims 1963). Since then high levels of a prolactin-like substance derived from the placenta have been detected in the blood of pregnant women (see chapter 4) and pregnant rats and it has been shown to be physiologically and antigenically similar to growth hormone (Contopoulos & Simpson 1956, Kaplan & Grumbach 1964, Welsh 1964, Kalousek et al 1968). Both clinically and experimentally growth hormone has a renotrophic effect and is associated with an increase in C_{PAH} and C_{IN} (White et al 1949, Heller et al 1955, Gershberg 1960, Morgen et al 1961). Increased levels of cortisol during pregnancy have been suggested as another factor. The secretory rate of cortisol in pregnancy is probably normal and the elevated levels of cortisol have been explained on the basis of an increase in steroid binding protein resulting from high oestrogen levels (Daughaday et al 1961). Doe et al (1967) and O'Connell & Welsh

(1969) have reported that during pregnancy levels of free cortisol are also elevated (see also chapter 9).

Pregnancy is associated with marked circulatory changes some of which are related to an increase in total blood volume, both red cell volume and plasma volume, and an increase in the resting cardiac output (Adams 1954, Bader *et al* 1955, Kerr 1968). Expansion of plasma volume is associated with an increase in GFR (Cargill 1948, Levinsky & Lalone 1963), which may be related to the release of a glomerular pressor substance from the liver (Uranga 1965, 1967). However, the hypervolaemia of pregnancy does not coincide in time with the changes in GFR, suggesting that the contribution of volume expansion to elevated levels of GFR and RBF is probably small (Berlin *et al* 1953).

TUBULAR FUNCTION

Tubular reabsorption

Glucose. There is an increased incidence of glycosuria in pregnancy. The overall reported incidence of glycosuria varies from 5 to 40% of subjects and it is clinically significant in 10% of women (Sims 1963). Fine (1967) reported that the excretion of greater than 100 mg of glucose per day was present in 70% of pregnant women compared to 20% in controls. Potentially, all pregnant women may develop glycosuria because of the increased filtered load of glucose (Donato & Turchetti 1955, Christensen 1958, Watanabe *et al* 1962). The increased delivery of glucose to the proximal tubule results in an augmentation of tubular reabsorption of glucose although the Tm_G does not rise in proportion to the GFR (Welsh & Sims 1960). Another important factor is the variation in the ability of the nephron population to reabsorb glucose. An impaired transport capacity in some nephrons is demonstrated by the widening of the splay in the titration curve of tubular glucose reabsorption. Reubi (1950) suggested that this was the major mechanism responsible for the glycosuria and was an important factor in the management of diabetes in pregnancy (see also chapter 11).

Phosphate. On the basis of the finding of enlarged parathyroid glands in pregnant animals it has been assumed that a physiological state of hyperparathyroidism exists in pregnancy. During pregnancy there is a heavy drain of calcium from mother to foetus especially in the last months. This has been postulated as a mechanism for the parathyroid glands to be in a hypersecretory state. Increased levels of parathyroid hormone cause a hyperphosphaturia by depressing the tubular reabsorption of phosphate. The percentage tubular reabsorption of phosphate in pregnancy is, however, within or greater than the normal range (Stoesser & Sims, quoted in Sims 1963) suggesting that parathyroid activity may in fact be normal. Growth hormone has an effect on tubular handling of phosphate opposite to that of parathyroid hormone (Gershberg 1960). It has therefore been suggested that the increase in the growth hormone-like substance during pregnancy may mask the effect of parathyroid hormone. Using a radio-immunoassay method, Lequin et al (1970) have found levels of parathyroid hormone in pregnancy ranging from 'non-detectable' to definitely elevated, the mean levels being higher than those for normal control subjects.

Amino acids. During pregnancy most amino acids are excreted in increased amounts in the urine but the net loss to the body is not great and is estimated as 0·7 g per day (Sims & Seldin 1949, Wallraff et al 1950, Christensen et al 1957, Pollak & Nettles 1960).

The amino-aciduria consists of histidine, tyrosine, serine, arginine, phenylalanine, threonine, tryptophan, glutamic acid, α-aminobutyric acid, asparagine, glycine, leucine, lysine, methionine, and guanidoacetic acid. Creatinine which has a common transport mechanism with amino acids is also excreted in increased amounts, the daily excretion in early pregnancy being 0·15 g, at 7 months of gestation 0·33 g and at term, 0·77 g (Pitts 1944, Clark et al 1951).

The main factors responsible for the augmented amino-aciduria in pregnancy are the increase in GFR and the proximal tubular reabsorption of amino acids being inappropriate in relation to the increased filtered load. This results in an increased clearance of the amino acids with a consequent decrease in their serum concentrations. The elevated levels of plasma cortisol in pregnancy have also been incriminated as a possible factor.

The histidinuria of pregnancy has been recognized for many years and the responsible mechanisms have been studied in greater depth (Page et al 1954, Peterson & Frank 1958). The clearance of histidine during pregnancy is 28 ml per min compared to 7 to 8 ml in the postpartum period. The rate of entry of histidine into the intracellular compartment is said to diminish, resulting in elevated plasma levels of histidine. Together with the hypertrophy of renal function in pregnancy, the filtered load of histidine is increased by 64%. The tubular reabsorption of histidine is also impaired and is responsible for 25% of the histidinuria.

Uric acid. In accord with the general augmented state of renal function, the clearance of uric acid is increased in pregnancy particularly in the first half (Chesley 1950, Christensen & Steenstrup 1958). The serum uric acid concentration and uric acid clearance in early pregnancy are 3.2 ± 0.8 mg/100 ml and 18.4 ± 4.9 ml/min respectively; in late pregnancy they are 3.0 ± 1.7 mg/100 ml and 13.0 ml/min respectively (Hayashi 1956, Pollak & Nettles 1960). Plasma levels of uric acid may be altered by changes in total body Na^+ and intravascular volume (Black *et al* 1950, Schwartz & Relman 1967, Hull *et al* 1967). Elevated levels of serum uric acid may occur in pregnancy with Na^+ depletion, mimicking the changes in toxaemia of pregnancy, and correction of the Na^+ depletion causes a fall in the serum uric acid level (Palomaki & Lindheimer 1970). This possibly results from proximal tubular rejection of uric acid with volume expansion. These changes may have relevance in the pathogenesis of eclampsia.

Tubular secretion

C_{PAH} is increased during pregnancy and the related changes have been discussed above.

Sodium and water

A marked tendency to retain Na^+ and water occurs in pregnancy, particularly in the weeks close to term (MacGillivray & Buchanan 1958, Lichton 1961, Hytten & Leitch 1964, Gray *et al* 1964). Balance studies in pregnant dogs reveal that 30 to 40% of ingested Na^+ is retained in the latter half of pregnancy (Robb *et al* 1970). Studies of total exchangeable Na^+ suggest that an increase of 460 to 570 mEq of Na^+ occurs in pregnancy and this is distributed between the Na^+ of the foetus and placenta, and the increase in blood volume during pregnancy (Plentl & Gray 1959, Davey *et al* 1961, Gray *et al* 1964). In the pregnant dog, the uterus and its contents contain 79.6% of the increase in salt and water (Robb *et al* 1970). The mean blood volume increases by 30 to 50%, peaks at 28 to 34 weeks of gestation and declines after the 36th week (Kerr 1968); an increase in total exchangeable Na^+ has been noted after the 12th week of pregnancy with a fall of 80 mEq of Na^+ by the 38th week (Plentl & Gray 1959, Davey et al 1961).

Oedema by itself without hypertension is common in pregnancy and in an extensive study, Thompson et al (1967) found significant oedema in 35% of pregnant women. During pregnancy the Na^+ balance is in a critical and finely balanced state and there are a number of factors which have opposing effects on the urinary excretion of Na^+.

Factors promoting Na^+ retention

Oestrogens. Dignam et al (1956) have shown that women given oestrogenic hormones over 3 to 6 days develop salt and water retention. The mechanism for this is not altogether clear but it has been attributed to a direct effect of the hormones on renal tubules. Na^+ retention appears to be greater when the hepatic inactivation of oestrogens is diminished as in cirrhosis of the liver and cardiac failure (Sims 1970). The secretion of oestrogens is greatly increased during pregnancy, the excretion of oestrone and oestradiol-17 increasing a hundred-fold and that of oestriol, a thousand-fold (Brown 1956) (see chapter 4).

Cortisol. During pregnancy there is an increase of both bound and unbound cortisol (Daughaday et al 1961, Doe et al 1967, O'Connell & Welsh 1969). Although the increase in GFR associated with elevated levels of cortisol would tend to cause natriuresis, its major effect is on the distal convoluted tubule where it promotes Na^+ reabsorption in exchange for H^+ and K^+.

Renin-angiotensin-aldosterone. Aldosterone is a potent Na^+ retaining steroid and there is now considerable evidence that the renin-angiotensin-aldosterone axis is greatly stimulated during pregnancy. Most investigators have found elevated levels of plasma renin concentration and plasma renin activity during pregnancy, the highest levels being in the earlier part of pregnancy (Brown et al 1963, Winer 1965, Brown et al 1966, Hodari et al 1967, Helmer & Judson 1967, Geelhoed & Vander 1968, Hodari & Hodgkinson 1968, Robb et al 1970). Some conflicting reports (Maebashi et al 1964, Gordon et al 1967) have been explained by the lack of evaluation and control of dietary Na^+ (Robb *et al* 1970). Apart from the maternal kidney, the high renin levels in pregnancy may also be derived from the foetal kidney (Hodari & Hodgkinson 1968). A renin-like substance has also been detected in the uterus in both pregnant and non-pregnant states (Gould *et al* 1964, Gross *et al* 1964, Bing & Faarup 1966, Hodari *et al* 1967, Ferris *et al* 1967, Skinner *et al* 1968), in the placenta (Stakemann 1960, Gould *et al* 1964, Gross *et al* 1964, Hodari *et al* 1967, Skinner *et al* 1968, Hodari *et al* 1969) and in amniotic fluid (Brown *et al* 1964). The greatest concentration of renin was found in the chorion (Skinner *et al* 1968) and tissue culture studies of renin production from the chorion, decidua and myometrium revealed the highest levels from the chorion and the lowest from the myometrium (Symonds *et al* 1968). The uterine enzyme has biochemical properties similar to renal renin, but appears to act on another substrate as well to produce a pressor substance (Carretero & Houle 1970). Foetal renin, however, is physiologically similar to maternal renin (Hodari 1968).

A number of other factors have also been postulated as being responsible for the high levels of renin in pregnancy. The administration of oral contraceptive agents which contain oestrogens, causes a rise of plasma renin substrate, plasma renin activity and aldosterone in some patients (Laragh *et al* 1967, Newton *et al* 1968) and dogs given β-oestradiol develop elevated levels of plasma renin activity (Robb *et al* 1970). The high levels of oestrogens during pregnancy have therefore been incriminated as a factor responsible for stimulation of renin production. The diminished catabolism of renin by the liver (Schneider *et al* 1969) and the elevated levels of plasma renin substrate itself (Helmer & Judson 1967, Robb *et al* 1970) may also contribute to the elevated levels of renin. However, Robb *et al* (1970) were unable to demonstrate that the sequestration of salt and water in the uterus played any role.

The increased production of renin will result in high levels of angiotensin. That this is so is suggested by a blunting of the pressor response and renal effects during angiotensin infusion in pregnancy (Abdul Karim & Assali 1961, Kaplan & Silah 1964, Hocken *et al* 1966, Talledo *et al* 1968). However, the modified response to the angiotensin infusion test may partly be explained by an increase in angiotensinase during pregnancy (Talledo *et al* 1967).

The renin-angiotensin mechanism is the most potent stimulator of aldosterone production and is responsible for the increased secretion and excretion of aldosterone in pregnancy (Jones et al 1959, Kumar & Barnes 1960, Brown et al 1963, Watanabe et al 1963, Winer 1965, Helmer & Judson 1967, Robb et al 1970). Although there are minor changes in the end products of aldosterone metabolism (Tait & Little 1968), the metabolic clearance rate of aldosterone is unaltered in pregnancy (Tait et al 1962). This suggests that the plasma concentration of aldosterone during pregnancy is increased (Sims 1970). The other factors responsible for the stimulation of aldosterone production are the complex natriuretic mechanisms in pregnancy (see below) and the high levels of oestrogens (Katz & Kappas 1967). In pregnancy aldosterone is responsive to changes in Na^+ loading (Watanabe et al 1963) and administration of aldosterone antagonists (Barnes & Buckingham 1958) and therefore plays an important role in the Na^+ balance.

Posture. In pregnancy the adoption of the supine posture causes a considerable translocation of fluid to the lower part of the body mainly by compression of the inferior vena cava, common iliac and femoral veins and the aorta by the gravid uterus (McLennan 1943, Howard et al 1953, Bieniarz et al 1966, Kerr 1968, Sims 1970). This causes a fall in GFR and RBF (approximately 20%) and salt and water retention with an accompanying decrease in urinary excretion of Na^+ and water and a decline in urinary flow rates (Walker et al 1934, Pritchard et al 1955, Assali et al 1959, Chesley & Sloan 1964, Kerr et al 1964, Kalousek et al 1968, Toback et al 1970). A similar effect of posture on renal function occurs even in the presence of volume expansion and a water induced diuresis, although to a modified extent (Hawker et al 1961, Lindheimer & Weston 1969). The possible physiological mechanism of the supine and erect antidiuresis is a decrease in the filtered load of Na^+ (Sims 1963), changes in humoral agents such as progesterone, oestrogens, angiotensin (Landau et al 1955, Landau & Lugibihl 1958, Landau et al 1960, Chesley et al 1963, Katz & Kappas 1967) and an increased drive in the sympathetic nervous system (Brigden et al 1950, Earley 1966, Gordon et al 1967). Similar changes in renal function in non-pregnant women may be produced by increasing intra-abdominal pressure, quiet standing and inferior vena caval compression (Bradley et al 1949, Faber et al 1953, Davis 1960). Altered permeabilities of capillaries and cellular membranes do not appear to contribute to the translocation of fluid in the interstitial space (Jones 1968, Chesley 1970). Physical factors such as pressure by the gravid uterus on the aorta, inferior vena cava and the ureters and changes in uterine blood flow have also been incriminated (Burwell 1938, Epstein *et al* 1953, Pritchard *et al* 1955, Heckel & Tobin 1956, Scott & Kerr 1963, Kerr *et al* 1964, Rector *et al* 1966, Wakim 1968, Rubi & Sala 1968).

Factors promoting natriuresis

GFR and RBF. The increase in GFR and RBF in pregnancy increases the filtered load of Na^+ by about 50% and this amounts to approximately 30,000 mEq of Na^+ a day. This is equivalent to ten to fifteen times the total exchangeable Na^+ of the body (Chesley 1960) and presents a very large extra load for the renal tubular reabsorptive mechanisms. It is, however, of interest that Watanabe *et al* (1963) were unable to find a correlation between the GFR as assessed by C_{CR} and the aldosterone secretion rate in pregnancy.

Progesterone. Progesterone is a potent natriuretic factor and has a strong spironolactone-like action (Landau & Lugibihl 1958, Gornall *et al* 1960). Progesterone given to non-gravid rabbits induces hypertension in the animals (Horobin & Lloyd 1970). The hypertensive effect of progesterone may be modified by simultaneous administration of saline suggesting that progesterone induces Na^+ loss and that this may be important in initiating hypertension. During pregnancy, particularly in the third trimester, progesterone secretion from the placenta is greatly increased and is ten-fold compared to that in the luteal phase of menstruation (Short & Eton 1959). There is also an associated rise in the unbound level of progesterone (Rosenthal *et al* 1969). Progesterone may be partly responsible for the stimulation of renin-angiotensin-aldosterone mechanism (Martin & Mills 1956, Chesley & Tepper 1967, Gray *et al* 1968).

Natriuretic or third factor. The expanded state of plasma volume in pregnancy (Berlin *et al* 1953, Adams 1954, Burwell & Metcalf 1958) may have considerable importance in the control of sodium balance because it exerts an effect not only on Na^+ transport in the proximal tubule and the loop of Henle but also on the GFR. As in the non-pregnant state, volume expansion with normal saline or hypotonic saline during pregnancy results in a marked natriuretic response (Black *et al* 1950, Crawford & Ludemann 1951, Strauss *et al* 1951, Diekmann et al 1952, Wesson 1957, Willson et al 1957, Chesley et al 1958, Auld et al 1968, Lindheimer & Weston 1969). Lindheimer & Weston (1969) have shown that the natriuresis results from a decrease in the fractional reabsorption of Na^+ in the proximal tubule.

In the latter part of pregnancy there is a tendency to hyponatraemia due to retention of water slightly in excess of Na^+ (Lichton 1961, 1963; Hytten et al 1966). The excess water at term amounts to 1–2 litres and in patients with oedema, up to 5 litres. Although the placenta produces vasopressinase (Zelinik 1965) the activity of ADH as measured by bioassay in rats is not increased in the serum and urine of pregnant women (Zuspan et al 1958, Friedberg et al 1960). The concentrating ability of the kidney as measured by the maximum quantity of free water extracted per unit of GFR ($Tm^C_{H_2O} \times 100/C_{IN}$) is within normal limits (McCartney et al 1964).

Potassium

Pregnancy is associated with a positive K^+ balance and diminished urinary excretion of K^+ particularly in the third trimester (Robb et al 1970). The total body K^+ is significantly increased even after deducting the total body content of K^+ in the foetus (Godfrey & Wadsworth 1970). The plasma concentration of K^+, however, tends to fall towards the end of pregnancy (Lucius et al 1970) because of the excess retention of water.

Acid-base changes in pregnancy

Changes in acid-base balance are not prominent during pregnancy. From the 10th week of gestation to term a fall in Pa^{CO_2} (35·5 mm Hg to 28 mm Hg) and standard serum bicarbonate (27 mEq/l to 21 mEq/l) may be explained on the basis of hyperventilation (Lucius et al 1970). Part of the fall in serum bicarbonate may be the result of plasma volume expansion which causes a decrease in the reabsorption of bicarbonate in the proximal tubule (Slatopolsky et al 1970). The most marked deviations in acid-base occur during labour when there develops a combination of hypocapnia from hormone induced pulmonary hyperventilation, and a metabolic acidosis (Fadl & Utting 1969). The latter occurs not as a result of a rise in plasma fixed acids as in renal insufficiency, but due to renal loss of bicarbonate to compensate for the hypocapnia.

ROLE OF PRESSOR AGENTS IN PREGNANCY

Although the urinary excretion of catecholamines does not alter during pregnancy (Zuspan 1970) an altered sensitivity to adrenaline and noradrenaline has been reported particularly when pregnancy is complicated by oedema, proteinuria or hypertension (Raab *et al* 1956, Talledo *et al* 1968).

Senior *et al* (1963) have reported increased concentrations of 5-hydroxytryptamine (Serotonin) in the plasma of pregnant women and also in the placentas when there is associated hypertension. This has been related to a monoamine oxidase (MAO) deficiency with reduced metabolism of 5-hydroxytryptamine (Sandler & Baldock 1963). These changes may influence the intra-renal distribution of blood flow and the resultant changes in nephron function.

EFFECTS OF POSTURE ON RENAL FUNCTION

In the supine and standing position, the gravid uterus compresses the inferior vena cava, aorta, common iliac veins, femoral veins and the ureters (McLennan 1943, Howard *et al* 1953, Kerr *et al* 1964, Bieniarz *et al* 1966, Kerr 1968, Sims 1970). These changes have a profound effect on various aspects of renal function. The alterations in renal blood flow, glomerular filtration rate, and salt and water excretion have been discussed above. In addition, the supine posture causes a reduction in urate excretion (Lindheimer 1969). Baird *et al* (1966) have studied the effect of the supine posture on ^{131}I-Hippuran renography. A delay in the renogram was noted only on the right side during the early months of pregnancy. After the 37th week of pregnancy the delay occurred on both sides. Other studies by Toback *et al* (1970) have not revealed any effect of posture and activity on the qualitative and quantitative excretion of protein.

CIRCADIAN RHYTHMS OF RENAL FUNCTION

A circadian rhythm exists for various aspects of renal function in the non-pregnant state, a daytime peak occurring with GFR and RBF and the urinary excretion of Na^+, K^+, Cl and HCO_3^-; nocturnal peaks occur for the urinary excretion of NH_4^+ and titrable acid (Kleitman 1923, Simpson 1924, 1926, 1929; Manchester 1933, Sirota *et al* 1950,

Stanbury & Thomson 1951, Wesson & Lauler 1961, Wesson 1964). In general the excretion of Na^+, K^+, Cl^- and HCO_3^- appears to correlate with the GFR (Sirota *et al* 1950, Stanbury & Thompson 1951, Wesson 1957, Wesson & Lauler 1961). The earlier studies to evaluate the diurnal variation of fluid and electrolyte excretion in pregnancy gave conflicting results (Horska & Vedra 1964, Pystynen & Pankamaa 1965). Kalousek *et al* (1969) have performed a careful study to re-evaluate the pattern of excretion of creatinine, Na^+, Cl^-, K^+, osmolar concentration and urine volume in eleven pregnant women at various intervals in the second and third trimesters. During recumbency, as in healthy non-pregnant women, the pattern of excretion was constant with daytime peaks. However, on days of activity the peaks for urine volume and Na^+ excretion were reversed, being high at night. These changes reverted back to normal on recumbency in the same patients suggesting that the alterations in the circadian rhythm were mainly the effect of posture.

The increase in urine volume and Na^+ excretion during recumbency in the non-pregnant state (Howard *et al* 1953, Kerr *et al* 1964) can be abolished by leg wrapping or standing in water (Pearce *et al* 1954, Gowenlock *et al* 1958) suggesting that hydrostatic pressures and venous return from the lower limbs are important contributing factors. During pregnancy, fluid accumulation in the lower limbs is enhanced. Kalousek *et al* (1969) have suggested that the nocturnal mobilization of fluid during recumbency is equivalent to an intravenous saline infusion resulting in volume expansion and consequently causing an increase in urine volume and Na^+ excretion.

REFERENCES

ABDUL KARIM R. & ASSALI N.S. (1961) Pressor response to angiotensin in pregnant and nonpregnant woman. *Amer. J. Obstet. Gynec.* **82**, 246–251.

ADAMS J.W. (1954) Cardiovascular physiology in normal pregnancy: studies with the dye dilution technique. *Amer. J. Obstet. Gynec.* **67**, 741–759.

ASSALI N.S., DIGNAM W.J. & DASGUPTA K. (1959) Renal function in human pregnancy. II. Effects of venous pooling on renal hemodynamics and water, electrolyte and aldosterone excretion during normal gestation. *J. Lab. clin. Med.* **54**, 394–408.

ASSAYKEEN T.A., GOLDFIEN A., OTSUKA K. & GANONG W.F. (1969) Effect of α and β blocking agents on the increase in plasma renin activity provoked in dogs by insulin-induced hypoglycaemia. IVth International Congress Nephrology, Stockholm, p. 137 (Abstracts).

AULD R.B., LALONE R.C. & LEVINSKY N.G. (1968) Regulation of sodium excretion during acute and chronic extracellular volume expansion in man. *J. clin. Invest.* **47**, 2a (Abstract).

BADER R.A., BADER M.E., ROSE D.J. & BRAUNWALD E. (1955) Haemodynamics at rest and during exercise in normal pregnancy as studied by cardiac catheterization. *J. clin. Invest.* **34**, 1524–1536.

BAIRD D.T., GASSON P.W. & DOIG A. (1966) The renogram in pregnancy, with particular reference to the changes produced by alteration in posture. *Amer. J. Obstet. Gynec.* **59**, 597–603.

BALINT P. & CHATEL R. (1967) Die Wirkung von Adrenalin und von Nonadrenalin auf die Nierenhamodynamik beim Hund. *Arch. Exp. Pathol. Pharmakol.* **258**, 24–36.

BARGER A.C. (1966) Renal haemodynamic factors in congestive heart failure. *Ann. New York Acad. Sci.* **139**, 276–284.

BARNES A.C. & BUCKINGHAM J.C. (1958) Electrolyte balance studies with the antihormones. *Amer. J. Obstet. Gynec.* **76**, 955–968.

BERLIN N.I., GOETSCH C., HYDE G.M. & PARSONS R.J. (1953) The blood volume in pregnancy as determined by p^{32}-labelled red blood cells. *Surg. Gynec. Obstet.* **97**, 173–176.

BIENIARZ J., MAQUEDA E. & CALDEYRO-BARCIA R. (1966) Compression of aorta by the uterus in late human pregnancy. I. Variations between femoral and brachial artery pressure with changes from hypertension to hypotension. *Amer. J. Obstet. Gynec.* **95**, 795–808.

BING J. & FAARUP P. (1966) A quantitative and qualitative study of renin in the different layers of rabbit uterus. *Acta Path. Microbiol. Scand.* **67**, 169–179.

BLACK D.A.K., PLATT R. & STANBURY S.W. (1950) Regulation of sodium excretion in normal and salt-depleted subjects. *Clin. Sci.* **9**, 205–221.

BONSNES R.W. & LANGE W.A. (1950) Inulin clearance during pregnancy. *Federation Proc.* **9**, 154.

BRADLEY S.E., MUDGE G.H., BLAKE W.D. & ALPHONSE P. (1949) The effect of increased intraabdominal pressure upon renal excretion of water, sodium and potassium in normal subjects and in patients with diabetes insipidus. *J. clin. Invest.* **28**, 772 (Abstract).

BRIGDEN W., HAWORTH S. & SHARPEY-SCHAFER E.P. (1950) Postural changes in the peripheral blood flow of normal subjects with observations on vasovagal fainting reactions as a result of tilting the lordotic posture, pregnancy and spinal anaesthesia. *Clin. Sci.* **9**, 79–91.

BROWN J.B. (1956) Urinary excretion of oestrogens during pregnancy, lactation, and the re-establishment of menstruation. *Lancet* **1**, 704–707.

BROWN J.J., DAVIES D.L., DOAK P.B., LEVER A.F. & ROBERTSON J.I.S. (1963) Plasma-renin in normal pregnancy. *Lancet* **2**, 900–902.

BROWN J.J., DAVIS D.L., LEVER A.F., ROBERTSON J.I.S. & TREE J. (1964) A renin-like enzyme in normal human urine. *Lancet* **2**, 709–711.

BROWN J.J., DAVIES D.L., DOAK P.B., LEVER A.F., ROBERTSON J.I.S. & TRUST P.

(1966) Plasma renin concentration in the hypertensive diseases of pregnancy. *J. Obstet. Gynaec. Brit. Cwlth.* **73**, 410–417.
BROWN J.J., DAVIES D.L., LEVER A.F., PARKER R.A. & ROBERTSON J.I.S. (1966) The assay of renin in single glomeruli and the appearance of the juxtaglomerular apparatus in the rabbit following renal artery constriction. *Clin. Sci.* **30**, 223–235.
BUCHT H. (1951) Studies on renal function in man. With special reference to glomerular filtration and renal plasma flow in pregnancy. *Scand. J. clin. Lab. Invest.* **3** (Suppl. 3), 1–64.
BURWELL C.S. (1938) The placenta as a modified arterio-venous fistula, considered in relation to circulatory adjustments of pregnancy. *Amer. J. Med. Sci.* **195**, 1–7.
BURWELL C.S. & METCALF J. (1958) *Heart Disease and Pregnancy: Physiology and Management*, p. 19. Boston, Little, Brown.
CARGILL W.R. (1948) Effect of intravenous administration of human serum albumin on renal function. *Proc. Soc. Exp. Biol. Med.* **68**, 189–192.
CARRETERO O.A. & HOULE J.A. (1970) A comparison of renin obtained from pregnant uterus and kidney of the dog. *Amer. J. Physiol.* **218**, 689–692.
CHESLEY L.C. (1950) Simultaneous renal clearances of urea and uric acid in differential diagnosis of late toxemias. *Amer. J. Obstet. Gynec.* **58**, 960–969.
CHESLEY, L.C. (1951) Kidney function in the normal and toxaemia pregnant woman. *Med. Clin. N. Amer.* **35**, 699–714.
CHESLEY L.C., VALENTI C. & REIN H. (1958) Excretion of sodium loads by non-pregnant and pregnant normal, hypertensive and pre-eclamptic women. *Metabolism* **7**, 575–588.
CHESLEY L.C. (1960) Renal functional changes in normal pregnancy. *Clin. Obstet. Gynec.* **3**, 349–363.
CHESLEY L.C., WYNN R.M. & SILVERMAN N.I. (1963) Renal effects of angiotensin II infusions in normotensive pregnant and nonpregnant women. *Circ. Res.* **13**, 232–238.
CHESLEY L.C. & SLOAN D.M. (1964) The effect of posture on renal function in late pregnancy. *Amer. J. Obstet. Gynec.* **89**, 754–759.
CHESLEY L.C. & TEPPER I.H. (1967) Effects of progesterone and estrogen on the sensitivity to angiotensin II. *J. clin. Endocr. & Metab.* **27**, 576–581.
CHESLEY L.C. (1970) The movement of radio active sodium in normal pregnant, non pregnant and pre-eclamptic women. *Amer. J. Obstet. Gynec.* **106**, 530–533.
CHRISTENSEN P.J., DATE J.W., SCHONHEYDER F. & VOLQUARTZ K. (1967) Amino-acids in blood plasma and urine during pregnancy. *Scand. K. Clin. Lab. Invest.* **9**, 54.
CHRISTENSEN P.J. (1958) Tubular reabsorption of glucose during pregnancy. *Scand. J. Clin. Lab. Invest.* **10**, 364–371.
CHRISTENSEN P.J. & STEENSTRUP O.R. (1958) Uric acid excretion with increasing plasma glucose concentration (pregnant and non-pregnant cases). *Scand. J. Clin. Lab. Invest.* **10**, 182–185.
CLAPP J.R. (1965) Urea reabsorption by the proximal tubule of the dog. *Proc. Soc. Exp. Biol. Med.* **120**, 521–525.
CLARK L.C., THOMPSON H. & BECK E.I. (1951) Excretion of creatine and creatinine during pregnancy. *Amer. J. Obstet. Gynec.* **62**, 576–583.

CONTOPOULOS A.N. & SIMPSON M.E. (1956) Increased growth hormone activity in plasma of pregnant rats. *Federation Proc.* **15**, 39 (Abstract).

CORTNEY M.A., NAGEL W. & THURAU K. (1965) A micropuncture study of the relationship between flow-rate through the loop of Henle and sodium concentration in the distal tubule. *Arch. Geo. Physiol.* **287**, 286–295.

CORVILAIN J. & ABROMOW M. (1962) Some effects of human growth hormone on renal haemodynamics and on tubular phosphate transport in man. *J. clin. Invest.* **41**, 1230–1235.

COTRAN R.S. & KASS E.H. (1958) Determination of volume of residual urine in the bladder without catheterization. *New Eng. J. Med.* **259**, 337–339.

CRABTREE E.G. (1942) *Urological Diseases of Pregnancy*. Boston, Little, Brown.

CRAWFORD B. & LUDEMANN H. (1951) The renal response to intravenous injection of sodium chloride solutions in man. *J. clin. Invest.* **30**, 1456–1462.

DANIELS E.G., HINMAN J.W., LEACH B.E. & MUIRHEAD E.E. (1967) Identification of prostaglandin E2 as the principal vasodepressor lipid of rabbit renal medulla. *Nature (Lond.)* **215**, 1298–1299.

DAUGHADAY W.H., HOLLOSZY J. & MARIZ I.K. (1961) Binding of corticosteroids by plasma proteins. VI. The binding of cortisol and aldosterone by corticosteroid binding globulin and by the oestrogen induced binding system of plasma. *J. clin. Endocr.* **21**, 53–61.

DAVEY D.A., O'SULLIVAN W.J. & MCCLURE BROWN J.C. (1961) Total exchangeable sodium in normal pregnancy and in pre-eclampsia. *Lancet* **1**, 519–523.

DAVIS, J.O. (1960) Mechanisms of salt and water retention in congestive heart failure. *Amer. J. Med.* **29**, 486–507.

DE ALVAREZ, R.R. (1958) Renal glomerulotubular mechanisms during normal pregnancy. I. Glomerular filtration, renal plasma flow and creatinine clearance. *Amer. J. Obstet. Gynec.* **75**, 931.

DE LUNA M.B. & HULET W.H. (1967) Urinary protein excretion in healthy infants, children and adults. *Proceedings of the American Society of Nephrology, Los Angeles*, p. 16 (Abstract).

DE WARDENER H.E., MILLS I.H., CLAPHAM W.F. & HAYTER C.J. (1961) Studies on efferent mechanism of sodium diuresis which follows administration of intravenous saline in dog. *Clin. Sci.* **21**, 249–258.

DE WARDENER H.E. (1969) Control of sodium reabsorption. *Brit. med. J.* **3**, 611–616, 676–683.

DIEKMANN W.J., POTTINGER R.E. & RYNKIEWICZ L.M. (1952) Etiology of pre-eclampsia. IV. Sodium chloride test for the diagnosis of pre-eclampsia. *Amer. J. Obstet. Gynec.* **63**, 783–791.

DIGNAM W.J., VOSKIAN J. & ASSALI N.S. (1956) Effects of oestrogens on renal haemodynamics and excretion of electrolytes in human subjects. *J. clin. Endocr.* **16**, 1032–1042.

DIGNAM W.J., TITUS P. & ASSALI N.S. (1958) Renal function in human pregnancy. I. Changes in glomerular filtration rate and renal plasma flow. *Proc. Soc. Exp. Biol. Med.* **97**, 512–514.

DIRKS J.H., CLAPP J.R. & BERLINER R.W. (1964) The protein concentration in the proximal tubule. *J. clin. Invest.* **43**, 916–921.

DOE R.P., VENNES J.A. & FLINK E.B. (1960) Diurnal variation of 17-hydroxycorticosteroids, sodium, potassium, magnesium and creatinine in normal subjects and in cases of treated adrenal insufficiency and Cushing's syndrome. *J. clin. Endocr.* **20**, 253–265.

DONATO L. & TURCHETTI G. (1955) Renal glycosuria in pregnancy. *Acta Med. Scand.* **152**, 223–230.

EARLEY L.E. (1964) Effect of renal arterial infusion of albumin on saline diuresis in the dog. *Proc. Soc. Exp. Biol. Med.* **116**, 262–265.

EARLEY L.E. & FRIEDLER R.M. (1965) Studies of the mechanism of natriuresis accompanying increased renal blood flow and its role in the renal response to extracellular volume expansion. *J. clin. Invest.* **44**, 1857–1865.

EARLEY L.E. (1966) Influence of haemodynamic factors on sodium reabsorption. *Ann. New York Acad. Sci.* **139**, 312–327.

EPSTEIN F.H., POST R.S. & McDOWELL M. (1953) The effect of an arteriovenous fistula on renal hemodynamics and electrolyte excretion. *J. clin. Invest.* **32**, 233–341.

EVANS T.N. (1955) Endogenous creatinine clearance as a measure of renal function in normal and toxaemic pregnancies. *Amer. J. Obstet. Gynec.* **70**, 122–134.

FABER S.J., BECKER W.H. & EICHNA L.W. (1953) Electrolyte and water excretions and renal hemodynamics during induced congestion of the superior and inferior vena cava of man. *J. clin. Invest.* **32**, 1145–1162.

FADL E.T. & UTTING J.E. (1969) Acid-base disturbance in obstetrics. *Proc. roy. Soc. Med.* **63**, 77–78.

FAINSTAT T. (1963) Ureteral dilatation in pregnancy: a review. *Obstet. Gynec. Survey* **18**, 845–860.

FERRIS T.F., GORDON P. & MULROW P.J. (1967) Rabbit uterus as a source of renin. *Amer. J. Physiol.* **212**, 698–702.

FINE J. (1967) Glycosuria of pregnancy. *Brit. med. J.* **1**, 205–210.

FRIEDBERG V., VORHERR H. & SCHULTE G. (1960) Adiuretin studies during normal pregnancy and pregnancy toxemias. *Arch. Gynäk.* **192**, 483–491. (Ger.).

GEELHOED G.W. & VANDER A.J. (1968) Plasma renin activities during pregnancy and parturition. *J. clin. Endocr. & Metab.* **28**, 412–415.

GERSHBERG H. (1960) Metabolic and renotropic effects of human growth hormone in disease. *J. clin. Endocr. & Metab.* **20**, 1107–1119.

GERTZ K.H., MANGOS J.A., BRAUN G. & PAGEL H.D. (1965) On the glomerular tubular balance in the rat kidney. *Arch. Ges. Physiol.* **285**, 360–372.

GIEBISCH G., KLOSE R.M. & MALNIC G. (1967) Renal tubular potassium transport. *Bull. Schweiz. Akad. Med. Wiss.* **23**, 287–312.

GODFREY B.E. & WADSWORTH G.R. (1970) Total body potassium in pregnant women. *J. Obstet. Gynaec. Brit. Cwlth.* **77**, 244–246.

GORDON R.D., KUCHEL O., LIDDLE G.W. & ISLAND D.P. (1967) Role of the sympathetic nervous system in regulating renin and aldosterone production in man. *J. clin. Invest.* **46**, 599–605.

GORDON R.D., FISHMAN L.M. & LIDDLE G.W. (1967) Plasma renin activity and aldosterone secretion in pregnant woman with primary aldosteronism. *J. clin. Endocr.* **27**, 385–388.

GORNALL A.G., ROBERTSON M.E. & TAIDLAW J.J. (1960) The influence of estrogen and progesterone on urinary sodium and aldosterone excretion, in E.Fuchs (ed.) *Proceedings of the First International Congress on Endocrinology*, p. 157. Copenhagen.

GOULD A.B., SKEGGS L.T. & KAHN J.R. (1964) The presence of renin activity in blood vessel wall. *J. Exp. Med.* **119**, 389–399.

GOWENLOCK A.H., MILLS J.N. & THOMAS S. (1959) Acute postural changes in aldosterone and electrolyte excretion in man. *J. Physiol.* **146**, 133–141.

GRAY M.J., MUNRO A.B., SIMS E.A.H., MEEKER C.I., SOLOMON S. & WATANABE M. (1964) Regulation of sodium and total body water metabolism in pregnancy. *Amer. J. Obstet. Gynec.* **89**, 760–765.

GROSS F., SCHAECHTELIN G., ZIEGLER M. & BERGER M. (1964) A renin-like substance in the placenta and uterus of the rabbit. *Lancet* **1**, 914–916.

GUTMAN A.B. & YU T.F. (1957) Renal function in gout. *Amer. J. Med.* **23**, 600–622.

HALVER B. (1966) The effect of parathyroid hormone on the tubular reabsorption of glucose. *Acta Med. Scand.* **179**, 427–432.

HARSING L., SZANTO G. & BARTHA J. (1967) Renal circulation during stop flow in the dog. *Amer. J. Physiol.* **213**, 935–938.

HAWKER R.W., WALMSEY C.F., ROBERT V.S., BLACKSHAW J.K. & DOWNES J.C. (1961) Oxytoxic activity of blood in parturient and lactating women. *J. clin. Endocr. & Metab.* **21**, 985–995.

HAYASHI T. (1956) Uric acid and endogenous creatinine clearance studies in normal pregnancy and toxemias of pregnancy. *Amer. J. Obstet. Gynec.* **71**, 859–870.

HAYASHI T. (1957) The effect of Benemid on uric acid excretion in normal pregnancy and in pre-eclampsia. *Amer. J. Obstet. Gynec.* **73**, 17–22.

HEACOX R., HARVEY A.M. & VANDER A.J. (1967) Hepatic inactivation of renin. *Circ. Res.* **21**, 149–152.

HEALY J.K., EDWARDS K.D.G. & WHYTE H.M. (1964) The phenol red clearance in normal man. *J. clin. Path.* **17**, 557–563.

HEALY J.K. (1968) Clinical assessment of glomerular filtration rate by different forms of creatinine clearance and a modified urinary phenolsulphonphthalein excretion test. *Amer. J. Med.* **44**, 348–368.

HECKEL G.P. & TOBIN C.E. (1956) Arteriovenous shunts in the myometrium. *Amer. J. Obstet. Gynec.* **71**, 199–205.

HELLER B.G., SMITH R.E. & LUBIN R.I. (1955) Renal function status in patients with acromegaly. *Clin. Res. Proc.* **3**, 13.

HELMER O.M. & JUDSON W.E. (1967) Influence of high renin substrate levels on renin-angiotensin system in pregnancy. *Amer. J. Obstet. Gynec.* **99**, 9–17.

HIERHOLZER K, WIEDERHOLT M., HOLZGREVE H., GIEBISCH G., KLOSE R.M. & WINDHAGER E.E. (1965) Micropuncture study of renal transtubular concentration gradients of sodium and potassium in adrenalectomized rats. *Arch. Ges. Physiol.* **285**, 193–210.

HIERHOLZER K., WIEDERHOLT M. & STOLTE H. (1966) Hemmung der natriumresorption im proximalen und distalen Konvolut adrenalektomierter Ratten. *Arch. Ges. Physiol.* **291**, 43–62.

HINSHAW L.B., DAY S.B. & CARLSON C.H. (1959) Tissue pressure as a causal factor in the autoregulation of blood flow in the isolated perfused kidney. *Amer. J. Physiol.* **197**, 309–312.

HINSHAW L.B., FLAIG R.D., LOGEMANN R.L. & CARLSON C.H. (1960) Intrarenal venous and tissue pressure and autoregulation of blood flow in the perfused kidney. *Amer. J. Physiol.* **198**, 891–894.

HOCKEN A.G., KARK R.M. & PASSOVOY M. (1966) The angiotensin-infusion test. *Lancet* **1**, 5–10.

HODARI A.A., BUMPUS F.M. & SMEBY R. (1967) Renin in experimental 'Toxaemia of Pregnancy'. *Obstet. & Gynec.* **30**, 8–15.

HODARI A.A., SMEBY R. & BUMPUS F.M. (1967) A renin-like substance in the human placenta. *Obstet. & Gynec.* **29**, 313–317.

HODARI A.A. & HODGKINSON C.P. (1968) Foetal kidney as a source of renin in the pregnant dogs. *Amer. J. Obstet. Gynec.* **102**, 691–701.

HODARI A.A. (1968) The contribution of the fetal kidney to experimental hypertensive disease of pregnancy. *Amer. J. Obstet. Gynec.* **101**, 17–22.

HODARI A.A., CARRETERO O.A. & HODGKINSON C.P. (1969) Uterine production of renin in normal and nephrectomized dogs. *Obstet. & Gynec.* **34**, 358–362.

HOROBIN D.F. & LLOYD I.J. (1970) Pre-eclamptic toxaemia: possible relevance of progesterone, salt and fursemide. *J. Obstet. Gynaec. Brit. Cwlth.* **77**, 253–258.

HORSKA S. & VEDRA B. (1964) Diurnal variations of glomerular filtration and renal handling of electrolytes in normal and toxemic pregnancies. *Amer. J. Obstet. Gynec.* **90**, 285–287.

HORSTER M. & THURAU K. (1968) Micropuncture studies on the filtration rate of single superficial and juxtamedullary glomeruli in the rat kidney. *Arch. Ges. Physiol.* **301**, 162–181.

HOWARD B.K., GOODSON J.H. & MENGERT W.F. (1953) Supine hypotensive syndrome in late pregnancy. *Obstet. & Gynec.* **1**, 371–377.

HULL A.R., SUKI W.N. & RECTOR F.C. JR. (1967) Mechanism of diuretic induced hyperuricemia. *Proceedings of the First Annual Meeting of the American Society of Nephrology, Los Angeles*, p. 31.

HYTTEN F.E. & LEITCH I. (1964) *The Physiology of Human Pregnancy*, p. 121. Philadelphia, F.A.Davis Co.

JAMISON R.L., BENNETT C.M. & BERLINER R.W. (1967) Countercurrent multiplication by the thin loops of Henle. *Amer. J. Physiol.* **212**, 352–366.

JAMISON R.L. (1969) The function of the thin loops of Henle in the urinary concentrating mechanism, in N. Alwall, F.Berglund & B.Josephson (eds.) *Proceedings of the Fourth International Congress on Nephrology, Stockholm*, **1**, 170–174. Basel, München, New York.

JONES E.M. (1968) Capillary permeability to plasma proteins during pregnancy. *J. Obstet. Gynaec. Brit. comm.* **75**, 295–299.

JONES K.M., LLOYD-JONES R., RIOEDEL A., TAIT J.F., TAIT S.A.S., BULBROOK R.D. & GREENWOOD F.C. (1959) Aldosterone secretion and metabolism in normal men and women and in pregnancy. *Acta Endocrinol.* **30**, 321–342.

KALOUSEK G., HLAVACEK C., NEDOSS B. & POLLAK V.E. (1969) Circadian rhythms of creatinine and electrolyte excretion in healthy pregnant women. *Amer. J. Obstet. Gynec.* **103**, 856–867.

KAPLAN S.L. & GRUMBACH M.M. (1964) Studies of a human and simian placental hormone with growth hormone-like and prolactin-like activities. *J. clin. Endocr. & Metab.* **24**, 80–100.

KAPLAN N.M. & SILAH J.G. (1964) The effect of angiotensin II on the blood pressure in humans in the hypertensive disease. *J. clin. Invest.* **43**, 659–669.

KATZ A.I. & EPSTEIN F.H. (1967) The physiological role of sodium-potassium activated adenosine triphosphatase in the active transport of cations across biological membranes. *Israel J. Med. Sci.* **3**, 155–166.

KATZ F.H. & KAPPAS A. (1967) The effects of oestradiol on plasma levels of cortisol and thyroid hormone-binding globulins and on aldosterone and cortisol secretion rates in man. *J. clin. Invest.* **46**, 1768–1777.

KERR M.G., SCOTT D.B. & SAMUELS E. (1964) Studies of the inferior vena cava in late pregnancy. *Brit. med. J.* **1**, 532–533.

KERR M.G. (1968) Cardiovascular dynamics in pregnancy and labour. *Brit. med. Bull.* **24**, 19–24.

KLEIT S., LEVIN D., PERENICH T. & CADE R. (1965) Renal excretion of ascorbic acid in dogs. *Amer. J. Physiol.* **209**, 195–198.

KLEITMAN N. (1923) Physiology of sleep: effects of prolonged sleeplessness on man. *Amer. J. Physiol.* **66**, 67–92.

KUMAR D. & BARNES A.C. (1960) Aldosterone in normal and abnormal pregnancy. *Obstet. Gynec. Surv.* **15**, 625–633.

LADEFOGED J. & PEDERSEN F. (1967) Renal blood flow, circulation times and vascular volume in normal man measured by the intra-arterial injection-external counting technique. *Acta Physiol. Scand.* **69**, 220–239.

LANDAU R.L., BARGENSTAL D.M., LUGIBIHL K. & KASCHUT M.E. (1955) The metabolic effects of progesterone in man. *J. clin. Endocr. & Metab.* **15**, 1194–1215.

LANDAU R.L. & LUGIBIHL K. (1958) Inhibition of the sodium-retaining influence of aldosterone by progesterone. *J. clin. Endocr. & Metab.* **18**, 1237–1245.

LANDAU R.L., PLOTZ E.J. & LUGIBIHL K. (1960) Effect of pregnancy on the metabolic influence of administered progesterone. *J. clin. Endocr. & Metab.* **20**, 1561–1575.

LARAGH J.H., SEALEY J.E., LEDINGHAM J.G.G. & NEWTON M.A. (1967) Oral contraceptives. Renin, aldosterone and high blood pressure. *J. Amer. med. ass.* **201**, 918–922.

LEQUIN R.M., HACKENG W.H.L. & SCHOPMAN W. (1970) A radio-immuno-assay for parathyroid hormone in man. *Acta Endocrinologica* **63**, 655–666.

LEVINSKY N.G. & LALONE R.C. (1963) The mechanism of sodium diuresis after saline infusion in the dog. *J. clin. Invest.* **42**, 1261–1271.

LEYSSAC P.P. (1966) The regulation of proximal tubular reabsorption in the mammalian kidney. *Acta Physiol. Scand.* **70**, (Suppl. 291), 1–152.

LICHTON I.J. (1961) Salt-saving in the pregnant rat. *Amer. J. Physiol.* **201**, 765–768.

LICHTON I.J. (1963) Urinary excretion of water, sodium and total solutes by the pregnant rat. *Amer. J. Physiol.* **204**, 563–567.

LINDHEIMER M.D., LALONE R.C. & LEVINSKY N.G. (1967) Evidence that an acute increase in glomerular filtration has little effect on sodium excretion in the dog unless extracellular volume is expanded. *J. clin. Invest.* **46**, 256–265.

LINDHEIMER M. (1969) Characterization of the effect of supine posture on renal function in late pregnancy, in *Proceedings of the Fourth International Congress of Nephrology, Stockholm*, p. 438. (Abstract).

LINDHEIMER M.D. & WESTON P.V. (1969) Effect of hypotonic expansion on sodium, water and urea excretion in late pregnancy: the influence of posture on these results. *J. clin. Invest.* **49**, 947–956.

LONGO L.D. & ASSALI N.S. (1960) Renal function in human pregnancy. IV. The urinary tract 'dead space' during normal gestation. *Amer. J. Obstet. Gynec.* **80**, 495–499.

LOUIS W.J. & DOYLE A.E. (1965) The effects of varying doses of angiotensin on renal function and blood pressure in man and dogs. *Clin. Sci.* **29**, 489–504.

LUCIUS H., GAHLENBECK H., KLEINE H.O., FABEL H. & BARTELS H. (1970) Respiratory functions, buffer system, and electrolyte concentrations of blood during human pregnancy. *Resp. Physiol.* **9**, 311–317.

MCCARTNEY C.P., SPARGO B., LORINCZ A.B., LEFEBVRE V. & NEWTON R.E. (1964) Renal structure and function in pregnant patients with acute hypertension; osmolar concentration. *Amer. J. Obstet. Gynec.* **90**, 579–592.

MCDONALD M., SCHRIER R.W. & LAULER D.P. (1967) Effect of acute extracellular volume expansion on cross-circulated dogs. *Nephron* **4**, 12.

MACGILLIVRAY I. & BUCHANAN T.J. (1958) Total exchangeable sodium and potassium in non-pregnant women and in normal and pre-eclamptic pregnancy. *Lancet* **2**, 1090–1093.

MCKENNA O.C. & ANGELAKOS R.T. (1958) The adrenergic innervation of the canine kidney. *Circ. Res.* **22**, 345–354.

MCLENNAN C.E. (1943) Antecubital and femoral venous pressure in normal and toxemic pregnancy. *Amer. J. Obstet. Gynec.* **45**, 568–591.

MAEBASHI M., AIDA M., YOSHINAGA K., ABE K., MIWA I. & WATANABE N. (1964) Estimation of circulating renin in normal and toxaemic pregnancy. *Tokohu J. Exp. Med.* **84**, 55–61.

MAHER F.T. & TAUXE W.N. (1969) Renal clearance in man of pharmaceuticals containing radio-active iodine. Influence of plasma binding. *J. Amer. med. ass.* **207**, 97–104.

MALNIC G., KLOSE R.M. & GIEBISCH G. (1966) Microperfusion study of distal tubular potassium and sodium transport in rat nephron. *Amer. J. Physiol.* **211**, 529–547.

MALNIC G., KLOSE R.M. & GIEBISCH G. (1966) Microperfusion study of distal tubular potassium and sodium transfer in rat kidney. *Amer. J. Physiol.* **211**, 548–559.

MANCHESTER R.C. (1933) Diurnal rhythm in water and mineral exchange. *J. clin. Invest.* **12**, 995–1008.

MARTIN J.D. & MILLS I.H. (1956) Aldosterone excretion in normal and toxaemic pregnancies. *Brit. med. J.* **2**, 571–573.

MARTINO J.A. & EARLEY L.E. (1967) Demonstration of a role of physical factors as determinants of the natriuretic response to volume expansion. *J. clin. Invest.* **46**, 1963–1978.

MATERSON B.J., JOHNSON A.E. & PEREZ-STABLE E.C. (1969) Inulin labeled with chromium 51 for determination of glomerular filtration rate. *J. Amer. med. ass.* **207**, 94–96.

MAUNSBACH A.B. (1966) Absorption of I^{125} labelled homologous albumin by rat kidney proximal tubule cells. A study of microperfused single proximal tubules by electron microscopic autoradiography and histochemistry. *J. Ultrastr. Res.* **15**, 197–241.

MILLS I.V. (1970) Renal regulation of sodium excretion. *Ann. Rev. Med.* **21**, 75–98.

MOFFAT D.B. (1967) The fine structure of the blood vessels of the renal medulla with particular reference to the control of the medullary circulation. *J. Ultrastr. Res.* **18**, 532–545.

MOFFAT D.B. (1968) Medullary blood flow during hydropoenia. *Nephron* **5**, 1–6.

MORGEN R.O., GONDA E. & BECK J.C. (1961) Renotropic effects of human growth hormone in man. *Clin. Res.* **9**, 206 (Abstract).

NANRA R.S., CLYNE D. & KINCAID-SMITH P. (1969) The use of the modified PSP excretion test in the clinical management of patients with renal allografts. *Med. J. Aust.* **1**, 1083–1086.

NEWTON M.A., SEALEY J.E., LEDINGHAM J.G.G. & LARAGH J.H. (1968) High blood pressure and oral contraceptives. Changes in plasma renin and renin substrate and in aldosterone excretion. *Amer. J. Obstet. Gynec.* **101**, 1037–1045.

O'CONNELL M. & WELSH G.W. 3rd (1969) Unbound plasma cortisol in pregnant and Enovid-E treated women as determined by ultra filtration. *J. clin. Endocr. & Metab.* **29**, 563–568.

ORLOFF J. & HANDLER J. (1967) The role of adenosine 3'5'-phosphate in the action of antidiuretic hormone. *Amer. J. Med.* **42**, 757–768.

PAGE E.W., GLENDENING M.B., DIGNAM W. & NARPER H.A. (1954) The causes of histidinuria in normal pregnancy. *Amer. J. Obstet. Gynec.* **68**, 110–118.

PALOMAKI J.F. & LINDHEIMER M.D. (1970) Sodium depletion simulating deterioration in a toxaemic pregnancy. *New Engl. J. Med.* **282**, 88–89.

PAPPENHEIMER J.R. & KINTER W.B. (1956) Hematocrit ratio of blood with mammalian kidney and its significance for renal haemodynamics. *Amer. J. Physiol.* **185**, 377–390.

PEARCE M.L., NEWMAN E.V. & BIRMINGHAM M.R. (1954) Some postural adjustments of salt and water excretion. *J. clin. Invest.* **33**, 1089–1094.

PEART W.S. (1965) Renin-angiotensin system. *Pharmacol Rev.* **17**, 143–182.

PETERSON H. & FRANK H. (1958) Untersuchugen uber die ruckresorption des histidins bei schwangeren, bei der renalen glykosurie und bei kranken mit eingeschrankter nierenfunktion. *Deutsche Arch. klin. Med.* **205**, 70–78.

PITTS R.F. (1944) Comparison of renal reabsorptive processes for several aminoacids. *Amer. J. Physiol.* **140**, 535–547.

PLENTL A.A. & GRAY M.J. (1959) Total body water, sodium space, and total exchangeable sodium in normal and toxaemic pregnant women. *Amer. J. Obstet. Gynec.* **78**, 472–478.

POLLAK V.E. & NETTLES J.B. (1960) The kidney in toxaemia of pregnancy: a clinical and pathological study based on renal biopsies. *Medicine* **39**, 469–526.

PRITCHARD J.A., BARNES A.C. & BRIGHT R.H. (1955) The effect of the supine position on renal function in the near-term pregnant woman. *J. clin. Invest.* **34**, 777–781.
PYSTYNEN P. & PANKAMAA P. (1965) Diurnal variations in urine excretion and sodium and potassium excretion in healthy and toxaemic women in late pregnancy. *Acta obstet. gynec. scand.* **44**, 408–415.
RAAB W., SCHROEHER C., WAGNER R. & GIGES W. (1956) Vascular reactivity and electrolytes in normal and toxemic pregnancy: pathogenic considerations and diagnostic pre-toxemia test. *J. clin. Endocr.* **16**, 1196–1216.
RECTOR F.C. Jr., BRUNNER F.P. & SELDIN D.W. (1966) Mechanism of glomerulotubular balance. I. Effect of aortic constriction and elevated ureteropelvic pressure on glomerular filtration rate, fractional reabsorption, transit time, and tubular size in the proximal tubule of the rat. *J. clin. Invest.* **45**, 590–602.
RECTOR F.C. Jr., SELLMAN J.C., MARTINEZ-MALDONADO M. & SELDIN D.W. (1967) The mechanism of suppression of proximal tubular reabsorption by saline infusion. *J. clin. Invest.* **46**, 47–56.
RECTOR F.C. Jr. & SELDIN D.W. quoted in E.E. Windhager (1969) Kidney, water and electrolytes. *Ann. Rev. Physiol.* **31**, 117–172.
REUBI F. (1950) Quelques aspects du diabéte rénal. *Helv. Méd. Acta* **17**, 493–497.
ROBB C.A., DAVIS J.O., JOHNSON J.A., BLAINE E.H., SCHNEIDER E.G. & BAUMBER J.S. (1970) Mechanisms regulating the renal excretion of sodium during pregnancy. *J. clin. Invest.* **49**, 871–880.
ROSENTHAL H.E., SLAUNWHITE W.R. Jr., & SANDBERG A.A. (1969) Transcortin: a corticosteroid-binding protein of plasma. X. Cortisol and progesterone interplay and unbound levels of these steroids in pregnancy. *J. clin. Endocr. & Metab.* **29**, 352–367.
RUBI R.A. & SALA N.L. (1968) Ureteral function in pregnant women. III. Effect of different positions and of fetal delivery on ureteral tonus. *Amer. J. Obstet. Gynec.* **101**, 230–237.
RUEDAS G. & WEISS C.L. (1967) Die Wirkung von Anderungen der Natriumkonzentration im Perfusionsmedium und von Strophanin auf die Glukoseresorption der isolierten Rattenniere. *Pflügers Arch.* **298**, 12–22.
SANDLER M. & BALDOCK E. (1963) *In vivo* monoamine oxidase activity in toxaemia of pregnancy. *J. Obstet. Brit. Cwlth.* **70**, 279–283.
SCHNEIDER E.G., ROSTORFER H.H. & NASH F.D. (1968) Distribution, volume and metabolic clearance rate of renin in anesthetized nephrectomized dogs. *Amer. J. Physiol.* **215**, 1115–1122.
SCHRIER R.W., MCDONALD K.M., JAGGER P.I. & LAULER D.P. (1967) The role of the adrenergic nervous response to acute extracellular fluid volume expansion. *Proc. Soc. Exp. Biol. Med.* **125**, 1157–1162.
SCHWARTZ W.B. & RELMAN A.S. (1967) Effects of electrolyte disorders on renal structure and function. *New Eng. J. Med.* **276**, 452–458.
SCOTT D.B. & KERR M.G. (1965) Inferior vena caval pressure in late pregnancy. *J. Obstet. Gynaec. Brit. Cwlth.* **70**, 1044–1049.
SELKURT E.E., WATHEN R.L. & SANTOS-MARTINEZ J. (1968) Creatinine excretion in the squirrel monkey. *Amer. J. Physiol.* **214**, 1363–1369.

SENIOR J.B., FAHIM I., SULLIVAN F.M. & ROBSON J.M. (1963) Possible role of 5-hydroxytryptamine in toxaemia of pregnancy. *Lancet* **2**, 553–554.

SHORT R.V. & ETON B. (1959) Progesterone in blood. III. Progesterone in the peripheral blood of pregnant women. *J. Endocr.* **18**, 418–425.

SIMPSON G.E. (1924) Diurnal variations in rate of urine excretion for 2 hour intervals: some associated factors. *J. biol. Chem.* **59**, 107–122.

SIMPSON G.E. (1926) Effect of sleep on urinary chlorides and pH. *J. biol. Chem.* **67**, 505–516.

SIMPSON G.E. (1929) Changes in composition of urine brought about by sleep and other factors. *J. biol. Chem.* **84**, 343–411.

SIMS E.A.H. & SELDIN D.W. (1949) Reabsorption of creatine and guanidoacetic acid by renal tubules. *Amer. J. Physiol.* **157**, 14–20.

SIMS E.A.H. & KRANTZ K.E. (1958) Serial studies of renal function during pregnancy and puerperium in normal women. *J. clin. Invest.* **37**, 1764–1774.

SIMS E.A.H. (1963) The kidney in pregnancy, in M.B.Strauss & L.G.Welt (eds.) in *Diseases of the Kidney*, p. 853. Boston, Little, Brown.

SIMS E.A.H. (1968) Renal function in normal pregnancy. *Clin. Obstet. Gynec.* **11**, 461–472.

SIMS E.A.H. (1970) Pre-eclampsia and related complications of pregnancy. *Amer. J. Obstet. Gynec.* **107**, 154–181.

SIROTA J.H., BALDWIN D.S. & VILLAREAL H. (1950) Diurnal variations of renal function in man. *J. clin. Invest.* **29**, 187–192.

SKINNER S.L., LUMBERS R.E. & SYMONDS E.M. (1968) Renin concentration in human foetal and maternal tissues. *Amer. J. Obstet. Gynec.* **101**, 529–533.

SLATOPOLSKY E., HOFFSTEN P., PURKERSON M. & BRICKER N.S. (1970) On the influence of extracellular fluid volume expansion and of uremia on bicarbonate reabsorption in man. *J. clin. Invest.* **49**, 988–998.

SMITH H. (1956) Summary interpretation of observations of renal haemodynamics in pre-eclampsia, in *Report of the First Ross Obstetric Research Conference*, pp. 75–76. Columbus, Ohio, Ross Laboratories.

SOHAR E., SCADRON E. & LEVITT M.F. (1956) Changes in renal haemodynamics during normal pregnancy. *Clin. Res. Proc.* **4**, 142 (Abstract).

STAKEMANN S. (1960) A renin-like pressor substance found in the placenta of the cat. *Acta Path. Microbiol. Scand.* **50**, 350–354.

STANBURY S.W. & THOMPSON A.E. (1951) Diurnal variation in electrolyte excretion. *Clin. Sci.* **10**, 267–293.

STOESSER S. & SIMS E.A.H. in SIMS E.A.H. (1963) The kidney in pregnancy, in M.B. Strauss & L.G.Welt (eds.) *Diseases of the Kidney*, p. 861. Boston, Little, Brown.

STRAUSS M.B., DAVIS R.K., ROSENBAUM J.D. & ROSSMEISL E.C. (1952) Production of increased renal sodium excretion by the hypotonic expansion of extracellular fluid volume in recumbent subjects. *J. clin. Invest.* **31**, 80–86.

SYMONDS E.M., STANLEY M.A. & SKINNER S.L. (1968) Production of renin by *in vitro* cultures of human chorion and uterine muscle. *Nature* **217**, 1152–1153.

TAIT J.F., LITTLE B., TAIT S.A.S. & FLOOD C. (1962) The metabolic clearance rate of aldosterone in pregnant and nonpregnant subjects estimated by both single-injection and constant-infusion methods. *J. clin. Invest.* **41**, 2093–2100.

TAIT J.F. & LITTLE B. (1968) The metabolism of orally and intravenously administered labeled aldosterone in pregnant subjects. *J. Clin. Invest.* **47**, 2423–2429.
TALLEDO O.E., RHODES K. & LIVINGSTON E. (1967) Renin-angiotensin system in normal and toxaemic pregnancies. II. Inactivation of angiotensin in normal pregnancy. *Amer. J. Obstet. Gynec.* **97**, 571–572.
TALLEDO O.E., CHESLEY L.C. & ZUSPAN F.P. (1968) Renin-angiotensin system in normal and toxemic pregnancies. 3. Differential sensitivity to angiotensin II and norepinephrine in toxemia of pregnancy. *Amer. J. Obstet. Gynec.* **100**, 218–221.
THOMPSON A.M., HYTTEN F.E. & BILLEWICS W.Z. (1967) The epidemiology of oedema during pregnancy. *J. Obstet. Gynaec. Brit. Cwlth.* **74**, 1–10.
THORBURN G.D., KOPALD H.H., HERD J.A., HOLLENBERG M., O'MORCHOE C.C.C. & BARGER A.C. (1963) Intrarenal distribution of nutrient blood flow determined with Krypton 85 in the unanesthetized dog. *Circ. Res.* **13**, 290–307.
THURAU K. (1964) Renal haemodynamics. *Amer. J. Med.* **36**, 698–719.
THURAU K. (1967) The nature of autoregulation of renal blood flow, in J.S.Handler (ed.) *Proceedings of the Third International Congress on Nephrology, Washington,* **1**, 162–173. Basel and New York, Hans Huber.
TOBACK F.G., HALL III, P.W. & LINDHEIMER M.D. (1970) Effect of posture on urinary protein patterns in non-pregnant, pregnant and toxaemic women. *Obstet. & Gynec.* **35**, 765–768.
TRAUT H.F. & MCLANE C.M. (1936) Physiological changes in the ureter associated with pregnancy. *Surg. Gynec. Obstet.* **62**, 65–72.
TRUNIGER B. & SCHMIDT-NIELSEN B. (1964) Intrarenal distribution of urea and related compounds, effects of nitrogen intake. *Amer. J. Physiol.* **207**, 971–978.
ULLRICH K.J. (1967) Renal transport of sodium, in J.S.Handler (ed.) *Proceedings of the Third International Congress on Nephrology, Washington,* **1**, 48–61. Basel and New York, Hans Huber.
URANGA J. (1965) The pressure in Bowman's capsule and the occluded ureter of the toad. *Acta Physiol. Latinoam.* **15**, 77–85.
URANGA J. (1967) Influence of the liver on regulation of glomerular pressure in the toad. *Amer. J. Physiol.* **213**, 1244–1248.
VANDER A.J. (1967) Control of renin release. *Physiol. Rev.* **47**, 359–382.
VAN WAGENAN G. & JENKINS R.H. (1939) Experimental examination of factors causing ureteral dilatation of pregnancy. *J. Urol.* **42**, 1010–1020.
VOGEL G., TERVOOREN U. & STOECKERT I. (1965) Untersuchugen zur Abhangigkeit des renal tubularen Glucose-Transportes vom Ionen- Angebot sowie des Na^+-Transportes vom Angebot an Glucose. *Arch. Ges. Physiol.* **288**, 359–368.
WAKIM K.G. (1968) How kidneys function in pregnancy. *Postgrad. Med.* **43**, 113–121.
WALKER E.W., MCMANUS M. & JANNEY J.C. (1934) Kidney function in pregnancy: II. Effects of posture on disease. *Proc. Soc. Exp. Biol. Med.* **31**, 392–397.
WALLRAFF E.B., BRUDIE E.C. & BORDEN A.L. (1950) Urinary excretion of amino acids in pregnancy. *J. clin. Invest.* **29**, 1542–1544.
WATANABE M., DOMINQUEZ O.U., MEEKER C.I., SIMS E.A.H., GRAY M.J. & SOLOMON S. (1962) Aldosterone and progesterone secretion in pregnancy. *J. clin. Invest.* **41**, 1408–1409 (Abstract).

WATANABE M., MEEKER C.I., GRAY M.J., SIMS E.A.H. & SOLOMON S. (1963) Secretion rate of aldosterone in normal pregnancy. *J. clin. Invest.* **42,** 1619–1631.

WEARRN J.T. & RICHARDS A.N. (1924) Observations on the composition of glomerular urine with particular reference to the problem of reabsorption in the renal tubules. *Amer. J. Physiol.* **71,** 209–227.

WELSH G.W. 3rd & SIMS E.A.H. (1960) The mechanism of renal glycosuria in pregnancy. *Diabetes* **9,** 363–369.

WELSH G.W. (1964) Abnormal glucose tolerance in pregnancy: studies of cortisol secretion and growth hormone. *Excerpta Med.* **74,** 164–165.

WESSON L.G. Jr. (1957) Glomerular and tubular factors in the renal excretion of sodium chloride. *Medicine* **36,** 281–396.

WESSON L.G. & LAULER D.P. (1961) Diurnal cycle of glomerular filtration rate and sodium and chloride excretion during responses to altered salt and water balance in man. *J. clin. Invest.* **40,** 1967–1977.

WESSON L.G. Jr. (1964) Electrolyte excretion in relation to diurnal cycles of renal function. *Medicine* **43,** 547–592.

WHITE H.L., HEINBECKER P. & ROLF D. (1949) Enhancing effects of growth hormone on renal function. *Amer. J. Physiol.* **157,** 47–51.

WIEDERHOLT M., HIERHOLZER K., WINDHAGER E.E. & GIEBISCH G. (1967) Microperfusion study of fluid reabsorption in the proximal tubules of rat kidneys. *Amer. J. Physiol.* **213,** 809–818.

WIEDERHOLT M. & WIEDERHOLT B. (1968) Der Einfluss von Dexamethason auf die Wasser- und Electrolytausscheidung adrenalektomierter Eatten. *Pflüger Arch.* **302,** 57–78.

WILLSON J.R., WILLIAMS JR. J.M. & HAYASHI T.T. (1957) Hypertonic saline infusions for the differential diagnosis of the toxemias of pregnancy. *Amer. J. Obstet. Gynec.* **73,** 30–36.

WILSON R.J., DE BONO E. & MILLS I.H. (1970) Quoted in MILLS I.V. (1970) Renal regulation of sodium excretion. *Ann. Rev. Med.* **21,** 75–98.

WINDHAGER E.E. (1968) Glomerulo-tubular balance of salt and water. *Physiologist* **11,** 103–114.

WINDHAGER E.E., LEWY J.E. & SPITZER A. (1969) Intrarenal control of proximal tubular reabsorption of sodium and water. *Nephron* **6,** 247–259.

WINER B.M. (1965) Renin in pregnancy and the menstrual cycle. *J. clin. Invest.* **44,** 1112 (Abstract).

ZELENIK J.S. (1965) Endocrine physiology of pregnancy. *Clin. Obstet. Gynec.* **8,** 528–549.

ZUSPAN F.P., BARNES A.C., & DILLHOFER J.R. (1958) The urinary excretion of antidiuretic substances by the obstetric and gynecologic patient. *Amer. J. Obstet. Gynec.* **76,** 619–625.

ZUSPAN F.P. (1970) Urinary excretion of epinephrine and norepinephrine during pregnancy. *J. clin. Endocr.* **30,** 357–360.

14

Respiratory Physiology in Pregnancy

ANN J. WOOLCOCK & JOHN READ

In this chapter the steps and mechanisms involved in normal respiration will be outlined, and methods used to study them briefly described. The limitations of such studies will be referred to. Finally, the effects of pregnancy on respiration will be described so far as these are known, and possible reasons for the changes that have been demonstrated will be discussed.

Though large numbers of pregnant women have been studied from the point of view of one or two respiratory parameters, systematic studies of the integrated pattern of pulmonary function during pregnancy are totally lacking. Again, many of the simple measurements that have been made suffer from technical defects which render the results of doubtful significance. An integrated study of pulmonary function during pregnancy would be rewarding. The authors have not undertaken such a study, but one of them (A.J.W.) has carried out a small series of detailed serial studies in pregnant women. Because of the paucity of available data in some areas, reference will be made to examples from these studies, which will be referred to in this chapter as 'the local series'.

In the outline of normal respiratory processes and the chief abnormalities which may arise in them, attention will not be restricted to those aspects where we can report corresponding studies in pregnancy. There are two reasons for this. First, a pregnant woman may develop any of the usual disorders of respiration, and these merit functional assessment in the usual way. Secondly, integrated studies of the respiratory changes accompanying normal pregnancy are long overdue, and the present chapter aims to provide some general background for their understanding when they do appear. Again, a broad view of respiration will be taken, rather than attempting to cover the detail commonly found in standard texts.

NORMAL RESPIRATION

The following represents a bird's eye view of the processes of normal respiration. Based on a variety of stimuli impinging on the respiratory centre, impulses are sent peripherally to the inspiratory muscles. Contraction of these muscles overcomes elastic forces in the chest wall, diaphragm, and lungs as well as flow-resistive forces generated by air flowing along the airways. As a result of enlargement of the thoracic cage, air flows into the lungs. When the inspiratory muscles cease to contract, recoil forces in the lungs are sufficient to return the thoracic cage to its resting level during the passive process of expiration. The air entering the lungs in inspiration mixes with gas already present in the air spaces and exchange of oxygen and carbon dioxide takes place continuously between the alveolar gas and the pulmonary capillary blood across the thin alveolar-capillary membrane by a process of simple diffusion. The distribution of ventilation and blood flow to the various parts of the lungs is fairly evenly matched.

The total process of respiration also requires an adequate haemoglobin content of the circulating blood for both oxygen and carbon dioxide carriage, as well as an adequate cardiac output to provide an adequate peripheral blood flow to subserve the metabolic needs of tissue respiration.

In considering the respiratory processes in more detail, it is necessary to break them up into several components, and this is done in succeeding sections. But understanding of respiration is often impeded by too much fragmentation, and the divisions adopted here are broader than usual: mechanics and ventilation of the lungs, gas exchange, and respiratory control.

LUNG MECHANICS, LUNG VOLUMES AND VENTILATION

The sequence in which the items in the above heading are placed is quite deliberate. It is commoner to consider them in the reverse order since, historically, knowledge of them developed in this way. However, an attempt will be made here to describe them as logically flowing one from the other.

During the process of breathing a certain quantum of work is done by the respiratory muscles each minute. During quiet breathing, the active work is done almost wholly during inspiration, and expiration is a passive event. The work energy is expended in two main ways:

(1) In overcoming elastic forces. During inspiration, the inward elastic forces of the lungs must be overcome. These elastic forces derive from two sources: (a) the elastic tissues contained within the lung substance; (b) the elastic forces generated by the presence of an air-liquid interface at the alveolar surface.

(2) In overcoming flow-resistive forces. Airways of normal calibre present some frictional resistance to air-flow though, compared with the situation which may arise in disease states, it is small.

An even smaller force is needed to overcome frictional resistance of the lung tissue; and this is usually ignored in considerations of respiratory work.

The balance of elastic forces in the total respiratory system should be considered further. If the chest wall is opened, the lungs retract inwards and the chest wall recoils outwards compared with their pre-existing positions. This is the simplest way of demonstrating that the normal resting position of the lungs and chest is determined by equality of the inward and outward elastic pull of the lungs and chest wall respectively. The resting position of the lungs (measured as the functional residual capacity or FRC*) is not then an arbitrary, fixed position; it is that position where the elastic recoils of the lungs and chest wall are equal and opposite in direction. It may change if there is any alteration in the elastic recoil properties of either the lungs or the chest wall, or both. Since the recoil properties of both lung and chest wall may change in either direction, it is not possible to infer from changes of FRC whether the primary change is within the lungs or chest wall, though this is often done in the literature.

To take two examples of change in FRC from lung disease: lung elasticity is reduced in the presence of the tissue destruction of emphysema and the resting position of the chest will be in an expanded position; in the presence of lung fibrosis, increased lung elasticity causes the opposite effect with a diminution of FRC.

With maximum voluntary inspiratory effort, additional air enters the lungs until they reach the position of total lung capacity (TLC). The TLC position is determined by the force that can be generated by the

* A glossary of the abbreviations used appears at the end of this chapter.

inspiratory muscles and the extent to which both lungs and chest wall (which both now have a recoil force inwards) can be distended by that inspiratory force.

When a maximum voluntary expiratory effort is made the lungs are reduced to residual volume (RV). This position is determined by the maximum force generated by the expiratory muscles and by the minimal size to which the chest cavity can be reduced by that force. An additional factor, of increasing importance over the age of 40 years, is closure of small airways as lung volume is reduced.

These lung volumes and the forces acting to determine them are summarized in Fig. 14.1.

Starting from FRC, the rhythmic expenditure of a certain quantity of inspiratory muscle work overcomes the elastic and flow-resistive forces to produce a certain change of chest volume and hence the entry of a certain volume of air at each breath (tidal volume or $\mathring{V}_T$). This tidal volume, repeated f times per minute generates a certain minute ventilation ($\mathring{V}_E$). To anticipate one aspect of respiratory control, the combination of tidal volume and respiratory frequency utilized for a given $\mathring{V}_E$ is that which will minimize inspiratory muscle work, allowing for elastic and flow-resistive factors. Expressed numerically, minute ventilation then becomes somewhat of an incident, rather than a precise value with some intrinsic biological meaning of its own. It represents merely the resultant of the respiratory muscle work generated on the one hand and the elastic and flow-resistive forces which must be overcome on the other.

Laboratory assessment

Tidal volume, respiratory frequency, and minute ventilation are easily measured by collecting expired gas over a given time interval and counting the respirations over that interval. In practice, the measurement of minute ventilation plays little part in the routine assessment of lung function, because the value varies widely between subjects of the same size and age, and because patients with advanced lung disease often have normal levels of $\mathring{V}_E$. However, where the measurement is made carefully with the subject at rest, and observations for a given subject are compared at different times, it may become of value in demonstrating changes.

The simplest test of the mechanical properties of the respiratory

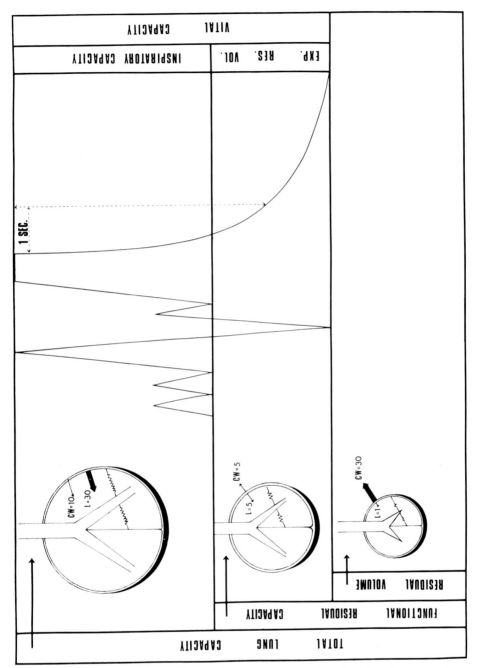

FIG. 14.1. Subdivisions of lung volume and forces acting to determine these lung volumes. The arrows labelled CW and L indicate the direction of elastic recoil of chest wall and lung respectively at three lung volumes, and the numbers are examples (in cm H_2O) of values in normal subjects. The breathing pattern on the right shows three tidal volumes, a maximum inspiration and expiration, and a forced expiratory volume in 1 second (FEV_1).

system is measurement of the vital capacity (VC). This is the amount of air which can be expired between TLC and RV (Fig. 14.1). It is important to emphasize that VC, which many people think of as a lung volume in its own right, is simply the difference between two primary lung volumes (TLC and RV), each of which is determined by various mechanical factors. It follows that reduction of vital capacity may result either from reduction of TLC, increase of RV, or from a combination of both.

Granted a co-operative subject, VC is a very reproducible value. It represents the measurement which has been made most frequently in normal subjects, in patients with respiratory disease, and also in pregnancy.

The forced expiratory volume in one second (FEV_1) is often measured at the same time as VC. VC is measured when the subject expires from TLC to RV at her own pace into a recording spirometer. For measurement of FEV_1 the expiration from TLC is made using maximal effort. Again, provided the subject is co-operative, this is a very reproducible measurement. A normal subject can expire more than 70% of her vital capacity in the first second (that is, FEV_1/VC is greater than 70%). In some young women the value may exceed 90%. If FEV_1 constitutes less than 70% of vital capacity, and there is no disorder of expiratory muscles, airways obstruction of some kind is present. The test may be repeated following inhalation of a rapidly acting bronchodilator to assess the acute reversibility of any airways obstruction.

If both VC and FEV_1 are normal by predicted standards, it is probable, but not certain, that the mechanical properties of the respiratory system are normal.

The significance of a reduced VC can only be explored by measurement of FRC. The subject breathes to and from a spirometer which initially contains some helium as a foreign indicator gas, and the change in concentration of helium is recorded after a new equilibrium concentration is obtained. By calculation, the resting volume of air in the lungs with which the helium equilibrated may then be derived. The volume between FRC and TLC (inspiratory capacity, IC) and between FRC and RV (expiratory reserve volume, ERV) may be determined while the patient is breathing into the spirometer. Given these values, TLC and RV may then be derived.

The significance of changes in FRC has already been discussed in terms of an altered balance of forces between lungs and chest wall. In the absence of changes in the chest wall, an increase in RV suggests that the airways are narrowed and are therefore becoming occluded at a higher lung volume than usual during expiration. A decrease of RV suggests that the airways are wider than normal or that there is a space-occupying lesion in the lungs. Reduction of TLC indicates some lesion interfering with the stretching of the lungs or the chest wall. An increase of TLC is usually due to the presence of airways obstruction, either temporary as in asthma or permanent as in bronchitis-emphysema.

Thus, a knowledge of the subdivisions of lung volume and of the FEV_1 usually allows assessment of the state of elasticity of lungs and chest wall and of the resistance of the airways. Sometimes, however, where the changes are small or where both airways obstruction and lung restriction are present, it is necessary to measure the elasticity of the lungs and chest wall and the airways resistance more directly.

Methods for doing this will not be described in detail. Essentially they depend on measuring the pressure drop across the lung by recording intra-oesophageal pressure from a balloon as an index of intrapleural pressure. Airways resistance follows from the simultaneous measurement of air flow at the mouth. Elasticity or lung compliance follows from the pressure change necessary to produce a given volume change (measured when flow has ceased). Compliance is the reciprocal of elasticity; compliance is high in conditions such as emphysema and low in conditions where the lungs are stiff, as in extensive pulmonary fibrosis. The compliance of the chest wall may be measured by inflating the respiratory system while the respiratory muscles are relaxed, measuring the pressure required to produce a given volume change, and subtracting from it the pressure required to produce this change in the lungs alone. Such measurements do not form part of a routine laboratory assessment, but they are of value in demonstrating underlying mechanisms.

Respiratory muscle work may be measured, but only with some difficulty. Broadly the methods fall into two groups. Changes in oxygen consumption may be measured at different levels of ventilation in the resting subject, to derive the oxygen cost of ventilation. Alternatively, the pressure-volume characteristics of the lungs and chest wall may be measured under dynamic conditions, and plotted on co-ordinate axes. The areas under the resulting curves are then measured and provide estimates of elastic and flow-resistive work. Both of these methods are strictly investigational procedures, not ones for routine use.

GAS EXCHANGE

The processes of gas exchange have been virtually ignored in pregnancy. Despite this, they will be described in some detail, especially from the point of view of gas and blood distribution in the lungs. It is intuitively likely that changes in this distribution will be demonstrated when they are sought in pregnancy.

Exchange of oxygen and carbon dioxide takes place by a process of simple diffusion across the alveolar-capillary membrane. Gas moves in either direction from the region of high to low partial pressure. Gas exchange has a finite time constant but, for practical purposes, the process is so rapid that there is equality of both O_2 and CO_2 partial pressures between alveolar gas and blood at the end of each individual capillary in the lung. If the alveolar-capillary membrane is thickened by disease, it is theoretically possible for an end-capillary oxygen tension gradient to persist. For all practical purposes, thickening of the membrane sufficient to interfere with CO_2 exchange is not compatible with life. Even then interference with oxygen transfer is only a secondary phenomenon in the condition of so-called alveolar-capillary block.

Since gaseous equilibrium is attained between alveolar gas and pulmonary capillary blood, then, apart from right-to-left shunting of blood, the factors which determine the partial pressures of oxygen and CO_2 (P_{O_2} and P_{CO_2}) in alveolar gas will be those which determine the partial pressures in the arterial blood. Of crucial importance in determining these alveolar gas tensions is the local ventilation-perfusion ratio ($\dot{V}a/\dot{Q}c$) in any individual alveolus.

It is obvious that, given an alveolus with a certain blood flow, the higher the ventilation of that alveolus the higher will be the P_{O_2} and the lower the P_{CO_2} within it. Again, for a fixed level of alveolar ventilation, the higher the flow of blood past the alveolus the more gas exchange will take place there, with movements of P_{O_2} and P_{CO_2} opposite in direction from those above. In the situation where both ventilation and blood flow of an alveolus may change independently, the final resulting alveolar P_{O_2} and P_{CO_2} are then determined by the ratio of local ventilation to local blood flow (perfusion). A high ventilation-perfusion ratio is associated with a high alveolar P_{O_2} and a low alveolar P_{CO_2}. A low ventilation-perfusion ratio is associated with gas tension changes in the opposite direction. This is a very precise relationship which can be quantitated. For a fixed composition of mixed venous blood entering the lung, any local combination of alveolar and capillary Po_2 and Pco_2 is precisely fixed by the local ventilation-perfusion ratio.

If the lungs are perfectly evenly ventilated and perfectly evenly perfused with blood, the same considerations that apply to a single alveolus will apply to the whole lungs. Where there is any maldistribution of ventilation or of blood flow (so that ventilation-perfusion ratios are not identical throughout the lungs), the composition of alveolar gas in various parts of the lungs will vary, as will the Po_2 and Pco_2 of end-capillary blood emerging from those regions.

Regional changes in ventilation may depend upon 'normal' mechanisms, such as the normal gravitational gradient of increasing ventilation as one proceeds from apex to base of the erect lung. In disease, maldistribution of ventilation depends primarily on the presence of regions with airways obstruction or reduced distensibility. Maldistribution of pulmonary blood flow may also be 'normal', as in the gravitational gradient of increasing blood flow from apex to base of the vertical lung. Of much greater magnitude is the maldistribution induced in disease by such factors as pulmonary vascular occlusion.

If both local ventilation and local blood flow to a region of the lung are reduced to the same extent (by an accident of disease or by some compensatory mechanism), local ventilation-perfusion ratio may be normal, though both components of it are reduced. Alveolar and capillary gas tensions in that region will then be normal. But a functional deficit remains in that, because of the reduction of ventilation and blood flow, the total gas exchange per minute in that region will be reduced.

Consideration of ventilation-perfusion ratio at the level of a single alveolus, while useful conceptually, does not help much in the overall situation. At the level of the whole lung two aspects are of importance: the ventilation-perfusion ratio of the whole lung, and maldistribution of ventilation-perfusion ratios throughout the lung.

At the level of the whole lung, and assuming that the lung is homogeneous, the ventilation-perfusion ratio may be either increased above or decreased below normal. The commonest circumstances producing these changes are hyperventilation and hypoventilation of the lungs respectively. These will produce changes in arterial Po_2 and Pco_2 in the directions to be expected from the resultant change in total ventilation-perfusion ratio.

Far more common is the situation of maldistribution of ventilation-blood flow ratios throughout the lung. An understanding of this situation is central to the understanding of most examples of disturbed arterial blood gas tensions.

When the ventilation-perfusion ratio is different in different regions of the lung, clearly the gas tensions in those regions and in the end-capillary blood leaving them cannot be the same. In addition, there is no combination of total ventilation and total blood flow which can, in the presence of this non-homogeneity, produce ideal levels of $P\mathrm{o}_2$ and $P\mathrm{co}_2$ in end-capillary blood in all parts of the lung. The level of total ventilation, for example, which produces an ideal ventilation-perfusion ratio in one part of the lung may leave the ventilation-perfusion ratio of another part unduly low. If total ventilation is increased to bring the ventilation-perfusion ratio of the second part towards more normal levels, ventilation and hence the ventilation-perfusion ratio of the first part will be unduly high. The first point which emerges, then, is that ventilation and gas exchange in a non-homogeneous lung must perforce be less efficient than in homogeneous lung.

So far as the final arterial blood gas tensions are concerned, the influence of maldistribution of ventilation-perfusion ratios is somewhat different for oxygen and CO_2. Because the CO_2 dissociation curve is fairly linear, regions of high ventilation-perfusion ratio can compensate for areas of low ventilation-perfusion ratio and the combination of these regions in the lungs can produce a normal arterial $P\mathrm{co}_2$. The situation with regard to oxygen is more complex. A region of low ventilation-perfusion ratio in the lungs will be associated with blood of a low oxygen tension and saturation leaving that region and contributing to the mixed arterial blood. Another region of high ventilation-perfusion ratio cannot really compensate significantly; even if it produces a higher than usual oxygen tension in its own component of the blood flow, this will not be associated with any significant increase in the oxygen content or saturation of that blood. This is because the upper part of the tension-saturation curve for oxygen is essentially flat. As a result, the hypoxaemic blood from the region of low ventilation-perfusion ratio will continue to have an effect in the mixed arterial blood, which will have a lower than normal oxygen tension.

An extreme example of ventilation-perfusion ratio abnormality is seen when the ratio reaches zero. The situation is then one of right-to-left shunting of blood. Some right-to-left shunts are based on congenital abnormality of blood vessels, but the commonest source in practice arises from non-ventilated regions of lung. These may result from complete airways closure or obstruction, alveolar collapse, or from organization of alveolar contents. The situation may be temporary or permanent. The functional hallmark is arterial hypoxaemia, with failure of Po_2 to reach predicted levels when breathing oxygen. Unless there is other coexistent lung disease, CO_2 retention almost never results from a right-to-left shunt, for hyperventilation of remaining healthy lung is more than adequate to remove CO_2 not removed from the shunted blood.

Laboratory assessment

Laboratory assessment of the efficiency of gas exchange involves measurement of the arterial blood gas tensions, often combined with other measurements. In all such assessments it is important that the subject be in a so-called 'steady state'. In this state, the metabolic requirements and the level of ventilation remain constant from minute to minute with no acute changes of hyper- or hypoventilation. Its presence is best judged by a respiratory quotient between about 0·7 and 0·9 which does not change from minute to minute. If the subject is not in a steady state, changes of arterial Po_2 and Pco_2 may simply reflect acute hyper- or hypoventilation and, taken in isolation, may give quite misleading information about gas exchange. In practical terms, arterial puncture itself is a common cause of acute hyperventilation.

Once the subject is in a steady condition, the most useful test of gas exchange is the determination of the Po_2, Pco_2 and pH of the arterial blood. Arterial blood is obtained by percutaneous puncture of the brachial, radial, or femoral artery, and withdrawal of blood anaerobically into a syringe containing anticoagulant.

If the arterial blood gas tensions are abnormal, then clearly gas exchange is abnormal, and further investigations may be necessary to determine the nature of the abnormality. Breathing room air, several patterns of blood gas disturbance are worthy of brief note, though few such patterns, in the absence of other evidence, are diagnostic of a single functional abnormality.

A rise of Pco_2 accompanying a fall of Po_2 is classically seen in pulmonary hypoventilation. A similar situation may, however, be seen in some patients with maldistribution of ventilation-blood flow ratios throughout the lungs. The two conditions can be distinguished by further analysis. The combination of a low Po_2 with a low Pco_2 is specifically an index of ventilation-perfusion maldistribution throughout the be expected to accompany pregnancy. In healthy women during normal pregnancy, arterial puncture has not been regarded as justified, and data concerning blood gases in normal pregnancy are very scant indeed. It is, however, possible to approach some aspects indirectly. The measurement of diffusing capacity can, for example, be carried out without performing arterial puncture, and such studies, as well as some related to ventilation and oxygen consumption, have been carried out during pregnancy.

RESPIRATORY CONTROL

Control of ventilation is a complex topic which is still poorly understood. In grossly oversimplified terms, one may conceive of a respiratory centre in the brain stem which is subject to a number of input and integrative influences, and the output of which is directed to the motor nerves of the respiratory muscles. A number of factors influence the respiratory centre. Important among these are the influences of arterial blood Po_2, Pco_2 and pH. CO_2 acts largely by a direct effect on the medullary chemoreceptors, with increases of CO_2 stimulating, and decreases of CO_2 depressing the respiratory centre. Hypoxia acts almost wholly via peripheral chemoreceptors, and little stimulus to ventilation takes place above an arterial Po_2 of 60 mm Hg. Changes of arterial blood pH also act largely via the peripheral chemoreceptors, with little central effect until pH becomes very low indeed. Other neural imput to the respiratory centre includes impulses from higher centres, from proprioceptive receptors in the lungs and chest wall, and from adjacent medullary centres. Apart from these well-known factors it is likely that a large number of other chemical and neural factors may influence the centre.

One of the puzzling aspects of respiratory control is the nature of the controlled variable or variables. In a normal subject an increasing arterial Pco_2 is associated with an increasing level of ventilation. Broadly then it might be thought that the output of the respiratory centre and its conversion into respiratory muscle work was determined to make an appropriate compromise between the level of arterial Pco_2 and the work being done in increased ventilation. If a restrictive load is placed upon the lungs of a normal subject (as, for instance, by tight chest strapping) the change of ventilation accompanying a given change of Pco_2 is exactly the same as in the unrestricted subject. It is achieved by an increase in rate and a diminution in depth of breathing compared with the normal circumstance, but the total level of ventilation is unchanged, and the work required to achieve it is obviously increased (Thompson & Read 1967). In this situation then, the same CO_2 stimulus appears to produce a quite different work output under two different circumstances.

The outcome is, however, different if an obstructive load is imposed on a normal subject (as by having him breathe through a narrow resistance). The increase in ventilation for a given CO_2 stimulus is then less in the obstructed than in the control test, and the amount of respiratory work done under the two circumstances is roughly equal (Eldridge & Davis 1959). Even in this apparently simple experimental comparison, then, there does not appear to be any simple model of ventilatory response to a single stimulus such as CO_2.

The response of ventilation to CO_2 stimuli may be readily measured in the laboratory, and, indeed, this is the most easily assessed of the respiratory stimuli. The process has been much simplified by the use of the rebreathing technique described by Read (1967).

CHANGES IN RESPIRATORY FUNCTION DURING PREGNANCY

LUNG VOLUMES AND MECHANICS

Vital capacity

Though not a primary lung volume, vital capacity has been the parameter of respiratory function most frequently measured during pregnancy. The earliest data were obtained in 1854 by Wintrich who found no change in VC during pregnancy. The numerous studies of VC since that time have been reviewed by Prowse & Gaensler (1965). Many of these studies have reported no change in VC (e.g. Cugell et al 1953, Ihrman 1960, Krumholz et al 1963), but several careful studies have shown VC to increase during pregnancy (Thomson & Cohen 1938, Widlund 1945, Gazioglu et al 1970). Of studies in the past 40 years, the only ones to have shown a decrease in VC have been those of Alward (1930) and of Rubin et al (1956). Each of these studies was based only on a comparison of measurements made in the last 2 weeks of pregnancy with measurements made post-partum. Despite this, and for what seem to be largely intuitive reasons, VC has been commonly thought to decrease during pregnancy.

Thomson & Cohen (1938) made a careful serial study of VC in thirty-one normal pregnant women from the 13th week to 7 weeks post-partum. VC increased in twenty-two women, and over the whole group the mean increase was 200 ml. There was a sharp mean fall of 300 ml in the early post-partum period. When VC increased, the increase began by the 20th week of pregnancy, and was seen when VC was measured in the sitting, the lying or the standing position.

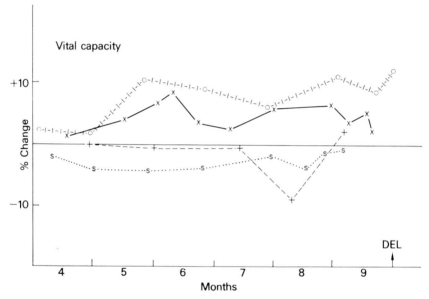

FIG. 14.2. Serial percentages changes in vital capacity in four subjects during pregnancy, compared with non-pregnant values on the same subjects. DEL = delivery.

The subcostal angle, as measured by the same authors, also showed a progressive increase in twenty-two out of twenty-five subjects. The mean changes were from 68° non-pregnant, to 80° by the 20th week, and to 104° by term. The increase in subcostal angle paralleled in time the increase in VC.

Serial studies of VC during pregnancy have been made in four normal subjects in the local study, and the results are shown in Fig. 14.2. VC showed no change in two subjects and an increase in two others. In one subject there was a slight decrease just before term, a decrease (in the last month) which has also been shown in other studies (Root & Root 1923, Enright *et al* 1935). This late fall may account for the change reported by Alward and by Rubin *et al*. In the supine position, VC, though lower than in the sitting position, showed the same changes.

Despite some apparent confusion, then, there seems little doubt that in a number of women there is a small but significant increase of VC during pregnancy. Possible reasons for this increase were discussed by Thomson & Cohen (1938). It is not simply an effect of training, for post-partum values have always been lower than those immediately before delivery. Thomson & Cohen suggested that the mobility of the thoracic cage is increased in pregnancy, and that this allows the larger VC, as well as the changes in subcostal angle. Since both these changes occurred well before any mechanical pressure from the enlarging uterus could play a role, they suggested that some hormonal factor might be responsible.

Functional residual capacity

Functional residual capacity (FRC) has been consistently shown to decrease in pregnancy. Cugell *et al* (1953) showed no significant change in FRC until the 6th month, followed by a mean decrease of 18%, maximal at term, in nineteen women in whom it was measured in the supine position. Similarly, in six subjects Gee *et al* (1967) found a decrease of 25% in FRC at the 8th month, and Gazioglu *et al* (1970) a fall of 12% in eight subjects at 36 weeks.

Fig. 14.3 shows the changes in FRC demonstrated by serial studies in four subjects in the local series. FRC decreased in all subjects, the maximum fall being 22%. The maximal decrease was not always at term but sometimes as early as the 7th month. The changes sometimes began as early as the 3rd month. Similar changes were obtained when FRC was measured in the supine position.

Total lung capacity and residual volume

Total lung capacity (TLC) has not been measured as frequently as VC or FRC, but in most cases it appears to remain unchanged or to decrease slightly in pregnancy. In one subject in the local series there was a definite increase in TLC at the time VC was maximally increased.

Residual volume (RV) has decreased during pregnancy in all studies where it has been measured (Krumholz *et al* 1964, Gazioglu *et al* 1970, Cugell *et al* 1953).

Reasons for the changes in the subdivisions of lung volume in pregnancy have not been investigated thoroughly. A decrease in FRC (which is fairly uniformly agreed upon) means that the elasticity of the chest wall, or of the lungs, or of both has changed. Gee *et al* (1967) measured

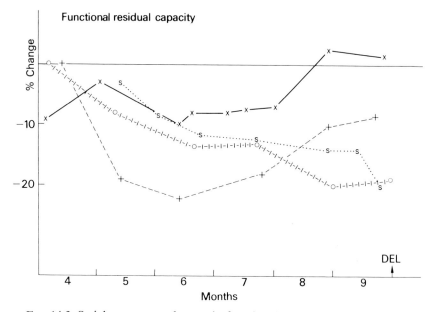

FIG. 14.3. Serial percentage changes in functional residual capacity in four subjects during pregnancy, compared with non-pregnant values in the same subjects. DEL, delivery.

the elasticity of the lungs during pregnancy and found it to be normal, and this was confirmed in two subjects in the local series. However, when the elasticity of the chest wall was measured in the local study, it had changed such that the outward recoil of the chest wall was reduced (Fig. 14.4) and the end-expiratory pressure in the lungs had become more positive. This seems to indicate that a change in the elasticity of the chest wall during pregnancy causes the fall of FRC.

The decrease in RV means either that the configuration of the chest wall has changed so that expiratory muscles can reduce the thoracic cavity to a smaller size, or that airways close at a lower lung volume than in the non-pregnant state.

Flow-resistive properties

Previous workers have shown no change in FEV_1 between the pregnant and the non-pregnant state (Cugell et al 1953, Gazioglu et al 1970,

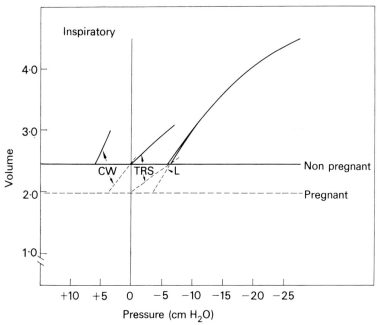

FIG. 14.4. Pressure volume curves for lung (L), chest wall (CW), and total respiratory system (TRS) in one subject, pregnant and non-pregnant. The continuous lines show non-pregnant values, the broken lines values during pregnancy (the two lung lines overlap). The two horizontal lines indicate the pregnant and non-pregnant levels of FRC. Volume in litres.

Rubin et al 1956). Of the four subjects in the local series, FEV_1 increased by about 6% in the two who showed a rise in vital capacity.

Gee et al (1967) showed a decrease in airways resistance from 1·8 to 0·9 cm of water/litre/second in the pregnant state and Rubin et al (1956) showed changes in pressure-flow relationships which also suggest a decrease in airways resistance. In the local series, airways resistance was measured in only two subjects, and found to decrease, the decrease being maximal by the end of the first trimester.

Goodland & Pommerenke (1952) have documented in detail the fall in alveolar $P\text{co}_2$ which follows ovulation. The size of the fall varies in different subjects, but among nine subjects these authors found an average fall of nearly 4 mm Hg. They and others (Goodland et al 1953) subsequently showed that a similar hyperventilation and fall of alveolar $P\text{co}_2$ could be induced even in males by the injection of 50 mg progesterone; and Lyons & Antonio (1959) found that the same drug reproduced in males the changes in respiratory centre threshold and sensitivity that they had demonstrated in pregnancy. Intramuscular progesterone also causes a fall of arterial $P\text{co}_2$ in patients with emphysema and hypercapnia (Tyler 1960), but analogues of progesterone do not show this activity. Whether progesterone accounts for the whole of the phenomenon of hyperventilation in pregnancy is still open to question.

DYSPNOEA IN PREGNANCY

Shortness of breath is a very common complaint of pregnant women. In orthodox terms no very convincing explanation of the phenomenon has been adduced. In modern terms dyspnoea is thought to arise when the work of the respiratory muscles during breathing reaches such a level as to be regarded by the subject as inappropriate to the circumstances. In the presence of the hyperventilation referred to above, the work of breathing in pregnancy is certainly increased. In addition, Bader et al (1959) have shown that the oxygen cost of each litre of ventilation is increased in pregnant compared with non-pregnant females. However, the increase is not of a magnitude that would lead one to expect dyspnoea on this basis alone.

So-called explanations, based on a change in the ratio of minute ventilation to maximum breathing capacity, provide no basis for the sensation of dyspnoea; and the mechanism of this common symptom in pregnancy must be regarded as remaining obscure.

CHANGES IN RESPIRATORY FUNCTION IN PREGNANCY ASSOCIATED WITH DISEASE

There are a limited number of studies of respiratory function in pregnancy among woman with pre-existing disease of heart or lungs. Gaensler et al (1953) studied seven patients with severe restrictive disorders due to preceding tuberculosis, or collapse or resection therapy.

There were two important differences compared with normal pregnant women. First, hyperventilation during pregnancy was less marked the greater the degree of lung restriction. Secondly, the increased ventilatory requirement was met in part by an increase in respiratory frequency, whilst in normal subjects it is met almost wholly by an increase in tidal volume.

Gazioglu et al (1970) studied eight patients with valvular heart disease and eight patients with chronic pulmonary disease during pregnancy. In the patients with valvular heart disease the increase in $\dot{V}_E$ late in pregnancy was greater than in normal subjects. The four patients with obstructive lung disease showed some increase in airways obstruction, though this was not associated with any clinical deterioration. The four patients with sarcoidosis showed no significant differences from the normal controls.

Though these studies are few in number, they and general experience seem to indicate that the respiratory system in general copes well with the stresses of pregnancy. Certainly the hazards do not appear comparable with those associated with disorders of the heart or kidneys.

GLOSSARY

A-a : alveolar-arterial (as applied to the difference in oxygen tensions between the two).
BSA : body surface area.
DL_{CO} : diffusing capacity of the lung measured with carbon monoxide.
ERV : expiratory reserve volume.
FEV_1 : forced expiratory volume in 1 second.
FRC : functional residual capacity.
IC : inspiratory capacity.
P_{CO_2} : partial pressure (tension) of carbon dioxide.
P_{O_2} : partial pressure (tension) of oxygen.
RV : residual volume.
TLC : total lung capacity.
VC : vital capacity.
$\dot{V}_A/\dot{Q}_c$: ventilation-blood flow ratio.
V_T : tidal volume.
V_D : dead space volume.
V_D/V_T : dead space/tidal volume ratio.
$\dot{V}_E$: minute ventilation.
$\dot{V}_{O_2}$: oxygen consumption per minute.

REFERENCES

ALWARD H.C. (1930) Observations on the vital capacity during the last month of pregnancy and the puerperium. *Amer. J. Obstet. Gynec.* **20**, 373–381.

BADER R.A., BADER M.E. & ROSE D.J. (1959) The oxygen cost of breathing in dyspnoeic subjects as studied in normal pregnant women. *Clin. Sci.* **18**, 223–235.

CUGELL D.W., FRANK R., GAENSLER E.A. & BADGER T.L. (1953) Pulmonary function in pregnancy. I. Serial observations in normal women. *Amer. Rev. Tuberc.* **67**, 568–597.

ELDRIDGE F. & DAVIS J.M. (1959) Effect of mechanical factors on respiratory work and ventilatory response to carbon dioxide. *J. Appl. Physiol.* **14**, 721–726.

ENRIGHT L., COLE V.V. & HITCHCOCK F.A. (1935) Basal metabolism and iodine excretion during pregnancy. *Amer. J. Physiol.* **113**, 221–228.

GAENSLER E.A., PATTON W.E., VERSTRAETEN J.M. & BADGER T.L. (1953) Pulmonary function in pregnancy. III. Serial observations in patients with pulmonary insufficiency. *Amer. Rev. Tuberc.* **67**, 779–797.

GAZIOGLU K., KALTREIDER N.L., ROSEN M. & YU P.N. (1970) Pulmonary function during pregnancy in normal women and in patients with cardiopulmonary disease. *Thorax* **25**, 445–450.

GEE J.B.L., PACKER B.S., MILLEN J.E. & ROBIN E.D. (1967) Pulmonary mechanics during pregnancy. *J. clin. Invest.* **46**, 945–952.

GOODLAND R.L. & POMMERENKE W.T. (1952) Cyclic fluctuations of the alveolar carbon dioxide tensions during the normal menstrual cycle. *Fertil. & Steril.* **3**, 394–401.

GOODLAND R.L., REYNOLDS J.G., MCCOORD A.B. & POMMERENKE W.T. (1953) Respiratory and electrolyte effects induced by estrogen and progesterone. *Fertil. & Steril.* **4**, 300–316.

GUZMAN C.A., CAPLAN R. & BECKLAKE M.R. (1969) Cardiorespiratory response to exercise during pregnancy. *Federation Proc.* **28**, 786.

HASSELBALCH K.A. (1912) Ein Beitrag zur Respirationphysiologie der Gravidität. *Skand. Arch. Physiol.* **27**, 1. Cited by PROWSE & GAENSLER (1965), q.v.

IHRMAN K. (1960) A clinical and physiological study of pregnancy in a material from Northern Sweden. III. Vital capacity and maximal breathing capacity during and after pregnancy. *Acta Soc. Med. Upsal.* **65**, 147–154.

KRUMHOLZ R.A., ECHT C.R. & ROSS J.C. (1963) Pulmonary diffusing capacity, capillary blood volume, lung volumes, and mechanics of ventilation in early and late pregnancy. *J. Lab. Clin. Med.* **63**, 648–655.

LYONS H.A. & ANTONIO R. (1959) The sensitivity of the respiratory center in pregnancy and after the administration of progesterone. *Assoc. Amer. Phys. Trans.* **72**, 173–180.

PLASS E.D. & OBERST F.W. (1938) Respiration and pulmonary ventilation in normal nonpregnant, pregnant, and puerperal women. *Amer. J. Obstet. Gynec.* **35**, 441–449.

PROWSE C.M. & GAENSLER E.A. (1965) Respiratory and acid-base changes during pregnancy. *Anaesthesiology* **26**, 351–392.

READ D.J.C. (1967) A clinical method for assessing the ventilatory response to carbon dioxide. *Aust. Ann. Med.* **16**, 20–32.

ROOT H.F. & ROOT H.K. (1923) The basal metabolism during pregnancy and the puerperium. *Arch. Int. Med.* **32**, 411–424.

RÖSSIER P.H. & HOTZ M. (1953) Respiratorische Funktion und Säurebasengleichgewicht in der Schwangerschaft. *Schweiz Med. Wschr.* **83**, 897–901.

RUBIN A., RUSSO N. & GOUCHER D. (1956) The effect of pregnancy upon pulmonary function in normal women. *Amer. J. Obstet. Gynec.* **72**, 963–969.

THOMPSON J.F. & READ D.J.C. (1967) Ventilatory responses to chest strapping and negative intrapulmonary pressures during progressive hypercapnia in conscious man. *Aust. J. Exp. Biol. Med. Sci.* **45**, 18P.

THOMPSON K.J. & COHEN M.E. (1938) Studies on the circulation in pregnancy. II. Vital capacity observations in normal pregnant women. *Surg. Gynec. Obstet.* **66**, 591–603.

TYLER J.M. (1960) The effect of progesterone on the respiration of patients with emphysema and hypercapnia. *J. clin. Invest.* **39**, 34–41.

WIDLUND G. (1945) The cardio-pulmonal function during pregnancy. *Acta obstet. gynec. scand.* **25**, Suppl. 1, 1–125.

WINTRICH M.A. (1854) Einleitung zur Darstellung der Krankheiten der Respirationsorgane. *Virchows Handb. Spec. Path. Ther.* **5**, 101. Cited by PROWSE & GAENSLER (1965) q.v.

15

Haemopoiesis and Coagulation

P.A.CASTALDI & D.R.HOCKING

ERYTHROPOIESIS

In the healthy adult, erythropoiesis takes place in the red bone marrow and studies by electron microscopy have shown that haemopoiesis takes place outside the marrow sinusoids which are filled with blood (Pease 1956). Red cells develop from the primitive reticulum cell through the stages of pronormoblast, basophilic, polychromatic and orthochromatic normoblasts to the reticulocyte and finally to mature red cells which gain access to the intravascular compartment through the endothelium of the sinusoids by a process of diapedesis (Stohlman 1960).

The major function of red cells is to provide a vehicle for the synthesis, transport and protection of haemoglobin and the fundamental stimulus to erythropoiesis is hypoxia (Grant & Root 1952). The rate of red cell production determines the size of the red cell mass, which in turn determines the haemoglobin concentration, the haemoglobin concentration determines the degree of tissue oxygenation and the degree of tissue oxygenation determines the rate of red cell production. The effect of tissue oxygen is mediated via erythropoietin which was described first by Carnot & Deflandre (1906). This substance is considered to be a hormone due to the fact that it is found in the circulating blood and acts on marrow while being formed in some other site.

Results of experimental work show that erythropoietin controls the normal production of red cells as well as the increased production in anaemia and in anoxia. The rate of erythropoietin production is determined by the relationship of availability of oxygen to required oxygen in with a predilection for Mediterranean and Negro subjects (Beutler 1970). In populations of this nature glucose 6-phosphate dehydrogenase deficiency is a common cause of anaemia in pregnancy and the neonate.

The function of red cells is to convey haemoglobin for the purpose of oxygen transport and the biconcave disc shape of the cell presents a large surface area for gaseous exchange. Oxygen combines with haemoglobin to form oxyhaemoglobin in situations of low Po_2 and each 1·0 g of haemoglobin combines with 1·34 ml of oxygen.

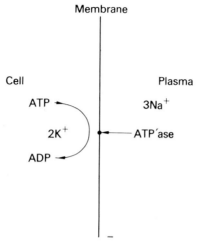

FIG. 15.1. Energy requirements for Na, K, differential across the red cell membrane.

Regulation of oxygen affinity for haemoglobin is through the action of 2,3-diphosphoglycerate (DPG) which is present in considerable amounts in red blood cells. Its action is by complexing haemoglobin in the deoxygenated state and this inhibits oxygen binding but facilitates its release at physiologic oxygen tensions (Benesch et al 1968). The level of DPG in the red cells thus influences oxygen release to tissues. It has been shown that DPG levels rise in high altitudes and that the oxygen–haemoglobin dissociation curve is shifted to the right. Further it has been shown that the affinity of DPG for haemoglobin F is less than for haemoglobin A thus facilitating the transport of oxygen from mother to foetus (Tyuma & Shimizu 1969). These observations are leading currently into investigation of the DPG state in several haemoglobins with high oxygen affinity producing polycythaemia. There are cases reported of inherited defects of red cell enzymes in which alterations of DPG level are found with consequent shifts of the oxygen–haemoglobin dissociation curves (Delivoria-Papadopoulos *et al* 1969).

The life span of the red cell in the peripheral circulation is approximately 120 days and the survival curves of random samples demonstrate an arithmetic type of disappearance over this period, the major determinant of red cell removal being senescence. There are many different methods employed to assess survival times (Berlin *et al* 1959), including Ashby's technique of differential agglutination and labelling *in vivo* and *in vitro* with a number of radio isotopes of which Cr^{51} is the most common (Dacie & Lewis 1968b).

Once the red cell is mature, no further protein or lipid synthesis occurs, and from this point the results of ageing may be divided into two categories. First a direct effect attributable to the decline in amount or activity of proteins, lipids and other substances not synthesized in the adult red cell; secondly, an indirect effect of these changes especially related to glycolysis. The first enzyme in the Pentose Phosphate pathway, glucose 6-phosphate dehydrogenase, declines as the cell ages (Marks *et al* 1958), and as the first two steps of this pathway provide the only sources of the coenzyme NADPH, the supply of this also reduces with ageing. The activity of the enzyme glyceraldehyde 3-phosphate dehydrogenase, necessary for the Embden-Meyerhof pathway of glycolysis, also diminished with ageing. As this enzyme is the only source of NADP and since it precedes the production of ATP in the cycle it has a profound effect in energy production. Other enzymes in the glycolytic pathways have been shown to decline with ageing and contribute to decreased energy production. Hence with the diminution of ATP and therefore energy production, it is probable that intracellular K^+ decreases, Na^+ increases, the cell takes in more water and becomes osmotically more fragile.

Ageing also affects the lipid of the red cell membrane and structural changes have been demonstrated by both conventional and scanning electron microscopy.

The actual point of disruption of the red cell at the end of its life span has never been demonstrated. Destruction may take place either intravascularly or far more commonly, extravascularly, within the confines of the spleen, liver or bone marrow. The sites and mechanism of destruction of red cells has been investigated extensively. It appears likely that the site depends on the degree rather than the mechanism of red cell damage, at least in haemolytic states (Wagner *et al* 1962). The spleen has been accepted as the major site of red cell destruction. Although this may be so in haemolytic states there is no evidence that in health it is any more active in this regard than other organs. Red blood cells which are abnormal, such as spherocytes or antibody treated cells, pool in the spleen and may be destroyed prematurely; diseases producing splenomegaly may result in splenic destruction of normal red cells. How this occurs is not clear (Prankerd 1965). There is evidence that red cells are removed at the end of their normal life span, *in vivo*, by phagocytosis within the bone marrow but the mechanism by which a phagocyte recognizes an aged cell is unknown (Bessis 1965).

The requirements for red cell production include a great variety of substances many of which are necessary for the maintenance of other body tissues as well, but there are many substances which are critical to erythropoiesis and without them detectable changes occur in the blood. These include protein and amino acids, iron and other minerals, folic acid, vitamin B_{12}, vitamin B_6, other B group vitamins, vitamin C and fat soluble vitamins.

Adequate dietary protein intake is required to provide amino acids for globin synthesis but deficiency must be gross to impair haemoglobin formation. This is commonly seen in some tropical zones, and in these areas is a frequent cause of anaemia of pregnancy (Menon & Rajan 1962). Maximal globin production occurs in the nucleated red cell in the bone marrow, a small amount takes place in the reticulocyte and none occurs in the mature red cell. Incubation of amino acids with reticulocytes has shown that they enter the cell, become bound to RNA granules and subsequently appear in peptide chains in the supporting medium (Dintzis *et al* 1958). The presumption is then, that per medium of the erythroblast RNA the amino acids are built into globin.

IRON

Iron is required for haem biosynthesis for incorporation into haemoglobin and haem enzymes (Bothwell 1968). Its absorption involves several stages. Ingested food iron is processed in the gut lumen to render it suitable for absorption. It enters the mucosal cells and is bound to a plasma β-globin called transferrin of which there are at least fifteen known genetic variants. All are identical in their iron binding function and each molecule is able to bind two atoms of iron until they are released at receptor sites on the surface of a red cell precursor. A second mechanism by which iron is made available is from senescent red cells. The phagocytosis of erythrocytes by reticulo-endothelial cells in the bone marrow has been described. Ferritin resulting from this breakdown collects in the reticulo-endothelial cells and may be transferred to developing normoblasts by a mechanism called ropheocytosis (Bessis & Breton-Gorius 1962).

The mucosal cells of the intestine take up iron (Bothwell & Finch 1962), and under normal circumstances absorption occurs in the duodenum and upper ileum probably related to the fact that pH and redox potential are optimal in those areas. Fe^{++} iron is absorbed better than Fe^{+++}. Phytates which form insoluble iron salts depress, and reducing agents, such as vitamin C, succinic acid and cysteine, increase iron absorption. Fructose potentiates absorption and this may relate to its ability to form a stable chelate with iron (Davis & Deller 1967).

Hydrochloric acid potentiates the absorption of ferric iron and iron deficient patients with histamine-fast achlorhydria absorb iron less efficiently (Jacobs *et al* 1964).

There is a factor described in normal gastric juice which binds iron and is markedly reduced in idiopathic haemochromatosis (Davis *et al* 1966), but there is no evidence that this plays a part in iron absorption. Pancreatic secretion was suggested to influence iron absorption when it was noted that patients with chronic pancreatitis may absorb iron excessively (Davis & Badenoch 1962).

Little information is available on the absorption of iron from various foods and there is great variation of availability; for example the iron in eggs and leaf vegetables is poorly absorbed, while muscle iron is well absorbed (Moore 1965). Haemoglobin is a good source of iron while that of ferritin is unavailable to the body (Hussain *et al* 1965).

Once iron is available for absorption it is taken up by the mucosal cells and the actual quantities passed to the plasma depend on several factors. Absorption is increased in iron deficient anaemia so that as iron stores are decreased absorption increases and as stores increase absorption decreases. Consequently iron absorption is regulated at least in part by the iron stores, and the rate of erythropoiesis also influences iron absorption (Bothwell *et al* 1958). Stimulation of erythropoiesis results in an increase of iron absorption and vice versa and this explains the iron overload often observed in chronic haemolytic anaemias. The actual mechanism of transfer of iron across the bowel mucosa is obscure but it is in part an active metabolic process (Jacobs *et al* 1966), a proportion of the iron being taken up by the mucosal cell and delivered rapidly to plasma transferrin and the remainder deposited as ferritin within the cell. In states requiring increased iron absorption little ferritin is deposited and the iron is transferred rapidly through the cell. Therefore it appears that ferritin acts as the mechanism modulating iron absorption.

The term 'mucosal block' was coined by Hahn in 1943 and used by Granich in the late 1940s to explain iron absorption regulation by intestinal mucosal cells. The hypothesis proposed the presence of an intestinal receptor which was physiologically saturated in normal situations. The degree of saturation of the iron receptor depended on the amount of iron in the mucosal cell which in turn depended on the overall pattern of iron storage in the body. Iron deficiency was therefore associated with desaturation of the receptor and more iron was absorbed. The hypothesis was criticized because large doses of iron failed to inhibit the absorption of a following test dose. The absorptive capacity of each mucosal cell is determined while it is being formed in the Crypt of Lieberkühn, after which they migrate out to the tips of the villi as they mature and are thence exfoliated. This explains the 24-hour lag in depression of iron absorption that occurs after the plasma iron level is raised artificially. The cells in the crypts are conditioned to absorb less iron by high plasma levels but this does not influence absorption until they have migrated to a functional position on the villi (Conrad *et al* 1964). The exfoliation of mucosal cells represent one avenue of iron excretion.

The normal daily diet contains 10–15 mg of iron of which 5–10% is absorbed. An adult requires 3–5 g of elemental iron for transporting oxygen from the lungs and carbon dioxide from the tissues, storing oxygen in muscle, utilizing oxygen in cellular oxidation and maintaining an adequate iron store. This is distributed as haemoglobin 65%, myoglobin 3%, storage-ferritin and haemosiderin 25%, plasma and extracellular fluids and in haem enzymes one percent and one percent unaccountable.

A normal adult male requires approximately 1 mg iron daily to remain in balance. The approximate daily loss is 0·6 mg from the gastrointestinal tract, 0·1–0·2 mg in the urine and 0·1–0·2 mg in sweat, desquamation of skin and in mucosal cells. In infancy and puberty and during menstruation an additional 1 mg per day is required (Pollycove

1966). Approximately 2·7 mg per day is required during the last two trimesters of pregnancy to meet foetal requirements and to compensate for blood loss at parturition (Kynes 1968), and 0·8 mg per day is required to balance loss during lactation.

If adequate iron is available during pregnancy the haemoglobin at term usually approximates the level in early pregnancy, but if iron intake is not adequate then the haemoglobin level is lower at term and may remain so for many months post-partum. This arises from increased demands for iron due to the normal iron loss, expansion of the red cell mass (approximately 25% as previously discussed), transfer of iron to the foetus (approximately 200–400 mg), the iron content of the cord and placenta (approximately 50–150 mg) and blood loss at delivery. In addition to these factors lactation over a period of approximately 6 months accounts for another 100–200 mg. On the other hand a saving of 250–450 mg occurs due to the period of amenorrhea. Adequate iron intake in pregnancy is not easy to attain, because as shown above, 700–1,400 mg of iron is required over the period of pregnancy and lactation additional to the normal requirement of 1·0 mg per day. This amounts to 2·5 mg of iron per day, when at best 4·0 mg is able to be absorbed from a daily diet containing 10–15 mg of iron (Chanarin *et al* 1968a). It is not surprising therefore that iron deficiency anaemia is extremely common in pregnant women not receiving iron supplements (Hocking & Castaldi 1971).

Other minerals, notably copper and cobalt have a role in erythropoiesis (Eluejem 1935, Marston 1959), but their role in man is ill-defined apart from the presence of cobalt as a component of vitamin B_{12}.

FOLIC ACID

Wills in 1932 demonstrated the haematological response to an autolysed yeast extract and to crude liver extracts, in Hindu women with megaloblastic anaemia. The response did not occur with purified liver extract and was therefore due to a factor other than vitamin B_{12}.

The first description of the clinical syndrome due to folic acid deficiency in pregnancy was by Channing in 1842. Snell & Peterson (1939) found an active factor in liver and green plants and named it folic acid.

Folic acid is obtained in the diet from fresh green vegetables as well as parenchymatous organs such as liver and kidneys. It is of interest that the Nutritional Education Section of the Health Department of

Australia gives the level of folic acid in raw fresh green peas and in fresh frozen green peas as 2×10^4 ng/100 g whereas canned green peas contain only 2×10^3 ng/100 g. There are widely different normal ranges reported by different authors for serum folate levels. Dacie and Lewis give a range 6–21 ng/ml (Dacie & Lewis 1968b), and Spray 2·1–28 ng/ml (Spray 1964).

The folic acid molecule consists of a pteridine ring, para-amino benzoic acid and L-glutamic acid and is known as pteroylglutamic acid. This is biochemically inactive and must be reduced to tetrahydrofolic acid to be enzymically active. Tetrahydrofolic acid functions as a co-enzyme in purine and pyrimidine metabolism and in the biosynthesis of thymine and methionine and hence DNA synthesis.

The principal biochemical reactions in which folate takes part include protein synthesis, in which the interconversion of serine to glycine, the interconversion of histidine to glutamic acid and methionine biosynthesis are of major importance, and nucleic acid synthesis, contributed to by the synthesis of purine, the catabolism of purine, and the synthesis of pyrimidines. There is also an association between iron and folate metabolisms which is poorly understood but has been shown to be important in iron deficiency anaemias of pregnancy in which there is a higher incidence of folate deficiency (Chanarin et al 1965).

Folic acid is absorbed from the proximal part of the small intestine. Deficiency of folic acid is commonly due to a poor diet and this is almost invariably the cause in megaloblastic anaemia associated with pregnancy, infancy and alcoholism. Deficiency may occur also due to lack of absorption from the gut in syndromes such as sprue, idiopathic steatorrhea and coeliac disease, as well as in diverticulae and blind loops of the small bowel. Excessive utilization and hence a relative lack may occur in chronic haemolytic anaemias. Some drugs act as folate antagonists and impair or destroy its action by blocking the enzyme folate reductase, for example Methotrexate. The megaloblastic anaemia resulting from long term use of anticonvulsant drugs is complex and may result from inhibition of some of the enzyme systems concerned with folate metabolism (Reynolds et al 1966), and from inhibition of intestinal mucosal conjugases (Rosenberg et al 1969). The classical findings are anaemia, with oval macrocytosis of the red cells, with anisocytosis and poikilocytosis. There may be neutropenia with the presence of hypersegmented neutrophils and thrombocytopenia. In the early stages the patient may not be anaemic and suspicion of the possibility of deficiency may arise by the presence of hypersegmented neutrophils as the only abnormality.

The bone marrow in pregnancy shows normally an increased cellularity but the presence of megaloblasts and 'giant' metamyelocytes is diagnostic of megaloblastic marrow.

The association of iron deficiency is not uncommon with the folate induced megaloblastic anaemia of pregnancy and in many cases the morphological abnormality does not become obvious until the iron deficiency is treated.

In the laboratory, folate levels are reduced below 3·5 ng/ml in the serum and 160 ng/ml in the red cells (Chanarin *et al* 1968b). Both of these techniques employ the micro-organism *Lactobacillus casei* as a biological assay.

In pregnancy there is a progressive decrease in serum folate levels towards term (Chanarin *et al* 1965) and in megaloblastic anaemia of pregnancy the level is generally lower than that in the folate deficient anaemias of the non-pregnant (Ball & Giles 1964). The association of a low serum folate and megaloblastic marrow must be demonstrated before a firm diagnosis can be made. Red cell folate activity accounts for 95% of the blood folate and in pregnancy the levels fall progressively from the first to the third trimester (Chanarin *et al* 1968b) and red cell folate levels of less than 80 ng/ml are frequently associated with megaloblastic bone marrow. But again, as in serum levels, low levels are not diagnostic of megaloblastic anaemia of pregnancy and must be associated with megaloblastic bone marrow for diagnosis.

Other laboratory techniques to demonstrate folate deficiency include assessment of the excretion of formiminoglutamic acid in the urine (FIGLU). Formiminoglutamic acid is formed during the metabolism of histidine. Fig. 15.2 shows the metabolic sequence. If either deficiency of folate occurs or there is some interference with the enzyme formimino transferase during treatment with folate antagonists, histidine is not metabolized completely to glutamic acid and formiminoglutamic acid collects in the urine. This can be assayed by many techniques, of which urine electrophoresis is the most common (Kohn *et al* 1961). Thin layer chromatography is frequently used in some laboratories (Roberts & Mohamed 1965). The value of the test is open to doubt as some series show the presence of FIGLU in the urine of patients with megaloblastic anaemia caused by vitamine B_{12} deficiency (Villamil & McCracken 1963), and this is associated with metabolic disturbance of folate due to the relationship with vitamin B_{12}. It is claimed that the basic cause of megaloblastic bone marrow is the lack of tetrahydro-pteroly-glutamic acid and that this may be brought about by either folate or B_{12} deficiency. The latter blocks the transformation of N5 tetrahydrofolate to tetrahydrofolate (Fig. 15.3). In addition, the test has been found to be unreliable in pregnancy due to altered absorption and excretion of histidine, as well as its transfer to the foetus (Chanarin *et al* 1963).

The body stores folate mainly in the liver to a total of 6–10 mg. Deficiency, with megaloblastic marrow, develops from a normal baseline

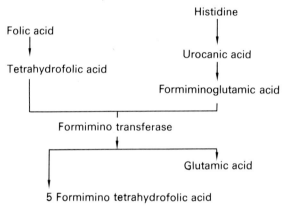

FIG. 15.2. Metabolic sequences in the FIGLU test.

level in approximately $4\frac{1}{2}$ months on a folate free diet (Herbert 1962a) and the minimal daily adult requirement is in the region of 50 μg (Herbert 1962b). Little is known about folate requirements in pregnancy but it has been found that red cell folate levels are maintained in women receiving supplementary folic acid, 100 μg daily, whereas in a group receiving no supplements the red cell folate levels fell slowly from 15 weeks to term (Chanarin *et al* 1968a). In practice the folate supplements used vary from 100 μg up to 15 mg per day. A possible reason for this gross variation may be the different populations surveyed with marked socio-economic differences. Another reason is that folic acid supplementation was first used before the physiological requirements were known.

The incidence of folic acid deficiency in pregnancy varies in different reports due to the type of population surveyed and also to the technique used to assess deficiency. One comprehensive survey of the literature gives a variation of incidence from 0·5% in a report in 1951 to 30% in a report in 1968 (Rothman 1970). The level of red cell folate is probably the most accurate method for determining deficiency. Serum folate level is the most popular technique but is not as reliable, due to the small amounts to be assayed when compared with the first method.

The incidence of megaloblastic anaemia of pregnancy due to folate deficiency again varies greatly with the type of population surveyed and appears to be higher in those countries where diet is poor. However there are conflicting reports from the same area. In Liverpool in 1957 one survey gave the incidence as 0·6% while another group from the same area in 1964 gave the incidence as 5·4% (Forshaw et al 1957, Hibbard 1964). In Nigeria in 1968 the incidence was given at 30% (Fleming et al 1968). Assessment of the many surveys available does show however, that using folate assays and bone marrow examination the incidence approaches 15–25%.

$$N5\text{-Methyl tetrahydrofolate} + \text{Homocysteine} \rightarrow \text{Methionine} + \text{Tetrahydrofolate}$$
$$B_{12} \text{ Coenzyme dependent}$$

FIG. 15.3. Inter-relationships of vitamin B_{12} and folic acid.

Many patients have no early symptoms and it has been reported that there is a slowly progressive anaemia which has failed to respond to iron therapy, and associated with low folate levels in the third trimester (Chanarin et al 1959). The major factor causing folate deficiency in pregnancy is dietary inadequacy. But foetal requirements are maintained even at the expense of the mother so that they are, at least, contributory. This is demonstrated by the observation that the folate level in cord blood is five times that in the maternal blood (Giles 1966).

There is some evidence to suggest that malabsorption of folic acid may occur in pregnancy (Chanarin et al 1959).

It has been suggested that folate deficiency is associated with an increased incidence of abnormal pregnancies including placental abruption, abortion (Streiff & Little 1965), foetal malformations (Scott et al 1970), prematurity, stillbirth and neonatal death (Rothman 1970). However, while the various authors quoted show no direct relationship, the association of hypertension, oedema and proteinuria with folate deficiency has been well demonstrated (Gatenby & Lillie 1960).

VITAMIN B_{12}

The form of vitamin B_{12} originally isolated was cyanocobalamin. The molecular structure of this compound consists of a corrinoid ring surrounding cobalt and a nucleotide group containing a base, 5,6-dimethyl benzimidazole. A cyanide group is carried by the trivalent cobalt atom which is also linked to the benzimidazole base. Recent work has suggested that naturally occurring compounds are chemically different from cyanocobalamin and that only these forms are biologically active. These compounds are termed cobamide coenzymes, three of which have been characterized in detail and are adenyl cobamide, benzimidazolyl cobamide and dimethyl benzimidazolyl cobamide.

Biochemically, the cobamide coenzymes participate with folic acid in methionine synthesis from homocysteine, in glutamate-isomerase reactions, in methyl malonyl-CoA-isomerase reactions and in the conversion of diols to desoxy aldehydes (Harris 1963). They are assumed, but with less evidence, to participate in the transformation of RNA to DNA (Weissbach & Dickerman 1965). Apart from the haematological manifestations, symptoms of deficiency may result from systemic, gastrointestinal and neurological disturbance. The cause of the deficiency may be dietary but this is extremely uncommon except in groups who exclude vitamin B_{12} from their diet, for example Vegans (Smith 1962).

The classic study of Castle established the nature of the gastric defect and he fed patients meat that had been predigested in a normal stomach (Castle 1960). This was followed by achieving a haematological response to a mixture of beef muscle and normal human gastric juice and led to the hypothesis of an extrinsic factor in the beef muscle and an intrinsic factor in the stomach. Liver was found to be a very active source of extrinsic factor and in 1948 vitamin B_{12} was isolated from crude liver extracts. This substance was found to be effective if given orally with normal human gastric juice as a source of intrinsic factor. Intrinsic factor has not been defined as accurately as extrinsic factor. It is a mucoprotein with a molecular weight of 50,000 and is stated to be produced by gastric parietal cells (Hoedmacker *et al* 1964). It has the ability to bind vitamin B_{12} and the IF-vitamin B_{12} complex is taken up by specific receptors in the microvilli of the distal ileum. Pathology of this region may be associated with vitamin B_{12} deficiency due to interference with absorption (Hocking & Carter 1967).

Little is known of the mechanism of access to the plasma across the mucosal cell but once in the plasma vitamin B_{12} is taken up by a β globulin, transcobalamin II and transported to the tissues (Hall & Finkler 1965). Transcobalamin I is another plasma vitamin B_{12} binding protein and is an α globulin. Its role is unknown, but vitamin B_{12} storage has been suggested as one function and the plasma level has been found to be raised in myeloproliferative disorders. Elevated transcobalamin II levels are found in pregnancy (Herbert 1968).

A third vitamin B_{12} binding protein is present in the cord blood of neonates and may relate to foetal or transplacental transport (Kumento *et al* 1967).

Vitamin B_{12} levels tend to be lower in pregnancy than in non-pregnant controls. The level falls to a minimum at 16 to 20 weeks but this does not indicate a true vitamin B_{12} deficiency (Temperly *et al* 1968). The low levels of B_{12} are generally associated with low serum folate and return to normal after treatment with folic acid (Giles 1966). There is no increased incidence of vitamin B_{12} deficiency causing megaloblastic anaemia of pregnancy, when compared with the non-pregnant population. In fact B_{12} deficiency in pregnancy is rare (Ball & Giles 1964). Vitamin B_{12} deficient women are unlikely to become pregnant (Jackson *et al* 1967) and there are many reported cases of infertile women who conceived after treatment of their deficiency (Adams 1956). Absorption of vitamin B_{12} is normal in pregnancy (Badenoch *et al* 1955), but deficiency may be associated with malabsorption syndromes (Baker *et al* 1962). Some women with mild malabsorption become pregnant and Baker (1962) studied seven such cases. All had moderate to severe anaemia with low vitamin B_{12} levels and breast milk levels were similar to serum levels. The infants born to these women had no anaemia and serum B_{12} levels were higher than in the mothers.

OTHER VITAMINS

Vitamin B_6 is the name applied to a group of naturally occurring compounds including pyridoxine, pyridoxal and pyridoxamine. The coenzyme is pyridoxal-5-phosphate and this is required in the biosynthesis of haemoglobin. Deficiency interferes with the synthesis of porphyrin and so iron is not used for erythropoiesis. Vitamin B_6 has many other functions, most of which involve the metabolism of amino acids. Pyridoxal-5-phosphate is a coenzyme for amino-acid decarboxylases,

HAEMOSTASIS

Current views about haemostasis assign a prim[...] emphasize the importance of the vessel, plasma [...] lytic factors. With growing emphasis on the s[...] atherosclerosis, it has become increasingly evi[...] factors can any longer be considered as isolate[...] been increasing interest in the study of the dyr[...] stasis with special reference to flow and the i[...] viscosity and thrombus formation (Dintenfass [...] search for an understanding of these variable [...] years in a clear definition of the nature of the m[...] acquired disorders of haemostasis and the ap[...] factory and reproducible methods to the study [...] logical and pathological states associated with [...] static mechanism or a tendency to thrombos[...] research effort are evident in a number of fiel[...] responsiveness of platelets in the aggregation a[...] more clearly defined; there has been better defir[...] between platelets and blood clotting; the coag[...] defined in terms now almost acceptable to th[...] with respect to the reactions involving fibrinoger[...] in thrombolytic therapy of thrombo-embolic di[...] impetus to the study of the fibrinolytic system. C[...] comparative species studies which are just beg[...] models for the study of haemostasis (Stokes &[...] widespread documentation employing repeated [...] factors involved in human haemostasis will a[...] standing of the mechanisms involved.

In pregnancy and the puerperium an increase [...] long been recognized and associated with incr[...] levels and platelet counts during pregnancy an[...] years more detailed studies have further docu[...] levels of a number of coagulation factors, as [...] fibrinolytic system to further explain these risks [...] ship of treatment with oral contraceptives ar[...] thrombotic disease has also stimulated work ir[...] hormonal changes on the factors involved. Preg[...] with an increased risk of excessive activation [...]

lysozymes including phosphatases, proteases and other enzymes are released. In this way a phagocytic vacuole is formed and the other cell contents protected. Archer (1966) has also demonstrated phagocytic activity of eosinophils but this is primarily for sensitized red cells and not directed against bacteria. It is related to an enzyme pattern in eosinophil granules different from that of neutrophils.

Less is known about the production and distribution of eosinophils and basophils. These are present in small numbers in normal blood and are much less accessible to study. Eosinophils are produced in the bone marrow but are believed to spend a much longer time in the tissues than neutrophils. A total life span of 8 to 10 days has been suggested for rat eosinophils with the majority spent in the lungs, skin and intestinal epithelium. The distribution of eosinophils is influenced by the adrenal and pituitary glands. Bassett (1962) has reported cyclic variation in the number of eosinophils present in the uterus of the rat in relation to the oestrus cycle. There does not appear to be any major variation in circulating eosinophils and basophils in humans in relation to the menstrual cycle or pregnancy but the major changes may in fact be occurring in the tissues and not detected by examination of the peripheral blood alone.

Detailed studies of the leucocyte count in relation to cyclic hormonal changes, pregnancy and parity have received sporadic attention, but the whole subject has recently been re-evaluated in the light of observations made in healthy individuals (Morley 1966) and in special detail in relation to pregnancy (Cruickshank & Alexander 1970, Cruickshank 1970). These authors examined the leucocyte counts in a large number of male and female blood donors and found that, while male counts did not vary significantly with age, in females the neutrophil and lymphocyte counts decreased significantly with age, the fall in the neutrophils occurring predominantly after the menopause. Decrease in the lymphocyte count in the females was more evenly distributed and remained at a lower level than the count in the males throughout. They also found that oral contraceptives were associated with higher neutrophil counts independently of the duration of taking these agents. Higher neutrophil counts were also associated with increasing parity. These observations point to an important influence of hormonal balance on the neutrophil count. Since the concentration of circulating neutrophils represents a balance between production in the bone marrow, release to the circulation, distribution in the peripheral blood between circulating and

marginated pools and finally escape
influences may operate at any of a
information presently available to
possibilities.

Oestrogens have been shown in cultu
late granulopoiesis while testosterone a
lating effect (Reisner 1966). On the othe
depress myeloproliferative activity (Kap
effects observed in relation to parity, or
may represent a balance between availa
is supported by changes in neutrophils
cycle and in pregnancy. Cruickshank (1
tions (Tysoe & Lowenstein 1950) that
sively during pregnancy up to the 30tl
level still above the normal mean in
Changes in lymphocytes and monocyte
pregnancy. The neutrophilia of pregn
creased oestrogen levels, placental produ
although much of the latter is in the
(chapter 9).

The neutrophilia of pregnancy is asso
of neutrophil alkaline phosphatase (NA
has been observed in relation to the ov
cycle and following oestrogen administra
demonstrated that this enzyme increase
pregnancy. They demonstrated that the
of pregnant women resulted from incre
with different substrate specificity from tl
or from that obtainable from the place
increased NAP of pregnancy was not c
related to oestrogen stimulation.

The immunological aspects of pregnan
volume (chapter 5). Little significant ch
circulating lymphocytes and monocytes
lymphocytes has been observed (Cruicks
first trimester but to attain a higher ste
Functional changes of great importanc
tissues must occur but these are not refle
and will not be discussed further in this

systems giving rise to complex coagulation disorders and defibrination, often of sudden onset. In this review it is intended to outline some of the mechanisms involved in the light of current knowledge and theory and to emphasize the changes that may be encountered in association with pregnancy.

PLATELETS

The platelet numbers may show little, if any change during pregnancy but it is common to observe an increase in platelet count after delivery. This increase has been associated, as in the post-operative period, with an increased risk of pelvic or peripheral venous thrombosis and a coincident increase in platelet adhesiveness first described by Wright (1942). Other factors, including stasis, alterations in coagulability and vascular endothelium—the triad of Virchow—must be at least of equal importance. Evidence for any real change in platelet function during pregnancy is scanty. Bolton *et al* (1968) showed a variable increase in platelet reactivity with adenosine diphosphate (ADP) and this may also be relevant to thrombotic episodes.

Under normal conditions there occur only mild degrees of variation in platelet numbers (Morley 1969) although exercise induced changes and cyclical variations have been described. The platelet has a relatively short life span of 7 to 10 days after production as a cytoplasmic fragment of megakaryocytes. The process of platelet production has now been well defined and results from a progressive maturation of the megakaryocyte, with release of platelets separately and in larger cytoplasmic fragments into the marrow sinusoids. Once released to the peripheral blood, a certain proportion is arrested briefly in the lungs and some modest reserve of functional platelets probably resides there. The majority, however, circulate and are ultimately removed by the normal spleen, by a random process largely dependent on age. There has been controversy about the determinants of platelet removal and a number of environmental and endogenous factors have been investigated (Murphy *et al* 1964, O'Neill & Firkin 1964) but it seems likely that a significant fraction is removed randomly in the maintenance of haemostasis and of the impermeability of vascular endothelium. It is possible that some are continuously removed in blood clotting although the evidence for continuous *in vivo* coagulation is inconclusive (Hjort & Hasselback 1961).

The rate of turnover of platelets may be calculated (Harker 1970) to be approximately 35,000/microlitre per day under resting conditions and a number of pathological states have been defined where immunological, coagulation and vascular changes lead to enhanced removal and increased turnover. It has recently been shown that turnover may be considerably influenced by pharmacological means with important implications in thrombosis control. Aspirin and other agents which affect connective tissues like phenylbutazone and dipyridamole, influence platelet turnover but also have important effects on platelet release reactions related to aggregation and thrombus formation (Harker & Slechter 1970, Zucker & Peterson 1970).

Platelet aggregation, or cohesion as it is sometimes called, is a fundamental reaction of this cell. It is associated with biochemical changes and energy release, morphological alterations suggesting a change in shape and contraction, as well as release of some soluble constituents and coincident activation of clot accelerating platelet lipid, probably located in membranes. Gaarder et al (1961) demonstrated that a factor from red cells caused aggregation and that this factor was ADP. Born (1962) and O'Brien (1962) defined methods of study and a variety of other compounds that would cause aggregation and in the subsequent years a large number of studies have been devoted to definition of the mechanism of the ADP effect. It is generally accepted that ADP is the most important agent, although others like adrenaline and 5 hydroxytryptamine (5HT) are of considerable significance, capable as they are of inducing aggregation, perhaps via a mechanism utilizing ADP. Platelets are rich in this nucleotide and also concentrate 5HT. Aggregation has been shown (Macmillan 1966) to be a diphasic reaction in the plasma of the majority of the subjects tested. The second, irreversible, phase of aggregation is associated with and mediated by the release of endogenous ADP, 5HT and other factors. It is this second phase which is inhibited by aspirin in a suppressive reaction that may be associated with prolongation of the bleeding time—a true disorder of haemostasis and of potential significance in the pharmacology of thrombosis.

Haemostasis is usually pictured as commencing with adhesion of platelets to the connective tissue fibres exposed by damage to the vessel wall. Adhesion may be duplicated experimentally with collagen or glass or other solids and may be shown to be associated with a release reaction providing ADP and other agents capable of propagating aggregation, and thus the haemostatic plug grows. Pederson et al (1967) drew attention also to the importance of ADP released from damaged red cells recalling the earliest reference to red cell ADP as a central agent in aggregation. These authors also demonstrated the appearance of fibrin within 30 seconds of wounding and so stressed the early availability of thrombin to consolidate plug formation.

The adhesion–aggregation reaction is associated with release of active platelet lipid to accelerate coagulation. There also occurs enhanced activation of surface active clotting factors XI and XII (Table 15.2) also concentrated on the platelet (Castaldi *et al* 1965) so that a propagated local reaction will rapidly lead to clotting and a consolidated thrombus. These reactions occur in flowing blood and the form and nature of the mass is greatly influenced by the flow stresses present (Dintenfass & Rozenberg 1965). Clotting may be initiated by any vascular injury or abnormality. When it occurs under conditions of limited flow or stasis, the coagulum is poorly structured and contains only scattered platelet aggregates intermingled with red cells and fibrin. When it occurs under conditions of considerable sheer stress, as may pertain in arteries and arterioles, and in the presence of distorted or diseased vessels the thrombus consists predominantly of masses of platelets densely aggregated with peripheral fibrin and red cells. These differences are reflected in the type of thrombus formed in arterioles and high pressure systems with a structure resembling that of the haemostatic plug and that formed in peripheral veins, the so-called red thrombus.

The viscosity of blood is related largely to the red cells and is that of a non-Newtonian fluid with increasing viscosity occurring in relation to increase in sheer stress. The theoretical possibility is that under appropriate conditions with high haematocrit levels, flowing blood may become extremely viscous and very liable to coagulation and thrombus formation. There have been studies relating blood viscosity to other parameters of haemostasis (Rozenberg 1970) which suggest that platelets also respond to increasing sheer stress by the formation of aggregates. This implies that blood viscosity is another parameter to be considered in haemostasis but is probably of even greater significance in thrombotic states.

It is evident that vascular integrity is fundamental to haemostasis. There are very few studies of the haemostatic properties of vessels and endothelium and it has long been accepted that platelets, in some ill-defined way, contributed to the vessel wall. Endothelial turnover has been shown to be markedly increased following endotoxin injection (Gaynor 1970) suggesting that a degree of self-replication operates to maintain normal vasculature. However, there is morphological evidence (Johnson et al 1965) that platelets are incorporated in the endothelium of small vessels and that this incorporation is associated with arrest of red cell escape after platelet transfusion in thrombocytopenic animals. This was the first direct evidence that platelets contributed to the endothelium. Other factors in vessels also involved in haemostasis include local contractility stimulated by amines such as 5HT released from platelets and possibly also by fibrinopeptides released during the clotting of fibrinogen (Laki 1951). Vascular endothelium also contains both activators and inhibition of plasminogen and plasmin and the former may be responsible for activator activity released by a number of stimuli. The vessels thus have a number of factors related to haemostasis but a great deal more needs to be learnt about their role.

Investigations of these aspects of haemostasis—platelet and vessel function and blood viscosity—are few in relation to hormonal changes in pregnancy. Dintenfass et al (1966) studied blood viscosity during the menstrual cycle and during ovulation suppression with oral contraceptives. Some cyclic increase was found with normal menstruation but studies of viscosity do not appear to have been performed during pregnancy. The platelet count has been found to decrease in the latter part of the menstrual cycle and rise again rapidly during menstruation (Pohle 1939). Similarly, platelet function has been poorly documented as a function of pregnancy although in the majority of cases there is no reason to suspect any abnormality. Quite extensive studies have been performed on platelets in the newborn and early infancy by Hrodek (1966) and he demonstrated some functional differences, especially in relation to adhesiveness and spreading of platelets on glass, in the newborn.

During pregnancy Bolton et al (1968) have shown enhanced sensitivity of platelets to ADP in a system employing changes in electrophoretic mobility of platelets, in three of six women, all in the latter stages of pregnancy. More widely used tests of platelet function such as bleeding times, glass bead adhesiveness studies and tests of aggregation with ADP, adrenaline and collagen in the photometric system described by Born (1962) do not appear to show any abnormality. It is also important that three of the group investigated by Bolton et al (1968) showed no increase in sensitivity to ADP. It is therefore difficult to attribute any increased thrombotic tendency to these findings. However, the latter authors related the changes observed in some patients taking oral contraceptives, where the evidence of increased sensitivity to ADP and abnormal platelet electrophoretic mobility was more convincing, to the presence of abnormal lipids, related to the phospholipid, lecithin. Since changes in plasma lipids also occur in pregnancy with increases in cholesterol, triglycerides and phospholipids (Oliver & Boyd 1955) there may be some connection with altered platelet function, although it seems that these changes are coincidental and more likely to be associated with alterations in plasma fibrinolytic activity than with platelet changes.

BLOOD COAGULATION

The effect of activators of the coagulation system is to convert the large protein molecule fibrinogen, circulating in solution in the plasma, into a gel with a characteristic bonded fibrillar structure that forms the basis of the haemostatic network and provides a scaffolding for subsequent healing of associated wounds. The means whereby this conversion occurs is the result of interaction of trace proteins known as coagulation factors which are identified by numerals (Table 15.2).

The activation process may employ either of two pathways and may proceed to completion over a period varying from a few seconds to several minutes, depending on the nature, extent and duration of the initiating stimulus. Coagulation may occur in a setting of vascular injury as a haemostatic and localized event, but the system is also subject to more widespread activity and it is the purpose of this section to discuss some of the factors, local and general, leading to activation of coagulation and those natural inhibitors and fibrinolytic activities that ordinarily balance the system.

The coagulation system in man and other mammals is complex in terms of the number of factors involved as coagulation proteins and in their interaction with one another, with initiating stimuli and with circulating cells. Coagulation in man occurs in an environment of rapid flow, under relatively high pressure in a complex vascular system and is, of necessity, capable of rapid response in protection against vascular injury in the maintenance of haemostasis. Although, for the sake of simplicity, it is necessary to consider the various components—cellular and plasmatic—excitatory and inhibitory—as separate entities, their close inter-relationship *in vivo* must be emphasized.

In an attempt to clarify both concepts and experimental interpretation a number of investigators have examined the haemostatic system in less evolved forms of life. In invertebrates such as *Homarus* the extracellular fluid contains few differentiated cell types and is incoagulable, although a coagulable protein similar to human fibrinogen can be demonstrated in the blood and in such forms may largely be derived from cells (Lorand et al 1963). The primitive coagulation system has been intensively studied in recent years in the hemolymph of the horseshoe crab, *Limulus Polythemus*, which belongs to the arthropod subphylum *chelicerates*, which also includes spiders. Their blood contains only one type of cell, the amaebocyte which is of approximately the same size as the human macrophage and contains many cytoplasmic granules. The cell-free plasma is incoagulable but the cells contain material which is coagulated by bacterial endotoxin. Coagulation in *Limulus* is associated with aggregation of amoebocytes together with degranulation and release of the clottable protein. The properties of this protein have been investigated by Solum (1970) who has shown that it is a low molecular weight protein not identical with mammalian fibrinogen. This work is of particular interest as it serves to provide a functional link between a primitive and evolved coagulation system, especially as fibrinogen is present in the platelets of man.

Other studies (Stokes & Firkin 1970) have demonstrated a more complex system already present in the blood of the cartilaginous fishes exemplified by the shark, *Heterodontus Portus Jacksoni*, estimated to have been in existence for 180 million years. There is present in this animal both an intrinsic plasma activator system and an extrinsic tissue activator system for the production of a clotting activity which converts shark fibrinogen to fibrin. Circulating cells are of several types including nucleated red and white cells and others, in low concentration resembling the platlet but nucleated and referred to as thrombocytes. Shark thrombin causes prompt coagulation of fibrinogen in shark plasma and this is accompanied by aggregation of the thrombocytes. These thrombocytes are not aggregated by ADP, 5HT or adrenaline which are aggregators of human platelets, but they do have some adhesive properties similar to human platelets and their response to thrombin does constitute a basic similarity to the more highly evolved form.

All of the features of the primitive forms exist in the human haemostatic factors. Platelets respond to an injurious stimulus by aggregation and a release reaction which participates in the coagulation sequence. Platelets also contain, and release, a clottable protein (Grette 1962, Castaldi & Caen 1965) which is probably identical to plasma fibrinogen, to assist in local haemostasis. The activation sequence and coagulation factors involved in the evolution of thrombin may follow one of two pathways and the individual factors and their reactions will be considered in this section.

Knowledge of the mechanisms of coagulation had accumulated in a progressive manner by careful observation of the effect of tissues on blood clotting until the so-called classic theory of Morawitz (1905) was proposed. The existence of at least four of the five reactants had been established, only prothrombin being entirely hypothetical and coagulation was proposed to consist of two stages. In the first, the hypothetical proenzyme prothrombin was converted to thrombin by the action of a tissue derivative. Thrombin then converted fibrinogen to fibrin. Haemophilia had long been recognized but it was not until the nature of this and other congenital bleeding disorders received detailed investigations in the last 40 years, that the complexity of the coagulation system became evident. Quick in 1935 introduced the prothrombin time test with a standardized extract of rabbit brain as tissue extract. This test absolutely confirmed the classic theory of Morawitz and lead to the better evaluation of a number of congenital and acquired disorders. There was still no recognition of any alternative activation sequence except that from the tissues and the nature of the deficiency in haemophilia remained an enigma.

In 1947 Owren detected a new coagulation defect, factor V deficiency, in which the tissue activation of prothrombin could not explain the abnormality and thus laid the basis for the subsequent discovery of the intrinsic system of prothrombin activation. The development of the prothrombin consumption test by Quick in 1947 and of the thromboplastin generation test in 1953, by Biggs and Douglas, permitted differentiation of the so-called haemophiloid states and established the existence of intrinsic activation of prothrombin from components peculiar to the blood itself. Seegers and his group, in a long series of experiments, have greatly contributed to knowledge about prothrombin and participating factors in its activation and this work has been summarized (Seegers 1969). The work of Margolis (1957) subsequently established the role of contact factors in the initiation of the intrinsic activation sequence confirming much earlier work (Bordet 1920) which had suggested the existence of contact-sensitive factors.

As may be expected in a period of rapid growth of knowledge, confusion in terminology developed and a numerical system was proposed and introduced by an international committee. Twelve factors were recognized and since then only one, factor VI has been removed and one, factor XIII (fibrin stabilizing factor) added. The last 15 years have seen further development in terms of purification of coagulation proteins although fine structural detail is still unknown for the majority and there has been more extensive definition of the nature of a number of coagulation deficiencies. Effective concentrates of semi-purified factors have been introduced and greatly facilitated the treatment of both congenital and acquired deficiencies.

Kinetic experiments conducted by Margolis & Bruce (1964) suggested that optimal activation in the intrinsic system was achieved by participation of all the recognized factors and that it was likely that no new participants would be detected. Experience has to date proved this prediction to be correct and allows some confidence in the presentation of the following table (Table 15.2).

TABLE 15.2. Blood coagulation factors

Factor		
I	Fibrinogen
II	Prothrombin
III	Tissue thromboplastin
IV	Calcium
V	Proaccelerin
VII	Proconvertin
VIII	Anti-haemophilic globulin
IX	Christmas factor
X	Stuart-Prower factor
XI	Plasma thromboplastin antecedent
XII	Hageman factor
XIII	Fibrin stabilizing factor

Fibrinogen has been purified (Blombäck & Blombäck 1956) and considerable information accumulated about its biochemical properties and reactions with enzymes. The molecule is the soluble precursor of fibrin and in plasma circulates in a negatively charged, mutually repulsive form that is rapidly converted to fibrin by the action of the enzyme thrombin. The intact molecule has a molecular weight of 340,000 and consists of three pairs of polypeptide chains in dimeric form. The three peptide chains designated α (A), β (B) and γ are linked by disulphide bonds at their N-terminal extremities (Blombäck et al 1968).

The concentration of fibrinogen in the plasma varies very little in health within a range for individuals of 150–400 mg per 100 ml. An increase during pregnancy is commonly observed (Ratnoff et al 1954), attributable to increased synthesis on the basis of turnover studies in pregnant monkeys (Regoeczi & Hobbs 1969). Fibrinogen is produced in the liver as has been demonstrated by Straub (1963) with human liver slices incubated *in vitro* and shown to produce a protein reacting specifically with rabbit anti-human fibrinogen antibody. Studies with ^{131}iodine labelled fibrinogen in normal recipients (McFarlane et al 1964) and in patients with congenital afibrinogenaemia (Gitlin & Borges 1953) have shown that the protein is distributed in both the plasma and extravascular space, equilibrium between the two being achieved shortly after injection. Labelled fibrinogen disappears from the plasma with a half-time of 80–100 hours and is catabolized at the rate of approximately 30% per day. These turnover data have been confirmed in humans with endogenously labelled fibrinogen in the studies of Brodsky et al (1970) using ^{75}Se selenomethionine incorporated into platelets and fibrinogen.

Apart from its distribution throughout the extracellular fluid, fibrinogen is also demonstrable in extracts of platelets (Ware et al 1948). There is no direct evidence of incorporation from the plasma and transfused labelled fibrinogen was not demonstrated in platelet extracts (Castaldi & Caen 1965) suggesting that it may also be produced by megakaryocytes. The close relationship of fibrinogen to platelets is of functional significance since there is good evidence that it is required as an essential co-factor with ADP and divalent cations in platelet aggregation (McLean et al 1964). Since the bleeding time is often prolonged in severe degrees of congenital afibrinogenaemia and is corrected by transfusion of purified fibrinogen (Castaldi & Caen 1965) there is additional evidence for the requirement of this protein in platelet reactions occurring early in haemostasis. Fibrinogen is also involved in the adhesive properties of other cells. In conditions where the fibrinogen level increases, such as pregnancy, there is a coincident increase in red cell aggregation and so also in the erythrocyte sedimentation rate.

The mechanisms leading to the relatively rapid removal of fibrinogen from plasma, as compared with other proteins have been the subject of considerable investigation. Since other proteins such as the immunoglobulins and albumin have a relatively long half-life in the circulation it is suggested that fibrinogen may be continuously removed in a process of coagulation. Other evidence for this concept, of an indirect nature, includes the observed rapid disappearance of other coagulation factors, the relatively short life span of platelets and the presence of degradation products of fibrin in normal blood. Much of this evidence has been reviewed by Hjort & Hasselback (1961). It seems possible that normal haemostasis may be associated with some degree of continuous coagulation. The tendency for disseminated coagulation to occur when a sufficiently strong stimulus, such as amniotic fluid embolism, retained dead foetus or widespread infection with release of large amounts of thromboplastic material into the circulation, is consistent with an alteration in an otherwise balanced state of activation and inhibition of coagulation. Under resting conditions the bulk of fibrinogen is probably removed by intracellular digestion and it is relevant that polymorphonuclear leucocytes have been shown by immunochemical means to contain fibrinogen and fibrin-related antigens (Barnhart 1965). They are believed thereby to perform a dual role in fibrinogen catabolism, both removing the circulating protein and also fibrin and its degradation products present in exudates. Barnhart stressed the occurrence of fibrin phagocytosis in thrombotic states where there may be an important contribution of neutrophils to thrombolysis, but did not discuss a possible participation in resting normal fibrinogen removal.

Fibrinogen is substrate for the two major proteolytic enzymes of plasma, thrombin and plasmin. Thrombin, produced by the coagulation system reacts specifically with fibrinogen and removes successively two peptide fragments from the α and β polypeptide chains respectively. This is a progressive reaction and involves limited hydrolysis of soluble fibrinogen followed by aggregation of the resulting monomer with reduced solubility and the formation of the extensive branched fibrin polymer characteristic of clotted plasma. Thrombin hydrolyses two arginine-glycine linkages, first the α chain and later the β chain. Hydrolysis of the α chain alone is sufficient to allow clotting of mammalian fibrinogen and is the mechanism of coagulation produced by some snake venoms, including reptilase from the Malayan Pit Viper. Thrombin, in the presence of calcium also activates factor XIII (fibrin stabilizing factor), a transpeptidase present in inactive form and a common contaminant of semi-purified fibrinogen. This enzyme is responsible for co-valent bonding of fibrin monomers in the production of the polymer necessary to maintain an intact haemostatic plug and resist, for a sufficient time, plasma proteolytic activity. These aspects of the reactions between thrombin and fibrinogen have been reviewed by Lorand (1965) (Fig. 15.4).

The fibrinopeptides released from the parent molecule by thrombin have molecular weights in the vicinity of 2000 and their amino-acid sequence has been determined (Blombäck et al 1962). They have been shown to have some properties that have received little attention and may be of some significance in haemostasis, since an antithrombin action has been found in a mixture of fibrinopeptides and isolated peptide B has powerful vasoconstrictor and cardiotonic properties (Bayley et al 1967). Other work has also demonstrated an interference with the aggregation of isolated human platelets with thrombin but not

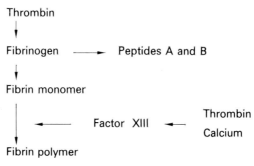

FIG. 15.4. The action of thrombin on fibrinogen.

with ADP or connective tissue by a mixture of fibrinopeptides (Gorman & Castaldi 1917). Since significant amounts of these substances may be present, especially in large clots, they may have important effects on local haemostasis.

Factor XIII, fibrin stabilizing factor, is a transpeptidase, activated by thrombin and responsible for the production of co-valent bonds between fibrin monomers. The existence of this enzyme was postulated by Laki & Lorand (1948) on the basis of solubility studies of purified fibrinogen. The subsequent demonstration by Duckert et al (1960) of congenital deficiency of this factor in a child with a bleeding tendency confirmed the physiological relevance of this factor. It has been found in close association with semi-purified fibrinogen and is also an important constituent of platelets (Nachman 1968). Stabilized fibrin is more resistant than monomer to the action of plasmin and is especially responsible for the effectiveness of the fibrin network in haemostasis. This factor has a relatively prolonged effect after transfusion with a half-life of 4 to 5 days, and very small quantities of normal plasma will correct the abnormality *in vitro* (Lorand et al 1969). An adequate quantitative assay has only recently been introduced, and there are, as yet, no studies to suggest changes in this factor in pregnancy.

The prothrombin complex of coagulation factors includes that group of factors II, VII, IX and X, which are produced by the liver with a requirement for vitamin K. This fact, together with biochemical properties of like behaviour, particularly in absorption by alkaline earths such as aluminium hydroxide and calcium phosphate, has led to controversy about their possible identity. Seegers (1969) has long postulated that they were prothrombin derivatives while others, and particularly the Oxford School, on the basis of the existence of congenital deficiency of one or other factor in the presence of normal activities of other constituents, have regarded them as separate entities. Current work has not resolved these possibilities and continued efforts aimed at separation and quantitation, particularly in congenital and acquired deficiencies will be required.

These four factors are particularly susceptible to vitamin K deficiency or antagonism such as by the coumarin and phenindione anticoagulants. Complete suppression of hepatic synthesis results in rapid depletion of their activities in the plasma. Factor VII disappears most rapidly and has a half-life in the vicinity of 6 hours. The other factors have slightly longer duration in the circulation and disappear more slowly. Factor IX for example, has a half-life of approximately 19 hours and that of prothrombin and factor X is similar. It is this group of factors in particular which change during pregnancy while levels of some increase with oral contraceptives.

Alexander *et al* (1956) examined ninety pregnant women at various stages during pregnancy. There was a tendency for factor VII levels to be highest in those subjects late in pregnancy. A few individuals had very high levels, with an average of 180 per cent as compared to 82 per cent in forty-one non-pregnant subjects. Factor VII levels declined after delivery but in seven of these women with high levels at the end of pregnancy decrease to the average normal did not occur before 6 weeks post-partum. Prothrombin (factor II) and factor V levels were not elevated in this group. Nossel *et al* (1966) similarly found no change in factor II and factor V levels but did find elevation of factor X and factor IX. High levels of factor X have been found in other studies (Hocking & Castaldi 1971) and it is of some interest that increases have been

Z

observed in patients with congenital deficiency of this factor, both during pregnancy and following oral contraceptives (Haber 1964).

Factor V, probably also produced in the liver, is not vitamin K dependent. It has now been obtained in relatively purified form and some reactions studied (Esnouf & Jobin 1967, Jobin & Esnouf 1967). These authors achieved a 6000-fold purification and established the nature of factor V interaction with factor X in the formation of the prothrombin converting complex. Factor V deficiency is a rare disorder first described by Owren (1947). There is controversy about changes of this factor during pregnancy and this has been reviewed by Ulutin (1969). In the writers' study (Hocking & Castaldi 1971) no change was observed, but tests were performed on frozen plasma so that some loss may have occurred during storage.

Factor VIII, anti-haemophilic globulin, has been extensively studied especially in relationship to the treatment of classic haemophilia and moderate success has been achieved in efforts at purification. Antibody neutralizing activity in normal and haemophilic plasma has been investigated. There is some evidence for heterogeneity of factor VIII cross-reacting material (Denson *et al* 1969) as had also been suggested for factor IX (Pfueller *et al* 1969). There have been a number of studies of factor VIII levels during pregnancy in normal women and in patients with Von Willebrand's disease and carriers of haemophilia. Nossel *et al* (1966) amongst others, have shown a progressive increase in factor VIII levels during pregnancy in normal women with return to normal being delayed for some weeks after delivery. Nillson *et al* (1959) observed the same changes in normal women and increase in factor VIII levels in a carrier woman whose initial level was 28% and rose to 69%. These findings were confirmed in carriers and patients with Von Willebrand's disease by Kasper *et al* (1964).

Of the contact factors Nossel *et al* (1966) found no increase during pregnancy and these factors have otherwise received little attention. The major quantative changes in coagulation factors during pregnancy, in the majority of studies, appear to involve components of the prothrombin complex and especially factors VII and X as well as factor VIII and fibrinogen. Some of these increases, especially of factors VII and X have also been observed during ovulation suppression with oral contraceptives and Poller & Thomson (1968) referred to high levels after protracted use of oral contraceptives while changes in other factors are much less constant than those observed in pregnancy. It is possible that these factor increases, involving both the intrinsic and extrinsic pathways, contribute to an increased tendency to coagulation and thrombo-embolic disease associated with pregnancy but it is difficult to correlate the changes in the individual patient.

Factor interactions in blood coagulation have been defined with considerable precision during the last decade. Macfarlane (1964) highlighted the concept of sequential reactions in the intrinsic system whereby coagulation factors present in very low concentrations could generate a prothrombin converting activity of sufficient magnitude to clot large amounts of fibrinogen. Further investigation by Hemker & Kahn (1967) and Jobin & Esnouf (1967) as well as Ratnoff (1966) and many other authors, appear to have confirmed these concepts and established that intermediate reactions occur with the formation of complexes with enzymatic activity. The current view is illustrated in Fig. 15.5 which depicts the reaction sequence in both the extrinsic system involving tissue thromboplastin and the intrinsic system following surface activation. It appears that in the intrinsic system the formation of two complexes, each involving phospholipid and calcium, occurs. These complexes have enzymatic activity on factor X and prothrombin respectively. The second complex may also be produced by a reaction with factor X involving tissue factor and factor VII. Activated factor X designated Xa then reacts with factor V, calcium and lipid to produce prothrombinase. Esnouf & Jobin (1966) showed that although activated factor X had prothrombin converting activity alone, optimal activity requires complex formation. Although most investigators agree with these concepts, the reactants are incompletely purified and much of the experimentation is indirect, so that for the present the results must be accepted with this limitation.

The phospholipid required in intrinsic factor interaction is provided by platelets and may be largely a product of membranes (Marcus *et al* 1966). It is suggested that phospholipid is required for the provision of a surface for reaction between the different clotting factors. Hyperlipidaemia has been associated with an increased tendency to thrombosis but the only direct association that appears to have been established is the occurrence of decreased fibrinolytic activity (Kwaan *et al* 1959). Serum cholesterol, triglycerides and phospholipids all rise progressively in pregnancy as the oestrogen and progestogen output increases (Oliver & Boyd 1955) and there is a coincident decrease in fibrinolytic activity (Macintosh & Kyobe 1966) that has been attributed to inhibitors of lytic activity (Brakman & Astrup 1963). The latter authors demonstrated a progressive increase of urokinase inhibitor during the second and third trimesters and Macintosh & Kyobe (1966) showed abrupt return of lysis to normal levels immediately after delivery suggesting a placental origin for the inhibitor. Changes in plasma lipids in pregnancy, or in any other condition, have not been shown to have any direct relationship to altered coagulability and any changes found in the latter are probably coincidental.

Natural inhibition of coagulation activation is provided by heparin and the so-called physiological antithrombins. Heparin antagonizes the activation of factor IX by activated contact product (Somer & Castaldi 1970) and blocks the action of prothrombinase. A co-factor present in

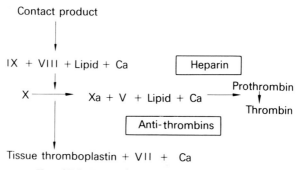

FIG. 15.5. Factor interaction in coagulation.

normal serum is required for the action of heparin and the latter is antagonized by a component of platelets, platelet factor 4, which may be of physiological importance in local haemostasis, and is of practical importance in increasing the dosage requirement of heparin in the presence of thrombocytosis.

The antithrombins have been reviewed by Von Kaulla & Von Kaulla (1967). A number of activities have been described and are numbered as follows: antithrombin I is an activity of fibrin in absorbing thrombin, antithrombin II is probably identical with heparin co-factor, antithrombin III is present in normal serum and probably includes more than one activity, as both immediate and delayed inhibitors occur. They are associated with α-1 antitrypsin of normal plasma with a wide spectrum of anti-protease activity including both thrombin and plasmin.

Antithrombin IV is ill defined and includes those activities resulting from the interplay of factors during clotting. To these have been added pathological activities, V a paraprotein and VI a fibrinolytic split product of fibrinogen, both of which interfere with fibrin polymerization. Antithrombin III is the activity usually measured in assessment of this system and this activity has been shown to be decreased in the third trimester of normal pregnancy (Peterson *et al* 1970). Since Egeberg (1965) has described familial deficiency of antithrombin III with an increased incidence of thrombosis in the affected families, this system may have considerable relevance to the balance of coagulation and decrease during pregnancy may contribute to the increased incidence of thrombosis in the puerperium.

The normal fibrinolytic system in plasma and extravascular fluids serves the function of removing fibrin and maintaining the patency of vessels and other channels. Extracellular fluids and plasma contain an inactive proenzyme, plasminogen, which may be activated by a number of stimuli to a powerful protease, plasmin, capable of digesting fibrin as well as a number of other proteins. The mechanism of activation of plasminogen is only partly known but both blood and a number of tissues contain activators. This tissue-dependent system bears a close resemblance to activation of prothrombin conversion and, indeed, sequential or coincident activation of two systems as well as other related peptides such as the plasma kinin system probably occurs. Some of these inter-relations are depicted in Fig. 15.6.

Tissue activators include those in vascular endothelium (Pugatch *et al* 1970, Todd 1964); veins are a more important source than arteries. Activator from this source is readily diffusible and may be responsible for the minor degree of fibrinolytic activity readily detectable in normal blood after minor stimuli such as exercise and adrenaline injection. Other tissues also contain less readily diffusible activators (Astrup 1966) that may only be released to the circulation when there is extensive tissue damage or necrosis. Activation of plasminogen to plasmin has been most intensively studied using streptokinase (Kline 1966) and urokinase (Lesuk *et al* 1965) as tissue activators have not been purified. These studies have led to much greater understanding of the fibrinolytic system and the introduction of thrombolytic therapy in the treatment of thrombo-embolic disease.

The activity of plasmin in the circulation is normally balanced by an inhibitor associated with α-2 antitrypsin which also has antithrombin properties. This latter, which is a progressively acting inhibitor forms non-dissociating complexes with plasmin and trypsin. Other anti-plasmin activity which is immediate acting as well as anti-activators also contribute to natural balance in this system. If activation occurs to a sufficient degree to overcome these inhibitors, plasmin may be detectable in the circulation or may be found in larger amounts in thrombin where it progressively degrades fibrin.

Plasmin hydrolyses a number of proteins but its preferred substrate is fibrin, although casein is commonly used in the laboratory assay. Plasminogen may be measured in this way and has been found in normal amounts during pregnancy even though fibrinolytic activity, as measured

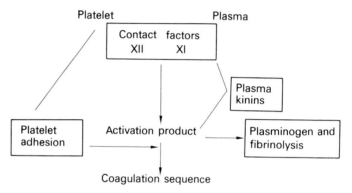

FIG. 15.6. Interactions in haemostasis.

with dilute clot lysis methods, is decreased (Shaper *et al* 1965). The action of plasmin on fibrinogen and fibrin proceeds at similar rates and results in progressive degradation with the ultimate formation of stable end products resistant to further digestion. These peptide fragments known as D and E with molecular weights of 80,000 and 30,000 retain immuno-reactivity with antifibrinogen antisera and are thus detectable in human serum. They have important effects on haemostasis and when present in increased amounts during enhanced fibrinolysis, interfere with fibrin polymerization and behave as antithrombins. Some earlier, transient products may also contribute to abnormal haemostasis by interfering with platelet function, but this latter aspect is less well documented (Kowalski 1968). Fibrin degradation products may be demonstrated in small amounts in normal blood and are also present during pregnancy in slightly increased amounts (Woodfield *et al* 1968) and especially during labour (Bonnar *et al* 1969). This may reflect fibrin deposition in the villi with local lysis and release of degradation products which is not of sufficient degree to give rise to haemostatic abnormalities. Normal labour has also been shown to be associated with a significant decrease of fibrinogen and factor VIII levels with enhanced fibrinolytic activity developing 2 to 4 hours after onset of labour (Kleiner *et al* 1970). These findings lend further support to the concept of enhanced coagulation activation with mild degrees of defibrination occurring in normal delivery and emphasize the precarious state of balance readily susceptible to excessive activation with enhanced stimuli in abnormal labour.

Abnormalities of pregnancy such as *abruptio placentae*, retained dead foetus, retroplacental bleeding or amniotic fluid embolism may be associated with fibrinolysis and a marked increase in degradation products (Bonnar *et al* 1969). This represents an extreme of activation in a haemostatic system which is normally balanced. It is possible that the overall increase in concentration of procoagulants, fibrinogen substrate and a possible relative decrease in plasma lytic activity may together be responsible for the increased incidence of thrombo-embolism associated with pregnancy. Other factors such as disturbance of flow and venous obstruction and alterations in platelets, as yet ill defined, may also be relevant.

REFERENCES

ADAMS E.B. (1956) Treatment of megaloblastic anaemia of pregnancy puerperium with vitamin B_{12}. *Brit. med. J.* **11**, 398.

ALEXANDER B., MEYERS L., KENNY J., GOCOSTEIN R., GUREWICH V. & GRINSPOON L. (1956) Blood coagulation in pregnancy. *New Eng. J. Med.* **254**, 358.

ALEXANIAN R. (1969) Erythropoietin and erythropoiesis in anaemic man following audiogene. *Blood* **33**, 564.

ALFREY C.P. JR & LAWE M. (1963) The anaemia of human riboflavin deficiency. *Blood* **28**, 811 (Abstract).

ALTMAN K.I. & MILLER G. (1963) A disturbance of tryptophan metabolism in congenital hypoplastic anaemia. *Nature* **172**, 868.

ARCHER G.T. (1966) The function of the eosinophil, in *Proceedings of the Eleventh International Congress on Blood Transfusion, Sydney*, p. 61.

ASTER R.H. (1966) Pooling of platelets in the spleen: role in the pathogenesis of hyperplasia thrombocytopenia. *J. clin. Invest.* **45**, 645.

ASTRUP T. (1966) Tissue activators of plasminogen. *Federation Proc.* **25**, 42.

ATHENS J.W., HAAB O.P., RAAB S.O., MAYER A.M., ASHENBRUCKER H., CARTWRIGHT G.E. & WINTROBE M.M. (1961) Leukokinetic studies. IV. The total

COOPER R.G., WEBSTER L.T. & HARRIS J.W. (1963) The role of mitochondria in ion metabolism in developing erythrocytes. *J. clin. Invest.* **42**, 926.

CRONKITE E.P. & FLIEDNER J.M. (1964) Granulocytopoiesis. *New Eng. J. Med.* **270**, 1347.

CRUICKSHANK J.M. (1970) The effects of parity on the leucocyte count in pregnant and non-pregnant women. *Brit. J. Haemat.* **18**, 531.

CRUICKSHANK J.M. & ALEXANDER M.K. (1970) The effect of age, sex, parity, haemoglobin level, and oral contraceptive preparation on the normal leucocyte count. *Brit. J. Haemat.* **18**, 541.

DACIE J.V. & LEWIS S.M. (1968a) *Practical Haematology*, 4th ed., pp. 12–14. London, J. & A. Churchill Ltd.

Ibid (1968b) *Practical Haematology*, 4th ed., p. 374. London, J. & A. Churchill Ltd.

Ibid (1968c) *Practical Haematology*, 4th ed., p. 409. London, J. & A. Churchill Ltd.

DAVIS A.E. & BADENOCH J. (1962) Iron absorption in pancreatic disease. *Lancet* **2**, 6.

DAVIS P.S. & DELLER D.J. (1967) The effect of orally administered chelating agents E.D.T.A., D.T.P.A. and fructose on radioiron absorption in man. *Aust. Ann. Med.* **16**, 70.

DAVIS P.S., LUKE C.G. & DELLER D.J. (1966) Reduction of iron binding protein in haemochromatosis. *Lancet* **2**, 1431.

DENSON K.W.E., BIGGS R., HADDOW M.E., BORRETT R. & COBB K. (1969) Two types of haemophilia (A^+ and A^-): A study of 48 cases. *Brit. J. Haemat.* **17**, 163.

DELIVORIA-PAPADOPOULUS M., OSKI F.A. & GOTTLIEB A.J. (1969) Oxygen-haemoglobin dissociation curves: effect of inherited enzyme defects of the red cell. *Science* **165**, 601.

DIAMANT Y.Z., THKO E., JADOUSKY E. & POLISUK W.Z. (1970) Leukocyte alkaline phosphate in pregnancy. Comparison with placental phosphatase activity. *Clin. Chim. Acta* **29**, 395.

DINNING J.S. (1962) Nutritional requirements for blood cell formation in experimental animals. *Physiol. Rev.* **42**, 169.

DINTENFASS L., JULIAN D.G. & MILLER G. (1966). Viscosity of blood in healthy young women. Effect of menstrual cycle. *Lancet* **1**, 234.

DINTENFASS L. & ROZENBERG M.C. (1965) The influence of the velocity gradient on *in vitro* blood coagulation and artificial thrombosis. *J. Atheroscler. Res.* **5**, 276.

DINTZIS H.M., BORSOOK M. & VINOGRAD J. (1958) *Microsomal Particles and Protein in Haemochromatosis*, p. 95. New York, Pergamon Press.

DUCKERT F., JUNG E. & SHMERLING D.H. (1960) A hitherto undescribed congenital haemorrhagic diathesis probably due to fibrin stabilizing factor deficiency. *Thromb. Diath. Haemorrh.* **5**, 179.

EGEBERG O. (1965) Inherited antithrombin deficiency causing thrombophilia. *Thromb. Diath. Haemorrh.* **13**, 516.

ESNOUF M.P. & JOBIN F. (1967) The isolation of factor V from bovine plasma. *Biochem. J.* **102**, 660.

ELUEHJEM C.A. (1935) The biological significance of copper and its relation to iron metabolism. *Physiol. Rev.* **15**, 471.

FISHER J.W., TAYLOR G. & PORTEUS D.D. (1965) Localization of erythropoietin in the glomeruli of sheep kidney by fluorescent antibody techniques. *Nature* **205**, 611.

FLEMING A.F., HENDRICKSE J.V. DEV. & ALLAN N.C. (1968) The prevention of megaloblastic anaemia of pregnancy in Nigeria. *J. Obstet. Gynaec. Brit. Cwlth.* **75**, 425.

FLIEDNER J.M., THOMAS E.D., MEYER L.M. & CRONKITE E.P. (1964) Fate of transfused H^3 thymidine labelled bone marrow cells in irradiated recipients: bone marrow conference. *Ann. New York Acad. Sci.* **114**, 510.

FORSHAW N.W.B., THELWELL-JONES A., CHISHOLM W.N. & McGINLEY W.K. (1957) Megaloblastic anaemia of pregnancy and the puerperium. *J. Obstet. Gynaec. Brit. Emp.* **64**, 255.

FOY H. & KONDI A. (1961) A pure red cell aplasia in marasmus and kwashiorkor treated with riboflavin. *Brit. med. J.* **1**, 937.

FRIED W., KILBRIDGE T., KRANTZ S., McDONALD T.P. & LANGE R.D. (1969) Studies on extrarenal erythropoietin. *J. Lab. clin. Med.* **73**, 244.

GAARDER A., JONSEN J., LALAND S., HELLEN A. & OWREN P. (1961) Adenosine diphosphate in red cells as a factor in the adhesiveness of human blood platelets. *Nature* **192**, 531.

GATENBY P.B.B. & LILLIE E.W. (1960) Clinical analysis of 100 cases of severe megaloblastic anaemia of pregnancy. *Brit. med. J.* **2**, 1111.

GAYNOR E. (1971) Increased mitotic activity on rabbit endothelium after endotoxin: an autoradiographic study. *Lab. Invest.* **24**, 318.

GILES C. (1966) An account of 335 cases of megaloblastic anaemia of pregnancy of the puerperium. *J. clin. Path.* **19**, 1.

GITLIN D. & BORGES W.H. (1953) Studies on the metabolism of fibrinogen in two patients with congenital afibrinogenaemia. *Blood* **8**, 679.

GOLDBERG A. (1963) The anaemia of scurvy. *Quart. J. Med.* **32**, 51.

GORMAN J.J. & CASTALDI, P.A. (1971) The anticoagulant action of fibrinopeptide A (Abstr.). *Proc. Aust. Soc. Med. Res.* **2**, 446.

GRANT W.C. & ROOT W.S. (1952) Fundamental stimulus for erythropoiesis. *Physiol. Rev.* **32**, 449.

GRETTE K. (1962) Studies on the mechanism of thrombin-catalysed hemostatic reactions in blood platelets. *Acta physiol. scand.* **56** (Suppl.), 195.

HABER S. (1964) Norethynodrel in the treatment of factor X deficiency. *Arch. intern. Med.* **114**, 89.

HALL C.A. & FINKLER A.E. (1965) The dynamics of transcobalamin II. A vitamin B_{12} binding substance in the plasma. *J. Lab. clin. Med.* **65**, 459.

HARKER L.A. (1970) Platelet production. *New Eng. J. Med.* **282**, 492.

HARKER L.A. & SLICHTER S.J. (1970) Studies of platelet and fibrinogen kinetics in patients with prosthetic heart valves. Personal communication.

HARRIS J.W. (1963) *The Red Cell*, p. 172. Harvard University Press.

HEMKER H.C. & KAHN M.J.P. (1967) Reaction sequence of blood coagulation. *Nature* **215**, 1201.

HERBERT V. (1962a) Experimental induction of folate deficiency in man. *Trans. Ass. Amer. Phycns.* **75**, 307.

HERBERT V. (1962b) Minimal daily adult folate requirement. *Arch. Int. Med.* **110**, 649.

HERBERT V. (1968) Diagnostic and prognostic values of measurement of serum vitamin B_{12} binding proteins. *Blood* **32**, 305..

HIBBARD B. (1964) The role of folic acid in pregnancy. *J. Obstet. Gynaec. Brit. Cwlth.* **71**, 529.

HILLMAN R.W., CABAUD P.G., MILSSON D.E., ARPIN P.D. & TUKANO R. (1963) Pyridoxine supplementation during pregnancy. *Amer. J. clin. Nutr.* **12**, 427.

HJORT P.F. & HASSELBACK R. (1961) A critical review of the evidence for a continuous hemostasis *in vivo*. *Thromb. Diath. Haemorrh.* **6**, 580.

HOCKING D.R. & CARTER N.G. (1967) Diverticula of ileum with megaloblastic anaemia. *Med. J. Aust.* **1**, 444.

HOCKING D.R. & CASTALDI P.A. (1971) Haemopoiesis and coagulation in pregnancy. *Proc. Second Congr. Asian-Pacific Div. Int. Soc. Haemat., Melbourne.*

HODGES R.E., OHLSON M.A. & BEAN W.B. (1958) Pantothenic acid deficiency in man. *J. clin. Invest.* **37**, 1642.

HODGSON G. & TOHÀ J. (1954) The erythropoietic effect of urine and plasma of repeatedly bled rabbits. *Blood* **9**, 299.

HOEDMACKER P.J., ABELS J, WASHTERS J.J., ARENDS A. & NIEWEG H.D. (1964) Observations about the site of production of Castle's gastric intrinsic factor. *Lab. Invest.* **13**, 1394.

HORWITT M.K., HARVEY C.C., ROTHWELL W.S., CUTLER J.L. & HAFRON D. (1956) Trylophan-niacin relationships in man. *J. Nutr.* **60** (Suppl.), 1.

HRODEK O. (1966) Blood platelets in the newborn. *Acta Univ. Carol. Med., Prague.*

HUSSAIN R., WALKER R.E., LAYRISSE M., CLARK P. & FINCH C.A. (1965) Nutritive value of food irons. *Amer. J. clin. Nutr.* **16**, 464.

JACKSON I.M.D., DOIG W.B. & MCDONALD G. (1967) P.A. as a cause of infertility. *Lancet* **11**, 1159.

JACOBS P., BOTHWELL T.H. & CHARLTON R.W. (1964) Role of hydrochloric acid in iron absorption. *J. appl. Physiol.* **19**, 187.

JACOBS, P., BOTHWELL T.H. & CHARLTON R.W. (1966) Intestinal iron transport: studies using a loop of gut with an artificial circulation. *Amer. J. Physiol.* **210**, 694.

JAFFE E.R. (1970) Hereditary haemolytic disorders and enzymatic deficiencies in human erythrocytes. *Blood* **35**, 116.

JANDL J.H. & GABUZDA G.J. (1953) Potentiation of pteroylglutamic acid by ascorbic acid in the anaemia of scurvy. *Proc. Soc. Exp. Biol. Med.* **84**, 452.

JEPSON J. & LOWENSTEIN L. (1968) Hormonal control of erythropoiesis during pregnancy in the mouse. *Brit. J. Haemat.* **14**, 555.

JOBIN F. & ESNOUF M.P. (1967) Studies on the formation of the prothrombin converting complex. *Biochem. J.* **102**, 666.

JOCUM N.O. (1970) Some characteristics of the clottable protein of limulus polythemus blood cells. *Thromb. Diath. Haemorrh.* **23**, 170.

JOHNSON S.A., BALBOA R.S., PEDERSON H.J. & BUCKLEY M. (1965) The ultrastructure of platelet participation in hemostasis. *Thromb. Diath. Haemorrh.* **13**, 65.

KAPPAS A. & PALMER R.H. (1963) Selected aspects of steroid pharmacology. *Pharm. Rev.* **15**, 123.
KASPER C.K., HOAG M.S., AGGELER P.M. & STONE S. (1964) Blood clotting factors in pregnancy: factor VIII concentrations in normal and AHF-deficient women. *J. Obstet. Gynec.* **24**, 242.
KLEINER G.J., MERSKEY C., JOHNSON A.J. & MARKUS W.B. (1970) Defibrination in normal and abnormal parturition. *Brit. J. Haemat.* **19**, 159.
KLINE D.L. (1966) Chemistry and biochemistry of the fibrinolytic system. *Federation Proc.* **25**, 31.
KOHN J., MOLLIN D.L. & ROSENBACH L.M. (1961) Conventional voltage electrophoresis in forminoglutamic acid determination in folic acid deficiency. *J. clin. Path.* **14**, 345.
KOWALSKI E. (1968) Fibrinogen derivatives and their biologic activities. *Seminar Haemat.* **5**, 45.
KUMENTO A., LOPEZ R., LUHBY A.L. & HALL C.A. (1967) B_{12} binders in human cord serum. *Clin. Rev.* **15**, 283.
KWAAN H.C., LO R. & MCFADZEAN A.J.I. (1967) On the production of plasma fibrinolytic activity within veins. *Clin. Sci.* **16**, 241.
KYNES M. (1968) Iron metabolism. *J. clin. Path.* **1**, 57.
LAKI K. (1951) The action of thrombin on fibrinogen. *Science* **114**, 435.
LAKI K. & LORAND L. (1948) On the solubility of fibrin clots. *Science* **108**, 280.
LANE M. & ALFREY C. JR. (1965) The anaemia of human riboflavin deficiency. *Blood* **25**, 432.
LANGE R.D., MCDONALD T.P. & JORDAN T. (1969) Antisera to erythropoietin. Partial characterization of two different antibodies. *J. Lab. clin. Med.* **73**, 78.
LAWRENCE A.C.K. (1962) Iron status in pregnancy. *J. Obstet. Gynaec. Brit. Cwlth.* **69**, 29.
LEHMANN H. & HUNTSMAN R.G. (1966) *Man's Haemoglobins*, p. 47. North-Holland Publishing Co.
LESUK A., TERMINILLO L. & TRAVER J.H. (1965) Crystalline human urokinase: some properties. *Science* **147**, 880.
LORAND L. (1965) Physiological roles of fibrinogen and fibrin. *Federation Proc.* **24**, 784.
LORAND L., DOOLITTLE R.F., KONISHI K. & RIGGS S.K. (1963) A new class of blood coagulation inhibitors. *Arch. Biochem. Biophys.* **102**, 171.
LORAND L., URAYAMA T., DEKIEWIET J.N.C. & NOSSEL H.C. (1969) Diagnostic and genetic studies on fibrin stabilizing factor with a new assay based on amine incorporation. *J. clin. Invest.* **48**, 1054.
LUND C.J. (1951) Studies on iron deficiency anaemia of pregnancy. *Amer. J. Obstet. Gynec.* **62**, 947.
MACFARLANE R.G. (1964) An enzyme cascade in the blood clotting mechanism, and its function as a biochemical amplifier. *Nature* **202**, 498.
MACINTOSH D.M. & KYOBE J. (1966) Fibrinolytic activity in pregnancy during parturition and in the puerperium. *Lancet* **2**, 874.
MACMILLAN D.C. (1966) Secondary clumping effect in human citrated platelet rich plasma produced by adenosine diphosphate and adrenaline. *Nature* **211**, 140.

McFarlane A.S., Todd D. & Cormwell S. (1964) Fibrinogen catabolism in humans. *Clin. Sci.* **26**, 415.

McLean J.R., Maxwell R.E. & Hertler D. (1964) Fibrinogen and adenosine diphosphate induced aggregation of platelets. *Nature* **202**, 605.

Magee H.E. & Milligan E.H.M. (1951) Haemoglobin levels before and after labour. *Brit. med. J.* **2**, 1307.

Majaj A.S. (1966) Vitamin E-responsive macrocytic anaemia in protein-calorie malnutrition. *Amer. J. clin. Nutr.* **18**, 362.

Marcus A.J., Zucker-Franklin D., Safier L.B. & Ullman H.L. (1966) Studies on human platelet granules and membranes. *J. clin. Invest.* **45**, 19.

Margolis J. (1957) Initiation of blood coagulation by glass and related surfaces. *J. Physiol.* **137**, 95.

Margolis J. & Bruce S. (1964) An experimental approach to the kinetics of blood coagulation. *Brit. J. Haemat.* **10**, 513.

Marks P.A., Johnson A.B. & Hirschberg E. (1958) Effect of age on enzyme activity of erythrocytes. *Proc. nat. Acad. Sci.* **44**, 529.

Medici P.T., Gordon A.S., Piliero S.B., Luhby A.L. & Yuceoglu P. (1957) Influence of transfusions on the erythropoietic stimulating factor (E.S.F.) of anaemic patients. *Acta Haemat.* **18**, 325.

Menon M.K.K. & Rajan L. (1962) Prophylaxis of anaemia of pregnancy. *J. Obstet. & Gynaec. of India* **12**, 9.

Mookerje A.S. & Hawkins W.W. (1960) Haemopoiesis in the rat in riboflavin deficiency. *Brit. J. Nutr.* **14**, 239.

Moore C.V. (1965) Iron nutrition and requirements. *Series Haemat.* **6**, 1.

Morawitz P. (1905) Die chemie der Blutgerinnung. *Ergebn Physiol.* **4**, 307.

Morley A.A. (1966) A neutrophil cycle in healthy individuals. *Lancet* **2**, 1220.

Morley A. (1969) Blood cell cycles in polycythaemia vera. *Aust. Ann. Med.* **18**, 124.

Murphy G.P., Kenny G.M. & Mirand E.A. (1970) Erythropoietin levels in patient with renal tumours or cysts. *Cancer* **26**, 191.

Murphy E.A., Robinson G.A., Russell H.C., Ozge A. & Crookston S.H. (1964) Blood platelet survival. *Thromb. Diath. Haemorrh.* **13** (Suppl.) 245.

Nachman R.L. (1968) Platelet proteins. *Seminars Haemat.* **5**, 18.

Nichol C.A. & Walsh A.D. (1955) Synthesis of citrovorum factor from folic acid by liver slices; augmentation by ascorbic acid. *Science* **121**, 275.

Nilsson I.M., Blombäck M., Thilen A. & Francken I.L. (1959) Carriers of haemophilia A. *Acta med. scand.* **165**, 357.

Nossel H.L., Lanzkowsky P., Cevy S., Mibaskan R.S. & Hansen J.D.L. (1966) Factor levels in pregnancy. *Thromb. Diath. Haemorrh.* **16**, 185.

O'Brien J.R. (1962) Platelet aggregation. II. Some results from a new method of study. *J. clin. Path.* **15**, 452.

Oliver M.F. & Boyd G.S. (1955) Plasma lipid and serum lipoprotein patterns during pregnancy and puerperium. *Clin. Sci.* **14**, 15.

O'Neill B. & Firkin B.G. (1964) Platelet survival studies in coagulation disorders, thrombocythaemia, and conditions associated with thrombocythaemia. *J. Lab. clin. Med.* **64**, 188.

Owren P.A. (1947) The coagulation of blood, investigation on a new clotting factor. *Acta med. scand.* (Suppl.), 194.
Pease D.C. (1956) Electron microscopic study of red bone marrow. *Blood* **11**, 501.
Pederson H.J., Tebo T.H. & Johnson S.A. (1967) Evidence of haemolysis in the initiation of haemostasis. *Amer. J. clin. Path.* **48**, 62.
Penny R., Galton D.A.G., Eisen V. & Scott J.T. (1966) Studies on neutrophil function. I. Physiological pharmacological aspects. *Brit. J. Haemat.* **12**, 623.
Penny R., Rozenberg M.C. & Firkin B.G. (1966) The splenic platelet pool. *Blood*, **27**, 1.
Perry S., Godwin H.A. & Zimmerman T.S. (1968) Physiology of the granulocyte. *J. Amer. med. ass.* **203**, 135.
Peterson R.A., Kurll P.E., Finely P. & Ettinger M.G. (1970) Changes in antithrombin III and plasminogen induced by oral contraceptives. *Amer. J. clin. Path.* **53**, 468.
Pfueller S., Somer J.B. & Castaldi P.A. (1969) Haemophilia B due to an abnormal factor IX. *Coagulation* **2**, 213.
Piliero S.J., Medici P.T., Pansky B., Luhby A.L. & Gordon A.S. (1956) Erythropoietic stimulating effects of plasma extracts from anaemic human subjects. *Proc. Soc. Exp. Biol.* **93**, 302.
Pisano J.J., Finlayson M. & Peyton P. (1968) Cross-link in fibrin polymeriza by factor XIII: -(-glutamyl) lysine. *Science* **160**, 892.
Plzak L.F., Fried W., Jacobson L.O. & Bethard W.F. (1955) Demonstration of stimulations of erythropoiesis by plasma from anaemic rats using Fe^{59}. *J. Lab. clin. Med.* **46**, 671.
Pohle F.J. (1939) The blood platelet count in relation to the menstrual cycle in normal women. *Amer. J. Med. Sci.* **197**, 40.
Poller L. & Thomson J.M. (1968) Effects of low-dose oral contraceptives on blood coagulation. *Brit. med. J.* **3**, 218.
Pollycove M. (1966) Iron metabolism and kinetics. *Seminars in Haemat.* **3**, 235.
Prankerd T.A.J. (1965) Non-immunological mechanisms of red cell destruction. *Series Haematology* **1**, 53.
Pugatch E.M.J., Forster E.A., Macfarlane D.E. & Poole J.C.F. (1970) The extraction and separation of activators and inhibitors of fibrinolysis from bovine endothelium and mesothelium. *Brit. J. Haemat.* **18**, 669.
Quick A.J. (1935) The prothrombin in haemophilia and in obstructive jaundice. *J. biol. Chem.* **109**, 73.
Rabinovitch M. (1968) Phagocytosis: the engulfment stage. *Seminars in Haemat.* **5**, 134.
Ratnoff O.D. The biology and pathology of the initial stages of blood coagulation, in E.Brown & C.V.Moore (eds.) *Progress in Haematology*, p. 204. New York, Grune & Stratton.
Ratnoff O.D., Colopy J.E. & Pritchard J.A. (1954) The blood clotting mechanism during normal parturition. *J. Lab. clin. Med.* **44**, 408.
Regoeczi E. & Hobbs K.R. (1969) Fibrinogen turnover in pregnancy. *Scand. J. Haemat.* **6**, 175.

Appendix A
Steroid Biosynthesis and Metabolism

BRYAN HUDSON

A comprehensive knowledge of all the biochemical events that occur in the synthesis and metabolism of steroid hormones may not be necessary for all those interested in reproductive physiology. The intention of this appendix is to provide the reader with some knowledge of the principles of steroid biosynthesis and metabolism but not to describe them in great detail. More comprehensive accounts can be found elsewhere (Klyne 1957, Dorfman & Ungar 1965).

The adrenal cortex, the ovary and the testis are the three important sites of steroid biosynthesis in the normal individual. Almost certainly the early events leading to the synthesis of cholesterol are common to all these tissues; thereafter, the formation of individual steroid hormones depends largely on the tissues in which the synthesis is occurring, so that the adrenal cortex forms mainly corticosteroids; the ovary, oestrogens and progesterone; and the testis, mainly testosterone. During pregnancy the placenta plays an additional role in steroid synthesis and metabolism.

Steroid chemistry

The basic carbon skeleton common to all steroids is the cyclophenanthrene nucleus in which carbon atoms in three cyclohexane and one cyclopentane ring are joined and can be considered to lie in a single plane. These atoms are numbered by convention as shown in Fig. A.1 and the rings are designated A to D as shown. The addition of a methyl group (CH_3) to carbon-13 gives a steroid nucleus containing 18 carbon atoms; this is the oestrane nucleus from which the oestrogens are

APPENDIX A

derived and the attached methyl group is described as the C-18 angular methyl grouping. The addition of another methyl group at C-10 gives a steroid nucleus containing 19 carbon atoms; this is called the androstane nucleus and the additional angular methyl group is C-19; this is the basic nucleus for the androgens. These added angular methyl groupings have been shown to lie above the plane of the carbon skeleton and serve as reference points to describe the relative positions of other substituent groupings. For instance, there are two hydrogen atoms attached to carbon atom 3, one of which lies above the plane of the molecule and the other below this plane. Thus, if the flat plane of this page could be

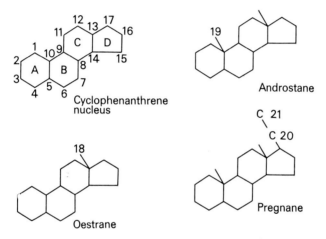

FIG. A.1. Reference nuclei in steroid nomenclature.

taken as the plane of the molecule, one hydrogen atom projects up from the page and the other down into the book; the C-18 and C-19 angular methyl groups both project upwards from the page. The different positions of the hydrogen atoms are by convention said to be *cis* or β if they lie above the plane of the ring and *trans* or α if they lie below the plane of the ring. When depicting the location of these atoms (or substituent groupings) relative to the ring, a solid line is drawn to indicate that the location of the atom is in the *cis* or β position, while a broken line indicates that the attachment is in the *trans* or α position. Thus, in the androstane nucleus, the hydrogen at C-5 may be either *cis* or *trans*, and the nucleus may be said to be 5β-androstane or 5α-androstane respectively.

The addition of a two carbon side-chain at C-17 yields a steroid nucleus that now contains 21 carbon atoms; this is the pregnane nucleus and is the basic structure on which progesterone and corticosteroids are built. In this nucleus the hydrogen atom attached to C-5 may, as in androstane, be either *cis* or *trans*, when the nucleus is said to be 5β-pregnane or 5α-pregnane (sometimes called *allo*pregnane). It is important to point out that these steroid nuclei—oestrane, androstane and pregnane—are not naturally occurring compounds and are devoid of hormonal action.

The nature of the modifications to these nuclei which are required before they become hormonally active will be described in the section on steroid biosynthesis. Several such modifications can occur; thus, two hydrogen atoms can be lost between two adjacent carbon atoms. Most commonly this occurs between carbon atoms 4 and 5 in ring A or between 5 and 6 in ring B. This unsaturation in ring A is a fundamental chemical property of the hormonally active corticosteroids and of progesterone and testosterone. The double bond in ring B between carbon atoms 5 and 6 is found in a number of naturally occurring products; cholesterol (which possesses an additional 6-carbon atoms on the side-chain), pregnenolone and dehydroepiandrosterone (DHEA).

Other important modifications to these nuclei include the addition of substituent groups of which the formation of alcohols and ketones most commonly occurs naturally. Thus, the molecule of cortisol has a basic pregnane structure, modified by the addition of three substituent alcoholic (or hydroxyl) groupings, two ketones and one double bond. This is depicted in Fig. A.2 in which it may be noted that the hydroxyl at C-11 is *cis* or β (and thus lies above the plane of the nucleus) while the hydroxyl at C-17 is *trans* or α and lies below the plane of the nucleus. Although the trivial name for this compound is cortisol or hydrocortisone, the official name is pregn-4-ene,11β,17α,21-triol-3,20 dione—suffixes 'ol' and 'one' being used to describe alcoholic ketonic groupings respectively. Likewise, with testosterone (Fig. A.2) the androstane nucleus is modified by the addition of one hydroxyl and one ketonic grouping with a double bond between carbon atoms 4 and 5. For this compound the trivial name is testosterone, while the official name is androst-4-ene-17βol-3-one.

An important modification to the A ring of oestrogens is the aromatization of the ring from its original cyclohexane structure to a benzene one. This, coupled with the addition of an hydroxyl substituent at C-3 makes the A ring phenolic in nature.

Steroids and their metabolites may undergo conjugation to form esters. Conjugation with glucuronic or sulphuric acids to form glucuronides or sulphates is the usual way in which steroid metabolites are *excreted*. Certain steroids, notably DHEA, may be *secreted* as sulphates.

Early steps in steroid biosynthesis

The basic building block for the biosynthesis of steroids is acetate—a two carbon compound. In 1953 Hechter *et al* showed that if isolated bovine adrenals were perfused with ^{14}C-acetate, radioactive cortisol and corticosterone were produced. These experiments have been con-

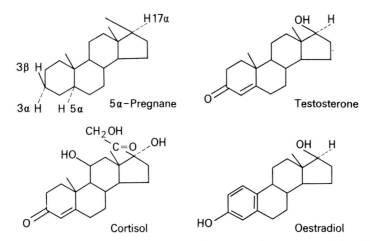

FIG. A.2. Modifications of pregnane, androstane and oestrane nuclei to give compounds with biological activity.

firmed in a variety of other mammalian adrenal cortical tissues. Experiments such as these also yield ^{14}C-cholesterol and there has been considerable debate as to whether cholesterol is an obligatory intermediate in steroid biosynthesis. Improved radiochemical techniques over the past 10 or more years have enabled some clarification of the role of cholesterol. It is now clear that significant amounts are formed locally in steroid producing tissues, and that it is normally an obligatory intermediate.

The starting point for cholesterol biosynthesis is acetyl-CoA which is derived from the catabolism of carbohydrate, fat and protein. Three molecules of acetyl CoA condense to form β-hydroxy-β-methyl glutaryl

CoA (HMG-CoA) which is then reduced to mevalonic acid, the formation of which is the first step that is unique to cholesterol synthesis. This irreversible step controls the rate of sterol synthesis and is catalysed by the enzyme HMG-CoA reductase.

Once formed, mevalonate is phosphorylated by ATP to form several active phosphorylated intermediates, and then decarboxylated to form isopentenyl pyrophosphate and its isomer dimethylallyl pyrophosphate which contain 5 carbon atoms. Six of these isoprenoid units then undergo a series of condensation reactions to form geranyl pyrophosphate (10 carbons), farnesyl pyrophosphate (15 carbons) and ultimately squalene (30 carbons).

Squalene has an open ring structure which resembles the steroid nucleus; this is converted to lanosterol by ring closure. Zymosterol is formed by the loss of three carbon atoms from demethylation. By rearrangement of one and the reduction of another double bond in this compound, cholesterol is formed.

Later stages in steroid biosynthesis

With the formation of cholesterol, the subsequent biosynthesis of steroids can occur by a number of different pathways. The first step in any of these is the loss of six carbon atoms from the cholesterol side-chain with the formation of pregnenolone. The reaction occurs within the mitochondria, involves the formation of 20α-hydroxycholesterol as an intermediate and is probably an important site of action of trophic hormones. The precise details of this side-chain cleavage and the nature of the reaction on which the formation of pregnenolone is based, have been reviewed by Hall (1970).

The first step in the *classical* pathway of steroid synthesis from pregnenolone is the formation of progesterone. Other pathways commence with the formation of 17α-hydroxypregnenolone or of pregnenolone sulphate. In the formation of progesterone two steps are involved; first, the conversion of the 3β-hydroxyl into a 3-oxo grouping; second, the isomerization of this intermediate which involves a shift of the double bond between carbon atoms 5 and 6, so that this is now found between carbon atoms 4 and 5. The two enzymes required for these transformations are the 3β-hydroxysteroid dehydrogenase which is microsomal and an isomerase which is cytoplasmic. This reaction is shown in Fig. A.5.

Once formed, progesterone may enter one of two biosynthetic

APPENDIX A

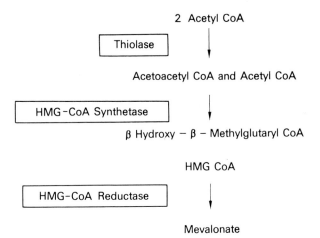

FIG. A.3. Initial steps in steroid synthesis.

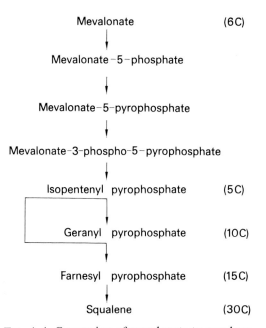

FIG. A.4. Conversion of mevalonate to squalene.

metabolism for corticosteroids and androgens is reductive; the first step is the reduction of the double bond between carbon atoms 4 and 5 with the formation of a *dihydro* compound; e.g. from cortisol this produces dihydrocortisol which is biologically inactive. This is the rate-limiting step, and is impaired in liver disease.

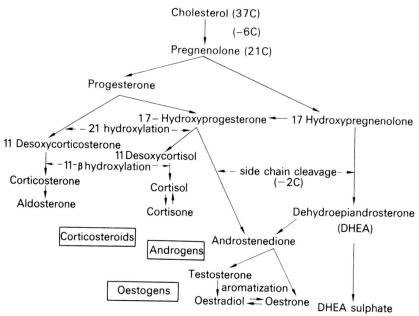

FIG. A.6. Pathways of steroid biosynthesis.

The next step is the reduction of the 3-oxo grouping to give a *tetrahydro* compound. Thus, in the metabolism of cortisol this yields tetrahydrocortisol, and in the metabolism of androstenedione gives androsterone (5α-androstane-3αol-17-one) or aetiocholanolone (5β-androstane-3αol-17-one). The next step is the conjugation with glucuronic or sulphuric acid to form glucuronide or sulphate conjugates. Corticosteroids are mainly excreted as glucuronides, while in the metabolism of androgens both glucuronides and sulphates are formed. The microsomal enzyme responsible for glucuronide formation is a glucuronyl transferase which catalyses the transfer of the glucuronide moiety from uridine diphosphoglucuronic acid to a hydroxyl group of the steroid. Glucuronides, which are readily water-soluble, are filtered at the glomerulus, and undergo very little tubular reabsorption. Sulphotransferases are soluble enzymes which transfer activated sulphate groups to a steroid hydroxyl grouping from 3'-phosphoadenosine 5'-phosphosulphate. The formation of tetrahydrocortisol glucuronide is shown in Fig. A.7.

Although the major pathway of corticosteroid catabolism is by the formation of tetrahydrocortisol or tetrahydrocortisone glucuronides,

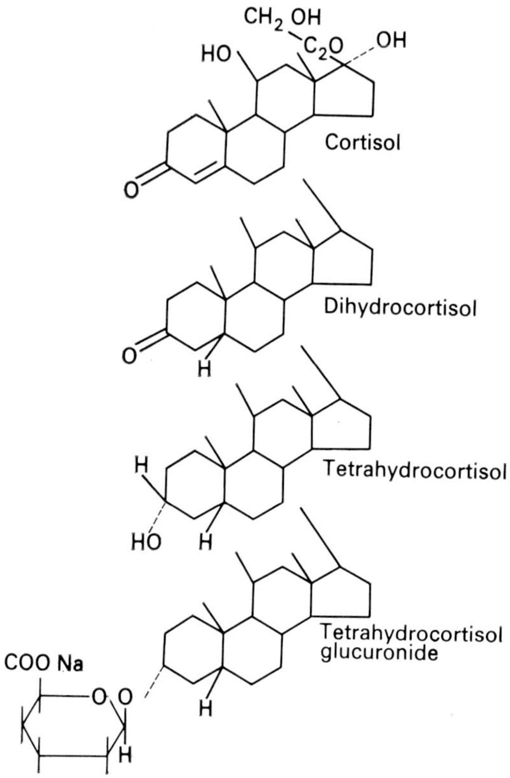

FIG. A.7. Catabolism of cortisol.

somewhere between 5 and 10% is further catabolized within the liver by the removal of the side-chain to form 11-hydroxy or 11-oxo-17-oxosteroid glucuronides such as 11-oxoaetiocholanolone and 11-hydroxyandrosterone. The further catabolism of testosterone principally involves the formation of tetrahydro 17-oxo compounds such as androsterone and aetiocholanolone which are conjugated with glucuronic acid (mainly) or sulphuric acid. Somewhat less than 1% of testosterone is directly conjugated with glucuronic acid to form testosterone glucuronide. The further metabolism of the three principal androgens is shown in Fig. A.8.

The catabolism of oestrogens is somewhat different from that of the corticosteroids, progesterone or androgens. Probably the most important metabolite is oestriol which is also formed in the liver. Many other metabolites of oestrone and oestradiol have been demonstrated such as: 2-hydroxyoestrone, 2-methoxyoestrone, 16α- and 16β hydroxyoestrone. These compounds, like corticosteroids and androgen metabolites are conjugated with glucuronic and sulphuric acids and excreted as glucuronides and sulphates; sometimes mixed sulphate and glucuronide conjugates are formed.

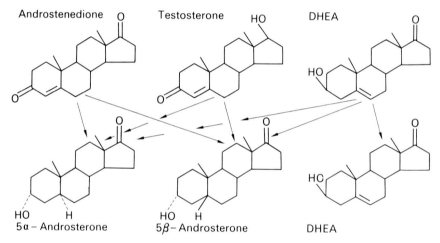

FIG. A.8. Catabolism of androgens.

Steroid secretion and metabolism by the placenta

During pregnancy the placenta assumes an important role in steroid metabolism. There is no direct evidence to suggest that the placenta is able to form steroids *de novo*—that is, from acetate. However, the foetal-placental unit converts steroids from proximate precursors both from mother and foetus. The important precursors from the maternal circulation are the sulphates of pregnenolone, 17α-hydroxypregnenolone and DHEA. These steroid sulphates, having crossed from the maternal circulation to the placenta are readily converted to free steroids by steroid sulphatases present in placental tissue. The placenta contains an abundance of the 3β-hydroxysteroid dehydrogenase and isomerase enzymes which enable an efficient conversion of Δ^5-3β hydroxysteroid precurors to progesterone, 17α-hydroxyprogesterone and androstenedione. Much of the progesterone formed is transferred to the foetus where it is further metabolized, usually by a series of hydroxylations in the foetal adrenal, testis and liver. In this way, cortisol is formed and further catabolized within the foetus. Progesterone formed by the placenta also passes into the maternal circulation, and is further metabolized to pregnanediol and pregnanolone by the mother. Maternal plasma levels of progesterone rise to a peak of about 15 μg/100 ml at term.

One of the striking changes in the pattern of urinary steroid excretion during pregnancy is the marked increase in the excretion of oestriol. This originates mainly from the foetus by aromatization of precursors—both maternal and foetal—such as DHEA and its sulphate, and from androstenedione and testosterone. The importance of foetal tissues in the formation of oestriol is shown by the fact that the urinary excretion rate of oestriol in a mother with an anencephalic foetus is considerably lower than normal. The excretion of oestriol throughout pregnancy slowly increases until term and is used as an index of foetal viability. More details concerning the role of the placenta in the synthesis and metabolism of steroid hormones may be found in reviews by Diczfalusy & Troen (1961), Ryan (1962) and Diczfalusy (1966). It is discussed further in chapter 4.

REFERENCES

BAULIEU E-E. (1962) Studies of conjugated 17-ketosteroids in a case of adrenal tumor. *J. clin. Endocr.* **22,** 501–510.

BAULIEU E-E., CORPÉCHOT C., DRAY F., EMILIOZZI R., LEBEAU M-C., MAUVIS-JARVIS P. & ROBEL P. (1965) An adrenal-secreted 'androgen': dehydroisoandrosterone sulfate. Its metabolism and a tentative generalization on the metabolism of other steroid conjugates in man. *Rec. Progr. Horm. Res.* **21,** 411–494.

DESPHANDE N., JENSEN V., CARSON P., BULBROOK R.D. & DOOUS T.W. (1970) Adrenal function in breast cancer: biogenesis of androgens by the human adrenal gland *in vivo. J. Endocr.* **47,** 231–242.

DICZFALUSY E. & TROEN P. (1961) Endocrine functions of the human placenta. *Vitamins & Hormones* **19,** 229–250.

DICZFALUSY E. (1966) Steroid metabolism in pregnancy, in *Proceedings of the Second International Congress on Hormonal Steroids*, pp. 82–93. International Congress Series 132. Amsterdam, Excerpta Medica Foundation.

DORFMAN R.I. & UNGAR F. (1965) *Metabolism of Steroid Hormones*, New York, Academic Press.
HALL P.F. (1970) Gonadotrophic regulation of testicular function, in K.B. Eik-Nes (ed.) *The Androgens of the Testis*, pp. 73–115. New York, Marcel Dekker Inc.
HECHTER O., SOLOMON M.M., ZAFFARONI A. & PINCUS G. (1953) Transformation of cholesterol and acetate of adrenal cortical hormones. *Arch. Biochem. Biophys.* **46**, 201–214.
KLYNE W. (1957) *Chemistry of Steroids*. London, Methuen.
RYAN K.J. (1962) Hormones of the placenta. *Amer. J. Obstet. Gynec.* **84**, 1695–1713.

Appendix B

Gonadotrophins—Chemistry and Measurement

HENRY G. BURGER

INTRODUCTION

The gonadotrophins, protein hormones which stimulate the functional activities of the gonads, are secreted by the pituitary and the placenta. Follicle stimulating hormone (FSH) promotes ovarian follicular growth and development, and acts on the testicular seminiferous epithelium; luteinizing hormone (LH) (alternatively known as interstitial cell stimulating hormone, ICSH) is concerned with ovarian and testicular steroid biosynthesis, ovulation and the function of the corpus luteum. Both of these gonadotrophins are secreted by the pituitary glands of both sexes, circulate in the peripheral blood, and are excreted in the urine. The generic term human pituitary gonadotrophin (HPG) is used to refer to pituitary extracts containing both FSH and LH; the term human menopausal gonadotrophin (HMG) refers to extracts prepared from the urine of post-menopausal females, again containing both biological activities.

The placental gonadotrophin in man is human chorionic gonadotrophin (HCG); it is secreted by the cells of the syncytio-trophoblast into the maternal circulation, and is excreted in large amounts in the urine. In its biological activities and chemical structure, it closely resembles LH.

Pituitary hormones with follicle stimulating and luteinizing properties are found in all other species of vertebrates: the nature of the placental hormones, however, shows species variation: one preparation which is used therapeutically and for biological assay is the equine placental hormone, pregnant mare's serum gonadotrophin (PMSG) which has follicle stimulating activity.

It is the purpose of this appendix to review the chemistry of the gonadotrophins and their assay both by biological and immunological methods. Normal values for the concentration of FSH and LH in biological fluids are given, and the question of appropriate reference standards for the various types of assay is discussed.

CHEMISTRY

The pituitary and placental gonadotrophins (together with thyroid stimulating hormone (TSH)) are glycoproteins i.e. proteins to which carbohydrate residues are co-valently linked. The carbohydrate moieties include hexose, hexosamine and sialic acid. Experience with experimentally produced antisera against FSH, LH and HCG indicates that these gonadotrophins and TSH share certain common antigenic sites (Odell et al 1968b, Ryan & Faiman 1968, Schlaff et al 1968) while other areas of the molecules are antigenically specific. Such observations suggest that these proteins share common structural features, and additional evidence for this concept has been reported by Papkoff & Ekblad (1970), who showed that ovine FSH and LH shared a common structural sub-unit (for further discussion of sub-unit structures—see below).

There is a considerable body of evidence which indicates that the structure of gonadotrophins extracted from the pituitary is not identical with that of the material excreted in urine: detailed studies of the nature of the circulating gonadotrophins have not been published. The differences between pituitary and urinary gonadotrophins will be discussed for each of the hormones individually.

Follicle stimulating hormone

Of the gonadotrophic hormones, FSH is the least stable, and has proved most difficult to purify, although a comparatively pure preparation was isolated in 1949 (Li et al). Preparations of increasing biological specific activity have been reported by Roos & Gemzell in 1964, Butt et al (1967), Roos (1967) and Saxena & Rathnam (1967), with a maximum of 14,000 I.U./mg. reported by Roos and confirmed by Peckham & Parlow (1969). For discussion of standards and assays, see below.

Yields of pure FSH from the pituitary are very low, e.g. 3–20 mg per kg wet weight of glands; the purified material has a molecular weight of approximately 30,000, and an isoelectric point at pH 4.2. The amino-acid composition includes a relatively high content of aspartic and glutamic acids, consistent with the acidic nature of the protein. The carbohydrate portion of the molecule is 17–18%, by weight, and must be taken into account when bioassay potency estimates of FSH are expressed in terms of a weight of material estimated by ultraviolet absorption spectrophotometry; the latter measurement estimates protein content but does not include the carbohydrate residues. The biological activity of FSH is destroyed by removal of sialic acid and neutral sugars; varying effects on biological activity of treatment with urea and with proteolytic enzymes have been reported; however, FSH is relatively stable to digestion with chymotrypsin, and advantage has been taken of the lability of LH to this enzyme in order to prepare FSH relatively free of LH.

Recent evidence indicates that FSH may consist of two sub-units of approximately equal molecular weight (Crooke & Gray 1968), and that dissociation into those sub-units leads to a large loss of biological activity.

The properties of urinary FSH differ somewhat from those of the material extracted from the pituitary: evidence for this was presented in detail by Roos (1967) who showed that the urinary material was of lower molecular weight, differed in its amino-acid composition and its carbohydrate residues, and was of lower biological specific activity (780 as compared with 14,000 I.U. per mg).

Luteinizing hormone

Purified LH preparations have been reported by a number of authors including Hartree *et al* 1964, Reichert & Parlow 1964, Parlow *et al* 1965, Hartree 1966. Specific biological activities of between 2500 and 4500 I.U./mg have been achieved.

Yields of purified LH from the pituitary are greater than those of FSH, e.g. a yield of approximately 140 mg per kg wet weight of pituitary glands has been reported by Hartree (1966).

The molecular weight has been reported to range from 28,000 to 42,000 (Reichert & Jiang 1965, Reichert *et al* 1968) but the former figure, obtained by calculation of the Stokes' radius and the sedimentation constant, is more likely to be accurate than figures derived from gel filtration on Sephadex columns. Squire *et al* (1962) obtained a value of 26,000. There is evidence that human (and ovine) LH dissociates into two approximately equal sub-units when diluted or exposed to acid conditions, 4 M guanidine hydrochloride or 8 M urea (Reichert & Midgley 1968, Hartree 1968, de la Llosa 1968, and Ryan 1968). Such dissociation is again accompanied by loss of biological activity. As indicated above, treatment of LH with chymotrypsin also results in loss of biological activity, although some immunological reactivity is retained.

Carbohydrate moieties make up 40% of the LH molecule by weight (Kathan *et al* 1967).

The properties of LH as it is excreted in urine differ somewhat from those in plasma; for instance, urinary LH is not inactivated by chymotrypsin treatment (Reichert 1968) and the relative biological and immunological potencies of the two materials also differ markedly (see below).

Human chorionic gonadotrophin

As already indicated, HCG resembles LH closely in its chemical composition, biological properties and immunological reactivity. However, the two proteins are not identical, and can be separated from each other, e.g. when they are both present in pregnancy plasma (Midgley *et al* 1967). Recent studies of the properties of HCG include those of van Hell *et al* (1968) and Bell *et al* (1969).

Because of the ready availability of HCG, it has been used diagnostically and therapeutically as a substitute for LH.

METHODS OF ASSAY

Introduction

Hormones may be measured by biological, chemical, immunological and radio-isotopic methods, or by combinations of these. Because the gonadotrophins are proteins, present in low concentrations (of the order of 10^{-10} M) in biological fluids which contain many proteins in much higher concentration (e.g. albumin approximately 10^{-3} M), chemical methods are unsuitable, lacking both sensitivity and specificity. Until the development of immunological methods of assay in the 1960s (e.g. Wide *et al* 1961), measurements were made by quantitative or qualitative observations of the biological effects of the gonadotrophins on their target tissues, and bioassays will therefore be discussed first. More recently, the use of the radioimmunoassay technique has permitted the specific measurement of FSH, LH, and HCG in small volumes of plasma and urine and these methods will be dealt with separately.

For both biological and immunological assays the activity of the unknown sample (e.g. plasma, urine) must be compared with an appropriate reference substance; because no highly purified reference gonadotrophin preparations are available in sufficient quantity for widespread use, a large and confusing number of standards has arisen. The situation is complicated further by the fact that the circulating forms of the gonadotrophic hormones may not be identical with those found in pituitary extracts, or with those excreted in the urine. The problem of standards will therefore be discussed after the assay methods have been reviewed.

The classical concepts regarding the roles of the gonadotrophins in reproductive physiology have been based entirely on the results of biological assays, particularly of urinary extracts. Radioimmunoassay, with its much greater sensitivity and precision has already allowed much clearer definition of hormone levels in circulating blood, and promises to clarify areas of current uncertainty, such as the disturbances which give rise to dysfunctional uterine bleeding.

General considerations affecting assay methods

Assessment of any assay procedure must include examination of its accuracy, precision, sensitivity and specificity, as well as its general practicability. *Accuracy* refers to the degree to which the assay result reflects the true hormone content of the sample and is customarily measured by recovery experiments in which known amounts of the hormone are added to biological samples, and the percentage recovery is estimated. *Precision* is a measure of the reproducibility of an assay result i.e. the range of the confidence limits. In a method of high precision, the coefficient of variation (standard deviation divided by the mean) may be as low as 2–3%. In bioassay methods, precision is assessed as an index, λ, which takes account of the biological variation of the assay animals, as well as the steepness of slope of the curve relating the dose of the preparation to the measured response. *Sensitivity* may be defined in various ways, but in general refers to the minimum quantity of the hormone which is detectable by the assay system, and which is distinguishable from measurements made in the absence of hormone ('the zero point') with a given degree of statistical probability. *Specificity* refers to the ability of an assay to measure a particular hormone to the exclusion of other hormones, and of other interfering substances. Specificity may therefore be considered as hormonal, e.g. a specific assay for FSH will not respond to LH, or to other pituitary hormones; it may also refer to the freedom from interference by non-hormonal factors—this consideration is important in regard to toxic effects of certain biological extracts on assay animals, and in regard to non-hormonal factors in samples of plasma as compared with the standards in immunoassay. Hormonal specificity has been a major problem in both types of gonadotrophin assay.

The concept of general practicability includes considerations of time, ease of performance, cost, necessary equipment and animals, and presence of specific requirements, such as isotope facilities or hypophysectomized animals. These matters are fully discussed, for example, by Loraine & Bell (1966).

BIOASSAY

Total gonadotrophin assay

The most widely used biological assay for clinical purposes is that of total gonadotrophin activity in urinary extracts; the test animal is the immature female mouse or rat, and the measured response is the increase in uterine or ovarian weight. Such assays are frequently, and quite incorrectly referred to as FSH assays: they are not specific for FSH but measure contributions from both gonadotrophins. Results of such assays are commonly given in terms of 'mouse units' or 'rat units', e.g. a urinary total gonadotrophin excretion of 75 mouse units daily means that 1/75 of a 24-hour urine extract will cause a doubling in weight of the uterus of the immature mouse, whereas 1/5 of a urine containing 5 mouse units must be injected to achieve a similar response.

This assay method is especially useful in the assessment of pituitary and gonadal function in clinical disorders such as hypopituitarism, ovarian dysgenesis and Klinefelter's syndrome. It is of less value in states of isolated gonadotrophin deficiency, when the total gonadotrophin assay may be negative despite the presence of normal amounts of at least one of the gonadotrophins, e.g. in selective FSH deficiency. Detailed physiological studies of pituitary-gonadal relationships, and definition of the hypogonadotrophic syndromes require specific assays for FSH and LH.

FSH assay

Detailed considerations of biological assays for FSH and LH are beyond the scope of this discussion and the reader is referred to the review by Loraine & Bell (1966).

The most widely used biological assay for FSH is the ovarian augmentation assay (Steelman & Pohley 1953, P.S. Brown 1955) which is based on the observation that HCG will augment the action of FSH on the ovary. In the assay, the sample is injected into the assay animal (intact female rat or mouse) together with an 'augmenting dose' of HCG and the ovarian weight is the assay end point. The assay is specific for FSH, and may be used to measure the content of pituitary extracts, and of plasma and urine samples. Sensitivity is such that the extract from a 48-hour collection of urine is required to measure FSH excretion in the normal adult male or menstruating female; 60–100 ml plasma are required for the assay of FSH in the blood of such subjects, and accordingly, very few studies of biological FSH activity in blood have been made. The mouse assay is more sensitive than that using rats. It is to be re-emphasized that the mouse uterus assay is *not* a specific assay for FSH.

LH assay

Two biological assays for LH are in common use. The ventral prostate (VPW) method (Greep *et al* 1942) depends on the stimulating effect of LH on testosterone biosynthesis in immature male rats which have been hypophysectomized to eliminate the effects of the endogenous gonadotrophin of the assay animal. The testosterone so produced causes growth of the prostate, and the increase in weight of the ventral lobe is measured as the assay end point. This is a specific but technically difficult and tedious assay. Sensitivity is broadly similar to that described above for FSH.

The second assay often used is the ovarian ascorbic acid depletion (OAAD) assay of Parlow (1961) which depends on the ability of LH to deplete the ascorbic acid content of the ovaries of immature rats rendered pseudopregnant by the injection of PMSG and HCG. This is a sensitive assay, but its specificity is open to some doubts when the test substances might contain LH releasing factor, which could release endogenous LH in the assay animal, and thus give an overestimate of the LH content of the test material. The toxicity of crude urinary extracts and the occurrence of non-specific responses have rendered this assay unsuitable for measurement of LH excretion in urine (Koed & Hamburger 1968).

An augmentation assay has been introduced recently by Donini *et al* (1968) and depends on the augmentation by LH of the stimulating effect of FSH on the growth of the mouse uterus. The sensitivity of this assay is superior to that of the OAAD method, but it has not yet been employed widely.

HCG assay

A large number of biological assays for HCG has been described, in which the effects of the hormone on the ovary, uterus or prostate are measured in rats and on the testes in amphibia. Commonly used methods include the ovarian hyperaemia test (Zondek *et al* 1945, and Albert 1948) and the increase in total prostatic weight in rats (Loraine 1950). Requirements for sensitivity are less stringent than for LH, as the amounts of HCG circulating and excreted in the urine in pregnancy and in trophoblastic tumours are relatively far greater than the amounts of LH encountered in biological fluids.

IMMUNOLOGICAL ASSAYS

General considerations

The method originally introduced by Wide *et al* (1961) was an assay for HCG in which the hormone present in the test sample caused inhibition of the agglutination of erythrocytes, coated with HCG, by rabbit anti-HCG serum. The assay was made quantitative by employing varying dilutions of the test sample and comparing the inhibition of agglutination with that produced by an appropriate reference standard. This technique can also be used for measuring LH because of the close immunological resemblance of the two hormones.

U.S.A., by the National Pituitary Agency, as an interim reference standard for plasma gonadotrophin immunoassay: while some laboratories report their results in terms of this standard, others have preferred to use the international urinary standard (2nd IRP-HMG). Table B.1 lists standards currently available or referred to in recent literature, and Table B.2 lists some potency estimates comparing certain of these preparations. A useful general review is by Rosemberg (1968).

TABLE B.2. Comparison of potencies of reference preparations

Preparation	Potency (I.U. 2nd IRP/mg)	Assay	Source
NIH FSH S1	26·5	OAR*	Rosemberg (1968)
NIH LH S1	51·3	VPW**	Rosemberg (1968)
	1538	OAAD	Reichert (1964b)
	588·0	OAAD***	Rosemberg (1968)
1st IRP	0·14 (FSH)	OAR	WHO
	0·5 (LH)	VPW	

* Ovarian augmentation reaction assay (Steelman-Pohley 1953).
** Ventral prostate weight assay (Greep *et al* 1942).
*** Ovarian ascorbic acid depletion assay (Parlow 1961).

Standards for biological assay and normal values

The most widely accepted standard for biological assays of pituitary extracts and plasma and urine samples, is the 2nd IRP-HMG, a preparation which, by definition, contains 40 I.U. of both FSH and LH in each ampoule. Its appropriate use as a standard therefore depends on the specificity of the biological assay method employed. This standard is unsuitable for use in assays for total gonadotrophin activity, as this will vary, depending on the proportions of FSH and LH in the test material. Although the 2nd IRP-HMG is a standard derived from post-menopausal urine, it is recommended for all types of biological assay of *human* gonadotrophins, whatever their source, in contrast to the situation with regard to immunoassay (see below).

Some representative normal values for the biological potency of various human samples are shown in Table B.3 together with the standard used and the method of bioassay. It is clear that very few estimates of biological potency have been made using human plasma; thus, the

TABLE B.3. Gonadotrophin concentrations in biological fluids—bioassay. VPW: Ventral prostate weight assay (Greep et al 1942); OAAD: Ovarian ascorbic acid depletion assay (Parlow 1961)

Subject	FSH (mean and range)	LH (mean and range)	Standard	Reference
I Plasma:	(mI.U./ml)	(mI.U./ml)		
Adult males	8·8	14·4 (VPW)	2nd IRP	Keller (1966)
		16·8 (OAAD)	2nd IRP	Keller (1966)
	11·5 (10·1–13·3)	6·1 (3·8–9·7) (VPW)	2nd IRP	Kulin (1968a)
Postmenopausal females	136–152	88 (VPW)	2nd IRP	Keller (1966)
		69·6 (OAAD)	2nd IRP	Keller (1966)
	64·6–78·0	74·6–84·4 (VPW)	2nd IRP	McArthur (1964)
II Urine	I.U.	I.U.		
Prepubertal children	2·2 ± 0·4 (sem)/litre	0·44 ± 0·07 (VPW) (sem)/litre	2nd IRP	Rifkind (1967)
Adult males	9·0 (3·5–30·6)/day		N.I.H. FSH S1	Becker (1965)
	5·6 ± 1·0 (sem)/litre	4·7 ± 0·6 (VPW) (sem)/litre	2nd IRP	Rifkind (1967)
Menstrual cycle				
Mid cycle peak	9·6/day	16/day (OAAD)	2nd IRP	Rosemberg (1965)
	no mid cycle peak	19·5/day (OAAD)	N.I.H. Stds.	Fukushima (1964)
Follicular phase		19·5/day (OAAD)	N.I.H. Stds.	Vorys (1965)
Luteal phase	3–14/day	4–12/day (OAAD)	2nd IRP	Stevens (1969a)

For discussion and references see Stevens (1969b)

APPENDIX B

interpretation of immunoassay data in biological terms awaits further study.

Standards for immunoassay, and normal values

In contrast to the *relative* uniformity which exists in the use of 2nd IRP-HMG (and the N.I.H. standard) as bioassay standards, various

TABLE B.4. Gonadotrophin concentrations in biological fluids—immunoassay. A: plasma of children and adult males

Subjects	FSH mI.U./ml (mean, range, SEM)	LH mI.U./ml (mean, range, SEM)	Standard*	Reference
PREPUBERTAL CHILDREN				
Age 3–8	10·1 (5·6–18·5)	10·2 (2·5–15·0)	2nd IRP	Saxena (1968)
Age < 7	< 5		2nd IRP	Odell (1968a)
	< 2·6 – 3·6		2nd IRP	Faiman (1967a)
	4·5 ± 0·9		2nd IRP	Raiti (1969a)
		0–11·0	LH	Schalch (1968)
		4·8 ± 1·2	2nd IRP	Johanson (1969)
		8·0(4·8 – 12·8)	2nd IRP	Odell (1967)
ADULT MALES				
	7·4 ± 1·9		2nd IRP	Raiti (1969a)
	6·0(3·5–8·6)		2nd IRP	Midgley (1967)
	4·9(< 2·6–11·8)		2nd IRP	Faiman (1967a)
	16·9(3·9 – 42·0)	14·0(2·5–31·5)	2nd IRP	Saxena (1968)
	4·14	13·12	2nd IRP	Franchimont (1968)
	1·26–8·2		2nd IRP	Rosselin (1967)
		10·9 ± 4·0	2nd IRP	Johanson (1969)
	5–25	14·4(6·4–25·6)	2nd IRP	Odell (1967, 1968a)
		22·5(5–68)	2nd IRP	Burger (1968)
		24·2(15–31)	2nd IRP	Catt (1968b)
Age 20–40		7·0 ± 2	LH	Schalch (1968)
Age 50–80		17·0 ± 4	LH	Schalch (1968)

* Results expressed in m.I.U. by conversion to 2nd IRP-HMG equivalents.

investigators have used a number of standards for immunoassay, and accordingly, results obtained in one laboratory cannot readily be correlated with those of another. The situation is further complicated by the fact that, even if the same standard is used different results are obtained for samples from normal subjects with different antisera, and even with the same antisera (Albert *et al* 1968). The range of values obtained may be seen in Tables B.4, 5 and 6 in which immunoassay data from a number of laboratories are shown. In general, the biological 'sense' of

TABLE B.5. Gonadotrophin concentrations in biological fluids—immunoassay. B: plasma of adult females

Subjects	FSH mI.U./ml (mean, range, SEM)	LH mI.U./ml (mean, range, SEM)	Standard*	Reference
FEMALE MENSTRUAL CYCLE				
Follicular phase	7–25	16·0	2nd IRP	Odell (1967, 1968a)
		12·0 ± 1	LH	Schalch (1968)
	19·4(7–27·2)	14·6(6–27·2)	2nd IRP	Saxena (1968)
Mid-cycle		102 ± 12	LH	Schalch (1968)
peak	51·8(28·2–93·5)	108·3(67–137)	2nd IRP	Saxena (1968)
	22·3	41·16(14·8–88·2)	2nd IRP	Franchimont (1968)
		27–54	2nd IRP	Midgley (1966)
		57·4	LH	Neill (1967)
	11–33	32·2–122·2	2nd IRP	Odell (1967, 1968a)
	5·3–20	42·9–105	LH, FSH	Faiman (1967b)
Luteal phase	5–15	12·0	2nd IRP	Odell (1967, 1968a)
		10 ± 1	LH	Schalch (1968)
	21·1(16·8–31·7)	16·3(7·4–21·3)	2nd IRP	Saxena (1968)
POSTMENOPAUSAL FEMALES:				
	40–250	48–96	2nd IRP	Odell (1967, 1968a)
	88	47·4	2nd IRP	Franchimont (1968)
Age 40–50		58	LH	Schalch (1968)
Age 50–70		105	LH	Schalch (1968)
	96(34·7–217)	65·8(0–90)	2nd IRP	Saxena (1968)
	66·4(5·3–245)		2nd IRP	Midgley (1967)
	51·0(32·8–73·2)		2nd IRP	Faiman (1967a)

* Results expressed in m.I.U. by conversion to 2nd IRP-HMG equivalents.

the values seen is similar: e.g. there is uniformity in the finding in normal females, of a short-lived mid-cycle peak in the plasma levels of LH, although the magnitude of the peak, as reported in mI.U./ml, varies widely. The interpretation of immunoassay data therefore requires a knowledge of the normal range for the particular laboratory.

There is lack of agreement over whether the standard for plasma (or serum) radioimmunoassay should be from the pituitary, the blood or the urine: as indicated above, an interim pituitary standard has been introduced through the National Pituitary Agency in the U.S.A.

CORRELATIONS BETWEEN BIOASSAY AND IMMUNOASSAY

It is of fundamental importance to recognize that biological and immunological assays may measure quite distinct portions of the

Table B.6. Gonadotrophin concentrations in biological fluids—immunoassay. C: urine of children, adult males and adult females

Subjects	FSH I.U. (mean, range, SEM)	LH I.U. (mean, range, SEM)	Standard*	Reference
PREPUBERTAL CHILDREN				
	$2 \cdot 2 \pm 1 \cdot 1$/litre		2nd IRP	Raiti (1969a)
		0·22–0·46/litre**	2nd IRP	Kulin (1968b)
ADULT MALES				
	$8 \cdot 5 \pm 3 \cdot 6$		2nd IRP	Raiti (1969a)
	2·2		2nd IRP	Franchimont (1968)
		4·3–7·1/litre**	2nd IRP	Kulin (1968b)
MENSTRUAL CYCLE				
Mid-cycle peak	46·7		2nd IRP	Franchimont (1968)
	4–12/day**	4–10/day**	2nd IRP	Stevens (1969a)

* Results expressed in mI.U. by conversion to 2nd IRP-HMG equivalents.
** On urinary extracts.

gonadotrophin molecules. Dissociation between the biological and immunological properties of several protein and polypeptide hormones has been described (Trenkle *et al* 1962, Hobson & Wide 1964, Imura *et al* 1965, Catt *et al* 1968a) and a similar situation applies to the gonadotrophins. Thus, if the biological and immunological potencies of urinary extracts are compared, using a urinary standard such as the 2nd IRP-HMG, there is excellent agreement; similarly, comparison of the potencies of pituitary extracts in terms of a common pituitary standard results in good agreement between bioassay and radioimmunoassay. However, comparison of these activities in a pituitary extract, using a urinary reference standard, results in a ratio of immunological to biological activity much greater than unity (e.g. Odell *et al* 1968c). It can be considered that a portion of the pituitary extract is immunologically active but biologically inactive, or that the urinary reference preparation has a biologically active, immunologically inactive component.

Furthermore, the fact that the gonadotrophins of a number of different species can be bioassayed in a single species of animal, e.g. the rat, implies that they may share a common structure responsible for biological activity; these preparations do not, however, share immunological determinants. It seems likely that during *in vivo* degradation, portions of the gonadotrophin molecule retaining their antigenic structure might be split off from the biologically active fragment of the protein. In all these circumstances, immunological assay would not reflect the biological activity present in the sample. It must therefore be emphasized that the interpretation of immunological assay data for gonadotrophins requires a knowledge of the normal range of values for the particular laboratory, and an awareness of the fact that the levels found do not necessarily give an accurate reflection of biological activity. Nevertheless, it can be stated confidently that the availability of radioimmunoassay procedures for the measurement of gonadotrophins has already led to marked advances in our knowledge of reproductive physiology, and will continue to contribute in the future. For further discussion, the reader is referred to Albert (1968).

Details of the physiology of the gonadotrophins, and their relationship to gonadal function, will be found in chapters 2 and 3.

REFERENCES

ALBERT A. (1948) A clinical bioassay for chorionic gonadotrophin. *J. clin. Endocr.* **8**, 619–620.

ALBERT A. (1968) Bioassay and radioimmunoassay of human gonadotrophins. *J. clin. Endocr.* **28**, 1683–1689.

ALBERT A., ROSEMBERG E., ROSS G.T., PAULSEN C.A. & RYAN R.J. (1968) Report of the National Pituitary Agency collaborative study on the radioimmunoassay of FSH and LH. *J. clin. Endocr.* **28**, 1214–1219.

BECKER K.L. & ALBERT A. (1965) Urinary excretion of follicle-stimulating and luteinizing hormones. *J. clin. Endocr.* **25**, 962–974.

BELL J.J., CANFIELD R.E. & SCIARRA J.J. (1969) Purification and characterization of human chorionic gonadotrophin. *Endocrinology* **84**, 298–307.

BERSON S.A. & YALOW R.S. (1968) Principles of immunoassay of peptide hormones in plasma, in E.B.Astwood & C.E.Cassidy (eds.) *Clinical Endocrinology*, II, pp. 699–720. New York, Grune & Stratton.

BROWN P.S. (1955) Assay of gonadotrophin from urine of non-pregnant human subjects. *J. Endocr.* **13**, 59–64.

BURGER H.G., OLIVER J.R., DAVIS JENNIFER & CATT K.J. (1968) Radioimmunoassay for human pituitary luteinizing hormone using paper chromatoelectrophoresis. *Aust. J. exp. Biol. med. Sci.* **46**, 541–553.

BUTT W.R., JENKINS J.F. & SOMERS P.J. (1967) Some observations on the chemical properties of human pituitary follicle stimulating hormone. *J. Endocr.* **38**, xi–xii.

CATT K.J., CAIN M.D. & COGHLAN J.P. (1968a) Radioimmunoassay of angiotensin in blood, in M. Margoulies (ed.) *Protein and Polypeptide Hormones*. pp. 77–80. Amsterdam, Excerpta Medica International Congress Series 161.

CATT K.J., NIALL H.D., TREGEAR G.W. & BURGER H.G. (1968b) Disc solid-phase radioimmunoassay of human luteinizing hormone. *J. clin. Endocr.* **28**, 121–126.

CROOKE A.C. & GRAY C.H. (1968) Some observations on the molecular weight and structure of human FSH, in E. Rosemberg (ed.) *Gonadotrophins*, pp. 33–36. Los Altos, California, Geron-X Inc.

DONINI P., PUZZUOLI D., D'ALESSIO I. & DONINI S. (1968) A new approach to the biological determination of the luteinizing hormone. *Acta endocr. (Kbh.)* **58**, 463–472.

FAIMAN C. & RYAN R.J. (1967a) Radioimmunoassay for human follicle stimulating hormone. *J. clin. Endocr.* **27**, 444–447.

FAIMAN C. & RYAN R.J. (1967b) Serum follicle-stimulating hormone and luteinizing hormone concentrations during the menstrual cycle as determined by radioimmunoassays. *J. clin. Endocr.* **27**, 1711–1716.

FRANCHIMONT P. (1968) Radioimmunoassay of gonadotropic hormones, in M. Margoulies (ed.) *Protein and Polypeptide Hormones*, pp. 99–116. Amsterdam, Excerpta Medica International Congress Series 161.

FUKUSHIMA M., STEVENS V.C., GANTT C.L. & VORYS N. (1964) Urinary FSH and LH excretion during the normal menstrual cycle. *J. clin. Endocr.* **24**, 205–213.

GREEP R.O., VAN DYKE H.B. & CHOW B.F. (1942) Gonadotrophins of the swine pituitary. *Endocrinology* **30**, 635–649.

HARTREE A.S., BUTT W.R. & KIRKHAM K.E. (1964) The separation and purification of human luteinizing and thyrotrophic hormones. *J. Endocr.* **29**, 61–69.

HARTREE A.S. (1966) Separation and partial purification of the protein hormones from human pituitary glands. *Biochem. J.* **100**, 754–761.

HARTREE A.S. (1968) In M. Margoulies (ed) *Protein and Polypeptide Hormones*, p. 799. Amsterdam, Excerpta Medica International Congress Series 161.

VAN HELL H., MATTHIJSEN R. & HOMAN J.D.H. (1968) Studies on human chorionic gonadotrophin. *Acta endocr. (Kbh.)* **59**, 89–104.

HOBSON B.M. & WIDE L. (1964) The immunological and biological activity of human chorionic gonadotropin in urine. *Acta endocr. (Kbh.)* **46**, 632–638.

IMURA H., SPARKS L.L., GRODSKY G.M. & FORSHAM P.H. (1965) Immunologic studies of adrenocorticotrophic hormone (ACTH): dissociation of biologic and immunologic activities. *J. clin. Endocr.* **25**, 1361–1369.

JAFFE R.B. & MIDGLEY A.R. JR. (1969) Current status of human gonadotrophin radio-immunoassay. *Obstet. Gynec. Surv.* **24**, 200–213.

JOHANSON A.J., GUYDA H., LIGHT C., MIGEON C.J. & BLIZZARD R.M. (1969) Serum luteinizing hormone by radioimmunoassay in normal children. *J. Pediat.* **74**, 416–424.

Karolinska Symposia on Research Methods in Reproductive Endocrinology (1969) *Acta endocr. (Kbh.)* Suppl. 142.

KATHAN R.H., REICHERT L.E. JR. & RYAN R.J. (1967) Comparison of the carbohydrate and amino acid composition of bovine, ovine and human luteinizing hormone. *Endocrinology* **81**, 45–48.

KELLER P.J. (1966) Studies on pituitary gonadotrophins in human plasma. II. FSH and LH in male and postmenopausal plasma. *Acta endocr. (Kbh.)* **52**, 348–356.

KOED H.J. & HAMBURGER C. (1968) The ovarian ascorbic acid depletion test for luteinizing hormone: lack of specificity. *Acta endocr. (Kbh.)* **59**, 629–635.

KULIN H.E., RIFKIND A.B. & ROSS G.T. (1968a) Bioassay determination of luteinizing hormone (LH) and follicle stimulating hormone in adult male plasma. *J. clin. Endocr.* **28**, 100–102.

KULIN H.E., RIFKIND A.B. & ROSS G.T. (1968b) Human luteinizing hormone (LH) activity in processed and unprocessed urine measured by radioimmunoassay and bioassay. *J. clin. Endocr.* **28**, 543–546.

LI C.H., SIMPSON MIRIAM E. & EVANS H.M. (1949) Isolation of pituitary follicle-stimulating hormone (FSH). *Science* **109**, 445–446.

LLOSA DE LA P. (1968) In M.Margoulies (ed.) *Protein and Polypeptide Hormones*, p. 798. Excerpta Medica International Congress Series 161.

LORAINE J.A. (1950) Estimation of chorionic gonadotrophin in urine of pregnant women. *J. Endocr.* **6**, 319–329.

LORAINE J.A. & BELL E.T. (1966) *Hormone Assays and their Clinical Application*, 2nd ed. Edinburgh, Livingstone.

MCARTHUR JANET W., ANTONIADES H.N., LARSON L.H., PENNELL R.B., INGERSOLL, F.M. & ULFELDER H. (1964) Follicle-stimulating hormone and luteinizing hormone content of pooled human menopausal plasma and of subfractions prepared by Cohn methods 6 and 9. *J. clin. Endocr.* **24**, 425–431.

MIDGLEY A.R. JR. (1966) Radioimmunoassay: a method for human chorionic gonadotrophin and human luteinizing hormone. *Endocrinology* **79**, 10–18.

MIDGLEY A.R. (1967) Radioimmunoassay of human follicle-stimulating hormone. *J. clin. Endocr.* **27**, 295–299.

MIDGLEY A.R. JR., FONG I.F. & JAFFE R.B. (1967) Gel filtration radioimmunoassay to distinguish human chorionic gonadotrophin from luteinizing hormone. *Nature (Lond.)* **213**, 733.

MIDGLEY A.R. JR. & JAFFE R.B. (1966) Human luteinizing hormone in serum during the menstrual cycle determination by radioimmunoassay. *J. clin. Endocr.* **26**, 1375–1381.

NEILL J.D., JOHANSSON E.D., DATTA J.K. & KNOBIL E. (1967) Relationship between the plasma levels of luteinizing hormone and progesterone during the normal menstrual cycle. *J. clin. Endocr.* **27**, 1167–1173.

ODELL W.D., PARLOW A.F., CARGILLE C.M. & ROSS G.T. (1968a) Radioimmunoassay for human follicle-stimulating hormone: physiological studies. *J. clin. Invest.* **47**, 2551–2562.

ODELL W.D., REICHERT L.E. & BATES R.W. (1968b) Pitfalls in the radioimmunoassay of carbohydrate containing polypeptide hormones, in M.Margoulies (ed.) *Protein and Polypeptide Hormones*, pp. 124–128. Excerpta Medica International Congress Series 161.

ODELL W.D., REICHERT L.E. & SWERDLOFF R.S. (1968c) Correlation between bioassay and immunoassay of human luteinizing hormone, in E. Rosemberg (ed.) *Gonadotrophins*, pp. 401–407. Los Altos, California, Geron-X Inc.

ODELL W.D., ROSS G.T. & RAYFORD P.L. (1967) Radioimmunoassay for luteinizing hormone in human plasma or serum: physiological studies. *J. clin. Invest.* **46**, 248–255.

PAPKOFF H. & EKBLAD M. (1970) Ovine follicle stimulating hormone: preparation and characterization of its subunits. *Biochem. Biophys. Res. Comm.* **40**, 614–621.

PARLOW A.F. (1961) Ovarian ascorbic acid depletion assay for LH, in A.Albert (ed.) *Human Pituitary Gonadotrophins: a Workshop Conference*, p. 301. Springfield, Ill. Charles C. Thomas.

PARLOW A.F., CONDLIFFE P.G., REICHERT L.E. Jr. & WILHELMI A.E. (1965) Recovery and partial purification of FSH and LH during the purification of TSH from human pituitary glands. *Endocrinology* **76**, 27–34.

PAUL W.E. & ODELL W.D. (1964) Radiation inactivation of the immunological and biological activities of human chorionic gonadotrophin. *Nature (Lond.)* **203**, 979–980.

PECKHAM W.D. & PARLOW A.F. (1969) On the isolation of human pituitary follicle-stimulating hormone. *Endocrinology* **84**, 953–957.

RAITI S. & DAVIS W.J. (1969) The principles and application of radioimmunoassay with special reference to the gonadotrophins. *Obstet. Gynec. Surv.* **24**, 289–310.

RAITI S., JOHANSON A., LIGHT C., MIGEON J. & BLIZZARD R.M. (1969a) Measurement of immunologically reactive follicle stimulating hormone in serum of normal male children and adults. *Metabolism* **18**, 234–240.

RAITI S.M., LIGHT C. & BLIZZARD R.M. (1969b) Urinary follicle-stimulating hormone excretion in boys and adult males as measured by radio-immunoassay. *J. clin. Endocr.* **29**, 884–890.

REICHER T.L.E. (1968) In E.Rosemberg (ed.) *Gonadotrophins*, p. 52. Los Altos, California, Geron-X Inc.

REICHERT L.E. & JIANG N.S. (1965) Comparative gel filtration and density gradient centrifugation studies on heterologous pituitary luteinizing hormones. *Endocrinology* **77**, 78–86.

REICHERT L.E. JR., KATHAN R.H. & RYAN R.J. (1968) Studies on the composition and properties of immunochemical grade human pituitary follicle-stimulating hormone (FSH): comparison with luteinizing hormone (LH). *Endocrinology* **82**, 109–114.

REICHERT L.E. Jr. & MIDGLEY A.R. Jr. (1968) In E. Rosemberg (ed.) *Gonadotrophins*, p. 25. Los Altos, California, Geron-X Inc.

REICHERT L.E. Jr. & PARLOW A.F. (1964a) Preparation of highly potent human pituitary gonadotrophins. *Proc. Soc. Exp. Biol. Med.* **115**, 286–288.

REICHERT L.E. & PARLOW A.F. (1964b) Partial purification and separation of human pituitary gonadotrophins. *Endocrinology* **74**, 236–243.

RIFKIND A.B., KULIN H.E. & ROSS G.T. (1967) Follicle-stimulating hormone (FSH) and luteinizing hormone (LH) in urine of prepubertal children. *J. clin. Invest.* **46**, 1925–1931.

ROOS P. (1967) *Human Follicle-Stimulating Hormone*. Uppsala, Almqvist & Wiksells.

ROOS P. & GEMZELL C.A. (1964) The isolation of human pituitary follicle-stimulating hormone. *Biochim. Biophys. Acta* **82**, 218–220.

ROSEMBERG E. (1968) Use of standards: General considerations, in E.Rosemberg (ed.) *Gonadotrophins*, pp. 383–391. Los Altos, California, Geron-X Inc.

ROSEMBERG E. & KELLER P. (1965) Studies on the urinary excretion of follicle stimulating hormone and luteinizing hormone activity during the menstrual cycle. *J. clin. Endocr.* **25**, 1262–1274.

ROSSELIN G. & DOLAIS J. (1967) Dosage de l'hormone folliculo-stimulante humaine (H.S.F.H.) par la méthode radio-immunologique. Premiers resultats obtenus chez l'hormone normal et hypogonadique; chez la femme au cours du cycle menstruel et après la menopause. Effet de différents frénateurs hypophysaires. *Presse méd.* **75**, 2027–2030.

RYAN R.J. (1968) In M.Margoulies (ed.) *Protein and Polypeptide Hormones*, p. 800. Excerpta Medica International Congress Series 161.

RYAN R.J. & FAIMAN C. (1968) Radioimmunoassay of human follicle stimulating hormone: a comparison of several FSH antisera, in M.Margoulies (ed.) *Protein and Polypeptide Hormones*, pp. 129–133 Amsterdam. Excerpta Medica International Congress Series 161.

SAXENA B.B., DEMURA H., GANDY H.M. & PETERSON R.E. (1968) Radioimmunoassay of human follicle-stimulating and luteinizing hormones in plasma. *J. clin. Endocr.* **28**, 519–534.

SAXENA B.B. & RATHNAM P. (1967) Purification of follicle-stimulating hormone from human pituitary glands. *J. biol. Chem.* **242**, 3769–3775.

SCHALCH D.S., PARLOW A.F., BOON R.C. & REICHLIN S. (1968) Measurement of human luteinizing hormone in plasma by radioimmunoassay. *J. clin. Invest.* **47**, 665–678.

SCHLAFF S., ROSEN S.W. & ROTH J. (1968) Antibody to human follicle-stimulating hormone: cross-reactivity with three other hormones. *J. clin. Invest.* **47**, 1722–1729.

SQUIRE P.G., LI C.H. & ANDERSEN R.N. (1962) Purification and characterization of human pituitary interstitial cell stimulating hormone. *Biochemistry (Wash.)* **1**, 412–418.

STEELMAN S.L. & POHLEY F.M. (1953) Assay of the follicle stimulating hormone based on the augmentation with human chorionic gonadotrophin. *Endocrinology* **53**, 604–616.

STEVENS V.C. (1969a) Comparison of FSH and LH patterns in plasma, urine and urinary extracts during the menstrual cycle. *J. clin. Endocr.* **29**, 904–910.

STEVENS V.C. (1969b) Discrepancies and similarities of urinary FSH and LH patterns as evaluated by different assay methods. *Acta endocr. (Kbh.)* Suppl. **142**, 338–356.

TRENKLE A., LI C.H., SADRI K. & ROBERTSON H. (1962) Effects of chemical treatment on immunochemical properties of human growth hormone and sheep interstitial cell stimulating hormone. *Arch. Biochem. Biophys.* **99**, 288–293.

VORYS N., ULLERY D. & STEVENS V. (1965) The effects of sex steroids on gonadotrophins. *Amer. J. Obstet. Gynec.* **93**, 641–654.

WIDE L., ROOS P. & GEMZELL C.A. (1961) Immunological determination of human pituitary luteinizing hormone (LH). *Acta endocr. (Kbh.)* **37**, 445–449.

WILDE C.E. (1969) The correlation between immunological and biological estimations of HCG in body fluids. *Acta endocr. (Kbh.)* Suppl. **142**, pp. 360–376.

ZONDEK B., BERNHARD A., SULMAN F. & BLACK R. (1945) The hyperemia effect of gonadotrophins on the ovary. *J. Amer. med. ass.* **128**, 939–944.

Appendix C

Measurement of Adrenal Steriods

A.W.STEINBECK

INTRODUCTION

Methodological problems are considerable in some areas of this field. The form of the steroid to be measured must be known, whether free or conjugated and, if so, whether glucuronide, sulphate or disulphate, or others. Non-recognition of sulphate conjugation in 'infant's' urine presented a picture of steroid excretion related only to glucuronides. Also, the appropriate conditions for intended enzymatic hydrolysis will need to be fulfilled with exclusion of inhibitors for the reaction, if the conjugate is to be disrupted prior to extraction of the steroid. Under other situations, steroid conjugates might be extracted and further purified as such, although this is unusual in most conditions of clinical investigations. At times, investigations will merit this approach with its attendant problems. Extraction procedures in all estimations should be quantitative and subsequent development of the extracts, whether by partition or chromatography in any of its varied forms, with or without derivative formation, should lead to adequately pure steroids and their resolution from impurities if maximum specificity is required. Internal standards within the procedure give improved accuracy and reproducibility; with high activity, radio-nucleide (radio-isotope) labelled steroids it is possible to gain considerable sensitivity in some steroid estimations, a high degree of specificity and accuracy, or reproducibility. Methods such as double-isotope-derivative-dilution techniques are essential for meaningful results with some steroid estimations; this applies to plasma aldosterone values, urinary free cortisol and aldosterone in infants among others. A competitive protein binding method would be simpler, nearly as accurate and valuable for the estimation of urinary free cortisol in normal adults. The 'blank' value in such methods should, theoretically, be zero for maximum confidence. The method for final measurement of steroids should be appropriate in terms of sensitivity of detection, discrimination and overall accuracy. At times, a less specific method is used because the information obtained is adequate for the study. At times, urine measurements only are shown and, at others, plasma or both.

The relative value of blood and urinary methods is a long-standing source of controversy. Urine excretions should reflect changes in plasma steroids if they are directly excreted over a significant time-interval but plasma levels can only reflect 'instantaneous' concentrations, shorter-lived changes or variations under selected conditions. Mostly, urinary hormones will give valuable information if the study is correctly undertaken. For instance, measurement of cortisol secretion rate may relate best to the actual amount of cortisol secreted but the total breakdown products of cortisol will give a better measure of its catabolism and may provide all the information required. The urinary free cortisol will give an overall expression of changes in plasma free cortisol levels over the 24 hours; it may provide more worthwhile information than the secretion rate and with less difficulty particularly if a competitive protein binding method is applicable to the urine. In new methods, it is essential that criteria of identification for steroids be applied critically (Brooks *et al* 1970).

In clinical medicine, many methods of steroid estimation are 'group' methods depending upon reaction of chemical groups in a colorimetric procedure, fluorimetric reaction or some similar method. With the adoption of competitive protein binding assays rather more specific methods have become available, although by the nature of the binding protein, steroids other than the one for which the estimate is given may influence the final figure, as for example cortisol and corticosterone. Also, they are not applicable in every situation nor are some of the simpler methods. Protein binding methods are particularly applicable to such individual steroids as cortisol, testosterone, progesterone, 17β-oestradiol but may not give adequate information under some circumstances. Allowing for lack of specificity, some methods will still provide good physiological information. The Porter-Silber method for '17-hydroxycorticosteroids', by the original Nelson & Samuels method (1952) or its modifications (Bayliss & Steinbeck 1953) probably overestimates the real value of cortisol by some 15–30% at normal levels, while the Mattingly (1962) method for '11-hydroxycorticosteroids', is very non-specific (Brooks et al 1970) with a 40% error. However, both have given pertinent physiological data.

It is possible to improve specificity of some methods by simple manœuvres, as extraction of Zimmermann chromogens prior to colorimetry in methods using this colorimetric stage. Use of some terms is confusing: accuracy is the inverse of the difference between an estimate and the true value, obtained by an ideal method, and as such depends upon specificity. Precision is a function of agreement between repeated estimates on fractions of the one sample, and is a major determinant of sensitivity. Sensitivity is considered the minimum value that can be distinguished significantly from a zero value. In this regard, an interlaboratory trial of a standardized method for 17-ketogenic steroids and one for 17-ketosteroids gave coefficients of variation between ± 4 and $\pm 14\%$ (Gray et al 1969), with larger variation for others.

There is good reason also not to use designations for an estimate that is not justified by the procedure. Frequently reference to plasma 'cortisol' is made, or else plasma '17-hydroxycorticosteroids' when Porter-Silber chromogens are measured. In this regard, it is possible that one of the later methods (Peterson et al 1957) will give a closer approximation to the true value, with less non-cortisol measured, than the earlier techniques. Likewise plasma '11-hydroxycorticosteroids' or 'cortisol' is used when the Mattingly (1962) method, or its modifications or other fluorimetric procedures, is followed. One designation used for these has been 'sulphuric acid induced fluorescence technique' as this is the analytical end point (Eik-Nes 1968). The designation Norymberski chromogens has been suggested (Eik-Nes 1968) for estimates more frequently known as 17-ketosteroids or 17-oxosteroids, and 17-ketogenic (17-hydroxycorticoids) or 17-oxogenic steroids (Appleby et al 1955), making allowance for chemical doubts. All these methods demand some dietary and therapeutic precautions if they are to have significance within their area of reference.

URINARY METHODS

17-oxosteroids (*17-ketosteroids*) (*17-OS or 17-KS*)

The classical procedure depends upon hydrolysis, some preliminary purification and development of the Zimmermann colour reaction. The steroids reacting do not have equal chromogenicity and a male, female difference in estimates gives a sex difference that is more virtual than real when corrected for individual steroids. Non-specific chromogens are present in large quantity in the urine of newborn infants.

The excretion of these steroids for healthy children seldom exceeds 2 mg/24 hr, once the newborn phase is passed, up to the age of about 8 years. The mean output rises rapidly at puberty and reaches a maximum at 20–25 years. According to the method adopted by the laboratory, the normal male range is 5–25 mg/24 hr, with lower levels in the aged, and the normal female range is 3–21 mg/24 hr. Values by the method in the first week of life (Zimmermann chromogens) reach 1·5–2·5 mg/24 hr, although higher values have been found (Bongiovanni & Root 1963), falling somewhat after the early weeks to lower values in the first year. Mean values of 11-oxy-17-oxosteroids rise progressively after the first year to adult values (Knorr 1965).

Mostly, the overlap between normal values and those found in abnormal adrenocortical states is so great that isolated values are of little value. In Cushing's syndrome normal values are possible but considerable elevation may occur when adrenocortical carcinoma is the cause. The values are important in virilization and are elevated in typical instances of congenital adrenal hyperplasia. Here the levels can exceed, even in a child, normal adult male values. In isosexual, male premature puberty the values will rise to those appropriate for pubertal development.

17-oxogenic (ketogenic) steroids, 17 hydroxycorticosteroids (total or direct ketogenic steroids) (17-*OGS* or 17*KGS* and 17-*OHCS*)

The earlier technique (Norymberski *et al* 1953) is seldom standard, the 'total' method (Appleby *et al* 1955) being performed, although modifications (Few 1961) have been widely adopted. In both methods pregnanetriol and related compounds can interfere and, in some circumstances, a greater than expected difference between the 17-ketogenic steroid value obtained by the indirect and direct method indicates the presence of such interfering compounds. The Few (1961) modification may be preferable for metyrapone studies, or in routine determinations.

These methods are derivatives in one sense of the original Zimmermann reaction for this color development remains the final analytic step. The side chains for the corticosteroids at C-17 in ring D are shown in Fig. C.1 and give a summary of the probable compounds estimated. As usually understood, the Norymberski procedures measure only those designated 17-hydroxycorticosteroids. The original 17-ketogenic method did not measure those compounds under heading A, called 21-deoxyketols. The direct method measures all under headings A–D, and both include dihydroxyacetones (heading C). These are more usually measured by a Porter-Silber technique. Extraction procedures are common to all but, in the earlier method, bismuthate oxidation is the preliminary step converting three of the possible four 17-hydroxycorticosteroid side-chains (B–D) to 17-ketosteroids. In the later method,

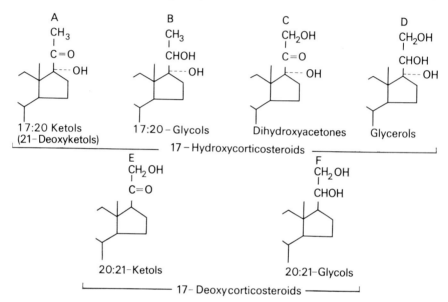

FIG C.1. The six types of corticosteroid side chain, according to Norymberski terminology for group chemical estimation.

borohydride reduction precedes the oxidation and all four side-chains (A–D) are measured. Neither method measures 17-deoxycorticosteroids (E and F), metabolites of corticosterone.

When pregnanetriol and related compounds are a large excretory component, as in congenital adrenal hyperplasia, there are methodological problems with both methods but more with the earlier technique. These problems have more recently been emphasized by Mitchell & Shackleton (1969). Large amounts of dehydroepiandrosterone, as with some tumours, may affect the accuracy of the final estimate. Also, some therapeutic substances prevent application of the method.

The two methods, except for the problems mentioned, give nearly similar results for normal subjects. In the first year, average excretion measures approximately 0·6 mg/24 hr, with a range 0·12–1·3 mg/24 hr, and in the second year about 1 mg/24 hr, with a wider scatter (Natoli & Natoli 1961). The scatter of results in normal children is a characteristic feature of the estimation, and often low values are obtained as an inherent feature of the method which needs to be understood. In this regard, ACTH stimulation tests are often needed to assess isolated and low values (Clayton *et al* 1953). From a study of mean values for different ages, there is a gradual increase in excretion values over the early years of life with a more obvious and significant increase during puberty. On occasions, individuals of both sexes will have values that are higher than the upper limit of an expected normal range, and this might be characteristic for them.

Values for adults, which may show seasonal variation are also subject to wide scatter, and often fluctuations, and may decrease with advanced age. Male values for 17-hydroxycorticosteroids have a range of 6–21 mg/24 hr and female values 3–18 mg/24 hr, although these will depend upon laboratory details and the criteria of exclusion for normality and representative estimates. The possibility of 'stress' increasing values cannot be discounted, and at time dietary factors are involved. Glucosuria will have been tested for before the estimation but is another source of error in a straight extraction procedure.

Porter-Silber chromogens

This method depends upon phenylhydrazine reaction with a dihydroxyacetone side-chain (C in Fig. C.1), and is often referred to as a 17-hydroxycorticosteroid value rather than the more descriptive Porter-Silber chromogens. The scatter of values can be considerable but, with standard and unchanging clinical conditions, results are reproducible. Adult values have a range of about 2–8 mg/24 hr. In children excretion is comparable or about 3 mg/m^2/24 hr (Wilkins 1965) but the method has inherent errors in newborn children (Mitchell & Shackleton 1969). Many non-steroidal substances can interfere with the estimation and patient supervision is important.

Cortisol

The term should only be used if cortisol, either as the free steroid or that released after the enzymatic hydrolysis, actually conjugated cortisol, is in fact measured. The actual excretion of the free steroid is a better guide to the free or unbound plasma level of cortisol occurring throughout the 24-hour period than 17-hydroxycorticosteroids (Norymberski chromogens) and Porter-Silber chromogens which measure metabolic products. As such the value is a guide to the plasma production or secretion rate of cortisol, provided there is no significant plasma-urine dissociation. Again there is a considerable scatter of values and adult excretions are mostly less than 60 µg/24 hr although many are lower, some occasionally higher. There is a progressive increase, on a weight basis, from 3 months of age to adult life, although excretion values from 1–15 years are not very dissimilar (Minick 1966). The estimation allows a better interpretation of the group methods particularly if high values have been obtained for them. Cortisol conjugates alter values but the normal ranges are not well studied and their significance is not sure.

The Mattingly (1962) method does not equate with cortisol in urine. The value includes cortisol, 6β-hydroxycortisol, some unconjugated metabolites of cortisol and equivalent unconjugated corticosterone and its derivatives. It may also be considerably interfered with by other substances. Cope (1965) regards this estimate as correlating better with cortisol secretion rates than 17-hydroxycorticosteroid (Norymberski) values; these latter are a measure of metabolic breakdown. However, there are a considerable number of vagaries in the urine method and high values are not unusual in abnormal situations, without a parallel increase in the free cortisol values measured by a satisfactory technique. A modification of the Mattingly (1962) method has been applied to studies in children (Cathro 1969) and in a small series from 3 months of age to puberty a positive correlation was found between the value and body weight, but less satisfactorily with surface area.

Pragnanetriol

This steroid is a marker for the common form of congenital adrenal hyperplasia, derived from 17α-hydroxy progesterone, and may also be found in excessive amounts with adrenal carcinomas, derived from another precursor.

Pregnanetriolone (11-ketopregnanetriol) is likewise a characteristic feature of early and late forms of congenital adrenal hyperplasia, Cushing's syndrome and has been found with Stein-Leventhal ovaries but there are problems in its estimation (Thomas & Steinbeck 1969). It should not be significantly present in normal urine.

The most accurate method for pregnanetriol is based upon the Peterson techniques and isotope dilution (New et al 1966); in this, values of <0·2 mg/24 hr are stated to be normal for all ages. With less specific methods, estimates of 0·8 mg/24 hr in infants to 2 mg/24 hr in adults are found (Bongiovanni & Eberlein 1958; Thomas & Steinbeck 1969) but these now need to be reassessed by isotope dilution or other more specific methods. In the luteal phase of the cycle higher values are found than in the follicular (Pickett et al 1959). ACTH stimulation will increase the excretion values.

11-deoxy/11-oxy ratio for 17-oxosteroids

These estimates depend upon urinary 11-deoxy-17-oxosteroid excretions being less than, or similar to, the 11-oxy-17-oxosteroid values to the age of 7–8 years for boys, and puberty for girls. An '11-oxygenation index' (Morris 1959, Hill 1960) gives an expression to the ratio of pregnanetriol and other 17-hydroxy steroids of the 11-deoxy form (cortisol precursors) to 11-oxy forms (cortisol and its metabolites). 17-oxosteroids derived from cortisol or 21-deoxycortisol are separated from those derived from 17α-hydroxyprogesterone. In 11β-hydroxylase deficiency, pregnanetriol excretion is not increased to the extent found in 21-hydroxylase deficiency, and may be normal; thus, differences are more obvious in this form of congenital adrenal hyperplasia. There appears to be no diurnal rhythm to alter the index (Hill 1960, Edwards et al 1964) and, in normal subjects, the index does not exceed 0·62, but in classical cases of congenital adrenal hyperplasia the index is greater than unity and higher with an 11β-hydroxylase deficiency.

Dehydroepiandrosterone

An increased excretion of this steroid occurs with some adrenocortical tumours in association with mixed varieties of Cushing's syndrome. It is also present in congenital adrenal hyperplasia with a Δ^5-3β-hydroxysteroid dehydrogenase deficiency. This compound can be measured adequately by gas liquid chromatography but most accurately by isotope dilution derivative techniques. Conjugates are significant, acid hydrolysis breaking these in group chemical methods, and the values need to be expressed as free and conjugated steroid.

Aldosterone

Free aldosterone can only be accurately estimated by double isotope derivative dilution techniques, and likewise the 18-glucuronide, which together give some measure of aldosterone production. In the neonatal period, where especial precautions are further necessary, steroid excretion is low and the conjugated excretion measures 3–13 $\mu g/m^2/24$ hr (New *et al* 1966). In children, probably excretion does not exceed 5 $\mu g/24$ hr (Visser 1966) provided that there is no sodium depletion. Values for an adequate sodium diet in normal males are 2·8–19·6 $\mu g/24$ hr and for females 3·3–23·6 $\mu g/24$ hr, in the follicular phase of the cycle. In general, the estimation is reserved for the facilities of a specialized laboratory.

PLASMA METHODS

Porter-Silber chromogens

Sampling for this estimation should take place at standard times, as for example 8–9 a.m. and 10 p.m.–midnight–2 a.m. for the presumed highest and lowest levels during the 24-hour period. However, for the best sampling more frequent estimations are required. Diurnal variation for the estimate appears established between the 1st and 3rd year (Franks 1966). The levels are much increased by 'stress', and this may occur quickly. Ordinarily, adult values have a morning range of 5–30 $\mu g/100$ ml plasma and the late night values are mostly below 8–10 $\mu g/100$ ml. However some morning values are exceptionally low, likewise some night values may be elevated, so that isolated determinations have limited value.

The method does not measure 21-deoxycortisol, as does the Mattingly (1962) method. 11-deoxycortisol (compound S), cortisone and tetrahydro-derivatives, that are not conjugated, will also be measured. The Peterson (Peterson *et al* 1957) technique significantly excludes 11-deoxycortisol. Cortisone is more prominent in plasma in early life than cortisol (Bro-Rasmussen *et al* 1962) but there are difficulties at this period (Hillman & Giroud 1965). The early morning plasma levels in infants show a wide scatter, compared with adults, have a lower mean at the 2nd to 5th day of 8·4 $\mu g/100$ ml plasma, but nevertheless higher than previously reported (Bertrand *et al* 1962). Methodological factors probably account for differences (Mitchell & Shackleton 1969). The

APPENDIX C 761

levels fall during the 2nd and 3rd week (mean morning value of 4·3 μg/100 ml) and adult values are reached at the end of the first month (Bertrand *et al* 1962).

Mattingly (1962) *method* (*sulphuric acid induced fluorescence*)

This method is relatively non-specific, but values appear to follow the clinical condition (Brooks *et al* 1970). About 40% (27–55%; mean 41%) is not cortisol (James *et al* 1967). It appears valuable for functional studies of the hypothalamic-pituitary-adrenal-system and is much utilized in Britain, whereas Porter-Silber methods are more widely adopted in the U.S.A. The plasma method has fewer vagaries than the urinary method but serum is preferable to plasma and heparin can interfere. It depends upon induced fluorescence of the steroids with sulphuric acid. Corticosterone fluoresces more intensely than cortisol but, normally, has only a low concentration. However, in childhood its contribution can be greater especially in the first 5 years, likewise in defective 17-hydroxylation. In children, the method has the advantage that smaller volumes of serum are required than for the Porter-Silber technique and that cortisone does not fluoresce. In all groups, synthetic steroids contribute little to total values. The range of values is similar to that of the Porter-Silber technique but the night value should not exceed 7 μg/100 ml plasma, although this is not sure (Jackson & Mowat 1970). The values obtained in infants are variable with a wide scatter and, although a semblance of diurnal rhythm may appear after the 1st month, it is not well established until the 1st to the 3rd year of life. The possible fallacies in the method need to be understood.

Specific steroids may be estimated by protein binding assay, isotope dilution and derivative techniques and radioimmunoassay. These methods are usually not yet adopted by routine clinical chemistry laboratories.

ADRENOCORTICAL STIMULATION TESTS

These are useful in excluding primary adrenocortical insufficiency and a variety of procedures is available all depending upon adequate ACTH administration. If a rapid screening test is wanted for outpatient conditions, 0·25 mg β^{1-24} corticotrophin injected intramuscularly allows comparison of the Mattingly (1962) SAFA value just before and 30 minutes after the injection. If the increase measures at least 7 µg/100 ml plasma and the 30 minute value is at least 18 µg/100 ml, there is an adequate response to stimulation to exclude insufficiency (Wood *et al* 1965). At times, the test may need to be prolonged to 2 hours, and especially in children (Cathro 1969). In children, a short ACTH stimulation test utilizing a single intravenous injection of 0·25 mg β^{1-24} corticotrophin should ordinarily have samples taken before and at 30 minutes, 1 and 2 hours after the injection. The intramuscular injection appears to fail in children from time to time, as it does in adults, although plasma responses are longer after intramuscular injection. With doubtful results from these tests, a full stimulation test with an intravenous infusion of 0·25 mg β^{1-24} corticotrophin over 8 hours following both the blood levels and urinary excretions of steroids before and after the infusion, is mandatory. In Cushing's disease, where the syndrome is due to bilateral adrenocortical hyperplasia or hyperfunction, a full stimulation test leads to a continued increased excretion of steroid beyond the day on which the infusion is given, several days in all, in most instances. Plasma responses will be generally excessive. Where the syndrome is due to a tumour, there is ordinarily little response to stimulation, although this may occur, measured by both plasma levels and urinary excretion of steroids; a carcinoma fails to respond. In Addison's disease, there is characteristically no response to stimulation although both plasma and urinary levels can be in the normal range ('partial Addison's disease').

In congenital adrenal hyperplasia of minor extent, there may also be a poor response and stimulation will need to be prolonged before the characteristic steroid pattern is disclosed; in classical examples with the typically featured steroid excretions, there may also be little response to ACTH stimulation but diagnostic problems are not troublesome; in fully treated examples, there may be little response to ACTH unless stimulation is prolonged but this situation seldom causes difficulty. Plasma levels of Porter-Silber chromogens may fail to increase with ACTH stimulation or increase to varying extents depending upon the analytical technique used and the form of the disease.

In ACTH stimulation tests it is essential that the ACTH be of a known potent batch, if synthetic β^{1-24} corticotrophin, or similar, is not used. None the less, human ACTH does show a prolonged action, assessed indirectly by estimation of peripheral plasma 'cortisol' levels compared with porcine ACTH on an equimolar basis (Upton *et al* 1970). There is probably also some difference in initial response between the polypeptides.

Metyrapone test

This test is less used than formerly and the results are only broadly quantitative. Reduction of 11β-hydroxylation leads to increased 11-deoxycortisol production and consequent endogenous corticotrophin release or increase, with subsequent adrenocortical stimulation. Plasma 'cortisol' levels will fall during the administration of the drug but plasma '11-deoxycortisol' values will rise. During the period of administration, but mostly in the succeeding 24 hours, the excretion of Porter-Silber or Norymberski chromogens rises. The actual increase is mostly less than that obtained with ACTH stimulation and any doubtful result such as a failed or poor response to inhibition should be checked by an ACTH stimulation test, if this has not been previously performed. Similar results are found in children (Lelong *et al* 1962).

Insulin hypoglycaemia test

An insulin induced hypoglycaemia tests the responsiveness of the hypothalamic-pituitary-adrenal system. Provided that blood glucose is reduced adequately within 20–30 minutes after the intravenous injection of purified porcine insulin (0·1 unit/kg body weight and up to 0·15–0·2 unit/kg body weight needs to be given if the lower dose is unsuccessful) failure of plasma 'cortisol' levels to rise suggests a defect in responsiveness of the system. The cause for this needs to be investigated subsequently. Plasma 'cortisol' is measured before injection and at 30 minutes, 1, $1\frac{1}{2}$, and 2 hours after injection; the maximum plasma increases are obtained at 1–2 hours. A failed response should be tested further by ensuring that the adrenal cortex responds to ACTH stimulation.

Dexamethasone suppression tests

These aim to reduce plasma levels and urinary excretion of steroids and thus assess the normality or otherwise of the hypothalamic-pituitary-adrenal system, particularly an apparent increase in adrenocortical secretion and its dependency (Liddle 1960). Mostly, in the first stage of the tests, 2 mg dexamethasone is given over 24 hours, as 0·5 mg at 6-hour intervals, and so spaced that appropriate blood samples in relation to a presumed day–night variation can be obtained; in the second stage of testing, dosage is increased to 8 mg over 24 hours, as 2 mg at 6-hour intervals. It may be necessary in some instances to arrange three levels of suppression to test suppressibility and dependency. Some paediatricians do not alter the dosage schedule for children (Cathro 1969) and similar doses, as well as metyrapone have been used in pregnancy (Maeyama & Nakagawa 1970). Ordinarily, normal adults will evidence suppression of 17-hydroxycorticosteroid (Norymberski chromogens) excretion with the lower dosage to below 40% of the control value or less than 5 mg/24 hr; Porter-Silber chromogens values become lower and at times are not detectable. By the 2nd or 3rd day, early morning plasma levels of Porter-Silber chromogens are below 5 μg/100 ml plasma. In obesity, values in both these estimations may not suppress so adequately, and with significant stress there will be some failure of suppression. The use of three levels of suppression can be valuable in some instances of obesity. There is no suppression in the usual instance of Cushing's syndrome but, in Cushing's disease with bilateral adrenocortical hyperplasia due to hypothalamic-pituitary overactivity, some suppression will occur with the high dosage. If adrenocortical carcinoma is the basis of the Cushing's syndrome, or ectopic ACTH production, there will be no suppression. With adrenocortical adenomas, there is mostly no suppression, although suggestive in others; adrenocortical rest tumours fail to suppress.

At times a simple screening test is adopted. Blood for steroid estimation is taken late at night (10 p.m.–midnight) and 1–2 mg dexamethasone given at 11 p.m. and a repeat blood estimation performed at 8–9 a.m. A value for Porter-Silber chromogens in the early morning blood of less than 5 μg/100 ml plasma excludes adrenocortical overactivity but failure to suppress has to be tested by a full dexamethasone suppression schedule. Occasionally, the spontaneous fluctuation in adrenocortical activity of Cushing's disease can be associated with early morning values scarcely different from zero.

SECRETION RATES

Secretion rates (or secretory rates) can be determined for steroids if a unique or near unique metabolite exists. If the metabolite comes from other sources production rates are estimated. The usual procedure is one in which an isotopically labelled steroid (with a stable lable and shown to be pure immediately preceding injection) is injected and the specific activity of appropriate urinary metabolites measured over a definite period, usually 24–72 hours. The specific activity is the dilution of the tracer label by the non-labelled steroid; both steroids of necessity need to have identical metabolic fates. This simple technique cannot be used to determine rigorously the rates of secretion of any hormone in pregnant women; as distinct from non-pregnant women or males, there will be several anatomical compartments (foetus, placenta and mother). Some hormones may be produced at more than one site and the tracer hormone may have a different metabolic or directional fate due to its injection into a maternal vein from that produced endogenously. With these provisions, the secretion rates of cortisol, using tetrahydrocortisone glucuronide for determining dilution of the tracer, and aldosterone using the 18-glucuronide conjugate for its dilution, may be measured. In the case of cortisol, other metabolites of quantitative significance generally give values similar to those for tetrahydrocortisone but sometimes do not, and the reason is not clear. If a urinary steroid is derived from more than one substance, secretory rates cannot be obtained by a single injection of the tracer only production rates and this applies particularly to C-19 steroids or androgens. Secretory rates of two hormones that participate in the metabolic products of one another may be obtained by simultaneous injection of both, as for example dehydroepiandrosterone and dehydroepiandrosterone sulphate in non-pregnant women. The theoretical concepts and derivation of parameters for secretion and production rates have been subject to much discussion (Tait & Burstein 1964, Lieberman & Gurpide 1966, Baird *et al* 1969).

Secretion rates for 24-hour periods have been expressed as mg/m^2 by investigators referring to children (New & Seaman 1970). A value for cortisol is 7·5 mg/m^2, which increases three-fold with ACTH, diminishing with metyrapone and dexamethasone suppression. With a 21-hydroxylase defect cortisol secretion will not increase with ACTH. The rate for 11-deoxycortisol (Compound S) is low, <0·5 mg/m^2, increasing eight-fold with ACTH, up to 100-fold with metyrapone. Corticosterone secretion measures about 2 mg/m^2 and shows similar relationships to cortisol for ACTH stimulation, 11β-hydroxylase inhibition and suppression of ACTH. Deoxycorticosterone rate is 55 $\mu g/m^2$, and similar to aldosterone. It increases considerably with ACTH, more so with 11β-hydroxylase inhibition and suppresses with dexamethasone, although not to the extent of cortisol. Aldosterone secretion doubles with ACTH stimulation, decreases with dexamethasone and to a greater extent with 11β-hydroxylase inhibition. Of all these steroids, only the secretion rate of aldosterone is related to sodium intake, and potassium. The values in the literature are, on the whole, similar although cortisol secretion rate is more variable; one normal series of $13\cdot5 \pm 4\cdot5$ mg/day rose to 71–278 mg/day after ACTH stimulation which suggests greater responsiveness than described above. Such secretion rates and their alteration in ACTH stimulation and suppression tests, as well as 11β-hydroxylase inhibition, are relevant in assessing enzymatic deficiencies as found in congenital adrenal hyperplasia.

METABOLIC CLEARANCE RATES

A metabolic clearance rate (MCR) is generally defined as the volume of blood irreversibly and completely cleared of steroid in unit time. It is a virtual volume. There are important presuppositions for this measurement and related ones; the calculation requires either an infusion of labelled steroid or, less satisfactorily, a single injection (plasma disappearance rate method). The concepts that form the basis of this measurement and related ones under steady-state conditions have been examined and their general usefulness described by Baird *et al* (1969). Clearance across an organ can also be measured. From time to time there has been a problem that secretion rates are not reproducible for different metabolites. This is discussed under aldosterone where tetrahydroaldosterone glucuronide and the 18-glucuronide conjugate are involved. A similar problem applies to cortisol in normal and abnormal situations. Secretion rates of cortisol calculated from different metabolites will differ in the range of 19–35% (New *et al* 1969), which may introduce problems in clinical situations. The different specific activities for *allo*-tetrahydrocortisol, tetrahydrocortisol and tetrahydrocortisone glucuronides are a problem (New *et al* 1969; Kowarski *et al* 1969). Also, the 'error' of such estimations is such that a cortisone secretion would not be estimated.

REFERENCES

Appleby J.I., Gibson G., Norymberski J.K. & Stubbs R.D. (1955) Indirect analysis of corticoserotids. 1. The determination of 17-hydroxycorticosteroids. *Biochem. J.* **60,** 453–460.

APPLEBY J.I. & NORYMBERSKI J.K. (1957) The urinary excretion of 17-hydroxycorticosteroids in human pregnancy. *J. Endocr.* **15**, 310.
BAIRD D.T., HORTON R., LONGCOPE C. & TAIT J.F. (1969) Steroid dynamics under steady-state conditions. *Rec. Progr. Horm. Res.* **25**, 611–656.
BAYLISS R.I.S. & STEINBECK A.W. (1953) A modified method for estimating 17-hydroxycorticosteroids in plasma. *Biochem. J.* **54**, 523–527.
BERTRAND J., GILLY R. & LORAS B. (1962) Free and conjugated plasma 17-hydroxycorticosteroids in the newborn during the first five days of life: effects of hydrocortisone and ACTH administration, in A.R.Currie, T.Symington & J.K.Grant (eds.) *The Human Adrenal Cortex*, pp. 608–623. Edinburgh, Livingstone.
BONGIOVANNI A.M. & EBERLEIN W.R. (1958) Critical analysis of methods for measurement of pregnane-3α, 17α, 20α-triol in human urine. *Analyt. Chem.* **30**, 388–393.
BONGIOVANNI A.M. & ROOT A.W. (1963) The adreno-genital syndrome. *New Eng. J. Med.* **268**, 1342–1351, 1391–1399.
BROOKS C.J.W., BROOKS R.V., FOTHERBY K., GRANT J.K., KLOPPER A. & KLYNE W. (1970) The identification of steroids. *J. Endocr.* **47**, 265–272.
BRO-RASMUSSEN F., BUUS O. & TROLLE D. (1962) Ratio cortisone/cortisol in mother and infant at birth. *Acta endocr. (Kbh.)* **40**, 579–583.
CATHRO D.M. (1969) Adrenal cortex and medulla, in D.Hubble (ed.) *Paediatric Endocrinology*, pp. 187–327. Oxford, Blackwell.
CLAYTON B.E. (1968) Assessment of pituitary-adrenal function in children. *Mem. Soc. Endocr.* **17**, 237–246.
DRAYER H.M. & GIROUD C.J.P. (1965) Corticosteroid sulfates in the urine of the human neonate. *Steroids* **5**, 289–317.
EDWARDS R.W.H., MAKIN H.L.J. & BARRATT T.M. (1964) The steroid 11-oxygenation index: a rapid method for use in the diagnosis of congenital adrenal hyperplasia. *J. Endocr.* **30**, 181–194.
EIK-NES K.B. (1968) Adrenocorticosteriods, in R.I.Dorfman (ed.) *Methods in Hormone Research*, 2nd ed., pp. 271–322. New York, Academic Press.
FEW J.D. (1961) A method for the analysis of urinary 17-hydroxycorticosteroids. *J. Endocr.* **22**, 31–46.
FRANKS R.C. (1967) Diurnal variation of plasma 17-hydroxycorticosteroids in children. *J. clin. Endocr.* **27**, 75–78.
GRAY C.H., BARON D.N., BROOKS R.V. & JAMES V.H.T. (1969) A critical appraisal of a method for estimating urinary 17-oxosteroids and total 17-oxogenic steroids. *Lancet* **i**, 124–127.
HILL E.E. (1960) Chromatography of the 17-ketogenic steroids in the diagnosis and control of congenital adrenal hyperplasia. *Acta endocr. (Kbh.)* **33**, 230–250.
HILLMAN D.A. & GIROUD C.J.P. (1965) Plasma cortisone and cortisol levels at birth and during the neonatal period. *J. clin. Endocr.* **25**, 243–248.
JACKSON I.M.D. & MOWAT J.I. (1970) Hypothalamic pituitary adrenal function in obesity and the cortisol secretion rate following prolonged starvation. *Acta endocr. (Kbh.)* **63**, 415–422.
JAMES V.H.T., TOWNSEND J. & FRASER R. (1967) Comparison of fluorimetric and

isotope procedures for the determination of plasma cortisol. *J. Endocr.* **37**, xxviii.

KNORR D. (1965) Age dependency of excretion of 17-hydroxycorticosteroids, androsterone, etiocholanolone and 11-oxy-17-ketosteroids. *Acta endocr. (Kbh.)* Suppl. 101, 33.

LELONG M., JAYLE M.F., JOSEPH R., CARLORBE P., JOB J.C., SCHOLLER R., BORNICHE P. & PASQUALINI J.R. (1962) L'épreuve à la Métopirone (SU-4885) chez l'enfant. *Annl. paediat.* **199**, 519–537.

LIDDLE G.W. (1960) Tests of pituitary-adrenal suppressibility in the diagnosis of Cushing's syndrome. *J. clin. Endocr.* **20**, 1539–1560.

LIEBERMAN S. & GURPIDE E. (1966) Isotope dilution methods for the estimation of rates of secretion of the steroid hormones, in G.Pincus, T.Nakao & J.F.Tait (eds.) *Steroid Dynamics*, pp. 531–547. New York, Academic Press.

MAEYAMA M. & NAKAGAWA T. (1970) Effects of ACTH, metopirone and dexamethasone on urinary steroid excretion in late pregnancy. *Steroids* **15**, 267–274.

MATTINGLY D. (1962) A simple fluorimetric method for the estimation of free 11-hydroxycorticoids in human plasma. *J. clin. Path.* **15**, 374–379.

MINICK M.C. (1966) Cortisol and cortisone excretion from infancy to adult life. *Metabolism* **15**, 359–363.

MITCHELL F.L. & SHACKLETON C.H. (1969) The investigation of steroid metabolism in early infancy, in O.Bodansky & C.P.Stewart (eds.) *Advances in Clinical Chemistry*, vol. 12, pp. 141–215. New York, Academic Press.

MORRIS R. (1959) Pregnanetriol in urine: determination as a 17-ketogenic steroid. *Acta endocr. (Kbh.)* **32**, 596–605.

NATOLI G. & NATOLI V. (1961) Studio dell'eliminazione urinaria degli steroidi 17-chetogenici nell'infanzia (secondo la metodica di Norymberski) 2. Ricerche in bambini normali appartenenti all prima infanzia. *Archo ital. Pediat. Pueric.* **21**, 380–392, quoted by Cathro (1969).

NELSON D.H. & SAMUELS L.T. (1952) A method for the determination of 17-hydroxycorticosteroids in blood: 17-hydroxycorticosterone in the peripheral circulation. *J. clin. Endocr.* **12**, 519–526.

NEW M.I., MILLER B. & PETERSON R.E. (1966) Aldosterone excretion in normal children and in children with adrenal hyperplasia. *J. clin. Invest.* **45**, 412–428.

NEW M.I. & SEAMAN M.P. (1970) Secretion rates of cortisol and aldosterone precursors in various forms of congenital adrenal hyperplasia. *J. clin. Endocr.* **30**, 361–371.

NEW M.I., SEAMAN M.P. & PETERSON R.E. (1969) A method for the simultaneous determination of the secretion rates of cortisol, 11-desoxycortisol, corticosterone, 11-desoxycorticosterone and aldosterone. *J. clin. Endocr.* **29**, 514–522.

NORYMBERSKI J.K., STUBBS R.D. & WEST H.F. (1953) Assessment of adrenocortical activity by assay of 17-ketogenic steroids in urine. *Lancet* **i**, 1276–1281.

PETERSON R.E., KARRER A. & GUERRA S.L. (1957) Evaluation of the Silber-Porter procedure for the determination of plasma hydrocortisone. *Analyt. Chem.* **29**, 144–149.

PICKETT M.T., KYRIAKIDES E.C., STERN M.I. & SOMMERVILLE I.F. (1959) Urinary pregnanetriol throughout the menstrual cycle. Observations on a healthy subject and an adrenalectomized woman. *Lancet* **ii**, 829–830.

TAIT J.F. & BURSTEIN S. (1964) *In vivo* studies of steroid dynamics in man, in G.Pincus, K.V.Thimann & E.B.Astwood (eds.) *The Hormones*, vol. V, pp. 441–557. New York, Academic Press.

THOMAS F.J. & STEINBECK A.W. (1969) Semiquantitative estimation of urinary pregnanetriol, pregnanetriolone and tetrahydro S in the investigation of adrenocortical function. *Acta endocr. (Kbh.)* **60,** 645–656.

THOMAS F.J. & STEINBECK A.W. (1969) Quantitative estimation of urinary pregnanetriol, pregnanetriolone, tetrahydro S and Δ^5-pregnenetriol in the investigation of adrenocortical function. *Acta endocr. (Kbh.)* **60,** 657–668.

UPTON G.V., HOLLINGSWORTH D.R., LANDE S., LERNER A.B. & AMATRUDA T.T. (1970) Comparison of purified human and porcine ACTH in man. *J. clin. Endocr.* **30,** 190–195.

VISSER H.K.A. (1966) The adrenal cortex in childhood. Part 1 Physiological aspects. Part 2 Pathological aspects. *Arch. Dis. Childh.* **41,** 2–16, 113–136.

WOOD J.B., FRANKLAND A.W., JAMES V.H.T. & LANDON J. (1965) A rapid test of adrenocortical function. *Lancet* **i,** 243–245.

Index

Abortion
  Abnormal karyotype and, 24, 25, 27
  Corpus luteum and, 337
  HCG levels in, 143
  Induction of with prostaglandins, 351
  Prostaglandin levels in, 312
Acid-base
  Changes in pregnancy, 623
  Foetal uterine contractions and, 349
  Liquor amnii and, 297
  Renal function and, 608
ACTH
  Administration of and foetal adrenal size, 179
  Adrenal cortex and, 416, 432
  Adrenocortical effects after birth, 416
  Aldosterone and, 436
  As a test of adrenocortical function, 761
  Circadian rhythm, 340
  Control of, 434
  DHAS secretion and, 431
  Drugs and, 435
  Extra-adrenal actions of, 436
  Foetal adrenal and, 162, 179
  Foetal hypothalamus and, 162
  Foetal pituitary and, 141
  Hypoglycaemia and, 433
  Induction of labor with, 186
  Maternal corticosteroids and, 166
  Placental isolation of, 148
  Surfactant and, 185
  Structure of, 417
  Theory of action, 56
Actomyosin
  Oestrogens and uterine content of, 334
Addison's disease, 449, 455, 459, 461, 462
Adenyl cyclase, 55
  Prostaglandins and, 57
  TSH and, 488

Adipose tissue—*see also* Fat
  Fatty acid synthesis in, 530
  Re-esterification and corticosteroids, 455
Adrenal cortex, 415—*see also* Adrenal hyperplasia
  ACTH and, 416, 432
  Anatomy, 415
  Anencephaly and, 160
  Congenital absence of, 167
  Embryology, 415
  Foetal ACTH and, 141, 163, 179
  Foetal adrenal, 177, 423
  Foetal adrenal and DHAS, 161
  Functional tests of, 761
  HCG and, 141, 179
  Histology, 416
  Maternal adrenalectomy and oestrogen levels, 158
  Neonatal function, 423
  Premature labour and, 336
  Puberty and, 419
  Steroid biosynthesis, 417
Adrenal hyperplasia
  DHAS secretion *in utero* and, 163
  Hypothalamic differentiation and, 108
  Induction with oestradiol, 180
  Pregnanetriol and, 425, 756
  Psychosexual development and, 71, 155
Adrenal medulla
  Foetal, 177, 388, 399
Adrenalin
  Foetal production, 178, 395
  Hypoglycaemia and, 433
  Myometrial response, 339
Ahumada-del Castillo syndrome, 45
Albumin
  Corticosteroid binding, 428
  Progesterone binding, 174
Aldosterone
  ACTH and, 436

Aldosterone (cont.)
  Assay of in urine, 760
  Control of, 436
  Effects on cardiovascular function, 583
  Excretion during pregnancy, 447
  Metabolism of, 422
  Pre-eclampsia and, 447
  Primary aldosteronism, 465
  Progesterone as antagonist, 70, 465
  Renal artery constriction and, 437
  Renal function and, 465
  Renin and angiotensin, relationships to, 437, 583, 619
  Secretion rate, 426
  Sodium retention in pregnancy and, 619
  Synthesis of, 418
  Transcortin binding, 428
Alkaline phosphatase
  As index of placental function, 164, 170
  Foetal, steroid induction of, 183
Allograft (syn. Homograft)
  Definition, 198
  Foetus as, 198, 232
  Histocompatibility antigens and, 221
  Rejection, immunological basis of, 221
  Rejection in pregnancy, 234
  Rejection in the foetus and newborn, 243
Alupent
  Suppression of labour with, 353
Amino acids
  In haemoglobin, 667
  Renal function and, 601, 617
Amnion
  Exchange across, 278
  Secretary function of, 283
  Ultrastructure of, 279
Amniotic fluid, 258—see also Liquor amnii
  Embolism, 705
  Lecithin in, 186, 312
  Meconium staining, 170
  Oestrogen concentration, 158, 304
  Progesterone in, 175, 302
Ammonia
  Renal function and, 609
Ampicillin
  Effect on urinary oestriol, 174

Anaemia
  Androgens in, 666
  Erythropoietin and, 664
Androgens—see also specific androgens
  ACTH and, 419
  Administration during pregnancy, 155
  As intermediary in oestrogen synthesis, 60
  Effect on foetal hypothalamus, 71
  Effect on HCG on, 141
  Erythropoiesis and, 665
  Hypothalamus, effect on, 107
  Inhibition of cortisol synthesis and, 418
  Metabolism of, 104
  Molecular actions, 108
  Neurological function and, 155
  Secretion from testis, 100
  17-oxosteroids and, 104
  Sexual differentiation and, 152, 154
  Thyroid function and, 486
  Transport of, 102
Androstenedione
  Metabolism of, 104
  Plasma levels, 100
  Renal clearance, 104
  Testicular secretion, 100
Anencephaly
  Cortisol production in, 181
  DHAS levels in, 159
  Foetal swallowing and, 267
  Liquor 17-oxosteroids and, 300
  Maternal metyrapone and, 179
  Oestrogen excretion and, 159, 165
  Onset of labour and, 186, 335
  Polyhydramnios and, 263
Aneuploidy—see Chromosomes—abnormality
Angiotensin, 612
  Aldosterone and, 437
  Oral contraceptives and, 583
  Pregnancy
    Effects on cardiovascular system, 583
    Sodium retention and, 619
Anosmia
  Effect on puberty, 46
Ante-partum haemorrhage
  Urinary oestriol and, 173
Antibodies
  Against seminal and testicular components, 226

Chromosomes (cont.)
  Mental development and, 29
  Mosaicism, 2
  Physical development and, 28
  Radiation damage and, 3, 15
  Sex chromosomes, 7, 8, 26
  Sex chromosome anomalies, 26
  Sexual differentiation and, 27, 148
  X chromosomes and female differentiation, 8, 27, 150
  X chromosomes and genetic activity, 11
  X chromosome and Lyon theory of inactivation, 11
  X chromosome inactivation, 10, 13, 30
  Y chromosome and infertility, 231
  Y chromosome and masculine differentiation, 8, 27, 150
  Y chromosome and quinacrine fluorescence, 9
Clonal selection theory, 216
Circulation—*see also* Blood flow
  Foetal, 386
    Control of, 388
    Hypoxia and, 390
  Maternal
    Thyrotoxicosis and, 484
    Time, 560
  Neonatal
    Changes after birth, 405
    Pulmonary circulation in acidosis, 401
  Placental, 376
Cistron—*see* Genes
Coagulation, 692—*see also* Haemostasis
  Coagulation factors, 691
Codon—*see* Genes
Colostrum
  Immunoglobulin A in, 205
Conception
  Immunological aspects, 222
  In vitro, 371
  Ovarian and pituitary hormones and, 84
  Tubal factors in, 363
Complement
  Fixation, 202
Congenital abnormalities
  Chromosomes and, 1, 8, 14, 15, 16, 18, 21, 25, 309
  Corticosteroids and, 468

Pre-natal diagnosis of, 210
Thymic aplasia, 213
Conn's syndrome, 465
Corpus luteum
  Control of, 58
  Cardiovascular function and, 584
  Conceptual cycle, 84
  HCG and, 140
  Hysterectomy and, 59
  Luteolysin and, 59
  Prostaglandins and, 59
  Removal of in pregnancy, 337
  Respiratory function and, 659
Cortexine, 150
Corticosteroids—*see also* Cortisol, etc.
  ACTH and, 416
  Biosynthesis, 417
  Catabolism, 419
  Excretion during pregnancy, 443
  Foetal liver glycogen and, 185
  Induction of labour with, 186
  Interrelationships with other hormones, 463
  Liquor amnii, levels of, 301
  Maternal administration and foetal hypothalamus, 162, 166
  Measurement of, 752
  Metabolic actions, 451
  Placental transfer of, 181, 450
  Plasma half-life of, 420
  Pregnancy, levels of, 349
  Prolonged suppression with, 435
  Secretion rates, 426, 764
  17-oxosteroids and, 420
  Surfactant and, 185
  Synthetic, 467
  Teratogenesis and, 468
  Umbilical cord levels, 440
Corticosteroid binding globulin (CBG) *see* Transcortin
Corticotrophin—*see* ACTH
Cortisol
  Administration of ACTH and foetal levels of, 179
  Assay of, 757
  Biosynthesis, 417
  Circadian rhythm, 429
  Effect on GFR and sodium in pregnancy, 619
  Effects of oestrogens on secretion and transport, 64, 428, 445

Cortisol (cont.)
   Excretion during pregnancy, 443
   Foetal metabolism of, 182, 444
   Functions of in foetus, 183
   HCS and, 144
   Immunosuppressive action in pregnancy, 235
   Inhibition by androgens, 418
   Interrelationship with cortisone, 420
   Liquor amnii levels of, 310
   Lymphopoenia in pregnancy and, 142, 446
   Metabolic actions, 451
   Metabolic clearance rate, 766
   Placental metabolism of, 163
   Placental transfer of, 310, 450
   Plasma binding of, 428
   Production rate in neonates, 181
   Secretion rate by foetal adrenal, 181
   Umbilical cord levels, 424, 440
Cortisone
   Convertibility to cortisol, 419
   Excretion during pregnancy, 443
   Umbilical cord levels, 440
Creatinine
   Liquor amnii, 291
   Relationship to blood urea and GFR, 600
Cri-du-chat syndrome
   Chromosome deletion and, 23
   Clinical features, 23, 26
Cushing's disease
   Glycosuria and, 455
   Pregnancy and, 450
   Resemblance to normal pregnancy, 446
   Treatment with corticosteroids, 462
Cyclic amp, 55
   ACTH and, 432
   Glycogen synthesis and, 530
   HCS and, 144
   LH and, 128
   Oestradiol and, 69
   Parathyroid hormones and, 494
   Thyroxine and, 488
Cyproterone acetate
   Effect on sexual differentiation, 105, 152
Cytogenetics, 1—*see also* Chromosomes and Genes

Deciduum
   Possible immunological role, 232
Dehydroepiandrosterone (DHA) and Sulphate (DHAS)
   ACTH and, 431
   Adrenal secretion rate, 426
   As a precursor of oestrogens, 160
   Assay of, 759
   DHAS and placental sulphatase deficiency, 163
   Effect of HCG in male infants on, 141
   Foetal adrenal and, 164, 166, 167
   Foetal hypothalamus and, 163
   Foetal suppression and maternal corticosteroids, 450
   Liquor amnii levels of, 300, 302
   Maternal administration and placental function, 170
   Placental utilization of, 166
$\Delta 5$-3β Hydroxysteroids—*see also* DHAS
   in liquor, 302
Deoxyribonucleic acid (DNA)
   Chromosomes, 2
   Composition, 31
   Duplication in cell division, 31
   Genetic code and, 31
   Inverted sequences in chromosome, 24
   Labelled DNA and chromosome incorporation, 3
   Messenger RNA and, 32
   Spermatogenesis and, 112
   Testosterone and, 110
Desoxyribonucleic acid—*see* Deoxyribonucleic acid
Diabetes insipidus, 339
Diabetes mellitus
   Assays of HCS in, 147
   Fatty acid oxidation in, 532
   Foetal effects, 545
   Gestational diabetes and HCS, 144
   HCG levels in pregnancy, 142
   Insulin levels in 538
   Oestriol assays in, 171
   Perinatal death rate in, 170, 547
   Polyhydramnios in, 263
   Pregnancy and, 543
   Thyrotoxicosis and, 544
Diamine oxidase
   As test of placental function, 164
Dihydrotestosterone

Relationship to action of testosterone, 109
Diphosphoglycerate (DPG)
  Oxygen affinity of haemoglobin and, 669
Down's syndrome—*see* Mongolism
Drosophila
  Gene dosage compensation in, 10, 11
Dyspnoea
  Pregnancy and, 660

Edward's syndrome
  Trisomy and, 25
Electrolytes—*see also* Sodium and Potassium
  in liquor amnii, 288
Embden-Meyerhof pathway
  Erythropoiesis and, 667
Enzyme
  Congenital defects in, diagnosis of, 309
  Corticosteroids and, 452
  Induction and genes, 34
  Induction of, and cortisol, 183
  Testicular feminism and, 28
Erythropoiesis, 664
  Androgens and, 665
  Erythropoietin and, 664
  Requirements for, 671
Erythropoietin, 664
  Assay, 665
  Sites of production, 665
  Structure, 665
Ethinyl oestradiol
  Metabolic effects, 64

Fabricius, bursa of, 212, 214
Fallopian tube, 361
  Cleavage in, 364
  Contractility control of, 367
  Fertilization in, 363
  Fluid movement, 371
  Ovum transport and, 365
  Prostaglandins and, 369
  Sperm transport and, 361
Fanconi's anaemia
  Genes and, 17
Fat—*see also* Adipose tissue
  Foetal metabolism of, 384
Fatty acids
  Diabetes mellitus and, 532

Oxidation, 532
Synthesis, 530
Fertilization—*see* Conception
Finger prints, 40
Foetus
  Absence of, and placental function, 156
  Acid-base, 297
  Adrenal function, 178, 425, 450
  Anencephaly—*see* Anencephaly
  Antigenic status of, 233
  As an allograft, 232
  Bradycardia, 170, 349
  Cardio-respiratory physiology, 376
  Cellular immunity in, 244
  Congenital absence of hypothalamus in, 162
  Cortisol function in, 183
  Cortisol metabolism in, 182, 425, 444
  Death of. Effect on oestrogen and pregnanediol excretion, 159, 169
  Diabetes and, 170, 545
  Dysmaturity of, 167
  Electrocardiography in, 396
  Exposure to foreign antigens, 250
  Foetal testosterone and differentiation 105
  Foeto-placental unit, 138, 156, 330
  Gut peristalsis, 267
  Haemolytic disease, 172 *and see* Rhesus incompatibility
  Haemorrhage, 407
  Immunoglobulin production in, 243
  Immunological development, 209, 242
  Involvement in maternal auto-immune disease, 248
  Labour, effects on, 349
  Maturity
    and liquor bilirubin, 293
    and liquor cells, 308
    and liquor creatinine, 291
  Oestriol assay as test for foeto-placental function, 165
  Pancreatic function with maternal diabetes, 546
  Passive immunology in, 242
  Progesterone metabolism and, 176
  Prognosis of related to urinary oestriol, 169
  Pulmonary secretion, 273
  Renal function, 271

Oogenesis
  Chromosome deletion during, 18
  Meiosis and, 6
  Non-disjunction during, 18
Oral contraceptives
  Effects on coagulation, 56
  Metabolic effects, 64, 533
  Myometrial effects, 347
  Renin angiotensin and, 583, 620
Orgasm
  Oxytocin and, 370
  Uterine contractility and, 346
Osmolality
  Liquor amnii, 288
  Renal function and, 603
Osteogenesis imperfecta, 36
Osteoporosis, 63
Ovarian function, 45
  Autoimmune disease and, 230
  Lipid metabolism and, 57
  Sexuality and, 71
  Steroidogenesis, 60
  Steroidogenesis FSH and, 53
    LH and, 56, 58
  Thymus and, 230
  Two cell theory of steroidogenesis, 60
Ovulation
  Hormonal changes associated with, 77
  Induction with FRF, 51
  Induction with LRF, 50
  Tubal secretion related to, 371
Ovum
  Fertilization, 636
  Transport, 365, 367
Oxygen—*see also* Po_2
  Consumption in neonates, 407, 408
  DPG, and 669
  Erythropoietin and, 664
  Foetus
    Arterial concentrations of, 386
    Capacity of foetal blood for, 381
    Dissociation curve of, 383
    Environment of, 376
    Uterine contractions and, 394
  Haemoglobin affinity, 669
  Hyperbaric, 403
  Maternal
    Breathing of and foetal PO_2, 396
    Exchange, 646
    Laboratory assessment, 649
    Measurement of consumption, 645

Oxytocin
  Induced labour and, 350
  Labour and, 337, 339
  Myometrial sensitivity, non-pregnant, 326
    pregnant, 329
  Orgasm and release of, 370
  Placental isolation of, 148
  Progesterone levels and sensitivity to, 175
  Water intoxication and, 351
Oxytocinase
  As index of placental function, 164, 170

Para-amniohippurate (PAH)
  Renal function and, 602
Parathyroid gland, 492
Parturition—*see* Labour
Patau's syndrome
  Trisomy and, 25
P_{CO_2}
  Foetal, 379
    Maternal hyperventilation and, 397
    Pulmonary circulation and, 389
    Uterine contractions and, 349
  Liquor and, 297
  Maternal
    Alveolar blood flow, and 646
    Laboratory assessment, 649
    Pulmonary ventilation and, 647
    Respiratory control and, 652
Pendred's syndrome, 481
Pentose shunt, 526
  Erythrocyte metabolism and, 667
pH
  Foetal, 379
    Brain survival and, 395
    Effect on oxygen release, 383
    Maternal hyperventilation and, 397
    Pulmonary circulation and, 389
    Uterine contractions and, 439
  of liquor, 298
Phenol red (PSP)
  Renal function, and 603
Phenotype
  Sex also specific syndromes such as Turner, *Cri-du-chat*
  Sex chromosome and, 10
Pheromones, 47
Philadelphia-chromosome, 23

Phylogenesis
  Immunological mechanisms in, 208
  —*see also* Immunology, developmental
Pineal gland, 47
  Melatonin and, 48
  Photosensitivity, 48
  Pineal tumours and sexual precocity, 48
Pituitary
  Foetal hypophysectomy and thyroid function, 509
  Foetal pituitary in anencephaly, 165 186
  Gonadotrophins—*see* LH, FSH
  Hypophysectomy, effect on testis, 111, 123
  Hypothalamus and, 49
  Immunological control of fertility and, 224
Placenta
  As an allograft, 232
  As an immunological barrier, 236, 247
  Circulation of, 376, 585
  Cortisol metabolism and, 163, 450, 726
  Dehydrogenase in, 181
  Evolution of, 198
  Foetal cotyledon, 377
  Foeto-placental complex, 156 *and see under* Foetus
  Functional tests of, 164, 170
  Growth, retardation of, 168
  Hormonal immunosuppression and, 235
  Hypertension and, 168
  Nitabuch's layer, possible immunological significance of, 238
  Oestrogen synthesis by, 158, 160, 166, 167, 726
  Oxygen consumption of, 381
  Pericellular fibrinoid layer in, 238
  Permeability to corticosteroids, 181, 301, 450
  Permeability to HGH, 546
  Permeability to thyroid hormones, 516
  Possible immunological competence of, 241
  Size of and ABO blood groups, 240
  Size of and immunological influences, 240
  Sulphatase in, 161, 163
  Transfer of iodine, 517
  Uterine contractility and, 333
Plasma cells
  Antibody production and, 207
Plasma volume, 555
  Oestrogen and, 584
Platelets, 688
Po_2
  Foetal, 380
    Maternal O_2 inhalation and, 396
    Pulmonary circulation and, 389
    Uterine contractions and, 349
  Liquor and, 297
  Maternal
    Alveolar blood flow and, 646
    Hyperventilation and, 396
    Laboratory assessment, 649
    Pulmonary ventilation, 647
    Respiratory control and, 652
Polar body, 6
Polyploid—*see* Chromosomes—abnormalities, numerical
Polyhydramnios—*see* Liquor amnii
Porter-Silber chromogens
  Assay in urine, 757
  Assay in plasma, 760
  Pregnancy and, 439
Potassium
  Aldosterone and, 347, 465
  Pregnancy and balance of, 623
  Renal function and, 608
Pre-eclampsia
  Aldosterone in, 447
  Corticosteroids in, 421, 448
  HCG levels in, 142, 147
  placental changes in, 168
Pregnancy—*see under* individual headings
Pregnanediol
  As index of placental function, 164, 170, 177
  Assays in pregnancy, relationships of HCS to, 147
  In normal cycle, 77
  Levels with placental sulphatase deficiency, 163
  Liquor amnii levels of, 302
  Maternally administered corticosteroids and, 166
  Umbilical cord ligation *in utero* and, 159

Pregnanetriol
  Adrenal hyperplasia and, 425, 757
  Assay of, 758
  Liquor amnii and, 307
  Normal cycle in, 682
Primordial follicles
  Turner's syndrome and, 27
Progesterone, 69, 174
  Aldosterone as an antogonist, 70, 465, 622
  Carboyhdrate and lipid metabolism effect on, 533
  Cardiovasuclar effects, 584
  Cellular actions, 70–1
  Effects on breasts, 70
  Effects on uterus, 69, 334
  General actions, 69
  HCG and, 141
  Immune reactions and, 235
  In normal cycle, 77
  Levels during pregnancy, 174
  Liquor amnii levels of, 302
  Metabolic effects, 70
  Metabolism of, 175, 176
  Onset of labour and, 175, 187
  Ovarian production of, 60
  Placental production, 156
  Protein binding of in plasma, 174, 429
  Relationship to LH peak, 77
  Renal function and, 70, 465, 622
  Respiratory function and, 659
  Venous distensibility and, 564
Prolactin, 51
  Clinical evidence for, 51
  Evidence for separate hormone, 52
  In rhesus monkey, 52
  PIF and, 52
  Relationship to HCS, 144
  Relationship to HGH, 51
  Theoretical control of, 52
Prolactin inhibitory factor (PIF), 49, 52
Prolonged pregnancy
  Foetal survival and, 172
  Urinary oestriol in, 173
Prostaglandins
  Abortion and, 60
  Co-action with adenyl cyclase, 57
  Contraception and, 59
  Labour induction, 351
  Liquor amnii levels of, 312
  Luteolysis and, 59, 188
  Onset of labour and, 187, 337
  Receptors for LH, 57
  Thyrocalcitonin and, 496
  Tubal response to, 369
Puberty, girls, 46
  Adrenocortical function and, 419
  Behavioural maturation and, 72
  Boys, 106
  Precocious puberty and pineal tumour, 78
  Sense of smell and, 46
  Testicular changes during, 95

Quinacrine fluorescence and Y chromosome, 9

Radioimmunoassay
  Principles of, 737
Red-cells—*see also* Erythropoiesis, 664
  Corticosteroids and, 459
  Life span, 670
  Relationship of shape to function, 669
  Sodium content and thyroid function, 503
  Volume in pregnancy, 558
Relaxin
  Placental isolation of, 148
Rem sleep, 430
Renal agenesis, 272
Renal clearance
  Aldosterone, 465
  Cortisol, 444
  Iodine, 478, 505
Renal tubules
  Changes in pregnancy, 616
  Function, 599
  Functional Anatomy, 594
Renin, 612
  Aldosterone and, 437
  Hepatic degradation, 612
  Liquor levels, 313
  Oral contraception and, 583
  Pregnancy and
    Relationship to cardiovascular changes, 583
    Sodium retention, 619
Renal artery stenosis, 437, 438

Respiration—*see also* Lung
  Maternal, 639
    Changes in pregnancy, 653
    Control, 652
    Control in pregnancy, 659
    Glossary of terminology, 661
    Laboratory assessment, 642
    Normal, 640
  Neonatal, 408
Respiratory distress syndrome (RDS)
  In babies of diabetic, 170
  Liquor lipids and, 312
  Prophylaxis with corticosteroids, 186
  Relationship to surfactant, 185
Resuscitation
  Neonatal, 400
Rhesus isoimmunization
  HCG assays and, 143
  Intra-uterine transfusion, immunological effects of, 245
  Liquor bilirubin and, 293
  Liquor protein and, 294
  Polyhydramnios and, 263
  Steroid excretion in, 172
Ribonucleic acid (RNA)
  ACTH and, 416
  Activity of X chromosome and, 150
  Corticosteroids and, 435
  Messenger RNA and gene transcription, 32
  Messenger RNA and gene translation, 32
  Oestradiol and RNA synthesis, 68
  Progesterone and RNA synthesis, 71
  Structure, 32
  Testosterone and RNA synthesis, 110
  Thyroxine and, 488
  Transfer RNA, 32
Ribosomes
  Messenger RNA and, 32
  Oestradiol and, 69
  Polypeptide synthesis, 32
Ritodrine
  Suppression of labour with, 352
Runt syndrome
  Experimental induction of by placental cells, 241
  Following intra-uterine transfusion, 246
  Graft-versus-host reaction and, 217

Sabbateir hypothesis, 388
Scaferrin—*see* Sperm—coating antigen
Sex-determination, 148
Seminal fluid
  Acute anaphylaxis and, 230
  Antibodies, 226
  Antigens, 225
  Chemistry of, 120
  Evaluation of, 117
  Sperm morphology, 119
Serotonin
  Oxytocic activity, 336
Sertoli cells
  Development of, 92
  During puberty, 95
  Possible role in steroidogenesis, 98
  Spermatogenesis and, 112
17-hydroxycorticosteroids
  Excretion during pregnancy, 443
  Liquor amnii levels of, 301
  Methods of assay in urine, 754
  Relationship to free cortisol, 428
17-oxosteroids
  Corticosteroids and, 420
  Effect of testosterone secretion on, 104
  Excretion during pregnancy, 448
  Liquor amnii levels of, 299
  Methods of assay, 730
  Relationship to androgen production, 104
Sex
  Chromosomal, 148
  External genital, 153
  Gonadal, 150
  Legal, 156
  Phenotypic, 154
  Psychological, 154
  Rearing, 155
  Sex-duct, 152
Sex-chromatin (Barr body), 8
  Drumsticks in leucocytes and, 8
  Lack of relationship to Y chromosomes, 9
  Liquor amnii cells and, 309
  Relationship to number of X chromosomes, 8
  Size of sex chromatin and abnormalities of X chromosome, 8
  Time of appearance in human foetus, 13

Sexual differentiation
  Chromosomes and, 27
  Thyroid and, 485
  Y chromosome deletion and, 28
Sodium
  Aldosterone and, 437, 465
  Factors promoting natriuresis in pregnancy, 622
  Factors promoting retention in pregnancy, 619
  Glucocorticoids and, 464
  Plasma volume in pregnancy and, 584
  Renal function and, 603, 606
  Retention during pregnancy, 618
  Thyrocalcitonin and, 494
  Thyroxine and, 488, 503
Sperm
  Antibodies, 226
    Immobilization test for and, infertility, 228
    Immunoglobulin classes, 230
    Micro-agglutination test for, and infertility, 227
  Antigen, 225, 230
  Capacity, 363
  Coating antigen(s), 225
  Fertilization, 363
  Transport, 361
Spermatogenesis, 111
  Control of, 128
  DNA synthesis and, 112
  Duration of, 115
  Evaluation of, 117
  Hypophysectomy and, 111
  Life cycle of seminiferous tubules and, 112
  Meiosis and, 6
  Morphology—*see* Seminal fluid
  Non-disjunction during, 18
  Puberty and, 95
  Testosterone and, 111
  Transport and storage, 115
  Xg blood group as marker, 20
Stature
  Adolescent growth spurt and, 106
  Oestradiol and, 63
  Permissive action of thyroxine, 484
  Phenotype and, 154
  Sex chromosomes and, 28, 29
Steroids—*see also* individual steroids
  Adrenal steroids, measurement of, 752

Biosynthesis and metabolism, 716
Catabolism, 723
Chemistry, 716
Placental, 726
Stroke volume
  Maternal during pregnancy, 572
Supine hypotension, 580
Surfactant
  Neonatal breathing and, 398
  Pulmonary maturity and, 185, 392
Systemic lupus erythematosus
  Placental transfer of antinuclear factor in, 248

Temperature
  Neonatal, 408
Testicular feminism
  Androgen insensitivity and, 153
  Enzymic defect and, 28
  Karyotype, 28
  Psycho-sexual development and, 155
Testicular hypoplasia—*see* Klinefelter's syndrome
Testis
  Antigenicity of, 225
  Biopsy of, 120
  Control of testicular function, 123
  Descent of in foetus, 93
  During childhood and puberty, 95
  Embryology and development, 92
    Time sequences compared with ovary, 93
  Foetal graft of and sexual differentiation, 152
  Foetal testis and testosterone synthesis, 94
  Oestrogen production from, 99
  Physiology and function, 91
  Steroid production from, 99
  Y chromosome and development, 149
Testosterone
  Adrenal venous levels, 426
  Biological actions, 105
  Biosynthesis, 96
    Flow sheet of, 98
  DNA and RNA, effect on, 110
  Effect of age on, 102
  HCG effect on, 125
  In normal menstrual cycle, 82
  Interconversion to oestradiol, 99
  Levels in homosexuality, 155

Testosterone (cont.)
  Liquor amnii levels of, 300
  Metabolic clearance rate, 101, 103
  Metabolism, of 104
  Molecular actions, 108
  Neonatal administration, 155
  Plasma binding of, 103
    Effect of oestrogens on, 103
  Plasma levels in males, 100
  Production by foetal testis, 95
  Relationship to dihydrotestosterone, 109
  Renal clearance, 104
  Secretion rate from testis, 100
  Sexual activity and, 127
  Spermatogenesis and, 111
Thalassaemia
  Controlling operon and, 35
Thymus
  As a primary lymphoid organ, 213
  Cell dynamics, 210
  Congenital absence, 213
  Corticosteroids and, 452
  Embryology, 210
  Hormones and, 235
  Hypoplasia in pregnancy, 235
  Immunological significance of, 213
  Ovarian function and, 230
  Relationship with secondary lymphoid organs, 210, 214
  Small lymphocytes and, 210
  Structure, 210
Thyrocalcitonin, 491
  Actions of, 494
  Disease and, 496
  Pregnancy and, 495
  Relationships to parathyroid, 492
  Structure of, 493
Thyroglobulin, 481
Thyroid, 478
  Blocking iodine uptake, 479
  Control of, 489
  Foetal, 509 development and HCT, 148
  Function in pregnancy, 504
  Functional tests, 497
  Glucocorticoids and, 463
  Iodine uptake, 478
  Neonatal thyroid function, 514
  Thyrocalcitonin, 491
Thyrotoxicosis, 483
  Menstrual function in, 485

  Transient neonatal form, 248
Thyrotrophin (TSH)
  Foetal pituitary and, 148
  Foetal thyroid and, 509
  Relationship to HCT, 147
  Relationship to placental, 730
  Relationship to TRF, 489
Thyrotrophin releasing factor (TRF)
  Relationship to TSH, 489
  Structure, 50, 489
  Tests and thyroid function and, 503
Thyroxine (T4), 480
  Action of, 486
  Biosynthesis, 480
  Cellular binding of, 487
  Control of, 489
  Genetic abnormalities and, 481
  Measurement of, 499
  Metabolism, 482, 483, 484
  Relationship to HGH, 484
  Transport, 482
Thyroxine binding globulin (TBG), 500
  Binding of T3 and T4, 482, 499
Thyroxine binding prealbumin
  Binding capacity of, 483
Transcortin (Corticosteroid binding globulin (CBG)), 427
  Aldosterone binding, 428
  Changes in pregnancy, 428
  Progesterone binding, 174
  Testosterone binding, 103
Transferrin
  Genes and, 35
  Ion transport and, 666
Transvestism, 155
Tricarboxylic acid cycle, 528
Triiodothyronine (T3)
  Biological activity, 483
  Biosynthesis, 480
  Measurement of, 501
  Metabolism, 482, 483, 484
  Placental transfer, 516
  Transport, 482
Triplo-X female, 14
  Fertility and, 27
  Frequency at birth, 15
  Non-disjunction and, 18
Trophoblast
  Antigenicity of, 238
  Antigens,

Trophoblast (*cont.*)
  Immunological enhancement and, 237
  Possible maternal iso-immunization and, 236, 250
  As an allograft, 233
  As an immunological barrier, 236
  Deportation, 236
  Deportation and immunological significance of, 241
  Experimental immunization with, 250
  Hormonal immunosuppression and, 236
  Hormonal impact in conceptual cycle, 84
  HCS and, 144
  HCT and, 147
  Pericellular fibrinoid layer, 239
  Proliferation of and ABO blood groups, 240
  Proliferation of, and immunological influences, 240
  Protective role in post-implantation embryo, 232
  Site of HCG production and, 138
  Tumours of and HCG levels, 142, 143
  Tumours of and HSC levels, 147
  Tumours of and immunological influences, 241
  Tumours of and molar thyrotrophin, 491
Turner's syndrome, 1, 12, 14, 27
  Abortion and, 27
  Frequency of birth, 15
  Mosaicism and, 20
  Stature and karyotype 28
Twins
  Anaphase lagging and, 21
  Genetic component and, 40
  Placental cross-circulation and chimaerism, 22
Polyhydramnios and, 263

Umbilical Cord
  Circulation of, 389
  Cortisol levels in, 424, 440
  Free fatty acid levels, 451
  Ligation of *in utero* and effect on oestrogens and pregnanediol, 159
  Neonatal ligation and placental transfusion, 406
  Progesterone in vein and artery, 175
  Ultrastructure of, 283
Urea
  In liquor amnii, 291
  Relationship to creatinine and GFR, 600
  Renal function and, 601, 610
Uric acid
  In liquor amnii, 291
  Renal function and, 618
Urogenital slit
  Sexual differentiation of, 93, 153
Uterus—*see also* Myometrium and Labour
  Blood flow, 573
  Blood flow and myometrium activity, 342, 581
  Gross effects of oestrogens, 62
  Immune responses and, 228, 237
  Progesterone and, 69
  Progesterone receptors in, 71
  Removal of, and luteolysis, 59
  Specific receptors for oestradiol, 66, 68
  Volume of, and activity in pregnancy, 330

Vagina
  Oestriol and, 62
Vasopressin
  Placental isolation of, 148
Venous pressure
  Maternal, during pregnancy, 564
Viruses
  Chromosomal changes and, 16
Vitamin B_{12}, 679
  Relationship to folic acid, 678
Vulva
  Oestradiol and maturation of, 61
  Vaginal cytology, 72

Water
  Cardiovascular function in pregnancy and, 584
  Intoxication and oxytocin, 351
  Renal function and, 603
  Retention during pregnancy, 518
  Turnover in amniotic fluid, 275

Wolffian ducts
  Foetal testis and, 152
  Foetal testosterone and, 105
  Sexual differentiation of, 93, 152

Yolk sac
  Migration of oogonia and spermatozoa from, 92, 151

Zona pellucida
  Possible immunological role, 232
Zuckerkandl, organs of, 178